METAPHYSICS AND THE PHILOSOPHY OF SCIENCE

METAPHYSICS AND THE PHILOSOPHY OF SCIENCE

The Classical Origins: Descartes to Kant

GERD BUCHDAHL

The MIT Press, Cambridge, Massachusetts

© Basil Blackwell 1969
SBN 262 02057 2 (Hardcover)

*Library of Congress Catalog
Card No.* 74–95579

Printed in
Great Britain

PREFACE

This book is meant to give philosophers of science as well as the general student some basic knowledge of philosophy, its problems and working methods, by a study of a crucial period in its history. Of necessity philosophy tends to become fragmented in proportion as its critical eye is focused on different fields, such as morals and politics, art, psychology, the physical and social sciences, and so on. Sometimes, as the subject becomes specialised and technical, particularly in the case of the philosophy of science, one loses sight of the general nature of philosophical thinking, and of the philosophical tradition that has accumulated over the centuries. Scientific detail as well as tendencies towards formalisation begin to take precedence over the more basic questions. This work is in part meant to redress the balance, and to make the contemporary technical philosophers of science aware that behind their subject lies a solid foundation of general philosophical thought.

Furthermore, this book also seeks to remind the students of general philosophy that their philosophical heritage is deeply grounded in, if not originating from, speculation about science, showing that the methods and results of scientific enquiry have had a profound direction on philosophical thought.

If we here proceed historically, this is because recent trends have ignored or omitted to mention the genuine core of earlier progress; and sometimes histories of the subject have not helped since they tend to bury the living core of the nature of philosophical *argument* under the detail of their reportage. Here I have tried to retain what is living in the thought and procedures of some of the great philosophers of the past, particularly by emphasising the nature of some of the logical manoeuvres which are permanently associated with philosophical thinking, and are an essential part of it. My objective has been to point to the growth of self-consciousness in philosophical thinking during recent times. In this, I have eschewed where possible the usual classifications of the history of philosophy, banishing some time-honoured prejudices by showing, for instance, that the 'rationalists' were not as rationalistic, nor the empiricists as empirical, as is commonly believed.

Owing to the emphasis on the philosophy of science, I hope that here and there I have put in new perspective some of the philosophical developments of the period discussed, as for instance, the conjectural and empiricist nature of much of Descartes' science, the significance of Locke's sceptical tone, Hume's double-approach to causality, the importance for Berkeley of the concept of the laws of nature, the relations between Leibniz's achievements in physics and his metaphysical doctrines. But, above all, I hope that by paying greater attention to Kant's views on scientific methodology, a more satisfying and accurate account of the teachings of this enigmatic philosopher has here been given.

Although there runs through the book a continuous theme, and the method of discussion used is meant to display something of the nature of philosophical argument and development, chapters—and sometimes even individual sections, especially in the chapter on Kant —are sufficiently self-contained to be read independently.

The breadth of this study, though still containing large gaps, forbids mention of many of the sources that have been used, and in any case I have tried to look at the main texts afresh through the spectacles of my principal interests, and I would want the reader of this book to do the same.

November 1, 1967

CONTENTS

I

INTRODUCTION: PHILOSOPHICAL APPRAISALS

In this book I want to trace the development of some of the philosophical doctrines and ideas of the classical period from Descartes to Kant. Since my principal interest is the philosophy of science, those aspects will be selected which illuminate most immediately topics currently debated by philosophers of science, such as the relative place and importance of deduction, induction and hypothesis; the formation of theoretical concepts; the relation of mathematics and physics; logical and extralogical presuppositions; the analysis of the concepts of causation and scientific law; the nature of scientific explanation in general. Of course, during the period under review, these matters formed a very intimate part of the general problems of philosophy; and much that goes under the name of general theory of knowledge arose out of the special problems of the methods and logical foundations of the sciences that were most rapidly developing during these centuries—mechanics, optics; and somewhat later, chemistry and biology. Their images and aims not only shape the general interests and trend of philosophical discussion, impressed as it is, for instance, by the importance of mathematics, or the emphasis on an observational and experimental ('empirical') approach; but also, many of the central scientific concepts come to function as the key-notions for philosophical systems which are developing concurrently. We shall study this influence below, through the importance of the foundations of dynamics for the philosophies of Descartes and of Leibniz; seventeenth-century corpuscularianism for the thinking of Locke; of Newtonian formulations, or again, biological systematics for the philosophy of Kant.

The key position which science occupies for the philosophers of this period is graphically underlined by the very titles of their major

1

works, and the general profession of their aims. Thus Descartes named his most popular work *A Discourse on the Method of Rightly Conducting the Reason, and seeking Truth in the Sciences*; Locke in his *Essay Concerning Human Understanding* tells his reader that he is setting out 'as an under-labourer' of the 'incomparable Mr Newton', his ambition being 'to be employed . . . in clearing the ground a little, and removing some of the rubbish that lies in the way of knowledge';[1] and one of Kant's major epistemological problems was formulated by him in the question 'How is Pure Natural Science Possible?'

For our understanding of the trend of philosophical speculation during the 150 years which span these three philosophical formulations, it is worth noting an interesting shift which occurs over this period. The shift is indicative of a very typical philosophical process, a peculiar sort of 'over-generalisation', of turning questions into 'meta-questions'. During these 150 years, the philosophical mind becomes increasingly self-conscious in its reflection on the idea of 'laying the foundations of scientific knowledge'. Thus whilst Descartes' *Discourse* still suggests that the philosopher's task is to search for methodological devices which will *produce reliable conclusions within* science, and even methods conducive to the *discovery* of new scientific truth, by the time of Kant philosophical aims have subtly shifted their ground, and one now searches after a justification for the whole *of* scientific knowledge, indeed, of knowledge as such. These two tasks are not always kept distinct. Special and general versions are frequently tangled in a tantalising fashion, as we shall see, for instance, in the work of Kant. But throughout, the lines of demarcation between these two enterprises are not drawn clearly; nor is it always possible to do so. To cite a more specific case, in order to answer the question how empirical generalisations are to be justified, one cannot always keep out considerations of scientific methodology, instead of fastening on the purely logical problem of the nature of empirical generalisations as such, or the meaning of justification. The question, 'how do we know that repeated tests under similar conditions will yield identical results?', is partly a question about the *concepts* of 'knowing', of 'uniformity', of 'law', and partly about information yielded by an extended investigation of nature. During the period under investigation, the shift towards the 'meta-question' is as much due to growing clarity concerning objects and limits of

[1] The Epistle to the Reader.

philosophical enquiry as to the successful development of some of the sciences. Thus whilst in the beginning Descartes' writings still constitute a composite attempt to discuss the logical foundations of science whilst simultaneously formulating some of the fundamental laws of dynamics and optics (at that time in a highly controversial state), Kant proceeds in the belief that there already exists an unassailable body of physical truth codified in more or less ultimate form, leaving the field clear for a solution of the generalised search for foundations.

At this point, a general warning is in place. I do not want to convey the impression that the history of philosophical ideas (doctrines, principles, methods) is an account of the progressive uncovering of errors and misconceptions, gradually to be replaced by more adequate and satisfactory notions, like so many items of philosophical truth. This would seriously misconstrue the whole tone and ethos of a philosophical enterprise, and not least the classical wisdom of the seventeenth and eighteenth centuries. These authors strike an immensely personal note. Each, more often than not, addresses himself to and argues against some central contention or other of his predecessors. It is like a living dialogue, a dynamic process of continuous debate—not at all some dead textbook formulation. At first sight, indeed, the tenor of this debate produces the misleading impression of bare statement and counterstatement, amounting to a series of progressive and mutual refutations. In fact, as we shall see, each of these thinkers is setting out from an entirely different basis, employing a different set of pictures or focal analogues; witness such key notions as 'idea', 'impression', 'perception', 'sensation'. By means of these each characterises what he considers to be the 'fundamental furniture' or 'basic inventory' of the world.

But such a basis (the use of a specific analogue to yield what is really a metaphysical foundation), being usually no more than a *partial* aspect of the whole, invariably and almost naturally is too narrow, and too unstable to function without further supports. Consequently, it requires enlargement in almost every case by appropriate metaphysical supplements that provide what we might term an additional metaphysical centre of gravity. Being metaphysical, and thus 'semantically opaque', such a centre will in turn need to be represented through some empirical or logical, or theological concept, making up what we shall call its 'analogical grammar'. Its true significance or function is, however, 'justificatory', for it acts as a sort of kingpin round which revolves the whole philosophical edifice. (This aspect of the metaphysical supplement we might call its 'ontological grammar'.)[1]

These remarks are purely declaratory, and clearly need explanation and defence. Since the rest of this book is devoted to their explication and justification, providing a 'point of view' for appraising the philosophical achievements of the writers whose doctrines I propose to examine, I shall not attempt to elaborate further for the present. But perhaps one brief example from among the following may be helpful.[2]

For Descartes, part of the basic inventory of the world is the individual and self-contained thinking-states of the self. These are regarded by him in a purely 'atomistic' fashion—a conception which clearly exceeds any straightforward empirical account; the states are, as it were, metaphysically distinct. On the other hand, a reconstruction of the world as it appears, or as Descartes' ideal requires, stands in need of substances and selves persisting in time. To restore therefore the broken chain of existence, Descartes needs to introduce a factor which will re-link the members of the chain, and this factor must have a metaphysical status in order to possess the required logical force. Now just as the discontinuous states of the self present something for which there are empirical and psychological models, so the additional factor is couched in terms of a model, namely Descartes' God; God here serving as the metaphysical construct, whose analogical grammar is taken from the realm of theology. Thus Descartes postulates that God preserves or re-creates at any moment those states of the self for which otherwise there would be no 'justification' to continue to exist.

In general, just as Descartes' God vouchsafes the continued existence of the external world, and—as we shall see—the reliability of sensory knowledge, so Locke's notion of 'simple ideas' will offer release from the 'dark closet' of the mind, 'ideas' in general being Locke's basic building blocks. In the same way, the God of Berkeley will serve as a foundation of inductive uniformity. And what custom and the imagination do for Hume in building up an external world of objects and causal relationships, the categorial forms of Aristotelian logic will provide for Kant.

[1] The application of the notions of 'analogical' and 'ontological grammar' to the case of the Aristotelian concept of the 'universal' has been worked out in my *Induction and Necessity in the Philosophy of Aristotle*, Aquinas Pamphlet No. 40, London, 1963. Cf. below, pp. 363–4.

[2] Cf. below, Ch. III, pp. 160ff., 175ff .

To the distinction between analogical and ontological grammar there corresponds another—as we shall find throughout what follows: the distinction between different 'levels' at which the argument of our philosophers proceeds; for instance, between an empirical or positive level, on the one hand, and a non-empirical, metaphysical, or even logical level on the other. Locke's theory of ideas, Hume's theory of causation, Leibniz's theory of forces will proceed at such different levels, although any connection that may exist between them will be construed—either consciously or unconsciously—by the different philosophers in different ways. An interesting feature of this process is that frequently these philosophers try to meet the deficiencies of their starting points by 'mixing' these levels, or sliding from one level to another, whether this be done wittingly or unwittingly, and although their logical entitlement to do so may be small.

Since it is one of the chief aims of this book to develop the themes implied in these initial remarks, I shall not try to develop them in any more detail for the present, confining myself instead for the moment to a general characterisation of the nature of the philosophical approach. I am suggesting, then, that the different forms of such an approach should not be thought of as either 'right' or 'wrong', let alone positions mutually refuting one other; but rather as sets of *alternative* philosophical responses to demands made by conflicting series of considerations. Progress in the development of philosophical history—in so far as this idea applies—will go on record as an increased degree of self-consciousness accompanying the construction of the different systems.

The manifest differences between the various schools will often reduce to differences of philosophical emphasis, rather than stand as genuinely contradictory statements. Let us illustrate this by an important example. It would normally be held that Descartes' estimate of the possibility of 'scientific knowledge' was more optimistic than that of Locke.[1] The former, as we shall see, claims not infrequently to possess certain unimpeachable principles supposed to govern some

---

[1] This difference is one of a group that has traditionally been designated by the respective labels of 'rationalism' and 'empiricism', 'schools' which have their most prominent representatives in Descartes and Locke respectively. But, as this example will illustrate, what really divides these two philosophers is not a genuine 'opposition' of viewpoints but rather differences of emphasis, and 'descriptive appraisals' of certain aspects in the various fields of 'knowledge'. The fluidity of the language in which these appraisals appear is most pronounced in the earliest of the thinkers of our classical period, though I do not think that this prevents them from representing types of the point of view here suggested.

system of scientific fact, for instance the phenomena of mechanics; whereas Locke's general position is that *all* scientific generalisations are more or less precarious and vulnerable. Now this is just the sort of difference in approach to scientific knowledge which people often associate, respectively, with 'dogmatic rationalism' and 'sober-minded empiricism'. A second look, however, soon shows that this difference is balanced by a compensating view about what really constitutes a proper object of 'knowledge', here: of 'scientific knowledge'. Disregarding for the moment a great number of possible interpretations of the complexities involved, and speaking only provisionally, Locke will be found prepared, in principle at least, to use the term 'knowledge' in the context of sensory experience, and to entertain the logical possibility of knowledge (doubtful though it may be) concerning the existence of things present to our senses. Now this Descartes does not allow. Instead he will invent a special terminology, referring to the result of sensory experience as 'the teaching of nature', a notion quite distinct from 'knowledge', and which—being one of those focal metaphysical key-concepts—must feed on a store of meaning borrowed from common everyday associations, whilst at the same time possessing a special 'philosophical' grammar as well.

The choice of such alternative appraisals has a curious consequence. Since Descartes never conceded even the logical possibility of experiential knowledge, in the sense just explained, he is never tempted to make much of a sceptical denial of our actual possession of it. Yet this is precisely the tenor of Locke's never-ending refrain.

But such appraisals should not of course be regarded as purely verbal manoeuvres. Thus Descartes' denial is linked in a complicated way with his more positive teaching also. This takes its starting point from the 'commonsense' doctrine that only a mathematical treatment of physical nature is capable of holding out any hope of scientific control. In its turn this comes to mean that only what can be pressed into a mathematical framework is knowable; something which eventually is rendered in the quasi-physical (or 'ontological') formulation: the world, as knowable, must be conceived under the guise of geometrical characteristics alone; a doctrine graphically emphasised in Descartes' famous doctrine that the essence of matter is extension.

On Locke's side likewise we meet with a compensating move of a more positive nature. When discussing the 'reality of human knowledge' (i.e. the question, as he puts it, of the 'conformity between our ideas and the reality of things'),[1] he remarks that without any doubt 'we have of mathematical truths . . . not only certain but real knowledge'.[2] This curious doctrine ('curious' if one expects him to hold in this matter views diametrically at variance with those of Descartes) is defended by saying that our mathematical conclusions will always be matched in nature because we simply ignore or dismiss what does not answer to the mathematical formulation. Now this may not seem so very different from Descartes' doctrine, in the form in which I stated it a moment ago. Both emphasise the idea that science involves the pressing of physical fact into mathematical form. But Locke is thinking of mathematical (and logical) 'truth' as 'knowledge' in a peculiar and somewhat Pickwickian sense, a kind of ceremonial label. Its status (by contrast with the 'rationalist' position) is diminished, for, as we have just seen, Locke is ready to *describe* the process of *sensation* (the possession of sensory 'ideas') as a kind of knowledge—indeed, in his eyes the very foundation of natural science.[3] But on the other hand, he adds the significant proviso that it is much 'narrower' than the kind of 'knowledge' involved in certain purely logical ('intuitive') judgments.[4]

Of course, it may be said that Locke here misdescribes a difference in kind as a difference of degree ('breadth of extent'). But such a view is itself tantamount to taking a philosophical stand in this matter. At best, Locke's position would seem to be a sort of half-way house. In fact, as we shall see in Chapter IV, Locke is really the one who strives to establish that there is a difference in kind, but fails to recognise this sufficiently, as evidenced by his complaint that we have 'very little' genuine scientific 'knowledge'. So the latter continues to linger as an ideal. At any rate, what is important for us to see is that this is a position which under the stresses and strains of several conflicting considerations attempts to *extend the dictionary* of 'knowledge', taking one of a number of possible views of the situation; in this case extending the grammar of 'knowledge' to a field which Descartes had regarded as being no more than the 'teaching of nature'; in his language something very different from, and inferior to, genuine 'knowledge'. Indeed, the emphasis is on 'difference', and it is not too much to say that in Descartes' use of this notion, 'the teaching of

---

[1] *Essay*, iv.4.3.
[2] Op. cit., iv.4.6.
[3] Cf. op. cit., iv.2.14; 3.5.
[4] For details, cf. below, ch. IV, sects. 1 and 3.

nature' in a confused way does express a more adequate recognition that we are dealing here with a difference in kind than can be found in Locke.

In general, neither side is at this stage clear on the distinction between mathematical or logical, and physical propositions; sometimes making it only a difference of degree; at others *sharing* the opinion that physical knowledge falls short of the stringent requirements of the former sort; whilst failing to see that this is not so much a deficiency as a difference in kind. That there is such a distinction in kind is itself, however, a philosophical point of view, not *clearly* erected into a fundamental principle until Kant. Descartes' and Locke's positions *might* then be viewed as members of a class of possible responses to the stresses and strains produced by a complex situation due to reflection on the interaction between mathematics and science; a fact which may help us to appreciate more clearly the function of the types of metaphysical supplementations alluded to already.[1]

Thus, Descartes clearly attaches greater pragmatic importance to certain ideals, for instance of mathematical knowledge, to what he calls the field of the 'clear and the distinct', and in consequence restricts the 'dictionary' of the term 'knowledge' more stringently than Locke. On the other hand, he supplements the domain, thus narrowed, by a number of 'opaque conceptions' (as I have called them), as witnessed in the suggestion that though sensory apprehension of the external world does not give us knowledge, it is not nothing, since 'God is no deceiver'. Which, in Descartes' language, is to say that though like all empirical, non-mathematical learning (the 'teaching of nature'), our perception of the external world does not reach intrinsically the standard of mathematical knowledge, it is nonetheless perfectly dependable. By contrast, Locke, when commenting on the question of our knowledge of the external world, expressly though grudgingly remarks that our 'assurance' of this 'deserves the name of *knowledge*'. 'Nobody', he writes, 'can, in earnest, be so sceptical as to be uncertain of the existence of those things which he sees and feels.'[2] One commentator on this passage[3] has remarked that it is 'question-begging'. I think, however, Locke's argument will be appreciated more clearly if we regard it as an

[1] Above, pp. 3–4.
[2] *Essay*, iv.11.3; Locke's italics.
[3] A. S. Pringle-Pattison, Abridged ed. of Locke's *Essay*, Oxford, 1934, p. 322.

interesting specimen of a central aspect of philosophising, the formation of a philosophical language, involving decisions concerning the grammar of certain key-terms of philosophy. Our example might be generalised. It indicates, at any rate in a preliminary way, the existence of alternative routes, the provision of alternative metaphysical centres, governed by the initial adoption of a starting point under the guidance of certain ideals, often suggested by the technical preoccupations of an historical period. The different metaphysical centres lead to alternative compensatory descriptions of identical material. Thus in the present case, omitting the finer details and provisos which our subsequent account will unfold, we find that both sides recognise mathematics and mathematical physics as an ideal of knowledge. Equally, both sides see that there is a difference of some sort or other between the domain of the statements of mathematics or logic and the domain of the realm of sensory fact.

But here the two schools diverge. For one side construes this difference as one of degree, the other as one in kind; a construction which since it involves matters *of principle*, 'hidden' so to speak beneath the logical and epistemological facts of the situation, must be regarded as being metaphysical. And it is these metaphysical differences which yield alternative 'weightings' of the identical logical situation as it presents itself to the contending schools.

Thus, basic for the rationalist approach, we shall find, is the central theme that, in Descartes' words, 'sensings are in truth merely confused modes of thinking'.[1] In plain language this means that the domain of sense is in some way (*though only in principle*) reducible to that of mathematical structure. But only, of course, in principle, since we have seen Descartes recognise the epistemic difference between mathematical knowledge and the 'teaching of nature'. Nevertheless, the metaphysical principle has the consequence that Descartes is more optimistic concerning the possibility of rational scientific knowledge, its significance and extent, and he is hence less troubled by the epistemic recalcitrance of sensory facts.

On the opposite side, Locke clearly assumes (though the principle appears as an explicit statement first only in Kant)[2] that the difference is one in kind, and no reduction in principle of one to the other is

[1] *Meditations*, VI; N. Kemp Smith, *Descartes' Philosophical Writings*, London, 1952, p. 257. (This edition will be referred to as K.S.) See also Leibniz, *New Essays*, iv.3.16, 6.7. For this doctrine in Descartes, cf. Ch. III, pp. 135, 166n.5, 179.
[2] Cf. *Prolegomena*, para. 13, Note III; ed. P. Lucas, Manchester, 1953, p. 47: 'Sensibility is not merely a confused kind of intellectual representation . . . it does

recognised. In consequence the epistemic limitations ('lack of certainty') of sensory knowledge constitute a greater worry for Locke and his followers; they are tempted to emphasise these limitations in a sceptical manner, whilst yet—paradoxically—their description, being fashioned by reference to the ideal of mathematical knowledge, makes Locke appear to construe the difference as one of degree, the very opposite of his true view. His belief in the generic distinctness of the two realms, coupled with the old and unconscious adherence to the ideal of rational knowledge, has the complex result of his attributing diminished importance to the pragmatic utility of the latter, whilst placing greater weight on the former; the 'rationalist' ideal thus sounding more and more like an empiricist scepticism.

The two opposing metaphysical principles thus result in the two sides describing their almost identical appraisals of the primitive logic of the situation in different ways. And yet, these two different metaphysics are themselves no more than quasi-axiomatic affirmations erected on the basis of some selected aspects of the logical and epistemological situation, itself not a matter of principle. So in the end, the outstanding feature is that there is here not so much straight opposition about matters of empirical fact or logical theory, but different responses, and resulting from this, different descriptions of the remaining and often conflicting features of the logic of the situation. There is thus certainly not just a gradual uncovering of philosophical 'truth', but rather *alternative* possibilities of response, taken up successively and in turn.

Still, though the idea of the gradual discovery of philosophical truth is too naive and misleading, this is not to say that one may not discern a *development* and even 'progress' in the history of philosophy. Clearly, there is development in the sense of a gradual exhaustion of possible alternative answers to a given problem, at least over a limited period; and furthermore, there is a gradual heightening of the degree of sophistication in the nature of the solutions offered. To take another example from the philosophy of science, the problem of the connection between the inductive foundation of natural laws and the

---

not consist in this logical difference of clarity or obscurity but in the genetic difference of the origin of knowledge itself.' Or, as it is put in the *Critique of Pure Reason*: the difference between 'sensible' and 'intelligible' is not that between 'clear and confused', but 'concerns their origin and content' (N. Kemp Smith translation, London, 1953, p. 84). For another Kantian passage in this vein, cf. below, Ch. III, p. 179n.3. See also Ch. VIII, sect. 5(b).

putative 'necessary connections' which such laws are often said to express. Here philosophy has almost exhausted all the possible solutions; some thinkers arguing that induction presupposes necessity (Aristotle, Ewing), and some that it does not (Braithwaite); some that there is necessity but no induction (Popper), others that there is induction but no necessity (Mill); and others again that there is neither (Schlick)—and so on. Here again, as remarked already (though the story extends beyond the chronological confines of our period), the actual logical nature of the very questions that are asked only gradually reveals itself.

Though the problem of induction emerged in its modern form during the classical period (in the writings of Hume), it does not at first play a very dominant part. Or rather, it is dressed in a different terminology, and occurs in a more general context, not unconnected with the problem discussed previously, that of the relation between sense and thought; it is the search for a satisfactory appraisal of the relation, or connection, or bond between the subject and the predicate of the ordinary S–P proposition.

Briefly and summarily, the topic emerges as follows. Scientific propositions, in particular those expressing law-like relations are, in the opinion of most of our classical philosophers, paradigm cases of the existence of special connections between the terms of these propositions. Speaking formally, the aim of science from the start was construed as being the establishment of connections whose degree of 'tightness' and 'indefeasibility' could be regarded as a measure of the success in arriving at the first principles of such a science—the models of mathematics and of logic on the one hand, and of theoretical reasoning on the other being of paramount importance in guiding the views concerning the nature of the connection involved.

The ideal existence of such connections is subsequently criticised by the 'British empiricists', at least in regard to the case of law-like propositions, e.g. those expressing causal relations. On the other side of the fence, some 'continental rationalists', Leibniz here being the central figure, taking the notion (and problem) of a propositional link more seriously, generalise the enquiry, by extending the problem beyond the case of 'universal' or 'law-like' propositions, to cover all propositions as such, whether general or particular. Finally, in Kant's 'critical purge', the empiricist criticism of causal connections is likewise extended to embrace the question of propositions of any kind, a special justification being demanded of the very concept of the 'propositional link'; and the Leibnizian postulation of the universal existence of this link (in the 'principle of sufficient reason') is subjected to a critical investigation and finally re-interpreted in an entirely novel way.

These are again the bare outlines of our guiding theme. In the next chapter I shall fill in the details to show how it arises and how it connects with many of the central ideas and problems of the philosophy of science. For the moment, we note that Descartes speaks of the 'conjunction' or 'union' of 'simple natures' which may be either 'necessary' or 'contingent'.[1] Locke's doctrine centres on the problem of the 'connection between our ideas',[2] and Berkeley and Hume echo this terminology. Leibniz, in a more 'grammatical vein', refers to the nature of the 'connection between subject and predicate' in many of his writings, e.g. *On the General Characteristic*,[3] and *Discourse on Metaphysics*;[4] elsewhere he speaks of a 'connection of the terms of a proposition'.[5] Finally, Kant, as is well known, investigates the question under the heading of the relation of the predicate to the subject;[6] or—more fundamentally, as we shall see—by discussing 'what is intended by the copula "is" '.[7]

---

[1] *Rules for the Guidance of our Mental Powers*, K.S., Rule XII, pp. 67–8. (Hereafter referred to as *Regulae*.)

[2] Cf. *Essay*, iv.3.10.

[3] G. W. Leibniz, *Philosophical Papers and Letters*, L. E. Loemker ed., Chicago, 1956, vol. i, ch. 24, p. 348.

[4] Op. cit., pp. 471ff.

[5] 'Letter to Arnauld', op. cit., p. 517.

[6] *Critique of Pure Reason*, trsl. Kemp Smith, p. 48. (This work will in future be referred to as K; thus here: K. 48.)

[7] Cf. K. 158–9.

II

LAWS OF NATURE AND THE PROPOSITIONAL LINK

I. PHILOSOPHICAL MODELS

I have suggested that one of the great central topics of classical philosophy is the discussion of the nature of the propositional link, in particular, the link that relates the subject and the predicate. It is a topic which will lead us to 'beat about all the neighbouring fields', yet serve as the thread guiding us through the labyrinth of the argument. When discussing the propositional link, we need not necessarily assume that there are privileged propositional forms, as for instance the subject–predicate form which most pre-twentieth-century philo-sophers singled out for discussion, to the exclusion of all others, such as forms involving relations. Since ultimately our concern will be with the 'ground' of the connection between propositional terms, a question involving primarily epistemological issues concerning the 'connection between our ideas' (in Locke's terminology), there is no reason to bar relational propositions from consideration. It is true that traditional philosophers excluded these, and indeed sought to reduce all propositional forms to the subject–predicate structure (S is P, or S–P for short), 'relations' being declared 'unreal'. *Per contra*, it may seem that by concentrating on the S–P form, we are creating artificial hurdles, and in general prejudice the discussion of the various issues involved.

I think that such fears are in general groundless, since the kind of logical problems generated by the S–P form tend to recur even when we take the less exclusive view. To cite just one example, no one used to be more insistent than Bertrand Russell that Leibniz's assumption of the primary status of the S–P form had been chiefly productive of his special metaphysical doctrines such as his theory of substantial

13

'monads' and his view of the 'analytical' nature of all propositions ascribing attributes to these substances.[1] Since Russell on general grounds held that it is wrong to attempt to reduce relational to subject–predicate forms, it was easy to infer that Leibniz's metaphysical formulations suffered *in toto* from this 'uncritical' assumption. Notwithstanding, Russell eventually had to face in his own later epistemological writings much the same problems as those encountered by Leibniz. Thus, in his *An Inquiry into Meaning and Truth*, the basic picture is that of 'a bundle of compresent qualities', regarded as a 'complex whole' (W), a proposal which is supposed to eliminate the need for any reference to the notion of 'substance', though surprisingly producing once again a doctrine which makes all judgments concerning the W's 'analytic', or—as Russell calls them—'judgments of analysis'.[2]

Now we are not concerned here with the details of Russell's theory, but only with the fact that the logical problems involved in explicating W are not unlike those encountered by Leibniz's notion of 'substance'.[3] Admittedly, this is only an *argumentum ad hominem*, but it may suffice for the present. It is at any rate sufficient for us to agree that the S–P form does accommodate itself rather easily to a representation of the problems surrounding the propositional link as a significant question, and it is advantageous from the point of view of the history which follows that it mirrors the conditions of what has been called 'substantival thinking',[4] illustrated by certain monistic philosophies. Thus F. H. Bradley argued that Reality was the ultimate subject of judgment.[5] More generally, when thinking of 'things' or even of 'corpuscles' as manifesting themselves through certain 'attributes' or 'qualities', it is particularly easy to 'imagine' the 'connection' between attribute and substance. For it does not need much imagination to think of the attribute as being 'linked' or 'attached' to the subject in virtue of the 'nature' of the subject.

It would be invidious at this stage to anticipate how we are to envisage this 'link', say for the case of the S–P proposition. Certainly

---

[1] Cf. B. Russell, *The Philosophy of Leibniz*, London, 1937, especially pp. 11ff. See also below, Ch. VII, sect. 4.

[2] Op. cit., London, 1940, p. 129.

[3] Cf. op. cit., pp. 98ff., 127ff.

[4] Cf. below, pp. 50–1.

[5] Cf. *Principles of Logic*, Chs. i and ii, Oxford, 1928, vol. i; and *Essays on Truth and Reality*, Ch. ix, Oxford, 1944, pp. 253–4.

one should not imagine that such a link necessarily corresponds to the 'is' of predication (in 'S is P') since the *verbal* expression of the copula 'is' can always be eliminated. (Cf. the expressions 'fx' and 'fa' of symbolic logic; or the employment of the device of inflection for predication, such as when we say 'The sun shines'.) On the other hand, it is much more difficult to indicate in positive terms what predication *intends*; for this depends very much on the view one holds on the logical (and ontological) nature of entities such as substance, attribute, quality, and the like. For that reason, any answer must await a discussion of these.

The same remarks hold, *mutatis mutandis*, if we wish to operate with relational forms. One cannot without further argument draw any 'ontological' conclusions from the grammatical structure of a relational proposition such as aRb, where a and b name certain 'things', and R might be taken to correspond to 'a relation', as though functioning like the 'substantives' a,b. This was, indeed, one of the basic denials of Wittgenstein's *Tractatus*: the complex sign aRb only *describes or shows* that a and b stand in a certain relation to one another; it does not *say* that a stands in the relation R to b. *A fortiori*, we cannot ask 'substantival' questions concerning R.[1] But a theory like Wittgenstein's (itself controversial) comes at the end of a long road of attempts to construe the propositional link in a variety of ways. And we must therefore proceed to consider the question in an unbiased way.

There are at any rate some further initial hurdles, concerning the appropriateness of the notion of a 'link' as such, and the corresponding one of a 'ground'. Take the proposition: 'The steel bar is fractured'; and suppose we asked what was the ground of the predication, in virtue of which 'being fractured' is asserted of the steel bar.

To this question several answers are possible, corresponding to the different logical sorts of 'ground'. First, one may give an *explanation* of the fact; and such an explanation may be of various kinds, which we need not explore for the moment. Common to them is that the fact in question 'follows' from the explanation (granted certain conditions). Secondly, one may state the observational or experimental *evidence* which supports the original assertion. For instance, one may point to the crack in the steel; or to the sudden lengthening in the bar as shown by the excessive movement of the indicator of some attached measuring instrument. This sort of situation led to the traditional

[1] Cf. *Tractatus Logico-Philosophicus*, London, 1961, 3.1432, 4022.

distinction between the '*ratio essendi*' and the '*ratio cognoscendi*'; or the corresponding and more recent distinction between 'constitutive' and 'epistemic grounds'. Thus the real ground of the bar fracturing might be a change in its chemical composition; whereas 'our ground' for the assertion would be the sensory evidence, i.e. the visible crack, or the movement of the needle.

The distinction referred to is not as obvious as it may appear at first sight; and it is particularly dangerous to generalise it. If it merely repeats the distinction between 'explanation of' and 'evidence for', it is pertinent to remember that the 'explanation' must in turn be a function of 'sensory evidence' at some stage or other, even if we do not hold that the game of evidence must end with the mentioning of a set of pure sense-data.[1]

As we shall see presently, the distinction between 'constitutive' and 'epistemic' grounds has a vital bearing on our story. A progressive penetration, criticism and purge of this distinction runs parallel with the development of philosophy during our period. Suppose, for instance, we take a case where the explanation model (or 'systemic model', as we shall call it) does not seem appropriate. There are two such cases. The first is what we may call propositions supplying basic evidence. Let such a proposition be 'this substance s is coloured green'. What are we to understand here by 'the ground of the proposition'? What more could be the ground for this assertion but the epistemic basis for the 'judgment' which couples the two terms, s and g? Which is to say simply that the speaker relates the terms in virtue of what he observes. One might now reply, however, that whilst it was true that we judge that s was g in virtue of the evidence, this is quite distinct from s being g in virtue of the nature of s, or the nature of things (depending on whether one is a pluralist or a monist).

But suppose we reply by insisting that there was no more than meets the eye? Now I am not concerned just now with the arguments by means of which one would support such a contention. We shall look at them when discussing the ideas of John Locke. But it is clear that the distinction between constitutive and epistemic itself depends upon some fairly far-reaching decisions concerning the questions just

---

[1] I add this proviso, because it has become in recent years difficult to maintain a rigid and absolute distinction between 'theoretical framework' on the one hand, and 'sensory data', or 'observational foundation' on the other. Nonetheless, so long as there is a difference here in kind, however narrow the gap, we may continue to retain a distinction which lies at the back of the whole of traditional classical philosophy.

broached; e.g. whether or not the universe is resolvable into a kind of 'logical atoms', as presented by the notion of 'simple idea' in the writings of Locke and Hume.

The second case which I want to consider here, though again only in preliminary fashion, is that of 'general propositions', including those expressing 'laws of nature'. Suppose we take 'All metals expand when heated'. Here it is not difficult to distinguish the explanation of the fact and the evidence for it. But quite apart from the significance (for ontology) of being able to offer an explanation, suppose we assumed for the moment either that there was no explanation, or that any explanation in the end reduces to the filling in of 'gaps' in our knowledge (or between the phenomena), so that we are after all left with a number of simple, not further explainable, states of affairs. The epistemic ground for the generalisation would then be simply the instances lying at the back of it, and which support it.[1] We might put this roughly by saying that our reason for asserting the generalisation is the instances, together with certain inductive principles involved in making the generalisation, e.g. that the sample was a fair one, etc.

Now however paradoxical it may appear at first sight, it is just at this point that many logicians feel strongly tempted to insist again on the constitutive-epistemic distinction—largely, it seems, for reasons concerning the inductive foundation of scientific knowledge.[2] 'There must *be* a ground, relating the facts in certain appropriate ways, so that we may consider ourselves entitled to assert the generalisation': this will be the contention. Now as we have just seen, this does have a straightforward sense in certain obvious cases. Thus, that the volume of a gas varies (approximately) inversely to its pressure (if moderate) at constant temperature, we *know* because we can make the experiment. Also, *the* ground, the 'reason for the fact', is supplied by the kinetic theory of gases. But once again, the game of 'reason-giving' must come to an end; or rather, the giving of reasons lives within the framework of some theory or other. Outside this, the notion of a 'constitutive ground' is as yet problematic. It is the exploration of a variety of possible responses to this problem, involving a number of metaphysical positions, which once more marks the development of classical philosophy.

[1] Not *just* the instances, of course, since the induction involved will bring in subsidiary considerations, such as that the instances are not all identical, but involve merely resemblances, i.e. identities and differences. But the complexities of induction may for the moment be ignored.

[2] Cf. below, our discussion of Johnson, pp. 27ff.

We may think of the cases where one speaks of the existence of a link, or a connection, and of grounds for such links or connections, as so many *models*. The models which we have mentioned, such as the explanation or systemic model, the substance-model, or the model suggested by the existence of generalisations or laws (let us call it the 'nomothetic model'), do not exhaust the list; thus there are more formal cases, where the existence of a connection between subject and predicate may be pictured by the propositions of logic, or by analytical propositions. Here, the history of the subject is illuminating because of the way in which some of these models throw a shadow on cases where they are not really appropriate. In such situations, an inappropriate model may prevent one from realising that actually there *is* a 'problem' when thinking in terms of a link at all. Or again, philosophers may be so dominated by certain models that they will banish the rest from the scene altogether. In this way, for instance, Hume refuses to speak of any 'objective' connections between terms (in any serious sense) except in the case of analytic and similar propositions. And on the other side, a philosopher like Leibniz will be so much under the domination of the explanation model that he will insist on the existence of 'links', and 'grounds' for these, in cases where the notion seems to be doubtfully applicable. In the end, we shall see how this double-sided situation will provide us with a key to unlock the real significance of the Kantian doctrine, which, whilst retaining the Leibnizian dogma of the universality of a link, will, through the 'critical method', as it were 'anaesthetise' that link.

We shall now explore in greater detail these three groups of models which have been introduced so far only in passing: (a) The formal, *a priori* group; (b) the nomothetic group; and finally, (c) the systemic or theoretical explanation group. The pursuit of my central topic thus leads us further afield, into the problems surrounding the logical nature of scientific law and scientific explanation. In this way, both central and peripheral, general and specific philosophical reflections will mutually illuminate each other.

In introducing, as I have done, this variety of models, I should explain first the choice of the term 'model' itself. To this end, consider a member of the second group (nomothetic), which includes the case of 'causal connections'. The strongest version of the causal link, that of 'necessary connection', was put forward explicitly (and of course criticised) by Hume. And indeed, in the course of this and the next chapter, we shall see that this concept of necessary connection has played, and continues to play, an important rôle in the philosophy of scientific explanation and law. But equally we shall find that very grave objections can be raised to this conception of a necessary connection. Very frequently such objections are couched in ontological language, when it is said, for instance, that there are no necessary connections in nature;[1] or that the notion is meaningless in the sense that there is nothing corresponding to it in nature, that it has at best only a purely psychological significance, etc.

It is, however, not at all certain whether adherents to the notion of necessary connection are compelled to embrace such ontological implications, for it is an open question what precisely such necessitarian claims imply. Instead of assuming forthwith any such simple-minded ideas of a 'correspondence' between necessity and nature, we may for the time being simply ask what logical *functions* a necessitarian interpretation is intended to fulfil. More generally, we simply investigate what sorts of situations, what kinds of 'pictures' have been put forward, and which of these have been 'pictures' of 'necessary connections'; and what reasons, or even motives, have been put forward for these 'pictures' as representatives of necessary connections. In this way, then, talk of 'models' is meant to take the 'ontological sting' out of a conception which might otherwise be dismissed with impatience,—for lack of an understanding of its true position in the logical thought of science and scientists. The notion of 'model' when taken in this sense will prevent any sort of 'ontological commitment' (whether positive or negative) from slipping too easily into the discussion; too hastily, that is, before the whole idea of an 'ontological position' has been discussed in detail; a discussion, indeed, which lies precisely at the back of the great argument that leads from Hume, *via* Leibniz, to Kant.

The use which I am proposing to make of this notion of 'a philosophical model' is like that suggested by a passage in Wittgenstein's *Philosophical Investigations*, a passage itself concerned with the notion of non-logical, or 'natural' necessity.

Wittgenstein, in his earlier writings, had himself held a view similar to that of some modern 'Humeans'. In the *Tractatus* he had expressed this when writing that 'there is no compulsion making one thing happen because another has happened. The only necessity that exists

---

[1] Cf. Braithwaite, *Scientific Explanation*, Cambridge, 1955, p. 294: 'In nature there is no extra element of necessary connection.'

is logical necessity'.[1] During his later period, in the *Philosophical Investigations*, he rhetorically represents a similar, perhaps more radical, view, in sections 371–2, as follows:

371. Essence is expressed by grammar.
372. Consider: 'The only correlate in language to a natural necessity is an arbitrary rule. It is the only thing which one can milk from this natural necessity into a proposition.'

But this is now followed by an objection which states his own later view:

374. The great difficulty here is not to represent the matter as if there were something one couldn't *do*. As though there really were perchance an object, from which I peel off its description, but then were unable to show it to anyone.—And the best I can propose is I suppose that we yield to the temptation to use this picture; but now investigate how the *application* of this picture goes.[2]

In this very important passage we have a clear indication of the way in which Wittgenstein draws the 'ontological sting' from one of the key metaphysical notions of philosophy, by telling the philosopher to 'yield to the temptation' to 'use the picture', by investigating the 'applications' which it finds in various contexts. In this spirit, we must now investigate the various models in terms of which the link binding the terms of a number of different propositional types has been envisaged.

2. THREE MODELS OF THE PROPOSITIONAL LINK

(a) *The formal and* a priori *group*

A well-known species of this group is 'analytic propositions'. First named in systematic fashion by Kant, they were recognised as such by earlier writers, though an acceptable account of this type of proposition is still a matter for debate. Such propositions are 'true' (formally true) in virtue of the existence of formally accepted definitions of the terms involved. 'All bachelors are unmarried' is a stock example. 'All bodies are extended' is Kant's example, and perhaps already slightly more controversial. (Such propositions are contrasted with so-called 'synthetic propositions' whose precise definition and significance we must defer until later.[3]) A second species is the

[1] Op. cit., 6.37.
[2] Trsl. G. E. M. Anscombe, Oxford, 1953; italics in text. I have slightly modified the translation.
[3] Cf. below, pp. 69f.

formally decidable axioms of truth-propositional logic, i.e. 'tautologies'. Yet another species are certain geometrical axioms, e.g. Euclid's axiom that things equal to the same thing are equal to one another.

Formal propositions of this kind are often said to be 'logically necessary', their contradictories being inconceivable; but as with the case of analytic propositions this is not a rigorous account of the term, if only because a distinguishing mark of the classical period of philosophy is that a precise connotation does not as yet exist. Precision at the start would hence defeat our object, quite apart from the matter still being controversial.

In the examples given, 'truth' is not in any direct way contingent upon the results of investigations into empirical matters of fact; it does not depend upon 'sensory experience'. What impressed the traditional philosophers was in fact a feature with certain psychological overtones (though one would not want to insist that it was *simply* a 'psychological matter'!): such propositions seemed instances of a more general group, all certifiable through sheer reflection on, or contemplation of, the terms involved; moreover, they possessed a peculiar kind of 'self-evidence'; often the term 'intuition' was used in this connection. From this there seems to have derived another label, that of the '*a priori*', as not being contingent upon sensory experience, contrasted with the *a posteriori* type of proposition.

Once again, we shall not give a rigorous analysis, let alone history, of this concept. Here, even more than in the previous case, it was precisely only as a result of the philosophical developments during our period that the concept of the *a priori* became in any way articulated; and here, also, the end is not yet in sight, though there was a temporary stopping-place in the philosophy of Kant. But it *is* important to realise that the term does not invariably have a rigorous meaning during the early years of classical philosophy. Thus, to cite just one example, when Descartes was attacked on the score of certain seemingly arbitrary assumptions made in his *Dioptrics* in the course of a proof of the law of refraction, he replied heatedly that he could hardly be at fault since he had after all 'demonstrated refraction geometrically and *a priori*'.[1] Nonetheless, Descartes had previously

---

[1] 1.3.1638. Cf. Ch. Adam and P. Tannery, *Œuvres de Descartes*, Paris, 1956, ii, p. 31. For the 'assumptions' and details of the 'demonstration', cf. below, ch. III, sect. 2(d), pp. 142–3.

made a special point of the fact that in the *Dioptrics* he had employed only a number of models ('comparisons') by whose aid he had given a successful theoretical derivation of the laws of geometrical optics, all independently and empirically verifiable. When subsequently questioned on the use of the term '*a priori*' in the above letter, he even insists in an emphatic manner that this is the only way to proceed in physics,[1] i.e. by way of the deductive development of *hypotheses*, for the purpose of explanation, the success of which could at the same time be counted a valid 'proof' of the hypotheses themselves.[2] Here it looks as though by an '*a priori* method' in science was meant the employment of the method of hypothesis, or hypothetico-deduction! And no doubt that was part of the meaning of the term: geometrical demonstration and theorising were here being contrasted both with purely observational ('empirical'—another thoroughly ambiguous term at this time) methods, or again, with methods of loose and *ad hoc* forms of theorising, not employing carefully contrived deductive formalisation.

But of course, this was not the sole meaning, if only because Descartes meant several things by 'hypothesis' and had special views of their relationship with the fundamental parts of mechanics on the one side and 'metaphysics' on the other. So we see again that any tensions in Descartes' methodological thinking would be mirrored in the logical grammar of a term like '*a priori*'. There is here a complicated relationship between the ambiguities of technical terms and their special philosophical meanings when tightened by philosophical enquiry.

Our discussion may have suggested that *a priori* propositions include only the group of formal (analytic, logical, etc.) propositions alluded to at the start. This is not the case. In the history of philosophy there are many propositions of an *a priori* nature which have been assented to on grounds other than their being analytic, or logical, or in some sense at least formally necessary. A celebrated case is the Euclidean axiom that two straight lines cannot enclose a finite space; or as an alternative version, the famous axiom of parallels. Again, Descartes, we shall find, will invoke a special intellectual capacity, in the 'light' of which (he speaks of 'the natural light'[3]) he claims to be able to see the unimpeachable truth (a truth

[1] 27.5.38. Ad. and Tan., ii, p. 142.
[2] Cf. *Discourse*, Ad. and Tan., vi, p. 76; K.S., p. 162.
[3] *Meditations*, III, K.S., p. 217, 219.

which we 'cannot anywise call in doubt'[1]) of certain propositions, such as 'that there must be at least as much reality in the efficient and total cause as in its effect'.[2] In an earlier work he had described the case of certain 'simple natures' where, so he claims, the concept of one entails the concept of the other, i.e. 'we could not conceive either distinctly, should we judge that the two are really apart from each other. Thus shape is united with extension, motion with duration or time, etc.'[3] Another 'necessary combination', cited as example, is that 4 and 3 are 7. It would be invidious to make a decision at this stage concerning the type of *a priori* proposition all these examples represent. Besides, this does not exhaust the list because there is another range yet to the notion of the *a priori*, as employed by Kant, where it overlaps to some degree with the concept of the 'transcendental', i.e. with the concepts and principles presupposed for the possibility of making certain kinds of cognitive judgments.

Whatever it is, both the notion of the formal, as well as of the *a priori*, lend themselves admirably as models in terms of which to conceive of the 'link' uniting the terms of the respective propositions. Indeed, Descartes himself already employs this very term when discussing certain axioms, such as 'things which are the same as a third thing are the same as one another'. Such axioms express (though according to Descartes, they are not required to *justify*) 'links (*vincula*) for the connecting together of the other simple natures'.[4] The notion of the *vinculum* is indeed a useful picture by which to summarise the various grounds on which the terms of *a priori* propositions of the type considered are united. Nor can there be any doubt of the significance here in using the notion of a link. Speaking still in picture fashion, the link represents the connection between the terms of the proposition; and it certifies our justification for propounding it.

Not surprisingly, early scientists and methodologists were fascinated and attracted by this type of proposition, for it has precisely some of those features with which they surrounded the most cherished bases of their scientific systems. As we shall see in Chapter IV, it was also the model that represented for Locke the ideal of truly scientific knowledge—albeit hardly ever reached. We have seen that the non-experiential and conceptual or intuitive character of the *a priori*

---

[1] K.S., pp. 217–18.   [2] K.S., p. 219.
[3] *Regulae*, XII, K.S., p. 67. At p. 68, more problematically, but important for later discussions, he gives the case that to doubt all things entails to know that one doubts.
[4] Ibid., pp. 65f.

propositions seemingly made them immune from the contagion of empirical falsifiability and thus guaranteed their unchallengeability as well as their 'universality', bestowing upon them a kind of 'necessity'; *both* characters which Kant was still to note as marks or 'criteria of *a priori* knowledge'.[1] It was clearly very tempting to seek the foundations and the basic statements of science among propositions possessing an *a priori* strength—a temptation to which not a few philosophers and quite a number of scientists yielded, and still yield. Nor, in the light of the admitted vagueness and ambiguity of the notion of the *a priori*, is this to be dismissed as a forlorn attempt *ab initio*.

Many efforts to utilise the *a priori* or rationalist model followed upon the merging of the Cartesian and Newtonian traditions during the eighteenth century.[2] Thus D'Alembert, a distinguished mathematical physicist of this period, in his *Traité de dynamique* asserts that 'the laws of equilibrium and of motion are necessary truths'.[3] Here, too, we must remember that terms like 'law' and 'necessary' do not have a fixed and invariable grammar. We have to see what grounds are given for the assignment of this terminology. The famous mathematician Euler, for instance, at about the same period (1760), declares Newton's First Law of motion, the so-called law of inertia, to be 'on equal footing with the truths of geometry'.[4] His point is that one may arrive at these laws 'by the path of reasoning' alone.[5] Concerning the 'observation' that the direction and speed of a body remain unchanged in the absence of interfering forces, he produces the following argument:

> In fact, one could not conceive why a body should turn aside from its path towards one side rather than another; hence, since nothing happens without reason, it follows that the body in question will always conserve the same direction, or that its motion will take place in a straight line.... In the same manner the speed of such a body could not change ... [for] there is no reason at all which could produce such a change.[6]

We shall discuss this kind of argument when considering Descartes' version of it. Here we only note the assimilation of a physical question

---

[1] Cf. K. 43–4.
[2] See my *The Image of Newton and Locke in the Age of Reason*, London, 1961, i.1.
[3] Op. cit., pp. 10, 60–1.
[4] *Lettres à une Princesse d'Allemagne*, Paris, 1845, p. 186.
[5] Op. cit., p. 187.   [6] Op. cit., p. 188.

to one of mathematics, logic and metaphysics; Euler's proof purports to employ purely conceptual considerations, both by the insistence that spontaneous change of velocity is *inconceivable*, and its invocation of the principle of sufficient reason.

As an aside, we may note in conclusion an interesting and much more recent case which occurs in the work of A. Landé, a physicist who has made contributions to the foundations of quantum mechanics. In his *From Dualism to Unity in Quantum Physics*, he maintains that the aim of a genuine '*explanation*' (e.g. of the wave-like statistical behaviour of particles) must be 'on the grounds of "principles nearly all so evident that one only needs to understand them in order to assent to them" '.[1] (The expression which he quotes is itself taken from Descartes.) Later in the same book this is explained to mean that wave-particle duality should be 'reducible to simple, almost self-evident physical ground-axioms, so that we can recognize it as a necessity rather than an oddity'.[2] The conception of 'self-evidence' here involved is of course odd. It is very likely that Landé's approach belongs more appropriately to the next group of models (nomothetic), and in particular, to the aspects mentioned on pp. 40ff. below.

Before proceeding to this, let us remind ourselves of the nature of our procedure. We are here only concerned (by way of a number of alternative models) with attempted characterisations of the propositional link. Thus, in the last example, we see that this link may be construed on the model of *a priori* propositions. In the next section, we shall find the link construed like the necessary connections putatively involved in scientific laws. It is, however, an entirely different question whether such a model does in fact rightfully apply; for instance, whether we are ever justified to construe the propositional link in an *a priori* fashion in any given case, a question which lies at the centre of the debate during the period we want to discuss. (Thus, Hume will argue that the *a priori*, or again, the analytic, model is not applicable to empirical generalisations.) There are in fact always three matters to be discussed: what are the models involved; how are such models to be construed or articulated; is a given model appropriate for a characterisation of the proposition in question? In this chapter, we are primarily though not wholly concerned with the first two questions; in the subsequent chapters, the last question will be more prominent.

Closely connected with the question: Does the model apply? is the

[1] Op. cit., Cambridge, 1960, p. xv.   [2] Op. cit., p. 41; cf. also p. 49.

question: How does the choice of model affect the *truth* of the proposition? This question is of importance in the case of the empirical generalisation of science; and it is again controversial, forming another part of the seventeenth- and eighteenth-century philosophical debate. In the course of this debate, moreover, the whole notion of scientific truth comes to be articulated in a very complex fashion, ending (in Kant) with the distinction between the probability on the one hand, and the rationality and possibility or intelligibility (let us refer to these jointly as 'plausibility') of scientific hypotheses on the other; a distinction which was subsequently lost, and which has been resurrected only in our own days. For the moment, it will be sufficient to say that the characterisation of a proposition through one model or another *may* be kept distinct from the question of its bearing on either the probability or the plausibility of hypotheses. There have in fact been philosophers who have kept these two aspects distinct quite deliberately. Thus, W. Kneale, in *Probability and Induction*,[1] construes scientific laws as principles of necessitation, and regards the concept of necessity involved on the lines of the *a priori* model. On the other hand, their inductive acceptance is made a function of certain inductive policies, formulated on semi-pragmatic grounds. Now it is of course controversial whether an *a priori* construction can be combined with an inductive foundation of this kind. But for us at present it is not necessary to come to any decision on this head, since we are only concerned with the neutral question of characterisation of alternative models. With these cautions in mind, let us turn to a discussion of the nomothetic model.

(b) *The nomothetic group*

If the previous group of models suggested a logical or conceptual bond uniting the terms of a proposition, the model to which we now turn involves (ultimately at least) something more in the nature of ontological considerations. We are here concerned with the semantic analysis of the concept of law, as used in the various natural sciences. Such an analysis has only been attempted in very recent years, and even here there is no final agreement; not very surprising since we shall discover the presence of metaphysical elements in the analysis. Recent work, however, is useful in enabling us to approach earlier writings.

[1] Oxford, 1949, *passim*. This is an important text which we shall use throughout as a modern equivalent of many of the classical views which we discuss.

We need not enter here into the details[1] but the following could be accepted as a reasonable shorthand account. Scientific laws, more specifically those that served as paradigms for the philosophers of the classical period, usually contain a high degree of conceptualisation. This means that their constituent terms carry a greater than usual content of theoretical superstructure, being frequently constructs of a highly complex nature. At the same time, such constructs make up the 'elementary scaffolding' of the physical system under consideration; witness components like the 'elements' and 'compounds' of chemistry; the 'atoms' and 'molecules' of physics; notions such as force, mass, energy, entropy, and so on. So much for structure. Functionally considered, propositions expressing laws, or being 'about' lawlike situations, whether of the qualitative or quantitative type (e.g. involving 'functional relations'), are intended to serve with unrestricted universality with respect to the class of objects involved, as well as with respect to spatial and temporal scope. They are no mere summaries of what has happened in the past, of the states of affairs that might for instance be offered as *evidence* in favour of such laws; for they claim predictive power, concerning what will happen on *any non-pre-assigned* 'future' occasion. (Hence many would hold that the proper logical form of such a law is best expressed through the hypothetical-conditional 'if-then', rather than the categorical 'all . . . are'.)

For a number of logicians, such an analysis does not however go far enough. For they would maintain that the 'compulsive nature' of statements of law is not sufficiently indicated by mere reference to, say, unrestricted universality. Mill drew attention to such an additional feature when he pointed out that the generality of 'causal' laws was not just that of invariability of succession but (in addition) of unconditionality.[2]

After Mill, the most influential English writer on this issue was W. E. Johnson.[3] I purposely choose Johnson because although many philosophers since his time have paid lipservice to the need to postulate a 'surplus-element' in the analysis of lawlike statements,

---

[1] See for instance, Braithwaite, op. cit., ch. ix; Ernest Nagel, *The Structure of Science*, New York, 1961, ch. iv; Arthur Pap, *An Introduction to the Philosophy of Science*, Glencoe, 1962, ch. xvi; Mario Bunge, *The Myth of Simplicity*, New York, 1963, ch. x.

[2] See *Auguste Comte and Positivism*, London, 1865, p. 57.

[3] *Logic*, Cambridge, 1924, vol. iii, ch. 1.

few have brought out so clearly the relevance of this for 'ontology'.[1]

Johnson distinguishes between 'universals of fact' and 'universals of law'. A 'logical formulation' of a law such as 'all metals expand when heated' would be—not: 'Anything, being a metal, *expands* if heated'; but: 'Anything, being a metal, *would expand* if it *were* heated.'[2] Such an analysis, it will be seen, extends the range of law into the modal realm 'beyond the actual into the range of the possible';[3] from what has been, is and will be the case, *as a matter of fact*, to what could be, might be or might have been. Now for Johnson at least, to this grammatical change there corresponds a point of view, involving what he calls 'an ontological distinction'.[4] The two points of view referred to at this place are 'the epistemic' and 'the constitutive points of view'.[5] Important as it may be, this is not an easy distinction to maintain; and though Johnson seems to have adopted it without question, its problematic nature becomes obvious when we consider that the outcome of the historical revolution which began with Hume, if not Berkeley or Locke, and ended with Kant had as a result precisely the denial of any 'dogmatic' distinction of this kind; a result (we note in passing) that has evidently been lost again in the sequel.

It is of course easy to bring out what Johnson *meant* to assert, even if it is difficult to make the distinction good. For the phrase 'empirical uniformity', used by Mill as supplying a proper rendering of 'law', involves (according to Johnson) a confusion, or ambiguity of points of view. If we understand it 'epistemically', 'empirical uniformity' connotes no more than the generalisation of the data, i.e. of the observed instances on which it is based. On the other hand, the extended modal analysis mentioned before makes it clear that the phrase 'empirical uniformity' has in addition to an epistemic, also a constitutive significance. That Johnson refers to this as an 'ontological distinction' implies that whilst 'generalisation of the data' is for him a

---

[1] But cf. R. Chisholm, in his classic paper on 'The Contrary-to-Fact Conditional', who after having failed to 'get rid of the subjunctive' in the analysis of law-like statements of the 'non-accidental' kind, concludes that this shows that the terms of such propositions are 'connected' in a special way: '*Connection* becomes an irreducible ontological category and a source of embarrassment for empiricism'. (In H. Feigl and W. Sellars, *Readings in Philosophical Analysis*, New York, 1949, p. 496. Chisholm's article is reprinted on pp. 482–97, from *Mind*, 55, 1946.)

[2] Cf. op. cit., p. 6.   [3] Ibid.

[4] Op. cit., p. 5.   [5] Cf. above, p. 16.

purely logical phrase, the existence of the modal function seems to point to something extra-logical; a claim, however, for which no separate argument is put forward. What we find instead is the contention that the existence of a constitutive aspect gives the generalisation 'validity in an objective sense'.[1] And indeed, when discussing the ontological status of causal relations in general, Johnson distinguishes 'the subjective or epistemic from the objective or constitutive relation'[2]—indicating thereby the way he views the constitutive relation.

We may conclude that the genuine motive for the postulation of a constitutive connection is the (supposed) need to supplement the logical grammar of lawlike statements, previously elucidated, by an ontological supplement, a constitutive addition, which is meant to bestow 'validity in an objective sense'. (At this point, hence, the logical analysis of laws merges into their epistemological justification!) But as already mentioned, the whole notion of the 'objective or constitutive' is precisely what requires a critical analysis. For this reason, if no other, the discussion of it by the classical philosophers makes their renewed reading a matter of topical value.

There is, however, a further aspect of Johnson's terminology which is of great historical interest. We have seen that his analysis of 'universals of law' is couched in the modal form 'If any thing of some given kind were characterised as p it would be characterised as q'. Now the fact indicated by such an expression Johnson calls 'nomically necessary' (to be distinguished from 'logically necessary'). This terminology shows that Johnson is here trying to give expression to the 'compulsive' character of statements of laws, or laws for short. It is the (putative) existence of such a constitutive necessitarian link which supplies the second model for our discussion of the nature and meaning of the link between the terms of propositions. The existence of such a close connection between Johnson's original modal formulation and its subsequent necessitarian interpretation may be questioned. On the other hand, the large body of recent discussions concerning the so-called 'counter-factual' aspect of statements of law would hardly have raised so much dust if the problem of necessity did not lie behind it.[3]

---

[1] Op. cit., p. 5.   [2] Op. cit., pp. 4–5.

[3] On the notion of the counterfactual, see Chisholm's paper referred to above, p. 28. n. 1; also by the same author, 'Law Statements and Counterfactual Inference', *Analysis*. **15**, 1955, 97–105, reprinted in E. H. Madden, *The Structure of Scientific*

Consider the remark, 'All the books in shelf A were green'. As G. E. Moore points out in his *Commonplace Book*, suppose that all the books on a certain shelf A are green, it would be absurd to produce a book bound in red and to say 'If this had been in shelf A, it would have been bound in green'. True, there is a sense in which I could make such an inference, viz. if I implied that the book in question had been identical with one of the books that were on shelf A. But, Moore adds, though this follows quite correctly from the original statement, the sense in which it follows is 'not the causal sense that being put in that shelf would have changed its colour'.[1] One feels here very strongly that there is something 'compulsive' about this causal sense, the sense of the 'transitivity of the action'; though how this 'compulsion' is to be analysed is quite another question. But it should not be thought that because it is difficult to produce a satisfactory analysis, therefore it is otiose to speak here of compulsiveness at all. There may be nothing wrong with this 'picture'; nor is it impossible that we might discover that there is no satisfactory substitute for it. Moore himself seems to feel this, for in another place, again discussing causal generalisations, and giving the example, 'being shot through the heart causes death', he adds the analytical note: 'If he was shot through the heart, he is dead ("must be")'.[2]

It is on the whole true, that a *lawlike* generalisation about certain factors putatively associated in universal fashion is an association not existing simply as a pure matter of fact, of, as it were, 'accidental fact', however extended the range of the generalisation may be held to be; but that there is involved some sort of *relationship* between the factors concerned, which is most adequately expressed by some such locution as 'Given A, then B *must* happen'.[3]

Now there are of course serious difficulties which this idea of a nomically necessary connection has to meet. These difficulties were first uncovered by the seventeenth and eighteenth century empiricist

Thought, London, 1960, pp. 229–35, and the bibliography there; A. Pap, op. cit., ch. 15, with bibliography; and the Symposium between P. Alexander and M. B. Hesse, 'Subjunctive Conditionals', *Arist. Soc. Proc.* Suppl. Vol. XXXVI, 1962, pp. 185–214.
[1] *Commonplace Book 1919–1953*, ed. C. Lewy, London, 1962, viii.6, pp. 348–9.
[2] Op. cit., vi, 11, p. 271.
[3] Some recent writers are equally emphatic. A very extreme pronouncement, for instance, occurs in N. Hartmann, *Philosophie der Natur*, Berlin, 1950, ch. 33 (a), p. 393: 'The lawlikeness of the law of nature is strict, unbreakable, one which possesses necessity.'

philosophers, who themselves were acquainted with the work of scientists and philosophers operating freely with the notion of necessity. Both sides, however, although necessity plays such a central rôle in the logical and ontological arguments or criticisms, seldom spent much time exploring the 'natural history' of the idea itself. This is a deficiency we must remedy; the task is facilitated by much recent work clarifying the 'logical status' of natural law.

It is true that recent periods have dealt less and less sympathetically with the idea of necessity. An age like the present which has seen a succession of theories, each replacing its predecessor, and which has adopted a manner of speech about science which continually stresses the provisional and tentative nature of its hypotheses, is ill attuned to a necessitarian position. In the light of this, some comment on the contemporary tone of the discussion is desirable. In the first place, it must not be forgotten that we have stated so far no more than the content of a certain semantic analysis—a statement of ideals by certain philosophers on what the logical grammar of lawlike statements is felt to be, or should be. Furthermore, the question of the justification of such an analysis is not to be dismissed simply with the remark that most scientific conclusions are invariably and eventually superseded. For we can always make the necessitarian aspect immune from any such empirical challenge, simply by the consideration that, in the face of contrary evidence, a putative law (even with a necessitarian logic) will be abandoned. All this shows that the assertion that a certain statement carries nomic necessity is not tantamount to the claim of perpetual immunity from potential falsification; it is certainly not claimed that one *knows* in a special way that the *statement* in question *is* immune. (There is thus a great difference between formal or logical and nomic necessity.)

What muddles this issue is that the adoption of a law proceeds in the light of supporting evidence. And with this, overtones of certainty and necessity, even in everyday commonsense thinking about science, creep in. For we naturally believe that scientifically well-attested conclusions carry far greater certainty and authority than opinions gathered haphazardly. Scientific generalisations which are deeply embedded within a given theoretical background, or which themselves have the place of primary principles employed to organise such a background, have usually been treated here with particular respect, the term 'Law of Nature' being sometimes especially reserved for them. And the more such principles embody the 'style' of scientific thinking of a period, employing, say, action-at-a-distance notions at one moment, or the field picture at another, atomistic modes of representation, or *per contra*, organismic or evolutionary ideas, the more they will be regarded as quasi-compulsive schemata, if not self-evident explanations of the fabric of nature.

Nevertheless, it does not *follow* from this that the corresponding laws carry necessity in the sense required, however much the facts just mentioned may be the *cause* of such a feeling of conviction. However, one might maintain that it is a *sign* of a law that it has the support of a complex background of knowledge; and though such signs do not *entail* necessity, one might count them as criteria. (At this point, we should remember that we are searching for 'pictures' of necessity!) Let us follow up this matter in a more formal way.

Comparing nomically necessary propositions with the logically necessary, and more generally, the *a priori* type of the first group, we note some interesting differences. First, the assertion of nomic necessity is not coupled with the claim that the denial of such statements would be inconceivable, as in the case of logical necessity.[1] Most logicians would hold that our ground for asserting logical necessity makes the assertion itself necessary; clearly this does not apply to nomic necessity. Nor would anyone nowadays be prepared to invoke here a kind of 'natural light' by which he 'sees' such a necessity—somewhat like the pre-Kantian *a priori* type. On the other hand, the assertion of nomic necessity seems to function as a sort of justificatory device, its presence bestowing validity upon a given generalisation. This indicates that this kind of necessity has a certain significance for 'ontology', is indeed intended to have an ontological status—as is clear from Johnson's argument. This aspect will be dealt with later. For it has been part of the critical attack of the classical British philosophers (Locke to an extent, but more so Berkeley and Hume) to impugn this whole conception of 'necessity in nature'; whilst Kant's work may be regarded as an attempt to preserve the idea without being committed thereby to the old, and putatively indefensible position. For Kant certainly did operate with the notion (though not the term) of nomic necessity. For him, the whole 'order of nature' requires it. This order, he writes in the *Critique of Judgment*, is known through certain 'particular rules' which despite the fact that they 'can only be empirically known' (and are hence 'contingent'), yet 'must be thought' by the understanding 'as laws (i.e. necessary), for otherwise they

---

[1] This will be taken up below. See especially Ch. VI, sects. 7–8, pp. 368ff.

would not constitute an order of nature, although their necessity can never be understood or comprehended by it'.[1]

Now though the idea of a nomically necessary link is not founded in any purely conceptual approach, on *a priori* lines, this is not to say that conceptual matters are irrelevant to it; still less that scientists wantonly adopt laws possessing (putatively) such a necessitarian grammar. What confuses the issue is that many of the statements called laws of nature in a primary sense (statements like Newton's laws of motion, the principle of conservation of energy, the law of conservation of mass, the chemical law of constant proportion, and so on), though well attested empirically, in addition frequently embody some very general conceptual considerations; furthermore, they usually contain one or more terms that are themselves 'implicitly defined'[2] by these and adjacent laws that act as basic explanatory principles of the science concerned. Such principles quite frequently formulate the basic language of a science, and it is that which gives them the appearance of being relatively impregnable. The abandonment of such principles introduces the necessity for modification and alteration of so many details at once that scientists are usually most unwilling to relinquish them, particularly when the principles in question have set up a common stock of 'pictures' in terms of which to 'understand', to 'grasp the significance of' the phenomena. The situation is therefore not so very different from that met in the discussion of the *a priori*,[3] and whilst such propositions are clearly not *a priori* in the sense of being intuitional, or perhaps analytic, only vaguely *resembling* the latter in certain respects,[4] they do have something in common with them; and some philosophers have in recent years adopted the term 'functional *a priori*' for just this sort of case.[5]

[1] Introduction, V, Bernard trsl., New York, 1961, p. 21. A similar conception under the label of 'natural necessity', has also been advocated by Kneale, op. cit. though Kneale claims Locke as a precedent (§17, p. 71).
[2] Hilary Putnam has aptly named them 'law-cluster terms'. See 'The Analytic and the Synthetic', in *Minnesota Studies in Philosophy of Science*, III, ed. H. Feigl and G. Maxwell, Minneapolis, 1962, pp. 376ff.
[3] Cf. above, pp. 21–4.
[4] That they do has been argued by W. V. Quine, in 'Two Dogmas of Empiricism', forming ch. 2 of *From a Logical Point of View*, Cambridge, 1961, pp. 20–46. For a reply, and modification, see Hilary Putnam, 'The Analytic and the Synthetic', op. cit., pp. 358–97. In the light of these arguments, the original 'confusion' of the various issues at the start of our period overlapping in such exasperating fashion is perhaps not surprising!
[5] Cf. A. Pap, *The A Priori in Physical Theory*, New York, 1946. Pap himself was

The whole situation, the mutual influences between the rational and the functional *a priori*, is in fact a matter of extreme fluidity.

We may, therefore, expect the earlier scientists and philosophers to be thoroughly muddled in these matters; and for this reason more recent clarifications may elucidate the earlier period for us. At the same time, let us hesitate before we dismiss the pronouncements and attitudes of the earlier scientists and philosophers simply as falsehoods or confusions. We must take care not to read into their terminology the hardened meanings of later generations. For the original language often concealed a variety of conceptions only later to be disentangled. Yet, any of the different ideas comprised under a single label may lend strength and significance to the others, which by themselves might have no sense at all.[1]

It is important to recognise that necessitarianism was not just the brain-child of philosophers; the general idea is buried deep in the world views of leading scientists, though for a variety of reasons this does not figure prominently in the histories of science of the epoch. Galileo, who in so many ways combines the leading intellectual facets of his age, is a pertinent example, particularly since many still regard him as no more than an early representative of the 'experimental approach', which later came to be associated with 'empiricism', and in turn calls up expectations of Locke and Hume. In fact the situation is much more complex, so it is necessary to enlarge on this a little.

Consider this matter in the context of an interesting passage in his *Dialogue Concerning the Two Chief World Systems* (1632) which crystallises some of these 'rationalist' tendencies alluded to, and which illustrates a whole stream of thought during the seventeenth and eighteenth centuries.[2] In this passage there is a criticism of the scientific approach of William Gilbert, the celebrated author of *De Magnete*, a work which combines painstaking empirical investigations with a certain amount of theoretical speculation—though the

---

influenced by the views of C. I. Lewis; cf. the latter's *Mind and the World-Order*, New York, 1929, chs. 8 and 9.

[1] Though I have emphasised the competing claims of uniformitarian and necessitarian interpretations of law, as being the two of the greatest importance for an understanding of the classical period, it should not be thought that they exhaust all possible analyses. But an exhaustive survey of the many different views lies beyond the confines of this volume; and anyway, definitions of law as, for instance, 'convention', or as 'description', will in the end come very close to one or the other of the uniformitarian-necessitarian pair.

[2] See also my *Image of Newton and Locke in the Age of Reason*, op. cit., and the bibliography there referred to.

two do not dwell entirely in such separate compartments as might appear at first sight. Galileo writes:

> He [Gilbert] seems to me worthy of great acclaim also for the many new and sound observations which he made ... [But] what I might have wished for in Gilbert would be a little more of the mathematician, and especially a thorough grounding in geometry, a discipline which would have rendered him less rash about accepting as rigorous proofs those reasons which he puts forward as *verae causae* for the correct conclusions he himself had observed. His reasons, candidly speaking, are not rigorous, and lack that force which must unquestionably be present in those adduced as necessary and eternal scientific conclusions.[1]

The tone of this passage is quite familiar to a present-day reader. It urges that explanations put forward by way of hypotheses should be developed deductively, step by step, so that thereby not only can it be shown that their consequences really do follow from them, but moreover that the consequences thus deduced factually do correspond to the observations. We should also agree that the deductive procedure would be at its most powerful and rigorous when carried out *more geometrico*, or more generally, in mathematical fashion. So far, this is a clear account of the so-called method of 'inverse deduction', or more properly 'retroduction', which has become such a prominent feature in more recent versions of the logic of science. There are however two further features in this passage which do not square so easily with this cosy 'modern' reading. They are, first, Galileo's reference to what is commonly known as the '*vera causa* doctrine'; secondly, and not altogether unconnected with the last, Galileo's 'necessitarianism'.

As to the former, Galileo implies that the explanatory hypotheses should have the status of '*verae causae*'; they must be explanations that are 'true'. It is not clear whether Galileo means more than that they should be *assumed* to be true, as in the 'inverse-deductivist' doctrine; or, that they should (on independent evidence) be also *known* to be true ('inductivist' doctrine). There are passages in other writings where the former interpretation is clearly suggested, e.g. in the *Dialogues concerning Two New Sciences* (1638); having propounded a certain basic axiom for the deductive development of theorems connected with the phenomena of freely falling bodies, Galileo thereupon proposes to regard it 'as a postulate, the absolute truth of which will be established when we find that all the inferences

[1] *Third Day*, tr. Stillman Drake, Berkeley, 1953, p. 406.

from it correspond to and agree perfectly with experiment'.[1] However, this would not agree with the normal interpretation of the *vera causa* doctrine, according to which what is required is an *independent support* for the truth of the hypotheses, although to the question, 'What is to count as such?', there is not always a clear answer.[2]

If we disregard 'historical hypotheses' and concentrate on the more prominent case of the so-called 'transcendent hypotheses' (e.g. those involving 'non-observables'), the most usual view would be that such hypotheses should, where possible, not conflict with the known theories and laws of nature, and more generally, with the whole contemporary 'style' of physical procedures and results. But clearly, even this is a relative matter: the decisive 'paradigm-situations' that are here involved, change over a period of time.[3] Nonetheless, this is not inconsistent with the consideration that constraints of *some* kind at least should be imposed upon the freedom to invent hypotheses *at any particular time*, and that the sheer desire for 'economy' and 'simplicity' will often be paramount in such a situation. No doubt this was what Newton himself had in mind in his classical formulation of the *vera causa* doctrine: 'We are to admit no more causes of natural things than such as are both true and sufficient to explain their appearances'; where he adds for further clarification:

> To this purpose the philosophers say that Nature does nothing in vain, and more is in vain when less will serve; for Nature is pleased with simplicity and affects not the pomp of superfluous causes.[4]

At this point the inductivist doctrine merges into another type, employing additional props for what we may call the 'supplementary consolidation' of hypotheses, props which are primarily relevant to what we previously termed 'the plausibility of hypothesis'.[5] Such a

---

[1] *Third Day*, tr. Crew and Salvio, Dover ed., n.d., p. 172.

[2] This problem of how to interpret the 'inductivist' *vera causa* doctrine has haunted the literature especially since it appeared in Newton's *Principia*, where the doctrine is stated in the first of the four 'Rules of Reasoning in Philosophy'. And the question got further tangled up with Newton's apparent hostility to the use of hypotheses, to which we shall return in later chapters.

   Whewell criticised Newton's views extensively in his *Philosophy of the Inductive Sciences*, 1840, vol. II, Part 2, Bk. 12, ch. 13, sects. 6–12, pp. 440ff. For a more sympathetic view cf. Mill, *Logic*, 7th ed., 1868, vol. 2. bk. 3, ch. 14, paras. 4–6, pp. 8ff.

[3] As has been extensively argued with exhaustive documentation in T. S. Kuhn, *The Structure of Scientific Revolutions*, Chicago, 1962.

[4] *Principia*, Bk. III, 'Rules of Reasoning in Philosophy'; Cajori ed., Berkeley, 1962, vol. II, p. 398.

[5] Cf. above, p. 26.

demand for additional supports seems to be at work in the treatise by Galileo last referred to (*Two New Sciences*), and which at first sight looked like an example of a purely inverse-deductivist approach. At least, there is a later interpolation in the main body of the text, added upon Galileo's suggestion by his pupil Viviani, which gives an intimation of Galileo's feeling that something more was needed. It is an addition, Galileo tells his reader, which is made

> for the better establishment on logical and experimental grounds, of the principle which we have above considered; and what is more important, for the purpose of deriving it geometrically, after demonstrating a single lemma which is fundamental in the science of motion.[1]

The 'lemma' mentioned involves in fact (though in a vague and tentative form) the 'principle of virtual work'—a principle whose position in statics is analogous to that enjoyed in dynamics by that of conservation of mechanical energy. It is however the tenor of the demonstration of the lemma which is of importance to us. For the use of the principle which it involves, Galileo implies, can be seen (as he says) to be completely and transparently 'clear'—almost (we might say) self-evident.[2]

The need for science to aim at further consolidation of its explanatory hypotheses connects closely with the second of the two features in the passage from Galileo's *Two Chief World Systems*, his 'necessitarianism', quoted before.[3] For he tells us that the grounds of an explanation must have a 'force' that will yield 'conclusions' that are 'necessary and eternal'. We are at once reminded that these two features were also criteria of the '*a priori*'.[4] This shows how easy it would be to confuse the two kinds of necessity involved, a confusion

---

[1] *Third Day*, p. 180.

[2] I say 'almost', for we are here concerned primarily with the case of 'nomic necessity', even though self-evidence may sometimes be difficult to keep out. Thus Kneale, in *Probability and Induction*, states that we are 'unable to see the intrinsic necessity of the law, and it seems clear that we shall never by any advance in natural science reach a point at which we can say that natural necessities have become self-evident.' (p. 92). Notwithstanding this, a few pages later we read: 'It is felt as an imperfection of a theory that it should assume any laws which cannot be seen to be intrinsically necessary.' This is a remark concerning the status of 'transcendent hypotheses', and he adds that 'the connections within the world of transcendent entities posited by a theory may all be self-evident . . .' (p. 97). Here, the use of 'self-evident' is unusual; but the whole doctrine is closely akin to Locke's, which we shall study in some detail later.

[3] Cf. above, p. 35.   [4] See above, p. 23.

almost inevitable. Nonetheless, there is a difference, though the difference does not necessarily reside in the 'meaning' of the term 'necessity'. As we have seen, nomic necessity is an aspect which involves attention to an 'ontological' relation. Certainly it is not postulated solely in virtue of purely logical features attaching to the lawlike proposition as such; quite the contrary: it is a necessity, which, in Kneale's happy phrase, remains 'opaque to the intellect'.[1] Rather, what is primary is that the statement, with which we are concerned, should (in the first instance) be an accepted scientific law. Such laws are statements that usually satisfy most if not all of the following criteria (our list is not complete).[2]

1. (a) The statement must satisfy certain conventional inductive criteria (using 'inductive' in a broad sense), e.g. absence of falsification under severe testing. Moreover, it must display a number of logical, functional, structural and systemic characteristics that have been mentioned briefly before.[3]

(b) The statement emerges after the deployment of certain 'consolidative devices', as in Galileo's proof of his 'lemma', involving additional methodological principles like those of 'simplicity' and 'symmetry'; frequently, it will display strong mathematical or physical analogies, and so on. As stated already, these criteria will jointly determine the 'probability' and 'plausibility' of the hypothesis of law.[4]

2. There is presupposed a certain analysis or appraisal of statements of law which hinges on considerations of which those outlined in our discussion of the views of W. E. Johnson are typical; they are viewed as 'universals of law'. In sum: that statements of law in general are regarded as nomically necessary is something which is a function of a certain philosophical analysis. But that some given statement is

---

[1] Op. cit., p. 97.

[2] I have discussed these criteria and their mutual relation in more detail in my 'Semantic Sources of the Concept of Law', in *Boston Studies in the Philosophy of Science*, vol. III, R. S. Cohen and M. W. Wartofsky, ed., Dortrecht, 1968, pp. 272–92. The topic will also be taken up again in ch. VIII, sect. 4 (c): iii–v, pp. 509ff.

A good discussion of the place of such principles in the consolidational establishment of hypotheses may be found in G. Schlesinger, *Method in the Physical Sciences*, London, 1963. See also E. Nagel, op. cit., 4,iii,4, pp. 64–7, for more general aspects of 'consolidation' or 'assimilation'. Nagel remarks in particular on the importance of the depth to which a certain 'law' is embedded within a theory, either as an axiom or a theorem, in order to count as law-like. In part this is measured by the degree of 'intellectual havoc that would ensue' from abandoning a fundamental explanatory hypothesis (op. cit., p. 66).

[3] Cf. above, pp. 26ff.   [4] Cf. above, pp. 26, 36f.

accepted as expressing a law (with such a necessitarian import) depends furthermore on considerations of the first kind (1), and in particular those mentioned under section (1b).

Obviously, a necessitarian analysis, like any other analysis of 'law', and perhaps even more so, has its difficulties. Since the history of classical philosophy is largely a criticism and articulation of these difficulties, we may reserve them for later. Yet it is noteworthy that the classical philosophers, although very much exercised by the problem of 'necessity', spent relatively little time on a detailed investigation of the grounds, motives and historical circumstances leading to this notion. Nor is this surprising since an investigation presupposes an advance of scientific sophistication not given at the beginning of our period. And for exactly the same reason, they were not always clear on the true relationship that existed between the inductive procedural situation sketched under (1a) and (1b) above, on the one hand, and the adoption of laws (with a necessitarian grammar) on the other. Nomic necessity, we have already seen, is clearly not a function of the rational *a priori*; in the same way, it does not emerge in any straightforward fashion as the accompaniment of some sort of inductive step. As the vast body of literature on induction in recent years has shown, the difficulties in the way of a satisfactory solution of the problems raised by the logic of inductive inference are considerable. The feelings of one philosopher on this problem are therefore easy to understand, when he says that the task

> difficult enough to justify our belief in scientific laws when they are regarded simply as generalisations . . . [will become still] more difficult if we are required to justify belief in propositions which are more than generalisations.[1]

A reaction from this is the view, already anticipated above, that the assignment of nomic necessity has nothing at all to do with the inductive establishment of the law as such, being entirely a matter of logical analysis (aspect (2) above).[2] This is a remarkable turning of the tables on the original (pre-Humean, Aristotelian) position, according to which the existence of a necessary connection between the terms of lawlike propositions was a very presupposition of inductive inference itself.[3] We shall see how the classical philosophers hammered their

---

[1] R. B. Braithwaite, op. cit., p. 11.
[2] This seems to be indeed Braithwaite's view; cf. op. cit., ch. 9.
[3] For an example in modern writings, cf. below, p. 42.

way out of this impasse of contradictions, and how their final views are indicative of possible alternatives—indeed not unlike the view I sketched a moment ago.

At any rate, I think that the 'consolidational' procedures mentioned under (2b) have a considerable bearing on the decisions of the scientist. Here really belong what is often referred to as the '*a priori* foundation' of scientific principles. The attempt by the sixteenth-century Stevinus to base the statical force-relations operative for inclined planes on the impossibility of perpetual motion is a case in point. We note how easy it is to regard this as an example of a rational *a priori* when a writer like Whewell says:

> [The impossibility of perpetual motion] is really *evident* in the case contemplated by Stevinus; for we cannot *conceive* a loop of chain to go on through all eternity . . . by the effect of its own weight.[1]

This oscillation between the rational and the functional *a priori* appears even more clearly in the thought of J. R. Mayer, one of the discoverers of the Principle of Conservation of Energy. Here was a law established by considerations of a large number of empirical factors, and capable of being tested empirically (as in the example of Galileo) by means of the deductive development of a number of empirically verifiable consequences. Yet, it is quite clear that Mayer regarded energy (or what he called 'force') as something which in a very special way was subject to the law of causation, in the form in which he asserts it, viz. 'the cause equals the effect' (*causa aequat effectum*). This is what Mayer has to say on the relationship between the empirical and the logical aspects of the matter:

> The proposition that a magnitude which does not arise out of nothing, cannot be destroyed either, is so simple and clear that against it one can object as little as against an axiom of geometry; and we may assume its truth so long as the opposite has not been shown to be an indubitably established fact.

---

[1] *Phil. Ind. Sc.*, 1840, Vol. I, Pt. I, Bk. iii, Sect. 6, p. 200; my italics. Such a rationalist interpretation was opposed by Mach, who insisted that Stevinus' principle was 'a *purely instinctive* cognition' (*The Science of Mechanics*, I.ii.2, 6th ed., Open Court, 1960, p. 34). But both Whewell and Mach share the correct perception that a principle like that of Stevinus is not a 'simple induction' based on a clear and limited set of experiments, but has a more deeply consolidated basis. Nor, in the light of the vagueness, and even ambiguity, of the '*a priori*', are rationalists like Whewell and positivists like Mach divided by much more than the associations of a seemingly contradictory terminology.

Or again:

> Either a given motion turns to nothing at its disappearance, or it will have an equivalent indestructible effect. If we decide unquestioningly for the latter, it is because we appeal both to the laws of thought and to experience.[1]

It is evident that Mayer does not claim that his principle *follows* from the law of causation; nor does he regard its opposite as inconceivable. It might be better to say that the relationship between the quantities involved and the concepts of logic are such as to express for Mayer the ideal of an adequate scientific explanation.[2]

So we see how confusions may easily arise. First, inductive considerations have *some* relevance to the conclusion, though the relation involved is not straightforward. Secondly, the modus of the functional *a priori* 'feeds' on the rational *a priori*; but they are not identical.[3] And thirdly, all this determines the nature and status of nomic necessity. There will be a particularly strong temptation to give a necessitarian analysis in a situation like Mayer's; nevertheless, in the end the necessitarian nexus is still only a function of the logical grammar of the lawlike proposition. Our appraisal of both the logical situation, the inductive backgrounds and the consolidational procedures will then provide the criteria by which we judge, and in virtue of which we assert, a given statement to be a law, regarded as a statement with necessitarian import. And whilst the claim to any special insight into the lawlike connection is no doubt misconceived, conceptual ramification of highly theoretical concepts involved in the statement of the law may play a part.

---

[1] *Die Mechanik der Wärme*, ed. J. J. Weyrauch, 3rd ed., 1893, pp. 247–8, 59.

[2] On this whole complex, the brilliant book by E. Meyerson, *Identity and Reality* (London, 1930), is still a *locus classicus*. And cf. his conclusion in this case that 'the conservation of energy, like inertia, like the conservation of matter, is neither empirical nor *a priori*; it is plausible' (p. 206).—We shall subsequently find that in Kant's opus such laws are also given a very special place.

[3] It is easy to confuse them. Descartes' attempt to reduce scientific knowledge to a mathematical basis, by the procrustean device of reducing matter to extension, was carried out under the assumption that it would be possible to arrive at an *a priori* foundation whose ideal status was very like 'logical necessity'. And Mayer's views might be thought in a similar fashion to produce the requisite 'clarity' by a 'reduction' of his principle to the law of causation. Nonetheless, I think the two enterprises can be logically distinguished, though I have tried to do this in a way which will reveal the very serious possibility of slides. We shall later show how this distinction is mirrored in the advance—in philosophers like Leibniz and Kant—from a view which regards scientific laws as dominated by the rational *a priori* (under the guise of Cartesian mechanism) to the much more adequate notion of the pressure of certain 'architectonic' or 'regulative principles'.

But there is still a fourth aspect that plays upon all this, making complexity more confounded. This is represented by latter-day Aristotelians, for whom nomic necessity has more than a mere 'logical' function and instead is turned into a justificatory device with an 'ontological' function. A. C. Ewing has given a very clear exposition of this sort of doctrine. There are, he says,

> strong arguments for the entailment theory. . . . We can after all make legitimate inferences from cause to effect. How could we do this if the cause did not in a very important sense entail the effect? The relation need not be exactly the same as the entailment which occurs in formal logical reasoning, but it must at least be analogous to it in the important respect that it justifies the conclusion.[1]

The difference between this and the approach I have suggested will be clear. For that makes a sharp distinction between the inductive foundation of the law, and its necessitarian analysis, the latter being an entirely separate matter, even where (as in the cases of Johnson and Chisholm) there was a claim that lawlike connections involve a peculiar 'ontological bond', bestowing 'objectivity' upon the causal relation.[2] I do not think, however, that these writers would maintain that the possibility of an inductive inferential step presupposes such a bond; and this possibility is explicitly rejected by others, like Braithwaite, Kneale and Popper.[3]

Nevertheless, it is only too tempting to take the further step, and employ the (putative) existence of a necessitarian bond precisely for the purpose of explaining the possibility (in principle) of an inductive inference. The argument being that only if a certain class of events *necessitates* another are we justified in inferring from the *observed* co-existence of members of the first class with members of the second class a probable coexistence in as yet unobserved instances. This was just the position first attacked by Hume who denied to necessitarian links the required 'objective' status. Hume's specific formulation of this attack may be open to criticism: this will be discussed in the sequel. But it does usefully bring out a flaw in the doctrine which invokes

---

[1] *The Fundamental Questions of Philosophy*, London, 1951, ch. 8, p. 163.
[2] Cf. above, pp. 28–9.
[3] For Braithwaite, see remark quoted above, p. 39. For Popper, see *The Logic of Scientific Discovery*, London, 1959, Appendix *x. Popper does assent to the doctrine that 'there are necessary laws of nature, in the sense of natural or physical necessity' and that they have a 'peculiar ontological status' (p. 438). He insists, however, that 'this doctrine is . . . incompatible with any theory of induction' (p. 432).

'entailment relations' for the justification of inductive inferences; it is that in the absence of further explanations the 'entailment' notion lacks *a meaning in the sense required*.

It is here that my account of the 'natural history' of the notion of necessity becomes relevant, for no doubt such an account helps to give the notion *a* meaning. What it does not do is to allow us to assume that such 'technical' meanings as we have educed are the ones required by the upholders of inductive justification through 'necessity'; though they may not be aware of this fact, as it is easy to borrow a meaning from one region of a concept and transfer it to another. Nor should such a procedure be condemned out of hand. It is just because the meanings of metaphysical concepts like necessity have 'ragged edges', carrying with them their never-ending systematic ambiguities, that positive doctrines like those of the Aristotelians and of Ewing, on the one hand, and proscriptive criticism like that of the Humeans, on the other, arise. It is in this way, indeed, that meanings first become 'fixed'. It is just this which makes a study of Hume's criticism of necessity interesting and rewarding; that it contributes in this way towards a 'fixation' of the meaning of necessity. Thus it was only owing to Hume's suggestion that the idea of necessity has no ontological counterpart in nature that the very notion of such an approach towards necessity first arose. Hume's critique of the possibility of a non-logical interpretation of necessity at the same time fixes the very meaning of the disputed concept, previously never formulated very clearly. It follows, moreover, that Hume's mode of talking is not necessarily unique: other 'pictures' may be experimented with. However, for the moment I am not concerned to evaluate (critically or otherwise) such points of view, but merely to indicate the intellectual map on which they occur; to 'investigate the application' (in Wittgenstein's phrase)[1] of the various models for necessity,—and of the links that relate the terms of various types of propositions.

All this is to speak with the knowledge of hindsight. And if we have given a sympathetic account of this picture of necessity, it is because it plays such a vital part in the thought of antique and classical periods of philosophy; only, the various strands and components that we have singled out are there still inextricably intertwined. For Galileo to claim 'necessity and eternity' for his scientific conclusions was no idle phrase; no more than when in a letter of 1615 he asserts that 'Nature

[1] Cf. above, p. 20.

acts through immutable laws which she never transgresses';[1] echoing older sentiments, already expressed in the utterings of a Heraclitus, that 'the sun will not overstep its measures . . . lest justice will find it out'; or of a Parmenides, that 'what is . . . is immovable in the bonds of mighty chains, without beginning and without end'.[2]

The thought that nature is 'subject to law', in a sense which makes law more than a mere description of regularities or uniformities, certainly became a commonplace during the seventeenth century; and the philosophers, or philosophically-minded scientists, echo and reflect it. In this way Descartes tells his reader that in dynamics he has

> discerned certain laws which have been so firmly established in nature by God, and the ideas of which He has implanted so firmly in our minds, that with sufficient reflection, we cannot doubt that they are strictly observed in everything that is or that happens in the world.[3]

In a similar fashion, Boyle after him speaks of 'nature' as an 'aggregate of bodies' which 'act and suffer, according to the laws of motion prescribed by the Author of things'.[4] And still later, Berkeley will speak of 'certain general laws that run through the whole chain of natural effects'.[5]

It is important to be aware that there was this attitude to nature during the whole of the seventeenth- and eighteenth-century period (and I think still later). For unless this is kept in mind we cannot understand a very paradoxical tension in the intellectual history of this period. On the one hand, the epistemological basis of our knowledge of these laws, or better (to be less confusing), of the presuppositions of a knowledge of laws understood in this necessitarian fashion, comes under close critical, and even sceptical, scrutiny. On the other hand, such criticism as exists is felt more deeply, than, say, in our own day, because the older views on the logical nature of laws, the idea of 'a nature of things', is retained—consciously or unconsciously. How otherwise understand the earnestness with which Hume pursues the question: 'Whence do we take this very idea of necessity?' He never really doubts that we *have* the idea, that it is clear and

---

[1] Quoted in E. A. Burtt, *The Metaphysical Foundations of Modern Physical Science*, London, 1932, p. 64.

[2] J. Burnet, *Early Greek Philosophy*, 4th ed., London, 1930, ch. 3, fragment 29. p. 135, ch. 4, (8), p. 175.

[3] *Discourse*, v, K.S., p. 147.

[4] *Works*, Birch ed., v, p. 177. For the complete passage, cf. below, p. 54.

[5] *Principles*, sect. 62. See also below, p. 55.

universally held, and that the concept of causal law practically spells necessity. And it is just because this view is so deeply entrenched, is never questioned, that the attacks on it are productive of such intellectual turmoil.

(c) *The systemic group*

The systemic model is not unrelated to the previous one; indeed, it is sometimes thought to be intimately connected with it. For many philosophers of science have held that an important criterion for regarding a statement as lawlike is that it should form part of a theory, i.e. a system of hypotheses or principles from which it is deducible, and by which it will thus be said to be 'explained'.[1]

Now it is easy to appreciate the need for such a criterion, if briefly we turn back to the Johnsonian difference between 'universals of fact' and 'universals of law', the latter involving 'nomic necessity'. If for the moment we ignore the syntactic and functional analysis of lawlike statements, and consider merely their epistemic basis, we see that such a basis is the same whether a law expresses a universal of fact or something more. Thus, the occurrence of an exceptionless class of observed ravens that are black, however far we imagine the evidential class to stretch into the future, is all the evidence that can be claimed by either school of logicians.

Now it is true that the proponent of the nomic view may have ready an objection. He will say that in the absence of the existence of a lawlike state of affairs, the series of black ravens, even if by hypothesis it stretched indefinitely and exceptionlessly into an indefinite future, would still be no more than an 'historical accident on a cosmic scale'.[2] This objection, however, still admits that there is no epistemic difference in the instantial *evidence* for the statement analysed in these two different ways. In other words, no 'factual' difference seems to correspond to the difference in the two analyses; and nothing in the 'evidence' seems to 'correspond' to the ascription of necessity.

Still, this deficiency need not deter us from our search. It simply

---

[1] It is not clear whether this is either a necessary or sufficient criterion of lawlikeness, nor of explanation. Some possible objections will be taken up presently. For the view outlined, cf. Nagel, op. cit., 4, iii, 2, pp. 59–62; Braithwaite, op. cit., ch. 9, pp. 301ff.

[2] Cf. W. Kneale, 'Natural Laws and Contrary to Fact Conditionals', *Analysis*, 10 (1950); reprinted in M. Macdonald (ed.), *Philosophy and Analysis*, Oxford, 1954, p. 229.

shows that nothing appertaining to 'counting of instances' yields a feature that explains, or yields motives for, this ascription of necessity. And we have indeed noted already in the last section that a far more powerful indicator of lawlikeness is the 'consolidational devices', referred to under the heading of 'simplicity' and 'symmetry'; the list could be enlarged but its discussion would lead too far afield.

Similarly, the motives provided by 'functional analysis' could be extended. Thus it has been suggested that when the necessitarian view is contrasted with the non-necessitarian, we find that adherents of the former are more definitely committed to give up a law in the face of non-explained exceptions. For if the uniformity is only one of 'fact', it is relatively easy to come to terms with the existence of some very occasional unexplainable exceptions, whereas a necessitarian must hold that such exceptions indicate an additional deficiency, viz. the non-existence of the necessitarian bond.[1] It is true that a white raven *formally* falsifies the generalisation of fact that all ravens are black, but we might be prepared to close an eye to a single unexplained exception, since it is only a matter of contingent fact that a white raven 'happened' to turn up. On the contrary, the proponent of the nomic view will find it difficult to retain the generalisation since he takes it to imply that a white raven *cannot* turn up.[2]

Now the fact that most scientists react very strongly against the suggestion 'to close an eye to exceptions' shows that they (at least unconsciously) take laws to possess a necessitarian logic. We need not go into their reasons for doing so; the explanation would have to be given in terms of pragmatic, psychological, evolutionary and more general ideological factors.[3] Such 'functional' considerations are, however, rather vague. And it is not surprising that a more formal and definitive model should have been desired; a request to which the systemic model at first sight seems the perfect answer.

What is suggested, then, is that if it can be shown that the statement in question is supported by a theory, it ceases to be an accidental uniformity, and is turned into a 'nomic necessity'. This is the view, for instance, taken by Braithwaite. Expounding Kneale, he writes:

---

[1] Cf. J. Bennett, 'The Status of Determinism', *Brit. J. Phil. Sc.*, **14** (1963), 115ff.

[2] 'Cannot', in the *nomic* sense of 'cannot': we are not of course concerned here with *logical* necessity.

[3] See for instance the history of Kepler's adoption of the concept of law, below, pp. 52–3.

The blackness of all ravens is surely 'accidental' if no reason can be given for such blackness; and this is equivalent to saying that there is no established scientific system in which the generalisation appears as a sequence.[1]

Let us now look at one rather obvious objection to this. It seems at first sight that the systemic criterion can hardly be relevant to the question of 'nomic necessity' at all, since surely propositions derived deductively from a set of antecedent hypotheses are 'necessary' only in the sense that they follow 'with necessity', i.e. if they are validly derived (in accordance with the rules of inference governing the system in question), the basic hypotheses being assumed true. But this is really just a species of formal necessity, and thus belongs to the group discussed under section (a) above. Besides, to counter the charge of logical triviality, we must presumably suppose that the basic hypotheses are themselves statements of law—so only removing our problem to another place!

On the last point, it is interesting to note that Braithwaite actually meets such a possible objection by allowing that a certain hypothesis will also be lawlike if it 'occurs in an established scientific deductive system as a higher-level hypothesis containing theoretical concepts'.[2] No special reason is given why this should be a criterion of lawlikeness, but perhaps there are here involved the sort of considerations already advanced as part of the previous 'nomothetic group of models', involving the case of hypotheses containing 'law-cluster terms'.[3]

The first criticism is harder to answer, but I think in the light of my general classification it is possible, at least, to understand the motive for and history of this criterion. For it is true that the explanatory hypotheses of a theory do supply a 'link' for the terms of the derived proposition. In the previous section we saw that the necessitarian model quite generally provides a picture for such a link. One may then easily say that the link provided by systemic explanation should also be regarded as bestowing necessity. For those who do not subscribe to the use of the necessitarian bond as a justificatory device,[4] but instead regard it as no more than what Braithwaite calls 'a honorific epithet',[5] this would not be a matter of much consequence, since the ascription of necessity in any case (on this view) depends on

[1] Op. cit., ch. 9, p. 304. [2] Op. cit., pp. 301–2. [3] See above, p. 33.
[4] See above, p. 42. [5] Op. cit., p. 302.

the nature of the 'knowledge or belief in the general proposition rather than in anything intrinsic to the general proposition itself'.[1] Without committing ourselves at this stage to such an 'anti-ontologist' position, we may adopt the neutral view that we are open to use the systemic aspect of science as one of the models for 'necessity', whether or not it ultimately makes sense to speak of 'necessity in nature'.

Let us now consider in greater detail the picture of the propositional link provided by the systemic or theoretical group of models. This will incidentally provide an opportunity to survey the broad outlines of the various types of explanations current at the outset of our period with which the average scientific and philosophical thinker would probably be familiar. We shall then be in a better position to understand the nature of the subsequent criticism. Earlier we mentioned Heraclitus and Parmenides;[2] and their sayings also embody some of the earliest theoretical explanation patterns: the reference to an underlying likeness; to an orderliness of the manifold of phenomena; an order not easy to define and hence often pictured through certain analogies, such as the 'justice' of Heraclitus, or the 'chains' of Parmenides.

These early models illustrate an essential aspect of 'explanation', viz. 'the reduction' of the properties of the *explanandum* to those of a more general class of things or of a 'fundamental object' of which they are, as it were, manifestations. One of the basic concepts here is that of an underlying continuum or substance, together with a definite and orderly class of elements 'arising' out of it, or 'tied down' to it. The inexorable nature of the process (corresponding to the 'prescriptive' force of a law) is likewise a prominent aspect; witness another of the early Greek cosmologists, Anaximander, who writes:

> And into that from which things take their rise they pass away once more, as is meet: for they make reparation and satisfaction to one another for their injustice according to the ordering of time.[3]

However, in the history of scientific thought, the process of explanation at this point divides and advances in different directions, related in some way to different emphases.

1. There is an emphasis on substance, and, competing with this, an emphasis on the existence of classes or 'substantial forms', related

---

[1] Op. cit., p. 301.   [2] See above, p. 44.   [3] Burnet, op. cit., ch. 1, sect. 13, p. 52.

together in systematic orderliness, as in a biological system of classification.

2. Subsequently, from an interest in substance, there is a shift to an
emphasis on relation, on function, on order of phenomena; something ultimately expressed through the concept of law. Nonetheless,
many of the logical features and problems of the first group apply also
to the second; a fact which makes an investigation of each relevant to
the other.

(i) *Substantial, and classificatory, analysis*

a. Emphasis on classification

The most interesting example is to be found in the writings of
Aristotle, and extends through the whole peripatetic tradition to the
seventeenth century. Its primary model is the zoological system. The
individual is thought of as belonging to a species, and the various
species present a system of stability and ordered harmony. What defines a species is a set of universal properties; and for Aristotelians the
'universal' is a sort of eternal form, having in some way 'more reality'
than the particular which it informs. The forms are 'nature' in one of
the many senses of this term, for Aristotle the most important. Moreover, in the Aristotelian conception of nature there is always a
reference to 'finality', to something teleological; the principle of order
or pattern tends towards optimum harmony. It is this principle of
harmony that keeps the individual species in some definite order.
There are not any 'quasi-physical' threads stretching across from one
to the other—each individual is 'fastened down' only in so far as it is
a member of a species; and indeed, even here the attachment is not
complete; the fixity of species is not absolute; the individual only *tends*
to realise the optimum perfection. (Cf. the importance and place
given to the existence of 'monstrosities' in the animal kingdom by the
earlier zoologists!)

It should be noted that the idea of the 'underlying continuum' is
here reduced to a minimum of importance; its shadow is the Aristotelian conception of nature as 'matter'; but matter is not something
existing by itself; it is reduced to 'pure potentiality', capable of
becoming 'informed', literally, and only thus emerging as an actual
individual, a real existent. Yet, the emphasis is on the 'universal', and
with this, on the species, and the ordered hierarchy of classes. The
individual rests snugly within the bosom of that species, left with a

relative degree of 'freedom', of 'accidental happening', each within its class, with no thought of connection with other classes, except indirectly, through the teleological ordering of the whole.

b. Emphasis on substance

As an alternative, explanation proceeds by reference to an underlying unchanging ground, the foundation on which the changing pattern of phenomena rests. The most obvious model for this is change of position or motion of a material particle, considered as itself unchanging. This is also the source of one of the most important of the earliest competitors of the peripatetic system of thought: the atomistic world-picture. Qualitative transformations are not here viewed as the actualisation of certain forms or universals, but as the expression of certain changes in the spatial configurations of unchanging 'substances', viz. the atoms. This viewpoint is more economical than the Aristotelian; yet, particularly in its early forms, it suffers from an inability to 'account' for a large number of characteristics of our universe, and so is usually accompanied by supplementary schemes, such as the seventeenth-century division of the world into the real 'primary qualities' (of atomistically conceived assemblies, corpuscularian structures), and the more subjective 'secondary qualities', colours, odours, and other 'sensations'.

The atomistic view has had a chequered history down to the present day. It has needed frequent modification to bring it in line with logical difficulties constantly discovered, and to make it adequate to further observational discoveries. Its most important logical feature, for the purpose of the present exposition, however, is the concept of the 'unchanging', of 'conservation', which it enshrines. Originally, this is expressed in Parmenides' saying, 'what is, is immovable'. Actually, this leaves the choice of an adequate candidate for immobility still open. For the early atomists, it was the atom itself, considered as substantial reality. With the advance of science, further powerful alternatives are added. Gradually, the notion of the 'unchanging nature of the atom' comes to be conceived 'attributively' rather than 'substantially'. The scientist primarily concentrates on the relevant *properties* of his substances rather than the substance itself. (Echoes of all this we shall find in Locke's writings.) Thus at the rise of the modern period of chemistry, we get the notion of the conservation of mass. Lavoisier refines it experimentally, but as a logical or metaphysical presupposition it was already present since the seventeenth century. Later, further candidates join the field: conservation of momentum, of energy, of electrical charge. We have seen already how these entities function as elements in the primary axioms of their respective branches of science, accompanied by subsidiary hypotheses. The axioms *define*, as it were, the subject-matter of their science; they are more akin to 'conventions', to 'custom' (remember: the early sense of law, as *nomos*, i.e. custom!) than to empirical generalisations, descriptions of fact. It is from this, as we saw, that springs, in part, the prescriptive, necessitarian, aspect of law.

(ii) *Emphasis on Law*

As time goes on, the concept of the state, and the changes of states of a substance, rather usurps the place that had been held by the substance itself in the scheme of things. When we discuss the foundations of dynamics, we are interested in the relationships between the velocity, time, distance, acceleration values of falling bodies. But there is no symbol for 'body' occurring in the equations of motion, e.g. the equation $v = gt$. As E. Cassirer showed with a wealth of illustration in his *Substance and Function*,[1] 'substance' becomes replaced by 'function', by the formulation of relations of invariance among groups of variables of state. Galileo, as usual, illustrates the trend in succinct fashion, witness one of his 'Letters on Sunspots':

> Although it may be vain to seek to determine the true substance of the sunspots, still it does not follow that we cannot know some properties of them, such as their location, motion, shape, size, opacity, mutability, generation and dissolution. These in turn may become the means by which we shall be able to philosophise better about other and more controversial substances.[2]

Locke in the seventeenth century graphically echoes this denigration of substance by his reference to it as an 'unknown I know not what'; and in contemporary quantum mechanics scientists are perfectly at ease with states that are not states of anything, i.e. of any substance. These are extremes; but the fact remains, that for the growth of science, from the logician's point of view, this replacement of substance by law is a central issue.

Of course, all this did not happen overnight, and it would be fascinating (if space permitted) to trace the growth of the concept of

---

[1] Chicago, 1923; Dover reprint, 1953.
[2] Cf. *Discoveries and Opinions of Galileo*, ed. Stillman Drake, New York, 1957, p. 124.

law (as an *explanatory* concept) at the threshold of the modern scientific period, during the seventeenth century. A few indications will have to suffice, so as to suggest some of the tensions in which the early scientists were involved. Kepler is one of the most interesting. As is well known, after years of effort he discovered the correct expressions for the orbit and velocities of Mars. In a letter of 1605, he writes of having successfully incorporated the inequalities of planetary motion into certain 'laws of nature', thus yielding a new type of astronomy free from 'hypotheses'.[1]

Behind this lies a whole revolution of thought. As long as the programme of astronomy proceeded on the assumption of the construction of 'hypotheses' involving the combination of *circles*, the astronomer could entertain the possibility of a substantival interpretation of these as being the largest circles of actual spheres of revolution. Once the spherical interpretation is made impossible (as the assumption of the ellipse as a true description, having *physical* significance, implies) the orbit of the planet for the time being loses its 'anchorage in reality'. Now the seventeenth-century scientists instinctively seem to have felt that they could make good this loss of substantial reality by according a more 'concrete' status to the concept of law as being itself an ultimate (a logical manoeuvre) introducing at the same time further causal factors capable of accounting for these new-found orbits. The operative concept involved in all this was force (e.g. gravitational force). But, as Newton realised very soon, what was relevant here again was only the *law of force*: as he puts it in one of his general remarks appended to the *Opticks*, principles such as that of gravity,

> I consider, not as occult qualities supposed to result from the specific forms of things, but as general *laws of nature*.[2]

All this will show how the systemic model, the explanatory function of laws, contributes further to their being conceived as prescriptive in nature: it helps us to see why the necessitarian logic of laws has had such a strong hold on the human imagination. The danger lies only in the tendency of the human mind to forget the origin of its 'pictures'. Thus, the purely 'logical' function of the concept of law in Kepler's thinking, a function which provides a background for its necessitarian interpretation, may be forgotten; the logical chains may come to be

---

[1] See E. Cassirer, *Das Erkenntnis-problem*, Berlin, 1922, i, p. 375 n. 1.
[2] Book 3, Pt. I, Qu. 31, Dover ed., 1952, p. 401, my italics.

viewed as quasi-physical, and ontological positions may result which then become an easy butt for later generations of philosophers.[1]

Not that such criticism should be regarded as otiose: far from it! It is a necessary follow-up to the almost inevitable hardening of the meanings of philosophical key-terms which are by nature systematically vague. But it is this very criticism which may itself be conducive to the process of 'hardening', 'pushing' the meaning of the term sometimes into an unwanted direction. The application of all this to the critique of necessity which we find in, say, Locke, or Berkeley, or Hume, will become apparent in the study of their writings. But already it will begin to become apparent that their mutual 'refutations' may well be missing their respective targets; that in their different 'languages' they may fail to be addressing each other. But if this is a fault, it is a necessary fault; it is what constitutes philosophy.[2]

To return to 'explanation': parallel with the tendency to replace substance by law we meet the changeover from an emphasis on 'natural classification', the hierarchy of a natural order of species, to a universe governed by forces and natural law. Bacon already foreshadows it, but the mechanistic alternative is worked out with relentless consistency in the work of Descartes. In his approach to biology, just as to physics, he attempts a thoroughgoing 'reduction' of the manifold of qualities, substantial forms and individual substances to a unitary starting point, in order to satisfy the methodological requisite of mathematisation, pictured in the doctrine of matter as extension. We must 'entirely reject' the search for 'final causes', he writes in the *Principles of Philosophy*; instead we must

> attempt solely to discover, through the faculty of reasoning . . . , *how* the things which we perceive by means of our senses could have been *produced*.[3]

---

[1] For a definitive and very concise recent comment, see e.g. Wittgenstein, *Tractatus*, op. cit., 6.371–2:
> The whole modern conception of the world is founded on the illusion that the so-called laws of nature are the explanations of natural phenomena. Thus people today stop at the laws of nature, treating them as something inviolable, just as God and Fate were treated in past ages. . . .

[2] Sometimes, a similar process may be discerned in the development of scientific thought proper. Thus, the meanings of such concepts as force, momentum, energy in dynamics, and phlogiston in chemistry, required a period of criticism which ended by 'hardening' the grammar of the respective terms in a way which either circumscribed their extensional domain in a definitive way, or—as in the case of phlogiston—ended by abandoning the concept.

[3] Bk. i, sect. 28, my italics; A. & T., ix, p. 37.

This 'how', as we saw earlier, is for Descartes expressed through 'certain laws', established by God, which are 'strictly observed in everything that is'.[1] So these laws, as with the Greeks, *govern* inexorably, though the point is made in theological language. But unlike the Greeks, Descartes thinks of *everything* that happens as being strictly subject to law. Nature is no longer an ordered 'cosmos', but has become a 'universe', a system all of whose parts are related to one another. These 'parts' are pictured as quasi-geometrical points, their supreme paradigm eventually becoming the point-particle of Newtonian dynamics. Each individual 'point', instead of being an 'instance' of a class or species, 'designed' to express some final purpose through the grand pattern of harmoniously related forms, outside which there is still room for freedom and accident—each such Newtonian particle *must* be conceived of as being a nodal point in a mesh of interlocking lines, each of which is the picture of natural law. 'Nature', from being a series of 'vertically' organised individual 'forms', takes on the appearance of a 'horizontal' mesh of laws.[2] Whilst the earlier picture emphasises the 'instance' of a 'uniformity', or a 'species', expressed through the class-form: 'All A are B', the modern version more properly fastens on the functional relation '$y = f(x,y,t)$', a conditional relation between variables of state.[3]

This notion of nature became fixed in the British tradition through the work of Boyle and other seventeenth-century writers. In '*A free Inquiry into the vulgar Notion of Nature*', Boyle expressed great distrust of the whole concept of nature because of its peripatetic origins and overtones. For him, nature, is

> the aggregate of the bodies that make up the world, framed as it is, considered as a principle, by virtue of which they act and suffer, according to the laws of motion prescribed by the Author of things . . . I shall express what I call general nature by cosmical mechanism, i.e. a comprisal of all the mechanical affections (figure, size, motion, etc.) that belong to the matter of the great system of the universe.[4]

---

[1] See above, p. 44.

[2] Compare the interesting logical version of the network analogy in Wittgenstein's *Tractatus*, 6.341.

[3] Cf. K. Lewin, 'The Conflict between Aristotelian and Galileian Modes of Thought in Contemporary Psychology', in *Journ. Gen. Psychol.*, 5 (1931), 141–77; subsequently published in *Dynamic Theory of Personality*, New York, 1935. Lewin overstresses, however, the 'statistical' aspect of the Aristotelian concept of class and shows insufficient understanding of Aristotle's concept of the 'universal'.

[4] *Works*, Birch ed., v, p. 177.

As to 'matter' Boyle, like everyone else during that century, is a corpuscularian. As Burtt[1] points out, the attempt to account for variety and change by means of one universal matter inevitably leads him to some form of atomic theory, and with this to the notion of 'point-particles' already mentioned. From here again the viewpoint passes into general philosophy. For Locke, the 'real essence' of any 'species' is 'the true internal constitution' of things from which 'flow' all its qualities, and 'which is the foundation of all those properties that are combined in it';[2] 'constitution' being of course framed on atomistic lines. Locke's terminology of 'essence' and 'species' has still about it an Aristotelian air, consonant with his conservatism. In Berkeley's theory of explanation, on the other hand, the concept of law has become central when he tells us that

> there are general laws that run through the whole chain of natural effects . . . and that explanation in science consists only in showing the conformity any particular phenomenon has to the general laws of nature.[3]

It was not until the work of Kant that all these conflicting strands, mechanistic and teleological explanation, the concepts of substance and law, the necessitarian aspect expressed through the idea of law as such, and in turn interpreted through the concept of system, were to be reconciled—or at any rate, uneasily united in a grand and unitary scheme.

Meanwhile, we must retrace our steps. We have seen how the systemic model, through the idea of an interlocking network of laws, aided and abetted by a very 'substantialised' conception of law itself, provides a powerful picture through which to conceive the link which unites the terms of some individual proposition; or in the case of empirical generalisations: the link which unites recurrent aspects of some given class of physical events. The primary pictures are either the underlying substance, conceived atomistically or otherwise, or the existence of some given 'species'; and as scientific thought becomes more sophisticated the central concept becomes that of a lawlike relation between certain fundamental aspects of nature itself.

The picture of a 'link', which is involved in the explanation model, may be represented graphically in several more or less formal ways.

---

[1] Op. cit., Ch. 6, p. 167, and ibid., n. 38.
[2] *Essay*, iii.6.6 and 9.
[3] *Principles of Human Knowledge*, sect. 62.

One of these is the very paradigm of a physical explanation during the classical period, viz. Newton's explanation of the motion of the planets. The second, and much older picture, is Aristotle's own syllogistic formalisation of the process of explanation. Let us briefly consider each of them, starting with the Aristotelian. As is well known, the explanation of, say, a universal proposition like 'all Greeks are mortal' is given in terms of two premises, a major and a minor, to the effect that 'all men are mortal', and that 'Greeks are men'. Put formally, and with abbreviation, we represent this as:

$$g \text{ is } M \quad \text{(minor)}$$
$$M \text{ is } m \quad \text{(major)}$$
$$\overline{g \text{ is } m} \quad \text{(conclusion)}.$$

In this scheme, Aristotle calls M the 'middle term'. It represents that which 'mediates' between m and g,—providing a link.

This scheme is admirably suited to the Aristotelian ideal of explanation sketched above, employing the notion of 'order', 'species' and 'universal'. The particular theory of the mediating link will therefore depend upon the general philosophical (and metaphysical) doctrine that underlies it. The 'explanatory power' as such is provided in the present case by the (putative) presence of a 'universal' (the 'essential' characterisation of manhood by mortality).

Let us now turn to the Newtonian example, although we shall again present only a highly schematic and simplified version. Kepler's well-known third law of planetary motion expresses a relationship between the average distance of the planet from the sun (R) and its period of rotation (T), given by the functional expression $T^2 = k.R^3$, where k is a constant. Now Newton showed that this law could be derived from a higher-level theory, which includes as premises the law of gravitation and the laws of motion.[1]

[1] Strictly speaking it is not Kepler's Law but some expression to which the former is an approximation that is derived from gravitational theory. It is rare that 'lower-level laws' stand in exact deductive relationship to their corresponding explanatory hypotheses. Each explanation to some extent affects its erstwhile explananda, modifying and re-articulating them in the process. Each achieved explanation must therefore include *certain implicit understandings*, adapting the original *empirical* expressions to the formal derivation. On this, see P. K. Feyerabend, 'Explanation, Reduction, and Empiricism', in *Minnesota Studies in Philosophy of Science*, 1962, vol. iii, pp. 28–97. Feyerabend takes the view that in the light of these and cognate considerations it is 'impossible to retain a formal theory of explanation, because these changes will introduce pragmatic or "subjective" considerations into the theory of explanation' (op. cit., p. 92). This seems too high a price to pay. Since all

As already suggested, it is these laws that provide the ultimate 'ground' relating the variables T and R. In detail: Using the laws of motion, one can derive an expression for the 'centripetal force' (F) which must be exerted for a small mass m to rotate round a larger mass M in a circle[1] of radius R with a period T, where $F = \dfrac{4\pi^2 mR}{T^2}$. Newton furthermore postulated that there was a universal force of attraction between any two such masses, whose magnitude is given by the expression $F = \dfrac{GMm}{R^2}$, where G is the universal constant of gravitation. If we now postulate the identity of the centripetal with the gravitational force, we can eliminate F, obtaining $\dfrac{4\pi^2 R}{T^2} = \dfrac{GM}{R^2}$, from which results $T^2 = k.R^3$, with $\dfrac{4\pi^2}{MG} = k$, a constant.

We note a formal analogy between this and the previous case: F may be regarded as a kind of 'middle term' which, upon being eliminated, yields the desired conclusion. However, F does not really link the terms of a proposition, but rather it is a kind of underlying ground, which supports the functional relation obtaining between T and R. There are two ways of looking at F, to see what it stands for. First, strictly speaking, F simply represents certain general relationships, 'rules', which—in so far as they hold—may be viewed as 'laws of nature'. And as such it was regarded (at least sometimes) by Newton, as we noted already in our quotation on p. 52, above. But secondly, one may make a 'model' of this F, using the suggestiveness of the term 'force', picturing what is involved as a kind of invisible thread or a field of action. And pictured thus, we say that a mass circles round a central body because there is acting upon it a certain attractive force. (This is at the same time a graphic representation of the 'prescriptive' power of laws!)

---

scientific thinking in the end is highly conceptualised, it may well be possible to continue operating with a formal account, keeping in mind the requirement that it is our views concerning the nature of the 'empirical basis' that require constant purging and criticism.

[1] To keep things simple, we have assumed the planetary path to approximate to that of a circle. The more sophisticated reader may also be referred to the account (itself a simplified version of the one occurring in the *Principia*) which Newton upon request sent to John Locke in 1690, the year of the publication of Locke's *Essay*. Cf. A. R. and M. B. Hall, *Unpublished Scientific Papers of Isaac Newton*, Cambridge, 1962, pt. iv, 2, pp. 293–301.

This second way amounts really to a re-substantialisation of a nomothetic kind of explanation. Some would say that without some model, a purely 'abstract' type of theory has little explanatory power.[1] There are of course a number of arguments for such a doctrine, but I think that the most powerful *motive* for it is that the 'lawfulness' of laws, their 'coercive power', is insufficiently represented by an abstract formula, and the relation which it represents; what is demanded is a kind of substantialisation of the explanation.[2] Not surprisingly, the battle was joined already during the classical period, for it was one of Berkeley's great achievements for the first time to have formulated the possibility of an 'abstract' and purely 'formal' type of explanation.[3]

In the sequel, this Newtonian picture becomes generalised. This is done by absorbing a number of traditional strands. We have seen how in the older scheme the variety of properties of a body was thought of as determined by corresponding 'substantial forms'—one of the oldest of explanatory devices. We have already mentioned Boyle's tract on the *Notion of Nature*. His *Origin of Forms and Qualities according to the Corpuscular Philosophy* (1667), deeply inspired by atomistic and Cartesian influences, like the previous work graphically represents the change-over. The underlying 'substance' is now envisaged in the manner of the corpuscularians. All elements (using this term here in the modern sense) are divided into 'insensible' corpuscles, distinguished by a 'determinate shape'. (The 'corpuscles' correspond somewhat to our modern atoms, the 'particles' to our protons or neutrons.) Finally, when 'these insensible corpuscles' are 'put into motion . . . that itself may produce great changes and new qualities in the body they compose'.[4]

To Boyle's scheme Newton adds the conception of force, whose universal explanatory power he anticipates as a generalisation from his gravitational theory. Thus, after speaking of his successful derivation of the celestial phenomena from the 'forces of gravity', he goes on:

I wish we could derive the rest of the phenomena of nature by the same kind of reasoning from mechanical principles, for I am induced by many reasons to suspect that they may all depend upon certain forces by which the particles of bodies, by some causes hitherto unknown, are

---

[1] See e.g. M. B. Hesse, *Forces and Fields*, London, 1961, ch. 1, pp. 21–28.
[2] Cf. p. 52, above.   [3] Cf. pp. 307ff., below.   [4] *Works*, iii, pp. 29f.

either mutually impelled toward one another and cohere in regular figures, or are repelled and recede from one another.[1]

It is perhaps not unimportant to realise that there was relatively little empirical evidence for the corpuscularian doctrine at the time when this general 'paradigm' was being developed. It was very likely the demand for some form of explanation itself, by way of a quasi-substantival link, that gave so much credence and such immense success to the only theory which offered these thinkers a reasonable alternative to the discredited Aristotelian doctrine. The corpuscularian doctrine possessed great imaginative power for those searching for a constitutive ground, a 'cause or reason', which might determine the uniform changes of behaviour of bodies ('substances') when subject to external influences. The favourite paradigm of this mechanistic picture was the watch. Thus Boyle, when considering the question of 'final causes', argues that it is not sufficient to know what a watch was made for, but that the scientist should 'make studious search' after the 'efficient causes', i.e. study the internal mechanism, the shape and constitution of the parts of the watch, its motions, the elastic forces that are involved, etc.[2]

I have referred to this mechanistic constitution as providing, or being, the 'constitutive ground' which is thought to explain the observable uniform behaviour of bodies. Now this reminds us of the term which we previously encountered in Johnson, where a distinction was made between the 'epistemic' and the 'constitutive' point of view; a distinction considered 'ontological' in kind, and mirrored in Johnson's two ways of regarding the causal relation, viz. 'subjective or epistemic, and objective or constitutive'.[3] Moreover, it was just this 'constitutive' ground which seemed to provide Johnson with an argument in favour of the distinction between the 'universal of fact' and the 'universal of law', the latter involving the feature of 'nomic necessity'. We see now how easily the systemic model provides us with something through which to imagine, and give a meaning to, the difference between the absence and the presence of a constitutive bond. By contrast, the nomothetic model does not quite so easily supply such an obvious picture—indeed, the account of the 'natural

---

[1] From the Preface to the 1st ed. of the *Principia*, op. cit., p. xviii.
[2] Cf. *A Disquisition about the Final Causes of Natural Things*, sect. iv, prop. V; *Works*, iv, p. 550.
[3] See above, pp. 28f.

history' of this concept will have shown already that much of its 'prescriptive' (and by implication: necessitarian) strength derives from certain logical (and syntactic) *demands*, and from statements of assumptions and presuppositions.

It is not surprising therefore that the nomothetic model should have come under attack during the seventeenth and eighteenth centuries, but in the course of this the power of the remaining models as explications of the constitutive aspect of the link comes to be questioned also, as does the whole notion of the distinction between an epistemic and a constitutive level.

Briefly, and in very rough outline—the subsequent chapters will endeavour to make the details comprehensible—the criticism takes the following form. Locke still works throughout implicitly (and unconsciously) with the distinction between epistemic and constitutive; compare his various dichotomies, such as real and nominal essence; primary and secondary qualities; things and ideas; necessary connection and coexistence. But he has already begun to doubt the usefulness of the necessitarian analysis since (as he puts it) no necessary connections can on the whole ever be 'perceived'. In consequence the systemic picture (interpreted substantivally) takes on the task of providing the essential links between the qualitative states of nature; Locke's corpuscularianism furnishes the required 'underlying connection', largely construed through the notion of the 'real essence'.

The critique is continued by Berkeley who, having realised (and sharpened) the metaphysical nature of Locke's substantive systemic model, rejects it. 'Explanation' in terms of 'attractive forces' and an underlying matter (as a system of insensible corpuscles) is rejected in so far as such an explanatory scheme proceeds by converting the explanatory entities into quasi-metaphysical objects. Berkeley's more formal approach construes explanation as a system of laws of nature regarded as empirical uniformities. For with the prop of Locke's substantive interpretation of the systemic model gone, the necessitarian analysis of laws comes under attack also, though Berkeley's readiness for this is in part due to his replacement of necessitarianism by the idea of God, who becomes the constitutive ground of the laws of nature. Hume completes the critique of necessitarianism by denying Berkeley's theological route of escape, whilst at the same time arguing the inapplicableness of the first (or *a priori*) model to the necessitarian analysis of natural laws.

Earlier, the systemic model, though shorn of its corpuscularianism, and converted into a thoroughly metaphysical construction, still holds sway in the philosophy of Leibniz, as expressed through the doctrines of monadic activity and pre-established harmony. It is not until Kant that we are made consciously aware of the existence of the epistemic-constitutive distinction as such, and above all, of the fact that in previous philosophies these notions had been construed both too narrowly and too widely. (Thus, for instance, the constitutive level had included both physical and metaphysical ingredients.) In consequence it becomes clear that a new and more refined classification is required, in which each member will have to be articulated through a more complex structure. Certain aspects of the constitutive level will be absorbed into the epistemic, both of which then generate a new dimension, called by Kant the 'phenomenal'. The remaining 'metaphysical' aspects of the constitutive level will reappear under a new label, the 'noumenal'.

One of the results, then, of the great epistemological criticism which begins with the work of Descartes, Locke, and especially Hume, and ends with Kant, is that it raises the question of the very significance of the distinction between the epistemic and the constitutive aspects. This is due largely to the greater emphasis put on the experiential basis of scientific knowledge, in particular, of the laws or principles of science. Indeed, anyone considering our survey of the notions of law and theory will have noticed that basically their philosophical grammar is governed by formal, or semantic or syntactic, rather than epistemic, considerations. And this reflects, of course, the historical sequence of events. The 'meaning' (or philosophical grammar) of these concepts developed first under the stimulus of considerations that ignored epistemological questions. Only gradually does the relevance of the latter come to be appreciated. To show in detail how this epistemological criticism developed and affected the earlier analyses will be the task of later chapters. In the remaining section of the present chapter I shall develop a little further (though still only in broad outline) the trend of these arguments, avoiding for the moment any of the finer distinctions, seeming contradictions and ambiguities which so frequently obscure the major issues. Such an outline will at the same time enable us to appreciate the bearing of the criticism of the different models for the propositional link on the problem of this link itself.

3. EPISTEMOLOGICAL CRITICISMS

Let us return, then, to the reaction of the great classical critics when confronted by the logical world-pictures which I have traced in the three groups of models through which the subject–predicate, and in general, the propositional, link is envisaged. I have already stressed that these models were taken seriously; they constituted ideals of scientific knowledge; and being very demanding ideals, since each in its own way pointed to very tight, or necessitarian, conceptions, the ensuing epistemological problems were felt all the more keenly. Modern philosophers are too prone to read the great classical philosophers as though they were solely concerned (in their investigations of scientific knowledge and its problems) with the notion and the inductive presuppositions of 'empirical generalisations'. To Locke's, Berkeley's, Leibniz's and Hume's as well as Kant's general underlying assumption, that scientific knowledge is couched in the framework of laws whose logic is of the necessitarian kind, has been attached too little significance.

Certainly, this was the assumption involved from the beginning in the language used by Galileo or Descartes, and of all those influenced by them: that scientific truths should embody 'necessary and eternal conclusions'.[1] Now any criticism of this point of view clearly involves a complex set of questions that are really quite distinct. First, is the necessitarian characterisation of scientific 'conclusions' a correct one? Secondly, even if its correctness were granted, in the pragmatic sense that it is a correct appraisal of scientific ideals, how would we 'justify' this characterisation in any philosophical sense? Or more specifically: how are we to 'define' the meaning of the relevant key terms, e.g. the meaning of the notion of 'necessity'? A question which in the criticism of Hume is interpreted as: 'what is it that corresponds to the element of necessity?' ('What is the "impression" corresponding to the "idea" of necessity?')[2] Thirdly, there is the question from both as to how we can establish scientific conclusions embodying a necessitarian logic.

Actually, behind the last question there lies a more general one, involving the problem of the justification of scientific conclusions, whatever their logical character, i.e. whether necessitarian or not.

---

[1] Cf. above, p. 35.

[2] It will be noted that at this point 'ontological' elements begin to insinuate themselves into the discussion of our model. We shall subsequently, especially in connection with Hume and Kant, subject the whole notion to a critical review.

Of these questions, the first was—as we have noted—not really ever asked at all during our period, for it is throughout assumed that in some sense or other all scientific statements have a necessitarian logic.[1] The other two questions are not usually clearly distinguished, and it is only Hume, and after him Kant, who explicitly do so.

At the very start, Descartes attempts an assimilation of the rational *a priori* model, in his attempt to lay the foundations of dynamics. At the same time, his writings already give evidence of certain empirical tendencies, positive views concerning the application of the inverse-deductive method, and the employment of purely heuristic models (e.g. in his optical writings), which set up a tension *vis-à-vis* the earlier, more deductive, tendencies.

Locke, too, is a thoroughgoing necessitarian. But, as was noted, for him such a view results in a pervasive scepticism. His ideals are so stringent that almost all 'knowledge' based on empirical observations becomes 'but judgment and opinion, not knowledge and certainty'.[2] 'Certainty and demonstration are things we must not in these matters pretend to', he writes;[3] and his terminology suggests (and his argument implies) that this is because information gathered through 'observation and experiment' does not come up to the ideal of yielding 'necessary connections', found for instance in the 'demonstrations' of the mathematicians. Necessity, which for Descartes is a 'methodological ideal' though realised perhaps only in certain special contexts, is for Locke mostly a shadowy ontological affair, which causes him to parade in the somewhat misleading guise of a sceptic.

Let us be quite clear, however, that Locke's was a 'philosophical complaint', and not a 'scientific' one. He was fully aware of the success of Newton's *Principia*. His preoccupation is with epistemological questions, though often as yet in a very specific and narrow way.[4] He is struggling to hammer out an adequate conception of 'scientific knowledge', but it is a very specialised struggle brought on by the conflict which he finds between the paradigms of mathematical and necessitarian forms on the one hand, and what he calls 'experience' on the other.

---

[1] Berkeley is perhaps an exception, but his doctrine of divine activity expresses much the same idea.

[2] *Essay*, iv. 12.10.   [3] *Essay*, iv. 3.26.

[4] He has not yet met in full Hume's problem of the justification of our knowledge of empirical *uniformities*, even though he repeatedly affirms that 'the certainty of universal propositions concerning substances is very narrow and scanty' (*Essay*, iv.6.13).

'Experience' is indeed the grand slogan, if not shibboleth, of our period, the second major guide-line for an appreciation of its intellectual tendencies—the first being the notion of system, law and mathematical form, allied to the notion of rational connection, which we met in our account of the three models. (As I have shown elsewhere, Newton's two major works, the *Principia* and the *Opticks*, each in turn represent admirably these two great currents of the seventeenth and eighteenth centuries.[1]) This is not the place to document the rising prestige of that all-important aspect and requisite of scientific progress, observation and experimentation (as seen by Locke's contemporaries); or the frequent abhorrence of 'the *a priori*', and of 'hypotheses'. Where the nomothetic and systemic models seemed to involve justification from 'above', and even 'beyond', the attempt to demand substantiation by sensory evidence drew men's eyes to the support from 'below', emphasising the most imposing aspect of empirical or *a posteriori* knowledge, its dependence for certification upon evidence which might at any moment be found to upset it. Of primary interest to us here are the echoes which these tendencies evoked in the minds of the major thinkers of the time; the critical responses which it drew forth from them; and upon the philosophical appraisals of the so-called 'empirical foundation'. When this foundation is investigated through the new philosophical spectacles, it is found (or seems) to run directly counter to the promise of the earlier ideals of 'necessity and eternity' of scientific knowledge, although this happens partly owing to the subtle transformation of the commonsense notion of 'empirical evidence' into private 'data of sensation'.

Man's knowledge comes from 'experience': thus Locke. This experience (for instance 'sensation'), yields 'ideas'. The place where these ideas are 'located' is the mind. Indeed, what the mind is 'applied about whilst thinking' are these 'ideas in the mind'. To know, observe, to be conscious of an external object, in the long run (and in a way to be considered later) is to be 'applied about', to contemplate, as it were, certain corresponding ideas, 'conveyed into the mind' by our senses.

We notice here already a sharpening of the older experientialist programme. The insistence that we should eschew hypotheses and 'vain constructions', that we should attend to the particular object or

---

[1] See my *Image of Newton and Locke in the Age of Reason*, i.1–2. It is the *Opticks* that emphasises the experiential strain.

occasion, the 'here-and-now' that comes under our observation, has unawares turned into the doctrine that *genuine knowledge* requires the locating of the 'object' *within the self* (a doctrine which is a heritage of the Cartesian standpoint). Only in this way can we be sure that the object of our knowledge exists, is real. (I ignore here for the moment the immense vagaries of Locke's notion of 'idea'; it is more important for my purpose to concentrate on what I take to be the essential aspects.) As an immediate consequence, the status of our 'knowledge of the external world' becomes a problem that will haunt all these philosophers down to Kant—and beyond. Locating the 'fundamental furniture of the world', as it were, in the mind, in turn produces unease with respect to everything that lacks this 'geographical' locus. Hence there results the search for additional ontological principles in order to enlarge the original basis that has now become too narrow for intellectual comfort. Nor do these difficulties concern only the problem of the objects of our external world; they embrace also that of the status of (our ideas of) the space and time in which such objects are located.

The 'idea', not surprisingly, shares many characteristics of the visual object. Above all, it is a 'particular existence'.[1]

We are now in a position to summarise the origins of the conflict in Locke between the demands of the 'semantics' of scientific propositions and their epistemic basis. It lies in his conception of 'experience' as being equivalent to the perception and combination of 'ideas'. These ideas he views each as being quasi-self-contained logical 'atoms'; they 'enter the mind by the senses simple and unmixed'.[2] Hence, complex ideas (and most of our ideas, including 'generalisations', are complex) must be compounded, the result of a 'subjective' synthesis.

Based on small psychological evidence, this story represents the following epistemological doctrine:

(a) All knowledge is based on experience. This entails that the character of the known is essentially determined by the character of

---

[1] Cf. *Essay*, iii.3.6: 'All things that exist are only particulars', with iv.17.8: 'Every-man's reasoning and knowledge is only about the ideas existing in his own mind, which are truly, every one of them, particular existences'. (Owing to the systematic ambiguities surrounding this term, this means not only that some corresponding psychical states are particular, but that what the idea 'means' or 'intends' must be particular also; which is not to deny, of course, that 'idea' at times connotes, for Locke, a concept.)

[2] Cf. *Essay*, ii.2.1.

the 'process' of knowing, an important mark of which is that the knower is able to consider the known as it is presented 'in itself', apart from all relations to other knowable objects (ideas). This doctrine, of what Whitehead called 'presentational immediacy' of the object of knowledge, together with the premises just stated, at once yields the 'atomicity' of all ideas as such. (A reflection perhaps of the prevalent physical corpuscularianism of the century.)

(b) As a consequence of (a), the combining of our simple 'ideas' (of objects) into complex ones of substances involves a subjective ('mental') factor, an element of spontaneity on our part. (Locke calls it 'voluntariness'.)

(c) The relations whereby we unite our ideas in experience do not exhibit the necessitarian character required by the argument from science; or better: required by the logical character of scientific constructions as defined through the logical appraisals we have considered, such as the concept of law, e.g. of causal law.

Nor is this appraisal limited solely to the case of those relations which are coextensive with laws; it extends further, to cover those which are putatively involved in our knowledge of objects, in so far as we mean by this a knowledge of the permanent coexistence of certain sets of properties which are said to be properties of what Locke calls 'particular substances'.

This is made clearer by considering an answer to a possible objection to (c) above. For it may be said that surely it is 'in experience' that the qualities of objects are united for us; therefore, the unity can hardly be of our own making. This is, however, not Locke's point. For he would certainly grant as much, but add that 'in experience' qualities are only united 'here and now', whereas the concept of a 'particular substance' requires that we conceive these qualities 'united' also at times transcending the present; but here lies the difficulty: experience cannot make us 'certain' that present coexistences will continue in cases not 'tried'.[1]

Now the 'certainty' for which Locke is here clamouring is of the rational kind. For in the passage in question, Locke contrasts our 'want of knowledge' with the unattainable ideal[2] of insight into the inner structure of objects, their 'real essence'. Elsewhere it is the ideal of necessity that is more clearly highlighted.

[1] *Essay*, iv.12.9.
[2] We shall discuss in Ch. IV the various senses in which for Locke this ideal is unattainable.

Let our complex idea of any species of substance be what it will, we can hardly, from the simple ideas contained in it, certainly determine the *necessary coexistence* of any other quality whatever. Our knowledge in all these enquiries reaches very little further than our experience.[1]

So we see that the experiential 'unity' is 'made voluntarily'[2] in a sense which contrasts with the metaphysical or near-metaphysical cases of the 'real essence', or of the 'necessary coexistence' of some supposed complex of qualities, and is thus itself metaphysical.

It follows that it is Locke's peculiar characterisation of the elements of 'experience' that clashes with his appraisal of the logical character of 'science' in the old sense of a structure of natural necessities, whether in the form of laws or of substances. The clash is presented in the form of a putative 'limitation' put upon the extent of our knowledge, an apparent scepticism which has not yet become sufficiently self-conscious about its own moorings. The notions of law and substance demand a connection or link which upon 'inspection' of the epistemic conditions of knowledge does not seem to be forthcoming. As we shall see in detail later, Berkeley and Hume will only extend and complete the consequences of Locke's metaphysical atomism.

Hume in particular was to shift Locke's position towards the general problem of the inductive foundation of science, interest in which had increased as a consequence of Newtonian pronouncements on induction, as in Newton's own insistence in the *Principia* that in 'experimental philosophy ... particular propositions are inferred from the phenomena and afterwards rendered general by induction'.[3] That this raised philosophical problems and not just scientific ones (of how to strengthen one's conclusions by further evidence) becomes clear if we consult Newton's fourth 'Rule of Reasoning', according to which

> we are to look upon propositions inferred by general induction from phenomena as accurately or very nearly true ... till such time as other phenomena occur by which they may either be made more accurate or liable to exceptions.[4]

If we consider these two passages carefully we see that there are two aspects involved, the first of which is the procedural one of

---

[1] *Essay*, iv.3.14; my italics. For the same point in respect of 'law', cf. iv.3.29.
[2] *Essay*, ii.12.2.
[3] *Principia*, Bk. iii General Scholium; op. cit., p. 547.
[4] *Principia*, op. cit., p. 400. I have discussed this formulation, and its relevance for the philosophy of induction in 'Inductive Inference and Inductive Process', *Aust. J. Phil.*, **34** (1956), 164–81.

formulating and testing a putative general hypothesis, leading to possible abandonment or modification; and the second, an inferential move, described as the 'rendering general' of the 'particular' propositions. It is important to distinguish these two aspects, in order to become clear about the nature of the philosophical (as distinct from methodological) problem that exercised the philosophers of our period.

Now it is possible to maintain that there is *no problem* in the scientist's determination of formulating general hypotheses; and that the processes of science are exhausted in the more or less complex testing and attempted falsification procedures devised from time to time.[1] Or, alternatively, that *no inference* from observed cases to law occurs, because laws live 'in a new dimension', which does not represent a mere extension of the realm of experience but is 'in a certain sense an abrogation of the whole realm';[2] or that because statements of law live in a different 'language-stratum', we are in need of an entirely new logic, 'still unexplored'.[3]

All such contentions are evidence of the feeling that laws are more than mere generalisations; indeed, we may note that the necessitarian analysis, regarded in this way, is just another, only stronger, alternative. But considering that the scientist's adoptions of laws cannot proceed in complete separation from the facts presented to us by nature; and considering further that the question of the 'applicability' of laws to future experience is after all still with us (requiring in turn *general* remarks concerning the limits of such application!);[4] this seems to me sufficient warrant for regarding the investigation of problems surrounding induction (and their history) as not totally misconceived. At any rate, none of the above considerations show that the discussion of the justification of the generalising inference, meaning by that, of the very notion of the adoption of a *general*

---

[1] They have been elaborated with much expertise, for instance, by Karl Popper, in *The Logic of Scientific Discovery*, op. cit., and elsewhere.

[2] E. Cassirer, *Determinism and Indeterminism in Modern Physics*, Yale, 1956, pp. 41–2.

[3] F. Waismann, 'Verifiability', *Proc. Arist. Soc.*, Supp. Vol. XIX (1945) p. 132; reprinted in A. G. N. Flew (Ed.), *Essays in Logic and Language*, First Series, Oxford, 1951, p. 129. I have discussed these positions critically in 'Convention, Falsification and Induction' (*Proc. Arist. Soc.*, Supp. Vol. XXXIV (1960)), pp. 113–30.

[4] Cf. Kneale, *Probability and Induction*, par. 17, pp. 76–7; 'a remark [*sc.* that mere rules will enable us to make successful predictions] which is in itself a universal statement about the world and contains all the difficulties . . . [found in] formulations of natural laws . . .'.

hypothesis as such, seen against the *particular* phenomena of our experience, is a senseless one.

Be that as it may, it is clearly *this* question, and not the methodological one, which exercised some of the great classical philosophers. And they sharpen it in a peculiar way: they give it a twist that amounts to a 'metaphysical decision'. We have already seen how the procedures of science emphasise the importance of the 'verdict of experience', and the central place held by the notion of 'the possibility of falsification'. Now it is easy to describe this situation as one where we say that 'we do not know' what the future may yet hold in store; that 'we do not yet know' whether a given hypothesis 'is true'. And it is equally easy to go on from this, and say that, no matter how much information is laid up concerning a certain hypothesis, we shall never know whether it is true. Or our sense of ignorance may be so great that we succumb to the temptation of saying that it is even 'nonsense' to speak of 'the truth' (however putative or however well founded) of an hypothesis;[1] that Newton was misguided to *speak* in terms of 'rendering general by induction' some particular proposition or other.

Now it must be noted that the sceptical move, when completed, has written into it already the kind of 'metaphysical decision' just referred to. And such a remark applies as much to our modern sceptics as to those of the time of Locke and Hume—a fact which after all constitutes the timeless interest attaching to critical history of philosophy. We can bring this out by an imaginary counter-objection to the sceptic.

We might say that we ourselves do not really 'know' either, (a) *whether we are entitled* to draw general conclusions from experience, however well founded and well consolidated, or, (b) *whether we are NOT so entitled!* The complaint that we *know*, just as much as that we do *not*, is equivalent to a general pronouncement about the logical character of the terms of the propositions in question.[2] What we are told is that the terms of the proposition are 'really separate': this is the true meaning after all of 'synthetic', when contrasted with 'analytic', in Kant's famous definition:

> Either the predicate B belongs to the subject A, or something which is (covertly) contained in this concept A; or B lies outside the concept A,

[1] Cf. above, p. 68.
[2] This is one of the general conclusions of this book which I hope a fresh look at Kant's 'critical standpoint' will suggest.

although it does indeed stand in connection with it. In the one case I entitle the judgment analytic, in the other synthetic.[1]

By themselves, the words of this definition do not tell us very much. Everything depends on their interpretation; on the way they work in context; on what the consequences are taken to be of the fact that a certain predicate B 'lies outside' the concept of the subject A; what the logical consequences are believed to be of this fact, and how we judge it to affect the consideration of the competing fact that A nonetheless in some sense stands 'in connection with' B. (To this I shall return.)

Now it is clear that the 'empiricist', just as much as the 'falsification-ist' interpretation of these terms is such as to imply that *no* genuine movement, *no* genuine conclusion from particular to general, is 'really possible'; or that it is *possible* only if we 'assume' certain very general postulates and principles which in turn cannot 'be proved', but must be 'presupposed', on pain of otherwise 'being unable to make any inference at all'.

Less dramatically, we may embark on a programme which seeks to 'explicate' the schemata of certain groups of rules which define 'good' inductive inference; or which are said to provide 'confirmation' for our hypotheses, in which case any notion of an actual inference (whereby we 'detach' the conclusion from the evidential premises) will be avoided. But such rules will be purely descriptive, mirroring beliefs concerning the strengthening of hypotheses that we hold otherwise on instinct, or which arise from an appraisal of the structure of the world as observed.

Now it is of the greatest importance not to consider such ways of describing the logic of the situation as either 'right', or (alternatively) as 'wrong'. For, as I have suggested, the descriptions in question amount to a kind of 'metaphysical appraisal', of what is to be counted a logically impeccable move and what is not. It involves the positive adoption of certain 'empirical analogues' (as we have wanted to call them), and the rejection of others that compete with them. Thus in the present case, the existence of 'evidence', however strong, is permitted no more than 'psychological relevance' for the 'putative truth' of the hypothesis. 'Logically', we are told, positive evidence is of no significance.

And just here lies the relevance of the first great example of a metaphysical appraisal which we meet in the writings of the classical

[1] K. 48. Cf. below, ch. VIII, sect. 6(c), p. 570, for a more detailed account.

empiricists. Their description of the situation in terms of 'ideas', each
—as we shall see in more detail later—'passively received', each
existing 'as particular', and in splendid 'private isolation', is a remarkably graphic embodiment of their appraisal of the logic of 'synthetic
propositions'. It is a logic which entails that in so far as we can speak
of a 'connection between terms' of a proposition at all, such a connection has a significance *only* 'in experience'—a fact which is enshrined
in the doctrine that all synthetic propositions are '*a posteriori*'. And
be it noted that the 'obviousness' of such a doctrine depends entirely
upon the adoption of suitable conceptions concerning the notion of
'experience', as well as one's appraisal of the logic of the 'terms'
involved; a situation which mirrors precisely the doctrinal fog surrounding the corresponding notions of the 'analytic' and the '*a
priori*'.

The important consideration that arises from this is once more that
the offering of clearly articulated definitions of these terms is only a
particular stage in the long-drawn-out process of trying to formulate
certain metaphysical appraisals. They are not a final answer to earlier
confusions, but rather, powerful stages in the development of philosophical doctrines. Such a reading will, of course, affect our evaluation of every stage of the story.

However much the apparatus of 'knowledge by way of ideas' may
be suggested by our ordinary ways of talking, at one level buttressed
by a vast set of logical and psychological considerations, there can be
no doubt that its ultimate significance lies at another level, in its
equivalence to a set of metaphysical appraisals, the key-term being a
metaphysical centre of gravity, which aims at supporting a structure
of otherwise 'physical dimensions'. As we said at the start, and shall
have frequent occasion to note again, the metaphysical entities appear
initially as elements in the 'ultimate furniture' of our (epistemological
or ontological) universe; and they turn up a second time, to make
good inadequacies in that 'furniture'. In this way, the God of Berkeley
will aim to supplement the initial elements of Berkeley's universe,
'ideas' and individual minds.

The appraisal of the logic of 'synthetic propositions', to which we
have referred, is conducted with particular severity by Hume. He
begins with and makes, as the very basis of his negative argument, the
'Aristotelian assumption' mentioned above,[1] that induction presupposes 'objective' necessary connection. He next proceeds to show that

[1] p. 42.

the conception of such a connection is otiose. The 'irrationality' of inductive inferences made by scientists in their everyday procedures follows then as an immediate consequence.

This conclusion is more far-reaching than it seems at first sight. It might be thought that one could counter it, or at least diminish its importance, by simply insisting that inductive inferences do not require or presuppose necessary connections; that they are inferences of a different *kind*; and that therefore the question of their justification is really quite unaffected by the rejection of the necessitarian position.

Such a reply, however, misses its aim, on at least two grounds: (1) We have already seen how systematically vague is the notion of necessity in the sense here involved. Indeed, we noted the extreme ('justificatory') case, in the passage from Ewing on p. 42 above, where it is so used that to reject it *is* equivalent to a rejection of inductive inference. (2) Hume's protest against the 'rationality' of inductive inference involves, as I suggested, a metaphysical appraisal of the logic of 'synthetic propositions'; and with this, of the nature of the propositional link. He seems to insist that in the light of the epistemic situation as envisaged by the philosophy of 'ideas', the various models which so far have been offered are inappropriate. As a result, what at first looked like an academic exercise (the question of the correct appraisal of the logical character of laws, i.e. uniformities *v.* nomic necessities) becomes a doctrine with potent consequences for the foundation of any knowledge whatever.

These difficulties are well known. A more far-reaching development, or generalisation, leading to what we might call a case of philosophical degeneracy, is, however, not so universally appreciated, and is less easily grasped. We meet it first in Leibniz, and then in Kant. (The nature of this development, it will be remembered, was meant to teach us something about the idea of philosophical 'progress' as such.)[1] We have seen how Locke's sceptical conclusions arise out of his metaphysical atomism. Although I have called it 'metaphysical', in Locke's formulation it parades of course in the guise of the consequences of the argument that knowledge is limited to experience, where this means the passage of separate and isolated ideas before the mind. Any connections between these ideas (excepting 'logical' connections), he implies, are almost totally beyond our comprehension; it being left unclear whether this is a matter of accidental or intrinsic incompetence! As I have suggested, he retains

[1] Cf. above, pp. 10ff.

the echo of the *a priorist* tradition in his contention that for science proper, such connections are a *sine qua non*. But we must not employ the conception of such a connection in any positive sense; it constitutes only an unrealisable and perhaps even somewhat misleading ideal: Locke is here, as we shall later show, in an ambivalent position. We have seen how Hume's criticism deepens this contention, taking the view that the notion of connection in the sense here required was strictly meaningless. Moreover, it was meaningless because we could not 'perceive it' in the way in which formal, or analytical, or logical, or conceptual connections can be perceived.

Now at this point Leibniz's reaction becomes of great interest. For Leibniz is inclined to dismiss such epistemological purity. It is true, he argues, that we are not always *conscious* of such connections. But the possibility of scientific knowledge, and indeed, knowledge in general, is not by this impugned. On the contrary, despite our epistemic incapacity to recognise necessary connections in the empirical, and hence contingent, i.e. non-logical, cases, we are nevertheless entitled to—and indeed must—postulate such connections. He recognises of course that there are different types of necessary connection, of the logical and the non-logical kind. But he holds that we are entitled to postulate the existence of connections between phenomena even if they are of the contingent kind—the case queried by Locke, and later by Hume. Leibniz, in other words, is inclined to dismiss the question of the relevance of the epistemological situation for ontology. It may be that at the 'empirical' level (let us call it 'phenomenal level', though not quite in the Kantian sense of this term) all is 'loose and separate', in a near-Humean sense of this expression; but why should we draw conclusions from this concerning what is 'really' the case? If at the phenomenal level things seem arranged in a quasi-atomistic fashion, why should we conclude that this must also hold at the subphenomenal level? Nor have we, so Leibniz seems to imply, a principle gainsaying the possibility of such a general connection at the metaphysical level. This being the case, might it not be said that in this way we can save the possibility of scientific knowledge, at least in principle, if not in fact?

In this last move, there lurks a dangerous confusion. We may be permitted to split off, in the way suggested, the phenomenal from the metaphysical level. But we cannot really be allowed—without further argument—to avail ourselves of the results of our metaphysical explorations for the purpose of pronouncements on phenomenal matters. This is a besetting sin of most philosophy. It is fair to say that, as Leibniz grows older, the separation between the two 'worlds' is increasingly realised in his writings, even though the origins of his search are never completely abandoned. But only Kant realises that the separation must be acknowledged in specific terms (by way of a new terminology), recognising that Leibniz's postulational technique needs replacement by a new method of proof—the transcendental method. The significance of such an acknowledgement still awaits recognition!

There is, however, yet another, even more noteworthy feature in Leibniz's argument. During the earlier stages, he limits his considerations to the case of the uniformities of nature, where his contention of the existence of an underlying connection 'which provides the means to predict future appearances successfully'[1] is equivalent to the provision of an 'ontological justification'—a conception noted before.[2] Leibniz is aware of the difficulties which face him here, and he unconsciously seeks to lessen the strain by introducing the systemic model, for he thinks of the required connection in terms of a 'reason which can always be given' for *any contingent* occurrence.[3] But in this last phrase (with its reference to *any* contingent happening) he is moving towards an astonishing generalisation: for we are now told that it must make sense to speak of connections or links *in every case*. There is to be postulated an underlying ground relating the subject and predicate of *all* propositions, particular or general, contingent or necessary (logically or nomically speaking). Thus, in a tract entitled *On Freedom* (and it is repeated many times elsewhere in his writings) he writes:

> In every true affirmative proposition, whether universal or particular, necessary or contingent, the predicate inheres in the subject.[4]

Of course, this language of 'inherence' imitates the formal and rational *a priori* model, but it is certainly not *meant* to be identical with it. Our main point is, however, that here we are at last offered the promise of a 'link' between the terms of *any* proposition, whatever be its grammatical structure.

As he writes in *The General Characteristic*: 'There must be some

[1] 'Letter to Foucher' (1675), Loemker, op. cit., p. 239.
[2] Cf. p. 42, above.
[3] *On the General Characteristic*, Loemker, op. cit., p. 349.
[4] Loemker, op cit., p. 405.

real connection between subject and predicate';[1] and this is meant to apply with total generality.

A terminological warning may be in place here. In the formalism of Kant, we should have to say that Leibniz's doctrine amounts to the claim that not only the formulae of logic and mathematics—which (together with the 'analytic' cases) may here be classed as 'logically necessary'—but also all propositions purporting to be about matters of fact, and hence for us surely 'synthetic' (Leibniz calls them 'contingent', contrasted with their opposite, 'logically necessary'), are in a certain sense 'analytic'. This shows that 'contingent' is not always tantamount to 'synthetic'. If, however, one wishes here to retain 'synthetic', then the Leibnizian bond between the propositional terms, instead of 'analytic', should be construed as *a priori*'; and this would be justified by the fact that the bond is postulated to exist not on 'experimental grounds' (in one or other of the many senses of this expression) but on the basis of some form of 'postulational ground' involving a general principle (viz., of 'sufficient reason'), whose source Leibniz asserts to lie 'in God'.

Here, again, we note a fluidity in our key-terminology which is sometimes not sufficiently appreciated in modern writings. On the other hand, the attempt of some philosophers to make do with fewer such terms has the effect of confounding the confusion and of doing less than justice to the complexities of the logical situation which we meet in the original sources. This in turn frustrates the contact between the latter and the modern writings. Moreover, such considerations will forewarn us that in Kant there are complexities lurking in a terminology which is only apparently clear-cut, despite the fact that Kant was the first writer to make the distinctions just referred to.

Leibniz's own procedure sheds an immense light upon the nature of philosophical argument. The demand for justification is stretched here beyond all limits; it consciously amounts to sheer postulation. The case is one of 'over-generalisation', until it assumes the air of paradox. Nonetheless, as we have learned in recent years, it is certainly possible that when philosophical extensions thus degenerate into almost literal falsehood, bordering on the paradoxical, such falsehoods may be highly illuminating and productive. Their very generality permits us to detect the queerness of a contention in a more obvious way, when, so long as it resided in a narrower compass, this

[1] Lomeker, op. cit., pp. 348–9.

might have escaped notice. But the queerness of Leibniz's position has an important sequel in the logic of Kant, for which it provides a most important clue.

Kant (as we shall show in the relevant chapter) responds in two ways to Leibniz's central contention (and a study of Kant's pre-critical writings will show that he always saw it to be a central question or problem): (1) He contends that although Leibniz discovered the principle of universal bondage (if I may so label his 'principle of sufficient reason'!), he only postulated it in a purely 'dogmatic' fashion. When Kant speaks here of 'dogma', he means that Leibniz did not provide a proof, which would show the necessity of such a bond as a fundamental presupposition of 'knowledge', or 'experience', expressed in statements concerning such 'experience'. (2) Positively, Kant accepts the basis of Leibniz's doctrine, that it is incumbent upon us to 'give an account' of the link which relates the terms of propositions (or judgments) *in whatever form*, i.e. particular as well as general. But the critique of the empiricists has persuaded him that none of the three models that we have discussed, i.e. the formal or *a priori*, the nomothetic and the systemic, are adequate as they stand to explicate the notion of the link. A new device, a new point of view, must first be initiated. This is provided by introducing a greater structural complexity into the notion of the 'phenomenal level', which is held necessarily to involve *a priori* as well as *a posteriori* elements; a sensory material basis as well as spatio-temporal and conceptual form. Kant will thus postulate that the notion of the empirical proposition, or rather judgment, requires the sensory basis to stand in need necessarily of 'form' (he calls it a 'transcendental' condition). And it is this form which provides the link or 'unity' of the material (sensory) elements of the judgment.

So we see that Locke's somewhat *psychologising* notion of 'unity', as well as Leibniz's *metaphysical* concept of the bond of predication, is here replaced by a '*transcendental*' construction. Finally, the nomothetic and systemic models, though rejected for the purpose of interpreting the link involved in the notion of the individual judgment, are subsequently reinstated for the articulation of the notion of the *system* of judgments (particularly judgments in general form, i.e. empirical generalisations), yielding thus an explanation of the concepts of scientific law and theory.

Once again we see that philosophical 'progress' consists in a re-distribution or re-location of the various key-notions at different levels, accompanied by a refinement if not additional creation of such levels. The nature of this process will be appreciated better after we have considered its form in greater detail.

I think the average reader of Kant is often hindered from properly grasping the central core of his work by an insufficient appreciation of —not only the general Leibnizian heritage which is worked into its very frame, but more importantly—the central aspect of this heritage, which is the doctrine that *in some sense at least* there must be an *a priori* link built into any proposition whatsoever, general or particular, necessary or contingent. Our average reader is too preoccupied (and too influenced by Kant's own 'guidance') with the problem of the foundations of scientific knowledge, e.g. of causation, substance, conservation, etc.[1] It is true that Kant saw himself as the heir and preserver of the Newtonian heritage—but not only that. The real significance of his doctrine lies in the fact that he secures these foundations by shifting the discussion on to such a wholly abstract ground that in the end it is drained of any genuine meaning whatever. Kant, so to speak, *anaesthetises* the metaphysics of both Hume and Leibniz. And though he claimed that for a proof of the possibility of scientific knowledge one had to supply transcendental foundations of experience in general, this feat was accomplished only by using a logical scaffolding on the queer lines laid down by Leibniz.

We shall later study this Kantian process of 'anaesthetisation' of the philosophical key-conceptions of his predecessors. It should not be thought that the result is frivolous. Quite the contrary: the Kantian philosophical procedure offers the philosopher a new 'autonomy', the significance of which has not yet been exploited. Conceptions such as 'external object', 'idea', 'perception'; the notion of 'voluntary combination' built by Locke into the concept of 'complex idea'; the 'metaphysical atomism' of these 'ideas'; all this in tension with the ideal of 'necessary connection'; objectivity of *a priori* conceptions; the foundations of dynamics; not to mention the search for an adequate foundation of ethics and religion (which we here ignore); the existence of an objective order of biological classification:—clearly such tensions could only be resolved by a complete critical purge of each of these conceptions; by a realisation that not only the notions of

---

[1] Just as he is misled by Kant into taking at face-value the 'exhaustive' division of propositions into analytic *a priori*, synthetic *a posteriori* and synthetic *a priori*, when —as we shall later see—the Kantian situation is in reality far more complex.

substance and law, but also of perception and experience, were in need of critical analysis.

Of course the attempt to provide such a synthesis has proved a failure—in a sense. 'Foundations of mechanics', particularly under the guise of special laws, such as those of Newton, make uneasy bedfellows with a general theory about the conditions of experience, supplemented by even more general schemes concerning 'psychology', cosmology and theology. Nonetheless, the positioning of the various elements, the conceptions of 'law', of 'system', of 'explanation' and 'hypothesis', is still—so we shall find—of significance for us today.

III

DESCARTES: METHOD AND METAPHYSICS

In the introductory pages of Chapter I attention was drawn to an important facet of philosophy as metaphysics: namely its selection of certain features or strands from the totality of experience which it then proceeds to regard as elements of 'the fundamental furniture of the universe'.[1] Such features then become the true 'foundations' of the universe, in terms of which all the rest is to be interpreted, and to which all else has in some way to be 'reduced'; what particular form such a reduction takes will naturally vary from case to case. It must not be supposed, however, that reduction is equivalent to elimination. On the contrary, it is an important aspect of metaphysical philosophy that it continues to take note, and indeed seeks to preserve, the total multitude of facts, however much it may seek to 'unify' them under the guidance of one or other of its leading conceptions. The degree of success in achieving integration of all this conflicting detail within an overall scheme without the introduction of obvious tensions varies; so does the extent to which a philosopher is forced to inject supplementary metaphysical conceptions in order to widen the basis of his fundamental starting point.

Students of philosophy usually limit their attention to the more obvious constructive features, with the effect that the philosophical system they are studying takes on a somewhat one-sided, not to say bizarre, appearance. They lose sight of the point that each of the great systems of philosophy seeks to accommodate *all* of the phenomena of our everyday world. The significance of this is best appreciated in the context of the specific concerns of some of the great philosophical systems. In the philosophy of Descartes, these are at least three-fold.

[1] Cf. above, pp. 3ff. See also my 'Science and Metaphysics', in *The Nature of Metaphysics*, D. F. Pears ed., London, 1957, pp. 61–82.

79

They connect with Descartes' disquisitions on the proper method suitable for the study of the sciences; with his investigations and discoveries in certain branches of science, such as optics and dynamics, together with the attendant framework of 'natural philosophy'; and finally, with his epistemological and metaphysical doctrines: epistemological, as regards the question, 'what is the proper basis of knowledge', behind which lies the more general search for an adequate conception of 'knowledge'; metaphysical, as regards the question, 'what are the proper objects of knowledge', considered as knowledge of 'the fundamental furniture of the world'.

Now at each level there arise in the epistemological situation conflicting claims which demand a hearing and whose relevance cannot be ignored. Each candidate for such a hearing claims to be 'fundamental' and seeks to usurp authority over the others. In this way, certain tensions are introduced, which a philosophical system seeks to eliminate by a process of redintegration. It is then that differential emphasis is given to one or other of the candidates, though this must not involve complete elimination of the alternatives. Whether the resulting strains and stresses can in the end be suitably released is another question. What interests us here is the way this is attempted.

Consider the contending candidates in the field of the logic and method of science. Here the claims of an empirical (experimental) basis of science are met by the demand for a 'theoretical justification', or basis, vouching for their integrity. The requirement to advance by the use of 'the method of hypothesis', the strength of a given set of hypotheses resting solely on their power to 'explain' a variety of observable consequences, is matched by the demand for an independent 'proof' of the former: a demand fed by a fascination with a deductive model of logic which regards the hypothetico-deductive approach as in some sense deficient, if not wholly 'fallacious'. Again, the demand that all the phenomena belonging to a given subject should be explained by an all-embracing systematic deduction from only a few primary principles is met by the consideration that a theoretical system requires for its application a reference to special boundary conditions, initial assumptions, and data of a specific and particular nature, themselves inevitably lacking the precision of the theoretical framework. And again, the requirement that the scientist should show the observed 'phenomena' in some given physical situation to be the resultant or product of an underlying 'internal structure' of elements and forces has to contend with the fact that the path of discovery inevitably is the reverse of this: going from the complex phenomena to the underlying structure, with the resultant conflicting demands concerning the basis of our knowledge of this structure.

Similar conflicts meet us at the other levels. In natural philosophy, the ideal of mathematics collides with the 'bruteness' of the phenomena of a sensory world which appears anything but mathematical. The model of 'necessary connections' suggested by the power of mathematics and logic is not obviously matched by the mere 'concatenation' of the qualitative phenomena of our universe. The requirement of an ideal of knowledge which demands the incorrigibility in principle of a proposition genuinely 'known' has to contend with the existence of 'opaque elements' which meet the enquiring scientist at every point. And just when the conditions for this ideal of 'knowledge' seem met, by locating its basis in the subject's 'self' and its 'ideas', in order to render it thus safe from 'external influences', it is confronted yet again by the need to 'explain' the 'existence of an external world' —of an 'other than the self'.

It is a measure of the greatness of Descartes' pioneering work (not for nothing has he been called 'the father of modern philosophy') that in his writings all these elements are still encompassed in their freshness, and with their resulting tensions. Here, these tensions are only half-resolved, for the outlines of the various philosophical concepts are as yet less definite, the resulting synthesis vaguer, but also less rigid than that in later philosophers. Moreover, despite appearances to the contrary, *all* the conflicting elements are preserved in Descartes' system. Only their ultimate 'weighting' misleadingly gives the impression that some vital elements, say some aspect of scientific methodology, have been omitted.[1] If this really were the case, a study of Descartes' methodological writings would be of little importance; in fact their interest lies precisely in the way in which the *conflicting* facets of method are integrated and *described* from a certain vantage point; one which has adopted as norm a selected part of that method in terms of which the whole is then to be interpreted.

The length of the present chapter is due to the necessity of indicating in some detail and by some textual analysis how Descartes' philosophical (and metaphysical) position arises easily and naturally

---

[1] Cf. a recent assessment: 'Descartes . . . remained unaware of the modern notion of a scientific hypothesis, and, indeed, would have rejected it.' A. Wollaston, *Descartes' Discourse on Method and other writings*, Harmondsworth, 1960, p. 26.

out of his scientific preoccupations. But there is a further reason for going into detail. It is with Descartes that the story of modern philosophy begins. The broad questions are asked; the lines of direction in which problems are to be explored are drawn for the first time. Above all, Descartes' peculiar brand of 'rationalism' will be shown to have roots in his science; nor is this just a mere historical curiosity, for many of Descartes' metaphysical starting-points, for instance the notion of 'clear and distinct' cognition, initially receive a *technical* meaning *via* certain scientific models of reasoning. This fact is important, because it will explain Descartes' optimism in expecting to achieve success with his programme of an ultimately deductive filiation of the whole range of phenomena and experience, starting from the *Cogito* and an apprehension of certain primary principles seen by 'the natural light'. For he believed that such a programme had been successful in some of the ranges of mathematics and science. These applications (witness the case of 'intuition'[1]) frequently then yield a technical meaning which is subsequently employed with a metaphysical intent. No mere analysis of the *philosophical* texts alone can therefore be sufficient for a grasp of Descartes' thought. Moreover, no philosopher requires closer attention to his actual writings because of their intellectual 'fluidity', and because of the need to see them in their specific contexts. To such a study this chapter must now proceed.

I. DESCARTES' 'METHOD', AND ITS PLACE IN HIS SCIENTIFIC AND PHILOSOPHICAL WRITINGS

(a) *'Universal Mathematics' and its objects*

When thinking of the great writers on scientific method in the seventeenth century, it is usual to pit Bacon against Descartes, and to contrast the inductive approach of the former with the preference for deduction by the latter. And with more recent views on scientific method in mind, many are inclined to dismiss Descartes' deductive method as indicative of an insufficient understanding of the experimental aspects of science, charging it with neglect of the requirement that scientific conclusions must ultimately be anchored in a 'solid basis of experience'. In fact, however, when we study Descartes' methodological views more carefully, they are found to exhibit a fascinating chiaroscuro, an interplay between both

[1] Cf. below, pp. 140ff.

'movements' of scientific reasoning, from data to explanation, from effect to cause; and (conversely) from theoretical principles to laws and from laws to the observations 'colligated' by such laws. Now, not only is the existence of both these 'movements' in science a fact, but Descartes takes cognisance of both, at least in his methodological pronouncements. In this, he is moreover influenced by actual models, serving as paradigm cases, and which appear to illustrate the various aspects of scientific reasoning. Descartes' estimation of the importance of such models will be appreciated if we remember that he himself made important contributions to mathematics, mechanics, dynamics and optics, sciences in which the different contrasting logical situations play their part.

From the start, as is well known, Descartes was fascinated by the mathematical model. No doubt, this stimulated the deductive emphasis in his thinking. Already in the *Regulae*, his earliest work, he comments that although 'there are two ways by which we arrive at the knowledge of things, viz. either by experience or by deduction', the former is 'often fallacious', because 'imperfectly understood'. And so

> arithmetic and geometry far surpass all the other known disciplines in certitude. They alone treat of an object so pure and simple as to admit of nothing that experience can render uncertain; they entirely consist in a sequence of consequences which are rationally deduced.[1]

Not that arithmetic and geometry alone are to be studied, Descartes adds; but we must insist

> that in our search for the direct road to truth we should not occupy ourselves with any object about which we are unable to have a certitude equal to that of arithmetical and geometrical demonstration.[2]

Clearly Descartes does not mean to stop at arithmetic and geometry. His point is that only those matters can receive 'scientific' treatment which are capable of being studied in the mathematical way, employing both the conceptual and the logical tools of the mathematician, the operative notions for Descartes being 'order and measure'. But this is universalisable; for

> in respect of measure it makes no difference whether it be in numbers, shapes, stars, sounds or any other object that such measure is sought,

[1] Op. cit., II, K.S., pp. 8–9.   [2] Ibid.

and ... there must therefore be some general science which explains all that can be enquired into respecting order and measure, without application to any one special matter.[1]

This 'general science' is Descartes' celebrated 'universal mathematics'. We need not spend much time here considering the precise connotation of this term. It is clear from the context that Descartes' central aim is to stress the need for a method (or logic) modelled on the best of the mathematicians, and to emphasise that the conceptual elements, too, should imitate the procedure of the mathematician; which is, to 'begin always with the things which are simplest and easiest'[2]—a fateful phrase, as we shall see, and one whose precise content may easily elude the unwary reader.

To understand the nature of Descartes' general approach as mediated by the mathematical model let us look in slightly more detail at the passage in Part II of the *Discourse*, just mentioned. Here, as before, he is found to emphasize the formal and orderly procedure of the geometers: their starting from what is 'simplest and easiest'. The passage is a classical one.

> Those long chains of reasonings, each step simple and easy, which geometers are wont to employ in arriving even at the most difficult of their demonstrations, have led me to surmise that all the things we human beings are competent to know are interconnected in the same manner, and that none are so remote as to be beyond our reach or so hidden that we cannot discover them—that is, provided we abstain from accepting as true what is not thus related, i.e. keep always to the order required for their deduction one from another.[3]

And what are the 'objects' i.e. the conceptual elements of such a science? The answer is that however different the subject-matter of the different branches of science may be, 'all agree in considering only the diverse relations or proportions to be found as holding between them'. Moreover, the simplest model in terms of which to conceive the relations involved is the geometrical model—the relations here 'holding between lines, there being nothing simpler and nothing that I can represent more distinctly by way of my imagination and senses'. But Descartes at once generalises this, for these lines may in their turn be 'expressed by certain symbols (Fr. 'chiffres')', a fact which enables us to collect 'all that is best in geometry and algebra'[4]—a reference to

---

[1] *Reg*. IV, K.S., p. 21.   [2] *Discourse*, ii, K.S., p. 129: third rule. Cf., p. 130.
[3] K.S., p. 130.   [4] K.S., p. 131, see below, pp. 12

Descartes' fourth 'Essay', the *Geometry*, where such a synthesis between algebra and geometry is attempted with considerable success, helping to lay the foundations of what is known as 'analytical geometry'.

A line participates of the genus of extension. The theme that extension, and comparisons between extensions, is the proper subject of science, had previously been broached more fully in *Regulae* XII and XIV. The logically simplest operation, according to Descartes, apart from 'the simple and naked intuition of one single thing' is 'comparison of two or more things with each other'. This is involved in any theoretical treatment of a scientific subject.

Consider for instance Descartes' discussion of 'the nature of the magnet' (one of his favourite subjects).[1] In trying to explain its attractive power, men like Gilbert had been wont to invoke the existence of 'sympathetic relations', and of a 'formate soul or animate form' exerting an action on other magnetic matter.[2] Descartes' objection to this procedure is that it involves conceptions which we do not understand and which are not 'evident'. Rather, having first made a requisite number of experiments, it will be necessary to adduce the basic elements (Descartes calls them 'simple natures') whose combined action might explain the experimental effects observed. The elements so postulated must not be far-fetched; rather, they should involve only what is a 'fundamental aspect' of all thought and experience: for

> if in the magnet there be some kind of entity the like of which our understanding has never yet apprehended, it is hopeless to expect that we shall ever be able to know it by way of reasoning; we should have to be furnished with some new sense . . .[3]

This will be Descartes' general refrain: the elimination of 'occult powers', 'spiritual substances' and the 'reduction' of the explanatory conceptual framework to a completely unitary nature which is matter in motion. Thus, Descartes demands, we must limit ourselves

> to discern *with all possible distinctness* that combination of *already known entities or natures* which gives rise to those effects which make their appearance in the magnet.[4]

[1] Cf. *Reg.* XII, XIII, XIV; K.S., pp. 73, 77, 85–6.
[2] Cf. Gilbert, *De Magnete*, ii.7, v.12, tr. Mottelay, Dover ed. 1958, pp. 123, 311.
[3] *Reg.* XIV, K.S., p. 85.   [4] Op. cit., p. 86; my italics.

4

The uninitiated reader might take this to be an instance of the kind of *a priori* method which he perhaps associates with the name of rationalism, particularly when he finds that Descartes holds that simple natures are 'already known', in the sense of 'intuited'.[1] Thus we are told that 'in the magnet there can be nothing to know which does not consist of certain simple natures, *known in and by themselves*'.[2] In a more formal mode of speech, we might be tempted to think of these simple natures as primary explanatory principles, particularly when we find Descartes speaking of 'that mixture of the simple natures' which produces 'all of the effects experienced in the magnet'.[3]

At this point it is important to beware of saddling Descartes (and other rationalists) with views they do not really hold. Descartes does not usually (and certainly not in the present case) take the view that scientific explanation requires the independent knowledge (through rational intuition) of explanatory hypotheses, whose deductive consequences describe observable phenomena of nature. For this reason, we must interpret Cartesian references to 'evidence' (or self-evidence) and 'knowledge'[4] very carefully, conscious of the fact that they bear a very fluid use, technically defined in some cases, though extended under the pressure of certain epistomological ideals and using external associations in other cases.

Thus, anyone acquainted with Descartes' very extensive account of the phenomena of magnetism[5] will be aware that his explanation invokes the *hypothesis* of certain screw-shaped particles corresponding to the pores in the magnetic material whose own shape determines whether these particles may pass through or not. The details of this theory need not detain us. What is important is that here already, in the middle of the discussion of the rational foundations of scientific knowledge, there enters a suppositional element, a fact which

---

[1] Cf. *Reg.* XII: 'We must make use of all the aids afforded by the understanding . . ; for the purpose of intuiting distinctly the simples which come before the mind' (K.S., p. 56).

[2] Op. cit., K.S., p. 73.   [3] Ibid.

[4] There is indeed a fatal ambiguity about the meaning of the French 'connaître' which means both 'to know' as well as 'to understand'. And we shall later see that Descartes does refer to 'simple natures' in terms of 'truth and falsity'. Cf. p. 105, below, and quotation there from *Reg.* XII, K.S. p. 66. This is a complex issue on which we shall be able to shed light only gradually.

[5] Cf. *Principles of Philosophy*, iv, sects. 144–87. For a brief summary of the theory, see J. F. Scott, *The Scientific Work of René Descartes*, London, 1952, ch. 12, pp. 188–93. Cf. also below, pp. 97f.

Descartes recognises quite explicitly, since his account in the *Principles* acknowledges that although by rational exploration we may discern the nature of matter and motion in general,

> we cannot determine by reason how big these pieces of matter are, how quickly they move, or what circles they describe. God might have arranged these things in countless different ways; which way he in fact chose rather than the rest is a thing we must learn from observation. Therefore, we are free to make any assumption we like about them, so long as all the consequences agree with experience.[1]

This is evidently a fairly accurate (if partial) description of the hypothetico-deductive form of reasoning, and we shall want to return to this topic in the sequel.[2] The question for the moment is 'How does this square with Descartes' demand that the basic elements of his theory should be "evident" and "already known"?'

If we return to the methodological account of the *Regulae*, we find the following. To be sure, we are to invoke in our explanatory account nothing but 'simple natures' which are 'known in and by themselves'. If the scientific enquirer sticks to this, there can be 'no doubt as to how he should proceed'.

> First of all, he diligently collects all the experiences to be had in regard to this stone [the loadstone, i.e. magnet], and from these he then endeavours to infer [*deducere conatur*] what the character of that mixture of the simple natures must be if it is to be effective in producing all of the effects thus experienced in the magnet. On determining this mixture, we can at once boldly assert that we have learned the true nature of the magnet, in so far as it can be discovered by man from his given experiences.[3]

It is clear from this that in Descartes' account of our theorising about the phenomena of the magnet we must distinguish two aspects. First, we have hypotheses about the actual shape and motions of certain elementary particles, formulated in the light of certain 'experiences', or experiments, i.e. governed by the fact that they explain the latter—a form of inference to which Descartes refers in the above passage as 'deduction', but clearly meaning 'hypothetico-deduction'.[4]

---

[1] *Principles*, iii. 46; transl. E. Anscombe and P. T. Geach, *Descartes' Philosophical Writings*, London, 1954, p. 225. (This will be referred to as Anscombe.)
[2] For 'hypothesis', cf. below, pp. 96, 102f., 118ff., 141ff.
[3] *Reg.* XII, K.S., p. 73.   [4] For this, cf. below, pp. 102ff.

The second aspect of the method concerns the prescription that the *concepts* employed in our theorising should be universally applicable, simple, clear, comparable as 'ratios', and expressible in mathematical form.[1] We should employ only what is common to all physical explanation; and what is common is the 'already known entities, viz. extension, shape, motion and the like . . .', as he tells us in a later rule.[2] Only this produces genuine 'understanding' ('knowledge').

We see: sometimes Descartes claims no more than rational *clarity* in respect of the 'intuitive' starting-points, i.e. *concepts*,—not rational *certainty* concerning *propositions*. He is not so much concerned here with inductive truth, as with what previously we have labelled the 'plausibility' of scientific principles.[3] But one cannot be sure whether at other times he does not want to extend the range of 'intuition' (what later he referred to as the domain of the 'clear and distinct') much further, especially in the realm of general epistemology.[4] But even in the narrower range of mathematical physics, the mathematical element may have led to more optimistic epistemic appraisals than the logic of the situation really allows. To this we shall return. For the moment we note that at the very start of Descartes' reflections on scientific method two extreme approaches, rational intuition and hypothetico-deduction, both hold sway. And we shall find that throughout Descartes' thinking these two aspects are productive of certain tensions that are never quite resolved, and which in fact get more pronounced as Descartes grows older.

Returning to Descartes' line of argument which leads him to the doctrine of a privileged set of 'simple natures' in science, this runs as follows. If we limit ourselves to ideas like extension, shape and motion, we can be sure that we 'understand' the nature of the objects which we are investigating. Just as the shape of a silver crown is conceived by us exactly like that of a golden crown, so, provided we concern ourselves with shape and similar 'natures' when introducing the explanatory entities of physics, we are assured that we may apply whatever reasoning we normally can in this context. Shape and motion are ideas which we 'discern with all possible distinctness'; they are capable of mathematical treatment; they allow most easily of the

[1] Cf. also the list of 'simple natures' given at *Reg*. VI: they are 'whatever is viewed as being independent, cause, simple, universal, one, equal, like, straight, and suchlike' (K.S., p. 25).

[2] *Reg*. XIV, K.S., p. 86.   [3] Cf. above, ch. II, pp. 26, 36, 38.

[4] This point will be taken up again below, pp. 104ff., 130ff.

operations of 'order and measure', of the method of 'proportions' by which we eventually arrive at the setting up of 'equations'.[1]

Of all these ideas, extension is the one 'most easily and distinctly depicted'. It is now not difficult to understand Descartes' formulation of the central aspect of his method: we must always ensure

> all our questions to have been so simplified that there is nothing else to be inquired into save only the knowing of a certain extension by the comparison of it with a certain other already known extension.

Be the subjects investigated ever so diverse, it is 'certain that whatever differences in ratio exist in other subjects can be found to hold also between two or more extensions.'[2]

Shape is another idea, related to extension and of almost as great an importance, for it is by various geometrical shapes that we represent to ourselves 'ideas of all things'. The twelfth Rule explains how perception of shape is involved in our experience of all qualities. Seeing that science finds difficulty in coping with such things as colour, Descartes suggests, let us abstract from every feature except shape, and let us think of the diversity existing between white, blue, red, etc., as one which exists between differently shaped sets of lines. Then whatever the nature of colour may be,

> it is certain that the infinite multiplicity of shapes suffices for the expression of all the differences in sensible things.[3]

We have arrived at the point where we see that the mathematical model suggests powerfully that all the objects of scientific knowledge should be considered under the aspect of extension—perhaps even, as it were, 'reduced' to extension! We shall not be surprised to find a corresponding metaphysical position in Descartes' later philosophical writings. Little more indeed was needed for Descartes to write in the *Principles*, i, sect. 53:

> Every substance has a principal property that constitutes its essential nature, and all others are reduced to this. Extension in length, breadth, and depth is what constitutes the very nature of corporeal substance. . . .[4]

We might say that, from holding that a successful scientific treatment of nature *presupposes* its being *considered* under the aspect of

[1] Cf. *Reg.* XIV, K.S., p. 88.   [2] K.S., p. 93; cf. below, p. 116.
[3] *Reg.* XII, K.S., p. 59.   [4] Anscombe, p. 192.

extension, Descartes slides into the assertion that (material) nature *is* essentially equivalent to extension, and that this alone *justifies* us in postulating the existence of genuine science.[1]

This is not to say that Descartes does not seek by means of argument to mediate between these two extremes. Thus the passage just mentioned goes on to say that

> any other possible attribute of body presupposes extension. . . . For example, shape is not conceivable except in an extended thing, nor motion except in an extended space.

In *Meditation* II a similar point occurs though considered more from the epistemological end in connection with the question: 'When can we be said to "comprehend distinctly", or to "apprehend adequately" a physical object such as a piece of wax?' The notion of 'distinctness' in general is a difficult one, as we shall see later. But here the context helps us. We cannot be said to comprehend *distinctly* any of those attributes of the wax that *change* whilst all the while the wax remains.[2] But if we abstract from all these changing attributes which we are normally capable of 'imaging' or 'sensing', all that remains is the extendedness of the body; and, not just this or that particular extension, but rather the whole 'variety of extensions' (the 'determinable', rather than the 'determinates') that a body is capable of and which clearly can never be pictured. Extension, in this sense, can therefore only be comprehended, not by sense or imagination, but 'by the mind alone', by 'thinking'. Descartes concludes that

> adequate apprehension is not a seeing, nor a touching, nor an imaging . . . but solely an inspection of the mind which may be imperfect and *confused*, as it formerly was, or *clear and distinct*, as it now is, according as my attention is directed less or more to the constituents composing the body.[3]

We should pay particular attention to the context here of the notion of the 'confused' and the 'clear and distinct'. Evidently when the mind focuses on the sensory aspect, it is 'confused'—we shall return to this fateful expression. To attain clarity and distinctness is

---

[1] Cf. above, pp. 6f. where this is compared with Locke's views on the 'reality of mathematical knowledge'; and below, pp. 116ff.

[2] Op. cit., K.S., pp. 207–9.

[3] Op. cit., p. 209; my italics. It will be noted that as for the case of the magnet, 'constituents' does not here refer to the elementary particles of which (according to Descartes' physics) a material body is composed but to those aspects of them which he designates by the terms 'shape' and 'extension'.

the function of 'thought', in fastening on the unchanging 'substance' beneath the changing attributes, distinguishing 'the wax from its external forms . . . stripped as it were of its vestments' and considered 'in complete nakedness'.[1] This substance has primarily a quantitative aspect—geometrical extension. Though in a sense we perceive physical objects by sensory observation,

> properly speaking, bodies are cognised not by the senses or by the imagination, but by the understanding alone. They are not thus known because seen or touched, but only in so far as they are apprehended by thought.[2]

Once again, the argument moves on two fronts, at two levels. First, though the world is given to us through sensory observation, a proper understanding of it is reserved for those aspects which result from the process of abstraction, and for those which unchangingly remain. And these are also the ones that yield to 'universal mathematics'. But, and secondly, that which remains after abstraction takes on an additional 'ontological' significance. What the mind 'considers in complete nakedness' is not *just* an abstraction; on the contrary, it is what 'really exists', 'properly speaking'. For Descartes argues that every other attribute of body *presupposes* extension; that all attributes are *reducible* to it.[3]

It is as though Descartes was passing from the contention that we are to study (and cannot but study) what is capable of order and measure, i.e. extension, to the assertion that matter or body *is* nothing but extension. Where earlier he had merely concentrated on the spatial and kinematical aspects of matter—as being the ones most 'clearly and distinctly' understood—he now reduces matter to these characteristics themselves!

The point is further reinforced by a doctrine which Descartes sometimes defends, viz. that we cannot really distinguish between corporeal substance and extension. True, in the passages quoted above, p. 89, from the *Principles* (cf. i. 51, 53), such a distinction is actually made and in the passages from the *Meditations* (on p. 91 above) this distinction is again implied, though this time between

---

[1] Op. cit., p. 210.
[2] Op. cit., p. 211. I have slightly modified the K.S. translation.
[3] Cf. *Principles*, ii. 4; Anscombe, p. 199: 'The nature of matter, or of a body considered in general does not consist in its being a thing that has hardness or weight, or colour, or any other sensible property, but simply in its being a thing that has extension in length, breadth, and depth.'

'substance' considered as a 'substratum' and the attributes which are said to 'inhere' in the substratum. On the other hand, in *Principles*, ii, it is implied that people wrongly distinguish the substance of a body from its extension; that the distinction is at best one 'only as regards our way of conceiving' the difference, 'just as number [differs] from what is numbered'; but that 'in actuality it is not possible to subtract the least bit of the quantity or extension without likewise removing just as much of the substance'.[1]

So it looks as though by 'reducing' all attributes to that of extension, and maintaining the equivalence of 'substance' and extension (or matter, or body, and extension[2]), Descartes has seemingly supplied himself with a handle whereby to fasten this 'abstract' conception of extension to a foundation in reality, to something that is more than merely 'abstract'.

But has he? To appreciate this question, it will be worth while (however briefly) to glance ahead and make a contrast with the teaching of one of the later 'empiricists', George Berkeley. Berkeley tells us that we cannot abstract from the idea of motion that of a body moving in some particular fashion, a body, moreover, possessing various further sensory attributes. True, he allows, there is *a sense* in which the process of abstraction can be performed. This applies when we '*consider*' the aspect of motion in isolation, as happens when the scientist performs certain demonstrations by the use of Newton's Second Law of motion ('The change of motion is proportional to the impressed force'). However,

> these propositions are to be understood of motion and extension in general. . . . It is only implied that whatever motion I *consider*, whether it be swift or slow, perpendicular, horizontal or oblique, or in whatever object, the axiom concerning it holds equally true.[3]

In just the same way,

> a man may *consider* a figure merely as triangular, without attending to the particular qualities of the angles, or relations of the sides.[4]

Here we see that evidently for Berkeley to 'consider' a figure, or a motion, in general, is a locution which entails a corresponding denial.

---

[1] Op. cit. ii, 5, 8, 9; Anscombe, pp. 199, 201.
[2] Cf. *Principles*, ii. 22: 'Matter, whose nature consists in being extended substance . . .' (Anscombe, p. 208).
[3] *Principles of Human Knowledge*, Introduction, 11; my italics.
[4] Op cit., 16; my italics.

In Berkeley's technical terminology, it is that we do not possess any 'abstract general ideas' of entities of this type, a doctrine which in turn is due to the 'received opinion' that 'every significant name stands for an idea':[1] which (according to Berkeley) misleads people into holding the mistaken opinion that a general name stands for a general abstract idea. Now in Berkeley's language, for anything to be an 'idea', it must be perceived by a mind; and that alone gives the 'idea' (and, as we shall see, for Berkeley this practically means, the corresponding 'object') an ontological standing.[2] It follows that there is something 'inconsistent' about the notion of an abstract general idea, for (Berkeley implies) all ideas which are *perceived* are 'particular' (and therefore 'concrete').

This objection teaches us something about the nature of philosophical controversy. Berkeley is as informed as was Descartes concerning the importance and use of key-conceptions of science such as extension, motion, force, etc. What he denies is that this has the consequences for ontology claimed by Descartes; a denial made possible by a particularly indulgent use of the notion of 'considering', and of course by Berkeley's peculiar theory of 'ideas'. For the moment let us note the general intentions of Berkeley's argument: if he is right then there is no sense in talking of 'reducing' sensory qualities to pure extension; nor of speaking as though the latter could in some sense exist, in and by itself, non-sensorily, only to be grasped by the mind. Berkeley lights up the point at which Descartes has supplemented perfectly straightforward methodological considerations by a metaphysical supplement. And Berkeley's method of argument already contains (in embryo) the requirement that such metaphysical supplementations may not be postulated 'dogmatically' but stand in need of a form of defence which relates them to 'experience'.

On the other hand, this is not a contention that would have impressed Descartes; he would not have admitted a 'refutation' which involved the plea that 'extension in general' cannot be perceived, since the core of his position is to lead us away from sensory

---

[1] Op. cit., 19; this doctrine is discussed in more detail in ch. V, sects. 2–3; cf. pp. 279ff.

[2] Cf. op. cit., 99: Extension and motion involve 'a twofold abstraction: first, it is supposed that extension, for example, may be abstracted from all other sensible qualities; and secondly, that the entity of extension may be abstracted from it being perceived. But ... all sensible qualities are alike sensations, and alike real: ... where the extension is, there is the colour too, to wit, *in his mind* ...' (my italics).

perception and to reflect on what is involved in the claims of thought. Still, whilst Berkeley's counter-arguments are hardly refutations, they do remind us that Descartes has not, so far, any very positive grounds for his metaphysics. Nor, if one of the main contentions of this book is accepted, would we expect him to offer us any clinching grounds.

(b) *Physical arguments for the doctrine of 'extension'*

But perhaps it is unjust to suggest that Descartes merely *postulated* the reducibility of all properties to extension? For he in fact believed himself to possess additional *physical* grounds for this doctrine. To study this it will be necessary to make a considerable excursion into Descartes' theory of matter and space in order to become clear as to just what is specifically physical, and what is conceptual or 'meta-physical' in his theory, and what relation exists between the two in Descartes' thinking. Physical considerations are hinted at in *Meditation* VI, for instance, where he harps on certain 'erroneous judgments' which we all pass continually, for instance,

> that in a hot body there is something similar to the idea of heat which is in my mind, that in a white or green body there is the very whiteness or greenness which I am sensing, that in a bitter or sweet body there are these very tastes, and so in other like instances . . .[1]

This passage embodies the celebrated doctrine which subsequently became known as the distinction between 'primary' and 'secondary' qualities' which, though already in the teaching of the Greek atomists, is to be found in Galileo, Gassendi and Hobbes, the terminology itself being probably due to Boyle, who employed it in his *Origin of Forms and Qualities* (1666). It was made prominent by Locke, according to whom the 'primary qualities' which are 'inseparable from body' include extension, figure, motion (equivalent to Descartes' list, though Locke adds 'solidity'). Secondary qualities, on the other hand, are in a different position.

> Secondary qualities [Locke writes] in truth are nothing in the objects themselves, but powers to produce various sensations in us by their primary qualities, i.e. by the bulk, figure, texture, and motion of their insensible parts, as colours, sounds, tastes, etc.[2]

Though I have quoted Locke's concise statement of the doctrine, it

[1] K.S., p. 258.   [2] *Essay*, ii. 8.9–10.

is—as we see—in outline already Descartes'. And since we wondered whether Descartes had *physical* grounds for his doctrine of the reducibility of sensory qualities to extension, let us consider the present distinction as a problem for physical theory. Now Descartes had indeed espoused such a theory already in his earliest writings, e.g. in his unpublished *Le Monde*. It appears more systematically in the *Dioptrics*, and the *Meteors*, as a theory of light and colour. The essential outlines of this connect in a curious way with his arguments concerning the primacy of extension and the nature of space, to which we must turn briefly.

Consider for instance *Principles*, ii. 2, where the argument resembles that about the piece of wax. Qualities like hardness, colour, heaviness, coldness and heat may all be abstracted, when 'thinking of a stone', nor does any change in them imply that the stone has lost the nature of body. For all that remains, and remains unchanging, is extension in length, breadth and depth; something which is just what is contained in the idea of space.

One might object that the extension of a body is the extension of *something*, whereas the idea of empty space does not imply the presence of anything. However, for Descartes this only proves that there is no sense in the idea of empty space, since this would suggest an 'extension which is the extension of nothing'; whereas, where there is extension, 'there must necessarily be substance as well'.[1] So, 'in reality the extension in length, breadth and depth that constitutes the space is absolutely the same as that which constitutes the body.'[2]

The conclusion is as subtle as it is extraordinary—anticipating, though only in a vague fashion, doctrines of some centuries later. It exacerbated seventeenth-century Newtonians such as Locke. For the moment, we can only note the result: all matter is continuous; there cannot be any truly empty space—Descartes speaks of 'the impossibility of a vacuum in a philosophical sense'.[3] The conclusion likewise entails a denial of atomism, at least in its ancient Greek form, for which the conception of empty space, in which the atoms are in continual motion, was essential. Instead we get a world of a single continuous matter of uniform density. Moreover, this matter (like space) is indefinitely *divisible*.

---

[1] *Principles*, ii. 16; Anscombe, p. 205.   [2] Op. cit., 10; Anscombe, p. 202.
[3] Op. cit., 16; Anscombe, p. 205.

So far, Descartes' conclusions are the outcome of purely conceptual arguments. In his language: all this can clearly and distinctly be grasped by thought. There is however a second concept, in a similar position. For there are to be found in the physical world a multitude of different bodies, endowed with a variety of qualities. Matter is not only divisible, but is *in fact divided*, and we require an additional *fundamental* concept (not contained in the concept of divisible matter as such), which will render intelligible the actual division into parts of different sizes. The required concept is motion. The reduction to a unitary concept like 'extension' turns out to be too radical. The construction of a physical plurality of particular 'elements' towards which Descartes is striving cannot be realised with the aid of extension alone but requires a concept in addition to that of extension. Its logical independence Descartes expresses by remarking that God created 'matter along with motion and rest in the beginning'.[1] Motion is not inherent in the concept of matter, though the laws of motion— as we shall later show—have a certain degree of '*a priori* rationality' about them. In this way, then, the existence of one unitary matter, divided into parts moving subject to certain laws of motion, can all be 'determined by reason'; an expression which we might explicate by saying that these results of Descartes' formulations concern the conceptual framework of physics, presupposed for any kind of physical investigation. No considerations of any empirical, i.e. observational, nature have so far been relevant.

All this must be sharply distinguished from the next step, which does involve *physical*, and indeed hypothetical, considerations. For, with this step it becomes necessary to make large numbers of additional *physical assumptions* concerning the actual sizes of bodies into which matter is divided, and their orbits and velocities.[2] What guarantees these 'suppositions' (hypotheses)? The answer, already anticipated, though perhaps astonishing for the reader who has been led traditionally to think of Descartes as a thoroughgoing 'rationalist', is: 'It is not by force of reason but by experience alone, that we can determine' all the variety of special conditions.[3] In what way does 'experience determine' this matter? Once more Descartes is astonishingly sanguine and 'modern': We are free to frame any suppositions,

---

[1] Op. cit., 36; Anscombe, p. 215. We shall see later that the reference to God's creative power is Descartes' way of saying that matter and motion have not just a logical but an ontological significance.
[2] Cf. above, p. 87.   [3] Cf. *Principles*, iii. 46; Anscombe, p. 225.

or hypotheses, we like, provided that all of their deduced consequences agree with observations.[1]

We shall return to the logical point that is here involved. For the
moment, look at the world Descartes constructs with his choice of
assumptions. It is guided by the need to give a plausible account of
the phenomena; but in addition Descartes is not unmindful of the
traditional classifications which had formed the stock-in-trade of an
older physics, such as the division into heat, light, the various states of
matter (aeriform, liquid, solid); magnetism, gravity, etc. In particular,
matter is conceived as divided into three 'elements', classes of
particles of increasing size, corresponding to the aspects of heat,
light, and the remaining states of matter. Though Descartes eschews
atomism, his hypothetical scheme is still particularistic, and this
helped considerably to reinforce the seventeenth-century fashion for
corpuscularianism.

With these assumptions, the great totality of physical phenomena
are claimed to be capable of explanation, such as heat, light, sound,
gravitation and inertia, magnetism, and a host of others. Descartes'
explanations in this field are, however, purely qualitative, and seldom
controlled by experimental tests; the sheer size of the undertaking
would have precluded this, as he indeed hints in both *Discourse* and
Principles.[2] But the explanations are often also vague, and the original
assumptions too narrow, avoiding (as they have to) all reference to
any kind of physical forces acting at a distance, and above all, taking
insufficient account of the all-important factor of 'mass' or 'inertia';
surely a *physical* process could hardly be derived from a purely
geometrical scheme. Nevertheless, we are concerned with intentions,
and not achievements. The point is that we have here a material
sense in which Descartes understands the notion of 'reduction' of all
attributes to 'extension'. However, this 'reduction' has also some
important consequences for Descartes' theory of perception; and to
study this, we must first briefly look at his theory of light and vision.

(c) *Light and the philosophy of perception*

Light and colour relate here to the 'second element', which is
assumed to consist of very small spherical particles, tightly packed—
the interstices between them filled by the yet smaller particles of the

---

[1] Ibid.
[2] *Discourse*, vi, K.S., pp. 154, 159; *Principles*, Preface, E. S. Haldane and G. R. T.
Ross, ed., Cambridge, 1967, i, p. 213. (Abbreviated as H. & R.)

'first element', which are relatively plastic and have no permanently fixed shape (to ensure that there will never be empty spaces).[1]

How, on this scheme, are we to explain for instance the nature of the light as emitted by the sun? A passage in the *Dioptrics* supplies an answer:

> Light which is the activation or motion of the sun or other luminous bodies, pushes a certain very subtle matter contained in all transparent bodies.[2]

It is in virtue of the sun's vortex motion that certain pressures are exerted at the centre. As Descartes explains in a letter to Mersenne, the way in which light presses the subtle matter in a straight line is like that in which a stone, whirled round in a sling, exerts pressure on the centre of that sling. One point is here worth noting: though light is said (in the first discourse of the *Dioptrics*) to be nothing but 'a certain very rapid and lively motion or activity', there is no actual motion, but only a 'tendency to move'.[3] In other words, Descartes' is not an 'emission theory of light'. But neither is it, strictly speaking, a wave-theory. Rather, the particles being tightly packed, as well as being 'infinitely hard', it follows that the pressure propagated from the sun towards any observer must travel with infinite speed. Descartes thought that the results of observations of lunar eclipses were actually confirmatory evidence for this conclusion (though in fact the negative result expected on his theory was obtained only because an observation of the actual retardation lies outside the limits of experimental error). He was only too ready to accept the result as supporting his whole physical theory, itself fitted to accommodate the conceptual ('metaphysical') scheme:

> If you could show me wrong [he writes to Beeckman], I should be prepared to confess that I know absolutely nothing in science, . . . [and that if any retardation] could be observed, I say that the whole stock of my science would be completely overthrown.[4]

The theory easily 'accounts' for colour: though the particles have no translational motion, they can rotate with different angular velocities;

[1] This is the 'element' involved in the explanation of the magnetic phenomena mentioned above, pp. 86f.

[2] Op. cit., v; A. & T., vi, p. 118. Cf. *Discourse on Method, Optics, Geometry, and Metereology*, Trans. P. J. Olscamp (Libr. of Lib. Arts, New York, 1965), p. 94. (Referred to as L. A.)

[3] Op. cit., i; A. & T., vi, p. 88; L.A., p. 70.

[4] Cf. A. & T., ii, Letter 57, p. 310; cf. also below, p. 140.

it is to these differential rotations that there *correspond* the different colours; if the motion is very rapid, the body appears red; if it is less, it will appear blue or green.[1]

The physical mechanism of the lines of pressure transmitted to our eyes is likened by Descartes to the case of the blind man who senses the existence of external bodies by the resistance they put up to pressure when poked with a stick. (This pressure also travels with 'infinite speed'!) The colours of variously coloured bodies are, in a similar fashion, only so many different 'tendencies to motion', set up in the nervous system.

> It is not necessary to assume the transmission of something material from the object to our eyes in order that we may see colours and light, nor even the occurrence in the object of anything resembling our ideas or sensations of it.[2]

All that is necessary is to assume 'that it is through the nerves that the impressions made by objects upon the external organs are transmitted to the soul in the brain'.[3] But they need not resemble them; nor do they, as we can see from Descartes' remarks in *Meditations* VI, quoted on p. 94 above, concerning the 'secondary qualities'.

The account that emerges from his *Dioptrics* iv and vi leads Descartes to speak of an 'effect' of external objects upon the sensory organs; the effect being 'to set in motion . . . the points of origin of the nerves in the brain'. Now it is this effect (a complex pattern of motions) which 'occasions the soul's perception of various qualities in the bodies'.[4]

We must be careful to distinguish two sorts of 'lack of resemblance': the one between the external quality and the brain-pattern—a lack of physical correspondence, involving no more than a minimum of 'structural similarity'; the other, between the brain pattern and the 'ideas' which we form of the external qualities, 'occasioned' by the presence of these brain-patterns. Now Descartes sometimes seems to imply that on those occasions when we normally talk of an observer viewing an external object, properly speaking it is 'the mind' that contemplates a 'corresponding' brain-pattern.[5] But this is not what he

---

[1] Cf. *Meteors*, VIII. Cf. L.A., pp. 337–8.
[2] *Dioptrics*, i, A. & T., vi, p. 85; Anscombe, p. 242; L.A., p. 68.
[3] Op. cit., iv, A. & T., loc. cit., p. 109; Anscombe, p. 242; L.A., p. 86.
[4] Cf. op. cit., A. & T., p. 114; Anscombe, p. 244; L.A., p. 90.
[5] Cf. *Reply to Objections* II, where images 'impressed in some part of the brain' appear as 'ideas . . . in so far as they [*sc.* the images] inform the mind, on its applying

means. For even though there may be some resemblance between brain-pattern, i.e. the 'image of our head' and the object, Descartes explicitly remarks that

> we must not think that it is by means of this resemblance that the picture makes us aware of the objects—as though we had another pair of eyes to see it, inside our brain; ... rather, we must hold that the movements by which the image is formed act directly on our soul *quâ* united to the body, and are ordained by Nature to give it such sensations. ... But there need be no resemblance here between the ideas conceived by the soul and the disturbances that cause them.[1]

Similarly, a few pages later: 'It is the soul that sees, not the eye; and only *by means of the brain* does the immediate act of seeing take place.'[2] Now if we put all this together, the following doctrine emerges. First, there is a radical difference between the perception, or rather 'the idea' (which is Descartes' technical term for this), and the 'corresponding' motions in the brain. Secondly, though it is simply a brute fact (if fact it be) that seeing takes place 'by means of the brain's motions', this is not something which itself is 'observed', nor understood. Mind and body must be *conceived* as utterly distinct; it is only an ordained fact, *uncomprehended*, that body and soul are 'intimately conjoined'.[3] To remind us of the theoretical distinctness, the lack of resemblance, Descartes employs—as will have been noted in the passage from the *Dioptrics*, A. & T., p. 114, quoted on p. 99— the fateful term 'occasion'. The point is well made in a little work of Descartes, entitled *Notes on a certain Programme* (1647), which comments on the case where we judge

> that such and such ideas, now present to the mind, refer to certain things

---

itself to that part of the brain' (A. & T., vii, p. 161, quoted in Kemp Smith, *New Studies in the Philosophy of Descartes*, London, 1952 (hereafter referred to as K.S.N.S.) p. 149, cf. H. & R., ii, p. 52).

[1] *Dioptrics*, vi, A. & T., p. 130; Anscombe, p. 246; L.A., p. 101. On the sense in which the soul is 'united' to the body by 'Nature', see below, pp. 178ff. This unity is in Descartes' view only a 'brute fact' which cannot be rationally grasped. Speaking rationally, they are distinct; and it is as an expression of this fact that Descartes insists on the logically possible lack of resemblance between 'image' and 'idea'.

[2] Op. cit., A. & T., p. 141; Anscombe, p. 253; L.A., p. 108; my italics.

[3] *Med.* VI; K.S., p. 257. For the same reason, Descartes can, as a scientist, sometimes speak of the 'motions that occur in the part of the brain by which the mind is immediately *affected*'. But as we shall see, the result of this, viz. sensations, and the whole connection between the motions and the sensations is simply 'brute fact'; it is—in Descartes' language—simply something that 'testifies to the power and goodness of God' (*Med.* VI; K.S., p. 263). Which is to say that the relation is not one that is comprehended as flowing 'rationally' from such of our *ideas* as we have

outside us. These things do not transmit the ideas to the mind through the sense-organs; but they transmitted something which gave to mind the *occasion* to form the ideas at that rather than at some other time.[1]

Descartes concludes that (at least in a sense) not only our ideas of motion and shapes, but also those of pain, colours, sounds, and the like must 'be *innate in us*' so 'that our mind may be able on the *occasion* of certain bodily movements to exhibit them to itself, inasmuch as there is no similarity between them and the bodily movements'.[2]

Here we have arrived at the end of the road on which we set out earlier to consider the objects of the 'universal mathematics'. It has led us to Descartes' 'dualistic position', the dichotomy between body and mind, extension and thought—though it is a doctrine which does allow a place for the idea of a factual union between body and mind at a certain level; but it is a union which the mind itself cannot grasp; it is only something we are 'taught by nature'.

One way to express the dualist notion is that the thought of an extended material object is not itself extended. As a reply to *Objections* V puts it:

> No corporeal species can be received into the mind; the conception or pure intellection of things, either corporeal or mental, is effected without any image or corporeal species.[3]

Let us retrace the steps towards the position by which all physical attributes of bodies are reducible to extension. As regards attributes like colour, hotness, etc., i.e. the 'secondary qualities', we must distinguish between the observer's experience of these, relating solely to the level of 'mind', having the status of 'ideas'; and the physical situations (particles in motion) which are 'the occasion' of these experiences. Now the latter are just so many modifications or specifications of 'extension', which they thus 'presuppose'; and only as such can they be grasped by the mathematical understanding. We at once see how such a position sets up a novel tension. For it tempts us to say, *either*, that the only things 'really given' are the 'ideas', all of

---

[1] A. & T., viii., pt. ii, p. 359, H. & R., i, p. 443; my italics.

[2] Ibid.; my italics. The problems surrounding the notion of 'innateness' are taken up below, pp. 110ff. For a critical discussion of this position, we must turn to Locke, cf. below, ch. IV, p. 235.

[3] A. & T., vii, p. 387; cf. H & R., ii, p. 231. 'Image' in this usage of Descartes is something which *is* extended, a pattern in the brain, indeed, in a special place of the brain, viz. the pineal gland.

which are 'innate in us'; *or*, that there are ultimately two real things, mind and extension, though entirely distinct. The last position, which is that adopted by Descartes, is also the one that gives rise to the greatest philosophical difficulties. The reductionist thesis has been made good only on condition that the 'qualitativeness' of secondary qualities be made a 'mental' matter. But why stop there? Why not go on and say that all our knowledge is exhausted in the stock of our ideas? In that way Descartes is driven to locate the epistemological centre of gravity in the thinking self, the self endowed with its ideas. But this is clearly too radical. New tensions will be produced by the claims of an 'external world' and a realm of the 'not-self'. The difficulty will be to give an acceptable 'account' of these, by building on the foundation of the 'self' and its 'ideas' alone. The attempt to do this leads Descartes to supplement his whole thesis by a number of constructive devices ('metaphysical constructions'), intended to define the notions of the 'real existence' of matter, *quâ* extension, and of mind.

It will be seen that the argument, so far, has been running at three different levels simultaneously: a logical, a physical and a meta-physical level. The logical argument insists that all physical processes shall be considered in their quantitative aspect alone, of order and measure; the candidate for which is extension. The physical argument involves the construction of a hypothetico-deductive scheme, whose elements are material particles endowed only with the attributes extension, shape and motion. I label this argument 'hypothetico-deductive', for three reasons, (1) because it involves explanations whose strength rests primarily on their power to lead to verifiable consequences;[1] (2) because the 'theoretical entities' involved in this scheme (the particles) are necessarily unobservable, since they do not possess any secondary qualities; (3) because there is a necessary gap between the theoretical hypotheses and the empirical conclusions which they are intended to explain. A theoretical scheme of this type requires certain 'coordinating propositions' or 'rules of correspondence'[2] which relate the hypothetical concepts to others with an empirical content. Such a correspondence rule is Descartes' assumption that a certain rate of rotation of spherical particles of the second

---

[1] Cf. pp. 96–7 above.

[2] Called a 'dictionary' in N. R. Campbell's *Physics: The Elements*, Cambridge, 1920, ch. 6, the *locus classicus* for a notion which has become a commonplace in writings on scientific theory construction in recent years. Cf. Nagel, op. cit., ch. 5 pp. 93ff.

element will, under certain conditions, be equivalent to, and an occasion for, the observation of a patch of red. This is the second sense in which Descartes understands the notion of the 'reduction' of qualities to extension. Suitable specifications of the shape, size and motion of particles are capable of yielding explanations of such other physical relationships as are observed to exist between the remaining qualities of physical bodies.

Now to both the physical and the logical arguments it could be objected that they move at only an 'ideal level'. Thus, the logical argument maintains only that the *ideas* of shape, motion, colour 'presuppose' that of extension. The physical argument only implies that, from certain statements about a number of assumed relationships between certain spatial entities, other statements can be deduced which when suitably interpreted yield observable states of affairs. A sufficiently 'sceptical' attitude towards 'the reality of scientific entities' might well maintain that such a system had no substantial implications for the *existence* of the entities postulated by the explanatory hypotheses. But clearly, this does not at all represent Descartes' intentions. For him, 'reduction' leads to something 'ontologically' fundamental, something requiring the aid of metaphysics.[1]

In outline, and anticipating, his move is two-pronged, and as follows. On the one side, we find straightforward ontological postulation, *via* a kind of causal principle, of a 'reality' transcending the range of 'a mere idea' of extension.[2] On the other, we are presented with the contention that sensory knowledge is unsafe, yields hasty judgments, etc.; that indeed ultimately, from the point of view of the thinking self, 'sensation is merely confused thought', the 'real' object of thought being extension. A mode of expression which, functioning in the writings of these rationalist philosophers (e.g. Descartes and Leibniz) as a *terminus technicus*, artfully conflates a dual appraisal: (1) sense is confused, in the ordinary meaning of this term; (2) thought contains as much 'reality' and of the same 'kind', as sense.

---

[1] One is of course free to object to the arbitrariness of such a move. Thus Nagel, op. cit., p. 152, after a discussion of the opposing 'instrumentalist' and 'realist' views of the logical status of theoretical entities, concludes that 'the opposition between these views is a conflict over preferred modes of speech', and Duhem's main objection to Descartes' physical procedure was that he attempted to give explanations of optical effects in terms of a particulate light-ether treated as a 'reality', when it was no more than a pictorial analogy. (*The Aim and Structure of Physical Theory*, tr. P. P. Wiener, Princeton, 1954, ch. 3, pp. 34–5.)

[2] This is the doctrine that 'God is no deceiver', which we shall discuss below, pp. 167–8.

This, roughly, is the way in which Descartes' views on science and method inexorably drive him towards a metaphysical construction. Later critiques of this scheme largely amount to a gradual whittling down of the contention that *metaphysical* constructions are a suitable answer to the epistemological difficulties raised by a logical and physical reductionism.

We need to trace now in more detail the tangled skein of Descartes' argument. To come more closely to grips with these epistemological difficulties, we must here turn to a preliminary consideration of Descartes' contrast between 'sensory' and 'rational' knowledge.

(d) *Sensory Knowledge and Rational Knowledge. The advance to metaphysics*

It will be remembered that the starting point of the whole exercise was to define an *object* of scientific knowledge whose nature would offer the possibility of a 'universal mathematics'. This object (extension) is the only one capable of yielding physical knowledge where 'knowledge' means 'apprehension of clear and distinct ideas'; in the language of the *Regulae*: of simple natures. Examples of natures that are 'materially simple' are 'shape, extension, motion, etc.'[1] Now these simple natures, according to Descartes, are related to 'complex natures' as cause to effect, the independent to the dependent, equal to the unequal, etc.[2] Simple natures have a certain logical priority; they are the logical elements pervading the complex phenomena. Thus, according to the sixth Rule, we should present the nexus between the elements of a scientific system in such a way that, although we start with what is observed first in the order of investigation, we finish by deducing it from that 'which is completely absolute'.[3] This reminds us of Descartes' methodological prescription that we should so analyse any complex whole that it can be shown to consist of elements, viz. 'simple natures' which can be 'intuited' in a way which vouchsafes truth.[4] Here it looks as though (translating into modern terms) we could have direct knowledge of the truth of the premises of a scientific

---

[1] *Reg.* XII, K.S., p. 65. Examples of 'mental simples' are knowledge, doubt, ignorance, and of those said to be common to either, existence, unity, duration. All these are 'innate absolutely', i.e. not even *correlated* with anything 'material', or 'external'. See below, pp. 110ff.

[2] *Reg.*, VI, K.S., pp. 25–6.

[3] Op. cit., p. 26.

[4] Cf. *Reg.* V; above, p. 86, and below, pp. 118ff, especially pp. 131ff.

demonstration, independently of its power to explain. This seems supported by a passage in the twelfth Rule.

Simple natures are one and all known *per se*, and never contain any falsity.[1]

On the other hand, we have noted when discussing the example of the magnet, that Descartes often operates with the conception of 'simple natures' in situations whose logic belongs to the 'hypothetico-deductive' field.[2] How far, then, is Descartes concerned here with methodological prescription, and how far with statements concerning the possibility (or metaphysical justification) of scientific truth?

Sometimes Descartes talks as though it made sense to speak of directly grasping the truth by intuition. His reference to the 'necessary conjunctions of simple natures',[3] and the *cogito* may be cited as examples. On the other hand, as we have already suspected—and we shall develop this point of view further in the sequel—such cases give a misleading impression of his views on scientific method. Furthermore, in so far as they *are* taken seriously, they presuppose the need for a special ontology. This need is indeed clearly recognised already in the *Regulae*, when Descartes discusses the nature of our knowledge of the external world. He insists 'that the understanding can never be deceived by anything experienced if it limits itself to intuiting the thing presented to it precisely as given'.[4]

But, we ask, what is it that is thus 'precisely as given'? Simply that it is an object which is grasped either by pure intellectual contemplation, or through sensory observation; in Descartes' words: 'either as it is in itself',[5] or (when perceived by sense) 'in some "image"', viz. of

---

[1] K.S., p. 66.

[2] Cf. above, pp. 86–7.

[3] *Reg*., XII, K.S., p. 67: 'If I say that 4 and 3 are 7, this combination is a necessary combination'.

[4] Op. cit., p. 69.

[5] Ibid. Kemp Smith's gloss on 'in itself' is 'as apprehended non-spatially, through the mind's direct contemplation of it'. This is the conception the meaning and possibility of which was later questioned by Kant. But in the present case, we cannot be sure of the examples Descartes has in mind; he may appeal to cases which involve what he calls 'necessary conjunctions', *Reg*. XII, K.S., p. 67. It is precisely the growing awareness that cases which look similar, are not, and may need differential treatment, that make up part of what we have called 'progress in the history of philosophy'. For instance, can we be sure that the case '4 + 3 = 7' is *like* the case 'I am, therefore God is', as Descartes claims; and that this is again *like* the case: 'I know, therefore I have a mind distinct from body'; or like yet another, which Descartes sometimes mentions: 'shape entails extension'? Decisions

some physical object. Now in relation to the second case Descartes already at this early stage stipulates provisos which will drive him towards ontology if his theory of knowledge is to be made good. It is true that we cannot tell here whether image means 'brain-pattern', associated with which the mind forms an 'idea', or whether the reference is to the 'idea' itself; these distinctions were not as yet clarified by Descartes. But on either interpretation, what follows in the passage under consideration is of the greatest interest: the understanding is not deceived as long as it does not

> proceed to judge that the imagination is thereby faithfully reporting the objects of the senses, or that the senses take over the true shapes of things or, in short, that external things are as they appear to be. For in all such judgments we are liable to error. . . .[1]

Here, with his demand for an object of 'universal mathematics' which would satisfy the conditions for the possibility of genuine knowledge, Descartes is inexorably driven 'inwards' (cf. the theory of 'innate ideas'), connected with the contrast here drawn between what is given in the imagination by the senses, and the world of external objects. Now the crucial question is: in what sense are we 'liable to error', when making the judgments referred to? In his answer to this question Descartes blurs the distinction between methodology, physics and ontology in a most interesting way. As regards methodology, we already know that Descartes' requirement to begin only with what is 'clear and distinct', or with what is 'simple', conceals a variety of moves that can only gradually be disentangled.[2] But at the physical level, likewise, though there are 'judgments', i.e. reasonings, which are liable to error, are 'bold' or 'hasty' in a straightforward sense and which are capable of cure, there are other cases which involve positing 'more than is given' in a far more radical and unavoidable way. Consider Descartes' example of the sun. I have two ideas of this, he says, one 'derived from the senses', a small disk the size of a penny; the other many times larger than the earth. This second sun 'has been arrived at by way of astronomical reasoning, i.e. elicited from certain notions innate in me, or formed by me in some other manner'.[3] Now

---

as to where the differences and similarities lie cannot be made lightly or haphazardly; they involve invariably a total philosophical point of view, a specific way of looking at things.

[1] Ibid.   [2] Cf. also below, pp. 130ff.
[3] *Med.* III, K.S., p. 218.

this is offered as a case where things differ greatly from our ideas. The reference to theoretical astronomy suggests that it is the untutored naïve mind that misunderstands the true size of the sun, judging hastily, and where more careful considerations would teach us the real state of affairs. On the other hand, the reference to 'innate' in this context misleadingly suggests that here we may expect to reach something which is true and indefeasible. This would add something with a methodological bite to the requirement to occupy ourselves with ideas that are clear and distinct (such as ratios of extensions).

Yet Descartes misleads; the situation is more complex. And its study may teach us a good deal about the nature of philosophical argument. In the same *Meditation*, only two pages earlier, he tells us that we must distinguish between *three* sorts of ideas, which (in a letter to Mersenne[1]) he calls 'adventitious' (e.g. our sensory 'vulgar' idea of the sun), factitious (that which the astronomers by their reasoning make of the sun), and finally, those that are innate (Descartes' examples here include the ideas of God, the mind, body, triangle).

On this several comments are necessary. First, it is not a case of sense judging hastily that the sun looks the size of a penny. Secondly, the conclusion that it is many times larger than the earth belongs to the field of theoretical reasoning which simply combines, in a judicious way, the deliverance of sense with the constructions of geometry and theoretical astronomy. Thirdly, it is not a conclusion to which attaches a superior power in virtue of there being involved 'innate ideas' (as misleadingly suggested by the first passage from the *Meditation* quoted before).[2] This is however what one might be tempted to conclude remembering that we have been told that 'simple natures are known *per se*, and never contain any falsity'.[3] What is true is that the notion of 'innateness', as here used, harps back (confusedly) to Descartes' theory of mathematical knowledge; the abstract and general nature of mathematics, its logical difference from knowledge pertaining to the empirical field, when treated as suggestive of an ideal of 'rational knowledge' *per se*, thus easily becoming once again a logical foundation for a complex epistemological situation.

The situation is fluid enough to conceal from Descartes the question which was to haunt his spiritual successors: 'How can "substantial

[1] June 14, 1641; A. & T. iii, pp. 382–3; quoted in K.S.N.S., p. 216.
[2] *Med.* III, K.S., p. 218; above, p. 106.
[3] Cf. above, p. 105, quotation from *Reg.* XII, K.S., p. 66.

knowledge" fail to be exposed to falsification?' Or shall we say, that introducing a greater 'rigidity' in our definitions of the key philosophical terms creates the seriousness of the problem? These questions will have to be taken up in the sequel. Certainly, I do not think that Descartes is here simply philosophically obtuse. Rather, it is an attempt to use a conception like 'innate idea' twice over, first, in a technical, mathematical, logical context; and then again, as a justificatory device.

We meet a different pair of contrasted cases in *Meditation* VI, where Descartes again warns his readers against 'judging inconsiderately'. He speaks of the 'erroneous' judgment that space, which to sensory inspection *appears* to be empty, is *really* a vacuum. This is a composite argument, partly belonging to the realm of sensory observation coupled with *scientific reasoning*, partly to the context of Descartes' *logical* arguments concerning the concept of space and extension. For, on the one hand, he is only saying that it would be hasty to infer a vacuum from the fact that there does not *appear* to be anything in a given space—seeing that there might well be very small unobservable bodies, invisible effluvia, or something on the lines of Faraday's later electric field. On the other hand, Descartes also wants to say that we have here a conflict between 'sense' and 'reason', since he purports to have shown that 'a vacuum in the philosophical sense' is impossible. It is only when interpreted in this way that the case is like the one (mentioned by him in the same breath) that we have already met, where we judge that the hotness or greenness of a body actually *is* in that body.[1]

Now the 'error' in this last judgment is clearly of a sort that is very different from the previous case in its 'scientific' version, let alone from the normal ones, as when I judge that a certain whitish powder is sugar when in fact it is salt! Here again, the queer cases ('queer' in the light of the whole way in which the theory of secondary qualities eventually emerges)[2] are 'normalised' by feeding on the grammar of the straightforward examples. Thus in the present case we could *only* be said to judge 'hastily' in the sense of not having sufficiently attended to Descartes' arguments concerning the primacy of extension

---

[1] See above, p. 94.

[2] A discussion of the distinction between primary and secondary qualities we shall defer to the next and later chapters, where it will be of more immediate importance, but we are here assuming the result that the distinction is not clear-cut. Certainly we shall find it extremely difficult to construe the contention, for instance, that the colour of a body is not one of its 'real' constituents.

and the 'reducibility' of all other qualities to it—clearly a highly Pickwickian sense of 'hasty judgment'! 'Conclusions regarding things located in the world outside us', Descartes tells us tantalisingly, are 'a task for the mind alone'.[1] The task is obviously composite, and belongs to a variety of levels, some capable of being handled by exercising sufficient care, others by using scientific theory. Yet others, however, are not tractable in this way at all, owing to the fact that Descartes has ended up by locating objects of 'infallible' knowledge among the realm of 'ideas' as such, be they 'ideas' like thinking, doubting; or like extension; or even of colour and touch. For when Descartes, in the passage cited above,[2] asks whether 'external things are always as they appear to be', from asking a question which can be tackled either by observations or theoretical science, he slides into quite another kind of question, viz. whether there 'is' anything corresponding to our 'ideas', e.g. of extension and of colour.

Now the various operative reasons for the intractability of this question by any ordinary means are these: first, for Descartes, *all* ideas have become in a sense 'innate', 'intra-mental'. Secondly, the notion of 'innateness' is still sufficiently tied to spatial notions to make it seem as though, with its ideas, the self was 'cut off' from the 'external world'. In particular, the 'real external world' is given to the understanding only through the *idea* of extension, which does not seem to exhaust the full notion of an external world. Yet, since that notion can clearly not be made good by processes of purely logical or scientific reasoning, Descartes already having exhausted these, he is driven towards some metaphysical construction which will supply him with the notion of a 'real world', as contrasted with an 'ideal' one.

To see that this is the correct reading, consider only the following. Take that extreme case of doubt concerning the existence of 'other things' lying 'beyond' those ideas of mine that represent external bodies. Descartes tells us that the possession of such ideas does not, by itself, supply any 'reason' for the assumption of the existence of any corresponding external bodies. Now he clearly recognises that this doubt is not susceptible of 'empirical' resolution, since he expressly tells us that we have 'no faculty' whereby we can discover the existence of any physical bodies transcending our ideas. His conclusion is that if we are nevertheless entitled to believe in their existence,

---

[1] *Med.* VI, K.S., p. 259.   [2] p. 106; *Reg.* XII, K.S., p. 69.

this is only because 'God is no deceiver'.[1] This shows that the 'metaphysical doubt' can be 'closed' only by a 'metaphysical construction'.

(e) *A note on 'innate ideas'*

In the passage quoted from the *Notes on a certain Programme*, Descartes speaks of the ideas of motion, and shape, colours, sounds, etc. as being 'innate in us'. Innate ideas are important in Descartes' thought, and this notion also supplements our understanding of the sense in which mind is distinct from body, however much suitable motions may come to be the 'occasion' for the formation of ideas in the mind.

In *Regulae* XII Descartes refers to certain things which 'we know as they really are'; and which we know 'so easily that for this it suffices that we be endowed with [the natural light of] reason'.[2]

The things referred to are concepts like knowledge, doubt, ignorance, volition. Their importance, as examples, for the present discussion lies in the fact that according to Descartes they are not borrowed from, nor due to, sensory impressions; that they are 'purely mental' and 'known by the understanding through a certain inborn light' alone, 'unaided by any corporeal image'.[3] As we saw, there is also a second group of 'simples' which are 'apprehended only in bodies, such as shape, extension, motion, etc.' It is not clear in this passage whether Descartes means that these concepts *are* 'derived from' perception, or whether rather they are *involved in* the perception of bodies; it is tempting to suggest an anticipation of the Kantian notion that they are 'presupposed for' perception, except that these sorts of distinctions are precisely the ones not as yet made by Descartes: the philosophical situation is still far more fluid. That he does mean something less than 'derived from' seems implied also by the nature of the third or 'mixed' group of concepts, attributable to either material or mental things, e.g. existence, unity, duration.[4] Finally, there is a fourth group, not mentioned in the *Regulae*, and which, as we saw, does not turn up until the *Notes on a certain Programme*: it includes the ideas of pain, colours, sounds and the like. These, we

[1] *Med.* VI, K.S., p. 256. See below, pp. 168ff.
[2] K.S., p. 65, trans. modified.   [3] Ibid.
[4] Ibid. It is interesting to note, in the light of Locke's subsequent opposition to the notion of 'innate idea', that existence and unity are classed among his 'simple ideas of reflection'. Cf. *Essay*, ii. 7. 7; see below, ch. IV, p. 182.

found, are formed 'on occasion' of the existence of certain patterns in the brain. But once more, the concept of 'occasion' displays an ambiguity. It may mean that the presence of the physical pattern produces an idea in the mind, in the way in which a physical seal produces its imprint in the wax (and we can never be certain that this model, though rejected by Descartes, does not retain some hold); or it may mean, that the physical disturbance in the brain is a 'sign' in the presence of which the mind forms a certain idea. Here, again, there are several possibilities. The 'sign' may set up some mental process in virtue of which the idea is formed; on the other hand, in the extreme case the 'causal chain' may be cut altogether, the physical change being the 'signal', which like a train-signal, *is* only the sign that a train is going to pass, but does not *cause* the train to pass.[1]

Here, as so often in philosophy, it is probably a somewhat unrewarding task to follow each of these arguments to 'innateness' in detail, in view of the vagueness or ambiguity of conceptions that play such plural rôles. The notion of innateness is required by Descartes' epistemology for several purposes. And the real nerve of his procedure (unlike his explicit reasoning) is likely to be affected by the fact that he has already settled for a favourable position in advance of his formal conclusions, since the conception of innateness is really the reflection of his ideal of knowledge. A basic element in these circumstances (as we shall find frequently throughout our story) is the existence of a technical analogue or paradigm which in the opinion of its defender exhibits the desired feature (here 'innateness') uncontroversially.[2]

For Descartes, a paradigm case of innateness occurs in the context of geometry. He holds that there is a very clear sense in which members of his second group of concepts, viz. shape and extension, are certainly not produced, and at most only 'occasioned', e.g. triangles, circles, etc. Consider the idea of a 'perfectly straight line', treated in the demonstrations of the geometer. Now sensory observation, says Descartes, *cannot* convey to the mind the idea of such a figure, since no physical line when closely examined is quite straight. On the other hand, we obviously do have the concept of a perfectly

---

[1] Cf. Berkeley, *Principles*, 65: 'The fire which I see is not the cause of the pain I suffer upon my approaching it, but the mark that forewarns me of it.'

[2] Other such cases include minimum principles in optics and dynamics, which we shall meet when we come to a study of Leibniz and Kant, and which will provide paradigmatic supports for a teleological approach.

straight line; and in fact when we see the physical line, 'we apprehend not it itself, but rather the authentic figure.'[1]

The argument purports to show that our idea of the straight line is in no sense a replica of either the physical line, or the pattern in the brain. Descartes here appears to make two assumptions: (1), that *we do have the idea of a perfectly straight line in the sense required*; (2), that *no actual physical line is straight in the sense required*. With these assumptions, we get a very clear idea of the way in which the notion of 'innateness' is meant to work. The essential consideration is that there is a lack of correspondence between the elements of the sensory world and the ideas which the mind possesses.

This kind of argument has a noble ancestry stretching back to Plato and is worthy of further scrutiny. To clarify issues, let us take the case of 'equality or sameness of length'.

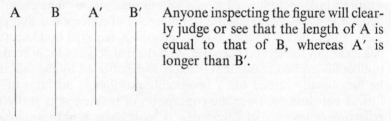

A   B   A'   B'   Anyone inspecting the figure will clearly judge or see that the length of A is equal to that of B, whereas A' is longer than B'.

If one now uses a measuring rod calibrated, say, to the nearest sixteenth inch, one will find that A is not equal to B. There is a temptation to draw the following two conclusions, (1) that A only looks equal to B, but is not really; (2) that the idea of equality of length which we have cannot be due to the *perception* of the 'equality' of *these* two lines, since they are not 'really' equal. There is a further conclusion, given the premise that we have the idea of 'perfect equality', viz. that we do not get this idea by perception but carry it around with us as part of our innate mental furniture. Many rationalists, in order to lessen the air of mystery, often add that the 'seeming equality' of our lines A, B 'suggests' or 'calls forth' this idea of the perfect case, which is only dormant in us prior to its perceptual evocation. Let us look at this with a critical eye and from the point of view of Kantian and post-Kantian approaches, in order to appreciate better the nature of the Cartesian standpoint.

Descartes' whole argument hinges partly on an unfortunate ambiguity surrounding the notion of two lengths 'looking equal'. In one of

[1] *Reply to Objections* v, A. & T., vii, pp. 381–2, H. & R., ii, p. 228.

its senses, this expression correctly contrasts with 'but are not really equal', where the context implies that the original is known to be modifiable at a subsequent measurement, the first being *knowingly* approximate. In a second sense, there is no such contrast, since here the *only* way in which we can discover the equality of two lines *is* by looking. Moreover, this second sense is more fundamental, since the former implies having first looked, and then looked again—both of these in the second and more fundamental sense of looking!

Consider now what is involved in the fundamental sense of looking, which lies at the basis of our judgments of equality. Here, whether you 'just look', or whether you use complicated manipulative instrumentation, there always must arrive a point at which we have to make 'a judgment of comparison'. For instance, we may lay off a given number of standard units of length against both A and B: then for any finite unit of length, there will always be some unit which will make the number of units when summed be either greater or less than A (or B). In other words, the end-point of A will not coincide with the end-point of the last unit, and its length will, to this extent, be 'indeterminate'. It follows that the end points of both A and B will lie 'somewhere', and we shall have to 'judge', by 'looking', whether A and B are equal. Furthermore, it is always conceivable that with a shorter unit (i.e. smaller scale) our 'judgment' might be 'erroneous'.

Two things emerge from this. First, the operational sense of equality of length is a function of judging, and thus 'looking' (in the fundamental sense). Secondly, the 'error' which emerges when we diminish the scale of the standard is a function of that standard and thus relative to it; it makes no sense to speak of 'error' when we abstract from this condition of measurement; and in the same way, the distinction between 'apparent equality' and 'real difference' of A and B (as when, upon further trials, one measures 'more carefully', or observes more minutely) is not unconditional, and certainly cannot be maintained at any given 'level' of observation ('looking', 'judging'). This corrects the second of the two conclusions mentioned above.

But it also corrects the first conclusion. For we find that in the 'deep sense' of 'looking', it is not the case that A and B only look equal but 'really' are not. Rather, their appearance is precisely such as would be taken to imply the judgment that they *are* equal: a fact backed up by the consideration that it is just by such cases that we learn to employ or apply the concept, and in this sense, acquire the concept of equality, including that of *correcting* the original judgment.

It remains to remark, however, that there is a 'pure' concept of equality of length which is itself exact, as contrasted with the necessary imprecision contained in the (applied) judgment of equality. A nominal definition of equality of length of two lines A″, B″, would be that they contain an equal number of equal unit lengths laid end to end. 'Equal number' would have to be defined liberally enough to include the case of two equal 'incommensurable' lengths, such as the diagonals of two squares whose sides were equal in the sense just defined. Now such a definition would be purely abstract, and related solely to the realm of number and pure geometry. And Descartes is right in thinking that the abstract cannot be 'derived' from the operational concept of equality; but this is simply because one belongs to pure mathematics and the other to mathematics applied to the field of perception.

There are two opposing positions traditionally held with regard to the relation between the pure and applied concepts. In terms of our example: according to the first, we begin with the perception of an approximate equality, and then by extended approximation, abstraction and idealisation arrive at the abstract notion of equality. ('Equality' as absence of 'inequality'.) The contrary opinion holds that one could not judge two lengths to be 'approximately' equal unless one already were in the possession of the pure idea. (By adding the condition that all actual pairs of length A, B are never equal, one then returns to Descartes' position.) Our argument shows that neither of these positions is adequate. What is true is that we presuppose counting and numbering as a condition of judgments of equality (we need only imagine our lines A and B several miles long!); but equally, such a judgment involves 'looking', in order to decide whether two physical lengths are equal, something (we have seen) involving an element of necessary imprecision, or rather: of decision-making. It means a 'seeing' (deciding) that the two lengths actually *are* equal, and our idea of equality of lengths is thus somehow connected with the perceptual situation. Moreover, it is natural (though misleading) to describe the process whereby we ignore the element of imprecision as one of 'abstraction'.

We can now see in this most interesting example how Descartes *describes* the various elements of this logical situation in a peculiar way. The conception of the length of a line as it 'really is in itself' is a curious amalgam of the pure number concept on the one hand, and the physical length on the other, that would result from a series of indefinitely repeated ever more 'accurate' measurements—with the conditions of measurement, however, removed! The assumption behind this is that the initial judgment, and indeed every successive judgment, is (because modifiable in subsequent trials, as mentioned) in some way 'inaccurate' or 'erroneous', which we have already seen to be a misleading description of the situation. Here, as in other cases, Descartes mixes straightforward paradigms of 'erroneous judgment' with other cases to which they do not really apply. For though in a *technical* sense, the 'real length' of a line may emerge only after repeated trial measurements, and hence the first and any subsequent measurements be termed 'provisional', nevertheless these are an *essential* element in any ascription of 'real length'; outside this process, 'real length' is undefined in Descartes' system and will have to be located (metaphysically) in a special 'world', or 'source'.[1] Descartes' lack of appreciation of the need for clearer distinctions of this kind tends to muddy the required division between the 'mathematical' and 'physical' world, with consequences which we must now study.

For him, the geometer's systems present us with an ideal of knowledge. It is for this reason that the physicist should likewise occupy himself with the 'object' of the geometer, with 'shape and extension'. In either case, we are attending to our 'innate ideas', which under the label of 'simple natures' are named in the *Regulae* as the basic elements of true science.[2] On the other hand, we have seen that the element of extension is reached by a process of abstraction, applied to the world of *material* bodies and their attributes. But why should the concepts of the geometer (and their mathematical relations) have any significance for the relationships of extensions studied by physics? Descartes must be saying that what the physicist apprehends in body (viz. extension) is precisely what the geometer apprehends in the relationships between his figures. But how can we be sure that the space and matter of physics are thus tractable by the geometer's systems? Descartes must be assuming that in seeing 'clearly and distinctly' what is involved in body, viz. extension, we *ipso facto* see also what is subject to the geometer's systematic constructions.

Consider Descartes' earliest defence of such an agreement, in *Regulae* XIV.[3] What is important both for physical science and geometry, are structural comparisons. All knowledge, apart from 'the simple and naked intuition of one single thing, is to be had by the

[1] Cf. below, p. 117. [2] Cf. above, p. 104. [3] Cf. above, p. 85.

comparison of two or more things with each other'. All qualitative comparisons are to be reduced to comparisons of 'the extension of a body that has shape'. The proportions which then result enable us to use the method of 'equations', as illustrated in the *Geometry*. We simplify a problem until it involves nothing further than the comparison of an 'unknown' extension with an 'already known extension'. And this method will universally apply to all quantifiable enquiry.

What is certain is that whatever differences in ratios exist in other subjects can be found to hold also between two or more extensions. Accordingly, our endeavour is sufficiently met if in extension itself we consider all the things that can aid us in the comprehension of differences in ratios.[1]

The 'things' here referred to include extensive magnitudes (not only length, but also weight, speed, etc.); purely numerical ratios; and finally, such 'shapes' as again enable us to represent 'differences of relations or proportions', as when points represent numbers, or when a 'genealogical tree' represents relations of descent in biology. All this leads Descartes to the following: success in scientific enquiry requires that one apply to it numerical methods which are generalisations of the kind used by the geometers and by those who study algebraic equations with one or more unknowns. Furthermore, the proper objects of science are those which are amenable to such a mathematical treatment. Only then can one hope to see the relationships involved as clearly as does the geometer in the purely geometrical case.

Now it is a perfectly respectable view that we find our way about in science only after having broken down our material into conceptual units that allow of the construction of deductive systems, where we know which of the complex relations depend on which other, fewer and simpler, ones.[2] If this were all that Descartes claims, then his 'method' would resolve itself into the requirement that all science must seek as great a conceptual clarity as possible; and that the primary tool for this is logic and mathematics.

However, woven into such a relatively trivial and straightforward thesis there is a more contentious one, brought about by the fact that Descartes slides from treating mathematics as a necessary *tool* for

---

[1] K.S., p. 93. Cf. above, p. 89, and below, pp. 127–8.
[2] As Descartes almost puts it himself; cf. *Reg.* VI, K.S., p. 26; see above, p. 104·

physics to postulating it as its *object*. For, not only is the world to be studied by means of the setting up of relations between 'extensions'; the world (in its material aspect) *is* extension. But clearly, such an identification requires defence. It might be thought that Descartes' thinking runs perhaps somewhat as follows: no other scientific treatment of the world is possible except that which considers it as a network of extensional relationships; but the world can thus be viewed; and further, there are a number of systematic mathematical constructions that have been successful. Such a 'pseudo-transcendental argument' (it would be exceedingly tenuous) is not what one finds in Descartes. On the contrary, he seems convinced that the assumption of an agreement between geometry and the world needs 'external' support. Now for Descartes there exists such a support; it is hinted at when he says that 'doubtless figures such as the geometers consider can *exist in reality*'. He does not tell us where, but he concludes:

> Indeed, we should not be able to recognise the geometrical triangle by looking at that which is drawn on paper, unless our mind possessed an idea of it *derived from some other source*.[1]

That source is presumably God. For as Descartes says in a letter to Mersenne,

> the mathematical truths which we entitle eternal have been established by God and are as dependent on him as are all other ceatures. ... · [Also] it is God who establishes the laws of nature. ... Now there is no one of the laws which is beyond our powers of apprehension if we apply the mind in the study of it: and all of them are inborn in our minds.[2]

We now see that the conception of the 'innateness' of an idea covers a considerable number of different situations, although there is sufficient resemblance or overlap to enable Descartes to slide (under cover of a single *term*) from one to the other. We can in fact distinguish at least four cases. An idea is 'innate', (a) if it is 'occasioned', in the way in which *all* ideas as such (including those involved in 'sensation') are quite distinct from the physical 'occasioning' situation (brain-pattern, etc.); (b) if there is no corresponding perceptual object, as in the case of the perfectly straight line, perfect equality, etc.; (c) in the case of certain abstract category concepts, such as knowledge, doubt, unity, etc.; and finally, (d) if the idea has a special metaphysical backing, as he claims for certain mathematical ideas, or

[1] *Reply to Objections* v, H. & R., ii, p. 228; my italics.
[2] A. & T., i, pp. 145–6, K.S.N.S., pp. 179–80.

for the ideas of God, perfection, etc. Such is his very fluid notion of innateness, and in the sequel it comes to be used in the following way.

To arrive at unimpeachable notions and relations, these must be considered innate, each and all. The world *quâ* extension is an innate idea, and so are the conceptions of mathematics. To guarantee an agreement between them, and to give the whole exercise any 'validity' and significance, Descartes has to maintain that this agreement is simply a fundamental (though 'contingent'[1]) fact about the universe. This solution (metaphysical in kind) we might compare with Kant's 'demonstration' of a similar agreement. Here the concepts of space are not 'innate' but 'transcendental', i.e. 'presupposed' for any cognition of extended objects as such. But Kant still postulates ('dogmatically', I think) that the specific forms or properties of that space which are thus presupposed, are 'mathematical truths' of a *unique* kind, viz. Euclidean. Nevertheless, Kant has begun at least with a critique which draws attention to and calls in question the nature of Descartes' blind assumption of his various metaphysical postulates.

Is Descartes entitled to claim the logical possibility of physical truth placed (at least in principle, if not in fact) on such an unimpeachable basis? After all, for the major parts of his cosmological theories in the *Principles*, Parts ii and iii, he claims no more than 'hypothetical' strength. Is this merely a deficiency, a second best? Or *are* there 'brute facts' and 'contingencies' in the Cartesian epistemological framework, which though sometimes suppressed, are nevertheless not to be ignored? Answers to this and related questions will require extensive exploration, to which we turn in the following section.

2. THE CONCEPT OF 'ANALYSIS', AND THE TENSIONS IN DESCARTES' CONCEPTION OF SCIENTIFIC METHODOLOGY BETWEEN DEDUCTION AND HYPOTHESIS

(a) *The need for hypotheses (the 'hypothetico-deductive' logic)*

That there are such 'tensions' will be obvious from what has preceded. On one hand we have met the contention that the epistemic

---

[1] According to Descartes, the 'eternal truths' are no more 'conjoined' to God's essence than any of His other productions. God is not 'necessitated' to create these truths. He is as free to 'bring it about that it should not be true that all lines drawn from the centre of a circle are equal, as He has been not to create the world'. Cf. letter to Mersenne, May 27, 1630; A. & T., i, pp. 151–3, K.S.N.S., p. 181.

edifice of science rests on a foundation of 'simple natures' (in the language of the *Regulae*), an 'intuition' of which alone can promise us achievement of 'truth': 'Simple natures . . . never contain any falsity.'[1] On the other hand, Descartes equally insists that the construction of his physics requires special hypotheses, the sole justification for which lies in the *explanatory* power which they have, with regard to the effects deducible from them.[2]

Superficially, the conflict seems due simply to the prevalence of alternative paradigms of actual methods for carrying out different scientific investigations. The method of 'simple natures' has deep roots in the studies which culminated in the *Geometry*, though being anchored at the same time in the metaphysics of extension. The belief in premises of scientific reasoning that can be 'seen' to be true is sustained, as we shall see later, by the example of the Cartesian laws of motion. The method of framing explanatory hypotheses has its natural footing in Descartes' cosmological theories; and it is employed more systemically in Descartes' studies in refraction and in the physics of the rainbow.

There are many other apparently contradictory statements of method. His physics starts, so Descartes often claims, 'from the simplest and most familiar principles which our minds know by their innate constitution'; they are such as we see to be 'self-evident';[3] 'clearer and more certain than the demonstrations of the geometers'.[4] The *Principles* in particular, however, tell also the other side of the story. 'If we find that from certain causes that we have assumed we can subsequently deduce all the other more particular phenomena not previously used to advance these causes, then this is a very strong argument assuring us that we are on the right path', Descartes tells us at iii, 42.[5] The passage at iv, 203 (quoted above) interestingly occurs in the context of the question, 'With what right does the physicist reason in terms of unobservable particles?' Descartes' reply is that he assumes that 'similar effects among sensible objects . . . arise from similar interactions of insensible bodies; especially as this seemed the only possible way of explaining them'.[6] And he adds:

This may give us an idea of the possible constitution of Nature; but we

---

[1] Above, pp. 86, 105. [2] Above, pp. 86f., 96, 141ff.
[3] *Principles*, iv, 203, iii, 43; Anscombe, pp. 223, 236.
[4] *Discourse*, v, K.S., p. 147.
[5] Anscombe, p. 223. I have slightly modified the translation so as to bring it in line with the wording of the French edition, which gives a better sense.
[6] Anscombe, p. 236. (For the earlier part of this passage, see below, p. 134.)

must not conclude that this is the actual constitution. . . . The supreme Craftsman might have produced all that we see in a variety of ways. I freely admit the truth of this; I shall think I have done enough if only what I have written is such as to accord accurately with all natural phenomena.[1]

The certainty which results from the explanatory power of hypotheses, regarded in this way, Descartes calls 'moral certainty' (what is 'certain for all practical purposes'), distinguishing it from the 'absolute certainty' of mathematical proofs and the knowledge that material objects exist (based for Descartes on the proof of the existence and nature of God).

Now such passages occur too frequently to be treated simply as a faulty vein in Descartes' putative enterprise of a unidirected explanatory scheme purely deductive in kind. He is certainly aware of the conflict, for he often tries to preserve the unitary scheme by a reduction of the range of logical power of his primary premises. For instance, *Principles* iv, 206 begins with an emphatic declaration that God's goodness guarantees that we cannot go wrong provided we use our intellectual faculties well and that the subject matter is of the kind which is self-evidently true, as with mathematical relationships, the doctrine of matter as extension, the nature of motion, and with whatever physical relationships are capable of mathematical treatment. However, Descartes goes on, this can only be claimed for 'the chief and most general principles', such as that all physical processes reduce to matter in motion; but here the deductive movement already stops. Consider for instance the following: the fixed stars cannot excite motion in our eyes unless all of matter between them and us is in motion also.

> From which it is perfectly *evident* that the skies must be fluid, i.e. composed of small parts separately in motion relative to one another; or at least, that there must be found in them such parts; for this is all one can say that I have *assumed* [by way of hypothesis] and which one may find in section 46 of Part iii, viz. that the skies are fluid. Thus, if you agree that this is sufficiently demonstrated by all the various optical effects as well as by all the other things that I have explained, I think one will likewise have to recognise that I have proved by mathematical demonstration (following the principles which I have established) all the matters that I have written of, *at least those most general* that concern the fabric of sky and earth.[2]

[1] Anscombe, p. 237.   [2] A. & T., ix, pp. 324–5; my italics.

An inspection of this passage will show how fluid is Descartes' methodological thought in this matter; how rapidly the connotation of such expressions as 'evident', 'demonstration', 'assumption', etc. varies from place to place. The most consistent interpretation is that Descartes is claiming only that his subsidiary assumptions are compatible with his more general principles, and subject to the same kind of mathematical treatment. This interpretation is borne out by a passage in *Discourse*, vi. Here, once again Descartes begins with his claim that the 'first causes' or 'principles' have been deduced 'from no other source than certain seeds of truth with which our souls are naturally endowed'—presumably an oblique reference to 'innate ideas'—, with the help of which we then 'deduce' certain very general principles common to most things—probably a reference to motion and its laws.[1] But here again the movement downward stops. For (Descartes admits) there are too many specific and particular 'forms and species of bodies' all compatible with our general principles and which God could have created by the introduction of suitable subsidiary causal schemes. The first principles are 'so simple and general' that of the particular effects there is hardly one that could not be accounted for 'in several different ways', i.e. by a variety of subsidiary causal schemes, clearly themselves again 'hypothetical'. The choice between the latter is determined by the result of certain 'crucial experiments' which will have to adjudicate between alternatives; in other words, we must here—as he explicitly remarks—'discover from the effects what are the causes'. And he adds:

> Thanks to this procedure . . . I can indeed venture to say that I have not observed any [effects] which I could not appropriately explain by the principles I had discovered.[2]

Passages like this show that Descartes is clearly not excluding observation and experiment, nor hypothesis; they play a vital part in the scheme. The 'seeds of truth' evidently yield principles with which the particular experimental effects they are meant to 'explain' are no more than *compatible*. They do not 'predict' these effects, since (if Descartes' contention is correct) there are too many 'special conditions' which would make any but the roughest guesses impossible.[3]

---

[1] Cf. below, pp. 147–55, for Descartes' laws of motion.   [2] K.S., pp. 153–4.
[3] Not that this situation is so unusual. 'Explanations' in history or any of the 'historical' sciences (such as geology) often suffer likewise from this assymetry

We may say that in this sense the relationship between the primary principles and the particular phenomena is 'epistemically opaque', and the resultant 'certainty' no more than 'moral'. For Descartes, ontologically, the effects do no doubt depend upon the causes, in the sense that 'God willed' them. Unlike Leibniz, however, Descartes does not possess a special principle of selection between alternatives, supplying us with a 'reason' for God's choice. We cannot comprehend a reason. All Descartes ultimately tells us is that (in regard to his secondary assumptions)

> those who notice how many deductions are here made from a few principles . . . even if they thought my assumption of these principles haphazard and groundless, would perhaps recognise that so many things could hardly hang together if they were false.[1]

Still Descartes is clearly no ordinary 'hypothetico-deductivist'. Any interpreter of his method must do justice to the interplay of the multiplicity of opposing forces in Descartes' scheme. Consider the following methodological passage from the *Principles*. After having remarked that a strong confirmation of hypotheses consists in their leading subsequently to additional verifiable deductions, he goes on, in the next section:

> But assuredly, if the only principles we use are such as to be quite evident, if we infer nothing from them except by mathematical deduction; and if these inferences agree accurately with all natural phenomena: then we should, I think, be wronging God if we were to suspect this discovery of the causes of things to be delusive.[2]

Now here, jointly with a statement of the principles of the hypothetico-deductive logic, we find the inclusion of the other, 'deductive' or '*a priori*' feature, with its harping back to principles that are 'quite evident'. We may interpret this, however, in a perfectly 'up-to-date' way. For our passage no doubt draws attention to an interesting aspect of scientific methodology. Descartes in effect claims that *three* conditions are required for a satisfactory scientific 'theory': principles that are evident; deductive inferences from these principles; agreement

---

between prediction and explanation. (For the logical problems involved, see B. Baumrin (ed.), *Philosophy of Science*, The Delaware Seminar, Vol. 1, New York, 1963, Part II.)

[1] *Principles*, iv, 205, Anscombe, pp. 237–8.
[2] Op. cit., iii, 43, Anscombe, pp. 223–4.

between these deductions and sensory experience (observation or experiment).[1] At this point, we must refer to what was said in Chapter II concerning the 'supplementary consolidation of hypotheses' and the interpretation of this notion by various scientists and philosophers of science.[2] As was noted there, though most people would subscribe to the last two on Descartes' list, there is considerable doubt concerning the first. What more do we need in addition to deductive power and experimental agreement? But, as we showed, there have always been considerable pressures towards basing an hypothesis on more than mere dependence on its verified consequences. Not only does the notion of 'explanation' as such appear to require more; there are also many occasions when the 'consequences' speak with an insufficiently distinct, and even divided, voice.

Again, as we saw, there is contained in the primary principles of science a 'direction-giving' force, determining the form of the investigation, and often indeed its very language. In any living science, the connection between explanatory hypotheses and the derivable consequences is far from complete, or even completable. The looser the bond, the greater the pressure towards additional foundations, reaching its maximum where the deductive development is at its most rudimentary. This is particularly evident in Descartes' biological theory. Here once again we meet with his admission that there is 'not as yet sufficient knowledge' to enable him to deduce 'effects from their causes', and that he has to follow once more the method of 'suppositions'.[3] He therefore proposes to treat the human body on the lines of the *model* of a 'machine'. Working out the 'consequences of this supposition' he claims to have explained all the relevant phenomena. Curiously, however, this explanation is said to possess 'the force of mathematical demonstrations' since 'the motion [of the blood] which I have now explained follows as necessarily from the very arrangement of the parts . . . as does the motions of a lock from the power, the situation, and the shape of its counterweights and wheels'. So it is by the aid of the mechanical model that we here 'deduce effects from their causes', by 'showing from what elements and in what manner nature *must* produce them'.[4]

---

[1] This is not unlike the account given of Galileo's method, analysed into the three steps of 'intuition or resolution, demonstration, and experiment', in E. A. Burtt, *The Metaphysical Foundations of Modern Physical Science*, op. cit., p. 70.

[2] Above, pp. 36–41.

[3] *Discourse*, v, A. & T., vi, pp. 45ff.; Everyman ed., pp. 36–46.

[4] Op. cit.; A. & T., p. 50; Everyman ed., p. 40; my italics.

Several aspects are at play. First, we have the deductive development of an hypothesis. Secondly, the hypothesis employs concepts of
a mechanical nature, with laws putatively applying to all matter. To
this end, the biological structure to which these laws are applied is
'modified' so as to assimilate the action to that of an inanimate
mechanism. And finally, the hypothesis leads to verifiable consequences. Naturally, Descartes cannot be certain that such a mechanistic approach is appropriate. His confidence is due to the trust which
he places in the universal power of his approach, based as it is on the
fundamental principle that matter is extension. In sum: whilst in this
case there is no deductive development of the basic principles, and
auxiliary suppositions have to be constantly injected, and the actual
physiological structure has to be replaced by a 'model', the true
significance of the mechanistic approach is here that of 'directing' and
'anticipating' method and results. It is for this that the basic approach
is propped up by an additional 'metaphysical' foundation.

The plethora of hypotheses with which Descartes fitted out his
physical and cosmological writings is well known; so much so that
when the Newtonians came to react against the *physical content* of
Cartesian theory they frequently did this by fastening on its purported
methodological weaknesses instead, enshrined in Newton's declaration
that in 'Experimental Philosophy' we must eschew hypotheses altogether, whether they be 'metaphysical or physical—whether of occult
qualities or mechanical',[1] the condemnation of 'mechanical' hypotheses being reserved (among others) for the Cartesians.

But probably it was not so much the formal logic of hypothetical
reasoning to which Newton was here objecting; as we shall see in its
proper place, he made ample use of this himself. Rather, he was
carping at its faulty application, its lack of mathematical development,
and insufficient methodological 'structure'. Thus one of the chief
faults of Descartes' approach to general physics (as distinct from his
brilliant excursions into optical theory)[2] was his inability to break
down the complex fabric of his subject-matter into subsidiary intermediate components in order to apply the kind of method which he
had sketched himself in the *Discourse*.[3] Besides, with our knowledge
of hindsight we can see that his physics lacked an appreciation of the
narrowness of its fundamental conceptual organisation; even though

---

[1] *Principia*, General Scholium, op. cit., p. 547. Cf. also below, ch. VI, sects. 1–2,
and Appendix, pp. 385ff.
[2] Cf. below, sect. 2(d).   [3] Cf. above, p. 121.

sometimes, as with the avoidance of 'force', this was deliberate. One cannot apply the conceptual scheme of any corpuscularian theory (whether of the Cartesian or Newtonian kind) until a far richer knowledge of the logically intermediate structure of the phenomena is forthcoming, and until at least the modicum of a set of empirical generalisations has been discovered. Prior to that, the general scheme can only give direction to research; and of course, just as often its anticipations may leap ahead and thereby lead the scientist astray.

Here, those components of an hypothesis which tend to consolidate its position within the body of a given science, and which contribute—in addition to the specific observational basis—to its plausibility, are precisely the ones which may also lead furthest astray. Now Descartes' 'intuitional' approach, amounting to the doctrine that the concepts of extension and motion should be given a privileged status in formulating hypotheses, was really just one such 'consolidative' feature. And it is here that Descartes' misreading of the logical status of consolidation causes it to interfere with an appreciation of the need for more complex and carefully contrived testing of his hypotheses.

It is as though he had believed that because the concepts of extension and motion are 'clear and distinct', intuitively evident, the hypotheses employing these concepts likewise possess a strength *independent of* their inductive basis. Instead, what such anticipatory concepts and principles do is to give the necessary *power to the scientist* to create hypotheses and theories, sufficiently strong to cope with a comprehensive and articulated interpretation of the phenomena.[1] Still, scientific successes may feed back into the conceptual scheme which has made their formulation possible, and to this extent strengthen our belief in the adequacy of the scheme itself. And here, interplay between the inductive (evidential) and the consolidative aspects of the scientific situation is fluid enough to explain the ease with which it is possible, by over-emphasising either one or the other of these two strands, to misconstrue the general logic of science. Descartes' scientific *practice* may suggest that he did. What is more important however is that a careful reading and inter-relating of his

[1] This view was later by Kant converted into the more rigid doctrine that the 'rationalist' component of science, the systematic and constructive activity of the scientist (called 'theoretical reason') not only defines 'the truth of its [methodological] rules', but even 'guarantees the correctness' of the whole scientific enterprise (what Kant calls 'the empirical employment of the understanding'). (Cf. K.535, 556; below, ch. VIII, sect. 4(c) iv).

various pronouncements suggest a more penetrating insight into this logic on his part than many later methodologists were to exhibit in the period immediately following.

(b) *Types of Analysis*

There are however more important paradigms which serve as signposts for the elucidation of Descartes' methodological principles. They connect on the one hand with Descartes' researches in optics, on the other with his mathematical studies as formulated in the *Geometry*. And though these subjects belong to very different logical fields, there is in Descartes' eyes a methodological concept which mediates between them, and which seems to have been a focal point for his attention from the very earliest writings: this is the concept of 'analysis'.[1] The interesting thing about analysis is that it acts for Descartes as a bridge between the different methodological approaches we have noted: at one extreme the 'intuition' of fundamental starting points; at the other, the assumption of hypotheses anchored solely in their explanatory power.

Both the *Regulae* and the *Discourse*[2] introduce the discussion of the primary rules of method by reference to certain 'ancient geometers', whom Descartes credits with having been in possession of 'certain vestiges of true mathematics' (found in Pappus and Diophantus) through the 'use of a certain "analysis"' which they applied in the resolution of all their problems'.[3] What is this method of Pappus? There is a statement in a work by Pappus (whose first Latin translation was published only a few years before Descartes[4]) which covers the two associated notions of 'Analysis' and 'Synthesis'. Both are said to be methods enabling mathematicians to *prove* theorems. 'Synthesis' is more or less equivalent to what is now called the 'axiomatic method' of proof, the most ancient model for which is Euclid's *Elements*. In this a theorem is proved by deriving it from axioms already known or assumed to be true; Pappus calls them

---

[1] Cf. L. J. Beck, *The Method of Descartes* (Oxford, 1952), pp. 278–84. E. J. Dijksterhuis, 'La Méthode et les Essais de Descartes', in *Descartes et le Cartesianisme Hollandais* (Paris, 1950), pp. 21–44. Also my chapter, 'Descartes' Anticipation of a Logic of Scientific Discovery', in A. C. Crombie (ed.), *Scientific Change* (London, 1963), pp. 399–417; and 'The Relevance of Descartes' Philosophy for Modern Philosophy of Science', *Brit. J. Hist. Sc.*, **1** (1963), pp. 227–49, especially section v.

[2] *Reg.* IV, K.S., pp. 17, 19; *Disc.* ii, K.S., p. 129.

[3] *Reg.* IV, K.S., p. 17.

[4] 'Synagoge' or *Collection*, Venice, 1589, reissued, 'Pisauri . . . 1602'. Cf. T. L. Heath, *A Manual of Greek Mathematics* (Oxford, 1931), pp. 435ff.

'first principles', and the Greek assumption was presumably that the truth of these should be 'self-evident'. The converse, analysis, starts by assuming the truth of the theorem to be proved, and deduces from the latter some other proposition, theorem or axiom, *already* known or assumed to be true.[1] In other words, here we move from the 'unknown' to the 'known', though it will be realised that this is not actually a method for *discovering*, but only for *proving* a proposition. The analytical proof is complete, if subsequently the theorem can also be proved 'synthetically'.

This 'analytical method' is as powerful as it is elegant. Its fascination for Descartes lay however in the possibility of its generalisation. Twice in *Regulae* IV he mentions 'rudimentary attempts' to extend the method to algebra; anyone reading his *Geometry* will find what use he makes there of the 'method of unknowns'. Already in *Regulae* XVII, when discussing 'complicated questions', dealing with the passage from 'known' to 'unknown', he 'adopts the device' of 'treating of the unknown intermediaries as if they were known'.[2]

An example may be helpful.[3] Consider the following 'problem': Given a line segment AB containing any point C, required to produce AB to D so that the rectangle AD. DB shall be equal to the square on CD.

$$\overset{\text{A}}{\vert} \quad \overset{a}{} \quad \overset{\text{C}}{\vert} \quad \overset{b}{} \quad \overset{\text{B}}{\vert} \quad \overset{x}{} \quad \overset{\text{D}}{\vert}$$

Now the method which Descartes proposes for the solution of such a problem involves two things: first, label the line segments AC, CB, by calling them a, b; treating a and b as numbers to be employed in algebraic calculations. Secondly, likewise treat BD = x as such a number, i.e. treat the unknown as though it were known, 'making no distinction between known and unknown lines'. Then we must set up an equation that will enable us to compute x. In the present case we should thus have:

$$(a + b + x) \cdot x = (b + x)^2, \text{ from which } x = \frac{b^2}{a - b}.\text{[4]}$$

---

[1] For an example of this in Greek geometry, see T. L. Heath, *The Thirteen Books of Euclid's Elements* (Cambridge, 1926), iii, pp. 440–4.

[2] K.S., p. 105.

[3] Due to Van Schooten, whose Latin edition of the *Geometry* appeared in 1683. See op. cit., p. 149, and *The Geometry of René Descartes* tr. D. E. Smith and M. L. Latham (Dover, Publ., 1954), pp. 6–9.

[4] Cf. *Geometry*, op. cit., pp. 6 and 9: 'If, then, we wish to solve any problem, we

No wonder Descartes attached such importance to the reduction of qualities to ratios between extensions and thus between numbers, since this yields such a powerful method for the solution of a variety of problems! No wonder he immediately asked why 'mathematics' should not 'also include astronomy, music, optics, mechanics and several other sciences'![1]

The species of analysis studied so far (let us call it 'analysis$_m$') by no means exhausts the genus, neither in Descartes' own writings nor those of his contemporaries.[2]

In his *Replies to Objections* II, for example, we find him elaborating on the theme of the two-fold method of proof.

> The manner of demonstration is twofold: one proceeds by an analysis *or resolution*, and the other by a synthesis *or composition*.
>
> Analysis shows the true way by which a thing has been methodically discovered [*inventée*] and allows us to see how the effects depend on the causes. . . .
>
> Synthesis, on the contrary, proceeds by quite a different route. It examines as it were the causes by their effects (though the effects are likewise proved here often by the causes). In this way it demonstrates clearly what is contained in its conclusions. It avails itself of a long series of definitions, postulates, axioms, theorems and problems, so that if anyone denies any of the consequences, it forces him to see how they are contained in the antecedents . . .
>
> The ancient geometers habitually only used this form of synthesis in their writings, not that they were totally ignorant of analysis, but (in my view) because they set so high a store on it that they kept it to themselves, like an important secret.[3]

The identification of analysis and synthesis with 'resolution' and 'composition', respectively, was of course not new. It can be found in

---

first suppose the solution already effected and give names to all the lines that seem needful for its construction—to those that are unknown as well as to those that are known. Then, making no distinction between known and unknown lines, we must unravel the difficulty in any way that shows most naturally the relations between these lines, until we find it possible to express a single quantity in two ways. This will constitute an equation, since the terms of one of these two expressions are together equal to the terms of the other. We must find as many such equations as there are supposed to be unknown lines.'

[1] *Reg.* IV, K.S., p. 20.

[2] We ignore the use of the term in modern mathematics, where it denotes such things as differential calculus, functions of a complex variable, etc.

[3] A. & T., ix, 121–2; H. & R., ii, pp. 148–9; my italics. For some mysterious reason, English translations omit the italicised words.

Galileo and his predecessors, and goes back at least as far as the writings of Grosseteste in the twelfth century.[1] 'Resolution' suggests that analysis involves a process of resolving a composite whole into its primitive parts, whose being 'explains' the whole. Moreover, this form of analysis (let us call it 'analysis$_r$') is thought of, not only as a 'method of proof' but also as constituting a 'way of discovery'; it mirrors the actual advance in science from effects to causes, from the complex phenomena to the elementary components: i.e. once again, from the known to the unknown! This need not necessarily be a substantive kind of analysis, as in chemistry. Still, that Descartes should think of the component elements of the phenomena as 'causes' is not surprising; it is a very common idea, still echoed in Newton's famous allusion to the method of analysis, in one of his *Queries*, appended to the *Opticks*:

> The investigation of difficult things by the method of analysis, ought ever to precede the method of composition. This analysis consists in making experiments and observations, and in drawing general conclusions from them by induction. . . . By this way of analysis we may proceed from compounds to ingredients, and from motions to the forces producing them; and in general, from effects to their causes. . . . And the synthesis consists in assuming the causes discovered, and established as principles, and by them explaining the phenomena proceeding from them, and proving the explanation.[2]

The idea of a sure and certain *method* which would lead us on to further discoveries suggests a somewhat naïve optimism which ignores the difficulties inherent in scientific analysis, the creation of new conceptions and the invention of explanatory possibilities. Descartes' own optimistic appraisal was partly guided by his successes in optics,[3] partly by his conception of 'method' as a very general and procedural set of regulative recommendations, connected with the idea of 'order' and 'measure'. Furthermore, 'analysis' may denote alternatively either 'method' or 'proof'. As a notion of proof, it has already been discussed under the label of 'analysis$_m$'. There is, however, yet another species of analytical proof, which likewise involves 'assuming what is to be proved as though it were known', and thereupon deducing from it certain consequences already known independently: this is clearly

---

[1] See A. C. Crombie, *Robert Grosseteste and the Origins of Experimental Science* (Oxford, 1953), pp. 63ff., 71–2, 75–9, 81–3, 193–4, 305, 311–12.

[2] Query 31, op. cit., pp. 404–5.

[3] See below, sect. 2(d), pp. 136ff.

the hypothetico-deductive order of proof (let us call it 'analysis$_h$').[1] Whatever power this form of reasoning possessed in Descartes' eyes would then automatically tend to devolve also on to his whole conception of analysis, whether considered as method or as proof. Finally, however, there is one other aspect of analysis—for Descartes perhaps the most important. Under the guise of 'analysis$_r$', and as an expansion of the idea of 'order and measure', it involves a demand for the regress to a special sort of starting point: a starting point which is viewed, sometimes, as a special set of particularly evident principles, and sometimes as the set of basic aspects which underlie the more 'derivative' set of ordinary phenomena—conceptions which (as regards science) are none other than shape, extension and motion. Analysis, when regarded in this way, amounts to the provision of a justification for and the revelation of the metaphysical basis of Descartes' scientific enterprise. The tantalising difficulties that face the interpreter of Descartes are precisely that he constantly moves from one of these levels to another, the trite and obvious methodological remark being used to bolster up the most complex and precarious ontological construction; some basic mathematical conception being used to highlight certain doubtful methodological viewpoints. In the next and succeeding sections we shall try to unravel some of these complexities.

(c) *Analysis as intuition of basic structure*

Let us first then take up one of the more obvious, methodological interpretations of analysis$_r$, or 'resolution'. This occurs in *Regulae* XIII: to trace the true relationships between causes and effects we must, on the one hand, not omit any operative conditions that might vitally affect the result; on the other hand, we must pare away all those conditions which are irrelevant to the solution of the problem. These rules are not unlike the methods of agreement and difference which since Bacon and Mill have formed an essential part of 'inductive logic'.

The aim of the process of paring away is, however (in Descartes' view), to arrive at the 'simplest elements'; and here matters become less obvious again! Consider *Regulae* V, which attacks the same

---

[1] See below, pp. 144–6; above, pp. 87, 118ff. Analysis$_h$ is clearly very different from analysis$_m$; and only the verbal resemblance of their respective 'definitions' would lead one to assume any important similarities.

aspect under the heading of 'order', another aspect of 'analysis'. 'Method consists entirely in the orderly handling of things', Descartes tells us; and this amounts to the 'resolution of involved and obscure data into those which are simpler'.[1] In science, the greatest degree of simplicity will be reached with the ideas of geometry and mathematics, though 'mathematics' need not be understood in any narrow and technical sense. Again the notion of simplicity! Now the more simple, as we saw before,[2] is also the more 'absolute'. But Descartes has in mind not just purely physical or mathematical elements, for the context of *Regulae* VI reveals that he really means to insist that our theoretical explanations should be in terms of 'logically primary' *conceptions* such as 'causes' (rather than 'effects'), 'extension' (rather than specific measurable things), 'length' (rather than extension), 'simple' (rather than complex), straight (rather than curved), like and equal (as against unlike and unequal), etc.[3] It is clear then that 'resolution into basic elements' ultimately refers us to the logical priority of certain concepts, either 'material' or logical in kind.[4] Now the importance of such concepts for Descartes lies in the fact that they are involved in what in the *Regulae* is called 'intuition', and later becomes 'clear and distinct perception'. Just as 'simple natures are one and all known *per se*, and never contain any falsity',[5] so 'there can be no falsity in the bare intuition of things, whether the things be simple or conjoined'.[6] The objects relative to 'intuition' are thus 'simple', or 'absolute', 'independent'. Once they yield to inspection, they are 'immediately disclosed to us either in this and that sense-experience, or by a light that is in us'.[7] Intuition is thus not relevant to every subject; and, on the other hand, when the logical nature of the subject is thus defined, then there is really no more to be done: intuition is here not a mysterious process or activity; it is more like the statement of a condition that must be satisfied if success is to be expected; that one may expect it is not guaranteed but only suggested by the paradigms which Descartes has in mind.[8] This is borne out also by the specific examples which Descartes enlists in the *Regulae*. For instance, sensory observation is rarely a candidate, for what is observed in this way is not

[1] K.S., p. 23.   [2] See the discussion above, pp. 104ff.
[3] Cf. K.S., pp. 26–7.   [4] Cf. above, pp. 104, 110, where examples are given.
[5] Cf. above, p. 105.
[6] *Reg.* XIII; K.S., p. 78. What is conjoined is still simples.
[7] *Reg.* VI; K.S., p. 27 (I omit Kemp Smith's 'native [in]' for 'insito'.)
[8] There is also a procedural interpretation of 'intuition': cf. below, pp. 140–1.

sufficiently 'independent' of surrounding physical conditions, quite apart from the fact that sensory qualities are hardly 'ultimate' or 'absolute' enough.[1] Thus, we judge all objects yellow when we suffer from jaundice, which shows the untrustworthiness of such judgments.[2] In the end, Descartes' examples of intuition come from different quarters. Consider *Regulae* III:

> Intuition is not the fluctuating reliance on the senses, nor the misleading judgment of a wrongly combining imagination, but the apprehension which the mind, pure and attentive, gives us so easily and so distinctly that we are thereby freed from all doubt as to what it is that we are apprehending. In other words, intuition is that non-dubious apprehension of a pure and attentive mind which is born in the sole light of reason. . . . Thus each of us can see by intuition that he exists, that he thinks, that the triangle is bounded by three lines only, the sphere by a single surface, and the like.[3]

Clearly, in his search for candidates, Descartes is being steadily driven towards more general epistemological considerations; to see by intuition that I exist, that I think, already foreshadows the *Cogito*.

The notion of intuition is however insufficiently sharp. The passive apprehension of what is simply given covers too much: not only abstract concepts, mathematical objects, the realm of extension, but also awareness of sensory data (provided all conditions of obvious error have been guarded against). Descartes in his later writings therefore replaced intuition by the notion of 'clear and distinct' perception of ideas. The most succinct statement of this occurs in the *Principles*, i, 45.

> The perception on which we can found an indubitable judgment must be not only clear but also distinct. I call *clear* that perception which is present and manifest to an attentive mind, just as we are said to see objects clearly when, being present to the gaze of our eyes, they operate on it sufficiently strongly. . . . But I call *distinct* that perception which is so precise and so different from all others that it contains within itself only that which appears plainly and evidently [*manifestement*] to whoever considers it properly.[4]

Descartes' illustration of the distinction is interesting for its own sake.

---

[1] Cf. above, pp. 106f., and below, pp. 155ff.
[2] For this example, cf. *Reg.* XII; K.S., p. 69.   [3] K.S., p. 12.
[4] A. & T. viii, p. 22, Anscombe, p. 190.

When, for instance, someone feels an intense pain, his awareness of it is so far clear, but not for that matter always distinct. For he normally confounds it with the false judgment which he makes concerning the nature of what he thinks exists in the injured part, which he believes to resemble the idea or the feeling of pain which is in his thought, in spite of the fact that he perceives clearly only the *feeling or the confused thought* which exists in him. In this way a cognition can sometimes be clear without being distinct, but it cannot be distinct without being also clear.[1]

The example reminds us of an earlier case[2] where we were said to judge 'over-hastily' that the secondary qualities reside in the bodies themselves. It shows at once that 'distinctness' has connotations which remove it from the 'ordinary' technical cases towards the realm of metaphysics. I can know something clearly as long as I attend to the 'here and now' of the 'immediately experienced'. (The incorrigibility of this 'immediate experience' itself involves an assumption which proved to be a basic starting point for a considerable number of empiricist theories in philosophy, from Locke to the twentieth century.) But whilst I may be said to perceive 'clearly' a 'bent stick' immersed in water, I shall 'judge falsely' if I imply that it *is* bent. And similarly with the other examples we discussed before, such as that of the 'two suns', the existence of a vacuum, etc.[3] The conception of the 'distinct', we may say, whilst being *introduced* through straightforward and quasi-technical examples, achieves a philosophical grip only through the contentious cases, e.g. the 'confused' judgment that there is in bodies something corresponding to the secondary qualities which we are normally said to perceive as their attributes. The insistence that we *understand* the meaning of distinctness in this extended (metaphysical) case is then further used as a lever jacking up the doctrine of reduction[4] and the idea of a special starting point of all knowledge in general. Only in this way can we understand the final result of the Cartesian programme of reduction when it insists, as in the last passage, that feelings are 'confused thought'.

The application of all this to 'the foundations of science' is expressed with great clarity and force at the end of Descartes' *Principles*, iv. 203. We have already come across the part of this passage where Descartes explains the need for using analogy in reasoning from the properties and structure of observable bodies to that of their

---

[1] Op. cit., i, 46; A. & T., viii, p. 22; my italics.   [2] Cf. above, pp. 94, 108.
[3] Cf. above, p. 106.   [4] Cf. above, pp. 109ff.

unobservable constituents.[1] The relation of this subsidiary logic to the general principles of science emerges in the earlier part:

> Since I assign determinate shapes, magnitudes and motions to the insensible particles of bodies, as if I had seen them, and yet admit that they do not fall under the senses, someone will perhaps demand how I have come to have knowledge of them. To him I reply that I first considered in general all the *clear and distinct notions of material things* to be found in our understanding, and that, finding no others except those of shapes, magnitudes and motions, and the rules whereby these three things can be diversified by each other—which rules are the principles of geometry and mechanics—I judged that all the knowledge man can have of nature must be derived from this source alone. The notions we have of sensible things, being all of them *confused and obscure*, cannot serve to give us any acquaintance with anything outside ourselves, but may on the other hand serve to impede it.[2]

The 'clear and distinct' applies to certain privileged *notions*, e.g. extension and motion. In science, only these must be allowed as the 'materials' for the construction of theories. In addition to 'notions', there are, however, geometry and mechanics which also contain 'universal relations' or 'rules'. In them likewise there 'can be no falsity', in so far as we perceive them 'clearly and distinctly'; as enshrined in Descartes' celebrated general meta-rule: 'The things we apprehend very clearly and distinctly are true'—although this rule in turn is subject to the condition of God's existence and nature.[3]

All these passages display a certain ambiguity. Are 'clear and distinct' notions merely a *condition* for 'seeking truth in the sciences'? Or is 'clear and distinct apprehension' a reference to an intellectual process? Or, finally, does this expression denote a statement of achievement, effected when truth has been reached? As usual, there is no clear-cut answer, nor are there any unequivocal examples. The nearest to both process and (putative) achievement is Descartes' theory of the 'rules' or laws of motion, a theory which itself requires the aid of the perfection of God![4] No doubt, in Descartes' view the whole doctrine was buttressed considerably by the radical contrast it drew between the 'confused and obscure' perception of 'sensible

---

[1] Above, p. 119.   [2] A. & T. ix, p. 321; H. & R., i, p. 299; my italics.
[3] Cf. *Discourse*, iv; K.S., pp. 141, 145; *Meditations* iii; K.S., pp. 214–5. See below, pp. 171ff.
[4] See below, sect. 6, pp. 147ff.

things', and the 'clear and distinct apprehension' of the geometrical and kinematical notions. The terms chosen suggest that we are here dealing with a difference of degree, whereas (with Hume and Kant and most modern philosophers) we may want to insist that the difference is generic. On the other hand, Descartes' belief that sensory knowledge was *inevitably* subject to error, and was thus *ineluctably* 'obscure', and his giving the conception of 'clear and distinct apprehension' a technical and metaphysical direction, in the end has much the same effect of restoring a difference in kind, denied only by the choice of terminology: a terminology which betrays the fluidity that still exists at this stage of philosophical development.

The practical intentions of Descartes' rationalist approach in the case of the unobservable particles of science are here of great significance. (See our last passage, *Principles*, iv, 203.) Descartes has reached the result that the general structure and laws of matter apply to *all* matter, macroscopic and microscopic. And though the specific hypotheses invoked for the latter are supported only 'hypothetically' and with the aid of 'analogy', as 'the only possible way of explaining' the phenomena,[1] the power of 'clear and distinct' foundations is invoked to provide the additional support for what is otherwise 'merely hypothetical'. Still, the support can clearly offer no more than 'practical' or 'moral certainty'.

We can now understand how the 'resolutive method' (analysis) covers two rather different things. On the one hand, it concerns the movement from the complex observed effects (the 'phenomena') to their component causes; or (to put it propositionally) from the statements of observed regularities of nature to their explanatory principles or hypotheses. This may be termed the methodological or formal sub-species of 'analysis$_r$'. But there is also a second or 'notional' sub-species, which consists in the move towards the logically and mathematically simplest elements of science as such. However—and Descartes is not always too clear on this—when the basic notions have been reached we are not, in virtue of this feat alone, also possessed of the primary explanatory principles; we have only regressed to the level of matter as extension and its geometrical relations, and in general, to the simplest and most absolute elements of experience. Though this guarantees success in principle, what the actual substantial relationships are to which the method may then be applied is still an open question.

[1] Ibid., p. 299.

This will explain better the place held by the hypothetical method: a method obviously far more prominent in Descartes' scientific thinking than appreciated by most of his commentators, and which needs therefore to be reconciled with his doctrine of the clear and distinct starting points. More than likely, the two sub-species of analysis, merge, providing for each other mutual support. The famous 'rules' of Descartes' method, enunciated so briefly and with tantalising insufficiency in the *Discourse*, and possessing only a very delusive clarity, themselves exhibit this double-spectrum.

(a) To accept nothing as true which I did not evidently know as such. . . .

(b) To divide each of the difficulties into as many parts as necessary for an adequate solution;

(c) an order, beginning with matters the simplest and easiest to know, and then ascending gradually, step by step, to the knowledge of the more complex. . . .

(d) In all cases make enumeration complete, i.e. make sure that there are no gaps in the chain of deductions, that nothing has been left out of account.[1]

It will be obvious that these rules are an amalgam of considerations actually belonging to totally separate fields, being partly a reference to general (methodological) forms of scientific procedure; partly a recommendation in favour of the theoretical regress towards the elementary explanatory principles of any subject matter; partly a reminder of the requirement that we must descend to the 'clear and the distinct' notions and relations of extension in general.

(d) *Resolutive Analysis and its application to optics*

Two gaps still remain before our account approaches even a modicum of completeness. One concerns the occurrence of analysis, in a genuine specimen of scientific enquiry, Descartes' *Dioptrics*, together with hints in *Regulae* VIII and IX. This shows what the method of regress to elementary constituents means in a concrete case. The second gap to be filled concerns Descartes' account of the principles of dynamics. It supplies material for the question whether and in what sense there are any laws in Cartesian science which are based on ideas that are 'clear and distinct'.

Cf. *Discourse*, ii; K.S., pp. 129–30. I have only paraphrased the rules, and my interpretation of (*d*), in particular, uses information and explanation supplied by *Reg*. VII. As noted, the rules of the *Discourse* are far too brief, and indeed almost trivial, to be of much use, and to appreciate their importance requires reference to the *Regulae*, and in fact to the whole of Descartes' teaching. This is indeed the reason for their appearance so late in this chapter.

The interest attaching to the example from optics is not only that it has all the appearance of a case involving regress to the 'intuition' of an 'absolute' starting point in the sense of 'primary principle', but also that it indicates the sense in which this method functions as a method of discovery. The example is an important law of geometrical optics, called 'Snell's Law', of which Descartes was a co-discoverer; and he certainly was the first to relate the quantities involved explicitly to the ratio of the velocities of light in different media. Put very briefly, according to this law, for any given pair of media the ratio of the sines of the angles of incidence and refraction is constant and equal to the ratio of the velocities in the respective media, for light of a given colour.[1] The constant is now called the 'index of refraction'.

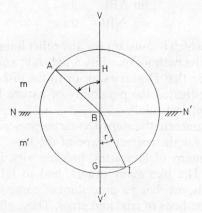

This law lies at the basis of the problem of constructing a lens whose shape is to be such that it will collect parallel rays at a point. As Descartes points out in the *Dioptrics*, this condition is satisfied for an elliptically shaped surface.

From his studies of conic sections Descartes was aware of the following geometrical property of an ellipse. If AB is parallel to the

[1] In the figure NN' represents the interface between the two media, m, m'; ABI is the light ray, refracted at B; VV' is the normal to NN'. ABH = i, and IBG = r are the angles of incidence and refraction, respectively. With AB = BI being unit radii of the circle, AH = sin i and GI = sin r. (Cf. L.A., p. 80.)

Descartes' expression for the ratio of the velocities is the inverse of the 'correct' value. The deeper reason for this is that his is really a particle-theory, and not a wave-theory, agreeing thus with the same expressions reached by Newton and Maupertuis, as against those of Huygens and Fermat.

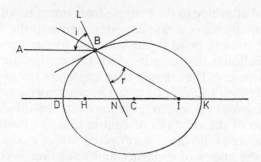

major axis DK, and if it meet the tangent to the ellipse at B, with LN
being the normal; and if BI meet DK at a focus I; then

$$\frac{\sin ABL}{\sin NBI} = \frac{\sin i}{\sin r}$$

will form a ratio which is constant for any other lines parallel to DK.[1]
Now this is also the path followed by a lightray; and it would not be
difficult to imagine that Descartes, prior to the discovery of his law,
should have investigated the possibility of nature following such a
convenient geometrical rule![2]

Let us now enumerate the steps Descartes prescribes for the scientist in his search for the correct shape of the lens. The problem had
been studied by many of Descartes' predecessors, including the for-
midable Kepler. He, like many others, had to fall back on rather
Baconian methods, involving a procedure of curve-fitting and several
other empirical methods of trial and error. These all failed to provide
an adequate solution; and they are quite unrelated to any of the
theoretical features of the situation.[3] But this is, Descartes tells us,
what is required. We must follow *Regulae* III, which says that in

---

[1] Cf. op. cit., ch. 8; A. & T., vi, pp. 168–71. L.A., 129–30.

[2] For the hypothesis concerning Descartes' actual discovery of Snell's Law, cf. G.
Milhaud, *Descartes Savant*, Paris, 1921, ch. 5, pp. 103–8. The procedure which
Descartes claims to have followed is of course quite different, but the derivation is
so far-fetched in detail, and involves such unlikely assumptions, that it is more than
probable that the above considerations did actually provide the necessary preliminary hints. (For the derivation, see p. 143, below.)

[3] For Kepler's solution, cf. *Ad Vitellionem Paralipomena. Works*, ch. 6, ed. Caspar,
ii, pp. 104–8. It is true that Kepler, whom Descartes on his own admission had
studied carefully, had also engaged in considerable theoretical discussion, especially
in the first chapter of this work (op. cit., pp. 20–8). But in the end theoretical
investigation and empirical curve-fitting are insufficiently related internally; and it
is here that Descartes' approach constitutes such a brilliant advance.

order to treat of any subject, we should examine 'what we can intuit clearly and evidently'. Now what does this mean in the present context?

Descartes' answer is as follows. We must not limit ourselves to purely mathematical aspects (e.g. the geometrical properties of the ellipse). Instead we must study the *physical conditions* of the problem empirically, e.g. note that the relation between i and r depends on the differences in the media through which the light passes; that this depends on the manner in which the ray penetrates the transparent body; that knowledge of this property of penetration in turn presupposes a knowledge of the action of light; and that finally in order to understand this we must have a general grasp of what constitutes a 'natural power': this being the last and 'most absolute' term in the whole series.[1] Moreover, it is this last term which must 'by intuition be clearly apprehended'.[2]

We see that in *this* example, 'resolution' really does involve descending to the actual physical conditions ('the causes') of the phenomena, whilst at the same time also indicating the logically most fundamental elements. A perception of what is involved simply and absolutely in the passage of light should give us an intelligible starting point.

There is a fundamental lesson implied in this procedure, a lesson which still has relevance for contemporary discussions. Logicians often say that hypotheses have to be 'guessed' in the light of the data. Now whilst there is some truth in this, it is usually added that the guessing is 'inspired' and 'informed'. It is carried out in the light of a complicated network of background knowledge which in our case connects with the physical basis of the passage of light through transparent media. Clearly, here we move from the data to the hypothesis (Snell's Law) *via* the study of the underlying mechanism involved in the problem, and subject to the initial guidance of qualitative observation. This is the reverse of the usual account of scientific method, according to which we pass from data to hypothesis and *then* only seek for an explanatory mechanism which might account for the hypothesis. What is true is that the physical theory injected into the situation is at this initial stage only *approximate*; more precise and accurate accounts will have to await adequate formulation of the laws for which we are still searching. In short, the procedure is one which employs the method of 'leap-frogging'.

[1] Cf. *Reg.* VIII; K.S., pp. 38–9.   [2] Ibid.

To return to Descartes: What (in this example) is meant by 'apprehending through intuition' the required 'natural power'? What is the semantic force of this phrase? The proposed answer occurs in the next rule, *Regulae* IX. Descartes now suddenly changes his method, and continues by proceeding under the guidance of certain results of his theory of light, which—as we have already noted—is itself a result of reasonings involving a curious mixture of the *logical* requirement that all space is full, together with the *hypothetical* (*physical*) *assumptions* which lead to the theory of the second element as the carrier of the light impulses![1]

The focal point of this theory was (it will be remembered) that light impulses were propagated with infinite speed. The status of this assumption is not too clear. Is it a 'mere hypothesis', or is it a conclusion of reasonings involving other hypotheses, seemingly confirmed by observation together with general considerations of a logical kind?[2] As a result of the 'resolution' of the optical problem, the principle of the speed of light ought to be a fundamental core of the theory; we may therefore understand Descartes' demand that we should show such a principle to constitute an elemental possibility, powerfully supported by general considerations of the motion of natural forces. And I think this is precisely what his mysterious 'intuitive apprehension' here amounts to; always remembering that it is all formulated in terms of 'distinct and perspicuous' notions of matter (extension) and motion. This is what he says:

We should devote our attention to the local motion of bodies, as being of all motions the most manifest. A stone, as being a body, is indeed, as I observe, unable to pass instantaneously from one place to another distant place. If, on the other hand, a power similar to that which moves the stone is to pass from one subject to another, and does so nakedly [i.e. without a material carrier] it must do so instantaneously. For instance, if I move one end of a staff of whatever length, I *easily conceive* the power by which that part of the staff is moved as *necessarily* moving at one and the same instant all its other parts, because the power is then communicated nakedly, and not as existing in some body (such as the stone) which carries it along.[3]

It will at once be seen that the 'intuitive act' which is involved in this example does *not* produce one of the results which we might have

---

[1] Cf. above, pp. 96–8.   [2] For the speed of light, cf. above, p. 98.
[3] Op. cit., K.S., pp. 46–7; my italics.

expected Descartes to desire. He cannot be certain that the process of the action of light is necessarily as here envisaged. The most he can say is that *if* it is in accordance with his account of one of the 'actions of natural powers', then it will be a process which he is capable of 'conceiving easily' to take place in a certain 'necessary' fashion. Secondly, it will be obvious that the logical notions shown in quotation marks in the previous sentence agree only *analogically* with the technical requirements of the 'clear and distinct', and of 'intuitive apprehension'.[1] The important point for our philosophical appraisal of the genesis and meaning of Descartes' rationalism is, however, that such cases supply the 'technical' or 'analogical grammar' for the key notions; and subsequently these come then to be employed 'justificatorily' for the purpose of 'laying foundations'.[2]

Descartes is indeed perfectly aware that he has no direct evidence for, and hence cannot be certain of, the actual nature of the fundamental physical process involved. He expressly remarks that it may be that when we are trying to develop the theory *deductively*, we are in fact 'unable to determine straightforwardly the nature of the action of light'. In that case, we should run over the various possibilities which earlier inspection of the variety of natural powers has already suggested. Now one of the difficulties of the present example is that one cannot work with infinite speeds; similarly, it is necessary to introduce a somewhat more concrete physical process than that offered by the rather vague notion of 'tendency to motion' of the minute particles of the 'second element', the ether of light. Descartes therefore suddenly switches to the remarkable position that a knowledge of one of the other alternatives may 'help us to understand it [*sc.* the action of light], *at least by analogy*'.[3] The intended analogy must be—as *Regulae* IX has hinted—that of a body in local motion, such as a stone, moving with a *finite* velocity.

We see, then, that at this point once again the deductive line is snapped and original restrictions are relaxed; the reasoning thereafter proceeds in accordance with a series of *models*. The models of the

---

[1] The possession of a strong analogy has more recently been held to be a *sine quâ non* of an hypothesis before it can be assigned the status of a genuine theory or explanation. (Cf. Campbell, *Physics. The Elements*, Ch. 6; and my 'Theory Construction. The Work of N. R. Campbell', in *Isis*, 55 (1964), 151–62.) It is easy to see how at an earlier stage analogy could function as a paradigm for the *a priori*-strength of an hypothesis.

[2] Cf. above, ch. I, pp. 3ff.

[3] *Reg.* VIII; K.S., p. 39; my italics.

Dioptrics are in fact on Descartes' own admission employed with the utmost abandon and disregard for mutual consistency in their actual physical action, though founded on the basic paradigm of the mechanical laws of matter and motion. We might describe all this as a procedural tightening of the qualitative hypothetical situation, which was initially introduced to begin the movement from the data to the colligating formula. Descartes is remarkably clear on the methodological basis of his procedure—witness the introductory remarks near the beginning of the *Dioptrics*:

> Now since I shall have to treat of light only in order to explain how its rays enter into the eye, and how they can be deflected by the various bodies which they encounter, there is no need for me to undertake to say anything about its true nature, and I think it will be sufficient if I avail myself of two or three comparisons, which will help us to understand it in the most convenient manner, in order to explain all those of its properties that experience allows us to know, and to deduce thereafter all the others which may not be so easily noticed. In all this I am using the method of the Astronomers who, even though their suppositions be nearly all of them false or uncertain, all the same draw from them many consequences which are perfectly correct and assured, since they agree with the diverse observations which they have made.[1]

Here we meet again the reference to support for hypotheses from consequences; it was an ancient discovery that descriptive astronomy is capable of reconciling its observations with a variety of possible hypotheses. Descartes' approach is however more revolutionary, for it employs 'models', some of which *deliberately* relax the conditions, such as those of infinite speed, which the *physical* account had demanded; and in doing this, Descartes is further compelled to abandon the original *hypothesis* of stationary ether particles in favour of an emission theory of light particles moving through space. His justification for this is once again a piece of analogical reasoning: light-rays seem to be reflected or absorbed much like the motion of a ball or a stone.

> And so it lies on hand to suppose that the activity or inclination towards motion, which as I said is the way in which we must think of light, follows here the same laws as does motion itself.[2]

The model in terms of which he goes on to 'deduce' Snell's Law employs the motion and impact of a tennis ball, when striking

[1] A. & T., vi, p. 83; L.A., p. 66.
[2] Op. cit., A.T., vi, p. 89; L.A., p. 70.

a variety of surfaces. The deduction, whilst simple in manner, is tortuous and full of arbitrary assumptions, including that of the *increased* velocity of the particles in the denser medium. This assumption is forced upon Descartes in order to obtain the qualitatively observed bending of the light in the denser medium *towards* the normal to the interface. (This in turn he tries to make comprehensible by additional illustrative devices.) The deduction does however lead to the brilliant result of relating the refractive index to the ratio of the speeds of light in the two media; it emerges in Descartes' demonstration of the law as follows.

He argues that the only circumstances requiring consideration are the component velocities of the particle at the moment of impact with the interface. In terms of the figure on p. 137,[1] if ABI represents the path of our particle, Descartes argues that the velocity-component parallel to NN' will not be affected but remain constant, since the forces on the particle when it strikes NN' act only in the direction normal to this.[2] In terms of the geometry of the figure this means that $V_r \sin r = V_i \sin i$, which—with $V_r/V_i = \text{const.}$—gives Snell's Law, where V_i, V_r are the velocities of light in m and m', in the direction of AB, BI, respectively.

It will be seen that Descartes' 'method' has here reached its goal. The 'question', after having been 'perfectly determined', has been reduced to 'proportions' specified through an 'equation'.[3] A proper scientific procedure now requires the testing of this 'conclusion', the test being equivalent to a demonstration that an elliptically shaped lens does in fact collect parallel rays at a point.[4] Accordingly, Descartes instructed a celebrated optician of his time (Mydorge) to make a lens of the appropriate shape and the experiment indeed brilliantly confirmed the prediction of the theory.[5]

In this example, 'method' turns out to be something little expected by the usual interpreter of Descartes' thought. It is through reasoning round a model in geometrical fashion and using the *laws* of motion of projectiles that Descartes manages to derive his formula. The logical form of the model is of some interest: there is putative identity of the physical *principles* governing the model and the theoretical

---

[1] Cf. ibid., n. 1.

[2] These forces are such as to *increase* the normal velocity component—clearly a very paradoxical assumption, and made solely in order to fit the qualitative observations.

[3] Cf. *Reg.* XIV; K.S., p. 88; cf. above, pp. 88–9.

[4] Cf. above, p. 137.   [5] See A. & T., i, p. 239, letter to Golius, 2.2.1632.

('hypothetical') structure; but the model is not meant to reproduce the theoretical *structure* as such: it is only a kind of formal
'analogue'.

How does this methodological procedure square with Descartes'
more rationalist pronouncements, and in particular, with the aspect
of analysis, which insists on resolution until we arrive at intuitive
starting points? (After all, it should be remembered that *Regulae*
VIII ostensibly seeks to illustrate this with its example from optics!)
So far we have noted three possible interpretations of this requirement:
the employment of concepts that have a relative logical priority; the
preference for mathematisation of a physical situation; finally, the
reductionist or 'mechanist' move to 'extension'. The question, however, still remains whether—at least in Descartes' *science*—there are
any *principles*, and not just *concepts*, that have an 'intuitive' basis; and
the answer which we have elucidated from our last example is that in
regard to the former, intuition has so far only an interpretation in
terms of the *analogical* power of certain fundamental generalities—
such as those of matter in motion.[1] This, however, leaves the explanatory principles of optics themselves still with a purely hypothetical
status. There is a famous passage in the last chapter of the *Discourse*,
where Descartes seems actually quite aware of this, and where he in
fact gives rather a subtle statement of the 'hypothetico-deductive'
logic:

> Some of the things of which I have spoken at the beginning of the
> *Dioptrics* and the *Meteors* may seem shocking at first sight, because I
> call them hypotheses and do not seem to have tried to prove them. I hope
> however that the reader will be satisfied if he has the patience to study
> the whole book attentively. For it seems to me that the reasonings are
> mutually connected in such a way that as the deductions from the
> hypotheses are demonstrated by the latter, these being the causes of the
> former, so the hypotheses themselves are in turn proved by the deduc-
> tions which are their effects. Nor should we imagine that thereby I
> commit the fallacy called by logicians reasoning in a circle. For since
> experience renders the majority of these effects quite *certain*, the causes
> from which I deduce them serve not so much to prove but to *explain*
> *them*; indeed, it is on the contrary the causes that are *proved* by the
> effects.[2]

We now see more clearly the formal analogy with 'analysis$_m$'. That,
too, claimed to prove a certain proposition by assuming its truth and

[1] Cf. above, pp. 140–1.   [2] Op. cit., vi; A. & T., vi, p. 76; cf. K.S., p. 162.

deducing from it consequences already and independently known to be true. However, this species of analysis, 'analysis$_h$', differs from 'analysis$_m$' in that our knowledge of the consequences is never 'axiomatic' but only 'empirically certain', as Descartes notes in the passage just quoted. Furthermore, unlike the case of analysis$_m$, though the consequences are deducible from the assumed 'hypothesis', the converse is never formally achievable in analysis$_h$. Nevertheless, we should not ignore the superficial analogy between the two species of analysis; and it is certainly a contributory factor in the grand synthetic picture of Descartes' logic. The situation is always fluid, enabling differential stresses to be placed, now at one point, now at the other.

And indeed, even though Descartes has been seen to have a fairly clear conception of the formal schema of hypothetico-deductive logic, his true view is more complex, for he has not really abandoned the requirement that every hypothesis (despite strong backing from its logical consequences) is in need of additional 'consolidative support'. It is the relation between these two that really constitutes Descartes' (and our) problem. And here it appears at first sight as though Descartes were still following a fairly traditional rationalist line—unless we remember the fluidity of his language. This can be confusing. Thus, even in the very section of the *Discourse* from which we have just quoted, and following immediately upon the passage given, we are suddenly told that his reason for having spoken of 'hypotheses' or 'suppositions' is really other than that first given. In explanation, Descartes slides into quite a different (and much older) use of the term 'hypothesis', going as far back as Plato, where an 'hypothesis' is a 'provisional axiom', held to be deducible from higher-level principles which (ultimately) are held to possess a superior epistemological status. Following this definition, Descartes tells us that he holds his hypothetical explanations to be really deducible 'from those primary truths which I have explained above'—we are not told whether this is a reference to the argument of the *Cogito*, or to the existence, goodness and conserving power of God, to the laws of motion,[1] or to concepts and principles of extension (geometry).

Such a deductivist account is partly due to attacks made on him for his sympathetic attitude to hypothetico-deductivism,[2] forcing him to

---

[1] These matters will be discussed in the subsequent sections of this chapter.
[2] Hostility to this form of inference was as common then as it has been since, because it is easy to regard it as an attempt to simulate a formal argument, in which case it

make an attempt to provide what we would now think of as a 'consolidative' device.

In a letter to Vatier,[1] who had commented with surprise on a method which, whilst it seemingly emphasised rational intuition, in the *Dioptrics* had instead gone so far as to offer explanations by way of models,[2] Descartes again shuffles by insisting that although in his treatise he had employed 'a *posteriori* demonstrations', this was only because to try and give *a priori* ones of every detail would have involved an exposition of the whole of his physics, and that the analytic method was sufficient for all practical purposes. More significantly, he adds that his omission of '*a priori*' physical proofs had been facilitated by his belief that he would certainly (if required) in principle be capable of deducing all his suppositions from 'the first principles' of his 'metaphysics'. We may understand this remark as being in part a reference to the background of physics and mechanics, and in part to the 'metaphysics' of extension and motion. On the other hand, it must be said that Descartes is not always too meticulous in remembering the strong use he has made of analogy and reasoning in terms of models, adapting his explanations to the desired conclusion, rather than the other way round.

The logical fluidity of the scene—witness the flexibility of the terminology—is illustrated by the ease with which Descartes, despite the logical nature of his theorising in the *Dioptrics*, could still claim, as we noted earlier, to have 'demonstrated refraction geometrically and *a priori*'.[3] Yet, when Mersenne enquires how in the light of the use made of his models, he can make such a claim, he responds with righteous indignation: 'to require from me geometrical demonstrations, in a question which concerns physics is to ask me to do the impossible'![4]

Clearly, there are innumerable stresses and strains which produce such differential emphases; and it is they which really make Descartes the 'father of modern philosophy'. In the case of Descartes, it is

---

is of course invalid. And even when regarded as a separate non-demonstrative form it still faces many difficulties, e.g. a plethora of alternative hypotheses capable of 'explaining' the known data—in fact the most prominent objection made to Descartes in his own day.

[1] 22.2.1638; A. & T., i, p. 563.
[2] Cf. the passage quoted from the Introduction, above, p. 142.
[3] Cf. above, ch. I, p. 21.   [4] A. & T., ii, p. 142.

necessary to consider all this technical detail because—so is my contention—it supplies the concrete 'paradigmatical' background for the direction of Descartes' metaphysics, and gives an initial meaning to metaphysical concepts and principles. In the later philosophers, the background is mostly lost, or no longer living; and the fluid terminology hardens, the positions are sharpened, the internal contradictions exposed.

(e) *The conceptual foundations of dynamics*

There is, however, more in Descartes' striving after an '*a priori* in physical theory' than a mere contention that his hypotheses should be *consistent* with a conceptual framework of the mathematics of extension; they also have to reckon (as we have noted) with certain 'rules of mechanics', with certain 'laws established in nature by God';[1] the various models and hypotheses are subject to the constraint of the laws of motion.[2] Again, in the passage from *Principles*, iv. 203[3] there is the claim that 'the clear and distinct notion of material things' does not only comprise *concepts* like shape, magnitude and motion, but also certain 'rules' which 'are the principles of geometry and mechanics'. Similarly, at iii. 46: self-evident principles assure us not only that the universe consists of one and the same matter, divisible and actually divided into many pieces with various circular motions, but also that 'the same quantity of motion is constantly preserved in the universe'. (It is only after this that subsidiary hypotheses are introduced.) In particular, there are certain specific 'rules or natural laws' in accordance with which the various specific motions take place.

Undoubtedly in Descartes' opinion these laws represented the supreme paradigm for the programme of a 'metaphysical foundation' of mechanics. The term 'metaphysical' is here used advisedly; for whilst the claim to the clarity and distinctness of the geometrised aspect of matter, viz. extension, is made by mere attention to the nature of our ideas, with motion the purely geometrical standpoint is insufficient. Not only is 'motion' not deducible from extension;[4] relative to extension, it is an idea that is 'opaque': not unnatural, in view of the difficulty Descartes would encounter in his day with any attempt to establish a mathematical calculus capable of dealing with it. At any rate, the idea that the total quantity of motion of all the

---

[1] Cf. *Principles*, iv, 203; *Discourse*, v, K.S., p. 147. See above, pp. 44, 121, 134.
[2] Above, p. 142. [3] Above, p. 134. [4] See above, p. 96.

parts of matter should remain constant is something which clearly needs independent support. Nor can such support be narrowly inductive, for the principle in question involves a generalisation of such astonishing breadth that we feel its true springs must lie elsewhere. I have touched on the nature and place of such higher-order principles in science in chapter II;[1] it is sufficient to note here that Descartes' mode of establishing the principle of conservation of the quantity of motion is an early and obscure recognition of it being neither purely conceptual nor a simple hypothesis nor yet an empirical generalisation, but rather—in the language of our earlier chapter—a statement whose logic is that of the 'functional *a priori*'.[2] Moreover, in the same place I sketched the way in which many of the earlier scientific systems of explanation involved the notion of substance, the prime operative aspect being an entity which persists through change, and in terms of which modes of change may themselves be correlated, and I noted the variety of 'specific candidates' by which the general notion of substance is interpreted, e.g. atom, energy, momentum, electric charge, etc.[3]

Descartes' candidate, though not actually that of Newtonian momentum, is related to the latter, though specialised, viz. quantity of motion ('Q_s'), defined as the product of speed into volume of a given quantity of matter. (Momentum 'Q_v', on the other hand, is defined as velocity (= speed in a given direction) times mass.) Now Descartes treats Q_s as though it were a thing or, indeed, a corporeal substance. For he considers it quite simply as 'created by God in the beginning', and thereafter 'preserved' by His 'ordinary cooperation'.[4]

We shall in the next section consider Descartes' arguments showing the need for such 'cooperation', with the aim of 'conserving' a corporeal or thinking substance (body and mind); a necessity which (Descartes holds) presupposes God's 'conservative', i.e. 're-creative' power and action at any instant.[5] This is itself a supplementary part of the main argument through which Descartes hopes to transcend the momentary 'idea' and reach some 'ontological ground' of 'continued existence': an 'other', contrasting with the 'self' which is all that the previous argument of the *Cogito* has been able to establish. If God exists, and if He exists in the mode in which our ideas have

---

[1] Above, pp. 33–4.   [2] Cf. above, pp. 33, 41.   [3] Above, pp. 50–1.
[4] *Principles*, ii. 36; cf. Anscombe, p. 215.
[5] *Meditations* III, K.S., pp. 227–8; see below, pp. 175–7.

reflected Him, then, given His creations—here matter and motion—we may expect that from the nature of His being there will result certain primary effects, viz. the continued existence of 'substance'.

It is in this fashion that Descartes argues in the section of the *Principles* where he introduces the principle of conservation of the total quantity of motion:

> We conceive it as belonging to God's perfection, not only that he should himself be unchangeable, but also that his operation should occur in a supremely constant and unchangeable manner. ... Consequently it is most reasonable to hold that, from the mere fact that God gave pieces of matter various motions at their first creation, and that he now preserves all this matter in being in the same way as he first created it, he must likewise preserve in it the same quantity of motion.[1]

The fascinating anticipatory leap in which Descartes here engages treats not only matter but also one of its states (or *modes*), namely motion, as a suitable receptacle for substantial existence—very much in contrast with the older Aristotelian definition of motion.[2] Similarly, Q_s is for him a fundamental magnitude. The notion of 'fundamental' is here a reference to the metaphysical ingredient, anchored as it is in God's recreative or conservative activity. On the other hand, the specification of Q_s as a suitable candidate in reality constitutes an empirical or hypothetical element. Indeed, it was pointed out very soon after Descartes that Q_s was in fact not invariably conserved, e.g. in certain instances of collisions between elastic bodies. Leibniz was one of those to expose the error, in his commentary on the *Principles*, and his remarks are interesting:

> Although the constancy of God may be supreme, and he may change nothing except in accordance with the laws of the series already laid down, we must still ask what it is, after all, that he has decreed should be conserved in the series—whether the quantity of motion, or something different such as the quantity of force.[3]

[1] Op. cit., ii. 36; Anscombe, pp. 215–6.

[2] Cf. the discussion of motion in *Principles*, ii. 24–29, as mere relative 'translation', something which 'can have no being outside the body', and which is simply a 'state of a body'. We may call this a novel conceptual exploration of a scientific conception.

[3] Op. cit., ed. Loemker, ii, 42, pp. 648–9, on Article 36 of the *Principles*. The reference to 'force' is to *vis viva*, defined as mass times the square of the velocity, mv^2, the conservation of which was discovered by Leibniz, partly to show that purely geometrical notions were insufficient to exhaust a description of the universe. Cf. below, ch. VII, sect. 3(a), (ii), pp. 416ff.

Malebranche, much affected by this criticism, subsequently tried to save the Cartesian notion of the 'quantity of motion', by replacing Q_s with Q_v, which is indeed conserved. And he concluded characteristically, that the proposition that God conserves a constant quantity of motion in the universe bears even more the stamp of the attribute of the Divine if it is understood 'in a correct sense, and in conformity with experience', i.e. if we use Q_v, instead of Q_s![1] In this way, the search for a conservative quantity can often act as a direction-giving device in the development of empirical science.

But Descartes has to his credit an even more astonishing achievement: the first published statement, and a fairly close version, of the so-called law of inertia, Newton's First Law of motion, which in Descartes' formulation, *Principles*, ii, 37, 39, reads:

> A moving body, as far as it can, goes on moving [uniformly], [and that] ... not in any oblique path, but only in straight lines.

This, it should be remembered, appeared forty years before the publication of Newton's *Principia*; it corrects the false principle of Galileo, according to which all bodies not subject to forces continue to move round a common centre, i.e. in a circle!

How does Descartes elucidate his principle? He mentions certain observational facts, which are intended to explain why we do *not* normally *observe* inertial motion, but soon turns to a 'rational' explanation. This, as expected, is based on his total world view embedded in the notion of matter as isotropic infinite space, distinguished only by purely 'passive' metrical properties.

The specific proof of the law presupposes his novel conception of 'motion' as a 'state' of matter. His 'first law' deals with the 'scalar' part of velocity, viz. speed. Just as the shape of a body does not change except under the action of external causes, he argues, so also its speed remains unchanged except for the action of some external cause. The detailed argument invokes a general principle:

> Every isolated thing continues in the same state so long as is possible and never changes it except when encountering another.[2]

This law, Descartes tells us, follows 'from the fact that God is not subject to change, and that he always acts in the same way'. God's action, it should be noted, is not here 'causal' in the sense in which one body 'causes' another to move through impact; the latter Descartes

---

[1] *Recherche de la Vérité*, Paris, 1946, vol. 3, p. 278.
[2] Op. cit., ii, 37; A. & T., ix, p. 84; Anscombe, p. 216.

calls 'a secondary or particular cause', and contrasts with the former, which is 'universal and primary'.

The second law (equivalent to the second part of the law of inertia) is the truly revolutionary extension of inertial motion to the rectilinear case. It is formulated as follows:

> Any given piece of matter considered by itself tends to go on moving, not in an oblique path, but only in straight lines.[1]

Once again, Descartes notes explicitly, 'this is confirmed by observation'; witness the behaviour of a stone whirled around in a sling. Also, as before, in addition to experience, there is 'a reason for this rule', which is once more 'the immutability and simplicity of the operation by which God preserves motion in matter'. Just as God preserves the speed, so He preserves the *direction* in which a body moves at any particular instant of time.

In *Le Monde* (ch. 7), Descartes is more explicit, and adds that straight-line motion is the simplest of all types of motion, being 'the only one whose nature is contained and realised in an instant'. His argument is that the direction of a body moving in a straight line is determined by considering just *one* instant, whereas circular motion requires at least two such instants. For Descartes to say that God conserves the speed and direction of a free particle is to say (I think) negatively that such conservative tendencies do not follow from the *concepts* of speed and direction, so that the resulting statement is not 'analytic' but 'synthetic'. Further, God's immutable action means for Descartes that there are conservation laws governing the material universe, and that finally this fact, though illustrated by experience, is not founded on it. Moreover, it had for him the status of an *a priori* truth.

The motives for such a procedure are not far to seek. We have previously noted a tendency to construe the conception of law of nature in such a way that it includes a referene to a 'constitutive ground'—a conception meant to bestow on the law an 'objective validity' which it lacks if we pay attention only to the epistemic aspect of empirical generalisation.[2] Now the conception of a constitutive ground is really equivalent to an *affirmation* that certain principles which would validate inductive inference are in fact 'true'. Our concept of 'law of nature' is therefore viewed as something which

---

[1] Op. cit., ii, 39; Anscombe, p. 217.   [2] Cf. above, ch. II, pp. 28ff.

though it may be triggered by what the empirical pointers supply, contains more; and for Descartes, this additional feature is something 'constitutive'.[1]

Now for Descartes, God plays the function of the source of such a constitutive link. The problem however then arises as to why there should be an agreement between what observation can only suggest or illustrate, and its constitutive ground. Some may be satisfied here with sheer postulation; others—like Leibniz—may whittle away further the importance of the realm of empirical phenomena and their laws.[2] The alternative which Descartes chooses is to make the constitutive link putatively a matter of rational insight.[3] In the case of the law of inertia, this is achieved by exhibiting it as a special case of a basic aspect of all experience, viz. causation, thus giving the impression that it, like causation, has the status of an *a priori* truth.[4] What is valuable in such a procedure is Descartes' recognition that the laws of motion, regarded as axioms of dynamics, possess a special status with respect to that science. (It is for instance difficult to 'test' such laws *within* the domain of dynamics without circularity.)

However, postponing for the moment the question of the status of the law of causation, it is somewhat implausible that the more specialised principles of science, however basic, should be derivable in

[1] Cf. Kant's celebrated dictum: 'Though all our cognitive grasp [*Erkenntnis*] begins with experience, it does not follow that it all arises out of experience' (K. 41). We shall later see that Kant of course differs from Descartes in construing the constitutive element as something which is a *condition* of experience, not something lying beyond it.

[2] Thus, for Leibniz, not only are the 'phenomena' merely semi-real, the laws of nature likewise are no more than 'subsidiary maxims', mere approximations to an underlying reality. Cf. ch. VII, sect. 5(b).

[3] Cf. above, ch. II, p. 44, for his statement of his general view on the status of the laws of nature. Descartes' natural theology will be taken up in section 3.

[4] This mode of procedure was aided by the fact that the notion of the *a priori* is, as I have already suggested, at this stage not yet very precise. This *a priori* is simply the contrary of *a posteriori* (founded on sensory experience); it is not identical with the kind of *a priori* of which examples are the 'clear and distinct' perception of some geometrical truth, or of the *Cogito*. The postulation of an *a priori* rule like the first law requires therefore a special defence. Kant's 'proof' of Newton's First Law in the *Metaphysical Principles of Natural Science*, iii, Theorem 3, involves the application of a very similar causal principle, which Kant 'borrows' from the 'Second Analogy', the whole apparatus of the *Critique of Pure Reason* being thus brought to bear on a (transcendental) proof of that principle—'transcendental', because (as we shall see) that is the only kind of proof, according to Kant, possible for a principle which is both synthetic as well as *a priori*. (Cf. ch. VIII, sect. 9(b).)

any simple way through an *a priori* procedure. True, there is no reason why one should *not* stumble by chance upon such principles by the use of the deductive method, particularly if they are very general; but there is always the suspicion that—as Kant once wrote ironically —in such cases

> the philosopher has so steered reason that by means of a very slight twisting of the grounds of his argument, aided and abetted by some surreptitious sideglances at certain observations and experimental results, he arrives miraculously where least expected by his trusting disciples, namely at the conclusion which everyone had already known previously to be the object of his proof.[1]

And there are indeed a whole battery of additional definitions and postulates and conceptual formulations which Descartes has presupposed, and which are either empirical matters, or built into the mathematical and conceptual framework of dynamics in general. Let us list some of them:

(1) The postulate that motion is a *state* of body.
(2) Translation is defined relativistically, but the implications of this fact are not faced, any more than they were by any other physicist prior to the time of Ernst Mach, i.e. end of the nineteenth century.
(3) Descartes tacitly assumes that the concept of 'motion' as occurring in the first law denotes 'speed', i.e. 'rate of change of displacement with respect to time'. Only if this is assumed does 'unchanged motion' become 'uniform speed'. As Poincaré remarked uniform motion (the motion with which to couple the application of the principle of causality) could equally well be defined as uniform acceleration.[2]
(4) There is the assumption that the direction of a body at a point P moving in a curved path is given by the tangent to the curve at P.
(5) Granted that 'direction' is conserved, it does not follow that a 'free' body will continue to move in a 'straight line' in the sense of Euclidean geometry, for the geometry of physical space may be non-Euclidean.

[1] *Dreams of a Spiritualist*, ii, p. 2; in *Works*, Academy ed., 1905, ii, pp. 358–9.
[2] Cf. 'The Principle of Inertia', *Science and Hypothesis*, iii, 6 (London, 1905), p. 91.

It cannot therefore be said that Descartes managed to deduce Newton's First Law of motion from the existence and immutability of God alone. Nevertheless, Descartes' procedure does contain an important hint for our views about the status of the law of inertia; it is a useful corrective to Newton's view that the law is derivable in a straightforward fashion from experiment and observation.[1] Thus, as just mentioned, it is extremely difficult to devise empirical tests for the law which do not in some way presuppose the law itself.[2] We may therefore look at this law, and others like it, as forming part of a theoretical framework in terms of which to interpret the phenomena and test and adjust other empirical laws of the theory. So if this is the true state of affairs, we may say that the core of Descartes' procedure is equivalent to an attempt to chart out the *conceptual foundations* of dynamics.

Unfortunately, such a charitable interpretation of Descartes' intentions is only possible by relaxing the strictness of a number of definitions and making due allowance for empirical features not expressly mentioned by him. No doubt he did not feel the need for this since the weight of the remainder of his system would tend to conceal from him any logical deficiencies or vaguenesses on the side of its individual members. With the collapse of the Cartesian system, each of the individual components had to be re-examined, the concepts sharpened, previous emphases repressed, new interpretations brought forward. Sometimes this has the result of making people forget certain important insights, for the sake of clarifying aspects of the older system. (Thus the Newtonian philosophers had little patience with Descartes' views on hypothesis, and certainly next to no understanding of the 'functional *a priori*' aspect of his rationalist approach to the foundations of dynamics.) At other times, an older notion may turn up in a later system, purged and interpreted in a novel way; an example of this is the change of status of the principles of simplicity, which in Descartes' system (as we have just seen) are described as one of God's attributes, whereas in the philosophy of

---

[1] 'These principles [*sc.* laws of motion] are deduced from the phenomena and made general by induction, which is the highest evidence that a proposition can have in this philosophy.' (From a letter to Cotes, in H. S. Thayer (ed.), *Newton's Philosophy of Nature* (New York, 1953), p. 6.)

[2] For this, cf. my 'Science and Logic: On Newton's second law of motion', *Br.J.Ph.Sc.*, 2 (1951), 217–35; M. B. Hesse, *Forces and Fields*, op. cit., pp. 134–43; B. Ellis, 'The Origin and Nature of Newton's Laws of Motion', in R. G. Colodny (ed.), *Beyond the Edge of Certainty* (Englewood Cliffs, 1965), pp. 29–67.

Leibniz and Kant they emerge as 'architectonic', or procedural, 'regulative' principles of scientific thought.

3. DESCARTES' METAPHYSICS. GOD AND THE EXTERNAL WORLD

(a) *Some motives for metaphysical construction*

Several results stand out from the discussions of this chapter. The various methodological elements of Descartes' science ultimately swing back towards a focal point, which is that of the 'intuitive' or 'clear and distinct apprehension' of certain privileged ideas. The Cartesian contrast between 'clear and distinct' and 'obscure and confused' is used as a lever, not only to emphasise the greater methodological power of the former, but also to insinuate the realisability (or indeed the significance) of a programme of 'reduction' of the field of 'sense' to 'thought'.[1]

The logic of the programme of maximum elimination of error by means of a retrenchment of everything except 'ideas' has also carried in its train a second consequence for Descartes' ontology. The desired avoidance of 'erroneous judgments', so we have found, is again and again interpreted as the requirement that our comprehension of ideas should exclude not only all consideration of the purely 'sensory' aspects, but further (and more importantly), all reference to 'external things'.[2] Safety from error comes to be interpreted as withdrawal of the mind from the world 'outside' and of its attending solely to its own ideas which it possesses 'innately' and of itself.[3] The problems that arise from this fateful development are as follows. No doubt, there is an external world endowed with sensory qualities, and comprehensible by the methods and investigations of science. But the programme of 'reduction' to something completely susceptible to 'clear and distinct apprehension' has resulted ultimately in the excision of everything except 'ideas', and indeed, of all but certain privileged cases of these ideas.

The problem is now whether it is possible to construct in an alternative way (by a process of 'reduplication') that external world of which the 'methodological doubt' seems for the moment to deprive us. It is important to understand the meaning of this question. There is *a* sense, of course, in which that world of external objects endowed with

[1] Cf. above, pp. 132ff.   [2] Cf. above, pp. 94f., 133f.   [3] Cf. above, pp. 99ff., 214–18.

secondary qualities is given for the observer. What is being demanded is, rather, a 'construction' (a 'proof' of existence) in which consideration is limited to 'ideas' as such alone, and to their nature and status, together with certain supplementary logical or metaphysical or moral principles: the minimum requirement being that ideas and principles should be 'apprehended' in a 'clear and distinct manner'. The proof of existence must not depend for support upon anything connecting with the realm of sensory phenomena, with 'experience', understood as sensory knowledge of and acquaintance with these phenomena. The question of our 'knowledge of the external world', as Descartes raises it, transcends that realm. This is asserted explicitly when he remarks that God has given us 'no faculty' whereby to 'discover' whether our *ideas* of external objects are 'produced' by the latter or by some other agent.[1]

Descartes' optimistic assessment of the range and power of 'clear and distinct perception', and of the 'natural light', will reinforce his belief in the success of a programme in which the proof of existence proceeds from the very nature and being of his 'ideas', of the process of 'ideation' (thinking), in quasi-deductive fashion. This programme meets, however, with peculiar difficulties. For Descartes' argument soon has to move from the realm of ideas to something *other* than ideas, the 'other' appearing in threefold succession as 'the self' considered as substantial possessor of ideas; as 'God'; and as the 'external world'. But does not this suggest that the deductive chain of the Cartesian argument will somewhere have to snap? We have seen that the standpoint of the methodological doubt demands the 'innateness' of ideas. On the other hand, already in the passage from the sixth Meditation just quoted we note the implicit contention that ideas are 'produced', have 'causes', and are viewed as 'effects' of such causes.[2] But how is this continuous movement from effect to a transcending cause to be defended?

Here there arises indeed a problem which Descartes had so far not considered very deeply, the problem of 'existence'. Whereas the *Regulae* had limited itself to exploring the conditions for a successful method of discovery of ideal *relations*, his later writings (*Meditations*, etc.) give evidence of an increasing awareness of the question of *existence*. The *Cogito* is the first step in this direction, the first attempt to *argue* from the nature and content of an idea to the existence of something other than the idea, such as the self, considered as a

[1] *Med.* VI, K.S., p. 256; cf. above pp., 109–10.　　[2] Cf. *Med.* III, K.S., pp. 219–20.

'thinking thing', or God, or 'actual existents', e.g. external objects. Where, before, the world of sense was transcended in favour of one of mathematical systematics, we now find the latter supplemented—as through a mirror procedure—by a world of ontological construction.

There is a further set of considerations which drives Descartes to 'metaphysics'. Most of the methodological requirements of the *Regulae* reduce to statements of *necessary conditions* which must be satisfied so that the 'understanding' will acquire a knowledge of propositions that are true; but it does not seem that Descartes had so far very clearly faced the question, whether there *are* any propositions which in this way can be 'proved' to be true. Of course, there are plenty of reasons why he should not have felt this need earlier; the examples from mathematics, optics and mechanics would tend to act as powerful inhibitors postponing such awkward questions. But they could not be repressed for ever. And so, in the first Meditation, Descartes embarks on a process of doubting far more radical and sustained than hitherto.[1]

So far, he tells us, mere sensory investigation has been rejected in favour of 'real existents', such as bodily shape, magnitude, number, location and time. And for this reason, too, arithmetic and geometry are 'certain and indubitable', compared with 'physics, astronomy, medicine' etc., which are 'of doubtful character'.[2] But now, even this very reliable body of knowledge will have to be questioned; for how do we know that we are not deceived even in respect of the truths of mathematics and geometry? Hitherto it has been assumed that if one 'sees' that $2 + 3 = 5$, then this proposition will be rightly judged to be *true*, proceeding in accordance with the 'general rule, that the things we apprehend very clearly and distinctly are true'.[3]

---

[1] It should not of course be supposed that the results of the previous period are abandoned; on the contrary, though the 'methodical' knife probes more deeply, many of the earlier assumptions and comparisons are retained. And in any case, it is not certain that Descartes ever really improves on his earlier position; cf. the assumption which he is forced to make in order to advance beyond the *Cogito*, and which amounts to a reiteration of his previous views. (See below, pp. 171f.)

[2] *Med.* I, K.S., pp. 198–9.

[3] *Discourse*, iv, K.S., p. 141. Descartes adds: 'bearing in mind, however, that there is some difficulty in rightly determining which are those we apprehend distinctly'! Likewise, *Med.* III, K.S., p. 214: 'I would now seem to be able to adopt as a general rule that everything I apprehend in a genuinely clear and distinct manner is true'; 'seem', because, as the argument goes on to maintain, the rule is perhaps only a necessary, and not a sufficient condition. Whether it is Descartes' final view that it is (or is not), is a problem with which Descartes scholars are still wrestling.

But could not some 'malignant demon'[1] deceive me even in these seemingly transparent cases? Descartes' appeal to such a possibility shows straight away that the latter would involve no ordinary doubt—he himself refers to the 'ground' of this doubt as 'metaphysical'[2]—and that—correspondingly—to safeguard the foundations by an attempted elimination of the power of the demon (through God's will) is no ordinary safeguarding, the truth thus saved no ordinary truth.

How could we be deceived in the mathematical cases? Perhaps we might make a mistake in our calculations, particularly if they are lengthy. But we must assume with Descartes that as part of the exercise of method, such calculations have been broken down into their smallest possible parts, concerning which everything is intuitively clear.[3] The kind of relation involved in such a proposition as $3 + 4 = 7$ Descartes (we saw) regards as necessary.[4] And his reason for this seems to be that in 'conceiving the number 7 distinctly' we already 'include in it in some confused manner the numbers 3 and 4'. It would be tempting to interpret this as implying that Descartes thought the proposition to be 'analytic'; but we have seen that such sharp distinctions do not as yet obtain at this stage of philosophical history. Descartes asks: 'How do I know that I am not myself deceived every time I add 2 and 3, or count the sides of a square . . . ?'[5] He may be expressing doubts here about the possibility of the application of our numerical concepts and operations to our actual world; or he may, more searchingly, do so with respect to our possession of 'distinct' concepts of such numbers and operations. The assumption that we have such concepts, yielding formalisable systems not leading to any contradictions, would itself then constitute a peculiar instance of a rationalist position.[6]

It is at any rate just possible that Descartes' view is more suitably explained by tightening it to give something like Kant's 'synthetic *a priori*' account of the propositions of arithmetic. Kant's argument is that such propositions involve a 'synthesis of intuition', a sort of universalised 'construction'. He, too, demands a justification of our assumption that such constructions or intuitive syntheses should be 'necessary' or 'universal' ('*a priori*') with respect to the world. And he

---

[1] *Med.* I, K.S., p. 201.   [2] *Med.* III, K.S., p. 215.
[3] Cf. *Reg.* VII, K.S., pp. 32–33, and cf. the proviso quoted on p. 157n.3 from *Discourse*, iv.
[4] See above, p. 105n.3.   [5] *Med.* I, K.S., p. 199.
[6] Cf. I. Lakatos, 'Proofs and Refutations', IV, *Br.J.Phil.Sc.*, **14** (1964), p. 335.

argues that the synthetic approach is presupposed as a condition of experience.[1] Such a view implies that the propositions in question are neither merely 'empirical' (since they are capable of acting as necessary 'norms'), nor simply the product of arbitrarily chosen definitions or axioms.

It is very difficult to attribute a sense to their possessing a third alternative status. What precisely can be meant by saying that though it appears necessary to me that $2 + 3 = 5$, and that though I cannot *logically* doubt this truth, nevertheless it may in fact not be true? Excluding trivial answers, the Kantian response would at least offer us a handle to approach Descartes' view of this question, the only difference—and it is of course a vital one—being that Descartes has no 'transcendental' apparatus or method. In default of this, he can only return to sheer 'metaphysical assertion'. We have already seen that for Descartes the 'innateness' of mathematical concepts carries with it the claim that mathematical relationships are founded in the 'nature of things', i.e. in God.[2] And he might have used the doctrine of clear and distinct perception to this end also. But though he at times toys with it, he does not, in the end, actually avail himself of it.[3] Not that he entirely relinquishes the standpoint of the clear and distinct! We shall presently find that this re-emerges under the guise of the 'natural light', whose deliverances, Descartes insists, we 'cannot anywise call in doubt': it is the ultimate arbiter between the 'true' and 'the false'.[4]

Let us first outline Descartes' formulation of his general move *via* the *Cogito* to God, and then proceed to consider some of the details. Descartes has often been accused here of being trapped in a logical circle.[5] If so, he did not see it, the reason probably being that he proceeds to interpolate a new and separate set of arguments. He embarks on an entirely new path, assuming that he is 'ignorant', and capable of 'metaphysically' doubting everything. He then argues for a proposition of which he can be certain, whose truth cannot (logically)

---

[1] He assumes, moreover (arbitrarily, it seems), that certain unique forms of this approach are selected in the same way, e.g. Euclidean geometry, the arithmetic of natural integers, etc.   [2] Cf. above, pp. 217f.

[3] Cf. *Med.* III, K.S., p. 215: 'Let who will deceive me, he will never be able to bring it about that . . . it will be true . . . that 2 and 3 could make more or less than 5. . . .' Nevertheless, he goes on to say, in order to eliminate the 'metaphysical' doubt that remains, it is necessary first to prove that there is a God, and that He is no deceiver.

[4] *Med.* III, K.S., p. 218. Cf. below, pp. 171f.

[5] For this 'circle', cf. also below, ibid.

be denied, and which he thinks to be the existential proposition that the self, *quâ* thinking being, exists. The truth of *this* proposition, and of this alone, though it too, as a matter of *fact*, is grasped clearly and distinctly, does not logically depend on this fact.

Having now proved that there is a thinking entity endowed with ideas, Descartes goes on to claim that at least one of these ideas demands the existence of a self-transcendent (and moreover, divine) complement. The proof employs a quasi-causal maxim, claimed to be certified by the 'natural light'. Descartes seems to have thought that the power of the 'natural light' when directed towards this idea and the supplementary existential maxim was not exposed to the force of the 'metaphysical doubt'. And it may be that this assumption was buttressed by a confidence engendered by the previous proof of the existence of the self, whose assumptions entailed that it was this self and its ideas (rather than any material external world) which could alone be a genuine basis or starting-point for a metaphysical reconstruction.[1] At any rate, the *aim* (and end) of the whole argument is to demonstrate, or really to 'postulate', that there is an 'ontological element' (the non-deceiving God) which entitles us to follow the dictate of reason in the case of those perceptions which are clear and distinct.[2]

There is a final motive which drives Descartes towards postulating a transcendent causal factor; this follows from the limited range of the *Cogito* argument. For this claims to demonstrate no more than that *if and in so far as* I am thinking at a certain instant of time, I am also an existent thing (a self or mind) at that time. But this does not entitle me *at that time* to *infer* categorically that *I am* a substance, let alone a substance *existing continuously through time*. Descartes' reason is that, first, each moment of time is distinct from every other, 'that the several parts of time are not derived from one another',[3] and that secondly, thinking (and its forms, i.e. ideas) does not disclose any powers of continuous creation.

This position which in an oblique fashion anticipates already later forms of 'metaphysical atomism', together with the other positions outlined so far in an introductory way, we must now consider in

---

[1] Cf. *Reply to Objections* II; Haldane & Ross, ii, p. 31: 'I have demonstrated that the mind, considered apart from what is customarily attributed to the body, is better known than the body viewed as separate from the mind; and this alone was what I intended to maintain [*sc.* in the *Cogito* argument of the second Meditation]'.
[2] Cf. *Med.* V, K.S., pp. 242, 247.   [3] *Reply to Objections* I; H. & R., ii, p. 15.

greater detail in order to relate them to the whole of Descartes' philosophy. But before doing so, and to emphasise its significance for what has gone before, let us ask how the trend of the Cartesian argument bears on the questions discussed in the first two chapters, and in particular, on the problem of what I called 'the propositional link'.

It will have become clear that from the start Descartes gives this matter a special twist, only those propositions being declared acceptable which achieve a peculiar sort of rational certainty. Judgments based on the field of sensory observation in particular are eliminated as possible candidates; and the problems which they pose are thereby —for the time being—shelved altogether; this being made all the easier because Descartes does not (as we have seen) recognise the sharp distinctions later drawn between the logical and the physical realm. But upon inspection most other propositions seem likewise to fail the test whose principle consists in the aim of arriving at something which (in the words of *Meditations* II) is 'necessarily true'.[1] On Descartes' own admission, there is in the end, and speaking strictly, only one such putative case, that of the *Cogito*. But the thinking self which this yields is regarded as standing in need of a necessary complement, the activity of God. Now admittedly such a God is assumed to offer the required guarantee that what is rationally intuited, *is* true. This fact, however, —if fact it be—is offered us as no more than an 'opaque' stipulation.[2]

And so we end up simply with the putative *existence* of a God, Whose nature will now provide the propositional link or foundation we were searching for. The link being 'opaque', it follows that not all, if indeed many, propositions of Descartes' science or epistemology are, strictly taken, 'transparently true', let alone 'logically necessary' or 'analytic'—if such terms are indeed applicable to Descartes' philosophy. The principle with which Descartes comes to operate increasingly, viz. that God is no deceiver, is an acknowledgement of this 'opacity'. At the level of those propositions which are apparently clear and distinct, it provides an assurance that what cannot thus be doubted is in fact true. And at the level of those propositions which lack this feature (e.g. those which we are 'taught

---

[1] Op. cit., K.S., pp. 203, 205.
[2] I call it 'opaque', as being non-intuitive, whilst yet serving as a postulate with a non-empirical, and non-logical, i.e. metaphysical, import.

by nature'[1]) or those which (as in the physical sciences) are, though based on the deliverance of our senses, further buttressed by such systematic support as they may derive from their hypothetico-deductive development, truth is once more certified by the same principle. In this way, the quest for transparency and certainty almost turns into its opposite, metaphysically opaque postulation, a result which will not surprise us after the general considerations of our previous chapters.

(b) *The 'Cogito, ergo sum'*

Since much Cartesian scholarship has been expended on investigations into the meaning and truth or validity of Descartes' proposition 'I think, therefore I am', we may be justified in considering it only very briefly.[2]

This is the 'one thing' which Descartes claims to be 'certain and indubitable', more so than the truths of the geometers and those dealing with the principles of extension and motion, let alone those belonging to the realm of 'sense'. Even if it *seems* only that I think, *if* I doubt that I think, yet I think. And in so thinking, and whilst (though only as long as) I am thinking, I am also existing; at least, existing as a thinking being. This much Descartes 'perceives clearly and distinctly'; and he perceives also that though his doubtings, affirmations, denials, imaginings and sensings may, apart from the self, be nothing at all, yet they are something *quâ* ways of thinking, in, and by pertaining to, the self.[3]

It is difficult to be sure how much Descartes is even claiming in this proposition and how far and in what sense the claim is justified. Kant, at the end of our period, will object that it is a mere tautology: *Cogito* = *sum cogitans*; and *sum cogitans* = *sum*; that I who think, whilst thinking, must be regarded always as *subject* (and not as a mere predicate, somehow attached to the thinking), is (he asserts) an absolutely certain ('apodictic') proposition, simply because it is 'identical'.[4] Now clearly, this cannot have been Descartes' intention. For we have seen that the whole point of the exercise is to see whether we

---

[1] Cf. below, pp. 178ff.
[2] For Descartes' arguments, see *Discourse* iv; K.S., pp. 140–1; *Med.*, II, K.S., pp. 202–7. It should be noted that in the latter work the proposition 'I think, therefore I am' does not occur as such. All that is argued there is that 'I am, I exist', *quâ* thinking being, 'necessarily', 'every time I propound it or mentally apprehend it'.
[3] Cf. *Med.* III, K.S., p. 213.
[4] Cf. 'The Paralogisms of Pure Reason', K.337 and K.369.

can *advance* from the position of pure thinking (as given through its 'forms', the 'ideas') to something extra, something in which 'thought resides immediately', viz. 'mind'; mind being for Descartes a substance.[1]

Now this Cartesian intention Kant also understood very well; but it is a claim which he rejects. Whilst the *Cogito* is valid as a tautology, considering the 'I' as subject, it does not show, he contends,

> that I, as *object*, am for myself a *self-subsistent* being or substance. The latter statement goes very far beyond the former, and demands for its proof data which are not to be met with in thought, and perhaps (in so far as I have regard to the thinking self merely as such) are much more than I shall ever find in it.[2]

Not that Kant's 'refutation' should be accepted uncritically![3] For that criticism not only assumes a much narrower interpretation of 'thinking' than Descartes himself had intended; above all it assumes the generic division between thought and sense which is precisely what the whole Cartesian philosophy is concerned to deny—at least at the ultimate level of metaphysics, with its contention of sensation as being confused thinking. On the other hand, it does not follow that the Cartesian standpoint can make the *Cogito* valid, though it may explain a little better Descartes' sanguine expectations.

Descartes' 'I think, *therefore* I am', should not be regarded as a kind of syllogism, with certain missing or suppressed premises. It is more like a shorthand summary of an argument; alternatively, it states an implication or entailment relation, holding between 'thinking' and 'being'. Whatever view we take raises intolerably complex issues, which (as I have said) we cannot discuss here. There is for example the famous dilemma (echoed already in the Kantian criticism): Descartes' proposition is necessarily true only if it says that 'I think' entails 'I think'. And it is unsupported if it says more, since the implication relation which it claims to hold between thinking and existing (being 'synthetic') requires an independent support which is not here forthcoming.

Another frequent criticism is the contention that 'to exist' is quite a different *sort* of predicate (as compared with 'to think'); is indeed not

---

[1] Cf. *Reply to Objections* II; H. & R., ii, p. 53.   [2] Op. cit., K.369.
[3] The reader should always remember what was said at the start about the nature of philosophical criticism; cf. above, ch. I, pp. 3–5, 10f.

a 'real predicate' at all. Unfortunately, this criticism also[1] presupposes a set of assumptions which completely fail to square with those of Descartes. (It certainly presupposes, as will become clear in later chapters, the whole gamut of the empiricist critique, ending in the contention that 'existence' is a concept necessarily tied up with 'perception', and not capable of simple 'dogmatic' postulation.[2])

So in the end these criticisms do no more than express very pithily their own fundamental assumptions; and they therefore almost always fail to act as decisive rebuttals of the philosophical point of view against which they are directed. On the other hand, they usually manage to put into focus a lack of clarity, of univocal meaning, in the position which they criticise. And so also here. Descartes operates with a set of traditional concepts, such as spirit, substance, matter, whose connotative boundary lines are very fluid indeed. And this makes it all the easier to assume as evident and uncriticisable what later generations will reject as incomprehensible or even senseless.

What offers itself thus to Descartes' 'intuitive eye' is a thinking being, an 'I' (a self) which thinks. Now in essence, the argument of the *Cogito* only considers the *notion* of a self which thinks. Such a self, capable only of forms of *thinking* (Descartes' liberal list includes doubting, affirming, understanding, denying, willing, imagining and —note!—sensing[3]) cannot—in so far as it engages at all in any of these 'activities'—but be thinking, always provided that it does engage in them. (That it does so is an assumption, and a contingent fact; and therefore Descartes (to preserve the rationalist account) subsequently must, and does, argue the need for a 'power' which is the 'author' or cause of this self regarded as 'a substance that thinks'.[4]) So we see that the minimum contention which this argument involves is that the notion of the self that thinks, of mind as substance,[5] is one least open to metaphysical doubting, and therefore one which should

---

[1] Once again it originates with Kant. Cf. K.504: ' *"Being"* is obviously not a real predicate; that is, it is not a concept of something which could be added to the concept of a thing.' The point figures prominently in much modern philosophical writing.

[2] Cf. Kant once more: '[Actual] existence [of a thing] has . . . to do only with the question whether such a thing be so given us that the perception of it can, if need be, precede the concept' (K.243).

[3] Cf. *Med.* II, K.S., p. 206.   [4] *Med.* III, K.S., pp. 227–8.

[5] 'I am myself a substance' is stated twice already in the third Meditation, without further argument. Cf. K.S., p. 224. A slightly more detailed account occurs in *Med.* VI; K.S., p. 254.

be made the kingpin of the epistemological exercise.[1] However, Descartes seems to be fairly clear that, for the purpose of 'ontological construction', it will be very necessary to supplement his argument about the *notion* of a thinking substance by bringing in the aid of additional causal maxims which will supply (what is suggested by previous 'commonsense') the complement to 'mere ideas', viz. what such ideas represent.

(c) *The self transcended: The Natural Light and the existence of God*

Descartes does not pass directly from the position which he has now reached—the existence of the self, or mental substance—to the existence of an external world of physical objects ('corporeal substance'); instead he advances by the circuitous route of first proving the existence of God. The reasons for this inversion of the more traditional route (to God *via* the external world) are briefly that a transcendence from the self 'side as to something other than these ideas requires two conditions. First, a principle postulating that all ideas are 'due' to something other than what possesses mere 'ideal' status; secondly, that there is only one idea whose 'cause' *could not* (logically speaking) be the self. This idea is the idea of perfection. As to the external world, Descartes argues that the idea of extension likewise points to a 'cause' other than the self (since the latter is a purely thinking substance whereas the former must belong to a substance which is unthinking). However, though the *idea* of an extended substance *seems* to point to a substance, that substance might be God (if God exists); in the end it is only God's goodness which will not allow us to be thus deceived and which guarantees that the causes of my bodily ideas are 'corporeal things'. We must now consider certain features of these arguments in somewhat greater detail.

We have seen in previous sections of this chapter how the search for what is 'clear and distinct' drives Descartes steadily 'inwards', leading him to a new problem, expressed as early as the *Regulae*, in the question whether 'external things are always as they appear to be'; or more radically, the question of the 'existence' (substantial beings) of anything regarded as transcending the realm of ideas.

This is mirrored also in the transition from the second Meditation to the third. The former has concluded that our knowledge of mind

---

[1] This is supported also by what Descartes says in *Reply to Objections* II, quoted on p. 160n.1, above.

and extension is far more clear and distinct than what we are 'aware of by way of the senses'.[1] In the third Meditation the conclusion is drawn that although we certainly do have the ideas which we are aware of, and that 'strictly speaking' none of them can be 'false', yet each and any of them may lead us into 'error'. The 'error' involved here is, however, of a deeper and more 'metaphysical' kind than that which concerned Descartes when laying the foundations of science, though it emerges from it, or, as it were, duplicates it. The adoption of 'extension' as the fundamental attribute of matter had led Descartes to the conclusion that many of our ideas, e.g. those of the sensory qualities, 'mislead us' into believing that these ideas 'represent what is nothing as if it were something'.[2] But behind this mode of talking there now emerges a more general (or 'degenerate') view which in turn leads to a fateful, 'metaphysical' generalisation of Descartes' problem. For he now describes our knowledge or awareness of things in our material environment as *'ideas or thoughts of* such things present to our mind'.[3] But his notion of the 'innateness' of all *ideas* (in the more extended sense of the *Notes on a Certain Programme*) is so construed as to cause Descartes to think of these *'things'* as 'outside ourselves',[4] thus in turn raising the question of our knowledge of the *existence* of things outside ourselves. Thus he writes:

Hitherto it has not been by any assured judgment, but only from a sort of blind impulse, that I have believed in the existence of things outside me and different from me, things which by way of the sense-organs or by whatever means they employ, have conveyed to me their ideas or images, and have thus impressed on me their similitudes.[5]

---

[1] *Med.* II, K.S. p. 208.

[2] *Med.* III, K.S., p. 222. 'Light and the colours, sounds, odours, tastes, heat and cold and the other tactual qualities, they present themselves to me so *confusedly and obscurely* that I cannot tell whether they are true or false, i.e. whether the ideas I have of them are ideas of real things . . .'; my italics. See also above, pp. 105–9.

[3] *Med.* III, K.S., p. 214.   [4] Cf. above, pp. 100f.; *Med.* III, K.S., p. 214.

[5] *Med.* III, K.S., p. 219. One should carefully note the way in which what is usually described as 'sensory knowledge' is redescribed by Descartes as 'judging by blind impulse'; in a similar way, our perception of secondary qualities, because it does not match up to the programme of clear and distinct conception, is called 'confused and obscure' (op. cit., K.S., p. 222). Though the rationalist cannot explain away the world of external things endowed with their secondary qualities, he will re-model the language in which he describes their occurrence—leaving open the question whether such re-descriptions are temporary expedients, theoretical or logical devices, or irreducible metaphysical modes of speech.

It should be appreciated that the doubt which is here implied concerning the existence of external objects is a very radical one, not to be countered by considerations of any empirical kind. When Descartes writes that the chief and most usual error to be met with in our judgments consists in holding that the ideas which are 'in me are similar to, conformed to, the things which are outside me',[1] the example advanced, viz. the sun being judged no larger than a ha'-penny, is not really appropriate.[2] And the inappropriateness is clearly demonstrated by the fact that the doubt is stilled, and the claims of knowledge restored, not by any empirical or scientifico-theoretical considerations, but by a further investigation of the nature of our ideas and by the introduction of additional principles that yield the required result through postulates whose claim to acceptance rests on the deliverances of the 'natural light of reason'.

Descartes' position here is complex and must be carefully considered, particularly since it was to exert such wide influence upon all subsequent theory of knowledge. Ideas concerned with the external world are said to be 'received by way of the senses'.[3] Nevertheless, for reasons which we have discussed,[4] they are, strictly speaking, 'innate', the sense in which they are so being best explained by remembering that *no* sensory information as *normally understood* by Descartes can give us an assurance that there are any external objects to which these ideas correspond.[5] (And, indeed, it follows from Descartes' teaching about secondary qualities that there *cannot* 'ultimately' be any such objects!)

Descartes now makes two assumptions: (a) the 'things beyond' the ideas 'resemble the ideas'; (b) these ideas 'must have been caused by other things'.[6] The first of these assumptions Descartes realises is precarious—i.e. philosophically precarious; a fact which we have already seen him express by insisting that our belief in the existence of things 'outside' is only due to 'blind impulse', or—to use Descartes' more technical expression—the 'teaching of nature',[7] 'nature' here being construed as 'a certain spontaneous impulse which constrains

---

[1] Op. cit., p. 216. [2] Cf. above, pp. 106–9.
[3] *Med.* VI, K.S., p. 252. [4] See above, pp. 100f and cf. p. 111.
[5] Cf. here Descartes' celebrated fable of the universal dream which perpetually simulates an external world (*Med.* VI, K.S., p. 253). To the unfalsifiability of this possibility there corresponds directly Descartes' ultimate 'solution' in terms of the metaphysics of a non-deceiving God.
[6] *Med.* VI, K.S., pp. 252, 256.
[7] For 'the teaching of nature', cf. below, pp. 177f. and *Med.* VI, K.S., pp. 252–3,

me to this belief, and not a natural light enabling me to know that the belief is true'.[1] Nor is it of much avail to object that ideas of external bodies are 'independent of my will', since being 'natural impulses' they may well have been 'produced' by myself.[2] On the other hand, it must be admitted that the appearance of their occurrence against my will requires some explanation, particularly in the light of the 'very strong inclination to believe that those ideas are conveyed to me by corporeal things'.[3]

It is as though Descartes wanted to say that a world that was only *apparently* 'represented' by our 'ideas' would be exceedingly bizarre, and faulty in design, containing an *ultimate* negative element, deficient, as it were, in a way in which a person would be morally deficient if proved to be totally unreliable in his promises—making a mockery of the notion of promise-giving itself.[4] *If* the world transcends the individual self at all, and thus transcending exhibits a drive towards what is 'positive', and indeed, towards 'perfection', then it would be wrong to question its most pervasive manifestations of transcendence through the ideas of external objects. 'God', as Descartes formulates the doctrine in a theological guise, 'is no deceiver'.[5] It will be seen that the core of this contention rests not on logical or empirical grounds but turns once again on 'metaphysical', not to say 'moral' considerations, and introduces thus an additional element into Descartes' epistemological armoury.

There is, however, still the second of the assumptions made before, an assumption involved also in the contention just mentioned, namely

---

257, 258-9. It should be noted how near this view comes to that of 'empiricism'. Thus Hume argued that our belief in the continued existence of external objects was due not to the senses or to reason but to 'the natural propensity of the imagination', to 'a kind of instinct or natural impulse' (*Treatise*, I.iv.2; Selby-Bigge ed., pp. 210, 214).

[1] *Med.* III, K.S., p. 217. For the 'natural light', cf. below, pp. 177ff.
[2] *Med.* III, K.S., pp. 217, 218.   [3] *Med.* VI, K.S., p. 256.
[4] Some present-day logicians hold that it would be almost a *social fault* to doubt the reliability of *individual* inductive procedures by questioning them through a *universal* inductive doubt.
[5] *Med.* VI, K.S., p. 256. It will be remembered that according to Descartes the principle that God is no deceiver is required also to certify even those truths that are seemingly known with clarity and distinctness. (Cf. *Med.* V, K.S., pp. 246-7, for a clear statement of this principle.) I must omit from consideration also the intermediate doctrine (to be found in *Med.* V and the fourth *Reply to Objections* II) that the principle is also required to guarantee the truth of those conclusions of which—whilst we believe them to be true—we have forgotten the demonstration.

that there is a world transcending not only 'ideas' but also the 'self', the possessor of those ideas; and that these ideas are 'caused by other things'; for otherwise an idea would 'have derived its origin from nothing'.[1] Now Descartes' conception of this key-term in his philosophical work, 'idea', is exceedingly complex, and it is not our purpose to follow the tortuous windings of Cartesian scholarship in pursuing this prey; we can only select some of the most central and obvious features. In one of its aspects an idea is said to be a 'mode of the mind', a 'manner' or 'way of thinking'. In a second aspect an idea is viewed as a 'form of thinking' or 'form of the thought'.[2] In this respect, it is sometimes viewed as an 'image', 'representing' an 'actual thing' or 'reality'; though sometimes, and in other contexts, it seems more properly described as a 'concept'. Consider a painter's reproduction of King's Chapel, Cambridge. The painting itself is a physical object, something 'actual'. In Descartes' scholastic terminology, this constitutes the 'formal reality' of the painting. Being a painting, it has however a certain 'representative power' also, it 'pictures' another physical object, the actual chapel. That aspect of the painting, in virtue of which it 'pictures' the actual chapel, its 'representative content', Descartes calls the 'objective reality' of the painting.[3] This model is then applied to the notion of an idea, where Descartes again distinguishes 'formal' (or actual) reality, i.e. its being an instance of the 'act' of 'thinking', and the 'objective' (or representational) reality of the thinking, or what is often called its 'content'.[4]

In terms of this model, Descartes' problem (the problem of 'transcendence') is to explain how the 'objective reality' of an idea can argue a corresponding 'formal reality', i.e. how far it follows from the fact of my thinking *of* a certain object that there *exists either* a corresponding object, *or* something to which the fact of my having the thought representing *that* object (i.e., with that 'content', with that 'objective reality') is due. Descartes seems to think of the *'content'* of the thought (its 'objective reality') as being *caused* in much

---

[1] *Med.* III, K.S., pp. 219, 221.

[2] An analogy (though as later philosophers have shown, very misleading) would be the distinction between the activity of singing, and the song.

[3] A modern reader might find it useful to compare this with the theory of Wittgenstein's *Tractatus*, though remembering that the operative notions there are fact and proposition, rather than object and concept (or idea). Cf. for instance *Tractatus* 2.151: 'Pictorial form is the possibility that things are related to one another in the same way as the elements of the picture.' The *'pictorial form'* corresponds to Descartes' 'objective reality'.

[4] Cf. *Med.* III, K.S., pp. 219–20. Also *Reply to Objections* II, H. & R., ii, pp. 52–3.

the way in which one might speak of a thought itself, or even a physical object, being caused.[1] Indeed, it is the 'content' ('objective reality') of the idea, rather than the idea viewed as a manner of thinking, for which the cause must primarily be sought; for in the latter respect, the idea is after all simply a state of the spiritual self, and thus (save for the *Cogito*) needs no special accounting for. But if any of its representational (or objective) reality 'be allowed as being met with in the idea and yet not in the cause of the idea, it must have derived its origin from nothing'.[2]

The kind of causal relation here involved is none too clear. We can however get a further hint from *Principles*, i, 17, where Descartes contends that

> when somebody possesses the idea of a highly complicated machine, we are justified in asking from what cause he derived it; did he somewhere see such a machine made by somebody else? Or is it that he has made such a careful study of mechanics, or is he so clever, that he could invent it on his own account, although he has never seen it anywhere? Any device that is found in the idea representatively, in a picture so to say, must occur in the cause (whatever this turns out to be) not just representatively or by way of reproduction, but actually.[3]

The drift of the argument seems to be this: any of the ideas which I have may 'have been produced by myself' or 'proceed from myself' (unless indeed, as regards those of our ideas which are 'obscure and confused', 'the light of reason' shows these to be 'ultimately nothing'); *or*, where an idea is *not* ultimately 'nothing', it must point to some other reality, in line with the case of the complex machine of which I can think only if there *is* such a machine (unless I have 'invented' the idea or 'compounded' it from other ideas.[4])

However, what shows us that the step from 'objective' to 'formal reality' (from thought to thing, from concept to object) is sanctioned?

---

[1] 'A stone which has not yet existed cannot now begin to be unless it be produced by some thing which possesses in itself, either formally or eminently, all that enters into the composition of the stone. . . . But neither can the idea of . . . the stone exist in me unless it too has been placed in me by a cause . . .'. (*Med.* III; K.S., p. 220.)

[2] Ibid.; pp. 220–1.   [3] Anscombe, p. 185.

[4] Cf. *Med.* III, K.S., pp. 222–3. Cf. Locke's doctrine that we cannot 'invent or frame one new simple idea in the mind, not taken in by the ways before mentioned' (*Essay*, ii.2.2). The difference is that the source of our 'ideas' is limited to 'sensation' and 'reflection', and excludes the source of 'innate ideas' in the Cartesian sense, i.e. God.

In answer, Descartes introduces—as the coping-stone of the whole system—a principle ('manifest by the natural light') that the ideas existing in me 'have been placed in me by a cause which contains in itself at least as much (formal) reality as I ascribe' to the ideas in respect of their 'objective reality'. This principle is apparently a special case of another more general one 'that there must be at least as much reality in the efficient and total cause as in its effect'.[1]

What is the status and basis of this principle? What is the power of 'the natural light', the 'light of reason'? Now we have seen that at one point Descartes allows his 'methodical doubt' to apply even in the case of 'clear and distinct perception', suggesting that it may at most be only a necessary, but not a sufficient guarantee of truth.[2] This would seem to suggest that the 'natural light' will provide us with additional guarantees. And surely, without this possibility, would not Descartes' reasoning here be involved in a gigantic circle, the verisimilitude of clear and distinct perception requiring the concurrence of God, whilst the proof of God's existence presupposes the former in turn? We have already suggested some sort of answer; let us press it further.[3]

Does the 'natural light' go beyond the principles of transparency? In answer to this question, Descartes proceeds with a curiously divided voice. According to *Med.* III,

> what the natural light shows me to be true (e.g. that inasmuch as I doubt, it follows that I am, and the like), I cannot anywise call in doubt, since I have in me no other faculty or power whereby to distinguish the true from the false, none as trustworthy as the natural light, and none that can teach me the falsity of what the natural light shows me to be true.[4]

This is really a negative argument, implying that the alternative to following the guidance of the 'natural light' would be a *pis aller*. No doubt the putative successes in science and mathematics, e.g. the 'intuition' of the *Regulae* and the conceptual explorations of dynamics, contributed much to Descartes' confidence; one is reminded of this in a letter to Mersenne, where he speaks of 'the natural light or the *intuitus mentis*',[5] implying an equation between the two.

---

[1] *Med.* III, K.S., pp. 219–20.   [2] Cf. above, p. 158.   [3] Above, p. 160.
[4] K.S., pp. 217–8.   [5] 16.10.1639. A. & T., ii, p. 599.

It is by the natural light that we perceive certain principles—
Descartes usually mentions only one, the causal principle already
mentioned;—'than which nothing can be more true or more evident':

> That there is nothing in the effect, that has not existed in a similar or in
> some more eminent form in the cause, is a first principle than which none
> clearer can be entertained.[1]

Yet, even so, Descartes still does not offer us anything to improve
on the previous situation, except (putatively) a very fundamental
principle. How could he have regarded the situation differently? Now
his point is that the application of the principle to the idea of a
perfect being is held to prove its existence—an existence of which we
indeed only then 'become aware'.[2] But given this conclusion, we also
see that to entertain the 'metaphysical doubt' with regard to clear and
distinct perceptions is to hold that God is a deceiver. Since we 'cannot
imagine' this, our rule is safe.[3]

Nevertheless, this mode of reasoning does not free us from the
logical circle. But it seems clear that Descartes is in fact working with
a much stronger principle. He really holds, not only that the natural
light entails a non-deceiving God, but that the former *is also entailed*
by the latter! And certainly this was the historical origin of the notion
of natural light, with its overtones of guidance by the divine spirit. As
a result, the notion of natural light is given a new sense, as becomes
clear when we study another passage in *Reply to Objections* II:

> In the case of our clearest and most accurate judgments which, if false,
> could not be corrected by any that are clearer, or by any other natural
> faculty, I clearly affirm that we cannot be deceived. For . . . it is contra-
> dictory that anything should proceed from Him [*sc.* God] that positively
> tends towards falsity. But since there is nothing real in us that is not
> given by God . . . and we have a real faculty of recognising truth, and
> distinguish it from falsehood, . . . unless this faculty tended towards
> truth, at least when properly employed (i.e. when we give assent to none
> but clear and distinct perceptions . . .), God, who has given it to us,
> would justly be held to be a deceiver.[4]

If we compare this with the passage last quoted from *Meditation*
III, we shall note the similarity in the wording, save for the additional
semantic force that has been provided *via* the introduction of the

---

[1] *Reply to Objections* II; H. & R., ii, p. 34.   [2] Op. cit., p. 41.
[3] Ibid.   [4] Op. cit., pp. 40–1.

notion of God. The natural light is a device whereby we are brought into harmony with the nature of things. True, Descartes is all the while insinuating that he has previously proved (independently) the existence of God. But he could not have done so (in any strong logical sense), unless he had assumed the extra power of the natural light.

What was it that concealed this from him? Perhaps he argued as follows: assuming its truth-injecting power, the natural light allows us to prove and thus become aware of, the existence of a non-deceitful God. And this conclusion subsequently adds the missing justificatory force to the original assumption. The erstwhile epistemic ignorance is rectified, *via* our 'proof', which induces us to become aware of the true situation. But the logic of the argument is peculiar: the special nature of the conclusion is used to inject additional strength into the premises! There was hardly need to go into the shadow-game of proof and inference; the whole process was simply employed to allow the emergence of those assumptions or principles which are required for the construction of an external world, and of a 'nature' transcending the realm of 'ideas'. Descartes is in fact saying that it is not enough to operate with the rule of clear and distinct perceptions; we must assume the truth-generating source of certain additional principles. What distinguishes the natural light from ordinary clear and distinct perception is, first, its special 'object', certain primary principles, such as that of 'causation'; secondly, that it has a special justificatory force, due to a natural harmony of mind and God. In the words of *Principles*, i. 30:

> The faculty which . . . we call the natural light, never perceives anything which is not true, in so far as it knows it clearly and distinctly; for we should have to believe that God is a deceiver, if He had so arranged it that we took what is false for what is true even though this faculty was being employed by us properly.

Once more we see how earlier successes of the methodological function of the concept of intuition are used subsequently to provide a justification for the purposes of enlargement of an epistemological basis, whose original extent has proved too small.

The application of his principle to the case of 'ideas', we have seen, leads Descartes to postulate that the content of any of our ideas must be matched by an actuality somewhere, viewed as 'cause'. To transcend the realm of the self's ideas, he must then be able to point to at least one idea that is 'representative' of a 'reality' which could not have 'proceeded from the self'. Such an idea, he claims, is the idea of God, by which Descartes means 'a substance that is infinite, immutable, independent, all-knowing, all-powerful'.[1] Just as the idea which I possess of a complicated machine (provided it is a 'positive' idea, and not just a word for 'something better than we have') demands the existence of such a machine, so does the 'positive' idea of perfection demand its actual source. But I myself am visibly limited and imperfect; to account for this idea I must explain it by a source beyond the self.

There is no need here to enter upon a minute criticism and appraisal of this argument, which would not find many defenders.[2] Its core is the step from idea to 'existence' (from 'objective' to 'formal' reality), from concept to object, from the possible to the actual. Now later philosophers, e.g. Kant, will object to this mode of reasoning (a kind of 'ontological argument') that such a move from idea to existence is quite impermissible, since we first have to establish the 'real possibility' of such an existence of anything. But to achieve this, we must, first, establish the ('transcendental') framework of conditions which would make the experience of such a thing possible; secondly, the thing would have to be given in the context of an actual experience as a whole, i.e. we would have to have cognitive grasp of it (either actually or hypothetically) *a posteriori*, a supposition which is of course precluded in Descartes' case.[3]

Still, as I have remarked before, objections of this nature, however powerful, do not necessarily constitute refutations, since their basis rests on such different starting points. We need only remember that the whole spirit of the Cartesian thinking is against the implicit Kantian assumption of the generic division between 'sense' and 'thought'. For Descartes the domain of sensory observation, being ultimately 'unreal', does not enter the argument at all, since it is something known only 'confusedly and obscurely', in some way 'reducible' to intellectual predicates such as extension. Still, if Kant's objections are not refutations, they do highlight eccentricities in the Cartesian modes of statement. Thus in Descartes, the sense in which there exists a field beyond the idea is left entirely opaque; we are left only with a 'moral certainty' of the existence of the external world;

[1] *Med.* III, K.S., p. 224.
[2] In fact, its plausibility depends, I think, on the putative success of Descartes' arguments concerning the innateness of geometrical ideas—arguments of which the one under discussion seems to be a special, if extreme, case. Cf. above, pp. 111–14.
[3] Cf. K.505–6; and below, Ch. VIII, sect. 4(a), pp. 488–90.

but on the face of it, morality and sensory reality seem uneasy partners
to join in any epistemological enterprise.

(d) The '*metaphysical atomism*' of the states of the self

Before considering this further we will look briefly at a second, and
supplementary argument of Descartes' for the existence of God,[1]
because this assigns to God some very interesting logical functions—
functions that we shall meet again in later philosophers, though
performed there by different agencies. (Indeed, looking at this from
the standpoint of later writers, will help considerably to appreciate
the logical place of God in Descartes' system.)

The central point of the argument—the atomism of the states of the
self—has been mentioned already.[2] Here we need to add the further
premise, which is Descartes' contention that the self, as thinking
substance, cannot arise out of nothing; recognising myself as limited,
I must postulate a cause of its existence. However, is it not possible to
suppose that the self has existed always? One could then simply
assume that it just goes on existing; there seems no need to postulate
creation. To this Descartes replies that continued existence needs as
much explanation as the beginning of any existence. The reason he
gives for this astonishing contention is as follows:

> The course of my life can be divided into innumerable parts, none of
> which is in any way dependent on the others. Accordingly *it does not
> follow* that because I was in existence a short time ago I must be in
> existence now, unless there be some cause which produces me, creates
> me as it were anew at this very instant, that is to say, conserves me.[3]

'It does not follow': *inductive* reasons are here plainly irrelevant. In
terms of the *Cogito*, we can contemplate a momentary thinking state
of the self, clearly and *distinctly*. Moreover, the 'distinctness' (logical
independence) is here absolute; unlike the case of the 'necessary
conjunction' between 3 + 4 and 7, here there is no 'confused relation'[4] or bond between the separate states of the self, whereas the

---

[1] The third, which actually bears the technical label, 'Ontological Argument', we
shall not discuss, being of rather less importance for our purposes. Cf. *Med.* V,
K.S., pp. 243–5.
[2] Above, p. 160.
[3] *Med.* III, K.S., p. 227. God, considered as an agent of conservation, was of course
a key-concept in Descartes' establishment of the law of conservation of quantity of
motion. Cf. above, pp. 148–50.
[4] Cf. above, p. 158.

concept of a substantial self apparently requires the existence of an entailment relation between the distinct states.

Of course, present-day philosophers would say that the search for an entailment relation here is misplaced since the proposition 'state A implies state B' is 'synthetic' and, being synthetic, is in need of an *a posteriori* basis; which—as we have just noted—Descartes' argument excludes. The modern view stems indeed from Hume, who took Descartes' point further: 'Reason can never satisfy us that the existence of any one object does ever imply that of another',[1] he writes, thus generalising Descartes' limitation of the argument to the states of the thinking self. Hume's criticism brings out with particular force something concerning the constructions of metaphysics. Descartes is clearly hankering after a bond whose strength would mirror that of logical entailment. On the other hand, logical entailment can offer him no more than a partial model, since his worry is about *existence*, which to him suggests the model of 'creative force'.

Nevertheless, Descartes' procedure here foreshadows important developments and responses in later philosophers. For they, too, wrestle with the problem of how to interpret the conceptions of substance and causation in a sense other than merely formal. What is for Descartes the activity of God will for Hume be custom and the imagination. More clearly still, Kant will face with renewed vigour the problems bequeathed by the Cartesian intellectualist atomism; but instead of an *a priori* logical relation, he will offer a 'transcendental synthesis', i.e. a relating of the separate data of sensation without which they could not become 'objects', and without which there could be no 'experience' of such objects.

A contrast of Descartes with Locke is similarly instructive. Descartes' ultimate turning away from the external sensory field, coupled with the doctrine of the 'atomism' of ideas, perceived 'distinctly', makes a solution by way of 'closing the gap' seem more urgent; for he will not be able to say—as Locke was to do later—that whilst (again for 'atomistic' reasons) there is little 'science' (in the sense of '*scientia*', i.e. 'rational knowledge'), yet we may 'content ourselves with probability' in the case of the 'universal propositions' of physics, since what matters is the casting of 'particulars' in the form of a theoretical system, the touchstone for which can only be 'those particular instances' by means of which 'the first discoverer found the truth'.[2]

---

[1] *Treatise*, I.iii. 7; op. cit., p. 97. Cf. also I. iii. 6, p. 86.
[2] *Essay*, iv. 6.15; iv. 7.11. For the atomism of man's existence, cf. iv. 11.9.

Descartes, when confronted by the logical consequences of his doctrines will neither be tempted, nor be able, to try Locke's method of pacification *via* a probabilistic approach; on the other hand, he really has no tools with which to master his problem. When coming to the 'substantial' principles of science, and of existence and substance in general, he must therefore opt for alternative 'descriptions', or 'demands'. Since perception of the clear and the distinct does not yield the continuity of a substantial existence, this must be supplied through postulation of an 'ontological' or 'opaque' ground, Descartes' God.

As an example, consider his celebrated dualistic argument, concerning the relation (or lack of relation) between mind and body. The principle invoked is similar to the one just encountered:

> Since I know that all the things I clearly and *distinctly apprehend* can be created by God exactly as I apprehend them, my being able to apprehend one thing apart from another is, in itself, sufficient to make me certain that the one is different from the other, or at least that it is within God's power to *posit them separately*; and even though I do not comprehend by what power this separation comes about, I shall have no option but to view them as different.[1]

In other words: *logical* distinctness creates a 'gap' which in Descartes' view can be closed only by an 'ontological' bridge. Perhaps *not always* 'ontological'! For—as Descartes goes on to argue in the paragraph from which we have just quoted—since I have clear and distinct ideas of myself as a thinking thing, as well as of my body, and though this makes it 'certain that I am truly distinct from my body, and can exist without it', yet it is just as 'certain' (though, we should note, 'certain' in another sense) that 'I have a body with which I am very closely conjoined'. Now what makes me certain in this second sense is 'the teaching of nature', Descartes' philosophical notion (or really *metaphysical*: for again this has an ordinary and a special grammar!) through which to deal with and place the field of sensory experience. To this conception we must now turn.

(e) *The 'Teaching of Nature' and the existence of the external world*
Granted Descartes' position, the language needed to describe the

---

[1] *Med.* VI; K.S., p. 254; my italics. Again compare Hume: 'All *distinct ideas are separable* from each other, and . . . 'twill be easy for us to conceive any object to be non-existent this moment, and existent the next, without conjoining to it the distinct idea of a cause or productive principle.' (*Treatise*, I.iii.3; op. cit., p. 79; italics mine.)

'deliverance of sense' will have to be readjusted also. The expression which Descartes employs is 'the teaching of nature'. This, we have already seen,¹ is contrasted with the 'natural light', as being 'only a certain spontaneous impulse' which constrains us to hold some given belief or other. This matter is discussed, significantly, in the last chapter of the *Meditations*, where Descartes sets out 'to examine the nature of sense' in order to discover whether sensory ideas can yield 'any certain proof of the existence of corporeal things'.² As we have seen, this they are not capable of achieving without the intervention of the principle of God's inability to deceive; and we need say no more about this. Rather, what is of interest now is to observe the precise place which 'sensing' receives in this scheme. And the view pressed by Descartes is that all judgments passed on objects of sense have only been taught me by nature.³ Now such judgments are deficient because they lack that insight into the connection of things which seemed guaranteed by the 'method'. Thus, hunger puts me in mind of taking food, and dryness of throat of drinking; this I experience. But I lack the 'reason': 'For assuredly *there is no affinity*, none at least that I can understand, between this twitching of the stomach and the desire to eat.'⁴

Similarly, concerning the connection between mind and body: though rationally 'it is certain that I am truly distinct from my body', *nature* teaches me *with equal certainty* that my body is in fact 'very closely conjoined' with myself—indeed, 'so intimately conjoined, and as it were intermingled with it, that with it I form a unitary whole'.⁵ Evidently the certainty of nature is of a kind different from that of reason. Descartes mentions in fact two explanations of the meaning of this term, nature. According to the first of these it is 'either God Himself or the order of created things'; secondly, it includes what is 'given me as a being composed of body as well as of mind', i.e. pains, pleasures, sensory apprehensions, as contrasted with any knowledge which 'the mind alone' might possess of rational or inferred truths, including that of external objects.⁶ These explanations yield two points: the certainty which the teaching of nature yields is of the 'moral' kind, a function of God's goodness; secondly, my sensory apprehensions are, *at the level of 'nature'*, ultimate: they are indeed

---

¹ Above, pp. 167–8.   ² *Med.* VI, K.S., p. 251.   ³ K.S., p. 253.
⁴ Ibid.; my italics. Like the empiricists later, Descartes is already complaining of the lack of any 'necessary connection'.
⁵ K.S., pp. 254, 257.   ⁶ K.S., pp. 257, 259.

(to borrow Leibniz's technical phrase for this notion, arising from roughly the same doctrine) *'phenomena bene fundata'*. It is only when we 'reflect'—if such reflection is possible—that they yield to a more transparent realm of structure and thought. This doctrine emerges at its clearest in the problem of the relation of mind and body and the relevance of bodily sensation. Here more than anywhere we have the final apotheosis of rationalist thinking, and—as I have maintained from the start—of that important philosophical re-description of the logical status of phenomena.

Consider a body depleted of nourishment. If all were rationally transparent, the mind would 'apprehend this expressly'; for instance, a structural configuration of one kind would be seen to entail that of a second kind. As it is, I am made aware of the deficiency only 'by confused sensings of hunger, thirst, pain, etc.',[1] which are due to the contingency that body and mind are *in fact* closely intermingled.

Is this fact 'ultimate'? There is no unequivocal answer to this question. In one sense, yes, for my 'sensuous apprehensions' are what they are; indeed, they are for this reason once even said to be 'sufficiently clear and distinct': namely (says Descartes), in order to testify 'to my mind what things are beneficial or harmful to the composite whole of which it is a part'.[2] Yet, and following the whole logic of his doctrine, Descartes holds that (ultimately) 'these sensings . . . are in truth merely confused modes of thinking, arising from and dependent on the union, and, as it were, the intermingling of mind and body'.[3]

Descartes fails in the end to make clear the position of the sensory and qualitative realm in his ontology; or rather, he leaves the matter

[1] K.S., p. 257.
[2] K.S., p. 259. Once more compare Hume: Only experience, 'nature' in Descartes' sense, and not 'reason', can inform us concerning the principles of connection between things; and these principles are no mere arbitrary or nominal constructions; they 'are the foundation of all our thoughts, and actions, so that upon their removal human nature must immediately perish and go to ruin'. (*Treatise*, I.iv.4; op. cit., p. 225.)
[3] Op. cit., 257. In order to relate it to what is to follow, it is worth comparing this with Kant's strictures on Leibniz, who held a similar though more systematic and generalised doctrine. Kant objects that such a view 'allowed sensibility no mode of intuition [Anschauung] peculiar to itself but sought for all representation of objects, even the empirical, in the understanding, and left to the senses nothing but the despicable task of confusing and distorting the representations of the former' ('Concepts of Reflection', K. 286). This of course constitutes no refutation. The insistence that sensibility is a mode of experience 'peculiar to itself' *is* simply denying that the sensory realm is ultimately resolvable into transparently rational perception. Since this idea itself is not 'clear', and relates so intimately to the whole vast complex of Descartes' scientific and methodological interpretations—as we have

opaque and dogmatic. God does not deceive me concerning this apparent order of created things: very well! But it does not follow that the realm of sense is ultimately analysable in the way Descartes suggests when he calls it 'confused thinking'. Here, as before in his scientific methodology, with its hard core of irreducible initial conditions and tentative hypotheses, Descartes has not been able to do more than suggest by a chosen language that the detail of 'phenomenal' particulars is amenable to the over-all constructions of reason. The tension which exists between his rational ideals and the claims of sensory experience is now handed on to his philosophical successors. If this chapter has been long and tortuous, it will nevertheless have set the scene, painted in the background, against which, to the time of Kant at least, these problems will present themselves to the philosophers.

---

followed them in this chapter—Kant's rebuttal can prove itself at best only by a more successful reshuffling of the various logical points of view, an appraisal which —as indicated in ch. I—is one of the central contentions of this book.

IV

LOCKE: NARROWING THE LIMITS OF SCIENTIFIC KNOWLEDGE

I. NEW DEMARCATIONS FOR THE PROVINCE OF KNOWLEDGE

We have seen that Descartes severely limits the extent of 'scientific knowledge'; yet, the tension set up by this limitation, which manifests itself on the one hand in the whole complex comprised under the notion of 'analysis', and on the other through 'the teaching of nature' and the doctrine fleetingly referred to as 'sensing = confused modes of thinking', is insufficient to fracture the rationalist venture; on the contrary, these forms of description have the effect of concealing the difficulties involved. 'Nature' is here, ultimately, 'God'.[1] And God is also the focal point round which Descartes' ontological construction of his universe is attempted, required as a consequence of the doctrine of the *Cogito* and the notion of 'innateness' and 'distinctness' of ideas. Just as the sensory realm can be grasped only 'obscurely and confusedly', through the guidance of 'nature' (but not 'reason'), so the realm of an external material world—indeed, the very conception of such a realm can be made good only through postulation of something 'other-than-self'; and though Descartes conceives this postulation as rationally certified, the 'object' so certified (God) cannot 'guarantee' that external world otherwise than as an 'opaque' fact, expressed metaphorically through the notion of God's unwillingness to deceive.

We have already seen something of Locke's readjustments to this 'picture of reality'.[2] The Cartesian notion of 'rational certification' of

[1] 'By nature, considered in general, I now understand no other than either God Himself or the order of created things as instituted by Him.' (K.S., p. 257; cf. above, p. 178.)
[2] Above, Ch. I, pp. 6–10; Ch. II, pp. 55, 63, 64–7.

181

concepts like those of the 'innateness' of our mathematical ideas and reason's guidance by the 'natural light', in a sense in which both are taken to carry with them the doctrine of divine *guarantees*,[1] is now given up; certainly this is one of the *intentions* of the first Book of Locke's *Essay*, directed against the doctrine of 'innate principles' and 'innate ideas', and issuing in the famous clarion call: 'Men must think and know for themselves.'[2] Indeed, the egocentricity of the Cartesian edifice in a certain respect might even be said to have been carried further. If, like Descartes, Locke wants to 'arrive at certainty', this must be possible without falling back on *'original notions and principles'*.[3] Locke's intention (with this use of 'original') is to protest against the doctrine that such of our ideas and principles as seem to us indubitable derive their logical force from some foundation located beyond 'the mind' in God, by whom they have been 'imprinted' in us. A principle like 'That it is impossible for the same thing to be, and not to be' must not be viewed like a 'native inscription', lodged in our minds by an outside source,[4] as though it were a proposition handed to us with a truth-voucher attached. Locke means to lay more securely 'the old foundations of knowledge and certainty'.[5] What is wrong with the picture of 'innate impression' is not so much its 'innateness'; we shall presently see that Locke assigns a considerable place to the mind's capacity, through 'reflection' upon its operations with its sensory ideas, to furnish us with an additional stock of non-sensory ideas, such as existence, unity, thinking, knowing, power.[6] Rather, the error lies in the misleading associations conveyed by the term 'impression', as though the 'truth' of these principles was somehow 'impressed' from without. On the contrary, Locke holds—surprisingly at first, if one has in mind the Cartesian implications of this term—that their 'self-evidence' is 'intuitive', depending on no proof, perceived directly.

> Thus the mind perceives that white is not black, that a circle is not a triangle, that three are more than two, and equal to one and two. Such

---

[1] For the mathematical case, see above, p. 117; for the 'natural light', see above, pp. 159–60.

[2] *Essay*, i.2.24. Locke's arguments there are however directed not only, or even primarily, against Descartes.

[3] i.2.1; italics mine.

[4] i.2.11–12; i.2.18 speaks of 'innate impression'.　　[5] Cf. i.4.24.

[6] Op. cit., ii.7.7. We have noted above, p. 110, that these are precisely examples of some of the Cartesian 'innate ideas'!

kind of truths the mind perceives at the first sight of the ideas together, by bare intuition, without the intervention of any other idea; and this kind of knowledge is the clearest and most certain that human frailty is capable of. . . . It is on this intuition that depends all the certainty and evidence of all our knowledge. . . . A man cannot conceive himself capable of a greater certainty, than to know that any idea in his mind is such as he perceives it to be. . . .[1]

This looks at first sight not unlike the ideal of knowledge painted by the Descartes of the *Regulae*.[2] But it is Descartes without the subsequent demon: Locke will not seek to shore up further the foundation of 'knowledge' by way of a transcendence of the self's intuitive cognitions through the notion of divine guarantees. And even though some men may call this 'pulling up of old foundations of knowledge and certainty', he tells his reader, his true intentions are to lay 'those foundations surer'.[3] How is this to be achieved? Clearly, it will have to be done by a number of readjustments, fresh delineations, re-interpretations. Descartes' system had contained, in an uneasy relationship, a variety of what, for want of a better term, we have called 'levels':[4] things that we perceive clearly and distinctly, such as mathematical relations and the existence of the self; things that we perceive only 'confusedly', through our 'ideas of sense', and the relations between such ideas; external objects regarded as the 'causes' of our ideas, whose existence is guaranteed 'demonstratively' only, through the mediation of God's goodness. And these various 'levels' are all uneasily held together, in the light of the paradigm of knowledge as clear and distinct cognition, through the notion of a divine guarantee. Locke will give up the latter, and radically re-interpret the former; nevertheless, the different 'levels' will reappear, though articulated in a different fashion, using a different nomenclature. The Cartesian divisions between the different 'levels' will become less fluid, the distinctions sharper. If 'clear and certain knowledge' is of the intuitive kind alone, Locke will begin the task of dividing it more radically from other kinds of intellectual functions colloquially designated 'knowledge' by ordinary men, such as our 'knowledge' of the behaviour of physical substances; or again, our 'knowledge' of an external world (as contrasted with our ideas); our 'knowledge' of

---

[1] iv.2.1. Cf. iv.7.19.
[2] Cf. above, p. 132, for Descartes' definition of 'intuition'.
[3] i.4.24.   [4] Cf. above, p. 5.

mathematical relationships; our 'knowledge' of sensory qualities in general.

Abandoning the Cartesian (divinely sanctioned) system of these various 'levels', it is as though Locke were going to allow them each to settle in a groove of their own, only here and there creating new comparisons and evaluations. We now find that by 'sensation' we can also be said to have a kind of knowledge ('sensitive knowledge') of the existence of external objects,[1] agreeing thus with the fact that this already 'passes under the name of knowledge' in general.[2] Again, the having of an idea 'received from an external object . . . [into] our mind' is said to be a kind of 'intuitive knowledge'.[3] The change in the attachment of the 'knowledge' label should not be underrated, even though it may appear at first sight as no more than a linguistic decision, particularly if it is realised that Locke will—like Descartes—apply 'ontological' restrictions to some of these ideas, by insisting that *ideas* of 'secondary' qualities' have no resemblance to corresponding *qualities* of external objects.[4] Nevertheless, the decision to speak here of 'knowledge' without hedging is an indication that the metaphysical centre of gravity has shifted; all those items called objects of knowledge are 'basic' and 'irreducible' epistemic elements; sense is here not reduced to (confused) thought; after all, in the end (as we shall see) 'all the materials of reason and knowledge'—according to Locke—come from 'experience', the source being either 'sensation' or 'reflection'.[5]

But if new centres of gravity thus arise in the absence of the Cartesian method of reduction to a unitary basis, we may also, note a contrary tendency, namely that of emphasising the gap between the different areas covered by sensation and thought. This expresses itself in the different ways in which Locke seeks to define the boundaries of knowledge, its 'degrees of certainty', its 'extent', and its 'reality'—the latter referring to the question of the 'agreement' of our ideas with and their relations to external objects. Here the division between the mathematical and the physical realm is deepened and

---

[1] Cf. iv.2.14. It is only a *sort* of knowledge because it is not 'demonstrative'; nor does it possess the kind of 'certainty' we get in logic and mathematics. Yet it would be 'folly' to reject it as a candidate for knowledge in virtue of this deficiency. (Cf. iv.11.10.)

[2] Ibid.

[3] Ibid. 'intuitive' here means 'direct', non-mediated; but it is not of the transparent kind referred to in the passage cited above, pp. 182–3.

[4] ii.8.15. Cf. below, pp. 226ff.   [5] ii.1.2–4.

made more absolute. There is no suggestion that physical reality 'in the nature of things' conforms to mathematics, one of Descartes' fundamental assumptions, with his 'ontological reduction' of matter to extension. To be sure, there is an agreement between mathematical propositions such as those of geometry and 'real things existing'. But this (according to Locke) is

> because real things are no farther concerned, nor intended to be meant by any such propositions, than as things really agree to those archetypes in the [geometer's] mind.[1]

In effect, we are given here a fresh demarcation of the place of mathematics in science. It concerns the discovery of relations between our 'abstract ideas', the method to be applied being that 'learned in the schools of mathematicians', who,

> from very plain and easy beginnings, by gentle degrees, and a continued chain of reasonings, proceed to the discovery and demonstration of truths that appear at first sight beyond human capacity.[2]

So mathematics does not give us 'substantial' knowledge, for instance of the properties of the magnet or of gold: '*Experience here must teach me* what reason cannot.'[3]

Has Locke recognised that he is dealing here with different *kinds* of knowledge? His reference to 'the degrees of knowledge', in the second chapter of Bk. iv, might suggest that he has not; but of the three 'degrees' mentioned, intuition, demonstration and perception of external objects, the last-named is (as already noted) recognised as being of a special sort. On the other hand, despite the desire to extend the denotation of knowledge, preoccupation with the intuitional model has the effect of describing the differences referred to in a curiously pejorative and scepticist language; there is a constantly recurring theme of the inferior nature of sensory knowledge, of its narrow extent, its small reach, its 'failure to conform' with 'things

---

[1] iv.4.6. This finds an echo in Newton: 'In mathematics we are to investigate the quantities of forces with their proportions consequent upon any conditions supposed; then, when we enter upon physics, we compare those proportions with the phenomena of nature, that we may know what conditions of those forces answer to several kinds of attractive bodies.' (*Principia*, Bk. I, Prop. 69, Scholium.)

[2] iv.12.7. Compare this with the passage in Descartes' *Discourse*, ii, cited above, p. 84, which has a similar message, yet leads to such a different 'metaphysics'!

[3] iv.12.9; italics in text. Admittedly, mathematics is not specifically in Locke's mind in this passage, but in view of Locke's theory of mathematical knowledge, it is covered by it.

themselves'.¹ Nevertheless, at least occasionally, there is an explicit recognition that a difference in kind is involved. Locke then *emphasises* (as in the mathematical example) the purely conceptual ('abstract' is his favourite term) nature of 'certain and evident' knowledge, implying really the avoidance of all ontological questions. He will then acknowledge that there are two extreme cases, two distinct epistemological types, as when at iv.11.14 he distinguishes between

> knowledge [which] is the consequence of the existence of things, producing ideas in our minds by our senses . . . [and] knowledge [which] is the consequence of the ideas (be they what they will) that are in our minds, producing there certain general propositions.

This is an anticipation of a dichotomy which Hume will make the basis of his doctrine, the distinction between 'relations of ideas' and 'matters of fact'.²

On the other hand, the note of scepticism becomes more insistent when we come to the case of 'general truths' or 'universal propositions', and to our 'knowledge concerning substances'.³ Here, our knowledge goes only 'a very little way'; 'our ignorance is great'.⁴ Indeed, Locke goes so far as to deny that there can be any genuine 'science'; the most we can hope for is 'judgment and opinion, not knowledge and certainty'.⁵ Locke is evidently experiencing some severe conflict; whereas in one place this is still knowledge, though of a very 'narrow' kind, in another this appellation is denied altogether. Yet, what looks like despair from ignorance seems to be generated at first sight by nothing but the sharpening of the division between different kinds of propositions. The reason given for our lack of a 'science of bodies' is that in this sphere of enquiry, 'certainty and demonstration are things we must not . . . pretend to'.⁶ This we should expect from the distinction between factual and logical knowledge Locke has himself already noted. As post-Humeans, we should not expect 'rational demonstrations' in questions that concern

---

¹ iv.2.14; 3.5; 3.9; 4.11. I have discussed the comparison between Locke and Descartes in preliminary fashion above, ch. I, pp. 5–8.
² Cf. *Enquiry*, iv.1, ed. Selby-Bigge, p. 35: 'All the objects of human reason or enquiry may naturally be divided into two kinds, to wit, *Relations of Ideas*, and *Matters of Fact*.' Cf. *Treatise*, iii.1.1; Selby-Bigge, p. 463.
³ Cf. iv.2.14; 2.6; 3.11.   ⁴ iv.3.9; 3.22.
⁵ iv.12.10. Cf. 3.26; and 2.14, where we find we can expect no better than 'faith or opinion'.
⁶ iv.3.26.

'matters of fact'; and we should not pretend to 'intuitive certainty' in a sphere of enquiry which was ultimately contingent upon a sensory basis. But this ought hardly to be described as a case of 'ignorance' and 'narrowness of knowledge'![1]

The reason for Locke's vacillating attitude must hence be sought in his failure to make a complete break with the Cartesian approach, and in his remaining in a position approximately half-way between the earlier period and that of Hume. Locke ascribes the fundamental cause of our ignorance to our inability to perceive the 'necessary connections' between each group of qualities of physical substances which the evidence of our senses shows us regularly to 'coexist'.[2] 'Experience' teaches us that certain things have 'a constant and regular connection' in the ordinary course of events; but we lack any sort of rational insight into these connections; being opaque to the intellect, they are just brute (*though permanent*) facts; as appears in Locke's theological version, when he writes: 'we can attribute their connection to nothing else but the arbitrary determination of that all-wise Agent who has made them. . . .'[3]

The dichotomy, necessity inherent in nature *v.* mere observable uniformity, is, as we saw earlier on, part of the intellectual tradition which Locke inherited. What is new is that it plays a more worrying part in Locke's philosophical thinking; our inability to perceive necessary connections begins to generate consequences of ever-growing importance. And in line with the fluidity and lack of univocal position in Locke's thinking, there is a parallel lack of definition regarding the significance to be attached to our lack of knowledge of necessary connections. At one extreme we seem to have a full-blooded 'uncritical' acceptance of natural necessity; at the other, lack of knowledge of this necessity appears to imply its absence in nature; and there are hints that in the absence of the 'epistemic' relation the very notion of a necessitarian bond regarded 'constitutively' loses its meaning.[4]

---

[1] As we go on, we shall discover, however, additional motives for Locke's sceptical tone.

[2] Cf. iv.3.10–14; 3.28–29; 6.13; 7.5; 11.9; ii.31.6. See also above, ch. II, p. 63.

[3] iv.3.28. Once again we are reminded of Descartes' 'teaching of nature'. Locke, if anything, feels the deficiency more deeply, because for him there is no doctrine according to which the sensory realm can ultimately be integrated into that of thought, and mathematics come to the rescue of physics.

[4] However, these *are* extremes, never clearly recognised till Hume. And Locke's doctrine of 'real essence', as well as of 'substance' (*quâ* substratum) will, as we shall see, complicate the issue further.

2. NATURAL NECESSITY: 'EPISTEMIC' AND 'CONSTITUTIVE' LEVELS

The fluidity of Locke's position is due to the fact that it is discussed in terms of, or proceeds at least under the shadow of, a number of additional dichotomous distinctions which pervade the whole logical framework of the *Essay*, where each in turn is infected with a similar lack of univocal expression. To bring this out into the open, the rest of this chapter will require a certain amount of textual criticism owing to the intricacy and ambiguity of Locke's thought. The first of these distinctions is that between the realm of 'ideas' and of an external reality behind these ideas, the 'external objects'. A second group concerns the contrast between 'real essence' and 'nominal essence'; between the internal corpuscularian structure of physical 'substances' and the complex of qualities which are sensorily observed generally to coexist in any given substance.[1]

A third pair, intimately associated in Locke's discussion with the previous one, though again distinct, is that of our ideas of primary and secondary qualities.[2] And lastly, we have the distinction between the notion of 'substance in general', considered as a substratum, and the qualities which may be thought of as inhering in it or as being supported by it.[3]

From the point of view of our present discussion, each of the last three dichotomies furnishes a context for Locke's conception of

---

[1] Cf. iii.6.2. To complicate matters, when speaking of the 'nominal essence', Locke usually refers to it as a 'complex *idea*', standing for the name of the substance, whereas the 'real essence' is considered as 'the *real* constitution of anything' (Cf. 6.6); this looks like only another version of the contrast between idea and external object, particularly since Locke insists frequently on our inability to have any knowledge of the 'real essence'. However, it is almost certain that the two dichotomies are entirely distinct. The considerations which led to Locke's notion of 'real essence' belong (as will be shown in greater detail below) to the realm of the contemporary physical science, and are neutral to any theory of 'representative perception'.

[2] ii.8.9–25. Primary qualities Locke holds to exist 'in bodies', but secondary qualities are 'nothing in the objects themselves' but only 'powers to produce various sensations in us by their primary qualities' (8.9), so that strictly he ought only to speak of '*ideas* of secondary qualities'. Still, at ii.8.15 and iv.3.13 he mentions 'secondary qualities' *simpliciter*.

[3] Cf. ii.23.2–6. Once more, Locke alternatively speaks of qualities and the *ideas* of qualities as inhering in the *substratum*. In an important article, Jonathan Bennett has shown that although these different dichotomies are quite distinct, it is easy to conflate them, particularly, since the first of them, that between 'idea' and 'external object', pervades the others. Cf. 'Substance, Reality, and Primary Qualities', *Am. Phil. Quart.*, **2** (1965), 1–17.

scientific knowledge; and in each, although Locke seems to adhere to a necessitarian position, he emphasises our lack of any knowledge of it; the concept of natural necessity is shown to be increasingly useless and inapplicable. And then, under the stress of these considerations, Locke begins (but only begins) to throw doubt on the ideal of scientific knowledge implicit in this notion (though only obliquely) by carping at the 'uselessness' of 'abstract maxims and principles' for the development of the natural sciences, and the essentially empirical background to the foundations of the latter in the work of his scientific contemporaries. As we shall see later, this complaint can be understood to be as much a hankering after the old rationalist gods as it is an attempt to provide new demarcations for scientific knowledge.[1]

The whole discussion gains considerably in interest if we remember that in recent years Locke's philosophy of natural law has exercised a certain influence on those philosophers of science who have been wanting to support a necessitarian interpretation. Among these, apart from Johnson, who was singled out earlier,[2] William Kneale in particular has argued in favour of the view that natural laws are principles of necessitation, despite their necessity being 'opaque to the intellect'.[3] Our inductive policies in terms of which we formulate principles of assent to scientific hypotheses can never establish, let alone yield '*a priori* insight' into their necessity. Nevertheless, unless such policies could be assumed at least in principle to lead to laws possessing necessitarian force, induction would not be a 'rational' process; scientists would have no reason for continuing their efforts.[4]

Evidently Kneale, like Johnson before him, operates again with the distinction between constitutive and epistemic levels. Lack of '*a priori* insight' into the necessity of natural laws does not militate against the plausibility of the assumption of necessary connections, although the only models offered are taken either from mathematics or the result of the 'intuitive induction' of such modal principles as 'a

---

[1] Cf. below, pp. 211–3.   [2] Cf. above, ch. II, pp. 27ff.
[3] *Probability and Induction*, p. 97. Cf. paras. 17 and 18, pp. 70–89, especially pp. 78–81, 88–89. Cf. above, ch. II, pp. 26, 37n.2, 42. Locke's necessitarian position with respect to scientific law finds nominal expression at iv.3.29: 'Things that, as far as our observation reaches, we constantly find to proceed regularly, we may conclude do act by a law set them: but yet by a law that we know not, [because] their connections and dependencies being not discoverable in our ideas, we can have but an experimental knowledge of them'.
[4] Op. cit., p. 258.

thing which is red cannot also be green at the same time'.[1] This is not the place for a separate critique of Kneale's position; our interest is solely in the relationship in which it stands to Locke, and through which each may be hoped to illuminate the other. Kneale rejects arguments like those of Hume, according to which the notion of natural necessary connections is otiose on the ground that they are compatible with conceivability of the opposite. But Kneale's rejection seems to be based largely on the contention that epistemic considerations are not decisive, and do not necessarily affect the question of such a bond considered from the constitutive point of view; and here he claims some backing from the authority of Locke. He understands Locke's doctrine to imply that

> we cannot have insight into the connections which we assert in natural science; but he [*sc.* Locke] holds nevertheless that they may be necessary in the same sense as connections which we are able to comprehend, and he implies that a sufficiently powerful mind which was furnished with the appropriate ideas of what he calls the real internal essences of things might see the necessity of the laws.[2]

But already we may see that Locke's position is by no means as clear-cut as is suggested by Kneale. On the contrary, Locke is precisely the philosopher at whose hands there begins, by implication, the process of erosion of this hitherto almost unquestioningly held distinction between the constitutive and the epistemic levels, culminating in the attempt at total demolition by Hume. For one thing, we note that, unlike Kneale, Locke regards the absence of intuitive insight into the alleged necessary connections implied by laws as a grave deficiency, leading to 'lack of certainty' in scientific knowledge. To be sure, through this language Locke may wish to convey only that although for the mind connections between most of its ideas are not 'visible',[3] they may nonetheless exist (just as invisible corpuscles exist!). But other passages suggest a stronger doctrine. Thus, iv.3.10 says that our knowledge of complex ideas of substances does not 'extend' very far because the 'simple ideas' of which they 'are made up' 'carry with them, *in their own nature*, no visible necessary connection or inconsistency with any other simple ideas'.[4] Some sections later Locke tells us that

although many of our ideas have a constant and regular connection in

[1] Op. cit., par. 9, p. 32. This example occurs also in Locke, *Essay*, iv.3.15.
[2] *Probability and Induction*, p. 71.
[3] iv.3.3.; 3.14.   [4] My italics.

the ordinary course of things; yet that connection being not discoverable in the ideas themselves, which appearing to have no necessary dependence one on another, we can attribute their connection to nothing else but the arbitrary determination of that all-wise Agent who has made them to be, and to operate as they do, in a way wholly above our weak understanding to conceive.[1]

This is a theological way of saying that the uniformity is one of 'brute fact', an 'historical accident on the cosmic scale'.[2] Perhaps it could be objected that we are here dealing after all only with connections between *ideas*. But in the next section, Locke repeats his position without specific emphasis upon ideas.

But the coherence and continuity of the parts of matter, the production of sensation in us of colours and sounds, &c., by impulse and motion; nay, the original rules and communication of motion, being such wherein we can discover no natural connection with any ideas we have, we cannot but ascribe them to the arbitrary will and good pleasure of the wise Architect.[3]

Taken together, passages such as these suggest that Locke is here tending to the view that lack of epistemic connection implies a lack of constitutive bond. Evidently, Locke's thinking is altogether (as so often with him) in a state of transition. From the 'uncritical' standpoint (attributed to him by Kneale without qualification) that there not only may, but must be necessary connections 'in nature', although by and large unknown, he moves to a position according to which such connections, being unknown, are denied to exist. Indeed, this suggests in turn a more extreme move: since knowledge of necessary connections always refers to ideas, Locke might well take the view, never stated expressly—it is only his lamentations deploring our 'ignorance of necessary connections' which suggest that the difficulty cuts more deeply for him than at first appears—that the notion of an extra-ideal connection of this kind was *senseless*. But it is only

---

[1] iv.3.28.   [2] For this phrase, cf. above, ch. II, p. 45.

[3] iv.3.29. Cf. ii.23.28, where 'communication of motion by impulse' is said to be 'obscure and inconceivable'. Similarly, at ii.21.4, communication of motion by impulse is declared to be too 'obscure' to yield a clear idea of 'power'. On the other hand, at iv.3.13, things like 'the change from rest to motion upon impulse . . . seem to us to have some connection one with another'. The more radical sceptical view fully anticipates the positions of Berkeley and Hume, except that Hume states his point in non-theistic language; cf. *Enquiry*, vii.2, ed. Selby-Bigge, p. 74: 'One event follows another; but we never can observe any tie between them. They seem *conjoined*, but never *connected*.'

suggested; there is no argument to this effect. The senselessness of the notion would only follow logically if a trancendence of ideal relations (to relations considered 'constitutively', as we have called them) became itself proscribed; and this step of course was taken explicitly only by Berkeley and Hume. For in their denial of the very possibility of any knowledge of—indeed, any significance of the conception of an external world, transcending our 'ideas', the notion of the existence of any external, non-ideal, relations is *ipso facto* denied as well.

Nor is this all. In the passage quoted already, in which Kneale characterises Locke's position, he refers to the 'real internal essences of things'; but this is quite a different matter, involving a separate dichotomous contrast (we shall turn to it a little later).[1] What is true is that it did supply Locke with a further model of the 'constitutive level'; which would conceal from him the true implications of his doctrine—or at any rate, prevent him from drawing the conclusion drawn later by Hume. But as we shall see, the status of that level (of 'real essences') is itself sufficiently ambiguous to allow Locke's doctrine to remain in the philosophical twilight which pervades it throughout. Anyway, Locke never squarely comes out with a denial (any more than an affirmation) of non-ideal, intrinsically necessary relations. Instead, the question of the 'reality' behind the 'ideas' (just as that of the relations between the members of the other dichotomies previously mentioned) is left in an extremely obscure and muddled state. And, this very vagueness may be taken to be equivalent to a move in the direction of a dissolution of the constitutive realm. For since Locke frequently employs phrases like 'qualities of substances' and 'ideas of qualities of substances' interchangeably, the impression is conveyed that lack of perception of necessary connections, equivalent to lack of the corresponding relations between ideas, is tantamount to a denial of the relation as such.[2]

I think it now becomes apparent that the revolutionary element in Locke's philosophy is the central emphasis on the notion of 'idea' as the fundamental object of thinking (including under this umbrella

[1] Below, pp. 204ff. In the passage referred to, Kneale unconsciously conflates them!
[2] Locke sometimes denies this equivocation between qualities and ideas of qualities: 'which ideas, if I speak of them sometimes as in the things themselves, I would be understood to mean those qualities in the objects which produce them in us (ii.8.8). Unfortunately, some of the arguments demand just this equivocation, if they are to succeed. Besides, it also echoes the vagueness of Locke's distinction between object as 'external cause of my perceptions', as the accusative element of my perceptual states, and as the 'scientific object' with its given 'internal constitution'. Cf. below, p. 248, for a summary of these various conceptions of object.

term, as Locke does, things like perceiving, reasoning, etc.). Moreover, with Locke's insistence that all ideas come ultimately from 'experience', i.e. what is yielded through 'sensation', as well as through 'reflection' on our mental 'operations' with the ideas of sensation, this 'idealism' manages admirably to absorb the incipient 'empiricism' of the contemporary scientific scene. On the other hand, Locke's conception, or rather ideal, of 'knowledge' is, as we have already seen, thoroughly 'rationalistic'.[1] And it is this fact which produces the peculiar tensions in Locke's theory of knowledge, though it is at the same time admirably adapted to a 'philosophy of transition'. Let us therefore explore in greater detail these two cornerstones of Locke's treatise, and particularly, his views on 'knowledge' in general—its 'degrees' and 'extent', in order to see more clearly the part played by 'ideas', and to get a clearer grasp (through textual analysis) of the variety of positions which Locke holds on the question of our knowledge of 'external objects'. This will enable us then to study the relevance of the various 'dichotomies' in his scheme already referred to, and thus finally to bring out Locke's position with respect to the foundations of scientific reasoning.

3. KNOWLEDGE: ITS 'DEGREES' AND 'EXTENT'

Among the various meanings given to the term 'knowledge' in the *Oxford English Dictionary* we find, among the more important ones, 'the fact of knowing a thing, state, etc. or person'; also, 'acquaintance with fact'; finally 'intellectual perception of fact or truth'. Locke's conception of 'knowledge' takes off primarily from the last of these, and that in a very special sense. The 'perception of the truth' involved in Locke's definition is of what he calls 'the agreement or disagreement of two ideas'.[2] 'Perception', again, is here of a very special sort. It occurs in fact in the third of three different contexts which Locke himself distinguishes. The first of these is the more common use, where one speaks of perceiving or noticing or seeing some motion or process or object. For reasons to be discussed later, the *object* of perception in this sense is (for Locke) always an *idea*, where 'idea'

---

[1] The true significance of this 'rationalism' for Locke's appraisal of the status of scientific knowledge (in the neutral sense of this term) will occupy us later. (See below, especially pp. 212ff.)

[2] iv.1.2.

itself occurs equivocally either as 'concept', as 'particular existent' of one sort or another, or as mental image. This is the first of the senses of 'perception' which Locke distinguishes at ii.21.5, as 'the perception of ideas in our minds'. The second connects with our understanding of meanings; it is 'the perception of the signification of signs'. But only the last of the three is the one relevant for his own, and more specific, definition of 'knowledge'; viz. 'the perception of the connection or repugnancy, agreement or disagreement, that there is between any of our ideas'. But although these three uses of 'perception' are quite distinct, the shadow cast by the first affects the relevance of the third, through the employment of the ubiquitous notion of 'idea'. Locke distinguishes four types of agreement or disagreement between ideas: '(1) Identity, or diversity, (2) relation, (3) coexistence, *or* necessary connection, (4) real existence'.[1] Examples which he gives of the four types are, respectively, 'Blue is not yellow'; 'Two triangles upon equal bases between two parallels are equal'; 'Iron is susceptible of magnetical impressions'; and, finally, 'God is'.[2]

This perception of agreement and disagreement of our ideas is itself of three kinds, depending on whether it involves 'intuition', 'reasoning' (also called 'demonstration') or 'sensation'.[3] As Locke develops his account, 'intuition' can be seen to correlate with the first of the four types (identity); reasoning with the second (relation). On the other hand, the third and fourth types are more complex. As to the former, to knowledge 'of coexistence' we come only by sensation, whereas our knowledge of 'necessary connection', where it exists (and Locke has some examples, as we shall see) is intuitive. Finally, of our knowledge of the existence of ourselves Locke gives a somewhat perfunctory account, as being 'intuitive'.[4] The existence of God is argued to be knowable by 'demonstration', and involves again a form of causal maxim; the argument, though reminiscent of Cartesian and

---

[1] iv.1.2, my italics. Coexistence and necessary connection are hardly synonyms, nor even alternatives. The 'or' admirably represents Locke's ambivalent attitude towards the status of necessary connections already noted.

[2] 1.7. The last class of 'agreement or disagreement between ideas' would again be rejected by those who hold that 'existence' is not an 'idea' commensurable with ideas (expressions) designating persons, objects, or even processes and qualities. Locke's transitional position frequently incorporates earlier views hardly compatible with his type of 'empiricism'. That the copula 'is' does not designate a 'real predicate' will be stated explicitly by Kant, but it is already, as we shall see, to be found in some of Hume's arguments. (Cf. below, ch. VI, pp. 354ff.; ch. VIII, pp. 487ff.).

[3] iv.3.2.   [4] iv.9.3; 10.2.

scholastic origins, is fairly perfunctory, and anyway its discussion does not belong here. What is of greater interest to us is that our knowledge of the existence of 'other things', i.e. of 'external objects', is said 'to be had only by sensation'.[1] Since, as we shall show in greater detail later, Locke, again like Descartes, holds that we have 'direct access' really only to our 'ideas', a reader may be puzzled by this phrase. For Locke does in fact employ *arguments* to 'prove' the existence of external objects although these arguments do not involve a reference to God's goodness except by the merest ceremonial formula: 'God has given me assurance enough of the existence of things without me.'[2] His more serious arguments are of a different kind. We shall come to them in their due place.[3]

For the moment, let us note that of the four sorts of perception, the first (of identity and diversity), as involving intuition, forms the primary paradigm of knowledge; for it is here that 'a man infallibly knows, as soon as ever he has them in his mind, that the ideas' must have the relation to each other which he perceives them to have.[4] These are the

> truths the mind perceives at the first sight of the ideas together, by bare intuition, without the intervention of any other idea; and this kind of knowledge is the clearest and most certain that human frailty is capable of. This part of knowledge is irresistible. . . . [5]

Evidently this is the case of 'intuitive induction' previously referred to in connection with Kneale.[6] One must be careful not to misdescribe Locke's intentions by insinuating modern interpretations, using terms like 'conceptual knowledge', 'synthetic *a priori* judgment', or 'analytic proposition', for once again Locke's thinking is far too fluid or ambiguous to permit identification through such classificatory devices—devices which indeed themselves constitute attempts to deal with the problems which Locke here raises by the introduction of alternative philosophical positions. His position is more like an amalgam of these; if we say that perception of agreement and disagreement with respect to identity is a purely conceptual matter, we overlook the ambiguity in Locke's use of idea as '*object* of perception' in the first of the senses distinguished above, and as pure *concept*. Certainly Locke wants to say that there is something about

---

[1] iv.11.1.   [2] iv.11.3.   [3] Cf. below, pp. 235ff.
[4] iv.1.4.   [5] iv.2.1. Cf. above, p. 183.   [6] Cf. above, pp. 189–90.

the idea, in the sense of objective quality such that when you have grasped, through, say, a process of learning by inspection of actual instances and under the guidance of a 'teacher', what it is for anything to be that quality (e.g. to be blue), you have also grasped what it is for anything to be *not* blue—and that this is not simply a *verbal* matter, connected with some 'arbitrary' decision concerning the use of words. Such propositions expressing intuitions of identity and diversity must not be confused with 'identical propositions' which are, to use Locke's term, 'trifling', i.e. more like the modern 'analytic' in one of its senses; or with those, 'analytic' in the Kantian sense, which affirm in the predicate what is already 'contained' in the 'complex idea' of the subject.[1] In other words, though Locke must in the intuitional case be concerned with meanings (concepts), such meanings are also taken to be 'tied down to' corresponding 'objects' (here, the qualities): a transition which is artfully concealed through the ambiguity of the term 'idea'. The fateful result of this we shall note presently, in the case of our perception of coexistence. Before doing so, we must briefly glance at what Locke calls agreement or disagreement in respect of 'relation', which corresponds also to what he calls 'demonstrative knowledge'. Here, we cannot directly intuit any agreement or disagreement, but must first interpose a process of 'reasoning', as when we prove that the sum of the internal angles of a triangle are equal to two right angles, for instance, by showing that our three angles are equal to 'some other angles' which themselves can be 'seen' to be equal to two right angles.[2]

This case is interesting because it bears on Locke's position

---

[1] Cf. iv.8.2–8.

[2] iv.2.2. Locke is not very specific on the detailed nature of the reasoning involved. Perhaps he had in mind something like the sketch which later appears in Kant's *Critique* (K.578), employing the following auxiliary construction: by drawing B–B'

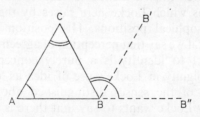

parallel to AC, and extending AB, to give BB", we obtain a new 'intervening set of angles', ABC, CBB', B'BB", the sum of which can be 'seen' to equal two right angles, whilst each of them equals, respectively, one of the internal angles.

regarding what he calls the different 'degrees of knowledge', and again, its differences in 'extent'. We have already seen that Locke has ambiguous views when he compares the fields of logic and mathematics with those of empirical science.[1] And whilst he does seem to say at times that we are here dealing with differences in kind, in the chapters on the 'degrees' and 'extent' of 'human knowledge' the discussion is conducted as though the differences were only of degree: with the result that the absence of features present in 'intuition' and 'demonstration' is construed as a 'deficiency'. The deficiency which Locke notes is of course the absence of necessary connections between our 'ideas' when we consider natural 'substances'.[2]

In his usual ambivalent way, Locke sometimes seems to suggest that we here *never* 'advance by contemplating our ideas', but that we employ observation solely in order to discover 'what qualities coexist',[3] whilst at other times he states that there are 'some few of the primary qualities that have a necessary dependence and visible connection one with another'.

Here, much would seem to depend on the examples Locke has on hand, but unfortunately those offered are not clear-cut. Thus iv.3.14 mentions 'figure necessarily supposes extension' and 'receiving or communicating motion by impulse supposes solidity'; 3.15 gives the 'incompatibility' of 'two smells or two colours at the same time'.[4] The first of these might be said to involve a relation of logical entailment, even though one might hesitate to designate it 'analytic'. The last example has frequently been held to involve 'intuitive induction'. But neither are of any relevance to the edifice of an empirical science. The second might be thought more promising were it not the case that—as Kant was later to point out—the concept of 'solidity' is one whose meaning is not very clear in such a context, being indeed thoroughly 'ambiguous'.[5] But it is interesting that Kant appears to have assumed that if the concept *were* univocal, the resulting proposition would be *a priori*; though only on condition that we first introduce the notion of force—something which involves the guidance of the category of quality and thus concerns a transcendental matter.[6]

---

[1] Cf. above, pp. 185–7.   [2] Cf. iii.6.6–9; iv.3.9–18; 6.12; 12.9–10.
[3] iv.12.9.; i.e. our 'perception' is by way of 'sensation' and not 'intuition'.
[4] Cf. also below, p. 202n.3 for another example to be found at iv.7.5.
[5] Cf. *Metaphysical Foundations of Natural Science*, sect. ii; ed. Bax, p. 170. Cf. below, ch. VIII, sect. 6(c), p. 566.
[6] Op. cit., pp. 170–1.

At any rate, Locke's vacillations are such that one gets the impression that he often believes our inability to *perceive* such connections to be a contingent, not a logical matter; that it is only an accidental inability on our part that conceals from us this bond. Now my main point for the moment is that much in such cases depends on the models under which we believe we can subsume our philosophical notions. And it is very probable that the mathematical cases (involved in 'demonstrative reasoning') offered just such a paradigm where one could sensibly speak of the existence of a necessary connection whose presence had so far not been made 'visible' owing to our inability to discover the relating bond, i.e. the 'intermediate ideas'. Thus, at iv.3, when discussing the differences in the 'extent' of 'intuitive', 'demonstrative' and 'sensitive' knowledge, he goes on to characterise the 'narrowness' of our knowledge by reference to just such a mathematical example:

> We have the ideas of a square, a circle, and equality: and yet, perhaps, shall never be able to find a circle equal to a square, and certainly know that it is so.[1]

It would therefore be easy for Locke to slide into the view that the absence of knowledge of necessary connections among our 'ideas of qualities', so far known to coexist only in virtue of sensory observation, was just such a contingent affair. And this view would be aided and abetted by the dichotomy in Locke's thinking (which lies never far beneath the surface) which distinguishes between relations between

---

[1] iv.3.6. A reference to the problem of 'squaring the circle', whose solution (not achieved until two centuries after Locke) consists in proving that π is transcendental, which shows that it is *impossible* to construct a square of which the area is equal to that enclosed by the circle.

A similar point is made at 3.18, where Locke remarks that in algebra it needs considerable 'sagacity in finding intermediate ideas', and that it is therefore difficult to say 'how far our knowledge reaches' (iv.3.7 and cf. 3.3–4). Kneale, op. cit., similarly argues that there are some mathematical theorems, such as Goldbach's, which are so far unproven, and whose contradictory we may therefore be able to 'conceive'.

Of course, to such examples one might object that if the theorem holds (or fails to hold), it does so necessarily. It is only our inability to discover the demonstration which conceals the connection from us. Yet Kneale, like Locke before him, uses this *model* to suggest the plausibility of the assumption of necessary connections in cases where no 'rational insight' is *ever* possible: assuming seemingly that although the latter cases are quite different from those of mathematics, this does not affect the possibility of the conception; and asserting simply that 'our rules for the use of symbols in the perceptual object terminology' are not necessarily inconsistent with this possibility! (Cf. op. cit., pp. 80, 88; and below, ch. VI, pp. 365ff.)

ideas, and the corresponding relations between the qualities of 'external objects'.[1]

At any rate, when we turn to the case of sensory perception of the coexistence of our ideas in 'particular substances', this is expressly said to be 'narrower' than the earlier cases. In the first instance, it is narrower in that it involves 'sensation' rather than 'intuition' or 'reasoning'. The inferiority is due to the fact that we have sensory knowledge only of things 'actually present to our senses';[2] what transcends the present moment, of that 'I cannot be certain'. Furthermore, there are the difficulties concerning my knowledge of the existence of 'external objects', as contrasted with my corresponding 'idea' of it, which the former has produced in me, and which alone is certainly 'in our minds'. Still, as already noted, though 'narrow', this is nevertheless, if grudgingly, conceded to be a case of 'knowledge'.[3]

However, perception of coexistence is also said not to 'extend very far', simply *because* it involves mere coexistence (rather than necessary connection). There is thus a discrepancy between what we actually know 'concerning substances', and what—in accordance with Locke's ideal—we ought to know.[4] Now what we do know is what qualities (especially 'secondary qualities') go together in each case of a natural substance, or rather, each 'species of substance'. Thus, our 'knowledge of the truth' that 'gold is fixed', 'amounts to no more than this':

> that fixedness, or a power to remain in the fire unconsumed, is an idea that always accompanies and is joined with that particular sort of yellowness, weight, fusibility, malleableness, and solubility in *aqua regia*, which makes our complex idea, signified by the word gold.[5]

In other words, the complex concept of gold comprises a number of others, regarded as being more elementary and more simple. Now our only reason for adopting this particular list is that it is 'based on

---

[1] Cf. above, pp. 191–2; and again, similarly, for the dichotomy between 'nominal' and 'real essence'.

[2] iv.3.5; the theme is taken up again at 11.9.

[3] iv.2.14; 3.2; 3.5; 11.3. Cf. above, pp. 183–4. Actually, at 2.14 and 11.3 this consideration is advanced to show that this knowledge is of an inferior '*degree*' of 'certainty'.

[4] And yet, as already hinted, we cannot be certain that Locke in the end believed it to be a realisable ideal, even a desirable one. His scepticism may just as well be interpreted as a warning to his reader not to search after false gods. Cf. below, pp. 212ff., concerning Locke's views on 'maxims', iv.7, and on the way to improve scientific knowledge, iv.12.

[5] iv.1.6.

experience',[1] i.e. on the observation of particular instances, of which it is a generalisation. But such inductive generalisations are highly problematic; we cannot

> be assured about them any farther than some few trials we make are able to reach. But whether they will succeed again another time, we cannot be certain. This hinders our certain knowledge of universal truths concerning natural bodies: and our reason carries us herein very little beyond particular matter of fact.[2]

There is a dual complaint here: (1) We cannot be certain that our list of qualities is necessary and sufficient for some given 'species'. This list is called by Locke the 'nominal essence' of the substance in question.[3] And although men in their efforts at classification are guided by what they observe to be correlated in any given species, the 'boundaries of the species' are 'made by men'; they are to this extent subjective, 'made by the mind', albeit 'not arbitrarily'.[4] Thus, knowing that such and such a property can be predicated of a given substance or species, we may want to know what other properties coexist with it in that substance or species. But this question can be answered *for the most part* (according to Locke) only by observation and experiment, and not by contemplating the erstwhile property (or rather, our idea of that property) alone. But we have already seen several times that this opens up the usual deficiency, viz. our answer cannot be obtained by a perception of the 'necessary connection or inconsistency' of the idea with any others.[5] (2) The lack of the connection carries over also into the logic of inductive generalisation. Evidently Locke assumes that in order to acquire the desired degree of knowledge yielded by generalisations, we must be able to presuppose the existence of necessary connections between the ideas in

---

[1] Cf. iv.3.14; 12.9.   [2] iv.3.25.   [3] Cf. iii.6.2; 3.15f.
[4] iii.6.25, 26, 28, 36. Cf. below, pp. 208–9; 214f.
[5] iv.3.10; 14.28. It is worth noting that the passage at iv.3.10 was especially singled out by Kant for comment in the *Prolegomena*, §3. But by the time of Kant, distinctions have been sharpened, and there is no question of allowing any 'intuitive' models to serve as aids towards the problem of necessary connections between our 'ideas', and any regress to a 'constitutive level' is likewise ruled out. Nevertheless, just like Locke, and under the influence of Leibniz and Hume, Kant conceives our knowledge of physical kinds as well as the laws governing them as involving something like necessary connections. He therefore interprets Locke's difficulty as the search for 'synthetic *a priori*' connections. However, notions like 'synthetic' and '*a priori*' now receive a rather different interpretation, expressing the peculiar nature of Kant's solution.

question. Their absence in Locke's view deprives empirical generalisations ('general truths' or 'universal propositions') of any claim to be
instances of genuine 'knowledge'; the title to be applied is more
properly that of 'faith or opinion'; of 'judgment'; or 'probability'.[1]

Locke goes on endlessly in this vein. But as mentioned, it is never
clear what his psychological attitude is to the deficiency which he
complains of, and what he considers to be its logical relevance. Thus,
my knowledge of the existence of some particular thing being described as my apprehension of a collection of simple ideas which go
together, such knowledge is said to extend no farther than 'the present
testimony of our senses'; to transcend such 'immediacy' once
more would presuppose the existence of a 'necessary connection'
between 'existence [of anything] a minute since' with its 'existence
now'.[2]

Now on one side we get the following reaction. Locke here realises
that the 'want of ideas' is a logical matter, which he expresses by
saying it is 'vain to expect demonstration and certainty in things not
capable of it'.[3] He realises that the kind of certainty of 'general
propositions' which he is here defining 'concerns not existence'—
which is his way of saying that they have no relevance to 'matters of
fact'—but is 'merely the consequence of the ideas . . . that are in our
minds'.[4] More radically still, elsewhere the paradigm of a 'universal
certain proposition' is given as 'All gold is malleable', on the ground
that it is analytic, or to use Locke's term, 'trifling'.[5] And we
have noted already that at iv.11.13–14, a clear distinction is drawn
between propositions concerning 'relations of ideas' and 'matters
of fact'.[6]

But on the other side, the alternative, namely 'getting and improv-
ing our knowledge' by turning to 'experience', is characterised as a
'weakness of our faculties in this state of mediocrity which . . . makes

---

[1] iv.2.14; 6.13; 15. This will later form also part of the basis for Hume's sceptical
conclusion. In the *Treatise*, i.3.6, he argues that inductive inference presupposes the
principle of uniformity; but that such a principle in turn rests upon the putative
existence of necessary connections, which of course (according to Hume) are ruled
out from the nature of the case. And cf. *Enquiry*, sect. iv.2, where the impossibility
of 'rational inference' is said to be due to the lack of a 'medium' or of 'interposing
ideas'. (ed. Selby-Bigge, pp. 34, 37.)

[2] iv.11.9. This incidentally shows that Locke appreciated that the difficulty concerning inductive generalisations extends to the problem of our knowledge of 'physical
objects', taken to exist continuously in time. Cf. Descartes' introduction of this
theme, above, Ch. III, pp. 175ff.

[3] iv.11.10.   [4] iv.11.14.   [5] iv.6.9.   [6] Above, p. 186.

me suspect that natural philosophy is not capable of being made a science'.[1]

Evidently one explanation of Locke's vacillation is that he slides from the paradigm of 'analytic proposition' to that of 'intuitive perception' of 'relations between abstract ideas', which includes the paradigm of mathematical relations[2] in which cases it had seemed to make sense to construe the 'want of ideas' as a contingent matter which might sometimes be rectified.[3] Again, as regards 'perception', it was easy to slide from the 'logical' to the 'empirical' level; from 'perceiving one quality to be associated with another' (in an experimental situation), *via* 'a contemplation of the corresponding ideas that are associated in the complex idea in my mind', to 'perceiving that the concept of one entails the concept of the other'.[4]

4. INTERPRETATIONS OF THE 'CONSTITUTIVE LEVEL'

Still, when everything is said and done, Locke has hardly faced the consequence of his various distinctions. With the knowledge of hindsight of Hume's subsequent arguments, one might protest that if any proposition was genuinely concerned with a perception of necessary connections between our ideas, then it could not at the same time have any bearing on the corresponding 'factual relations'. Or to put it the other way round: if you could conceive of any relation being severed, as you must with matters of empirical fact, then surely this proves the relation not to be a necessary one.[5] It would follow from this that the *absence* of *perception* of necessary connections in the factual cases should not be construed as a deficiency. There must be further explanations for Locke's refusal to follow this turn of the argument. The first of these is a distinction which he makes between

---

[1] 12.10. Similarly at 3.26: whilst we lack the requisite ideal of certainty, 'we are not capable of scientifical knowledge'.

[2] Cf. iv.12.7.

[3] Cf. above, pp. 196f. Besides, Locke certainly thought that he had some genuine examples of non-analytic, non-mathematical intuitively perceived necessary connections, which had the appearance of basic scientific principles. Thus at iv.7.5, he mentions the proposition 'two bodies cannot be in the same place'; here 'the idea of filling a place equal to the contents of its superficies, being annexed to our idea of body' makes such a proposition 'self-evident'.

[4] But it must be said that Locke never suggests that 'experience' *can legitimise* an *inference* from one to the other, such as Hume's critical comments on this argument seem to imply. (Cf. *Enquiry*, iv.2, op. cit., pp. 36–7.)

[5] Cf. Hume's argument, *Enquiry*, iv.2, op. cit., p. 35; *Treatise*, i.3.14, Selby-Bigge ed., pp. 161–2.

the realm of ideas and the realm of 'external objects' as the causes of these ideas. Although it is true, according to Locke, that 'the mind' has no 'immediate' knowledge of things, 'but only by the intervention of the ideas it has of them',[1] he nevertheless continues to treat the distinction between the 'epistemic' and 'constitutive' realm as perfectly significant,—unlike Hume, who was to argue that

> since nothing is ever present to the mind but perceptions, and since all ideas are deriv'd from something antecedently present to the mind; it follows, that 'tis impossible for us so much as to conceive or form an idea of any thing specifically different from ideas and impressions.[2]

At any rate, as long as one operates with the distinction between these two realms, it seems significant that whilst we have not so far discovered particular relations between certain given *ideas*, these relations may nevertheless *exist* between the 'objects' at the constitutive level.[3] We shall presently consider in greater detail Locke's 'theory of perception' that lies behind the distinction, and which presumably—in Locke's eyes—supplies further substance to the contention that our lack of knowledge of necessary connections is not simply a 'logical' matter, but denotes a real deficiency. But before turning to this, we must note that there plays into the epistemological argument yet another one, which though fed partly by epistemological considerations, primarily hinges on scientific and methodological matters, and which causes Locke to regard the deficiency just referred to as being of an empirical nature. The distinction involved is indeed itself a composite affair; it concerns, first, the dichotomous pair 'real

---

[1] iv.4.3.

[2] *Treatise*, i.2.6, op. cit., p. 67. Phenomenalists have wrestled with the problem of an 'external existence' ev(r since. The most common solution is to sanction the significance of the distinction between object and idea, by 'constructing' the former from the realm of the latter, in one form or another, whilst preserving certain features which do not allow of 'reduction', through various metaphysical entities, such as Berkeley's God, or Kant's 'thing-in-itself'. Alternative methods either adopt the distinction between the 'epistemic' and the 'constitutive' realm as irreducibly basic, unaffected by epistemological criticism (in the manner of Locke, and, after him Kneale); or, in the manner of the modern phenomenologists, treat the idea of the external object as an irreducible item of one's cognitive situation. We shall return to the question in Locke's theory of the external world.

[3] This would be on the lines of Kneale's argument in *Probability and Induction*, where it is held that while it is not possible to speak of necessary connections between 'sensa', such connections may exist at the 'perceptual object' level, where a 'perceptual object' is the kind of thing of which Kneale—with some hesitation— seems to deny our having any knowledge. (Cf. op. cit., §18, p. 88; and below, ch. VI, pp. 365ff.)

essence'—'nominal essence'; secondly, the 'primary—secondary qualities' distinction. Finally, 'in the wings', so to speak, we cannot leave out of account Locke's teaching concerning 'substance', in the sense of 'substratum'.

(a) *Atomic structure and 'real essence'*

The scientific doctrine which Locke takes over from his contemporaries is a corpuscularianism of the atomistic kind, particularly as it appears in the writings of Boyle; what the latter called 'the corpuscularian doctrine of qualities'.[1] According to Boyle's theory, the various 'real phenomena of nature', are due to, and can be explained by reference to, the great variety of particles of which all matter consists, and which are 'too small . . . to be sensible'. The properties of these particles which are alone required for this explanatory scheme are size, shape, order, texture and solidity, as well as motion; they are 'so simple, clear and comprehensive' as being 'applicable to all the real phenomena of nature'.[2] Boyle thinks of these particles, when 'associated into one visible body' and being 'put into motion', as *producing* 'new qualities in the body they compose', examples of which are 'heat', 'cohesion, fluidity, firmness', 'corrosiveness', 'gravity', and 'colour'.[3] Thus, heat is explained as rapid motion of the constituent particles; fluidity and firmness, by the shape of the constituent particles; 'whiteness . . . considered as a quality in the object', by reference to the surface texture of the body which reflects the rays of light that fall on it.[4]

Several points must be noted about this theory. Matter and motion are once again the 'primary principles', together with the small set of the other mechanical properties, already mentioned. The rest of the 'real phenomena' or 'real qualities' are derivative, in the sense that they can, according to Boyle, be 'explicated' by the former.[5] The main reasons given for the choice of his primary principles is that they are

[1] Cf. Marie Boas Hall, *Robert Boyle on Natural Philosophy. An Essay with Selections from his Writings.* Indiana, 1965, p. 239. Cf. above, ch. II.2, pp. 50f. This doctrine is not only Boyle's; his is, however, a good example of seventeenth century corpuscularianism.

[2] The Cartesian tone of all this is unmistakable, except for the inclusion of solidity.

[3] Op. cit., pp. 208, 213, 215. The sting of Boyle's thesis is directed against the Aristotelian doctrine of 'substantial forms', according to which these qualities 'proceed from' such forms, or from 'any other causes but mechanical' (p. 234).

[4] Op. cit., pp. 243, 246, 256.

[5] It is in this sense, as he explains, that he tries to 'demonstrate' how 'the qualities of bodies . . . proceed' from mechanical causes. (Op. cit., p. 234.)

the 'fewest possible', the 'simplest', 'universal and mathematical', and that they are 'intelligible', in the sense that we can see 'how they operate' in producing the derivative phenomena. Secondly, the theory unblushingly involves reference to 'unobservables', viz. the 'minute and insensible corpuscles', since they are 'too small' to be seen by the naked eye.[1] As a result—and Boyle is perfectly clear on this—the theory is hypothetico-deductive;[2] the explanation is only a 'possible' one; but it is, says Boyle, 'sufficient'. And though the 'corpuscularian doctrine of qualities' is only an 'hypothesis', it is an hypothesis which renders 'intelligible' a vast number of effects 'without crossing the laws of nature, or other phenomena'; nor are there any which visibly disagree with it. And finally, there are many experiments which 'although they be not direct proofs of the preferableness of our doctrine, yet they may serve for confirmation of it'.[3]

However, as we saw in our discussion of Descartes, a hypothetico-deductive schema is compatible with a number of philosophical positions respecting the status of the 'theoretical entities' involved. In recent years, theoretical terms have been taken frequently to have the sole function of entering as partially interpreted symbols into the axioms of some given theory, from which, with suitable interpretations, empirically testable statements can be derived.[4] But the seventeenth-century scientists almost certainly treated them as denoting 'real, and existing objects'; and indeed, in the Cartesian vein, regarded them, by a sort of 'metaphysical extension' as forming the fundamental material furniture of the world.[5] In other words, from saying that geometrical properties are the most suitable for giving an explanation of the majority of physical phenomena (qualitatively and quantitatively), they turned to regard the explanatory entities involved as, in some sense, the only 'real' or 'primary' ones, from which the rest, or 'secondary' ones, might 'proceed', or be 'produced'. On the former view, the difference between the 'primary' and the 'secondary' properties is no more than a formal, or logical matter; on the latter, it becomes an ontological question. For the problem which here arises —in the eyes of the seventeenth century thinkers—is where precisely

---

[1] Op. cit., pp. 208, 213, 191–7.   [2] For this, see above, ch. III, pp. 102–3.
[3] Op. cit., pp. 234, 239.
[4] Cf. E. Nagel, *The Structure of Science*, op. cit., chs. 5–6.
[5] A good historical account of the various stages of atomism,—metaphysical, realist, positivist and instrumentalist, may be found in G. Bachelard, *Les Intuitions atomistiques*, Paris, 1933. For Locke's 'realism', cf. also M. Mandelbaum, *Philosophy, Science and Sense Perception*, Baltimore, 1964, ch. 1.

the 'secondary' properties are to be located—to what 'ontological domain' they belong. Indeed, in a sense, the 'emergence' of the secondary properties seemed something of a 'miracle'. Having banished 'substantial forms' which precisely had furnished a response to the requirement of 'causal continuity', the philosopher was now faced with the embarrassing question of the 'origin' of all the secondary properties—or at least most of them.

It is important to realise, however, that the resultant problems are (at least partly) due to the 'metaphysical interpretation' of the hypothetico-deductive scientific theory. Consider, by contrast, the 'functional' or 'instrumentalist' approach to explanation. Here, too, we may begin with entities specified solely through their 'primary' qualities, e.g. mass, velocity, acceleration, etc. The derivation of laws concerning other than these 'fundamental quantities', e.g. the relation between pressure and temperature in a gas, is made possible then through the addition of certain 'rules of correspondence',[1] which *correlate* the theoretical quantities (e.g. mass, velocity, etc., and functions of these) with empirical quantities, such as temperature. Such a scheme asks no questions concerning the 'status' of the entities concerned. If it is silent about properties of its explanatory entities other than the fundamental ones, this need not necessarily be construed as a *lack* of such properties. Only the 'metaphysical approach' converts the lack of mention of other than primary qualities into a lack of possession.[2] Again, as regards some of these 'derivative' properties, for instance temperature, such a method is neutral towards the question of its 'status'. It assumes this property as 'given', or rather, as defined in an experimental context. Nor is this correlational procedure limited to 'quantitative' properties like 'temperature'. It is perfectly possible to apply it to more purely 'qualitative' aspects, such as colour. Thus, an optical theory may contain theoretical statements concerning frequencies, velocities, optical media, wave lengths. By adding certain correspondence rules correlating frequencies of a certain sort with specific shades of colours (as 'seen') of a certain sort, the required empirical relationships can again be derived.[3] It only remains to remark that the correspondence rules are neither definitions, nor '*a priori*' in any usual sense; rather, they are 'functional', for they are inserted in order to make the

---

[1] Cf. above, ch. III, p. 102.   [2] Cf. above, ch. II, pp. 50f.
[3] For a useful example, cf. N. R. Campbell, *The Principles of Electricity*, London, 1912 ch. 4, pp. 71–4.

deductions possible. On the other hand, one would usually require that the rules should not be purely arbitrary; they will be expected to be an integral part of the complex web of empirical scientific experience; more importantly, they will be expected to exhibit relevant analogies.[1] Therefore, although the presence of analogy may seem to make the correspondence rules relatively 'obvious', they certainly do not exhibit the feature of intuitive transparency on which we find Locke insisting as a requirement of our perception of necessary connection, and with it, of scientific knowledge in his sense.

Altogether this 'functionalist' approach to theory, together with a 'realist' interpretation of the explanatory entities, admirably represents one of the strains in Locke's thinking which we sought when trying to understand his motives for treating absence of 'genuine scientific knowledge' not just as an 'epistemological' matter, but as an empirical, or at best, methodological deficiency. For Locke works throughout with this corpuscularian scheme; we shall therefore expect his scepticism with regard to the possibility of scientific knowledge partly to be interpretable, first, in terms of his views concerning the epistemic status of the hypothetical 'axioms' of our theory; and secondly, by reference to some of the logical features present in any such explanation schema as we have just noted.

Now there is a sense in which the 'derived' laws of any hypothetico-deductive theory may be said to involve 'necessary connections': they will hold with necessity, if they are validly derived from true explanatory axioms ('hypotheses'). This case occurs explicitly in Locke, when he writes that if we knew the 'real essence', i.e. the atomic constitution, of a body, 'the properties we discover in that body would depend on . . . and be deducible from it, *and their necessary connection with it* be known'.[2] But it is clear that one might want to stipulate a variety of conditions before reaching a satisfactory explanation. The minimum condition is that the 'explanandum' should be validly derivable from the 'explanans'. A stronger, additional condition is that the explanans should be known to be true independently of its explanatory strength with respect to the explanandum in question.

[1] On this whole question, Cf. M. B. Hesse, *Models and Analogies in Science* (London, 1963), ch. 1. For Locke's specific reaction, cf. below, pp. 225ff.

[2] ii.31.6; my italics. This is also the first of the two criteria for 'necessity' of laws, which Braithwaite stipulates in his *Scientific Explanation*, p. 302, where a generalisation is said to be lawlike if it 'occurs in an established scientific deductive system as a deduction from higher-level hypotheses. . . .' Cf. above, ch. II, p. 46f.

The strongest condition is that the explanatory axioms should be known to be necessarily true.[1]

If we bear these simple logical points in mind, many of Locke's remarks throughout the *Essay* receive an obvious interpretation; and they will, moreover, be seen to be quite independent of the epistemological questions—though it would be too much to expect that they will not, at the same time, also throw their shadow on these.

One of Locke's most frequent remarks is that if we knew the atomic constitution of a body or physical substance, then we should know why the observable ('sensible') qualities coexist in the way they do: in short, we could see them coexist necessarily in the way they are now merely observed to do in experiment. Let one passage stand for many.

> I doubt not but if we could discover the figure, size, texture, and motion of the minute constituent parts of any two bodies, we should *know without trial* several of their operations one upon another, *as we do now the properties of a square or a triangle.* Did we know the mechanical affections of the particles of rhubarb, hemlock, opium, and a man, as a watchmaker does those of a watch, whereby it performs its operations, and of a file, which by rubbing on them will alter the figure of any of the wheels, we should be able to tell beforehand that rhubarb will purge, hemlock kill, and opium make a man sleep, as well as a watchmaker can, that a little piece of paper laid on the balance will keep the watch from going, till it be removed. . . . The dissolving of silver in *aqua fortis,* and gold in *aqua regia,* and not *vice versa,* would be then perhaps no more difficult to know, than it is to a smith to understand why the turning of one key will open a lock, and not the turning of another.[2]

The internal corpuscular structure of a given class ('sort or species') of bodies Locke usually refers to as the 'real essence' of that class. And so 'the real essence is that constitution of the parts of matter on which these qualities and their union depend';[3] i.e. the union of

---

[1] Cf. above, ch. II, pp. 35–41. And we have seen in ch. III how variably Descartes interprets these conditions in different parts of the sciences.

[2] iv.3.25; italics mine.

[3] iii.6.6. Locke recognises also a second signification of the term 'real essence', which is that of a 'form' or 'universal', in which—as Locke puts it, echoing the Platonic formulation—'natural things', i.e. a class of particulars agreeing in a given number of properties, such as an animal species, 'every one of them partake', and 'according to which . . . [they] are made' (iii.3.17). Here, again, Locke emphasises that such 'essences' cannot be known, and are useless for the task of defining species, but it will be seen that this question raises quite different, more metaphysical issues, connecting, as it does, with Locke's doctrine that 'universality belongs not to things themselves, which are all of them particular in their existence', from which he infers that 'general terms', and the corresponding 'abstract ideas' are

sensory qualities which observation, experiment and induction teach us normally to coexist. And here lies a source of our ignorance: we lack the required 'knowledge' of this real essence.

> It is evident the internal constitution, whereon their properties depend, is unknown to us. For to go no farther than the grossest and most obvious we can imagine amongst them, what is that texture of parts, that real essence, that makes lead and antimony fusible; wood and stones, not? What makes lead and iron malleable; antimony and stones, not?[1]

Or again, at iv.3.11:

> Not knowing what size, figure, and texture of parts they are, on which depend, and from which result, those qualities which make our complex idea of gold, it is impossible we should know what other qualities result from, or are incompatible with, the same constitution of the insensible parts of gold. . . .

Here, the necessitarian ideal is again very plain, but it is visualised in terms of the model of deducibility of observable qualities from atomic theory.[2] Evidently, Locke is opting for the view here that a genuine science would have to possess *knowledge* at least of the *factual truth* of its explanatory hypotheses; it is not enough merely to *assume* them, however much they may be capable of 'accounting for the phenomena'.[3] And indeed, as will be found presently, he subscribes for the most part to the stronger requirement, that we need to know the necessary connections involved in such hypotheses. At which point of course the ideal of intuitive knowledge insinuates itself confusedly into the discussion. In the chapter on the 'extent of human knowledge' these deficiencies are in fact listed together, described there as 'want of ideas' and as 'want of a discoverable connection between the ideas we have'.[4]

'creatures of our own making', albeit under the guidance of the observed resemblances between the particulars (3.11–13). Here, again, the logical nature of our 'ignorance' is quite different from that of our ignorance of the corpuscular structure of substances, which is an empirical, scientific, question.

[1] iii.6.9.

[2] I harp on this fact because Locke's scepticism has mostly been interpreted solely in the light of his theory of ideas. The true account is far more complex, being an uneasy fusion of the scientifico-methodological with the epistemological problems. *Per contra*, the methodological difficulties will give the appearance of creating difficulties for the epistemological account which are far from what they seem.

[3] I say that this is his view *here*; we shall presently see that he has more than one conception of the function of hypotheses in science.

[4] iv.3.22.

It is in part because of the 'want of ideas' that—as Locke exclaims —'our ignorance [is] great'. But the reasons given for this want of ideas are perfectly straightforward empirical ones. Although we operate with the notions of 'bulk, figure, and motion' in general, we are ignorant of the 'particular' information required in order to specify 'ways of operation, whereby the effects which we daily see are produced'.[1] But not only do we lack knowledge of 'particular motions': the main reason we cannot determine the mode of operation of the constituents of matter (and their 'primary qualities') is that they are too 'minute':

> Whilst we are destitute of senses acute enough to discover the minute particles of bodies, and to use ideas of their mechanical affections, we must be content to be ignorant of their properties and ways of operation.[2]

And in the next section he repeats that since we are ignorant 'of the primary qualities of the minute parts' of plants and animals, 'we cannot tell what effects they will produce; nor when we see those effects can we so much as guess, much less know, their manner of production'.[3]

It seems fairly clear, then, that the 'ignorance' with which we are concerned here is of an empirical kind; it is due both to the complexity of the universe and our inability to acquire sensory evidence of the atomic constituents of matter. Locke does not say—as has sometimes been argued—that we lack knowledge of the 'real essence' of bodies *in principle*;[4] the most that can be admitted is that in so far as

---

[1] iv.3.24. With this should be compared the more sanguine attitude of Descartes. Cf. above, ch. III, pp. 96f. 118f.

[2] iv.3.25.   [3] iv.3.26. The same point is repeated at 6.11.

[4] The reference to 'the supposition of essences that cannot be known' (iii.3.17) is hardly decisive, being probably directed against the notion of 'essence' as 'substantial form', which involves considerations belonging to the philosophical problem of universals. In his *First Letter* to Stillingfleet, Locke says that 'the real constitutions or essences of particular things existing do not depend on the idea of men, but on the will of the Creator', which is as much as to say that they are part of the fundamental furniture of the universe, and as such, brute or contingent fact. This would seem to suggest a further reason for our ignorance, since there plays into this whole question once again Locke's theory of ideas. For any essence to become an object of knowledge it must then be interpreted as being perceived as *ideas*. But quite apart from the usual difficulties with the perception of connections between ideas, it may well be that Locke thought it was of the very nature of 'real essence' as he regarded it, viz. a particular structural configuration, involving God's will, that it *could* not become an object at the level of ideas, since anything that does would and 'does depend [as he tells us in the same letter], upon the ideas of men', and is thus once more 'subjective' or 'arbitrary'.

such knowledge involves 'necessary connections' there *might* be a logical deficiency; but we have seen already that, even on this matter, Locke has no decided univocal position. The chief object of the present complaint must therefore be an insistence that scientific knowledge, in Locke's sense of that term, presupposes an independent knowledge of the factual truth of explanatory principles. It may be taken as a word of warning issued on the enthusiastic claims made by the contemporary corpuscularians.

Does this mean that he is opposed to the use of hypotheses? This can hardly be the case, if only because otherwise his unquestioning employment of the corpuscularian scheme would make no sense. In the *Essay* Locke makes, in fact, some explicit pronouncements on the function of hypotheses in science. (This occurs, significantly, in the chapter headed 'Of the Improvement of our Knowledge'.) It is perfectly acceptable, he there tells us, that in order 'to explain any phenomena of nature, [one should] make use of any probable hypothesis whatsoever'.[1] And such hypotheses are particularly important in that they 'often direct us to new discoveries'. However, what Locke censures is the attitude which misconstrues the status of such hypotheses. They stand or fall by the phenomena which they explain; and we cannot know whether they hold, apart from the organised experiments by which we test them.

Moreover, Locke goes on to say, this does not procure us genuine knowledge, or 'science'[2] (in the archaic sense of *scientia*). And the reason he gives is that we depend here totally on the making of particular experiments, instead of being able (as we ought to) to discover the 'real essences' of bodies.[3] Through this reference we are therefore once again thrown back on to the confused criterion of 'genuine' knowledge: direct observation of the corpuscular structure, so that we may know not only the truth of our premises—an empirical matter—but rather perceive the necessary connections: something which (as already shown) Locke at times seems to realise must *logically* elude us. But more: Locke in a confused fashion here interprets the ideal of knowledge as the need (*per impossibile*) for the corpuscular structure to function on lines of the older scholastic substantial forms. To insist on our 'lack of knowledge' is then a misleading way of saying that the atomic constitution necessarily lacks the metaphysical power attributed by the Aristotelians (albeit in Locke's eyes quite impermissibly) to substantial forms, as effecting

[1] iv.12.13. [2] iv.12.10. [3] iv.12.12.

'real' connections, and 'real' separations, between the different classes of natural substances.[1] It is because Locke is becoming aware that the old rationalist approaches to science are inapplicable[2] that he is more prone to emphasise the limitations imposed on the extent of our knowledge; he is then concerned to scotch *false claims*; and in the attempt to do so he tries to constrain both the hypothetico-deductive use of hypotheses, as well as nailing what he takes to be the views of the 'schools'.

Take care that the name of *principles* deceive us not, nor impose on us, by making us receive that for an unquestionable truth which is really, at best, but a very doubtful conjecture, such as are most (I had almost said all) of the hypotheses in natural philosophy.[3]

Commentators on Locke have often held the view that this constitutes a disparagement of hypothesis in science, but this is obviously not Locke's prime intention. His tone is as much an admonition not to misconstrue these indispensable logical tools of science, as it is the result of his own often misconceived comparisons with the intuitionist ideal of knowledge. Indeed, in the end it appears that Locke's placing of this ideal in the arena of his discussions becomes at the same time a means of emphasising its logical inapplicability. If this is not fully brought out, it is because logical and empirical deficiencies are constantly mixed up in the debate. However, even in our own day, when it is said that science 'only' employs the method of hypothesis, this is after all a mode of expression which itself labours under equivocation; it can be taken as a characterisation of the nature of scientific knowledge, by contrast with, say, mathematics and logic; or it can be taken to be an expression of a real deficiency.

Locke similarly employs these insights in his appraisal of Newton's method as it emerges in the latter's *Principia*, in remarks added to the fourth edition of the *Essay*. Here, his insistence is not so much that we have no knowledge of 'real essences' but that 'general maxims', i.e. general logical principles (such as 'what is, is', or 'the whole is equal to all its parts') are not, as claimed by 'Scholastic men', 'the foundations whereon any science hath been built'. Nor are they of any use in making 'new discoveries of yet unknown truths'. On the contrary, the

---

[1] Cf. above, p. 210n.4; below, p. 214n.4.
[2] As his remarks a few sections earlier have indeed already made perfectly clear, with their distinction between the 'two sorts of propositions', factual and conceptual. (Cf. iv.11.13–14; and above, pp. 186, 197, 201.)
[3] iv.12.13.

basic axioms, in such works as Newton's 'never enough to be admired book', really amount to no more than the 'finding out intermediate ideas, that showed the agreement or disagreement of the ideas, as expressed in the propositions he demonstrated'.[1] And the context shows that Locke here thinks of Newton's 'axioms' as generalisations of 'particular instances'—in which he echoes Newton himself.[2] 'For in *particulars* our knowledge begins, and so spreads itself, by degrees to *generals*.'[3] Which is not to say (Locke continues) that afterwards men may not cast their discoveries in the form of a demonstrative treatise (like Newton's), proceeding by the 'synthetic method'. Danger begins when the major axioms selected for this purpose take on a look of familiarity amounting to 'self-evidence'. For, what makes these axioms useful (apart from their systemic functions) is not this feature of self-evidence, but the fact that they are based on previous experience. And it is just because the 'maxims' are useless for this purpose, being of a different logical kind, that it is pointless for the scientific axioms to borrow fine feathers, so to speak, from the former. Once more, then, Locke's scepticism can here be understood to be more like an appeal to forswear false gods than a mere searching after an elusive ideal of rationality. Once free from this, his positive account comes to the fore.

If the method of hypothesis does not live up to, and should not be confused with, the logic of rational insight, its processes are nonetheless not simply arbitrary. If, as we have seen Locke note, hypotheses depend on confirming evidence, and thus are exposed to the vagaries of induction,[4] this can also once more be taken as a remark about what is alone appropriate in 'natural philosophy'. But then we may turn at once to the natural methods which the scientist here employs, among the foremost of which is 'analogy'. For 'in things which sense cannot discover, analogy is the great rule of probability'.[5] Thus we reason from analogy, says Locke, when we argue from the production

---

[1] iv.7.11.
[2] For an example of Newtonian 'axioms' employed as 'intermediate ideas', cf. above, ch. II.2, pp. 56f.
[3] Ibid.
[4] Cf. iv.12.9: 'The want of ideas of their real essences sends us from our own thoughts to the things themselves as they exist. Experience here must teach us what reason cannot: and it is by trying alone that I can certainly know what other qualities coexist with those' in my experience; '. . . which experience (which way ever it prove in that particular body I examine) makes me not certain that it is so in all, or any other . . . but that which I have tried.'
[5] iv.16.12.

8

of heat by rubbing together of two bodies violently, to the hypothesis that 'heat and fire consists in a violent agitation of the imperceptible minute parts of the burning matter'. Similarly, we argue from the reflecting and refracting properties of observable bodies that 'produce in our eyes the different appearances of several colours' to the assumption that the colour of bodies 'is in them nothing but the different arrangement and refraction of their minute and insensible parts'.[1]

Locke has no doubt of the powerful extent of analogy in nature, for, like so many writers of his and preceding centuries, he is a firm believer in the 'great chain of being', 'finding', as he does, 'in all parts of the creation that fall under human observation, that there is a gradual connection of one with another, without any great or discernible gaps betwixt them'.[2]

It may seem surprising that Locke could suddenly here 'discover connections'—has he not laboured our inability to do so? But in fact this denial *at one level* (the field of intuitive knowledge) seems to liberate him to operate all the more with the idea of connection at another, empirical, hypothetical and positive level. Let us note in passing that this bears a certain similarity to his philosophy of natural classification. Here, too, 'real essences' are of no avail; and such classifications of species which we manage to effect, carry on under the guidance of empirical pointers, which manifest indeed that there is a 'constant production of particular beings . . . very much alike and of kin one to another'. Yet the boundary lines which we draw are drawn in virtue of what we discover, and not of any 'insight into', or knowledge of, 'real essences'.[3] From the point of view which takes 'real essence' as its standard, our classifications are quite arbitrary, in the sense of having about them 'nothing essential'.[4] But if we attend to

---

[1] Ibid.

[2] Ibid. For the concept of 'the great chain of being', cf. Arthur Lovejoy's book of this title, especially chs. 6 and 8. Lovejoy quotes from Locke's *Essay*, iii.6.12, but the more impressive section iv.16.12 he passes by. The idea is central in the philosophy of Leibniz.

[3] iii.6.37. See above, pp. 397f.

[4] Cf.iii.6.5. Here, again, there is—as remarked above, p. 210n.4—a mixture of philosophical (metaphysical) and scientific discussion. For the ascription of 'arbitrariness' to our classifications is more radical than mere absence of knowledge of the atomic constitution of classes of substances or species could account for. It is due to Locke's apparent 'conceptualist' position, according to which 'universality belongs not to things themselves', since these are all 'particular in their existence'. (Cf. iii.3.6: 'All things that exist are only particular'; and 3.11 and iv.17.8 contend in the same vein for 'ideas'). Now the metaphysician seems to believe that 'the

the qualities which we *observe* to be 'always joined and existing together', we may use such hints to fashion a classification which is (in another sense) not at all 'arbitrary'.[1] In other words, Locke is here working with three levels at which to view classification: 'essential', 'natural', and 'arbitrary' or 'artificial' (the latter in the technical sense of the term, employed by botanists and zoologists).[2] If we forget the fact that he is all the time operating at these different levels, we shall continually meet with seemingly contradictory statements. Instead, for Locke the different levels hold one another in check; and it is only if one or the other is abandoned—as happens in the philosophies of his successors—that the need for further constructions becomes obvious.

(b) *Substance as substratum*

The longer we study Locke, the more we see that the notes of scepticism and of empiricism go hand in hand, and that behind this there lies the variety of dichotomous levels with which he is still operating. Some of these we have now studied in their relevant contexts: the levels of 'real essence' and 'nominal essence'; theoretical entities and the hypotheses concerning their mode of functioning, and the derivative sensory phenomena which they are supposed to explain;

---

essence of a species rests safe and entire' only in virtue of the 'real existence' of such an essence, holding it to be 'ingenerable and incorruptible'. And only with such real universals would the boundaries of a species be eternally fixed. But, argues Locke, such incorruptible 'essences' can only be 'taken for ideas established in the mind'—they are nothing but 'abstract ideas', garnered essentially from our inspection of the 'nominal essences'. Nor is there anything 'eternal' about the 'real constitution of things' (in the scientific sense), for this dwells entirely in the field of the 'particular', with which it stands or falls; at best, it is a result of God's will.

As we shall see, it is not until Kant's work that the scientific and the metaphysical strands in the argument concerning 'real essences', and hence the 'constitutive level', are disentangled. Locke, as usual, whilst undermining the metaphysical status of the notion, still continues to use it as expressing an epistemological ideal, underpinning his own conceptualist position.

[1] iii.6.28; cf. 6.36–37.

[2] 'Arbitrary', that is to say once more, by comparison with 'natural' classification, just as the latter is 'arbitrary' by comparison with 'essential' classification. There is a parallel here to the philosophy of induction: Humean uniformities are called 'accidental' by reference to a standard which insists that genuine uniformities of law must involve necessary connections. On the other hand, there is clearly also a sense of 'accidental uniformity', which contrasts with those uniform associations that seem to be perfectly invariable and unconditional. And just as in the taxonomic case, associations and meanings from one level easily insinuate themselves into the others. The progress of philosophical self-consciousness may be identified with increasing awareness of the existence of such levels.

the primary qualities viewed as standing in some form of causal relationship to secondary qualities. Others we have yet to consider in more detail, e.g. the substance–attribute distinction; and the juxtaposition of external object with idea. Some of these stand in close relationship to others; thus the real essence is in some way a sort of metaphysical extension of the theoretical entities level. And the distinction between the external object and the idea, as well as that between real and nominal essence, mirrors aspects of the division between what we have called the constitutive and epistemic levels, at least in so far as Locke regards the nominal essence as 'complex *idea*'.[1]

I have spoken of the 'metaphysical extension' involved in the notion of 'real essence'. Locke's scepticism with regard to our knowledge of the latter is further illuminated by his views on 'substance', regarded as substratum. Our problem was, it will be remembered, why, if we *cannot perceive* any necessary connections between sensible qualities, Locke should continue as though it made sense to assume that there might nevertheless *exist* such connections. Mathematics offered us one model; physical theorising, and the notion of real essence connected therewith, another. Now substance (*quâ* substratum) gives Locke yet another handle in terms of which to attach significance to his contention.

Sensory qualities, as far as observation goes, merely 'coexist' in any species or 'sort' of physical substance or body.[2] We have seen that Locke thinks of these qualities as 'proceeding' from the 'real essence', the corpuscularian constitution of the body, or the 'insensible particles' with their essential or primary qualities of extension, shape, size and motion. Ultimately, it therefore seems, we are always concerned with qualities. It is these which—in Locke's words—we 'observe to exist united together',[3] or which we should be able to observe in the case of the now unobservable corpuscles, were we

---

[1] Which is by no means always the case, for Locke, as we have already seen, slips easily from *idea* of quality to *quality*, and *vice versa*; and as we note from the passage at iv.12.9, p. 213n.4 above, he sometimes *contrasts* 'our *ideas* of their real essences' with 'things themselves as they exist'.

[2] Their enumeration will supply the 'nominal essence'. This use of 'substance' must be carefully distinguished from that of substance as substratum. The former is exemplified by the chemist's use, when he speaks of a substance like carbon-tetrachloride. Locke frequently cites gold as an example. To avoid confusion, where this might arise, I shall write substance with the subscripts $_p$ and $_m$ for the physical and the substratum sense, respectively.

[3] ii.23.6. Alternatively, Locke refers to such 'sensible qualities' also as 'observable ideas' (e.g. 23.3).

possessed of 'microscopical eyes'.[1] At this point Locke begins to consider the whole matter by way of a certain amount of 'logical analysis'. Although we 'observe qualities',

> when we speak of any sort of substance, we say it is a thing having such and such qualities; as body is a thing that is extended, figured, and capable of motion; a spirit, a thing capable of thinking.[2]

Now this form of speech, and of the grammar of our language, suggests that there is something besides the qualities, viz. substance$_m$.[3] The qualities are variously described as inhering in, being supported by, existing in, resting in, the substance$_m$. Moreover, thinking of qualities in the way we do, the notion of substance is evidently needed; and its main logical function is clearly brought out when Locke says that without substance$_m$ 'we cannot conceive how they should subsist alone'.[4] This 'substance in general' (substance$_m$) is what 'makes the whole combination of simple ideas subsist of itself'.[5] 'Every one who understands the language'[6] must think in this way: Locke is not too concerned about the ontological justification of the implications of our ways of talking and thinking. Most commentators seem curiously blind to Locke's intentions here; debates as to whether he subscribes to substance$_m$, or whether he dismisses it, have been endless. Now it is true that he continually insists that we 'do not know what it is', when we refer to substance$_m$.[7] This is because it is supposed to be 'something besides the extension, figure, solidity, motion, thinking or observable ideas'.[8] But Locke's account of knowledge and the understanding, based, as it is, solely on a foundation of 'ideas of sensations' (which correspond to sensory qualities), and whatever emerges through 'reflection' on such ideas, cannot easily cope with such category concepts as substance, causation and the

---

[1] ii.23.12. In these contexts, Locke's 'realism' with respect to the theoretical entities of his corpuscularianism is evident.

[2] ii.23.3.

[3] It is interesting to see how, in the passage just cited, we slide from substance$_p$ to substance$_m$!

[4] ii.23.4.   [5] ii.23.6.   [6] Ibid.

[7] His references to 'the word substance . . . an uncertain supposition of we know not what' (i.4.19), and similarly at ii.23.2, are as classical as they are misleading.

[8] ii.23.3. Similarly at 23.1, 'not imagining how these simple ideas can subsist by themselves, we accustom ourselves to *suppose* some *substratum* . . .' (first italics mine). Is Locke implying by his language that we 'suppose' this falsely, or confusedly; or do we do so necessarily; or perhaps necessarily though by way of a kind of 'transcendental illusion'?

like: a fact which he describes by asserting that 'we have no clear or distinct idea of that thing we suppose a support.'[1]

This last remark is a fascinating specimen of a head-on collision between Descartes' and Locke's philosophical thinking, and it illustrates well those shifts in position, which—as I have contended throughout—make up an important part of the history of philosophy. For, contrary to Locke, Descartes had held that we do have 'clear and distinct notions or ideas ... of conscious substance ... and of corporeal substance'.[2] This seems a straightforward, and almost arbitrary clash, for Descartes gives much the same account of substance$_m$ as does Locke. *Reply to Objections* II defines 'substance' as 'a subject' in which 'there reside' certain properties, that we perceive, and as anything 'by means of which there exist' these properties.[3] So we perceive as such only the qualities; and indeed Descartes admits that 'substance ... in itself has no effect on us'; 'we have no immediate knowledge of it';[4] it is only from 'the mere fact that we perceive certain forms or attributes which must inhere in something in order to have existence, [that] we name the things in which they exist a substance'.[5]

But how in that case can we be said to have a clear idea of substance? Descartes' answer is simply:

from any attribute we readily apprehend substance, because of the axiom that a nonentity can have no attributes, properties, or qualities. From perceiving the presence of an attribute we conclude to the necessary presence also of some existing thing or substance to which it may be attributed.[6]

It is not obvious how clarity and distinctness can follow from this argument, which resembles very closely that of Locke. But Descartes goes on to emphasise that the 'nature and essence' of body is extension; and the context seems to suggest that it is in virtue of this that bodily substance is a clear and distinct idea. If so, this must be due to his occasional identification of material substance and extension.[7] In the end, therefore, Descartes' different conclusion is due again to his

---

[1] ii.23.4.   [2] *Principles*, i.54; Anscombe, p. 193.   [3] H. & R., p. 53.
[4] *Principles*, i.52.   [5] *Reply to Objections* IV; H. & R., p. 98.
[6] *Principles*, i.52; Anscombe, p. 192. It should be noted that according to the teaching of the *Meditations*, this conclusion depends on the *Cogito* and on the argument for the existence of God and the external world.
[7] As remarked above, ch. III, p. 91.

preoccupation with the place held by mathematics, and his consequent 'confusion' of sense and thought. *Per contra*, Lockes' denial
of the clarity of the idea of substance must be his way of insisting on
keeping these things distinct—a fact we have already noted—and on
focusing with greater insistence on the 'empirical', observational,
side of the basis of our knowledge. And though his ideals of genuine
scientific knowledge are as elevated as those of Descartes, he will
refuse to seek their satisfaction from the mathematical aspects of
knowledge.

However, it does not follow from all this that Locke has to hold
that there is no such thing as substance$_m$ for the simple reason that
to say 'substance$_m$ exists', or 'substance$_m$ is observed', is nonsense.

Locke's notion of substance$_m$ is of interest because it provides a
possible model for the 'bond' or 'tie' involved in the subject–predicate
proposition, and which in the second chapter I marked out as an
important guiding thread for this history. At the same time, we shall
see how the development of Locke's 'idealism' creates growing difficulties for such an interpretation.

There is an ambiguity about the notion of 'thing'. In the sense in
which a thing is taken to *be* a material object or a person, say, it is
characterised by the S–P proposition as a whole. On the other hand,
in the sense in which Locke refers to thing at 23.3, viz. 'a body is a
thing that is extended . . .', thing is more like the variable x, in the
locution of modern logic when we say 'there is an x such that x is extended'. What Locke implies is that in this sense the S, as well as the P,
in S–P, is an incomplete expression when taken by itself. No wonder
that Locke describes 'substance', when taken in the sense of S,
as something known only confusedly or not at all. But it is queer that
he should have gone on to talk as though the predicate-quality P is
'knowable' without any further ado. What is knowable is always only
'some*thing* (S) which *is* such and such (P)'. Likewise, Locke is silent
about the ontological status of P, in the sense of uninstantiated
quality. The reason may well be that, when reflecting on this matter,
he takes the 'particular substance' as the fundamental existential unit.
To say that substance$_m$ makes it possible for the qualities to 'subsist
by themselves', and 'independently', is not something to which he
attributes much 'ontological significance', nor something raising the
kind of ontological questions which Leibniz was involved in when
discussing the problem of the transition from the 'possible' to the
'actual'. All he limits himself to is a reference to substance$_m$ as the

'cause' of the 'union' of the group of qualities making up a particular substance$_p$.[1]

It should be noted that it was slightly misleading to identify 'substance$_m$' with 'thing' in the sense of 'S', since the notion of 'substance$_m$' stands for the fact that a certain conjoined number of qualities are instantiated, i.e. that there *is* an S which is P_1, P_2 ... etc. If for a moment we think of qualities as non-substantial universals (leaving aside the question of their logical or ontological status), we might then think of substance$_m$ as a kind of 'active force'[2] or 'creative ground', responsible for the 'actualisation' of the qualities.[3]

This is a somewhat Leibnizian interpretation; and Locke's sceptical tone suggests at the very least that he realises that this is only a logician's way of regarding the matter, although with the analysis given it is certainly senseless to ask whether substance$_m$ 'exists' or 'is knowable'. Locke may well have intended to imply that this was no more than a way of *thinking* about qualities in relation as characterising a particular substance, and that the distinction between S and P was only a grammarian's abstraction. However, in that case (as already remarked) Locke should have drawn the consequence and acknowledged an equal scepticism with regard to the notion of quality. The reason he does not is to be sought in his habitual fundamental slide from 'quality' to 'idea of quality', where the basic axiom is that 'the ideas existing in his own mind, ... are truly every one of them particular existences'.[4] When Locke thinks of it in this way, 'particular substances' are no longer part of the 'ultimate ontological furniture', and are replaced by 'ideas', themselves composed of 'simple ideas', as we shall see later. But with this, the older notion of substance$_m$ becomes inapplicable. True, at first 'ideas' are quasi-adjectival and mind-dependent (as in Berkeley), but as the story unfolds (e.g. with Hume),[5] 'ideas' (or their equivalents) grow ever more 'independent' and 'substantial' causing the older notions of

---

[1] ii.23.6. Cf. below, p. 221, for the complete passage.

[2] As Locke's use of the term 'cause' suggests. This model is used by Leibniz. Cf. below, ch. VII, sect. 3(a): ii, pp. 409. 419–21, 424.

[3] Locke does not seem to have asked himself whether substance$_m$ ensured at the same time the 'individuality' of the actualised entity, or whether there could be two actual entities differing numerically although characterised by identical qualities.

[4] iv.17.8; and cf. iii.3.6: 'All things that exist are only particulars', suggesting implicit similarities between the 'philosophical grammar' of 'thing' and 'idea'.

[5] Similar tendencies may be found in later philosophers such as Ernst Mach and some aspects of Bertrand Russell's doctrine of Logical Atomism.

'substance$_m$' to become correspondingly shadowy,[1] and with it, this particular way of interpreting the notion of a 'bond'.

On the whole, then, Locke is simply drawing attention to the fact that in respect of material 'sorts of substances$_p$' as *things* or *bodies* existing, the logic of our language demands that we think of the corresponding qualities as embodied or instantiated;[2] whilst at the same time he is pointing out that the concept of substance$_m$ which this involves does not denote an entity which can be used epistemically for the delineation of 'particular sorts of substances$_p$'. For this can be done only through an inductive study of those groups of qualities which we find in nature. On the other hand, there is nothing which suggests that Locke had made up his mind on the ontological implications of his operating with the concept of substance$_m$ at all. Such phrases as he employs certainly suggest that he was quite prepared to *talk* as though he agreed to the metaphysical extension involved, as when he writes:

> Whatever therefore be the secret and abstract nature of substance in general, all the ideas we have of particular distinct sorts of substances are nothing but several combinations of simple ideas co-existing in such, though unknown, cause of their union, as makes the whole subsist of itself. [But] it is [only] by such combinations of simple ideas, and nothing else, that we represent particular sorts of substances to ourselves.[3]

So we may *think* one particular species (substance$_p$) distinct from another in virtue of that which *causes* it to manifest itself as just that combination (union, unity) of qualities, but the substantial$_m$ ground is not available to supply us with *knowledge* of how to define the particular group of qualities (the 'nominal essence') in question. It is sufficient to remark that here is another context by which to grasp Locke's sceptical complaint about our 'mere' knowledge of the 'nominal essences' of particular substances$_p$, for in these implications of his thinking about the logic of 'quality' he finds another means of introducing additional epistemological levels, through which he can attach some meaning to what he conceives the 'deficiency' to be. If

---

[1] Thus for Kant, substance$_m$ is not the '*cause*' of the unity of 'appearances' (here corresponding to the sensory element in Locke's 'ideas of qualities'), but the '*condition*' of our 'experience', i.e. of 'the possibility of all synthetic unity of perceptions'; something 'in relation to which alone all time-relations of appearances can be determined'. Cf. K.214, 213.

[2] Lewis Carroll's 'grin of the Cheshire cat' can *not* remain behind after the cat has disappeared!

[3] ii.23.6. Cf. also iii.6.21, with its recurring phrase '. . . cause of their union'.

there *were* such a substantial ground of the combination of qualities as this analysis suggests, and if we could have an independent insight into it, then we should possess that superior source of certainty after which a 'true science' is searching.

Nevertheless, Locke offers no indication of his actual position. He may be saying that there is such a ground but we have no insight into it; or he may be saying that the *thought* of such a ground, and what it would give us, is a pointer to what a true science would be like, although it is *only* a thought. And just because of this, anyone who claims through some verbal manoeuvres, employing the doctrine of 'substantial forms',[1] any special kind of knowledge is sailing under false flags. Once more Locke's scepticism would be employed to demarcate the true processes and necessary limitations of scientific knowledge. He would be saying that we are not to be misled by any extensions of logical (and even metaphysical) considerations into the field of empirical science.

Locke, as the passage from 23.6 shows, thinks of substance$_m$ as the 'cause of the union' of the ideas of qualities (or just qualities) definitory of any particular species of things (substance$_p$); it is a 'cause', 'as makes the whole subsist of itself'. The emphasis is here on the 'existential aspect' of things, on their 'that', as against their 'what'. This point is worth mentioning, because the question is often asked whether 'substance$_m$' and 'real essence' are not really identified in Locke's scheme.[2] This is all the more tempting an assumption because he sometimes refers to the real essence as an underlying 'union' and 'root' from which 'spring' or proceed the sensory qualities. Thus at iv.3.11, after having made his usual point that our complex ideas of substances$_p$ are for the most part made up of their secondary qualities which depend 'upon the primary qualities of their minute

---

[1] 23.3. This expression artfully straddles both 'substance$_m$' and 'real essence'! Cf. next footnote.

[2] It must be admitted that on at least one occasion Locke speaks of the *nominal* essence of body as being not just 'bare extension but an extended solid *thing*' (iii.6.21, my italics). Similarly, the real essence is 'the real internal constitution of *things*', though at other times it is said to be the 'internal constitution' of the 'insensible parts' of things. There is an ambiguity here which is also implied in more modern usage, as when we speak of the 'internal structure' of an element. In each case one may refer either to the properties and relations of the elementary particles alone, or to the latter as well as the former. It will be remembered that it was this ambiguity which enabled Descartes to slide from body to bare extension. The ambiguity likewise gives Locke's definition of real essence more plausibility than it might otherwise possess.

and insensible parts', he contends that we are ignorant of the latter because we cannot know the 'necessary union or inconsistency one with another' of these secondary qualities, since we do not know 'the root they spring from, not knowing what size, figure, and texture of parts they are'. This certainly makes the two look rather similar. Against this, in his *First Letter* to Stillingfleet, Locke distinguishes explicitly between substance and real essence, when he writes:

> I do not take them [*sc.* real essences] to flow from the substance in any created being, but to be in everything that internal constitution or frame or modification of the substance, which God ... thinks fit to give to every particular creature, when he gives a being.[1]

But this may be a reference to substance$_p$, and is therefore inconclusive. Rather, the real confusion in their identification lies in its ignoring the totally different contexts and logical positions assigned by Locke to these two notions. For the 'real essence' does refer us to something possessing (hypothetical) existence, at least when it is regarded as the structure of unobservable particles. Whereas the question of the *existence* of substance$_m$ cannot even arise. Through the conception of real essence Locke means to refer to the atomic constitution of a material body, to *what* that constitution really is. But it is a 'logical' fact that where there is such a body, with such and such a constitution possessing such and such primary qualities, the fact of such a body's existence may be construed in terms of the substance$_m$-attribute logic. Nevertheless, for Locke the two concepts, though distinct, do often appear closely related, for through reference to the unknown 'cause of their union as makes the whole subsist of itself' (at 23.6) he obliquely refers to the ultimacy of the fact *that* there *is* an internal constitution. Put in Locke's theological language: that a particular species or class of substances$_p$ possesses such and such a real essence depends on the will of the Creator; but this has also an existential implication: for it makes the qualities 'subsist' of themselves. To the differences and similarities between these conceptions there correspond the different shades of meaning which we must attach to Locke's sceptical demarcation of the limits of physical knowledge; which is our excuse for having embarked briefly here on a certain amount of verbal 'textual criticism'.[2]

[1] Quoted in A. S. Pringle-Pattison's edition of the *Essay* (Oxford, 1934), p. 233.
[2] As a final caution let us add that the dichotomy of 'substance$_m$ behind its attributes' must on no account be confused with the quite different distinction concerning the 'external object' or 'external qualities' behind the ideas which are

5. IDEAS AND OBJECTS

Let us briefly take stock, by reflecting on Locke's position as far as we have examined it. We noted that the level of our empirical knowledge of matter of fact is conceived by Locke to be extremely limited in extent, and inferior in degree of certainty. His various discussions of the deficiency which this implies tell us at the same time what he conceives as necessary for the link or bond relating the terms of these propositions. The models with which Locke operates are various and of many different kinds, ranging from 'intuitive induction', and even 'analytic propositions' (the formal or *a priori* models of chapter II) *via* mathematical reasoning, and the systemic model of the hypothetico-deductive explanation schema, to the more metaphysically extended models of the ultimate particles (specified through primary qualities alone) and, finally, substance as substratum, with the fleeting suggestion that the 'real essence' of body is determined by 'the will of God'.[1]

It will be seen that—identifying the level of 'ideas' as the epistemic level—the 'constitutive level' is not univocally determined through these different models but is itself differentiated in a variety of ways. It is a measure of the narrowness of Locke's conception of the epistemic basis that he requires such a complex and differentiated set of 'constitutive levels', in terms of which to supply the notion of a 'medium' or 'bond'. And, as we anticipated before,[2] it is the problems raised by these models, and particularly their appropriateness, that supply subsequent criticism with its text, requiring ever greater sophistication in its treatment of the epistemic level, in proportion as it abandons the employment of the constitutive models as an appropriate means of mending the epistemological deficiencies.

With these broader questions in mind, we must now turn to Locke's conception of the level of 'ideas'; for we can no longer

---

sometimes by Locke said to 'represent' the former (e.g. at iv.4.4), but, more usually, to be caused by them. To this we shall return. (Cf. below, pp. 251ff.)

[1] This last suggestion (the very model of an 'ontological postulation') should not be taken too lightly, for we shall find something like it playing a focal role in the philosophy of Leibniz, for whom those determinate happenings which are not necessitated in a mathematical or logical sense, and which he calls 'contingent', are explained by reference to a 'sufficient reason', which ultimately resides in God, on Whose 'will' depend both existence and essence, though it is a will that acts in accordance with a principle of 'fitness'. (Cf. *Monadology*, articles 36, 38, 43, 46, 48. Below, p. 269.)

[2] Cf. above, ch. II, sect. 2(c), pp. 59–61.

continue, as we have, to accept his easy identification of 'quality' and 'substance$_p$' with 'idea of quality' and 'idea of substance$_p$'. Moreover, in this contrast between 'idea' and 'external world' which is here involved, we shall meet a final model through which to construe a 'constitutive element' in Locke's epistemological scheme.

(a) *The gap between theoretical structure and the empirical realm*

We may usefully introduce this topic by a reference to further 'deficiencies' (as Locke sees it) in our scientific knowledge. To this end, let us revert to our earlier characterisation of the logical nature of the hypothetico-deductive system.[1] We found there, that in order to give what is called an 'explanation' of the observable uniformities studied by the scientist, through hypotheses that involve a reference to attributes and relations of unobservable ('theoretical') entities, we require certain 'correspondence rules' which correlate the latter with the former. This is a perfectly general matter; it applies to 'operational' concepts such as temperature (which involve a great deal of sophisticated construction) as much as to colour, odour and sound, and—indeed—to the particular 'sensible quality' which temperature was invented to measure, namely 'degree of hotness'. From this logical situation there follows an interesting consequence: even if it were the case that in some way or other we had direct intuitive knowledge of the theoretical hypotheses of our corpuscularian theory, and could thus 'see' them to be intrinsically necessary; and even if it were the case (as it must be, if we are to have a valid explanation) that the derivative laws follow logically from the hypotheses, we still could not claim to have arrived at a demonstration of any necessary connections concerning the derivative laws, because (as we have seen) the deduction presupposes correspondence rules which are far from transparent.[2]

Now here we have obviously a new 'deficiency'; and we shall not be surprised that it adds further grist to Locke's 'sceptical mill'. For evidently the need for correspondence rules denotes the existence of

---

[1] Above, pp. 206f.

[2] Cf. above, p. 206. Kneale, op. cit., §29, p. 97, makes the same point, although he does not explicitly state the reason for the lack of transparency involved in our 'explanation of natural laws'. He writes: 'Although the connections *within* the world of transcendent entities posited by a theory may all be self-evident, the relations *between* this world and the world of perceptual objects remain opaque to the intellect, and it is only by assuming these relations that we can explain our laws about observables'.

an intuitional gap between the theoretical structure and the observable phenomena. So, for Locke, not only is there ignorance of the former, but a like ignorance of the connection between it and the latter. This is expressed with admirable clarity at iv.3.12:

> Besides this ignorance of the primary qualities of the insensible parts of bodies, on which depend all their secondary qualities, there is yet another and more incurable part of ignorance, ... and that is, that there is no discoverable connection between any secondary quality and those primary qualities that it depends on.

A little later he writes again:

> We neither know the real constitution of the minute parts on which their qualities do depend; nor, did we know them, could we discover any necessary connection between them and any of the secondary qualities.[1]

Why is this 'ignorance' even 'more incurable'? Might we not, at least in principle, be able to overcome it? But this is not the case; for as we shall see immediately, the methodological gap between theoretical structure and empirical world is mirrored by a corresponding gap, though of an ontological kind, between the external material world and the mind's ideas of such a world. We noted before that Locke does on occasion give some examples of our perception of necessary connections between primary qualities.[2] But from the absence of examples of transparent connections between primary and secondary qualities, we may deduce that he had a more serious reason than any merely quasi-contingent absence. The explanation is that the gap which now opens is one (according to Locke) between the insensible particles and their primary *qualities* considered as 'mechanical affections of bodies' which exist as '*things themselves*' (i.e. independently of any cognitive reference) on the one side, and the *ideas* of the secondary qualities (colour, taste, sound), which the former 'produce' in our minds.[3] That is to say, instead of speaking of secondary *qualities* being produced by, or proceeding from, primary qualities, the productive relation is now envisaged as holding between the primary qualities and our *ideas* of these secondary qualities. Locke writes:

> Our minds not being able to discover any connectionbe twixt these primary qualities of bodies and the *sensations* that are produced in us by them, we can never be able to establish certain and undoubted rules of

---

[1] iv.3.14.   [2] Cf. iv.3.14–15, 7.5, above, pp. 197, 202.
[3] iv.3.13, 3.28; and for 'things themselves', cf. iv.4.4.

the consequence or coexistence of any *secondary qualities*, though we could discover the size, figure, or motion of those invisible parts which immediately produce them. We are so far from knowing what figure, size, or motion of parts produce a yellow colour, a sweet taste, or a sharp sound, that we *can by no means conceive* how any size, figure, or motion of any particles can possibly produce in us the *idea of any colour*, taste, or sound whatsoever; *there is no conceivable connection* betwixt the one and the other.[1]

It is no longer a merely contingent fact that we lack insight into the connections between primary qualities. Much more seriously, there are no '*conceivable* connections' between qualities and our ideas of qualities. Evidently, the background to this is the Cartesian dualism between 'mind' and 'matter'. For

> these mechanical affections of bodies having no affinity at all with those ideas they produce in us (there being no conceivable connection between any impulse of any sort of body, and any perception or a colour or smell which we find in our minds), we can have no distinct knowledge of such operations beyond our experience; and can reason no otherwise about them than as effects produced by the appointment of an infinitely wise Agent.[2]

In these passages, it looks again as though what was primarily denied is 'rational insight', preventing us from 'deducing' our ideas of secondary qualities from 'bodily causes'. This relation is asserted to be no more than a 'fundamental fact'—as the reference to the 'infinitely wise agent' is meant to imply.[3] But more interesting, and surprising, it is a fact of which Locke maintains that we actually have 'experience': it is 'experience [that] shows us' that 'primary qualities' produce 'ideas of sensible secondary qualities . . . in us',[4] so this is evidently a 'dualism' placed '*within* experience', at the level of 'ideas'. It concerns a distinction between the concepts of matter and motion on the one hand, and the concepts of 'thinking about' matter and motion (e.g. imagining, perceiving, etc.), on the other. It is these that lack 'affinity'.

---

[1] iv.3.13; my italics.

[2] iv.3.28. Cf. the parallel argument in Descartes, above ch. III, pp. 99ff. At ii.8.13, Locke similarly talks of God 'annexing' such and such 'ideas of colour' to such and such 'motions with which they have no similitude'.

[3] Cf. also iv.3.29, where 'the production in us of colours and sounds, &c., by impulse and motion, . . . wherein we can discover no natural connection with any ideas we have' is ascribed 'to the arbitrary will and good pleasure of the wise Architect.' Note again the similarity with Descartes' position, above, ch. III, 99–101, 178ff.

[4] iv.3.28. But it is *only* experience, Locke means to say, not intuitive insight.

Nevertheless, Locke's language, with its suggestions of a more profound epistemological deficiency, aims at the same time in another direction. For it appears to involve a slide from the *intra*-experiential dualism[1] so far considered, to a dualism holding *between* matter (regarded as 'external' in the sense of non-experienceable) and 'mind', regarded as the 'locus' of our 'ideas', including those of matter and motion. But on that interpretation, we do not just lack 'distinct knowledge' of the *connections* between these two 'realms', rather, we are deprived of any knowledge of the 'material' realm altogether. (This slide presupposes certain views about the logical status of ideas on Locke's part, to which we shall return. Certainly, the hardening into the second type of dualism can only be avoided by preventing 'ideas' from being treated as a kind of pseudo-object—albeit 'mental' —as a sort of 'ghost in the machine'.[2])

Locke's claim, therefore, that we have 'experience' of the productive relation between atomic structure and our 'sensations' or 'ideas' involves some fundamental ambiguities. For he does after all hold in other parts of the *Essay* to the basic tenet that to have empirical knowledge, or 'experience', of objects, i.e. to 'perceive' these, is to be 'exercised' about the corresponding 'ideas'.[3] And that surely ought to imply that the contrast between 'external qualities' and 'internal ideas' can hardly be a matter of 'experience'. At any rate, Locke's position is clearly caught in an ambiguity concerning the notions 'external' and 'internal'.[4] One of the reasons for Locke's seeming inconsistencies is that the contrast between primary and secondary qualities can be (and by Locke often is) construed at the epistemic level (of 'ideas') alone.[5] It is just one of Locke's fascinating muddles that distinctions developed at one level are used to make plausible the existence of a multiplicity of additional levels.

Of course, we have seen throughout that Locke, over a large part of the *Essay*, is never very meticulous in distinguishing clearly between 'qualities' and 'ideas of qualities', using these terms pretty interchangeably, so that it is only too understandable that the present difficulty should have concealed itself. Indeed, the whole presentation of Locke's various doctrines so far discussed makes it evident how much of the discussion is carried on at the 'realist level'. Or perhaps

---

[1] Fed no doubt by the *physical* account of the interaction between other objects and the sense organs of my body.
[2] Cf. below, pp. 234-7.   [3] Cf. ii.1.5; 9.1; iv.4.3.
[4] Cf. below, pp. 241ff, for a similar ambiguity concerning the concept of object.
[5] Cf. below, pp. 233ff.

better, the very ambivalence of quality *versus* idea of quality, which has puzzled so many commentators, suggests that the logical status of the concepts that involve this ambiguity is fairly neutral to the mind–matter question. To cite another instance, it would be a great mistake to think that Locke's concept of 'substance' is to be identified with the concept of 'external object', in the sense in which the latter contrasts with 'idea', for we have noted that Locke uses interchangeably the phrases 'substance' and 'idea of substance' (subscript p).[1] Nevertheless, in the light of his explicit reminder quoted several times already, viz. that 'it is evident the mind knows not things immediately, but only by the intervention of the ideas it has of them',[2] there is at least the question, how Locke manages to construe the contrast between 'external' and 'internal', and still more, how he can speak of 'experience of external qualities', when, after all, that very phrase needs translation, in accordance with his own canons, into 'being exercised about ideas of external qualities'.

One thing at least seems fairly clear: it is precisely when the view is taken that though it is *true* (as Locke clearly believes it to be) that 'impulse and motion' produce 'ideas of secondary qualities in us', we can never have *knowledge* of this productive relation (since empirical knowledge is confined to the level of ideas alone), that Locke's point concerning the inconceivability of any necessary connection between the primary and secondary qualities gets additional teeth.[3] The levels of the material and the mental realms are now totally distinct; 'empirical knowledge' of any relation between them non-existent.

(b) *'Ideas' as concepts and as perceptual objects*

We must dig further therefore into the problem by asking how this whole story of Locke could ever have started? How can he give a quasi-empirical account of the commerce between the realm of

---

[1] For the same reason, 'substance$_p$' is misleadingly rendered as 'phenomenal object', in, say, Kant's sense, as though it contrasted with 'noumenal object'.

[2] iv.4.3; this has served as a 'motto' for the so-called 'Theory of Representative Perception'.

[3] At iv.11.2 it is made quite clear that Locke believed in the 'constitutive fact' whilst at the same time holding to our 'epistemic ignorance' of the relation: 'It is therefore the actual receiving of ideas from without that gives us notice of the existence of other things, and makes us know that something does exist at that time without us which causes that idea in us, though perhaps we neither know nor consider how it does it.' Does Locke want to say that we know *that* it does it? Remember, that at iv.3.28 he had maintained that 'experience shows' this to be the case. More likely, this is again an incurable inconsistency due to a mixing of levels.

'external' primary qualities and our *ideas* of secondary qualities (*and* —it turns out—primary as well), when as the result of that account he ought not to be in possession of any information concerning the primaries, when conceived as 'outside the mind'?

A full discussion of the complex questions thus raised ('the problem of perception') transcends the limits of this book. Nevertheless, I shall try to outline some of the points involved, since this is vital for an understanding of Locke's immediate critics. Besides, it is crucial for our purposes, since this alone can help us to grasp some of the complexities and ambiguities that surround the notion of 'idea'.

It is one of Locke's basic assumptions that the extent of our knowledge depends, as a prime condition, on the range of our *ideas*,[1] all of which are 'in our minds'.[2] This is also the Cartesian tradition: ideas are our starting points; what is not an idea, or entailed by the ideas that we have, is not to be accepted.[3] But Locke, echoing the cry of the British scientists of his time, adds further constraints; he does not allow that any of the ideas with which the mind is 'furnished' are 'innate to it'. Following the slogan that all 'knowledge is founded on experience' of the external world, there is the somewhat 'technical' parallel, according to which either all our 'ideas' are 'conveyed into the mind' by our senses, or through our reflection on the mind's operation with these sensory ideas they result in a second set of ideas, e.g. perception, thinking, etc.[4]

'Idea' is here used in a quite technical sense, in some respects broader, in others narrower than is normal. It is what a man's 'mind is applied about whilst thinking';[5] but 'thinking' includes, and is sometimes by Locke even equated with, 'perception'.[6] Since there are no ideas 'innate to mind', Locke sometimes likens the latter to something that initially is like a piece of 'white paper, void of all

[1] iv.3.23.   [2] ii.8.8; cf. 1.1.

[3] The tensions and necessary supplementations which result from this in the philosophy of both Descartes and Locke will by now be obvious.

[4] ii.1.2–4. Locke spends very little time on the detailed way in which the latter can really be said to be part of our 'experience' in the same sense as are the ideas of sensation. For an instance of how this vagueness opens the way for attack, by means of a narrower interpretation of the concept, cf. below, pp. 264f., Hume's view on 'power'. There is also the difficulty of the distinction, here slipped in, between us (or our 'understanding', cf. ii.1.4) and our 'minds', which Locke never faces, and which does not become prominent until we meet Kant's distinction between 'pure understanding' and 'empirical consciousness'.

[5] ii.1.1.

[6] ii.6.2. But cf. 9.1, where he notes an eccentricity in this usage. Cf. above, pp. 193ff., for Locke's uses of 'perception'.

characters',[1] or like an 'empty cabinet', 'a closet wholly shut from light', a 'dark room'.[2] He regards the process of perception as one in which there are 'let in ideas of things without'. Now in their normal connotation, we should be inclined to think of ideas as *concepts*. Quite frequently, of course, Locke does regard them in this way, and his meaning is roughly that in virtue of perceiving certain objects and qualities we come to acquire the corresponding concepts. He then seems to distinguish between perceptions and ideas.[3] At other times, he speaks of 'ideas *or* perceptions',[4] and refers to the example of a snowball which has the 'powers to produce in us the ideas of white, cold, and round'.[5] At other times, these powers are said to produce in us 'sensations',[6] thus suggesting that Locke here thought of ideas as 'particular occurrences', e.g. instances of colour perception.[7] This slide from idea as concept to idea as sensory particular is of the utmost importance for the development of Locke's doctrine of perception. Let us suppose I am looking at a snowball; to know that it is a snowball, I must have the requisite concept; and seeing it, I may say that the object before me answers very well to my idea of a snowball. But Locke also wants to say that the snowball (the external object) is causally responsible for certain sensory occurrences, viz. 'ideas', which are the actual particular objects of my visual perception. Now it is true that in perceiving the snowball I may *also be aware* (though by no means always) that what I perceive is the kind of thing I know to be a snowball. But it clearly does not follow from this that *what* I am then perceiving, i.e. the object of my perception, is ideas rather than the external object and its qualities.[8]

Here we again see how the ambiguity surrounding the notion of 'idea' marks the latter out as a privileged servant for a metaphysical doctrine. It has to do double duty: on the one hand it expresses the decision to define the realm of knowledge in terms of ideas alone;[9] on

---

[1] ii.1.2. [2] i.2.15; ii.11.17. [3] E.g. ii.1.3. [4] ii.8.7; my italics.
[5] ii.8.8. These 'powers' are here defined as the 'qualities' of the snowball.
[6] ii.8.10; 8.13; 8.15. Sometimes Locke is not consistent in his terminology. Sensation is often taken to be the physiological aspect which is a condition of perception, e.g. ii.1.23; but at 1.3, 9.2 and 19.1 perception and sensation are more or less identified, though contrasted with the 'impression made on the body . . . by an external object' (19.1).
[7] But we should avoid identifying them with twentieth-century philosophers' 'sense-data'.
[8] The latter doctrine is certainly asserted explicitly at ii.1.25 and 9.1 jointly, as well as at iv.4.3, and is besides implied by innumerable other passages.
[9] We now grasp the deeper significance of the definition of knowledge as perception of the agreement and disagreement between our ideas!

the other hand, its place in Locke's theory is so adjusted as to give the impression of maintaining communication with the realm of the external world. 'Idea of sensation' is constantly viewed as the *effect* of some external cause.[1]

(c) *Primary and Secondary Qualities*

But we must enquire further into the grounds which made it possible for Locke to maintain a position which appeared to deny 'direct access' to external things whilst continuing confidently to describe their relationship with our ideas. One of these is what I have called the 'metaphysical extension' of the hypothetico-deductive explanation pattern of seventeenth-century corpuscularianism.[2]

Beginning with an explanation of the variety of sensory phenomena by means of the 'insensible particles' to which the scientist had ascribed no more than the attributes of size, figure and motion, and this for the sake of convenience, simplicity, intelligibility,[3] we end by being told that the explanatory particles 'in reality' possess *only* these attributes, and that they lack all others especially 'the secondary' ones. Evidently this does not strictly follow from the theory, for from the standpoint of theorising we can only say that the secondary qualities are simply not required for purposes of explanation; it is only the application of the law of the excluded middle which forces us to decide that they either have or lack these qualities. Indeed, if we assert the lack positively, then the particles become unobservable in principle, since—as Berkeley was to point out[4]—we can hardly be capable of observing extension and shape without some at least of the

---

[1] Even where this feature is reduced to a minimum, as in Hume and Kant, the chosen terminology still betrays origins and concealed pressures and intentions; cf. Hume's term 'impression', and Kant's 'appearance'.

[2] Cf. above, p. 205.   [3] Cf. above, on Boyle, pp. 204f.

[4] *Principles of Human Knowledge*, x: 'I desire any one to reflect and try, whether he can, by an abstraction of thought, conceive the extension and motion of a body, without all other sensible qualities'. Cf. *First Dialogue*, Everyman ed., p. 224, where Berkeley maintains that although we can 'consider or treat' the two groups of qualities separated 'abstractedly', we cannot form the corresponding 'idea'. This is precisely Berkeley's way of proscribing the *metaphysical* extension of the notion of insensible particles and their primary qualities. What he means to imply is that between the realms of formal abstract theory and particular sensory reality there is no third 'metaphysical dimension'. And of course, Locke would have been unhappy to be convicted of the accusation of admitting such a realm, but it is only the sham empirical argument mentioned in the text which conceals from him the true implications of his thought. (For Berkeley's opposition, cf. below, ch. V, pp. 280ff.)

attendant qualities.[1] Surprisingly, this is precisely what Locke suggests at one point,[2] but the actual examples he gives of observations through microscopes, by his own admission, are only cases of *change* in the secondary qualities, and not their disappearance. We must conclude that Locke's realist interpretation of his theoretical entities is of the genuinely metaphysical kind. Moreover, all this leads to a sort of metaphysical sharpening of the distinction between 'external world', the 'realm of matter', and mind. For since Locke must now hold that only primary qualities 'do really exist in bodies themselves'[3] whereas secondary qualities do not, the latter must necessarily be located in a different 'realm', viz. that of the mind. Here, again, we note a kind of metaphysical extension, or slide: in the scientific, hypothetico-deductive schema, the secondary qualities are 'derivative'; but in the extended scheme, this derivativeness is interpreted as giving them an 'ideal' status.[4] This may in part explain Locke's curious conviction that there is a resemblance between our ideas of primary qualities and the qualities themselves, which does not exist in the case of secondary qualities.[5] For there can of course be no resemblance if one of the terms of the relation is non-existent.[6] Clearly, Locke's argument is simply a *conclusion* from the 'metaphysical extension' accorded to the status of primary qualities.

Indeed, there follows a more general consequence from this observation. Locke puts forward a number of arguments to show that there is an intrinsic difference between primary and secondary qualities, and intends to prove that the former alone are 'inseparable from the body',[7] whereas the latter are not. For they may fail to be inseparable in different ways: First, they may in some sense have a peculiar 'mental status'; there being nothing like the ideas existing in the

---

[1] Of course, if with Berkeley we say that to exist *is* to be perceived, our 'insensible particles' would vanish *ipso facto* except as 'theoretical entities'. But we need not make this decision concerning the meaning of 'existence', and certainly Locke would not wish to do so.

[2] ii.23.11–12. [3] ii.8.15.

[4] Boyle already holds that 'colour is not . . . an inherent quality of the object', unlike 'local motion'. (Boas Hall, op. cit., pp. 255–6). And in *The Origin of Forms and Qualities, According to the Corpuscular Philosophy*, he declares that the secondary qualities only exist as the consequence of the existence of 'certain sensible and rational beings that we call men' (*Works*, op. cit., ii, p. 466).

[5] Ibid.

[6] It may be objected that resemblance still does not follow even for the primary qualities case, but much depends on how this relation is construed, and what 'topological' transformations are allowed.

[7] ii.8.9.

bodies themselves'.[1] Secondly, the *specific* way in which they manifest themselves depends not only on the state of the body but is, further, relative to the presence of physical media (e.g. light, air, etc.), as well as a physiological mechanism on the part of an observer. Many of Locke's examples demonstrating these facts are relevant only to the second of these; and obviously this amounts to no more than a difference of degree between two sets of physical predicates, a difference which is located at a single (viz. epistemic) level. It assumes an acquaintance with both primary and secondary qualities at this level, and is thus quite neutral to the first set of differences just noted. Thus, colour, taste and odour depend upon the presence of sensory mechanisms;[2] bodies do not manifest their colour when not exposed to light;[3] specific degrees of heat sensations are relative to the thermal state of the body,[4] and so on. All this splits up again in two different ways: a difference which is blurred by the ambiguous use of the term 'idea'. Secondary qualities may not be 'in the external body' in the sense that they are 'derivative' upon its atomic or molecular state. (Cf. the physicist's definition of absolute temperature of a gas as a function of the average kinetic energy of its gaseous particles.) The second way is through Locke's emphasis on the putative *actual non-existence* of the secondary qualities in the external object. Our 'observing' such qualities can hence only be described as the having of certain 'ideas', totally differing in kind from anything in the object. As an example, a certain poison can produce the 'sensation of sickness' (Locke says, alternatively, 'idea of sickness') and of pain. But clearly the sickness or pain are not in the poison; the poison only operates on my body 'by the motion, size, and figure of its particles'; . . . pain and sickness are only 'the effects of its operations on the stomach'.[5]

Of course, the example does not at all prove Locke's contention; as has been abundantly shown in recent writings, the kind of sensations illustrated by sensations (i.e. feelings) of pain cannot be identified logically with sensations (i.e. perceptions) of, say, colour; their identification hides the eccentricity of Locke's perception of qualities as the 'having of ideas' of such qualities. But his metaphysical assumption of the non-existence of secondary qualities 'in the material bodies themselves' forces on him this identification, as much as it

---

[1] 8.15.   [2] ii.8.17–18.   [3] 8.19.   [4] 8.21.

[5] 8.18. It will be noted that the 'stomach' is just as much an external object as is the poison, relative to the 'mind'; we are here concerned only with the 'primary qualities' constitution of the stomach, as it is 'related' to the 'sensation of sickness'.

explains his continued confidence in the existence of such bodies as though it were non-controversial.

(d) *Realist and idealist alternatives*

The last example, with its talk of the poison acting through 'the motion, size and figure of its particles' on the sensory organs serves as a graphic illustration of the general direction taken by Locke's causal account of perception.[1] And this account has really far more radical consequences for Locke's theory of knowledge than the distinctions between primary and secondary qualities made so far. It involves us also in a more general critical consideration of the whole doctrine of perception—something which we postponed when we traced Descartes' response to the difficulties which it raises.[2] Such a critique cannot be delayed any longer if we are to have a balanced perspective on both the doctrine itself, and the influence which it wielded over Locke's successors.

The most important discussion of this occurs at ii.8.12. The 'external object' and its qualities are extrinsic to the mind. This is analogous to the way in which the object is also external to my body, including my brain. We must therefore assume that certain 'imperceptible bodies' (light particles) are transmitted from the object to my eyes, 'and thereby convey to the brain some motion which *produces* these ideas we have of them in us'.[3]

This sounds harmless enough, but it will be obvious from what has gone before, and particularly in the light of the ambiguous nature of the term 'idea', that the process indicated in the last passage bears a number of possible interpretations; the different interpretations making up a large part of the subsequent philosophical debate. For this reason it is necessary to stand back and contrast Locke's doctrine with critical alternatives which have been developed since his time. Developments in our own day, which amount to important criticisms of the model of the 'mind' as a separate quasi-physical entity, and of the doctrine of the interposition between observer and object of a realm of 'sense-data', as well as phenomenological approaches, all show that Locke's assumption stands in need of a far-reaching

---

[1] Although Locke elsewhere discusses the nature of causal action, particularly in his analysis of the notion of 'power' (ii.21), it never receives the critical questioning bestowed on it by his successors. Cf. below, pp. 261–5.

[2] Cf. above, ch. III, p. 101.

[3] My italics. Cf. also 8.13, 8.15; 23.7; iv.6.11, for similar accounts; and cf. Descartes' account, above, ch. III, pp. 99–101.

critique. History of philosophy, by presenting us with the beginnings as well as the apotheosis of philosophical doctrines, can contribute vitally towards their proper evaluation.

At the outset, let us note that—like Descartes before him—Locke here artfully blends a scientific and an epistemological account. Thus it is quite true that unless our bodies (bodily senses, brain, etc.) are causally acted on by external objects through the relevant media, no physiological processes will occur of the kind which we associate with 'consciousness' or 'awareness' of the external world—including our bodies. This much is a matter of empirical knowledge, based on observation, theory and inference. But Locke goes on from there to describe this situation by saying that certain given brain activities ('motions') furthermore *produce* 'ideas in our minds', of just these self-same external objects; ideas that are said to 'represent to us things under those appearances which they are fitted to produce in us'.[1]

The trouble with this development of the theory is that the 'production of ideas' by what is said to be non-ideal cannot be an 'empirical' matter. Locke himself half-recognises this when he writes that 'we neither know nor consider how this production is effected.[2] What *is* empirical is that when I close my eyes I cease to have some given 'perceptual idea' (to use Lockean terminology) of which I previously was apprised, where however—again in the language of Locke's system—the closing of my eyes (if empirical) ought itself to be an 'ideal' matter.[3] But with this, the 'external world' becomes itself 'problematic'; concerning it, we have—in Descartes' words—'no information', and can only make 'suppositions'.[4]

To appreciate the bearing of Locke's doctrine, let us ask how one might bypass these sceptical conclusions. One way is not to regard the object perceived as a 'perceptual *idea*'. It is true that to opt for

---

[1] iv.4.4.

[2] iv.9.2. Remember that iv.3.28 had made the causal relation between primary qualities and *ideas* of secondary qualities a matter of 'experience' (cf. above, p. 227). But we should not automatically assume that 'primary qualities' as such belong into the realm of 'external objects'.

[3] Berkeley and Hume of course drew one consequence of this account. Thus Hume writes that 'nothing can ever be present to the mind but an image or perception', and that there is no 'immediate intercourse between the mind and the object.' (*Enquiry*, xii.1, op. cit., p. 152.) But as can be seen from what follows, other answers are possible.

[4] *Med.* VI, K.S. p. 252. Cf. above, ch. III, sect. 3(c), pp. 322ff., for Descartes' method of tackling this problem.

such a locution is extremely tempting, for a number of reasons. Thus one may draw on the natural grammar of 'idea' in the context in which we are rightly said to 'imagine an object'—such as when we close our eyes and imagine a table which we may have been viewing a moment ago.[1] But it does not follow that 'idea' is the correct model for the case of perception. More bewitching, however, is that the production story leads Locke to treat the mind as though it were a kind of physical *camera obscura*, and as though the images ('ideas') inside it were *produced* on the lines of physiological changes that are produced by external objects, the 'self' viewing only these 'images'.[2] Here, 'mind', and 'self' seem to be quasi-separate 'objects'. Now against this one may urge that the 'subject' which is involved in all 'experience' can never *at the same time* be also an 'object', as which, one among others, it would have to be presentable in perceptual experience.[3] And similar remarks must go for 'our minds within us',[4] in which the ideas are supposed to be located. Indeed, it is clear that the Lockean locution of ideas being 'in us' or 'in our minds' is again a misleading spatial metaphor.[5] Altogether, if the epistemological notion of mind and self *as separate 'objects'* is dropped, the use of 'idea' for 'object of perception' becomes tenuous, as involving needless puzzles.[6] On the other hand, we can see how easy it was for Locke to switch from one use to the other.

What would the Lockean causal account be like when we adopt an 'objectivist' or 'realist' language? We might interpret ii.8.12 as

---

[1] It is remarkable that Locke succumbed to this model, despite the fact that he uses the 'qualitative' *difference* between 'imagining' and 'perceiving' in one of his many arguments intended to show that sometimes we are entitled to 'infer the existence' of an external object corresponding to our idea of it. For he writes that we are 'invincibly conscious' of the difference between looking on the sun by day and thinking on it by night. Cf. iv.2.14, and below, p. 241f. And note Berkeley's identical example, *Principles*, sect. 36; cf. below, ch. V, p. 295n.2.

[2] Descartes already had warned his readers of the misleading nature of this model. (Cf. *Dioptrics*, VI, above, ch. III, p. 100.)

[3] Wittgenstein, in this connection, used the simile of the eye and the visual field, pointing out that the eye itself must never be counted as part of the field. Cf. *Tractatus*, 5.631 to 5.641. The denial of the *existence* of such a 'transcendental' self (*quâ* object) goes back to Kant; cf. K.333–34, 377–78.

[4] ii.1.4.

[5] Cf. ii.3.1: 'These sensory organs, or the nerves which are the conduits to convey them [*sc.* sensory impressions] from without to their audience in the brain, the mind's presence-room (as I may so call it) . . .'.

[6] But it does not follow uniquely that, as Wittgenstein thought, 'sollipsism . . . coincides with pure realism' (*Tractatus*, 5.64). The Kantian treatment shows that a great variety of intermediate positions are still possible. (Cf. K.344–52).

follows: Our awareness of external objects presupposes certain physiological happenings in our body (brain, etc.). In virtue of these we,
as observers, are put in a state of 'consciousness', rendered in our
language by a locution according to which we 'perceive the object'.
To 'have an *idea* [of the object] in us' would be equivalent to *our*
'being apprised of the object,' to attend to it, to take notice of it.
Here, the world of external objects, including my body with its sensory
organs, together with any causal relations that hold between them,
are to be understood as things (rather than 'ideas') that we can be
aware of. (This would represent the position of many modern 'realists', as well as 'phenomenologists'. The Kantian doctrine, as we shall
see, is more complex; whilst it holds that what, experientially speaking, we are aware of is an 'external object' rather than an 'internal
idea', the 'externality' of the object presupposes the 'representation'
of the 'forms' of space (and time), which is a 'transcendental' matter,
and, in this sense, 'subjective'.)[1]

Such a 'subject–object' relation would be basic. 'Subject' would
stand here simply for the fact that *I* (as a person) am aware of (or
perhaps even only reflecting on) such an *object*—where 'object', in
suitable circumstances, e.g. in the context of perception—is a physical
member of the external world. Of course, my 'being-aware-of-an-
object' involves a 'mental fact'. And particularly if I self-consciously
report on the latter, I may describe myself as being—in Locke's
words—'applied about the idea' which is in my mind. Nevertheless,
the 'object' itself does not thereby acquire the status of an 'idea'. It
would be more correct to say that in the *general* case the item at which
'I' direct my attention as such has not yet *either* a 'mental' *or* a
'material' status.[2] What actual status it acquires depends on further
considerations. (Obviously the object which I imagine, or which I
hallucinate, cannot be 'material' in any straightforward sense.)[3]

---

[1] Cf. K.348; 245; 71.

[2] Indeed, the phrase from ii.8.15, '. . . produces these ideas we have of them [*sc.* the
objects] in us', is still as such neutral to the status of these objects. It is only Locke's
slide from this to the doctrine that we have direct 'access' only to objects in the
sense of 'ideas' which hardens the logic of the situation in a definite direction. But
it must be admitted that Locke was seldom aware of the full consequences of this
slide, or indeed that he was guilty of it. This was left to his immediate successors, as
when Hume tells us that 'nothing can ever be present to the mind but an image or
perception', and that there is no 'immediate intercourse between the mind and the
object'. (*Enquiry*, xii.1, op. cit., p. 152.)

[3] This shows again that Locke's insistence (at iv.2.14, quoted above, p. 237n.1) on
the basic difference between 'imagining' and 'perceiving' an object (e.g. the sun) does

This construction avoids some of the difficulties previously noted, where we asked how Locke can *know* that there are any external objects that produce ideas of themselves in us, when 'knowledge' is circumscribed within the realm of ideas. For Locke's account suggests that I ought to be somehow aware of the external object producing an idea of itself in my mind. Of course, we can see some obvious analogies which misled Locke into his way of describing the situation.[1] It certainly makes sense to say that one can observe (or at least, make an inference concerning) external objects acting on our bodies and their sensory organs. Again, it makes sense to say that one can observe an external object, say a happening, producing a certain *sensation* in us, e.g. of fright or anger. And again, it makes sense (if slightly strained) to say that we can observe the hammer of a bell producing a sound in us. But the first case concerns an account of physical and scientific relationships; the second is about sensations in the ordinary sense of this term (and not Locke's extraordinary one, whereby sensations are identified with ideas); and the third is not an account which is relevant here, involving as it does the relation between a visual and an auditory phenomenon. But there can be little doubt that these analogies, together with the slides mentioned in the preceding pages, contributed towards Locke's doctrine, which involves the conclusion that what we are directly acquainted with is only 'the idea we receive from an external object' into 'our minds', and that we can only '*infer* the existence of anything without us which corresponds to that idea'; which, as with Descartes, is for him a problematic matter.[2]

(e) *The 'projective model' of perception, and the question of correspondence*

There must, however, be more serious reasons which suggest to Locke his interpretation of the place of 'ideas' in the epistemological situation, and which prevent him from treating objects of perception in a realist manner. For what plays into all this is the combination of the causal account with the metaphysical extension of the scientific

---

not catch the question of the status of the object, but presupposes a prior decision. Or we might say that this *defines* the distinction, rather than *proves* it.

[1] Remember once again his contention, mentioned before, that 'experience shows us' that primary qualities produce secondary quality sensations in us. Cf. above, pp. 227f.

[2] iv.2.14; my italics. But as we shall see, Locke's way of dealing with the problem is quite different from Descartes'.

doctrine of the existence of 'insensible particles' possessing primary qualities alone. Body and the external world are exhaustively characterised through these primary qualities. Therefore, so Locke must argue, although what I *seem* to see is a world of colour as well as shape, sound as well as motion, I can surely not 'really' see or hear what is not there, viz. colour and sound. And this supplies him with an excellent reason for holding that we must translate 'observing an object' into 'observing an idea'.[1] The model resulting is the following: any knowledge of the external world ultimately is conditional upon certain 'motions' in the brain. As a matter of brute fact (by a divine fiat) these motions are correlated with 'corresponding' ideas, which will not *necessarily* 'resemble' anything in the physical situation. This leads (as remarked already) to the model of the idea as the photographic image that emerges where the photo-sensitive paper is acted on by certain chemical substances. The relation of the image to the paper is like that of the idea to the brain. And the 'self', or what Locke calls the 'understanding', might then be thought of as inspecting these image-like ideas.

Perhaps the model of the photographic 'images' is too naïve, but even if he avoided it, Locke must still have retained some of its logical features. It is then as though the self, stimulated *quâ* body, *projected* certain images beyond itself. Such a model fits certain cases fairly naturally—as when we are *imagining* an object, e.g. a scene; and so perhaps characterises certain limited aspects of the cognitive situation. As a result of an extended application of this model, 'perceiving' comes to be treated as a kind of 'imagining'. 'Imagining an object' is regarded as a unitary act, suggesting to us that the grammatical accusative need not necessarily denote anything independently existing—not even 'ideally'; a little like saying that someone sleeps the sleep of the just, or is singing a song, or dreaming (or having) a dream.[2]

One of Locke's motives for treating the 'seeing an object' as the unitary projective act of the 'having of an idea', where the idea then becomes the 'object of inspection', is that this is supposed to cope with the case that troubled Locke so much, and which he describes as

---

[1] For a clear case of such a translation, cf. ii.23.6: '. . . those qualities or simple ideas, which he has observed to exist united together.'
[2] This ignores competing alternative cases, e.g. 'grasping an object', 'shooting a bird' or 'hitting a target', in which there is both an act and an object. Why should not 'perceiving' or 'seeing' be more like these cases?

our perception 'of anything without us' when 'no such thing exists'.[1] The problem which this apparently poses is that grammatically, 'seeing' stands putatively in an accusative relation to an existing object; it has the existential entailment of an 'achievement-verb'.[2] And Locke attempts to meet the case of the non-existing object by the very tempting device of interpolating the 'idea' between the 'self' and the 'object', because it removes the logical clash between the existential implication of 'seeing' and the non-existence of the 'object', since here—as Locke phrases it—there is nothing that 'corresponds to that idea'.[3]

Unfortunately, we are now saddled with Descartes' old question of how, if ever, we can 'infer the existence of anything without us which corresponds to that idea'.[4] And this question is here the more troublesome because Locke never makes it clear in what sense (and at what level) there is supposed to be this failure on the part of the 'object corresponding to the idea' to exist. Evidently, Locke's problem arises in at least two quite different contexts, though he fails to distinguish them explicitly. The first context arises out of the 'metaphysical extension' of the scientific story, according to which there is an external world of matter in motion that 'produces' our 'ideas'. Here it must be noted that since it is a result of this account that we lack 'direct access' to anything except our 'ideas', obviously any inference to the 'cause' of these ideas is impermissible; a fact I have noted by speaking of the 'metaphysical extension' of the corpuscularian account. It was for this reason that Descartes had attempted to settle this question by falling back on the power and goodness of God,[5] meeting more clearheadedly than Locke a metaphysical problem through a metaphysical solution.

If Locke does not properly appreciate the logical nature of his problem, it is because in the sections of the *Essay* at present under review he unconsciously switches to a second and quite different context, involving considerations at the *'epistemic' level*, i.e. the level of 'ideas' *alone*. For whilst there are here sometimes fleeting hints of a more fundamental appreciation of the existence of *two* levels, with his reference to our 'invincible consciousness' of the difference

---

[1] iv.2.14.
[2] For this term, cf. G. Ryle, *The Concept of Mind* (London, 1949), pp. 151, 223.
[3] iv.2.14.
[4] Ibid. Cf. Descartes, *Med.* III and VI; K.S., pp. 218, 252.
[5] Cf. above, ch. III, pp. 166–8, 178.

between 'imagining' and 'perceiving',[1] by and large he answers his question simply by reference to the purely empirical and pragmatic differences that exist between those cases where an 'object' does, and where it does not exist when we have the corresponding 'idea': all the while describing a situation which clearly demands that the 'object' itself be regarded as possessing the status of a 'perception' (= 'idea').[2] But clearly this will not do, since no arguments which involve only a single given level (here, that of 'ideas') can relevantly bear on a distinction which involves its transcendence. This is easy to see when we realise that comparisons of perceptual situations either presuppose, or are at least neutral to, questions concerning the status of the 'object' occurring therein.

(f) The 'problem of perception': Varieties of the 'concept of object'

There are really several quite different cases involved here, to which there correspond different senses of 'object'. (1) The first concerns the contrast between object as 'idea' (O_I) and object as 'external cause of my idea' (O_C). Locke's question how we know that anything 'corresponds to the idea' might imply that the idea could logically have been generated, either by some Cartesian 'evil demon', or simply spontaneously—though no doubt Locke would not have subscribed to such an extreme position, involving the rejection of the principle of causation.[3] Closely related to O_C are two further senses of 'external object', which for Locke support the plausibility of O_C. (2) First of all there is the case of what we may call the 'theoretical object' (O_T) whose hypothetical postulation we have already seen to be involved in Locke's corpuscularianism. O_T further merges into O_M, equivalent to what I have called the 'metaphysical extension' of O_T: the corpuscles construed as existing in themselves endowed with primary qualities alone.[4]

(3) We meet quite a different case if we imagine that a certain perceptual situation is due to a change in our own bodies rather than

---

[1] Cf. the passage from iv.2.14, referred to above, p. 237n.1. The only hint that he is aware of the presence of the 'metaphysical' context occurs at iv.4.4., where he attempts to meet it by reference to the 'atomic' nature, i.e. the uncreatibility and indestructibility of 'simple ideas'. (Cf. below, pp. 251ff.)

[2] At iv.11.4–7 further criteria are adduced to characterise the difference: perception presupposes sense-organs, imagination does not; sensory ideas we 'cannot avoid' having; there is a coherence between the testimony of the different senses.

[3] As we shall see, 'simple ideas' are due to the operation of the external world; 'complex ideas' may be subjectively conditioned.

[4] Cf. above, pp. 206f., 233, 240f.

to something going on in the rest of the external world. Boyle had laid particular stress on this when he remarked that we often believe we 'see a colour' when this is due only to certain motions in the brain having been produced artificially, and not through the action of light.[1] Here the existence of an external physical object is simulated, though none is present—a fact (it is important to note) that can in principle always be discovered subsequently by myself or others through empirical means. Let us call this sense of physical object O_P. (This is more or less also the sense in which Locke speaks of 'particular substances'.) An absence of O_P will here have to be construed as a qualitative difference in respect of O_I. This case is similar to another which involves again O_P; only, here the object seemingly presented is due to sensory deception, the manipulation of mirrors, mirages, etc. In these cases there *is* an external object, though it is not what it pretends to be.

Comparing these different cases, we can see that Locke tries to settle the problems raised by (1) through references to (2) and (3). In particular, using (3) he may maintain that in the absence of any epistemic foundation ('knowledge') of O_C, it is still possible *to give a sense* to the contrasted possibilities of the presence and absence (respectively) of 'external objects', (O_P), when these *seem* to be 'presented' or 'given in a perceptual situation, in the manner of Locke's immediate philosophical successors, through criteria like compulsoriness of our ideas (O_I), or their vividness, and mutual coherence.

It will be seen that the idealist position by and large operates primarily with a concept of external object, which is interpreted in a special way (in the sense of O_I, or rather its special case, O_P). But might one not then ask why it was worth the bother to continue speaking as though the object to which we had direct access was merely an 'idea'?[2] The answer to this—and Berkeley was to give one not unlike it—would have to be that the use of the term 'idea' reminds us that the 'existential entailment' of 'to see' does not entitle us to

---

[1] Boas Hall, op. cit., pp. 255–6. Descartes had already mentioned this example, and a similar doctrine may be found in other contemporary writings, e.g. in Hobbes. Cf. *Human Nature* (Molesworth ed. 1839, Vol. IV of *English Works*), ch. 2, sect. 5. Similar considerations we found already to bear on the primary-secondary qualities distinction; a distinction which in the present context we must understand to operate at the epistemic level.

[2] We shall in fact see later that Berkeley has no objection to permitting the *synonymous* use of 'thing' and 'idea'.

infer that the grammatical relation involved in the perceptual process between a cognitive self and an object is one between the former and an object construed in the sense of O_C or O_M. (To indicate this, let us represent the object involved in the perceptual relation by O_R.) But what is not so clear is whether O_R is correctly characterised as 'idea' —however much the 'causation' and 'projection' models may suggest this.

To get the measure of Locke's theory here, let us first suggest an alternative with a 'realist' bias. To do this, we first of all reject Locke's account of O_C, with its confusion of the physical process of interaction between environment and bodily-states on the one hand, and the body–mind relation on the other. Further, we retain that part of Locke's 'projection' model which postulates that the fundamental starting point of any epistemological account must be the notion of the bodily self (the 'person') whose perceptual situation (when it finds itself confronted by an object) is to be rendered by the locution 'I see O'. At this point, there is however still the old temptation to treat 'O' as a projected image. But here remember how we *learn* the concept of 'external object' (in the sense of O_R): namely, very much through situations which Locke has used to characterise the existence of objects in the sense of O_P.[1] In other words, we shall say that through the locution 'I see O' we basically *intend* to refer to a real external object O_R, but that the *signs* of our being successful in 'reaching' such an object are supplied by just these 'criteria'. Now what bedevils this account (a difficulty fed by considerations of 'perceptual failure' (case (3) above, p. 242) is our knowledge that we may fail in our intention, and that our 'intended' object O_R can hence not be identical with any 'real' physical object. If none the less we want to retain a realist interpretation, we shall have to build the notion of 'failure' into the very account of perception itself.

Suppose I claim to 'see an object O', and suppose further that this claim is successfully challenged by a second observer or group of observers, where such counter-claims will, however, equally be understood in a 'realist' sense—differing only in content. Unlike what is suggested by a theory like Locke's, I shall then *not* say that what I saw was an object, though possessing an 'ideal' nature, but instead shall simply describe my claim as *having been unsuccessful*. This alternative locution seems perfectly possible if we remember that

---

[1] And which are later refined by phenomenalists and idealists to serve as reduction accounts for the *meaning* of 'object'.

although each individual's perceptual statements entail an achievement claim, and thus a claim to the existence of an object (O_R), the claim does not entail its own success—*is indeed neutral to that question.* Knowledge of the possibility of a lack of success no more invalidates our justification for regarding the object as an object in the sense of O_R (rather than O_I) than does the fact that our knowledge of the possibility of a truth-claim being falsifiable impugns the logical propriety of expressing some truth claim on a given occasion.

A considerable part of Locke's motivation for his theory of ideas is his view that a foundation for 'knowledge' presupposes 'certainty' of the truth of the respective propositions which are supposed to supply that foundation. Since normally he regards (like Descartes) propositions concerning external objects as inferential, they lack the required degree of certainty; by contrast, 'ideas' are 'in our minds', 'united to our minds', in consequence of which—as he says at iv.2.14—'there can be nothing more certain than that the idea we receive from an external object is in our minds'.[1] Now as we saw, the achievement-imputation of the language of perception, as given in the proposition 'I see O' falsely imputes the requirement of certainty. A 'realist' theory of perception seeks to avoid this implication. Certainly, in a perceptual situation which I believe to be veridical I *intend* an object existing. But where none exists (or one, displaying differences from that intended) one does not have to conclude that our judgment concerns a non-existing object, viz. an 'idea', thus making the object of my perception an idea;[2]—an 'idea' which is believed to be 'projectively thrown' by (Locke says 'produced by') the relevant bodily process, and coming 'between' the real object and the observer. (In which case, of course, we shall have to say that we have to do with 'ideas' (or 'sense-data') also in the veridical cases.) Instead, we may equally well use a locution according to which we have simply *failed in our claim* of such-and-such an object as existing. (In other words, perception-language must be regarded as containing an in-built proviso of possibility of failure.)

It must be realised, of course, that on the realist interpretation, to say that our claim is always to a real object existing, does not

---

[1] Cf. ii.8.12.
[2] Modern phenomenalism, although employing the terminology of 'sense-impressions' instead of 'ideas' employs similar arguments. For this, and a 'realist' criticism, cf. R. Grossman, *The Structure of Mind* (Univ. of Wisconsin, 1965), ch. 6, especially pp. 224ff.

9

guarantee the existence of the objects. This is simply an empirical question, one that can be settled only *after* an agreement on the logical analysis of the perceptual propositions has been arrived at.[1] And the latter involves an ontological question.

However, to suggest that the 'realist–idealist' dispute revolves round 'ontology' can lead to misunderstandings. For it calls forth the false expectation that the claim expressed as 'I see O', understood as it now is as referring to an object existing, can somehow, in virtue of the claim alone, 'give' us an existing object. The fact is that the ontological difference between an analysis of the object as 'idea' and as 'object existing' is never more than a difference *intended*; it is not one that can be 'guaranteed', much less 'pointed to', in any empirical sense. It is just because of this that the existential or realist analysis manages to be immune from, or neutral to, the question of 'achievement' or 'success', in respect of veridical perceptual situations. And it is for this reason that the question of 'certainty', or 'guaranteed truth', in any empirical sense, cannot come up in the individual judgment 'I see O'. The question whether O *does* exist, to repeat, can only be answered in the concatenated context of experience.

It will be clear, therefore, that the two alternative views on the status of the 'object of perception' concern strictly nothing more than competing logical analyses, involving competing ontological points of view. Thus the 'idealist'—as Berkeley later never wearies of repeating—does not really 'argue against the *existence* of any one thing that we can apprehend',[2] nor does he wish to discredit the 'evidence of the senses'.[3] *Per contra*, 'objects' become not one whit more 'real' on the 'realist' analysis. On the other hand, all such analyses involve the adoption of one notion (model) or another as in some sense 'privileged'; and the philosopher has then the task of explaining certain contradictory features once he has adopted that model. As a consequence of this, the model becomes 'strained': the idealist becomes embarrassed by having to introduce additional conceptions into his ontology, which in the case of Locke are often no more than simple empiricist ingressions—as is evidenced by his confusion of 'levels'—or the strains may manifest themselves by the presence of theological

---

[1] With an 'idealist' analysis, short of sheer 'postulation' the empirical object can be 'recaptured' only in terms of such things as 'coherence' between our ideas, or between the testimony of the different senses, as already mentioned above. (Cf. pp. 241f. and p. 242n.2, reference to iv.11.4–7.)

[2] *Princ.*, sect. 34; my italics.   [3] Op. cit., sect. 39.

notions, such as Berkeley's God. The realist (as in our proposed model) has to introduce a technical notion such as 'possible failure', which is built 'intrinsically' into the analysis of the perceptual process: a clear sign that some 'metaphysical stand' is being taken. (All of which is not to say that one model may not be less confusing than, and hence preferable to, another.)

Locke never clearly settled on one side or the other of the fence; he never held clearly, as Hume for instance was to do later, that 'nothing can ever be present to the mind but an *image or* perception', rather than an 'object'.[1] Sometimes, the reference to ideas was meant as no more than a reminder that all empirical cognition presupposes a relation to a perceiving subject. Thus Locke instances the fact that whenever we do not take any 'notice of' the 'impressions' that are made on our sensory organs, there will be 'no perception',[2] where he goes on to describe this want of perception by saying that in such cases 'no idea will be imprinted on the mind'. And as we have already noted, at least some of our ideas (viz. 'simple ideas')[3] 'represent to us things under those appearances which they are fitted to produce in us'. The very term 'appearance' here almost foreshadows a doctrine which finds its full articulation only in the philosophy of Kant. According to this, 'knowledge of objects' must be interpreted inevitably in terms of conditions relative to the perceptual subject; and this defines also the sense in which there can be no possibility of knowing a 'thing as it is in itself'. But here this does not imply that the object '*quâ* appearance' is an 'idea'. For, as Kant puts it in an important passage in the *Prolegomena*,

the existence of the thing that appears is not thereby cancelled as with real idealism, but it is only shown that we cannot know it at all through the senses as it is in itself.[4]

[1] Cf. *Enquiry*, xii.1, op. cit., p. 152; my italics. Cf. above, pp. 236n., 238n.
[2] ii.9.2. Is there involved here also an oblique reference to the aspect of idea which corresponds to the notion of 'seeing' as a kind of 'psychic act' or 'process'? (Cf. Descartes' aspect of idea as a 'manner or way of thinking' in *Med.* III, K.S., p. 220.) But the theory that seeing is a 'process' has been severely criticised more recently (e.g. in G. Ryle, *Dilemmas* (Cambridge, 1954), p. 102), and anyway we need not impute distinctions to Locke which he does not explicitly make.
[3] iv.4.4. Cf. above, p. 236; below, pp. 251ff.
[4] §13, Note II; op. cit., 1953, p. 46. Kant is thus not a straightforward 'realist'. His non-realist approach to space and time has the effect of introducing a further classification of 'realism', what he calls 'transcendental' and 'empirical' respectively, and the same for 'idealism'. And Kant characterises his own doctrine as being both 'empirical realist' and 'transcendental idealist'. (Cf. K.72,78, 345–6.)

6. CONSEQUENCES OF THE THEORY OF IDEAS

The difficulties of a realist interpretation are as multitudinous as are those on an idealist basis. To treat the subject thoroughly would require a separate volume. If I have sketched a rough alternative to the 'idealist' position, it was only in order to obtain some sort of perspective on the nature of Locke's own theory. But it must be admitted that the concept of the object in Locke's philosophy appears in a bewildering number of guises, with its object as 'idea' (O_I), as 'cause of ideas' (O_C), theoretical object (O_T), the latter's 'metaphysical extension' (O_M)—itself confusedly borrowing from the 'realist' position (O_R)—and the positive object as 'particular substance' (O_P).[1] All these versions answer to certain aspects of the logic of different situations, and Locke has neither integrated them, nor decided which are basic and which are derivative. And perhaps to attempt this is a task forlorn. Locke's famous 'inconsistencies' may simply be an unconscious admission of an impossibility of integration. Still, this is what Locke's empiricist successors will attempt.

(a) *The distinction between 'simple' and 'complex' ideas*

However, Locke himself already did set out on this path; there is certainly to be found here the view that our world is, as it were, built up from 'ideas'; and that we cannot—at least epistemically speaking —break out of the circle of ideas, even though (partly by virtue of his sprawling 'inconsistencies') Locke managed to retain the notion of the constitutive realm (O_C; O_M). At this point, the 'grammar of knowledge' is re-drawn still further. When embracing an extreme idealist position, the central place held by the notion of 'idea' will give new form to the old problem of the 'subjectivity' of knowledge; and will furthermore be used to create a new, and quite metaphysical, way in which Locke seeks to 'break the circle of ideas' in ordert or each 'reality'.[2] Here, the logical properties of 'ideas' as such are used to enable Locke both to describe his difficulties and supply attempted

---

[1] A further member ought to be added here, viz. substance $_m$, i.e. substance in the sense of substratum, because later writers, particularly Berkeley, tended to slide from 'external bodies' (regarded as causes of our ideas), *via* 'matter or material substance', to substance in the sense of 'support of accidents', arguments against the former being treated as though they were relevant to the latter. (Cf. Berkeley, *Principles*, sects. 33, 35, 37.) The slide (though misguided) is understandable, being based on what *seems* common to both cases, viz. 'something' behind our 'ideas'.

[2] Cf. Locke's conception of the 'reality of our knowledge' (iv.4).

solutions, one of the main distinctions for this purpose being that between 'simple' and 'complex' ideas. To this we must now turn.

When Locke talks as an 'idealist', although he cannot *know* what lies behind the 'idea', the causal account of perception, according to which a sensory idea is *'produced'* by the corresponding 'impression on the body' due to 'an external object',[1] still throws its shadow. Now corresponding to this causal doctrine there is also a logical claim. In the acquisition of many of our ideas—creatures (it will be remembered) of 'sensation' and 'reflection'—we are entirely 'passive'.[2] Primarily, this must mean that the ideas in question are not affected by any 'activity' on the part of the 'self' or the 'mind'; for that, it will be remembered, is originally a *tabula rasa*, an 'empty cabinet'; conceived as purely 'receptive' with respect to these ideas.[3] Such ideas Locke calls 'simple'; and they are contrasted with 'complex' ones, concerning which the mind 'exerts several acts of its own'.[4] These 'acts' include not only the 'combining' or 'joining' or 'uniting' of simple ideas resulting in 'complex ideas', but also 'comparing', and 'separating' or 'abstracting', giving us respectively 'relations' and 'general ideas'. Locke speaks of these acts being 'made voluntarily', which is to emphasise that the mind is not here 'constrained' as it is in the case of the 'simple ideas'; only in respect of the latter can the mind 'have no more nor other ideas than what are suggested to it'; as to the complex ideas, we can pass 'infinitely beyond what sensation or reflection' give us.[5] Since in respect of its 'complex ideas' (e.g. of

---

[1] ii.19.1.

[2] ii.1.25; cf. 9.1; 12.1. This is clearly a purely logical point. Locke is perfectly aware —as already noted—that 'perception' presupposes a 'taking notice of' our 'impressions of sensation'. And in a lengthy section (ii.9.8–10) he explains the important part played by 'judgment' in altering our 'bare perceptions'. The 'logical point' on the other hand, is—as explained in the text—a response to the assumptive requirement of certitude—of having to provide an anchor for 'the reality of human knowledge' (as discussed in iv.4.1).

[3] This will turn up again in Berkeley, Hume and Kant; note, for instance, Kant's definition of 'sensibility', as 'the capacity (receptivity) for receiving representations through the mode in which we are affected by objects' (K.65). But for our understanding of the contrast between Kant and Locke it is vital to see already now that what is so 'received' does not have the status of a Lockean 'idea', for *as* it is received it does not occur (or must not be regarded as occurring) *with* its spatio-temporal forms. It is for this reason that these representations (what Kant calls the 'matter of sensation') can never even begin to 'duplicate' external objects, as do Locke's 'ideas'. Kant's procedure, we shall want to say, amounts to a 'critical purge', or 'critical constraint', of that notion. (Cf. below, ch. VIII, especially sect. 8, pp. 621f., 628.)

[4] ii.12.1.

[5] 12.2 For our purposes, the most important example of 'complex ideas' are those

'particular substances') the mind is not constrained by what is 'actually given', the 'limits' which it draws are not subject to the kind of checks exercised in the case of 'simple ideas'. They are, as we saw when discussing Locke's notions of classification and species, at best carried out under the guidance of the teaching of nature.[1] But the more intrinsic deficiency, we can now see, is viewed by Locke against the background of his conception of 'simple ideas'. Unfortunately, which of our ideas are simple and which complex we are given no special criteria to decide, barring the irrelevant reference to the physical process of the corresponding sensation, and a denotative reference to a number of examples of physical qualities of objects.[2] Nor need we enter on a particular criticism of the logical difficulties which arise if we ask in what sense Locke's simple ideas of qualities can be said to be 'simple'. (Is it a *concept* which is unanalysable, indefinable? Is the area of a sensed colour-patch indivisible? There are an indefinite number of possibilities here.) I think it is more fruitful instead to fasten on the metaphysical function of this whole doctrine.

The notion that the fundamental material constituents of the world are simple has always attracted thinkers and scientists, it being argued that what is thus simple can be neither naturally created nor destroyed.[3] Locke echoes this when he writes that man's power

---

of 'particular substances', and, as we have seen in detail already, it is just because these are 'made by the mind' that there attaches to them this peculiar 'uncertainty' and 'arbitrariness' which we have noted above (Cf. p. 214). *Per contra*, the true nature of this 'uncertainty' can be understood only by reference to the special part played by the concept of 'idea'.

[1] Cf. iii.6.28; and above, pp. 214f. Quite clearly, the lack of 'certainty' involved is not an empirical matter—no more than was Locke's more specialised worry concerning the 'boundaries of species'.

[2] Here, again, it is important to remember that Kant later was to hold that the 'synthetic activity' is presupposed for *all* concepts, including such qualities as, e.g., 'red in general' (his example; K.154a). The question 'how can the concept be made to agree with the object' (Kant's central question) becomes therefore at once not only more important but also more 'generalised', in the way in which many philosophical questions are typically generalised. This peculiar kind of generalisation will also facilitate the Kantian solution, which amounts to saying that any such 'object' *must* conform to our 'concept'. (Cf. K.22.)

[3] Newton's *Opticks* (though appearing later in time than the *Essay*) is a *locus classicus* for the physical case: 'That nature may be lasting, the changes of corporeal things are to be placed only in the various separations and new associations and motions of these permanent particles'; particles, which are indivisible, for 'no ordinary power' is 'able to divide what God himself made one in the first creation'. (p. 400.)

reaches no farther than to compound and divide the materials that are made to his hand, but can do nothing towards the making the least particle of new matter, or destroying one atom of what is already in being.[1]

Interestingly, he regards the position of 'ideas' in much the same light, since he expressly uses this physical example as an analogy for man's 'little world of his own understanding'. For, he tells his reader,

> it is not in the power of the most exalted wit or enlarged understanding, by any quickness or variety of thought, to invent or frame one new simple idea in the mind, not taken in by the ways before mentioned; nor can any force of the understanding destroy those that are there.[2]

(b) *The 'metaphysical' argument for the 'reality' of the external world*

Let us then assume that the mind is, as it were, shut up with its own ideas.[3] But, so Locke says, although the mind has as its 'immediate object' only ideas, 'there is something farther *intended*'.[4] For knowledge to be what Locke calls 'real', there must be a 'conformity between our ideas and the reality of things'.[5] Now we know already that there is no guarantee of such conformity with respect to complex ideas.[6] They can go 'infinitely beyond what sensation or reflection furnished it with'. However, in so far as we take the 'idealist' or

---

[1] ii.2.2.　　[2] Ibid.

[3] This is the radical position taken up at iv.1. Cf.: 'Since the mind, in all its thoughts and reasonings, hath no other immediate object but its own ideas, which it alone does or can contemplate, it is evident that our knowledge is only conversant about them' (iv.1.1).

[4] iv.4.2. Here we see that Locke was perfectly aware of the 'intentional' nature of the perceptual process. But instead of bringing this fact into the analysis of that process, which we have seen to be the 'realist' move, Locke takes account of this by way of a 'metaphysical' hypothesis. Here we have once more a clear example of these alternative adjustments of different philosophical schools to identical phenomenological situations.

[5] iv.4.3.

[6] Cf. ii.12.2: When the mind 'has once got these simple ideas, it is not confined barely to observation, and what offers itself from without; it can, by its own power, put together those ideas it has, and make new complex ones which it never received so united.' It should be noted that at ii. 31.2-3, simple ideas are held to be always 'adequate', unlike complex ideas of substances which cannot for certain 'copy things as they really do exist', owing to our ignorance of their inner constitution, from which alone we can deduce with necessity their various properties. But the distinction between 'nominal' and 'real' essence, which is here brought in, is not really relevant to the question of how our ideas, *quâ* 'percepts', fail to 'conform to reality', but only, how the *concept* we have of a substance may fail to include the 'correct' bundle of qualities by which to define the substance. Here we see again how the contrast between 'reality' and 'idea' gets off the ground only *via* the intrusion of considerations which refer to quite a different set of levels—a fact no doubt

'representationalist' position of iv.1.1 seriously, Locke does not have available the use of those constant references to 'observation', 'sensation', etc., and indeed altogether his language referring to a 'realm without', from which ideas are 'taken in', borrows quite invalidly from the epistemic distinction between external objects and our sensory constitution.¹ Actually, the only *logical* premises which Locke has available are the following: (1) The mind is an 'empty cabinet' and does not contain initially any of its own ideas; nor can it cause any of them unaided. (2) The mind can only join or separate the ideas which it does contain. From these two premises Locke wants to infer that *if the mind contains* any ideas that are not thus joined or separated (and are hence 'simple ideas'), they will necessarily 'conform' to external objects, in the sense of qualities and similar 'simple' constituents.² That the mind does contain 'simple ideas' is therefore a third premise which Locke requires for his proof of the external world. Simple ideas are thus seen to be one of those metaphysical centres which I have suggested as lying at the heart of philosophical doctrines. The purely assumptive nature of all this is, however, concealed by being blended with the empirical apparatus connected with perception, especially with vision.³

It is thus fairly clear that the metaphysical core of Locke's theory can only maintain itself by importing throughout considerations belonging to the scientifico-theoretical and empirical levels of the situation. Although the whole doctrine is immensely muddled, this is offset by the more honest insight we have into the confusing tensions produced by the different aspects with which Locke is dealing in the *Essay*.

---

concealed because of the epistemological ambiguities which surround the notion of 'real essence'.

¹ Cf. passage from ii.2.2 quoted on p. 251. And note also the question-begging language of ii.1.25, where Locke speaks of 'the objects of our senses' that 'obtrude their particular ideas upon our minds'. The bewitching analogy with the scientific doctrine of 'motions' or 'particles' impinging on our senses is really now quite unavailable.

² Cf. iv.4.4. Locke wants to say that the ideas are 'taken in', though this expression is itself now systematically misleading.

³ We have noted already an earlier and quite separate metaphysical assumption, viz. the 'resemblance' between primary qualities and our ideas of them. But the grounds for this assumption are quite different, as can be seen from the fact that not all these ideas are necessarily simple. Locke is here subject to some conflicting tensions, whereby he calls such things as space and extension 'simple' at ii.5, but 'complex' (i.e. 'simple modes', which are 'complex ideas' according to his own teaching) at ii.13–15. (Cf. below, p. 254.)

(c) *Ideas and Relations: Space, Time and Causation*

To return to our 'simple ideas': they are now the ultimate 'epistemic' constituents of Locke's world—'the whole materials of our knowledge', as he puts it at ii.25.9. As a matter of logic (though not of empirical fact), these 'ideas' are not only simple, but also 'distinct' from all other ideas; indeed, says Locke,

> nothing can be plainer to a man than the clear and distinct perception he has of those simple ideas.[1]

Each simple idea, then, is in a sense self-contained and cut off from any others. And here again is a further source of the scepticism expressed in Locke's contention that our knowledge of 'particular substances' concerns for the most part only the 'mere coexistence of our ideas'. He holds without question that any 'relations' existing between ideas have a less basic status than that possessed by the ideas themselves; relations being a function of the mind's 'voluntary' activity of 'comparison'. And for Locke this means that—as he states in a passage fateful for the subsequent history of the subject—a 'relation . . . is something extraneous and superinduced'; in the end all relations 'terminate in simple ideas'.[2] Locke is here, of course, not just making a remark applying to the level of 'ideas'. Rather, he is following the doctrine according to which only substances and their attributes are 'fully real', whereas—as he expresses it at ii.30.5— 'relations . . . have no other reality but what they have in the minds of men'. They are 'not contained in the real existence of things, but something extraneous and superinduced',[3] not touching the 'real essence' of anything which, it will be remembered, is the corpuscular constitution which anything 'has within itself, without relation to anything'.[4]

I cannot go into the historical background of this doctrine, except to note that it receives considerable support in Locke's 'idealist' form, which makes 'simple ideas' basic. Since the latter are the only constituents of the mind's furniture 'conforming' with 'certainty' to

---

[1] ii.2.1. The title of 'clear and distinct perception' evidently attaches to a set of candidates which differs from that of Descartes. But Locke's definition of these terms is not helpful (cf. ii.29.2–3): it has 'honorific' rather than 'metaphysical' intent.

[2] ii.25.8–9. How he reconciles this with the search for 'objective' necessary connections between such ideas, he leaves as obscure as this question had been left by Descartes (cf. ch. III, pp. 103, 131f.). It will be for Hume to pronounce on the demise of such connections for this type of empiricism.

[3] ii.25.8.   [4] ii.25.8.

anything in 'external reality', it follows that any 'relating' (just as any 'combining') of these ideas will have to be a function of the mind's 'activity' alone, and be in this sense 'mind-dependent'. Now such a doctrine has a particular interest since it bears very intimately on the question of the status of space and time, to which it will be worth while to devote some attention both for its intrinsic interest and because of its relevance for an understanding of later developments.

Locke, as usual, explores and partially straddles a number of different positions, often seemingly contradictory to one another and defended separately by different and opposing later writers. Thus, if—as Leibniz was later to hold—space is regarded as some kind of relational order, it followed that it could have the status of 'an ideal thing' only.[1] But although this will presently be seen to be one of Locke's views also, he holds, as usual, many additional and apparently mutually inconsistent positions at once; which, though it may be confusing for his reader, has the advantage of displaying more openly the many conflicting aspects which the concept of space involves. First of all, let us note that at ii.13.2 he speaks of space as a 'simple idea'; and its 'reality (in Locke's sense) would therefore seem to follow from this alone. What is more, space is even said to be a sensory idea; it, like extension, figure, rest and motion, are said to 'make perceivable impressions both on the eyes and touch',[2] a contention which no doubt is intended to establish that our knowledge of spatial and dynamical facts is not located in some realm of the 'pure understanding', as maintained by Descartes, and reiterated by Leibniz when commenting on this passage.[3] Against this, Locke explicitly claims that

> those even large and abstract ideas of space, time, and infinity *are derived from sensation or reflection*, being no other than what the mind, by the ordinary use of its faculties, employed about ideas received from the objects of sense, or from the operations it observes in itself about them, may and does attain to.[4]

Such a phraseology is sufficiently vague and open to allow almost any position and interpretation. Thus one might, like Kant, arrive at the conclusion after 'reflection', that the idea of space, though in one sense 'given in experience' by way of extensive qualities, is really a 'subjective form', which must be regarded as a 'presupposition' of

---

[1] Cf. *New Essays*, ii.13.17; *Clarke-Leibniz Correspondence*, L.V.47; ed. by H. G. Alexander (Manchester, 1956), pp. 70–71.
[2] ii.5. [3] *New Essays*, ii.5. [4] ii.12.8; italics in text.

sensory experience—remember here the celebrated phrase: 'Though all knowledge begins with experience, it does not follow that it all arises out of experience'![1] This consideration is all the more apposite when we find Locke admitting that ideas like space and time are really 'modifications' ('simple modes') of certain 'simple ideas', which 'the mind either finds in things existing, or is able to make within itself, *without the help of any extrinsical object, or any foreign suggestion*'.[2]

The metaphysical intentions of Locke's use of 'simple idea' in the *Essay* are obvious. When he forgets this special use of 'idea', or when it is not yet fully established, as in his earlier *Journals*, where he is still arguing explicitly against the 'reality of space', he in fact makes the opposite point, contending that 'our having ideas of space in our head proves not the existence of anything without us'.[3]

This earlier position is at once also a reminder that Locke's doctrine of space is not, any more than any of his other doctrines, in any way simple, consistent and straightforward; and if there are no straight contradictions between his earlier and later views, nevertheless there are interesting changes of emphasis, a study of which, *via* the *Journals*, can help us to appreciate his views more clearly. It then becomes evident that Locke does not stick consistently to any 'realist' position.

First of all, as just noted, the idea of space (like duration) is not strictly simple at all; when speaking more carefully, Locke calls it a 'simple mode'—modes being a species of 'complex ideas': a simple mode is a combination of simple ideas of the same kind.[4] Furthermore, a difficulty here is that in the case of space and time the 'simple' components are, as Locke realises, certainly not simple in the sense of being indivisible—a further indication of the supplementary metaphysical use Locke makes of 'simple'. Rather, he takes as simple here —in the technical sense—the smallest portion of space that is still visible—the celebrated notion of the *minimum visibile*.[5] By so doing, Locke will have it both ways: the component parts, being 'simple', must answer to *something* 'real'; on the other hand, our having the complex idea of a composite space, including infinite, or rather: 'boundless space', is not necessarily a 'proof' of 'the existence' of such

---

[1] K.41.   [2] ii.13.1.
[3] 20.1.1678. Cf. *An Early Draft of Locke's Essay* (ed. R. I. Aaron & J. Gibb, Oxford, 1936), p. 101.
[4] ii.12.5.   [5] ii.15.9.

a 'thing'.[1] For the latter amounts to no more than our 'joining' together the component 'ideas' (minimum spaces) without letting such a process come to an end. Indeed, it is actually—Locke insists— 'impossible the mind should be stopped anywhere in its progress in this space, how farsoever it extends its thoughts',[2] but he does not make it clear what kind of 'impossibility' this is. In this way Locke attempts to latch the idea of infinite space back on to 'experience', i.e. sensation and reflection,[3] and to counter the Cartesian epistemology.

Still, careful reading of Locke's various discussions shows that he felt that the force of the notion of 'simplicity' (in its 'metaphysical' sense) was not sufficient to decide the issue concerning the nature of space. From the beginning, he was not prepared to shelve the question in the manner of Descartes, for throughout he insists that space (and extension) are quite distinct from the notion of body. The distinctness of the *ideas* is sufficient, so he contends, to guarantee their separateness; although it must be admitted that one of the premises of the argument, viz. that 'body *includes* . . . solidity or resistance to motion' is not so much defended as simply asserted.[4] At this one should not be surprised; it constitutes Locke's way of stating the basic principles of his anti-Cartesian philosophy, with its resistance to Descartes' reduction of matter to extension, and through this, to the purely 'rational' principles of 'mathesis universalis'.[5]

This is not to say that Locke does not *seem* to argue for the distinction between body and extension and space in a number of ways.[6] But of many of these arguments the Cartesians were perfectly aware. Thus they would not have been perturbed to be told that 'motion proves a vacuum'—hardly imagining that such an argument was fatal to their position. Other arguments implicitly presuppose that we can distinguish between 'individual space', regarded as occupied by body, and empty space, unoccupied. This is a distinction expressly mentioned by Descartes,[7] only to be dismissed as not being 'essential'. Similarly Locke's contention that men have at least the 'idea' of a

---

[1] Cf. ii.17.4.   [2] Ibid.

[3] That this is his intention, he explicitly points out at ii.17.22.

[4] ii.13.12. Nor could Locke have claimed more, since he had expressly stated tha the question of what the 'idea of body' *essentially* entails, cannot be argued on abstract ground since we lack knowledge of 'essence' in the sense required. (Cf. iii.6.5.)

[5] Cf. above, ch. III, pp. 81–4.

[6] Cf. ii.13.11–27. Journals, 16.9.1677; 20.1.1678. Op. cit., pp. 95–96 100–3.

[7] *Principles*, ii, 10–11. Anscombe, pp. 202–3.

vacuum,[1] or that we must allow that God *could* create such a vacuum,[2] would be rejected by the Cartesians as involving ideas and suppositions which are 'obscure and confused', and inconsistent with their fundamental claim that the complete 'reduction' to bare extension is both possible and necessary. Sometimes the rejection of Locke's position amounts simply to postulation of contrary premises. Thus Leibniz, commenting on ii.13.22, asserts that such a supposition would be inconsistent with 'divine perfection'.[3]

We should, of course, not think that strength and plausibility are not often on Locke's side. But if, like Locke, we settle for the distinction between body and extension and space as being essential, we are, as Locke perceived very clearly, saddled with the well-known difficulties surrounding the status of empty space. For Locke it seemed vital to retain this concept since it answered well to the atomistic notions current in the scientific picture of his time, particularly after the appearance of Newton's *Principia*, three years before the publication of the *Essay*, and from which Locke's position sought support.

Broadly speaking, Locke considers the idea of space in two ways. In the first of these, space is somehow connected with the concept of distance between bodies. In the second way, it is more the idea of an empty volume of space, a vacuum—sometimes spoken of as the 'boundless ocean . . . of immensity', or the 'uniform infinite ocean of . . . space'.[4] Now as regards distance, Locke at first seems to assign to it, too, the status of 'simple idea of sense' (i.e. 'sight' and 'touch'); and he clinches this by the (at first sight) disarming contention that

> it is needless to go to prove that men *perceive* by their sight a *distance* between bodies of different colours . . .[5]

That this sort of perception involves, however, at least a certain amount of abstraction becomes immediately obvious in the next section where it is made clear that we have the idea of space only by 'considering' it *quâ* 'length between any two beings', and 'without *considering* anything else between them'—which gives us 'distance'.[6]

This suggests at once that we are not here 'merely passive', as we are in the 'reception' of our other 'ideas of sensation', and the object so 'perceived' is no ordinary object. As the *Journal* shows more explicitly, Locke was mostly of the opinion that distance is not to be

---

[1] ii.13.24.   [2] ii.13.22.   [3] *New Essays*, ii.13.22, op. cit., p. 155.
[4] ii.13.3; 15.5. The other 'ocean' is that of 'eternity' or 'duration'.
[5] ii.13.2; my italics.   [6] ii.13.3; my italics.

likened to a thing, not even 'an imaginary *thing*', being really 'a *relation* between two separate things'.[1] And since, as we have seen, the later Locke certainly assigns a more 'subjective' status to relations, we ought to expect this to mirror itself also in Locke's views concerning space. In 1677 we find him saying that since distance is to be regarded as a 'relation', it follows that 'were there no beings [*sc.* things] at all we might truly say there were no distance', and that for the same reason we are mistaken 'to consider . . . space as some positive real being existing without them'.[2]

This conclusion at once affects also Locke's second way of regarding space, viz. as 'empty ocean'. He cannot reject the concept, since he has argued for the conceptual difference between body and space. And so he concludes that space is 'imaginary', and that it 'signifies no more but a bare *possibility* that body *may* exist where now there is none.[3] In other words, it has a kind of modal existence only. Once again the conclusion would seem to be that space is not 'a positive real being'.

Locke evidently came to feel intensely uncomfortable about this conclusion. And the different moves which he employs subsequently to lessen the discomfort are themselves of great interest, teaching us, as they do, a great deal about the nature of philosophical responses to such difficulties. The first of these responses we have already noted: it is just the *decision* to call space a 'simple idea'. The second connects with an argument about the relevance of category considerations. We saw above that Locke declares relations to be 'mental' creatures rather than possessing full 'reality'.[4] Now he has further argued that space cannot be substantial, since he wishes to distinguish between bodily substance and mere extension. The conclusion drawn from this by his opponents is that this seems to make space 'nothing'—a conclusion with which the younger Locke had himself originally flirted, only to decide against it under the pressure of Newton's notions of space (and time). His answer in the *Essay* is that our understanding of 'substance' is so obscure as not to be able to settle the issue at all. Indeed he goes farther; he contends that, since the notion of substance is used in so many different senses (for God, finite minds, matter), we do not know

---

[1] 20.6.1676; my italics, op. cit., p. 78. Cf. pp. 95, 100. One of the purposes of the distinction Locke here makes is to indicate also the difference between body and extension.

[2] 16.9.1677. Op. cit., p. 95.

[3] Op. cit., p. 96; my italics. This is repeated in the next year, cf. op. cit., p. 100.

[4] pp. 253f.

whether this classification is exhaustive. 'What', he asks, 'hinders why another may not make a fourth?'[1] This is a truly visionary anticipation of Kant's addition of the 'forms of space and time' (as 'forms of intuition') to the accepted 'ultimate furniture' of epistemology, supplementing matter ('matter' *quâ* 'sensation') and mind (*quâ* conceptual framework of knowledge). Once again we meet in these philosophical 'expansions' those readjustments in the history of thought which we have claimed are among its essential traits.[2]

The third kind of response is found in the idea of God. Already in the *Journal* of 1677 Locke hints at this, writing in his usual tentative way that although 'imaginary space is just no thing', and thus no more than 'a bare possibility that body may exist where now there is none', yet 'if there be a necessity to suppose a being there, it must be God whose being we thus make, i.e. suppose extended but not inpenetrable'.[3] In the *Essay*, this theme recurs with greater emphasis, although Locke's hypothetical and tentative language makes it clear that we have to do here with conscious attempts at alternative conceptual analysis. When he writes—the language is that of the seventeenth-century Cambridge Platonists, above all of Henry Moore—that no one should say 'that beyond the bounds of body there is nothing at all, unless he will confine God within the limits of matter'; or that 'God, every one easily allows, fills eternity; and ... likewise fills immensity',[4] Locke uses the authority of a theological concept to underline his claim to the freedom of extending the limits of his ontology. Nevertheless, the hesitancy of his voice is obvious; and in the end he clearly leaves the issue open between those who think of 'space' as 'only a relation' and those who regard it as something which, though not matter, is yet something 'positive', viz. 'filled by God':[5] a doctrine which is echoed in the 'General Scholium' added by Newton to the second edition of the *Principia*, Bk. III (1713), where we read of God that, 'by existing always and everywhere, he constitutes duration and space'.[6] Similarly, in the 28th Query of the *Opticks* (added in 1706), Newton had already referred to 'infinite space' as the 'sensory' of God,—one of those remarks which soon after were to become the seeds for the celebrated controversy between Leibniz and the Newtonians, enshrined in the Clarke–Leibniz correspondence.[7] On

---

[1] ii 13.16–17.   [2] Cf. below, ch. VIII, sect. 7(c), especially pp. 594ff.
[3] 16.9.1677. Op. cit., p. 96.   [4] ii.15.2–3.   [5] ii.13.27.   [6] Op. cit., p. 545.
[7] Cf. *Opticks*, op. cit., p. 370; also Query 31, op. cit., p. 403. For the *Correspondence*, cf. op. cit., pp. 167, 174.

the whole, then, Locke's doctrine of space is uneasily balanced on the shoulders of the notion of 'simple ideas', and in particular, sensory ideas; and a forced expansion of this framework in the direction of another metaphysical centre of gravity, the concept of God, all supported by brushing aside older prejudices in favour of certain categories like substance, and by the invocation of scientific concepts like the vacuum which had only recently become more prominent. But too much mere assertion, and the lack of any attempt to explicate these notions, once again opens the way for his successors to sharpen Locke's doctrines, if not oppose them.

The ideality of relations also has consequences for Locke's theory of causation, although here again the issues are hardly as yet worked out. Unfortunately, for an 'idealist' epistemology this is a serious matter, since the 'relation of cause and effect' is a very important one —indeed Locke believes it to be 'the most comprehensive relation, wherein all things that do or can exist are concerned'.[1]

Obviously a large intellectual gap divides 'real essence' and 'substance' *quâ* 'substratum', in which all qualities of a body 'have their union', from the atomistically conceived ideas related only by some function of the mind's powers. But it must be realised that this 'subjectivist' approach is strictly tied to Locke's philosophical analysis of the meaning of 'ideas'. It is only by contrast with the simple ideas that relations like those of cause-and-effect will be characterised as 'subjective'; for Locke such a characterisation does not affect relations in the field of 'positive knowledge'. Yet here again the intrusion from this level appears (at least in Locke's discussion) as a kind of *deus ex machina*, to extricate him from difficulties originally connected with the metaphysics of ideas. As in the case of his discussion of 'external objects', the different levels become mixed up. Let us consider this in more detail.

Locke's analysis concerning the atomicity of simple ideas is, we have already seen, a logical, or more properly, a metaphysical thesis. One might, at the *empirical* or positive level, say that things behave the way they behave; and Locke himself, at this level, has (we have already seen) no compunction in declaring that qualities are not at all mutually independent, and that 'for aught we know'

> the great parts and wheels, as I may so say, of this stupendous structure of the universe, may . . . have such a connection and dependence in their

[1] ii.25.11.

influences and operations one upon another, that perhaps things in this our mansion would ... cease to be what they are, if some one of the stars or great bodies incomprehensibly remote from us should cease to be or move as it does.[1]

These facets of philosophical thinking, their constant operation at several different levels or dimensions simultaneously, we must never lose sight of, if we wish to gain an understanding of its nature. Philosophy's great complexity is precisely due to the fact that whilst, in order to maintain its positions, it needs empirical analogies as much as logical analysis, it must yet keep distinct the various levels (logical, empirical, metaphysical) over which it ranges. This also explains why some explications of important notions, like the concept of causal relation, will satisfy one philosopher, yet seem quite irrelevant to another.

Locke's discussion of causation is a case in point. It occurs mostly in the famous chapter entitled 'Of Power'.[2] From the start it is clear that he does not question at all the 'existence' of causal activity. (1) He does not question the principle of causality:

Whatever change is observed, the mind *must* collect a power somewhere, able to make that change. ...[3]

(2) Similarly, he appears to adhere to some form of the principle 'same cause—same effect' for, as he tells us, it has been '*constantly observed*' that

the like changes will for the future be made in the same things by like agents, and by the like ways.[4]

It will be noted that whilst (1) is for Locke apparently an inferential

[1] iv.6.11. Cf. above, pp. 214f. for our reference to the 'great chain of being'.
[2] ii.21. Strictly speaking, Locke seems to *distinguish* between a *cause*, which has the *power* to act, and the *effect*, which has a *power* to be acted upon. This might explain why the cause-effect idea is called a *complex* 'idea of relation' (ii.26), whereas power (although Locke admits that it 'includes in it some kind of relation') is asserted to be an idea which 'I think, may well have a place amongst other *simple* ideas'. (ii.21.3; my italics.) At ii.7.8 also, power is said to be a simple idea received from sensation and reflection. But no good reason is given for this piece of philosophical legislation; the true motive is no doubt that Locke wants power to be an ultimate constituent. But this is just what has to be contended for, and not simply asserted. Hume, who admitted that we had the idea, denied that it named an ultimate constituent and held it to be a function of some kind of mental activity.
[3] 21.4; my italics.
[4] ii.21.1 Hume later, and despite his sceptical conclusions, will say similarly that 'this principle [*sc.* the same cause always produces the same effect] we derive from experience' (*Treatise*, i.3.15; op. cit., p. 173); and again this will be a remark made at the 'positive level'.

matter, as is evident from the use of the word 'must', (2) is said to be a matter of what is 'observed'. So is another case (3): particular causal instances, such as the production of motion, both through the exercise of our will, and by virtue of the impulse of bodies on one another, are also observed to occur. Of these, he writes, 'we have by daily experience clear evidence'.[1] Indeed, he is emphatic on this: 'Do we not every moment experiment its [voluntary motion] in ourselves; and therefore can it be doubted?'[2]

However, the real question is the relevance of these empirical matters for epistemological analysis; how far the doctrine of ideas on the one hand and the tacit assumption of various constitutive levels play into this analysis. Here, Locke is obviously subject to some strains. Assumption (1) is never questioned. And as to the constitutive question, Locke glances at it only to dismiss it: 'my present business being not to search into the original of power . . .'.[3] Instead, what *is* his business is to enquire 'how we come by the idea of it'.[4] Why is this important? In the first place, because unless we have the corresponding idea (and Locke, though with some hesitation, as we have noted, holds that we have such a *simple* idea) we can have no 'knowledge' of it.[5] Unless we have the idea, transactions which, according to the experiential mode of speech without question involve power, i.e. causal activity, will be 'obscure and inconceivable'.[6] But more, there is Locke's metaphysical assumption that simple ideas 'conform' to 'reality'.[7]

So not only can we, lacking the idea, not understand 'the manner how' one thing operates on another; there is a strong hint that there might 'ultimately' be something missing. But it is as yet no more than a hint; and not until Hume do we find the thesis clearly developed that where you cannot explain or present a certain idea (what Hume calls 'impression'), you lack 'justification' for what the idea seemingly

---

[1] ii.23.28.   [2] ii.23.25.   [3] ii.21.2.   [4] Ibid.
[5] Remember the pregnant saying that 'the simple ideas we receive from sensation and reflection are the boundaries of our thoughts' (ii.23.29).
[6] ii.23.28. Cf. 21.4; iv.3.29.
[7] Cf. iv.4.3–4. Of course, Locke is there mostly thinking of ideas of sensation, whereas power is said to be 'a simple idea of both sensation and reflection' (ii.5.7). He never worked out in detail the nature of this process of reflection, and Hume's point that 'reason alone can never give rise to' the idea of power seems to be a specific and sharpened interpretation of this concept of reflection which in its original context is still very vague. We shall take this point up again presently. (Cf. below, pp. 264f.)

denotes. The realm of objective reality hovering behind the Lockean epistemological scene prevents him from drawing this conclusion explicity.

The demand that the idea should be exhibited is really a request for 'justification'. The question, does causation exist, or is it real, amounts (as do so many philosophical questions) to the request of showing how the concept is possible.[1] After all, Locke had maintained that the *experience* of causal activity is plain for everyone to see. But when he proceeds by saying that it is only 'the manner how' the causal agent operates that is unknown, although this makes it appear as if it were a request for further information about some hypothetical mechanism,[2] his real purpose is to search for an 'objective' feature which would *entitle* us to speak of causal activity—exhibit its possibility.

But in the case of Locke it is never made *clear* what we are searching for; and the intrusion of empirical analogues makes the situation confusing. Locke mentions two such analogues of causal activity: the first is one billiard ball setting another ball in motion 'by impulse', the second, motion consequent upon our willing to move an object such as part of our own body, e.g. our arm. Though with hesitation, he rejects the former as a relevant example. The specific reason given is that this case involves only a loss of 'motion' by the first ball, and a similar gain by the other.[3] But, as Locke notes, although this phenomenon is undoubtedly in accordance with a law, it must be judged as irrelevant for the present purpose, since we are here face to face once again with our general difficulty, which is that of mechanical action being 'opaque to the intellect', no 'connections and dependences' being 'discoverable in our ideas'. (This is simply again the intrusion of the intuitive model of knowledge, which we have discussed at length.)[4]

Locke, true to his ambivalent approach, is not quite certain that

---

[1] That this *is* the question actually asked was however not realised before Kant, who for the first time states this clearly. (Cf. *Critique*, Introduction, VI; K.55–58; and *Prol.*, §5.)
[2] With which demand such a request is often confused; and indeed, the request for a mechanism is what often gives sense to the metaphysical puzzle!
[3] ii.21.4.
[4] iv.3.29. The 'possibility' of impact-action is later one of Kant's chief objectives to demonstrate. But here, 'possibility' will be demonstrated not through the conception of intuitive insight, but through 'transcendental' construction. (Cf. ch. VIII, sect. 9(c), pp. 680f.)

this argument is conclusive; after having mentioned his doubts, he sanguinely concludes that

> if, from the impulse bodies are observed to make one upon another, any one thinks he has a clear idea of power, it serves as well to my purpose . . .[1]

Commentators normally think this a muddle; it is better to say that one can never be certain of the relevance of empirical cases. On the other hand, Locke is more convinced of the relevance of the second model, for 'the mind every day affords us ideas of an active power of moving of bodies'.[2] There is no need to criticise this argument in detail. It is usual to object here that this example cannot be generalised to include any purported transeunt action operating between inanimate bodies as required for instance by Locke's doctrine that 'sensible qualities' are nothing 'but the powers of different bodies [insensible particles] in relation to our perception'.[3] But perhaps Locke is only concerned to exhibit an empirical *analogue* capable of furnishing the *idea* of power that we make for ourselves in interpreting causal action? Still, there is something in the objection, which will also emphasise the treacherousness of the intrusion of the empirical cases. What Locke wants to offer, after all, through his talk about the 'simple idea of power', obtained by 'reflecting' on cases of *soi-disant* causal action is (as has been suggested) something providing a genuine causal connection between ideas which are as such 'distinct'; one in virtue of which we can make an inference, from the 'cause' to the 'effect'.[4] But can we be certain that the volition analogue is satisfactory for this purpose?

Now, as is well known, to this question Hume replied with a decisive no, and his opposition is explicitly and by name directed against Locke.[5] His general argument will occupy a later chapter, but it is important to set his specific contention in the present context since this helps to place Locke's position in a proper perspective. The important point is that Hume certainly *admits* that we *experience* such a thing as a 'force', an 'animal *nisus*'. What he insists, however,

---

[1] ii.21.4; a better way being the example of volition.

[2] ii.23.28.   [3] ii.21.3.

[4] Cf. Hume, *Treatise*, i.3.6 (op. cit., p. 94): 'Tho' causation be a *philosophical* relation, as implying contiguity, succession, and constant conjunction, yet 'tis only so far as it is a *natural* relation, and produces an union among our ideas, that we are able to reason upon it, or draw any inference from it.' (Italics in the text.)

[5] Cf. *Treatise*, i.3.14, op. cit., pp. 156–7, and Appendix, p. 632. *Enquiry*, vii.1, op. cit., pp. 63–69. Cf. below, ch. VI, pp. 330–2, 375f.

is that this goes nowhere to 'afford . . . an accurate precise idea of power', even though it 'enters very much into that vulgar, inaccurate idea, which is formed of it'.[1] What Hume has done here is to sharpen the idea of power, by identifying it with 'necessary connection'.[2] And with his reasoning, as will be shown in greater detail later, it is not difficult for him to argue that experiencing 'an endeavour to move, followed by motion' is not a proof that there is a necessary connection; from which he concludes that 'reason' or 'reflection' cannot yield the corresponding idea.

Whether we accept this conclusion or not, the argument does emphasise a duplicity in Locke's employment of the notion of 'reflection' (corresponding to that contained in the notion of 'idea'). In a popular, straightforward sense, reflection on certain phenomena (observational, introspective) might quite well yield ideas like 'power' or 'cause'—indeed, how else could we come by them?![3] But those who like Descartes designate them as 'innate', or Kant (later) as *a priori*, mean to imply thereby that such concepts are intended for a certain privileged function; to bring out this fact, they reject the intrusion of 'empirical' considerations, as muddling the issue. Hume, in a sense, follows the same path in his criticism of Locke, even though in the end he makes the causal relation a complex empirical function of the operations of the mind.[4]

7. LOCKE'S PHILOSOPHICAL POSITION ON THE QUESTION OF THE STATUS OF SCIENTIFIC LAWS

Let us now, after so much tortuous, but necessary, textual criticism, give a general estimate of Locke's redrawing of the map of knowledge, and particularly of the various aspects of his apparent 'scepticism', which, as we have seen, covers a very large variety of things. Above all, frequently its tone is misleading. We have no knowledge of 'real essence': partly this is only to spotlight the need for hypothetico-deductive reasoning, coupled with the requirement that 'hypotheses'

---

[1] *Enquiry*, op. cit., p. 67n.1. Cf. also below, p. 331n.3.

[2] Of which it is, according to him, a near synonym; cf. *Treatise*, op. cit., p. 157.

[3] This is irrespective of whether we hold that certain uniformities *produce* the concept of causality (something like Hume's view), or whether they make us aware of it, or even whether causality is already presupposed in our perception of these uniformities (Kant's view).

[4] He confusingly describes this as 'deriving the idea from experience'. But this—as we shall find—is misleading, and requires careful analysis. Cf. below, ch. VI, sect. 9, pp. 374ff.

ought to be founded 'inductively', like empirical generalisations, though in the corpuscularian theory of his time they were not. 'There can be no such thing as a science of matter': partly this is only saying that people have had wrong ideas about the logical nature of 'natural philosophy' (what we now call empirical science). Its hypotheses have falsely paraded as self-evident principles; the deductive nature of its presentation has been misunderstood. On the other hand, this is still a 'deficiency'. For natural philosophy is after all a search for the truth of its laws,[1] which ought to be a search after the 'eternal constitution of nature'—'law-likeness' being interpreted in necessitarian fashion. But Locke, unlike Descartes, does not admit the Cartesian claims of mathematics, nor has he much inkling of some of the sophisticated views about the 'conceptual foundations of science', of the kind we have found anticipated in Descartes.[2] Instead the counterclaim of the need for an 'empirical basis' ends up by hardening into a doctrine where we are appraised only of our 'ideas'. This in turn supplies a necessitarian model for scientific knowledge of the laws and properties of substances, making it tantamount to the intuition of necessary connections between these ideas. It is not clear—as we have seen— whether Locke is here again saying that empirical science is really misconstrued as a search after such intuitions, or whether—with few exceptions—it just does not possess them. (Nor is it clear whether Locke thinks that we shall ever discover any more relevant cases than those he has mentioned.)[3] What is certain is that he wants to say that there can be no question of *discovering* any necessary connections except by way of our perception of this 'agreement or disagreement' between our ideas. *And it is sufficient to inspect our ideas to know this.*

But we now see the far-reaching implications of this for theory of knowledge, since—it will be remembered—'idea' stands, among other

---

[1] It must be remembered that propositions about the 'nominal essence' of particular substances are only a special case of such laws. Cf. above, p. 198n.3.

[2] Cf. above, ch. III, pp. 147–54.

[3] The fact that on occasion we have found him to remind us that empirical propositions 'lack the certainty' of 'trifling', i.e. analytic, propositions, suggests that, sometimes at least, he is pointing to a difference in the logic of scientific statements, rather than bemoaning our inability to reach a certain ideal. Nevertheless, there *was* this ideal; and the search for 'certainty', apparently vouchsafed by our possession of 'simple ideas', 'passively taken in', like the similar search for intuition of necessary connections (as the apparently only acceptable analysis of the laws of science), was still powerful enough to make purely inductive generalisations seem a mere second-best.

things, confusedly for perceptual object ('O_I' in our symbolism).[1] Now it is part of the philosophical grammar of 'ideas' that they present themselves directly, nakedly as it were, to the introspective gaze of the 'understanding'. As Locke puts it very graphically:

> For let any idea be as it will, it can be no other but such as the mind perceives it to be; and that very perception sufficiently distinguishes it from all other ideas.[2]

As this passage shows, ideas can have no other relations or mutual connections except those that manifest themselves to the 'intuitive' understanding. There can be no 'hidden connections'; and any 'necessary connections', in order to fall within the realm of significant discourse, must be, at least potentially, transparent.

This enables us now to deal further with Kneale's claim, discussed in earlier sections of this chapter, that the conception of law-like necessary connections can gain support from Locke. Kneale, so we saw, in order to make good his theory, has to fall back on to the notion of additional constitutive levels, what *he* describes as the level of 'perceptual objects'.[3] But his claim to Locke's support in this is questionable, since it involves a univocal verdict on the latter's equivocal position. Locke holds, as I have shown at length, a highly ambiguous position regarding the notion of physical object. *Quâ* object in sense O_I this is clearly unavailable to a necessitarian of Kneale's kind, since this occurs only at the epistemic level; nor is there anything in Locke to give support to the suggestion that the external object, regarded as cause of our ideas, (O_C), contains such connections. What about the 'theoretical object', O_T? There is a suggestion in Kneale that this might yield a model for a necessitarian interpretation, for he writes:

> It is felt as an imperfection of a theory that it should assume any laws which cannot be seen to be intrinsically necessary, and attempts are therefore made to specify the hypothetical entities of the system in such a way that any connections between them required for the purposes of the theory are intrinsically necessary.[4]

---

[1] In Locke the question of synthesis or construction of 'objects' out of sense-data does not as yet arise. For this reason 'perceptual object' stands interchangeably for percepts of qualities as well as the objects (in a neutral sense of this term) possessing these qualities; cf. examples given of ideas at ii.1.1.

[2] ii.29.5.   [3] See above, pp. 189–91.

[4] Op. cit., §29, p. 97. For the context of this passage, cf. above, p. 225 and footnote n. 2.

Well, the imperfection alluded to in this passage is keenly felt by Locke. But there is nothing in the *Essay* to show that he thought it possible to mend it; for we have seen him to be adamant in his insistence on the hypothetical nature of the corpuscularian principles. The nearest to a constitutive foundation of necessity in Locke is his doctrine of 'real essence', and no doubt his view that this is the result of the action of God's will[1] does imply the existence of an object with relations that are both factual and 'contingent', and yet 'eternal' and 'determinate', in a sense which comes close to that demanded by the necessitarians. But the 'real essence' is not a 'perceptual object'.[2] If it did become an object of perception it would be subject to Locke's rules governing the 'epistemic' relations of ideas. And Locke's own epistemic conception of 'necessary connection' does in his theory clearly not apply to the constitutive structure 'willed' by the divine fiat. To affirm the contrary amounts in fact—as it does in Kneale's theory—to no more than 'dogmatic' assertion. And such assertions can easily be converted into their converse, straight denial, which is the path taken by Hume.

At any rate, Locke has not fully worked out these implications, partly because he does not quite say that the model of intuitive perception of agreement and disagreement is irrelevant[3] for the question of 'nomic necessity', as Hume was to do later; partly because he has not quite worked out the implications of his tenet of the 'simplicity' (i.e. 'atomicity') of his 'simple ideas'; and partly because he was lulled into a false sense of metaphysical security by his assumption of the various kinds of 'constitutive levels' which we have found beneath the surface of his epistemic realm.

We can develop the significance of this further, by reference to a fundamental comment made by Leibniz on Locke's definition of knowledge. Leibniz, when discussing Locke's definition of 'knowledge of truth' as our *perception of* the agreement or disagreement between our ideas, makes an objection to it, which, though seemingly innocuous and almost trite, yet by what it implies is important and basic. He contends that when we 'know truth only empirically' we have no such

---

[1] Above, p. 210.

[2] But cf. Kneale's reference, above, p. 190, to the 'internal constitution' of the 'perceptual object', which—so he implies—may quite well be subject to necessitarian relations.

[3] Nor is it, when properly understood: cf. what we have said concerning the 'conceptual exploration' aspect of science, in connection with our discussion of the 'functional *a priori*'. (Cf. ch. II.)

perception, since we do not then know 'the connection of things and the reason there is in what we have experienced'.[1] Now partly this is a disagreement about where to draw the boundary lines of the concept of knowledge, which Locke had purposely drawn narrowly, leaving 'empirical truth' to the mercy of 'judgment' or 'opinion'. Leibniz at first sight *seems* more tolerant, for he apparently allows us to talk of knowledge even where we do not have an explanation of (i.e. insight into the necessary connections implied by) the generalisations of science. However, as we shall see in more detail later, Leibniz's more sanguine approach is due to the fact that he has himself 'metaphysically extended' this realm of 'explanation' since he holds that even where we *can never* apprehend it, there nevertheless *always exists* a ground ('sufficient reason') for every such empirical proposition, the ground ultimately 'lying in God'.[2] Now against this, Locke is evidently on the way to making more prominent the relevance of the epistemic aspect of things. True, there is the level of 'real essence' from which 'flow' the properties of a particular substance. But this is not worked very intimately into his scheme; it is beginning to be an excrescence and, as we have seen, much is made of our total ignorance of it.

The peculiar blending of the voices of scepticism and emerging empiricism can be gleaned with especial clarity from examples where there is a clash between new discoveries and traditional viewpoints in science; for here Locke's plea of 'ignorance' can also be seen to act as a weapon in the hands of the new science against established modes of thought, as a kind of liberator. All this is very prominent in Locke's reaction to the Newtonian hypothesis of gravitational attraction.[3] Locke, at one with the Cartesians, with Leibniz, and many of his contemporaries, finds 'action-at-a-distance' unintelligible. The reasons for this judgment were as varied as they were confused. A frequent contention was that such an 'action' was 'against nature', although this might not mean more than that the phenomenon was (apparently) inconsistent with the fundamental *laws* of the contemporary mechanics; a system which involved action by impulse only; and a conceptual scheme which, in the case of Leibniz, was further propped up by being woven into his general metaphysical doctrines.[4]

---

[1] *New Essays*, iv.1.2; op. cit., p. 400.

[2] Cf. *Monadology*, sect. 36, 38. Cf. above, p. 224n.1; below, ch. VII, sect. 4(a), pp. 445f.

[3] For Newton's own attitude, cf. below, Appendix to ch. VI.

[4] This will be studied in more detail in ch. VII, but one or two passages may be helpful here. (Cf. also, below, p. 440n.5.) Thus in the Preface to the *New Essays*,

Locke's own opposition to action-at-a-distance, in the days prior to the publication of Newton's *Principia*, was for reasons less clearly stated.[1] His general attitude, in the early editions of the *Essay*, is expressed by the contention that not only is this kind of action, unlike action by impulse, 'inconceivable', but also that no body *can* act on *another* in any way but through contact. In the fourth edition of the *Essay*, however, Locke introduced a subtle change.[2] He still maintains that action-at-a-distance is inconceivable, but he deleted the claim that one body could not (as a matter of physical reality) act on another at a distance; the reason being—as he explained to Stillingfleet—that

> I have been convinced by the judicious Mr. Newton's incomparable book ... that the gravitation of matter towards matter in ways *inconceivable* to me is not only a demonstration that God ... can put into bodies powers and modes of acting beyond what can be derived from our idea of body or explained by what we know of matter; but it is furthermore an incontestable instance that he really does so.[3]

Here is a revolutionary change: the 'inconceivability' of a process is not an objection—in Locke's eyes—to the *contingent fact* of its happening, even if it be an immutable fact, as here implied by the theological metaphor. But further: it is important to remember here also a result of our earlier discussions,[4] where we found that Locke does not really in the end hold action by impulse to be more intelligible than any other kind; in fact, it is just as unintelligible. Thus, in the passage

---

Leibniz writes: 'Matter will not naturally possess the attraction mentioned above, ... because it is impossible to conceive how this takes place there, i.e. to explain it mechanically' (p. 61). Also, Letter to Conti: 'It is not sufficient to say: God has made such a law of Nature, therefore the thing is natural. It is necessary that the law should be capable of being fulfilled by the nature of created things', i.e. through the 'impulsion' from 'other bodies'. (Quoted in A. Koyré, *Newtonian Studies*, (London, 1965), p. 144.)

[1] Thus, the impossibility of action-at-a-distance in the sense of self-contradictoriness of the conception was averred by Locke as well as by Clarke, both holding that for one body to act on another 'without intermediate means' (*Clarke-Leibniz Correspondence*, Clarke's Fourth Reply, sect. 45), or 'where it is not' (Locke's *Essay*, 1st to 3rd eds., ii.8.11) was either 'a contradiction' (Clarke) or 'impossible to conceive' (Locke). Clarke inferred from this that one *does* necessarily require a medium, though for him one that was 'invisible and intangible' (ibid). Locke's conclusion, on the other hand, was quite different, making—as we shall see presently—a virtue out of the supposed deficiency. In his reply to Stillingfleet, he sometimes also voices a more Cartesian or Leibnizian attitude, as when he writes that the 'attraction of matter by matter ... [we cannot] derive from the essence of matter or body in general' (quoted in Koyré, op. cit., p. 155).

[2] Cf. ii.8.11.   [3] Cf. Koyré, op. cit., p. 155; my italics.   [4] Cf. above, pp. 190f.

from ii.23.28 in particular, we found Locke expressing his conviction that 'communication of motion by impulse' is altogether 'inconceivable', though he had again added, significantly, that this lack of intelligibility is quite distinct from the fact itself which is of course observed 'by daily experience'.[1]

Now I have suggested in the present chapter that this complaint of lack of intelligibility can be understood in two ways. On the one hand, it harps back to the ideal of intuitive knowledge, and of a 'science' in the old scholastic sense of *scientia*. On the other hand, it may also be understood as an attempt at a novel characterisation of the kind of inquiry of which Newton's was a crowning achievement, affirming that statements of contingent matter of fact were not capable of displaying the older intuitive links, being of a different logical type. Viewed from this angle, the lack of intelligibility of action-at-a-distance, far from being a deficiency, evidently was for Locke a genuine virtue.

It must be admitted that this method of delineating the realm of contingency, the domain of statements of matter of fact, by emphasising their 'inconceivability', was slightly eccentric. It still mirrored the ingredient of the misleading model of intuitive insight. Nor did Locke completely shake off the belief that some kind of (constitutive) link between these contingent events, though epistemically 'opaque', was still required. For both in the letter to Stillingfleet, concerning the phenomenon of gravity, as well as in the passage from iv.3.29, concerning 'communication of motion',[2] there is an insistence that these phenomena have their underlying link ascribed to God's action. There is here an obscure recognition that the foundation of a scientific theoretical concept requires two elements: first, inductive support, based on observation and theory; secondly, something, of which both 'conceivability' and 'divine action' are an obscure recognition, although the first is still misconstrued as 'rational intuition' and the second as a peculiar kind of ontological ground. And both will be developed in later philosophers through a variety of models, from Berkeley's instrumentalism, coupled with the causal efficacy of God, and Hume's psychological account, to Kant's notion of 'construction', understood as providing a transcendental account of the real

---

[1] It will be noted that Locke's assertion of 'inconceivability' involves two quite different grounds, the first having to do with the insufficiency of a concept (body cannot act *in distans*), the second with the lack of necessary connection between ideas that for us only 'coexist'.

[2] Cf. above, p. 191.

'possibility' of a concept.[1] All these function as explications of that element which is meant to provide additional 'consolidative' supports, particularly for those hypotheses of science which purport to treat of fundamental laws of nature.

Locke's own position is here very much still a half-way house. It uses the 'unintelligibility' of a phenomenon as a part-characterisation of it as a contingent fact. But at the same time, its true significance for the contingency of any fact it obscurely ploughs into the conception of God's activity.

We can test the claim that for Locke 'inconceivability' was an instrument for delineating 'contingency'. For if the latter should subsequently receive independent recognition, one might expect lack of intelligibility to become once more a worry. Now this is precisely what we shall find when we turn to Hume,[2] where action-at-a-distance comes to be altogether proscribed. The contingent status of 'matters of fact' being explicated in terms of Hume's 'metaphysical atomism', the apparent logical gap presented by action-at-a-distance is given once again quite an independent significance, Hume's own response being the result of certain conceptual views on space, together with an adoption of Newton's hypothetical ether as theoretically accounting for gravitational action—a purely *physical* matter.

Throughout, then, we find the Lockean philosophy in a state of transition. It needed only the cutting away, so to speak, of the 'constitutive ground' to bring the problem of the 'relation between our ideas' (including that between the components of the simple constituents of our 'complex ideas') into sharp focus. (As we saw, Descartes' contrast between 'knowledge' and the 'teaching of nature' suffered from the same uneasy relationship.)

Similar considerations apply to Locke's scepticism with regard to our knowledge of an external reality lying behind our ideas. Here we have found no 'proofs' of the Cartesian kind through which to break through the 'circle of our ideas'. Locke's hostility to 'innate ideas' is perhaps sufficient to assure this, and *his* proofs of the existence of the self and of God, though both asserted to be 'demonstrable', are mere vestigial structures in the system.[3]

On the whole, we see that the total fabric of Locke's philosophy is far more than the usual label of 'empiricism' or 'idealism' would lead us to suppose. Quite apart from his 'intuitionism', he is certainly no

---

[1] Cf. also below, ch. VIII, sect. 4(c): iv, for Kant's criteria of hypotheses.
[2] Cf. especially below, p. 330.   [3] Cf. iv.9.3; 10.1–11.

straightforward inductivist, either in the sense of dissociating himself from the employment of hypotheses, or of possessing 'positivist' tendencies leading him to deny the existence of a structure beneath the 'sensory realm'; or, finally, in the sense of putting any simple-minded trust in the deliverance of the senses. It is true, all this is an obstacle to his ideal of 'certainty' with respect to the simple ideas actually located in our minds—yielding yet another form of 'intuitive knowledge',[1] different from the intuition of necessary connections. But he is quite clear on the considerable rôle of 'complex ideas' and the 'uncertainties' which these introduce into the scene; and anyway, the major part of the doctrine of simple ideas is intended more as a logical myth: it is Locke's way of breaking through the circle of ideas, of being able to postulate a 'reality' to which they conform. This doctrine which in the end involves the postulation of a privileged *basis* for knowledge, is nowadays not fashionable, although a version, in the form of a theory of sense-data, enjoyed a considerable vogue not long ago. In the immediate sequel to Locke, however, the doctrine of the 'simple ideas' (or in Hume's version: of 'impressions' and 'ideas') was to become an important focus, until transformed in the philosophy of Kant.

I have spent much time on Descartes, and more still on Locke. This is inescapable, for it is here that the main themes arise, and that we can observe specifically the shaping of a great philosophical enterprise, something very different from the intellectual activities involved in pure science or applied technology. In none of these do we get the peculiar mixture of 'dimensions' or 'levels' which we have discerned in philosophical writing; something which the apparent 'contradictoriness' or 'confusions' of a Locke or a Descartes highlight with particular clarity. At this early stage, it is not possible to separate the narrower concerns of philosophy of science from their broader philosophical entanglements. In Locke's writings in particular we note the 'openness' of the conclusions, and his successors will often do no more than 'tighten' these in one direction or another, although it is not suggested that ambiguities of the kind we have met do not lurk in their writings also. In order to confine this work to a reasonable length, it will be necessary in subsequent chapters to limit attention to those areas where this tightening process takes place, and even then only in connection with special topics such as those of law and causation, space, time and explanation, and the idea of system.

[1] iv.2.14.

We shall follow the progressive critique of the implied notion of 'constitutive realm', with its connected problem of how the 'link' between 'ideas' is conceived by the various philosophers we shall study; or again, the problems that result from the attempted identification of idea and object; and we shall trace the various arguments as they involve theological, ontological, logical, physical, systemic and methodological conceptions in bewildering succession and variety.

V

BERKELEY: NEW CONCEPTIONS OF
SCIENTIFIC LAW AND EXPLANATION

I. CLOSER ATTENTION TO LANGUAGE AND MEANING

Descartes and Locke have been dealt with in some detail in the previous two chapters because it seemed vital to indicate the *growth* and *complexity* of their ideas, and to see how these were kept only uneasily in a balanced tension or equilibrium, an equilibrium which might be upset by the slightest of readjustments in emphasis. An awareness of this will help us to appreciate the basic presuppositions and problems that beset the thought of their philosophical successors. Above all this is obvious from a contemplation of Locke's tentative and ambiguous approach, with its attempt to straddle the various demands made on his epistemology from the different sides of the contemporary intellectual scene; evidently, many alternative solutions were still possible, through a mere change of emphasis, a mere sharpening of the key concepts involved.

In the next few chapters we shall be able to follow only briefly the directions of the considerable criticism which Locke's doctrines were to meet, acting as they did as a philosophical watershed. Perhaps such brevity may be excusable if only because the attentive reader will by now be in a position to attempt for himself certain predictions of the directions which this criticism was to take. At any rate, for the larger part I shall concentrate on some of the issues in the logic of science, centred on the lines indicated in the second chapter. If this means that no detailed critical commentary can be offered on the more purely philosophical issues found in writers like Berkeley and Hume and Leibniz, the excuse must be that this is not possible within the confines of a single volume, seeking—as it does—merely to relate some

275

connected themes which run through the history of this period.[1] At any rate, as far as Berkeley and Hume are concerned, we shall see how new ideas on the place of causation and law, and the nature of explanation, will emerge as a natural concomitant of this more specialised approach.

We may start by going back to Locke's persistently sceptical note. His scepticism, we saw, had really looked in several directions at once. It related to the rationalist idea of scientific knowledge as perception of necessary connections, and if at one moment it had said that the enquiries of the natural scientists cannot in the nature of things concern such connections, since the domain of their enquiries forms a logically autonomous realm, the next moment it was still harping back to the old Aristotelian and Cartesian demand that the principles of science ought really to be transparent. Again, the weight of the corpuscular philosophy, forcing as it does upon the scientist the logic of hypothesis, seemed to collide with the ideal of certain and unimpeachable sensory knowledge, as possessed in the immediate ('intuitive') perception of the 'idea'; from this there arose the complaint of the unknowability of the real essences of things (taken in the physical sense of 'real essence' as corpuscular constitution). And finally, the whole theory of perception which revolves round the notion of 'idea' led to the ultimate sceptical complaint concerning our knowledge of the external world of 'objects', considered as causes of these ideas. Now in each of these cases, Locke had tried to convey the impression that the deficiencies in question were an actual limitation upon the extent of our knowledge. Insensible corpuscles exist; only we have no access to them; and the same is true of the external objects. Necessary connections between our ideas exist; and again Locke (we saw) commonly talks as though it was merely a contingent deficiency that deprives us of a sufficient number of such connections in order to build up a proper *scientia*.

Against such a background, Berkeley's philosophy may be seen as an attempt to banish many of these sceptical notes. 'Scepticism extirpated': this is the important result claimed for his enquiries.[2] To be sure, from the outset he adopts Locke's theory of 'ideas' as a basic tenet: all the 'objects of human knowledge' are 'ideas', either of

---

[1] Similarly, Berkeley's critique of the foundations of the differential calculus has not been dealt with, as it adds little that would be relevant to our central themes.
[2] *A Treatise concerning the Principles of Human Knowledge* (1710), section 85 (in this chapter abbreviated as P, e.g. P.85) Cf. P.87, P.96.

sensation, or reflection, or memory and the imagination;[1] with 'idea' here having much the same meaning and weight as it had borne in Locke's philosophy—at least in its ultimate issue. But what is missing is the tentativeness of Locke's doctrines, if also their inconsistencies. Nor is there any longer this pronounced sense of a philosophy emerging as a response to conflicting considerations—with conflict built into the very philosophy itself. Instead, the consequences of the doctrine of ideas are here worked out with a more relentless logic; and where it is too narrow to bear the weight of the completed edifice, it is shored up more openly by supplementary notions, instead of being made at least apparently consistent by an ambiguous use of terms.

Thus, we noted Locke's habit of using the notion of 'material substance' ('substance$_p$') in a neutral sense that lent plausibility to the story of external objects affecting our minds, forgetting that, given Locke's theory of ideas, in the end such substances could live really only an 'ideal life'. Here Berkeley is more forthright, more narrowly doctrinaire. He will insist that if such 'substances' are to be anything for us, they *must* be 'ideas'. And just as Locke's ambiguous attitudes towards 'external objects', 'substance$_m$', and the like, had resulted in the furtherance of a seemingly total 'scepticism', so Berkeley's greater clarity and single-valued interpretation of the epistemological situation will present itself as an attempt to eradicate such sceptical conclusions. As to the substances[2] considered as external objects, supposed to 'cause' our corresponding ideas: Berkeley's attitude is no longer that we have only very little or no knowledge of such things; instead, the notions of the existence of such 'unthinking things are *words without meaning*'; they involve logical 'repugnancy'; are 'useless'; indeed, even 'altogether *inconceivable*'.[3]

So we see: from the start Locke's response to the epistemological problems which he had encountered is transformed and sharpened. Operating from the basis of the Lockean theory of ideas, Berkeley takes for granted the conclusion that there can be no knowledge of external objects (in the sense of O_C), of substance (in the sense of S_m), without any of the vacillation and tentativeness which had characterised Locke's procedure and which had allowed

---

[1] P.1.

[2] Here, and throughout, Berkeley's use of the term 'material substance' frequently conflates the concepts of 'external object' (O_C), 'substance $_m$', not to say that of 'real essence', with the latter further involving the notion of 'primary qualities'.

[3] P.24, 23; 19; 17. Cf. P.65; (my italics).

him (at least as regards external objects) occasionally to offer *empirical* arguments in favour of their existence, with the consequent confusion of the various senses of the concept of object (O_I, O_C, O_P).[1] The important point is that this sharpening of the argument in Berkeley's hands is achieved by an initial concentration on the question of *meaning*, with the result that—as we shall see presently—he will not be content with Locke's half-way house, whose sceptical implications had seemingly made possible a subtle straddling of realist and idealist positions at one and the same time. Berkeley's result is obtained by asking, not just more persistently: 'How do we know that such and such objects exist?'; nor even: 'What do we *mean* by such and such an object?'; but more radically: 'What is meant by the term exist?'[2] The whole procedure is expressed very explicitly at P.81:

> Nothing seems of more importance, towards erecting a firm system of sound and real knowledge, which may be proof against the assaults of scepticism, than to lay the beginning in a distinct explication of what is meant by thing, reality, existence: for in vain shall we dispute concerning the real existence of things, or pretend to any knowledge thereof, so long as we have not fixed the meaning of those words.

Nor is this approach restricted solely to the concepts mentioned so far; when considering objections to his system we shall quite generally find Berkeley employing this method of the determination of 'proper meanings'. Thus (as we shall see) one immediate objection to his general epistemological doctrine ('*esse = percipi*') is that it apparently provides no place for the unobservable atomic entities of the contemporary physical scientists; entities in terms of which the explanations of physical and chemical phenomena were being framed. Similarly, given Berkeley's basic position, the conception of a universal gravitational force will *prima facie* have to be treated with suspicion, if not positive hostility. Now in all these cases Berkeley will seek to evade the resulting difficulties by adding to his general position a further reconsideration of what we can properly *mean* by 'giving an explanation' in science,[3] where the proposed definition seeks to avoid any reference to the assumption of unobservable corpuscles.

The trend of the argument is thus the reverse of that found in Locke. There the significance of the epistemological results of the

---

[1] Cf. above, pp. 242ff.   [2] P.3.
[3] Cf. P.50, 58, 62, 104–5. Also *De Motu* (1721), sections 37, 69 (in this chapter abbreviated as M). Cf. below, pp. 307–17.

theory of ideas was not faced in all its peculiar consequences, partly perhaps because the scientists' theoretical enquiries seemed to imply the propriety of continuing to work with such concepts, and (as it were) simulating a hidden reality. Against this, Berkeley will give a 'reductionist' account of scientific explanation whose result will be to bolster up his reductionist epistemology in general.[1]

Still, we should not make too much of this shift to concentration on meaning, for almost always to the philosopher's quest for definitions and meanings there corresponds a search for 'reality'; the question 'What does so and so mean?' being a concealed form of asking 'Is so and so real?' And in Berkeley's philosophy the intimate relationship between these two questions, and the corresponding shift from the level of epistemology (not to say psychology) to that of ontology, is particularly pronounced. But at first sight at least this is not obvious, for Berkeley seems to put great emphasis on considerations of a linguistic nature. He introduces his *Principles* by urging 'caution in the use of language',[2] since—so he tells his reader—'a chief source of error in all parts of knowledge' is 'the nature and abuse of language':[3] all of which seems to indicate a fairly semantic approach. Nevertheless, as soon as we enquire more deeply, we find that at the heart of Berkeley's doctrine of language and the meaning of terms there lurk the more usual metaphysical positions.

2. THE SIGNIFICANCE OF GENERAL TERMS

To see this, we must briefly consider Berkeley' streatment of 'abstract general ideas', which Hume was later to praise as 'one of the greatest and most valuable discoveries' made by any philosopher of his period.[4] Berkeley, referring to Locke's discussion of this topic in the *Essay*,[5] points out that the theory of the meaning of terms frequently makes certain assumptions, the most important of which is

---

[1] We should be warned, of course, that in fact considerations affecting the epistemological status of physical objects are relatively independent from those concerning the theoretical entities of the natural sciences. A 'reductionist' account of the latter may live amicably with a realist account of the former. But Berkeley does not usually see this.

[2] Introduction, sect. 21, (abbr. as I.21).   [3] I.6.

[4] Cf. *Treatise*, i.1.7; op. cit., p. 17. Berkeley's discussion occurs in the introductory sections to the *Principles*, in particular I.6 to 25.

[5] ii.11.9–11; 12.1; iii.3.1–14; iv.7.9.

'that every significant name stands for an idea'.[1] According to Berkeley, this is rather a hasty assumption, leading to some dangerous consequences. To be sure, there is no difficulty concerning those names which are terms for particulars or individuals. But grave problems arise with abstract and general terms, expressions such as extension, colour, motion, man, humanity. It is by no means certain, Berkeley wants to say, that these terms can possess a similarly straightforward and clear meaning. For, using the principle that to every term there corresponds an idea, the meaning of an abstract general term would have to be a corresponding abstract general idea. Now Berkeley strenuously denies that we have any abstract general ideas.

The nature of this denial is, however, by no means clear. Is this a psychological thesis? If it is, it is curious that Locke should have maintained just such a thesis, when he writes that general words are words 'which are used for signs of general ideas'.[2] When there is a clash of this kind between two philosophers, it is seldom based purely on considerations of empirical (including in this, psychological) matters of fact. However much such a controversy may *appear* to be about an empirical matter (as suggested by Berkeley's appeal to introspection, to convince us that we have no abstract general *images*), the philosophical ambiguity or 'softness' of the controversial concept ensures that the argument will really turn on an attempt to clarify the consequences of certain deep-lying philosophical assumptions.

Consider Locke's account. Ideas, he had said, become general by means of a process of 'abstraction', which 'separates' from particular ideas the special circumstances of space and time and any other particularities which may 'determine' them.[3] Thus we may form 'the general idea of a triangle' by abstracting from its being either oblique or rectangular, equilateral or isoceles, etc.[4] At first sight, surely, it is difficult to see what could be wrong with such an account, taken in its ordinary sense. Its denial would seem to imply that we cannot meaningfully talk about triangles, or mammals, or colours in general.

Yet this is the account against which Berkeley is apparently moving in the *Principles*. Consider for instance the example already mentioned

---

[1] I.19. Cf. *Essay*, iii.2.1–5; also 2.4: '. . . words, . . . can properly and immediately signify nothing but the ideas that are in the mind of the speaker . . .'.
[2] *Essay*, iii.3.11.   [3] Cf. iii.3.6.
[4] Cf. iv.7.9 for this example.

in his criticism of the Cartesian and Lockean argument to 'extension'.[1] He takes Locke to mean that we are in the possession of 'general ideas' in precisely the sense in which we are said to have 'particular ideas'.[2] Now we could (he writes) come by such ideas (*per impossibile*) only by 'abstraction'. Taking for example a particular object which is coloured, extended and moving, we first abstract two of these qualities, retaining only its particular extension. We thereupon abstract further all the particular determinations of extension, and finally arrive at the abstract idea of extension.[3] However, Berkeley contends, it is impossible that we should ever have such 'abstract general ideas'; the only ideas which we can 'represent' to ourselves (he sometimes says 'imagine') are 'ideas of those *particular* things' that we have previously 'perceived'.[4] Thus, if I imagine a hand or an eye, these must always be of some 'particular shape and colour'.[5]

Perhaps Berkeley is merely harping on a putative psychological inability to have 'general *images*'? He is sometimes supposed to have interpreted Locke in this way, but we shall presently see that his disagreement cuts deeper, for his thesis has almost certainly 'ontological' implications.[6] If Berkeley were literally only concerned with what can be imagined, it would not even be empirically true to claim for instance that a certain table which we imagine must always be imagined by us with a particular and determinate length. What is true is that an actually existing table is properly said to possess such a determinate length.[7] If one does not know its actual length, it makes sense to try to determine it by measurement—which is not true

---

[1] Cf. above, ch. III, pp. 92f.; and note the way Berkeley's position affects Locke's views about the atomic constitution of matter (above, ch. IV, p. 233).
[2] Of course, how unclear *this* sense is I have laboured at length to show in the previous chapter.   [3] Cf. I.7–8.   [4] I.10; my italics.
[5] Ibid. The phraseology suggests that just as the phrase 'particular ideas' (this expression occurs e.g. at I.19) means 'ideas of particular things', so 'abstract general ideas' means 'ideas of abstract general things'. Naturally, for Berkeley this must not be taken to entail that there '*are* things', apart from 'ideas'; we are concerned so far only with the representative function of ideas; with what Descartes had called their 'objective reality' (cf. above, ch. III, pp. 169f).
[6] The postulate involved occurs explicitly in the *First Dialogue*: 'It is a universally received maxim, that *every thing which exists is particular* (Everyman ed., p. 223; Berkeley's italics). This leaves the choice of candidates that are 'particular' still relatively open. P.1, 38, 91 imply (as one might expect) that Berkeley is thinking of 'sensory qualities', a 'combination' of which is said to constitute an object. There are many subtle reasons, of both a psychological and logical nature, that make for the philosophical desire to view the world as ultimately 'reducible' to such 'elements'. In this book we shall not criticise this approach further in any systematic manner.
[7] We ignore complications due to the modern 'Uncertainty Principle'.

of the imagined table and ought hence not be true of the 'ideal table', as long as the term 'idea' is used in anything like its conventional sense. But then, of course, it is *not* used in this way; as with Locke,[1] the grammar of 'thing' and 'idea' are often interchanged, in this way making the philosopher's position sound more defensible than it actually is.

Berkeley's intentions emerge further in the argument that although we can imagine or consider complex objects possessing qualities never found united in fact, and similarly imagine objects lacking such qualities of parts, what we cannot do is to separate or abstract, and indeed, '*conceive* separately' those qualities one from the other 'which it is *impossible* should *exist* so separated'.[2]

But how does Berkeley determine here what is possible and what impossible? And what sort of impossibility is this? His examples of what *is* imaginable (such as of a body without eyes, or nose or any parts) suggest that it is not *physical* possibility which he has in mind. 'Logical possibility' suggests itself perhaps; but this notion is not really available until we are clearer on what a crucial instance, like the denial of the abstract idea of extension, actually implies. However this may be, his reference to 'existence' in the passage just cited shows that he is concerned with the question of what is ultimately 'given' in the 'nature of things'. What no doubt he wishes to say is that 'extension' as such, triangles in general, humanity (and other similar 'universals') are not 'given' (unlike 'particular ideas'); are not part of the 'ultimate furniture of things'.

There is, however, a more serious complication, which will further illustrate the way in which 'ideas' have taken over a heavy metaphysical task in Berkeley's philosophy. This will also explain the nature of his opposition to Locke's views. The complication is Berkeley's well-known doctrine (to which we shall turn in more detail presently) that the distinction between 'thing' (considered as external object) and 'idea' is epistemologically otiose, and that to speak of the 'existence' of a material thing is 'ultimately' to speak of 'ideas'. His contention concerning the 'particularity' and 'concreteness' of all ideas becomes itself an ultimate and basic axiom.[3] And it is for this

---

[1] Cf. above, ch. IV, p. 220 and n.4.

[2] I.10; my italics. Cf. also P.5: Abstraction properly so called 'extends only to the conceiving separately such objects as it is possible may really exist or be actually perceived asunder'.

[3] As with Locke, the grammar of 'idea' and 'thing' are fused also as regards the question of 'particularity'. Thus whilst in the *Principles* Berkeley's language refers

reason that Berkeley's inability to conceive extension, say, apart from certain secondary qualities like colour, or odour, or tactile sensations, is not just a simple 'logical' matter, in any purely definitional sense, but has more to do with the general conditions of communication.

Now let us compare all this with Locke's theory of general terms. Locke likewise works with an 'anti-universals' axiom,[1] when he writes 'that all things that exist are only particulars';[2] that 'general and universal belong not to the real existence of things'.[3] We may expect Locke to infer that like so many of his 'complex ideas' and 'relations', such general ideas also (unlike the 'simple particular ideas') do not automatically 'conform' to any real things.[4] They were, as he had put it at iii.3.11, mere 'inventions and creatures of the understanding'; they had merely conceptual significance. On the other hand, it is important to note that when Locke in this way contrasts 'things' and 'creatures of the understanding', he writes as a metaphysical realist. His basic epistemological inventory contains three levels: 'words', 'ideas' and 'things'—each of them certainly highly complex and ambiguous notions, as the discussions of our last chapter will have shown. And it is, I think, because of this that Locke sees no difficulty in speaking of 'general ideas', as when he declares that general words are words 'which are used for signs of general ideas'.[5]

But this at once helps us to understand the intended sting of Berkeley's *opposition* to 'general ideas'. This is *two-fold*: it is directed, first, against the ontological notion of 'universals *in rebus*'; secondly, against the implication that besides the realm of ideas there is a second realm of 'things'. Even the very *expression* 'abstract general idea', which carried the possibility of such Lockean implications, had to be resisted. For since he lacks the distinction between 'things' and 'concepts', or 'ideas', which Locke—perhaps inconsistently—had allowed in this context, the tasks set for ideas in Berkeley's philosophy becomes further charged with ontological significance, and this

---

mostly to the particularity and concreteness of our *ideas*, or at least speaks of our ideas of particulars, and even 'particular ideas', in *De Motu* we find him saying that 'the natures of things . . . are in fact singulars and concrete'. (Cf. M.7, and above, p. 281n.6.)

[1] Such a position has indeed considerable ancestry.   [2] iii. 3.6; cf. 3.1.

[3] iii.3.11. This, it will be remembered, lay also at the back of the philosophical doctrine which was involved in Locke's hostility to the notion of 'real essence' in the sense of the universal 'species' or 'substantial form' (Cf. 3.17, and above, ch. IV, pp. 210f., 214f.)

[4] For this sense of 'conform', cf. above, ch. IV, sect. 6(b), pp. 251f.

[5] iii.3.11.

consideration will extend also to ideas that are abstract and general. The situation is indeed very confusing: for Berkeley to allow the notion of 'abstract general idea' would *be* to admit the existence of universals in the sense in which, as we saw, Locke had already denied it. Berkeley will therefore have to say (in *his* language) that the very notion is 'inconceivable', and that the corresponding term has no meaning.

It is certainly important to place the emphasis of Berkeley's opposition to Locke at the point of Berkeley's proper intentions. For quite apart from the already-mentioned lightweight objection (which Locke might easily have rebutted) that we cannot frame general images, Locke does ultimately subscribe to a positive account which is not unlike Berkeley's, for he says expressly that 'ideas are general when they are set up as the representative of many particular things'.[1] Moreover, as Berkeley expressly remarks, he is not opposed to the notion of 'general ideas', but only to that of '*abstract* general ideas'.[2] What he means to emphasise by this is that so-called 'general ideas' are 'reducible'; that they are 'in themselves particular', and that their 'generality' is explained, as it was for Locke, by the fact that (a little like words when employed as symbols) they 'represent or stand for all other particular ideas of *the same sort*'.[3] Thus, to revert to our example of the 'triangle in general', such a concept involves no more than the *use* of some particular triangle (such as might be employed by a geometer in a drawing) to 'denote indifferently' a whole group of triangles.[4] To operate with the 'universal idea of a triangle' in such

---

[1] iii.3.11. Locke means: particular things which have a certain 'similitude' to one another. (Cf. 3.13) Berkeley will say that they 'represent or stand for' a given set of 'particular ideas of the same sort' (I.12).   [2] I.12.

[3] Ibid. It is not our purpose to follow up the endless discussions that have been occasioned by the words italicised by Berkeley, for in this book we are not concerned with such questions as the so-called 'problem of universals', except in so far as this might help to characterise the nature of Berkeley's notion of 'particular idea'. I will only say this much: The main problem is whether Berkeley did allow 'universals' in by the back-door when speaking of particulars 'of the same sort'. Perhaps Berkeley might insist on a *basic capacity* to 'intuit' the fact that a certain group of particulars resemble each other in a certain respect. His chief intention is surely to maintain that our assertion of the existence of a similarity must not be understood to be *sanctioned* (in the sense of being peculiarly justified) by an additional feature (called 'the real universal'), believed in some sense to *exist* in the way (or at least in a way similar to that) in which there exists the group of particulars. Whether universals are presupposed as a psychological or logical matter is then a less exciting question, to which the answer may well be in the affirmative. Unfortunately many realists seem to have believed that they had made a case in favour of the ontological contention when they have only shown the need for a satisfaction of certain logical or psychological requirements—a conclusion which hardly follows.

[4] Cf. I.12.

a demonstration is *not* to 'frame an idea' of a triangle which lacks all particular determinations, but amounts to the use of some 'particular triangle' being 'considered' as representing a *class* of particular triangles of a certain sort (e.g. rectilinear).[1]

> A man may *consider* a figure merely as triangular, *without attending* to the particular qualities of the angles, or relations of the sides. So far he may abstract: but this will never prove that he can frame an abstract general *inconsistent* idea of a triangle.[2]

No doubt, Berkeley's method of shifting attention away from 'meaning' to 'use' is very important, intending, as it does, to eliminate misleading ontological inferences from the locutions of our language, e.g. the language of geometers. When a geometer appears 'to reason round a triangle on a blackboard', he is not drawing general conclusions from a particular triangle; but neither is he 'really thinking' of some 'abstract universal triangle'. Rather, he is reasoning about those properties of his triangle which are held in common with the class of triangles concerned in the demonstration, that is to say, with all those triangles which are given the properties mentioned in the definitions, postulates and axioms.[3]

3. FROM 'MEANING' TO 'USE': INSTRUMENTALISM AND SCIENTIFIC CONCEPTS

Berkeley's method of dealing with geometrical concepts involves, we see, above all a shift from 'meaning' to 'use'. In the pursuit of this doctrine, he comes to develop the notion of 'use' primarily as the employment of the relevant terms or symbols within some deductive system of theory, by contrast with his earlier approach which had sought to meet the difficulty presented by general abstract terms through emphasis on their 'representative character'. Germs of this 'instrumentalist' approach may be discerned already in the *Principles*, though the theme is developed further and applied to a consideration of the theoretical notions of physical science in Berkeley's later *De Motu*. At I.19, where Berkeley seeks to purge the doctrine 'that every

---

[1] Cf. I.13.   [2] I.16; my italics.

[3] Of course, 'being concerned' with a particular triangle *quâ* representative symbol, the geometer may still be held to refer to, and in this sense, to think of, 'triangles in general'. But, as we have just seen, Berkeley would not object to this, provided the phrase 'thinking of', is not taken as equivalent to 'having the idea of', where this is then given the misleading ontological interpretation rejected by him.

significant name stands for an idea' of its dangerous implications, he refers to the practice of algebra, where we frequently have occasion to use 'letters' which, though they *may* represent particular quantities, yet 'it is not requisite that in every step each letter suggest to your thoughts that particular quantity it was appointed to stand for'. There is thus no need that 'significant names . . . every time they are *used*, excite in the understanding the ideas they are made to stand for'.[1] The mistake, as Berkeley explains later on, in connection with his opposition to the idea of 'pure space' ('distinct from, or conceivable without body and motion'), lies in the fact that 'we are apt to think every noun substantive stands for a distinct idea'.[2]

A similar shift of emphasis from the representative character of general abstract terms to their employment as 'implicitly defined terms' within some deductive system occurs in Berkeley's considera-tion of certain theoretical terms of physics.[3] Thus (to take one of his own examples) in Newton's *Principia* we find general propositions like 'the change of motion is proportional to the impressed force'. Such a locution might again suggest 'motion in general', not to mention 'force'. Here again, Berkeley at first offers us a reductive analysis which hinges on the general application of the term in question. Such an expression, he tells his reader, amounts to no more than this:

> Whatever motion I consider, whether it be swift or slow, perpendicular, horizontal, or oblique, or in whatever object, the axiom concerning it holds equally true.[4]

On the other hand, in *De Motu*, where Berkeley tackles once more the difficulty presented by the 'abstract terms' of physical science for a philosophy whose ideal of knowledge is the mind's 'meditation . . . on the particular and the concrete',[5] Berkeley seeks a solution more by reference to the systematic, hypothetico-deductive development of scientific theories.

The kind of concepts here called into question are force, attraction, corpuscles (atoms), action and reaction, space, extension, time and motion. We should not imagine that these all raise similar and unique difficulties—notwithstanding that Berkeley sometimes gives

---

[1] Op. cit.; my italics.   [2] P.116. Cf. below, pp. 321–2.
[3] For this, cf. above, ch. III, pp. 102–3; ch. IV, pp. 205–7.
[4] I.11. This passage was referred to above, in our discussion of Berkeley's reaction to Descartes, ch. III, p. 92.   [5] M.4.

this impression. 'Unobservable entities' like atoms will raise one sort of problem for a philosophy which (as we shall find presently) identifies the 'existence' of things with their 'being perceived'. On the other hand, force, extension and space can hardly be called 'unobservable' in the same sense. They are problematic, rather, because not only are they 'general', but they involve 'abstraction' from concrete contexts, whilst yet we continue to regard them as existing in some sense independently from that which is moved and extended, exerts force, acts and is acted upon.[1] 'Force', as well as 'gravity', moreover, is impugned also because it is intended for situations where the 'efficient causality' of matter is being invoked, a notion (as we shall see presently) here trenchantly opposed.[2] Common to all these cases is their postulation of entities for which there is no room in Berkeley's basic inventory. Therefore he tries to banish the life of such terms to the realm of hypothetico-deductive logic. This can, however, be understood in two ways, both of which are hinted at in *De Motu*. The first regards theoretical concepts as 'logical constructions out of the observed events and objects' by which we discover their presence. Like the theories in which they are embedded, we think of them as 'concise, abridged descriptions', either (on one view) of certain sensory elements like colours, pressures, odours, or (on another), of the realm of observable objects and their relations.[3]

There are hints of such a view in the *Principles* but the clearest statement occurs in *De Motu* (sect. 7), where Berkeley writes that the purpose of 'general and abstract terms' is 'to abbreviate speech' and to serve 'for instructional purposes'. In the *Principles* also, terms like 'attraction' or 'gravity' are said to be 'reducible' to 'general rules' (or laws) of nature.[4]

There are, however, hints also of a second and more sophisticated doctrine. Even at P.105, the reductionist account is referred to in a context where natural philosophy is said to be an explanation of 'particular effects' by attention to certain 'analogies, harmonies, and

---

[1] Cf. P.99, quoted above, ch. III, p. 93n.2, where 'extension' is said to involve a 'twofold abstraction', from 'sensible qualities' and from 'being perceived'.

[2] Cf. especially below, pp. 310f.

[3] Cf. R. B. Braithwaite, *Scientific Explanation*, ch. 3, pp. 50–53; Nagel, op. cit., ch. 6.II. Nagel credits Berkeley with being a forerunner of such a view in its first version. But it will be clear from our account that he was steeped too deeply in the great seventeenth and eighteenth century systems of physics to have consistently held to this doctrine, which pays too little attention to the deductive complexities of instrumentalist methodology.   [4] P.104, 105.

agreements . . . in the works of nature'. Commentators, taken up with the reductionist position, have seldom seen this for what it is, namely, a reference to the *systematic* structure of science. The *De Motu* shows, however, that this is the more likely meaning intended. 'Attraction', Berkeley there maintains (and he refers to the authority of Newton), was 'introduced . . . not as a true, physical quality, but only as a mathematical hypothesis'.[1]

The expression 'mathematical hypothesis' had long been used to denote those scientific assumptions which (for one reason or another) were believed to lack physical significance or reference to any physical reality, whilst at the same time being capable of 'accounting for' observable phenomena.[2] The view, which we have already met once or twice, which regards theories as purely formal or abstract instruments for prediction, i.e. deduction of testable propositions, is particularly suitable for incorporating such hypotheses into its methodology. For on this view not only does the question of the independent physical truth of the 'explanatory hypotheses' not arise, but according to the more extreme version of such an 'instrumentalist approach' the hypotheses may well even lack any meaning—for instance when appearing as purely formal, symbolic structures, with which the mathematician can operate in accordance with well-defined 'rules of play'. The 'meaning' of the constituent terms of such hypotheses may at best then be said to lie in their use.[3] And their use consists in helping us to

---

[1] M.17 Cf. below, pp. 312ff. The same point, using this expression, is made at M.28 and 66. Newton's own 'mathematical' approach is due to considerations mentioned already in connection with Locke, above, ch. IV, pp. 269ff. In the *Principia* Newton had indeed insisted that he was using words like 'attraction', 'impulse', etc., to denote 'forces not physically but mathematically', and that they were not to be taken 'to define the kind or the manner of any action'; and that he was not attributing 'forces, in a true and physical sense, to certain centres' when speaking of the latter 'as attracting or as endued with attractive powers' (pp. 5–6, Def. VIII). In his *Siris*, Berkeley again echoes this: 'What is said of forces residing in bodies, whether attracting or repelling, is to be regarded only as a mathematical hypothesis, and not as any thing really existing in nature.' (Op. cit., sect. 234.) Cf. also above, ch. II, pp. 52ff., 58ff.; and below, ch. VI, Appendix.

[2] The most celebrated instance of an hypothesis treated in this way was the heliocentric theory of Copernicus. This could account economically and elegantly for a great variety of astronomical observations, known in Copernicus' time, but was inconsistent with any of the then known, or at least widely accepted, principles of dynamics. Present-day scientists often use the term 'model' (in one of the many senses of that word) for such explanatory assumptions. We have already seen something of the use of the methodology of models in Descartes' theory of optics. (Cf. above, ch. III, pp. 141ff.)

[3] For a brief account of 'instrumentalism' cf. Nagel, op. cit., ch. 6, III. Also above, ch. III, pp. 102f.; ch. IV, pp. 205f.

articulate a theoretical framework, which through the provision of a kind of symbolic network, links together those concrete and particular objects and events which are open to observation or experimental inspection. Clearly, such an approach would give Berkeley a powerful tool through which to reconcile the use of abstract scientific terms with his view that they could not represent anything real 'in the nature of things'. Terms like force, gravity, attraction turn out to be part of a functional device, entering into the axioms of a given science like dynamics or theoretical astronomy. As an instance, Berkeley mentions the resolution of a force by means of the 'parallelogram of forces'. The resolved forces do not 'really' exist; they only

> serve the purpose of mechanical science and reckoning; but to be of service to reckoning and mathematical demonstrations is one thing, to set forth the nature of things is another.[1]

Similarly at M.39 he notes that the abstract and general terms of mechanical science 'are of first utility for theories and formulations, as also for computations about motion'; but once again, such 'devices' or 'fictions' do not exist 'in the nature of things'. And, as we have already abundantly noted, for Berkeley in 'nature' there exist only things that are 'concrete and particular'. His account of these 'abstract' terms was after all so adjusted as to show that they do possess a meaning without the importation of a corresponding 'reality' or 'object'.

Terms like 'reality' and 'object' are, however, systematically vague; for instance, are we talking of the realm of physical reality, or is there intended in Berkeley's argument the rejection only of some philosophical 'metaphysical reality'?[2] To clinch the matter, and reinforce his position on what can rightfully be said to 'exist', Berkeley develops further the doctrine that the meaning of any significant term is some corresponding idea—now that he has purged this doctrine of those cases involving general and abstract terms, merging it with the basic principle that what exists, exists as concrete and particular. It is obvious that his intentions could be realised most satisfactorily by linking all this up with a Lockean theory of

---

[1] M.18.

[2] We shall see in a moment that sometimes Berkeley insists explicitly that he is only rejecting entities belonging to the 'metaphysical' or 'philosopher's' realm; cf. P.35, where (concerning his rejection of 'matter') he remarks: 'The only thing whose existence we deny is that which philosophers call matter or corporeal substance.' It is, alas, not certain that he has not denied rather more!

perception, in the version which says that in the perceptual situation we are concerned with nothing but 'ideas'. Many of the logical and meta-physical contentions which Berkeley wants to make, for instance the denial to physical nature of genuine transeunt causality, could then be rendered in terms of an account of the logical grammar of 'ideas'. Berkeley's important contribution lies in sharpening some of the consequences of Locke's theory by bringing to light the ambiguities in the latter. One result of this will be the elimination of additional entities from the basic inventory of 'nature', particularly those already treated with some suspicion by Locke, e.g. substratum, external object, real essences, transeunt causality ('power'). After discussing the problems arising out of Berkeley's theory of 'existence' and 'law', we shall review his use of the concept of explanation in dealing with those 'entities' which stand in need of questioning in the light of his doctrines.[1]

4. THE ESSE = PERCIPI DOCTRINE

We have already seen that one of the objectives of Berkeley's harangue against the notion of general abstract ideas concerns the implied contention that there is no second realm of external objects (or qualities) which 'conform' to our ideas. We must now look, at least briefly, into this contention, central for Berkeley. The philosophy of Locke obviously contained a number of (what we may call) non-ideal regions: External objects (O_C); primary qualities which were said to 'resemble' our ideas of them; the substantial substratum (S_m) in which adhered or were 'united' the observable qualities (or our 'ideas of qualities', as Locke had sometimes put it); the insensible particles that provided the internal constitution of physical sub-stances (S_p), and that corresponded to some 'theoretical object' (O_T), which—when extended 'realistically'—yielded an object 'in itself', O_M.[2]

Locke had offered a great variety of arguments for all these different kinds of objects; and he had displayed different degrees of sceptical doubt concerning each of them. Nevertheless, common to his conclusion is always the refrain: we lack (or mostly lack) any knowledge of these entities, since they are not objects of perception; where in Locke's terminology this is eventually translated into the point that we possess no sensory 'ideas' of these entities. And it is

[1] Cf. below, pp. 310–16.   [2] Cf. above, pp. 242ff.

here that Berkeley asks his momentous question: Does it *mean* anything to speak of existence in such cases where no perception of an object is *possible*? Already, of course, the whole tenor of empiricist philosophy had been to drive home the importance of sensory perception. Natural philosophers had been enjoined *ad nauseam* to eschew assumptions and hypotheses which could not be tested empirically. Berkeley goes farther. According to him, not only does it not *mean* anything to speak of an object which cannot become an object of perception; but the condition that whatever we postulate as existing shall be perceptible is now also written into the very *meaning of the concept of existence*, in the famous phrase: the *esse* of things is *percipi*;[1] their being *is* to be perceived. Berkeley has here taken Locke to an extreme position. The move was not unexpected, considering the pressures which had impelled Locke to label the 'object of perception' an idea. Once this had been done, sooner or later someone would point out that there was little room left for a second layer of objects.[2] 'What are all those sensible objects which are said to have an existence natural or real', he tells us, 'but things we perceive by sense, and what do we perceive besides our own ideas or sensations?'[3] But by definition 'ideas' are things that cannot live 'without the mind'. No wonder that if things = ideas, things must necessarily be either perceived or be nothing: a conclusion which strongly suggests that their very 'being' or 'existence' is their *percipi*.

We must desist here from a lengthy critique of Berkeley's analysis of 'existence'; many commentaries have amply provided it. No doubt one wants to protest that although an object which exists will be perceived under suitable circumstances by some observer; and

[1] P.3. Cf. P.7, 22, 23. Let us note at once that we should beware of the conclusion which philosophers often draw from this development, which is to assume that Berkeley must be an 'inductivist Newtonian', who is vocal in his opposition to 'hypothesis'. As we have already seen in the previous section (and as will become clearer later) this is by no means the case. Contrary to what is usually supposed, the '*esse = percipi*' doctrine in the end liberates the logical situation, and indeed drives Berkeley to a more sustained employment of a hypothetico-deductive approach.

[2] Cf. Hume, *Treatise*: '. . . mankind who as they perceive only one being, can never assent to the opinion of a double existence and representation'; and he refers to the 'hypothesis' of 'the double existence of perceptions and objects' (pp. 202, 215). But, as we shall see, Berkeley's theory has not entirely emancipated itself from such an 'hypothesis'.

[3] Cf. P.4. There is an interesting slide in this locution. Strictly, Berkeley should translate 'perceiving an object' into '*having* an idea of this object', whereas he converts it into '*perceiving* our ideas or sensations'. This change conceals the eccentric use of 'idea', and eases the way towards the identification of 'idea' and 'thing'. (Cf. P.38, where he acknowledges the eccentricity.)

although, if it cannot or even could never be perceived, it might be said to be nothing to us; yet surely it does not follow that existence is what we *mean* when we say that the object is being perceived. On the other hand, such and similar criticisms are not very useful in guiding us along the main developmental outlines of our history. What we can say is that the Lockean analysis of the perceptual situation ends by regarding the object of perception as 'idea', where this was a relatively arbitrary choice—a kind of 'philosophical decision'—; the argument could equally well have gone the other way, ending up by treating the object of perception as *the* external object.[1] Berkeley, having settled for the 'idealist' alternative, develops the implications of the grammar of 'idea': having denominated the object an 'idea', it follows that its being is that of 'being perceived'—an inversion of the way in which he actually presents his reasoning! Let us, however, proceed along Berkeleyan lines, and see how he will tackle the concept of the object with this interpretation.

And here it seems that Berkeley has surely gone too far, in his desire to banish 'external bodies' or 'inward essences' and 'insensible particles'.[2] He wanted to establish the view that what cannot be observed *in principle* does not exist. But like modern philosophers of science who have sometimes operated with this programme, only to find that they cannot apply it without at least additional rules telling us where to 'draw the line', Berkeley calls in question too much—not only the list of things just mentioned, but also all those objects which are normally said to exist in a commonsense mode of existence. Not only atoms are banished but so also are the stars beyond the telescope, and small organisms beyond the limits of observation by the microscope![3] Or to put it differently: if you try to keep out a troublesome entity by applying the criterion: '*Could* it become an idea?', then most of the philosophical labour will have to be spent on clarifying the logical nature of this 'could'; and the idealist postulate is a relatively crude method of determining the basic inventory of things.

Berkeley therefore is forced very soon to enlarge the narrow basis on which he has placed objective existence. 'The table exists' does not

---

[1] No wonder Berkeley could protest that to call things *ideas* 'matters little' and does not deprive us 'of any one thing in nature' (*Third Dialogue*, ed. cit., p. 289, and P.34).　　[2] Cf. P.18, 102, for these examples.

[3] These do not *seem* capable of the treatment which Berkeley has given to some of the abstract concepts of science like force, space, time, motion, although we shall see presently that he does sometimes (but *only* sometimes) try to do this. And it should be said that if this is a confusion, it is still with us.

only mean that I see and feel it, but also that 'if I was in my study I might perceive it . . .' (although neither I nor anyone else may ever be in it). Nor is this subjunctive analysis sufficient, for Berkeley adds significantly, '. . . or that some other spirit actually does perceive it',[1] thereby restoring 'categorical symmetry'. He clearly feels that the reduction of an 'actually existing object' to what *would be* perceived *if* certain conditions *were* satisfied is unsatisfactory.[2] As P.48 makes clear, Berkeley in this way seeks to ensure the *continued* existence of the object; moreover, nothing less than 'the mind of some eternal spirit'[3] will do: God. Unfortunately, even neglecting the arbitrariness of the conclusion, this will not work even on its own terms. As the *Third Dialogue* admits, the things 'eternally in the mind of God' have, relative to finite observers, only some sort of 'hypothetical existence', and so the old deficiency breaks out again. Besides, these eternal or 'archetypal' ideas are not really in time, whereas physical objects (even if not perceived) would normally be thought to exist in an objective time.[4] Moreover, the argument assumes that whilst my ideas of sensation are 'perceived' but not caused by myself, God's ideas are 'caused' by Himself (a little like my ideas of imagination) though at the same time perceived by me, as though God possessed hypnotic powers. Thus, at P.29 and P.72, Berkeley remarks that our 'ideas of sensation' (unlike those of 'reflection or memory') are 'produced' by God, who 'excites' these ideas 'in our minds'.[5] All this comes close to replacing the words 'physical external object' by 'God's power'.

Digressing for a moment, we may wonder at this procedure; for had not Berkeley's main contention been to proscribe any inference from our 'ideas' to 'external objects'; and with this, from 'effects' to

---

[1] P.3. Cf. *Third Dialogue*, op. cit., p. 290.

[2] One of Berkeley's spiritual inheritors, J. S. Mill, did however adopt the reduction, defining the external object (also called 'matter') as 'a Permanent Possibility of Sensation' (*Examination of Sir William Hamilton*, 6th ed. (1889), ch. 11, p. 233).

[3] P.6. As P.27 explains, although we have no 'idea' of minds (since a mind is that which *perceives* and is never perceived) yet we may be said to have a 'notion' of it. Berkeley does little to develop his grounds for this permissive extension of 'reality'. With Hume, 'mind' will be dissolved further into something 'ideal', i.e. 'perceptions'; Kant, whilst retaining mind as a basic concept of his epistemology, will articulate it in dual fashion, as empirical, and as transcendental consciousness. Berkeley's 'notion' of 'mind' has more resemblance to the latter than the former.

[4] *Dialogues*, op. cit., pp. 290–3. A denial of this—which we find in Kant—would on the other hand require a far more extensive reconstruction of the argument than Berkeley ever envisaged.

[5] That this is a straightforward application of the causal axiom emerges from the fact that Berkeley notes that we 'collect' this inference 'from the light of reason'.

'causes'? Berkeley's reply suggests that his objection is not so much to this form of inference, but to the construal of the cause of our ideas as 'matter or corporeal substance'. It is no wonder, Berkeley points out, echoing Locke's very phrase, that the materialists admit their inability to comprehend the '*manner*' in which such ideas are supposed to be produced by 'external bodies',[1] for anything so utterly dissociated from the linguistic context of mind and ideas has really 'no meaning', and indeed (Berkeley suggests) 'involves a contradiction'.[2] By contrast, we can comprehend the relation between God (or infinite spirit) and the world of ideas, because the concept of such a spirit has a meaning *on analogy with* the concept of finite mind—of which concept we are said to have a 'notion'.[3] So if we assume that our ideas are produced at all, then the inference to a spiritual cause is— according to Berkeley—at least not logically otiose.

I shall not discuss the many weaknesses that can be found in Berkeley's theological argument. Let us return, rather, to the core of his philosophical position: things, regarded as 'combinations of sensible qualities',[4] possess 'ideal' status only. As such, the only relation they can have is to some 'mind'. The sole way in which we can move from them (regarded as objects or contents of *our* 'minds') to an 'other' on which they supposedly depend is *via* a model with 'mental' characteristics.[5] The latter thus becomes the

---

[1] P.19; cf. Locke's *Essay*, iv.11.2; above p. 229.

[2] P.54. Does this mean more than that it contradicts the doctrine of the *esse* = *percipi*? At P.73, Berkeley supports this by impermissibly sliding from the use of 'material substance' as 'external body' to that of 'substratum'. Cf. below, p. 297.

[3] P.140 affirms that 'our idea of spirit' is meaningful on the ground of its 'resemblance' to our 'own mind'. In the *Third Dialogue* 'my soul' is said to 'furnish me with . . . an image, or likeness of God'.

[4] P.38; cf. P.1, P.91. At P.99 Berkeley defines 'objects of sense' as 'nothing but . . . sensations combined, blended, or (if one may so speak) concreted together'. This 'blending' process foreshadows Kant's later concept of 'synthesis', except that the latter has the dual significance of a transcendental as well as psychological process, partly owing to the different logical treatment given to Berkeley's 'ideas'.

[5] The insinuation of such a quasi-causal dependence at this point is arbitrary, as Hume will point out. And in Kant's doctrine, it will be replaced by the notion of 'transcendental condition', a condition which there becomes 'absolute' because that which occurs in experience *quâ* sensational content is not allowed to exist in some quasi-objective spatio-temporal framework of its own, and is not allowed to have some spurious 'objective reality', which is then denied immediately (as in Berkeley) through insistence on its 'ideal status'. (Cf. P.43–44, where 'outness in space' is implied to *belong* to the sensory ideas only *via* certain 'visible ideas and sensations attending vision' in the sense that these can 'suggest' things existing at a distance; a reductionist account which implies that these things do not 'actually exist' in 'external space'. Cf. below, ch. VIII, sect. 5(c), for this interpretation, and its criti-

metaphysical centre of gravity for Berkeley's notion of 'reality', necessarily supplementing the original basis, the realm of 'ideas'.

This is not an aspect which is commonly emphasised by Berkeley's commentators, who more normally, and understandably, fasten on his negative, restrictionist tendencies. For Berkeley clearly vacillates in his attempts to reconstruct his universe on the narrow basis of ideas, witness the dispositional analysis of existence as contrasted with the introduction of God; or his attempt to make the explication of the 'real' (as against the imaginary) lie in its 'steadiness, order and coherence' and relative 'strength',[1] on the one hand, and its being produced 'by another and more powerful spirit', on the other. But the former definition (using criteria 'intrinsic to experience') is insufficient to characterise the difference intended (some 'sensations' are 'non-veridical' for instance; and there are many more difficulties of this sort), for which reason Berkeley falls back on the second alternative, which uses an 'extrinsic' criterion. Perhaps the latter in the end reduces to something that is simply 'basic' and non-inferential. A hint of this is provided at P.36 where he says that the difference between perception and imagination is phenomenologically 'given'.[2] But this is not developed further.

5. THE EXPULSION OF EXTERNAL OBJECTS, SUBSTANCE AND REAL ESSENCE

The most prominent of Berkeley's negative contentions is the attempted proscription of 'external bodies', 'material substance', 'internal essence', and the distinction between primary and secondary qualities—at least in the way in which such a distinction would lend

---

cism.) I harp on these comparisons because only in this way can the critical purges by later philosophers be seen as illuminating the doctrines of their predecessors.

[1] We shall see presently that this 'order and coherence' is defined by reference to the concept of 'the laws of nature'. (Cf. P.30, 36.)

[2] Ideas of sense occur in accordance with the laws of nature, and thus 'speak themselves the effects of a mind more powerful . . . than human'; and it is 'in this sense, the sun that I see by day is the real sun, and that which I imagine by night is the idea of the former'. Now this example had occurred in Locke's *Essay*, iv.2.14, where he asserts that we are 'invincibly conscious' of the difference intended. (Above, p. 237n.1) And I have myself claimed that some such assumption is required if we want to construct a non-idealist theory of external objectivity. If this is accepted, we can at once grasp the real nerve of Berkeley's reference to 'God' in this context.

itself (as it had done in Locke) to a theory of the two-layered universe: the 'ideal' and the 'non-ideal'.

Berkeley has no difficulty in demolishing Locke's contention that the difference between primary and secondary qualities is a difference in kind. First, primary qualities are as relative to points of view as are secondary, even if there are differences in degree between the two cases. Secondly, it is not possible to 'abstract' (in the sense involved in the doctrine of abstract general ideas) primary from secondary qualities, which shows it to be impossible to conceive of the former existing as 'external qualities' and unperceived. If we are to speak meaningfully of primary qualities at all, they must be acknowledged to be 'ideal'; the Lockean notion that primary qualities 'resemble' our ideas of them supposes the former 'unperceived', a 'contradictory' supposition: 'An idea can be like nothing but an idea.'[1] All this shows again a sharpening of what was vague in Locke; the slide from 'quality' to 'idea of quality' is brought out in the open. It does not, however, necessarily support Berkeley's *general* idealist contention (as he indeed seems to sense at P.15); it was perfectly possible for Kant later to agree to the differences between primary and secondary qualities but to maintain that both were real in the sense of 'empirical reality'—not just 'ideas'.[2]

Berkeley's argument against 'material substance' is rather more complex, largely because he uses terms like 'external body', 'matter', 'corporeal substance', etc., interchangeably,[3] although they do not always denote the same entity. Also, like Locke, he speaks of 'substance' sometimes as 'the combination of sensory qualities' (Locke's S_p), sometimes as 'substratum' or 'support' or 'subject' (Locke's S_m).[4]

Berkeley has, of course, little difficulty in demolishing the celebrated 'inference' from our ideas to the existence of 'external bodies'. He merely repeats the point of his predecessors that it might be possible for us to have all our ideas without their being produced by external bodies—at least: *material* bodies.[5] (We have already seen that the objection is not—as was done by Hume later—extended to the case of 'spiritual causes'. Nor was he as yet prepared to surrender that part of the original logical grammar of 'ideas' which required that they

---

[1] P.8, 9, 14, 90. Cf. *First Dialogue*, op. cit., pp. 218–24.
[2] Cf. *Prol.*, par. 13, Note II; K.73–74, 1st ed.
[3] Cf. P.8, P.9, P.19 and *First Dialogue*, op. cit., p. 218.
[4] Cf. P. 37, 11, 12, 49, 73.   [5] Cf. e.g. P.57.

stand in some relation to a subject or mind.) But naturally, the whole of Berkeley's position would be irreconcilable with the notion of the external object (O_C).

Because of the interchangeable terminology just mentioned, Berkeley regards the last argument, directed 'against the existence of matter', as strengthened by his various contentions against 'material' substance', in the sense of 'substratum' (S_m)—for good measure seeking further support from the failure to make good the distinction between primary and secondary qualities already mentioned.[1]

Concerning substance as a 'substratum' or 'subject', Berkeley adds little to Locke's characterisation,[2] but refuses to remain in Locke's position of neutral and detached analysis concerning the logical grammar of category concepts like quality, substance, etc., by turning Locke's confused 'unknown I know not what' into an unqualified 'I do not know nor have any understanding of' such a substratum.[3] But when we ask for Berkeley's reasons for this reaction, we are not really offered anything except the original position which limits the basic inventory of reality to 'ideas' and 'minds'. Berkeley certainly seems to subscribe sometimes to Locke's characterisation of qualities which need to 'exist in a substance'.[4] His modifications of this basic assumption consist first, in denying significance to any other substance than 'mind'; secondly, in asserting that the relation in question is therefore that of 'being in a mind' or 'being perceived'. It will be seen that with this, the predicates, having been prised off a background of substance (as understood by Descartes and Locke), are now related solely through a 'mental bond'. The relatively uncritical analysis of this bond emerges in Berkeley's vacillating attitudes towards the meaning of the concepts of 'existence' and 'reality', as already indicated. And this will hereafter constitute the problem for Berkeley's successors: to offer a rationale of this concept which will avoid sheer 'dogmatic postulation', and above all, the theological model. As this emerges in

[1] Thus, the argument against O_C is described at P.19 as being directed against 'matter or corporeal substances' (though note the plural!), whilst his contention against abstract extension, regarded in the manner of Descartes as a 'mode of matter', and as 'substratum', is likewise called an 'indeterminate description' of 'matter or corporeal substance' (P.11). But though this may produce unwarranted consent on the part of a careless reader, I do not think that Berkeley's arguments are vitiated on *these* grounds.   [2] Cf. above, ch. IV, pp. 215ff.

[3] P.73 declares it to be 'utterly impossible'. In a similar way, Berkeley draws attention to the metaphorical nature of the concept of 'support' which is here invoked though he does not show in detail *why* the metaphor is inadmissible. (Cf. P.37 and 49.)   [4] Cf. P.73.

Hume, it will be constituted through the notion of the 'imagination'; in Kant, through the logic of the transcendental apparatus of concepts, and the forms of space and time.

6. THE ATOMISM OF IDEAS

We have already seen something of Berkeley's efforts to cope with the contemporary concepts of science in the light of his restricted ontology. In the rest of this chapter we shall sketch further the way in which certain scientific and methodological concepts are integrated within this philosophy—playing indeed a major role in its articulation. This will concern in particular the notions of causation, explanation, law of nature, induction. Here Berkeley, like the other figures in our story so far, will prove to be transitional. The theory of ideas will purge these concepts of older and misleading meanings, and indeed use them in lieu of less powerful philosophical paradigms.

Berkeley, we noted, sometimes insists that it makes little difference whether we refer to the objects of perception as things or as ideas; and we should expect him to say this, since he has destroyed the notion of the 'external object' (O_c), which had enabled Locke and others to draw a contrast between thing and idea. On the other hand, this manoeuvre supplies Berkeley with an instrument through which to express a number of logical claims concerning the relations between things or qualities simply as insights which he derives from an inspection of our ideas. And first and foremost, we here meet again a development of the Lockean approach towards 'coexistence'; the claim that the perception of coexistence involves a deficiency, a lack of 'necessary connection'.

We should, of course, expect Berkeley to come to such a view. He has after all 'prised' qualities ('ideas of qualities', or 'ideas' for short) from their foundation in a putative 'substance'; similarly, his opposition to 'inward essence' has made it impossible to think of the secondary qualities as having their 'union' or 'root' in such an essence. Where the Lockean atomism of our ideas in relation to sense-experience had still been contrasted with some deeply buried realm in which the connections are perhaps restored, now this is no longer available: there *are* only ideas; and their atomistic character becomes a more pressing preoccupation. Berkeley refers again and again in the *Principles* to the central passage (P.25) where this is discussed, indicating the importance he attaches to it. Remarkably, this is headed

'Refutation of Locke', because it purports to prove the falsehood of the doctrine that our 'sensations' (and the secondary qualities) are the effects of 'powers resulting from the configuration, number, motion, and size of corpuscles'.[1] The intention here is not so much to question the existence of the 'insensible particles' but to contend that even *with* such particles we have no 'explanation' (understood as 'causation' or 'production') of our sensations. The question of the 'reality' of the particles receives independent consideration.

In this passage Berkeley simply takes Locke's atomism to a more relentless conclusion. We have seen how incessantly Locke already had stressed our inability to *see* any 'necessary connections' between primary and secondary qualities. Berkeley converts this into the contention that no such qualities, that is to say, ideas, *can* have any 'power to act' on one another.[2] And his conclusion is based on what he conceives their epistemological status to be: 'whoever shall attend to his ideas ... will not perceive in them any power or activity; there is therefore no such thing contained in them.'[3]

We should be clear that Berkeley's argument here is concerned with a logical characterisation of ideas, not with their 'factual content' (so to speak) which inspection might reveal. So when he says that we cannot perceive any power or activity in ideas, he really means (as he puts it) that an inspection of 'the very being of an idea' is sufficient to prove that it 'contains' nothing 'active', in the sense that the idea as such is 'passive and inert'. This reference to 'passive' reminds us of Locke's remark that 'in the reception of simple ideas, the understanding is most of all passive',[4] although he had added that our perception of ideas presupposes their being 'taken notice of in the understanding'.[5] The latter reappears in Berkeley as the point

---

[1] Cf. P.102, where Berkeley similarly singles out for censure 'inward essences', taken as equivalent to primary qualities of 'insensible particles'.

[2] P.25 speaks of one idea falsely believed to be 'producing' alterations in another; or 'to be the cause' of anything. P.31 mentions that we know that fire warms 'not by discovering any necessary connection between our ideas'. P.60 says that 'ideas have nothing powerful or operative in them, nor have any necessary connection with the effects ascribed to them'. P.102 opposes the theory of secondary qualities 'depending' on or 'flowing from' primary ones.

Clearly, in Berkeley's mind, the notion of 'necessary connection' is very closely related to that of 'power' or 'causation', and of 'production', whereas Locke had still treated them under separate headings; not until Hume is their close affinity made explicit. (Cf. *Treatise*, I.3.2, op. cit., p. 77.)

[3] P.25.   [4] *Essay*, ii.1.25. Cf. above, p. 249.

[5] ii.9.2. This seemed to make the understanding *both* passive or receptive *and* active. Kant will later construct the notion of 'receptivity' (belonging to sensibility, cf.

that the understanding (and the will—and only these) represent the notion of 'agent' or activity.[1] On the other hand, he seems to be saying that it is their 'being perceived' that makes us describe ideas as 'passive'.[2] (We have already noted the eccentricity of speaking of *ideas* as 'being perceived'.[3] But then this is due to the *decision* to identify the grammar of 'thing' and 'idea' for at least the present purpose.) They are 'passive' as standing in an accusative relation to the perceptual subject ('mind'); but they are also passive in that they are necessarily in need of 'support', of being sustained, by some mind; my own, as regards ideas of the imagination, God's mind, as regards those of sensation.[4] Moreover, in either case they must be regarded as *effects*.

In consequence Berkeley gives us a quasi-psychological model through which to formulate his atomistic position, which thereby becomes a radical version of Locke's; for here—it will be remembered —one had still envisaged the logical *possibility* of discovering necessary connections between those ideas which seemed so far to be merely 'coexisting'.[5] Where Hume will presently employ the test which asks 'Is it conceivable that the ideas might be separated?', in order to determine whether or not one can speak of necessary connections, Berkeley's procedure reminds us of the metaphysical sting

---

above, p. 249n.2) to take care of the former, reserving for the understanding the 'act of spontaneity' which consists in the 'awareness' of the present idea. (As he puts it: The 'I am aware of this' must accompany all my representations. Cf. K.65, 151). It is of course always a question whether such further classificational stratifications are possible; at any rate, their lack permits both Locke and Berkeley to move in a different direction.

[1] Certainly in the case of imagination, the mind is active; but in sensory perception 'some other will' produces the ideas. (Cf. P.28–9.) Berkeley is here silent on the question whether in the latter case our minds are active also, or purely passive. At M.25 he says that 'the mind can be called, correctly enough, a principle of motion'.

[2] 'Since they . . . [our ideas] exist only in the mind, it follows that there is nothing in them [i.e. no activity] but what is perceived' (P.25). This is clearly a misleading phrase. It is not *that we perceive* no activity in our ideas (e.g. the action of a pile-driver) but *that they are perceived* (perceptual states) which deprives them of this. It is thus the analysis of the status of ideas which removes all 'causal efficacy' and related notions from Berkeley's world; and no empirical counter-instances (so often urged by incomprehending critics) can affect this position. If we want to criticise it, we need quite different tools.

[3] Cf. above, p. 291n.3.   [4] P.28–9. Cf. n.1 above.

[5] Let us remember that we are not concerned here with the case of 'analytic propositions' but rather the dark field where we seek to extend the intuitional powers of the mind to yield a model of necessary connection between ideas apparently logically distinct.

contained in this notion itself. In a left-handed way he makes it clear to us—rightly or wrongly—that we are not necessarily concerned here with *logical* necessity, something too easily suggested by Hume's later approach. As we shall see in more detail presently, he feels that for causal connections to serve as genuine explanations, as genuine connections, they have to be set apart from the 'phenomena', moved into a different 'realm'. The paradigm of 'ideas' (in the sense of 'images', passively moved about) is meant to represent this deficiency graphically. But it is of course only an analogy. What the missing 'power or activity' is to the 'ideas', transeunt causality is to 'things'.

All this should make it clear that the 'atomistic' position here espoused cannot simply be attacked by arguments rooted in the consideration that many of the terms of our language, and in particular of scientific language, are not *in fact* 'cut off' from their context in the way Berkeley's doctrine (and Locke's before him) suggests. For the 'theory-laden' aspect of that language[1] in no way detracts from the attempts of our philosophers to reach greater clarity on the notion of the 'synthetic' or the 'contingent'.[2] If the development of scientific theory, which becomes progressively built into our language, seems to diminish increasingly the 'purely synthetic' character of propositions, it may be that all that is really involved is that the connections between the terms of our language have to be construed on an 'empirical basis' *leavened by* '*theory*'.[3] But this does not touch the problem of the thinkers of our period who are trying to elicit the significance of that feature of 'connectivity' which for so long had been felt to dwell (rightly or wrongly), 'over and above' the purely empirical and theoretical basis. *Our* account is solely concerned with *this* feature; with the advance from 'purely dogmatic' assertion *via* 'theological' and 'psychological' models, to the eventual Kantian attempt to construe the connection in terms of a transcendental logic, or some underlying 'logic of our language'. But it is of course true that the model which was used by Berkeley suggests too easy a solution; for Berkeley seeks to restore the broken link between the terms of the proposition (between the 'ideas') by the obvious device of referring us to the 'mind's activity' (and in particular: to the mind of God).

---

[1] Cf. above, ch. II, 33ff.

[2] Cf. above, ch. II, p. 75, and the caution mentioned there.

[3] For the way this will affect our appraisal of Hume, cf. below, ch. VI, pp. 369–71.

7. GOD, THE LAWS OF NATURE AND THE CRITIQUE OF 'CAUSAL ACTIVITY'

The theological model is employed in two or three different ways. We have already mentioned that, according to Berkeley, 'ideas of sensation' differ from 'ideas of the imagination' in being 'more strong, lively and distinct', and possessing more 'steadiness, order and coherence'.[1] Further, their occurrence in my mind is not subject to my will. As to the latter, Berkeley adopts one of the suggestions canvassed by Descartes in the Third and Sixth Meditations[2] that these ideas are 'excited by the will of another and more powerful spirit', a 'spirit that produces them'.[3] We see that this solution gives us a transcendent causal link, a link of which the world of ideas had been deprived.[4] This is particularly prominent when Berkeley deals with the method of explanation by way of the putative action of forces in mechanics. His positive instrumentalist doctrine we have already discussed:[5] such concepts must 'be regarded only as a mathematical hypothesis, and not as any thing really existing in nature'.[6] The reason is plain: things *quâ* ideas contain no transeunt causality:

'It is impossible for an idea to do any thing, or, strictly speaking, to be the cause of any thing.'[7]

Applying this to the special case of force in mechanics, Berkeley

---

[1] P.30, 33. Constancy and coherence are later adopted by Hume as important elements in his account of the causation of our concept of external objects. On the other hand he rejects Berkeley's doctrine of ideas of sensation not depending on our will, let alone depending on the will of God. (Cf. *Treatise*, i.4.2, op. cit., pp. 194–5.)
[2] Cf. above, ch. III, pp. 167–8.   [3] P.33; 29.
[4] We need not remind the reader that such an employment of the concept and principle of causation is entirely non-critical. Hume will presently dismiss it as impermissible; Kant will attempt to demonstrate more fully that any 'application' of the concept (and principle) of causality beyond the context of 'experience' (i.e. of the cognitive judgment) is, to say the least, 'problematic', and is permissible only when regarded as a 'demand of reason'.   [5] Above, pp. 287–9.
[6] *Siris*, sect. 234. Cf. my 'Science and Metaphysics', in D. F. Pears (ed.), *The Nature of Metaphysics*, op. cit., especially section iv.
[7] P.25. The 'strictly speaking' makes plain that this is a matter of logical analysis, and not empirical fact, or rather, not in accordance with plain speech. Berkeley does admit that for practical purposes, 'we must think with the learned, and speak with the vulgar'. In recent times, such a cavalier attitude to the implications of 'vulgar speech' has been questioned; cf. J. L. Austin, *Sense and Sensibilia*, ch. 11, which singles out for attack this approach of Berkeley, drawing particular attention to his use of such terms as 'strictness, precision, accuracy' etc. (p. 133n.). We shall return to this below, p. 305.

writes: 'Nothing mechanical is or really can be a cause.' But he then significantly goes on to say:

> We cannot make even one single step in accounting for the phenomena, without admitting the immediate presence and . . . action of an incorporeal agent, who connects, moves and disposes all things according to such rules . . . as seems good to him.[1]

Here the instrumentalist method of operating with such explanatory concepts is supplemented by an additional feature. Berkeley is clearly not satisfied with his account—sufficient as it may be for the scientist. He seems to feel that it has deprived us of the feature of causal efficacy, which must therefore be located elsewhere; here: in God's activity. However, this solution is a very radical one; it now claims that *every* happening is caused, activated, 'moved', albeit by God.[2]

There is, however, a further way in which God's 'wisdom, goodness and power' manifests itself.[3] The first of Berkeley's ways of distinguishing sensation from imagination seemed to involve what we may call 'immanent' criteria, e.g. those of order and coherence; might not these criteria perhaps be sufficient for the distinction without invoking God? But just as in the last case, we shall see that what at first presents itself as a novel solution, on inspection is insufficient and requires the notion of Berkeley's God. We shall have to approach this in a roundabout way, through a discussion of Berkeley's notion of 'laws of nature'. In any case this is itself of intrinsic interest to us, for Berkeley's doctrine here provides an interesting shift away from older notions of causation and explanation.

Whenever Berkeley speaks in such contexts of 'order and coherence', he makes it clear that he is referring to the 'connection' of ideas in accordance with certain 'set rules or established methods', which (as he says) are 'called the laws of nature'.[4] It appears that the

---

[1] *Siris*, sects. 247, 237.

[2] This doctrine resembles (as Berkeley admits, though with reservations; cf. P.53, 67), the views most prominently defended by Malebranche (*Recherche de la Vérité* (1675), vi.2.3, and 'elucidation 15' of that chapter) that God alone is a 'true cause', and that all 'natural causes' are merely 'occasions' for the happening of that which they would *normally speaking* be said to cause or bring about. Berkeley however rejects even 'occasions' as substitutes for secondary causes.   [3] Cf. P.72.

[4] P.30. Kant was later to define the realm of the 'actual' as 'whatever is connected with perception in accordance with empirical laws' (K.250). In neither case does it seem that this criterion distinguishes sufficiently clearly between imagination, reality, sensory illusions, etc. But the point is not important for us.

conception of 'law of nature' is here introduced as a new (and additional) metaphysical prop for the notion of 'reality', to restore what had seemed lost by the decision to convert things into 'ideas'.[1] I say 'metaphysical', although at first sight the notion is introduced as though it were empirical: Berkeley speaks of our *observation of the settled laws of nature*,[2] and he tells us that 'these we learn by experience, which teaches us that such and such ideas are attended with such and such other ideas, in the ordinary course of things'.[3] And again at P.62:

> There are certain general laws of nature that run through the whole chain of natural effects: these are learned by the observation and study of nature . . . .

Now note that the conception of 'laws of nature' does not only define more specifically what Berkeley intended by 'order and coherence'; it is introduced even more importantly as a translation or reduction of the notions of causation and explanation. Let us consider these briefly in turn.

We have already discussed Berkeley's denial of transeunt causality. Ideas being 'inefficacious perceptions', he tells us explicitly, 'one idea cannot be the cause of another'.[4] What then is his positive explication of this term? His answer is that what we call a cause is in reality no more than 'a mark or sign with the thing signified'.[5] The fire which I see is not the cause of the pain I feel when I come close to it, but simply a mark which forewarns me of the pain to come. Of course, a 'sign' can be used in this way only if accompanied by a 'rule' in accordance with which I can predict those ideas which usually follow upon certain other ideas.

> That food nourishes, sleep refreshes, and fire warms us . . . , all this we know, not by discovering any necessary connection between our ideas, but only by the observation of the settled laws of nature.[6]

This passage shows that Berkeley's 'doctrine of signs' must be taken with a pinch of salt. It is of course obvious that normal usage[7]

---

[1] For the place and function of the conception of law (the 'nomothetic model') for the 'link 'between the term sof propositions, cf. above, ch. II, sect. 2(b), pp. 26ff, especially, pp. 43–5.   [2] P.31; my italics.
[3] P.30.   [4] P.64.   [5] P.65.
[6] P.31; cf. P.65. Notice the similarity of this example to that of Descartes, mentioned above, ch. III, p. 178, where Descartes speaks of 'lack of affinity', resulting in our having to fall back on the 'teaching of nature'.
[7] As already suggested, above, p. 302n.7.

does not sanction the substitution of 'sign' for 'cause'. My friend's hat in the lobby may be *sign*, and forewarn me, of his presence in my study, but it is not the *cause* of his presence. A fever may be the sign or indicator of a disease, and not its cause; at most, in cases like the last we might say that both sign and what it 'signified' are tied together by some underlying causal situation.

However, reference to 'usage' in these cases is evidence of a misunderstanding of the purport of these philosophical 'redescriptions'. The intention of the philosopher is to remove misleading associations implicit in the common terminology, and to claim that the 'true' situation is more appropriately described by the new term. What he wants to maintain is that the situation is more properly regarded as being *like* the one *normally* described by the proposed substitution. Thus Berkeley wants to say that the situation which holds in regard to causal connections is *less misleadingly* described by the 'sign' terminology. True, strictly speaking it seems absurd, if not downright false, to describe it in this way—though only when considered 'empirically'. But as we have seen, Berkeley's concern is with 'philosophers'.[1] The substitution terminology is meant as a protest against the insinuation of the existence of what Berkeley believes to be 'nonexisting' entities of a *metaphysical* kind, here: 'causal links', taken in this way. Nor is his merely a straight denial, for the notion of 'sign' is only the first shot; it is followed immediately (as P.31; cf. p. 303 shows) by the proposal of the institution of a new philosophical concept, viz. the 'laws of nature'.

So we learn here a great deal about the general nature of philosophical thinking. To a large extent it consists in the employment of arguments by which certain concepts are shifted around whose 'normal' meanings express something about the 'metaphysical' situation. Whenever we hear such pronouncements, as that things are 'really only ideas', that empirical generalisations are 'really only hypotheses', that 'abstract general ideas' are *really* only general names, we can be sure that the true intentions of these bold philosophical slogans are as members of a class of philosophical appraisals which this last from Berkeley typifies. Nor does the argument stop there. For similar strategies become mandatory again as soon as the philosopher has adopted the novel point of view, manifesting itself in the new

---

[1] Just as when (concerning his argument about 'substance') he writes that only 'the philosophic, not the vulgar substance' is 'taken away' (P.37).

terminology. This is also the case in what follows, concerning the conception of the laws of nature.

If the law of nature is clearly intended to supply part of the missing furniture which has been lost owing to the abandonment of 'substance', 'causation', and the like, we should expect the difficulties in the analysis of the concept of law itself, noted in our second chapter, to arise with renewed force.[1] It is unlikely that Berkeley would be unconscious of these, but we may expect that the conception of God's activity will do duty here also. This appears in two ways.

First, the 'law-likeness of laws'. Although I have stressed so far Berkeley's empirical approach towards laws, their discovery by observation, and although momentarily, at P.31, Berkeley had *contrasted* 'laws of nature' with 'necessary connections', suggesting thereby a non-necessitarian analysis of law, he also holds that 'this consistent, uniform working' of which laws give us evidence, is due to 'that governing Spirit whose will constitutes the laws of nature'.[2] Secondly, we have the question of the inductive justification. Sometimes Berkeley simply asserts that on the basis of our 'experience' of 'the train and succession of ideas in our minds' we can make not 'uncertain conjectures, but sure and well-grounded predictions'.[3] At P.105, on the other hand, he tells us that the 'rules' based on the observation of analogies and uniformities enable us to make only 'very probable conjectures'. Why only 'very probable'? The answer is given two sections later, where we encounter at the same time Berkeley's postulate of induction:

> By diligent observation of the phenomena within our view, we may discover the general laws of nature, and from them deduce the other phenomena, I do not say demonstrate; for all deductions of that kind depend on a supposition that the Author of nature always operates uniformly, and in a constant observance of those rules we take for principles: which we cannot evidently know.[4]

The problem which Hume was presently to tackle is here bypassed through the reference to the 'Author of Nature', providing us with

---

[1] Cf. above, pp. 26–35.
[2] P.32. Cf. P.106, which speaks of the 'governing spirit who causes certain ...
[phenomena] according to various laws'.   [3] P.59.
[4] P.107. Cf. Newton, *Opticks*, Qu.31: 'And although the arguing from Experiments and Observations by Induction be no Demonstration of general Conclusions; yet it is the best way of arguing which the Nature of Things admits of ...' (op. cit., p. 404). Newton does not here refer to God explicitly.

the paradigm of a metaphysical basis, as we promised to show. Moreover, note that we meet here the third of the ways in which this God plays His part: supplying us with a justificational backing for 'uniformity'; what was called, in chapter II, its 'constitutive ground'.[1] (The other two ways had been: imposition of the specific laws of nature, and production of the individual phenomena. It will be seen that each of these differs from the other in respect of increasing degrees of generality.)

8. THE NEW CONCEPT OF EXPLANATION

Berkeley's elimination of transeunt or 'efficient' causality from the world of phenomena results in pressures leading to the introduction of new supplementary models for reality; the conception of 'laws of nature' is one such analogue. More interesting even is the effect which this phenomenalist pruning has on Berkeley's account of explanation. The latter is, indeed, not due solely to the attempt to offer us substitutes for transeunt causation; it turns up also in the context of Berkeley's answers to objections arising from an apparent consequence of the theory of ideas which makes it impossible to cope with the unobservable entities of the science of his time: atoms on the one hand; the microscopic organisms and structural components of organisms on the other; and again, such phenomena as the motion of the earth, itself surely 'unobservable' in any straightforward way.

In an earlier chapter we noted the older explanation models which had come down to the seventeenth and eighteenth centuries, and which by that time had, if anything, hardened into an unquestioned methodology.[2] I refer in particular to the emphasis on 'substance',

---

The language here used by Berkeley intimately resembles that used by Newton. Thus, in the next section Berkeley refers to scientists 'who frame general rules from the phenomena, and afterwards derive the phenomena from those rules' (whom he thereby takes to 'consider signs rather than causes'). This might be compared with Newton's words in the 31st Query, appended to the 2nd ed. (1717) of the *Opticks*, where he says that although he has not discovered the 'causes of those principles', what he has done is 'to derive two or three general principles of motion from phenomena, and afterward . . . [shown how] all corporeal things follow from those manifest principles.' Similar remarks occur in the General Scholium of the *Principia*, added to the second edition. The point I am stressing is that by this time Newton's *Principia*, and the form in which it is cast, had become a paradigm not only for scientists but for philosophers, when reflecting on the foundations of science and on the logic of phenomena in general.

[1] Cf. pp. 28ff., 61ff.
[2] Cf. ch. II, sect. 2(c), especially pp. 50f.

exemplified *par excellence* by the corpuscularian system, where 'explanation of the phenomena' is often taken to be equivalent to 'production of sensory phenomena' by some 'underlying structure of insensible particles'. Moreover, a useful, and common way of picturing such a model was the image of the watch of clockwork.[1] Here, the 'movements' of the structure of spring and wheels 'produce' the motions of the hands on the dial. For Berkeley, clearly the two crucial difficulties will be, first, that the 'particles' are unobservable, and, secondly, that whatever the analysis of the concept of an unobservable material entity may be, it cannot—on his principles as already seen— exhibit 'productive power'.

Now, contrary to what one might expect, the fact that an explanatory entity is not observable does not invariably cause Berkeley anguish, for he often interprets this to mean that it is not being perceived simply as an accidental matter of fact. Important examples are the motion of the earth, the internal organs and tissue-structure of plants and animals, and the whole world of minute organisms, so small 'as scare to be discerned by the best microscope'.[2] But we have already seen that Berkeley is not averse to employing conditional language to allow for such cases; the unobserved table exists in the sense that it would be perceived *if* someone *were* to direct his gaze at it.[3] This is also the reply he offers to the objection of the unobservability of the earth's motion; and once more the concept of 'the established rules of nature' is employed to give the assumption the stability it requires:

> The question whether the earth moves or no, amounts in reality to no more than this, to wit, whether we have reason to conclude from what hath been observed by astronomers, that if we were placed in such and such circumstances, and such or such a position and distance, both from the earth and sun, we should perceive the former to move among the choir of the planets . . .: and this, by the established rules of nature . . . is reasonably collected from the phenomena.[4]

We see that Berkeley is here moving towards the view that the assumption of physical reality is perfectly possible provided the entity or process in question is part of the condition of perceptibility.

---

[1] Cf. Locke, *Essay*, iv.3.25, for just one among many instances in the contemporary literature. Naturally, it turns up in Berkeley, P.60, where we read of 'the clock-work of nature'.
[2] P.60.   [3] P.3. Cf. above, p. 293.   [4] P.58.

(Although we must not forget that the latter concept is itself capable of manifold interpretations.[1]) So long as such an entity or process has the status of 'idea', it is acceptable; only, since Berkeley never makes it clear what *that* status really is (obviously 'being perceived' is much too narrow as well as too general), cases like those here under consideration come in turn to define that very status.

A similar situation surrounds the question of 'unobserved' body-tissue and microscopic organisms. Here, the relevant sections in the *Principles* require very careful study, for they are written in a very tentative and double-edged way; we can only state the results. At P.64, Berkeley describes them as

> so many instruments in the hand of nature, that being hid, *as it were, behind the scenes,* have a secret operation in producing those appearances which are seen on the theatre of the world, being themselves discernible only to the curious eye of the philosopher [*sc.* natural scientist].[2]

Now this passage shows that Berkeley does *not* deny the existence of micro-entities in principle, provided they are regarded as 'ideas'.[3] His point is quite a different one: supposing we do so regard them, they would at once (*quâ* ideas) be deprived thereby of causal efficacy, since the latter is reserved, as we have seen, for the Creator.

There is, however, a puzzle here. For one might ask (as Berkeley points out) why in that case God should use such complicated (invisible) means to produce the visible appearances? This question Berkeley admits to present a genuine 'difficulty',[4] to which he has only the somewhat lame answer that it is merely through such a complex organisation that God can 'act agreeably to the rules of mechanism, by him for wise ends established'[5]—once more 'laws of nature' defining the extent of nature; together with a bare hint of a teleological causality: the structure of phenomena, visible as well as

---

[1] Let us always remember that in Kant's hands it becomes (as 'conditions of the possibility of experience') nothing less than the framework of space and time, as well as the system of principles generated by the categories.   [2] My italics.

[3] Cf. the locution: '. . . what reason can be assigned why God should make us . . . behold so great variety of *ideas* . . .' (ibid.; my italics).   [4] P.61.

[5] P.62. 'A particular size, figure, motion and disposition of parts are necessary, though not absolutely to the producing any effect, yet to the producing it according to the standing mechanical laws of nature'. (Ibid.)

This passage shows clearly that Berkeley is not opposed to the corpuscularian doctrine provided it is taken *in a certain sense*, nor to speaking of 'production' and 'necessity', again *in that sense*; only, they must not cross his metaphysical proscriptions.

invisible, which 'exists' is that which is conducive to the best state of affairs.[1]

Here again is the contention that the concept of explanation must not be construed in terms of 'production'.[2] And this is, indeed, also the most powerful motive for Berkeley's rejection of the corpuscularian doctrine. Both P.25 and P.102 assert this explicitly. 'To say . . . that [our sensations] are the effects of powers resulting from the configuration, number, motion, and size of corpuscles, must certainly be false',[3] because this is to assign causal efficacy to the corpuscles.

However, there is a further reason for Berkeley's hostility to this doctrine. For he implies that it is possible to interpret the corpuscles in the sense of 'corporeal substance' or 'matter', 'in which extension, figure and motion do actually subsist'.[4] Interpreted in this way, they fall of course under Berkeley's hostile arguments against the use of the distinction between primary and secondary qualities for the purpose of shoring up the doctrine of a non-ideal realm of external reality[5] (an argument which, it will be remembered, Berkeley had further sought to clinch by identifying material substance with 'substratum'). In other words, Berkeley brings out in the open the procedure which uses scientific methodology in order to support a certain metaphysical construction, e.g. the theory of representative perception.[6] This explains why at times he is quite prepared to *speak of* 'size, figure, motion and disposition of the parts' of bodies 'producing' their 'effects'.[7] What he will not allow is the postulation of some 'reality', which would have to subsist at the ontological level of 'ideas', and which might correspond (*per impossibile*) to the term 'produce' (or cause). He does of course in fact believe (as already seen)

[1] 'Actuality' is defined in these terms also by Leibniz. (Cf. *Discourse on Metaphysics*, sects. 13, 19.)

[2] I.e. 'production' in Berkeley's special sense. (See above, p. 309n.5.)

[3] P.25. This 'falsehood', it will be remembered, was Locke's view. But once more we must realise that Berkeley is using his terms in a special sense.

[4] P.9. This corresponds to what in the previous chapter I called the Lockean 'metaphysical extension' of the concept of theoretical particles, the move from O_T to O_M. (Cf. above, pp. 295–7.)   [5] Cf. P.10.

[6] Our argument here takes up the subject where we left it in an earlier section, above, pp. 287–90, where we considered it in the light of the problems raised by the doctrine of 'abstract ideas', whereas now we are more concerned with the use Berkeley makes of the concept of explanation in his attempts to banish certain *metaphysical* positions which he considers mistaken. Ultimately, of course, the two aspects coalesce. Berkeley's points, and the whole approach, repeat themselves with remarkable similarity—but with greater confusion—in the writings of Mach and Duhem. (Cf. in particular the latter's *Aim and Structure of Physical Theory*, Pt. I, chs. 1 and 2.)   [7] P.62. Cf. p. 309n.5.

that genuine causal efficacy has to be a feature which subsists *somewhere*; though for him it must be 'over and above' the level of empirical reality (i.e., of 'ideas'); at any rate, the level has to be different from that assigned to the empirical realm.[1] But it does not follow that despite these metaphysical qualms (and corresponding decisions) Berkeley (and other phenomenalists like him) could not make room for concepts (and 'entities') like atoms, forces and causal or productive capacities *in some sense*.[2] Too many philosophers (even nowadays) think that a philosophical analysis which 'reduces' certain concepts by moving them to some other level must thereby be involved in corresponding *empirical* denials also. All that such misunderstandings show is the difficulty of a demarcation of the realm of 'metaphysics'.

As a result, Berkeley will therefore have to formulate a notion of 'explanation'[3] that will eliminate reference, not so much to unobservable entities, as to a metaphysical level with which the former have become confused.[4] Sometimes, the new proposals are indeed couched in reductionist language;[5] they then appear to eliminate completely the realm of 'unobservables', here: corpuscles. At P.50 Berkeley tackles the objection that on his principles 'the whole corpuscular philosophy' is 'destroyed': and his reply is

that there is not any one *phenomenon* explained on that [corpuscularian] supposition, which may not as well be explained without it, as might easily be made appear by an *induction of particulars*.[6]

[1] All the philosophers of our period share this view. Descartes, like Berkeley, had reserved primary causal efficacy for God. Hume will find its model in some function of the mind; Kant will locate it in the transcendental framework which is a condition of the 'possibility of experience'.

[2] How well a denial of transeunt causality and active force, at one level, may coexist in perfect harmony with their admission at another (positive) level, will be shown to constitute a key feature for our understanding of Hume.

[3] Cf. above, pp. 287–90.

[4] This amounted to a sharpening and making explicit of a metaphysical notion of explanation which had never been clearly grasped. Once understood, it was easy to confute it, except that the new formulation in turn denies more than it intends. To achieve a proper clarification of the issues involved would require a far more sustained investigation, and a more sophisticated definition, of the realm of 'phenomena' (phenomenal objects). This we do not meet until Kant, where only those 'entities' are proscribed which escape the network of the formal (intuitive and conceptual) conditions of experience (space, time and the categories). (Cf. particularly Kant's comments on the place of the employment of hypotheses, below, pp. 516ff.; also pp. 547–8.)

[5] For the contrast between 'reductionist' and 'instrumentalist' approaches, cf. above, pp. 287ff.   [6] Berkeley's italics.

For, he goes on, when we consider what we *mean* by explanation, it amounts to no more than 'to show, *why* upon such and such occasions we are affected with such and such ideas'.[1]

The choice of the word 'why' is ambiguous, and points in two directions. The first corresponds rather to 'that': Berkeley wants to say that explanation of a phenomenon is tantamount to showing that it is a member of a law-like regularity. This is the account he gives at P.62:

> [To explain a phenomenon] consists only in showing the conformity any particular phenomenon hath to the general laws of nature, or which is the same thing, in discovering the *uniformity* there is in the production of natural effects.[2]

This approach emphasises once again the notion of law. But there is a second: quite often Berkeley realises that there is a more appropriate model by which we may analyse the notion of explanation, viz. the concept of a scientific system of laws regarded as a deductive development of certain chosen axioms. Berkeley, writing twenty-three years after the publication of Newton's *Principia*, could hardly fail to be influenced by the form in which the latter was presented. And although the emphasis is on the misleading metaphysical conclusions drawn from corpuscularianism and the Newtonian theory of universal gravitation, he avails himself nevertheless of relevant *formal* features of that approach. This is particularly marked when he deals with the concept of gravitational attraction. We have already mentioned the doctrine of 'mathematical hypotheses' which was meant as an account of the significance of 'abstract' scientific terms.[3] Here we shall only stress the systemic aspect. When we explain the phenomena of dynamics, what is the function of 'attraction'? It clearly cannot be the name for anything like an 'efficient cause'.[4] Nor does it inform us concerning the mechanism lying behind the phenomenon.[5] Indeed, what we mostly mean by this term is a certain 'effect', viz. accelerated motion.[6] And further, 'attraction' does not designate just *any* motion, but rather, motion in accordance with a law.[7]

---

[1] Op. cit.; italics mine.   [2] Berkeley's italics.   [3] Cf. above, pp. 287ff.
[4] P.105. Like Locke before him (cf. above, ch. IV, pp. 269ff.), Berkeley slides from 'we cannot understand how matter acts across a distance' to 'we cannot understand how matter acts causally under any circumstances'.
[5] P.103–4.   [6] Ibid. Cf. *Siris*, 231, 234.
[7] Cf. Newton: 'Active principles, such as is that of gravity ... I consider, not as occult qualities ... but as general laws of nature' (*Opticks*, Qu.31, op. cit., p. 401).

So far, however, we are still only dealing with the law-like aspect of the concept, and not that of its forming a basic element in a deductive theory. We get a hint at P.104, where Berkeley notes that certain *analogies* existing between terrestrial and celestial motions suggest that they have 'a mutual tendency towards each other', for which we use the 'general name attraction'. 'Accounting' for such phenomena is equivalent to letting the suggestive analogies 'reduce' these to attraction, in the sense that they are 'a particular example of the general law'. Now here, however, it is clear that Berkeley means by 'reduction' something effected by the construction of an axiomatic system like that of Newton's *Principia*.[1]

The point is important. Earlier, *De Motu* had repeated a common complaint at that time against Newton, that 'gravity' was an attempt to provide an explanation by means of an 'occult quality'; and 'occult explains nothing'.[2] Now opponents of Newton, like Leibniz, object against 'gravity' mainly that it involves the postulation of 'action-at-a-distance', which is 'against the nature of things',[3] meaning, for Leibniz: in conflict with the programme of mechanics which limits itself to explanation in terms of impact, and the laws of impact, between material bodies. This is, however, not Berkeley's objection, nor is his response to Newton like that of Leibniz. For his point is that *no forces* (regarded as providing models for transeunt causality) can provide us with explanations in science. His objection to 'occult qualities' is directed therefore only against them as 'causal agents'. If, on the other hand, we 'reduce' their meaning to that of terms in a deductive system, they are perfectly acceptable. And within such a system, the question of whether we should employ impact notions only, or whether we should allow 'distance forces', is unsubstantial. His emphasis on systematic explanation thus amounts to an important advance in clarifying the methodological structure of science. Berkeley is more explicit in *De Motu*:

A thing can be said to be explained mechanically . . . when it is reduced

---

[1] Though the aspect of 'analogy' involved in this form of reasoning has only been noted in detail in very recent times.   [2] M.6. Cf. M.4–5.

[3] Cf. *New Essays*, Preface, quoted above, ch. IV, p. 269n.4. This Newt on increasingly agreed to, defending his position by insisting that though the mechanism that lay behind the phenomena of attraction was unknown, and at best no more than a subject of speculation at that time, in the systematic sections of his *Principia* he was concerned only with the laws of the phenomenon. (Cf. above, p. 287.) And this is Berkeley's position, except that he takes it further by insisting that explanation by way of a systematic formulation of laws is the *only* explanation of which science is capable.

to those most simple and universal principles, and shown by accurate reasoning to be in agreement and connection with them.[1]

M.39 notes that *terms* like force, action, etc. are of 'first utility for theories and formulations'. But we need not concern ourselves with the seat of forces:

> The traditional formulations of rules and laws of motions, along with the theorems thence deduced remain unshaken, provided that sensible effects and the reasoning grounded in them are granted.[2]

A clearer example of the 'hypothetico-deductive method' could not be found. But the clearest statement occurs perhaps at P.69:

> Physically . . . a thing is explained not by assigning its truly active and incorporeal cause, but by showing its connection with mechanical principles, such as action and reaction are always opposite and equal.

The conclusion to which Berkeley comes again and again is that we are dealing here with the mathematical expression of dynamical phenomena, a conclusion reinforced in his mind by certain contradictory views held concerning the 'nature of force' in his time.[3] But the important thing is that the expression 'mathematical hypothesis' is one which now directs us to this new 'centre of gravity', the axiomatic system. The 'principles of experimental philosophy' which 'explain the phenomena' are not to be confused with transeunt causal agents; they are 'foundations of knowledge' rather than 'of existence'. Thus the principles of mechanics are

> the primary laws of motions which have been proved by experiments, elaborated by reason, and rendered universal. [They are principles in the sense that] from them are derived both general mechanical theorems and particular explanations of the phenomena.[4]

As usual, the language does not quite correspond to Berkeley's true intentions. He talks as though the scientists had been confused when offering, as explanations in science, things, relations or processes, instead of—so he seems to say—*propositions* concerning these things, relations or processes. But if this be an error to which writers have been subject, it can only be due to hasty language; no one would

---

[1] M.37.   [2] M.28.

[3] Cf. M.67-8, which discusses the conflicting views of Newton, Borelli, Torricelli; alas, a very unsafe ground on which to base such a general logical conclusion.

[4] M.36. The similarity to Newton's phraseology is again remarkable.

seriously be misled or confused in this way. Whereas, what Berkeley is really suggesting is—as already indicated—that we should replace physical forces by dynamical laws of motion; and lest this once more look like a contrast between a physical entity and a linguistic one, we hasten to add that he meant: existing uniformities of motion.

But this still does not fully catch Berkeley's true intentions, which are surely metaphysical (as already suggested in my remarks concerning his theory of causation in general). Transeunt causal agents materially acting *between* things ('ideas') are only opposed in so far as they dwell in a region set over and above that of the 'phenomena' ('ideas'). But explanatory entities which offer themselves as 'ultimate explainers' (so Berkeley wants to say) *must* be removed from the realm of phenomena, to another level of 'reality'; this was after all a presupposition inherent in earlier speculative natural philosophy stretching back into Greek times. Now one half of Berkeley still accepts this, but then insists that such entities must be 'spiritual'. Yet, the other half sees quite clearly that an explanation schema *can* be constructed which avoids falling back on extra-phenomenal entities except for the incidence of logical form; here lies the significance of the introduction of the notion of empirical laws, and their systematic integration. And yet, even here—as we have seen—the break is not made complete; for when Berkeley faces the problem of justification of these laws, and of the law-likeness of nature, he falls back once more upon the goodness and power of God.

Using the example from gravitational attraction we might sum this up by saying that whilst superficially it appears that Berkeley is replacing a *physical* explanation by a *formal* derivation, in reality he is more concerned to argue that explanation in science should not involve a reference to *metaphysical* entities; whilst at the same time confusedly continuing in the belief that *genuine* explanation *ought* to involve metaphysical efficacy. But since 'force' also plays the rôle of an empirical analogue for such a metaphysical agency, he continues to speak as though he was concerned with eliminating *physical* agencies from natural science as well.

The 'principle of existence', referred to previously,[1] clearly denotes a reference to God. This shows that Berkeley is still subject to a number of opposing pulls, and that older concepts have not as yet been integrated with his basic position. This can be appreciated best if we consider subsequent developments. For gradually, as the concept

---

[1] Above, p. 314, in the passage from M.36.

of God as 'creative force' is whittled away, the significance of 'systematisation' increases, reaching its apogée in Kantian philosophy, where God has become simply the 'idea' (or analogical image) of the 'systematic unity' of theoretical science.[1] Moreover, in respect of the phenomenal world, the notion of God as the ground of all *possible* existence, will be replaced by the conception of 'possible experience', as comprising 'all empirical reality as the condition of its possibility'.[2] And as to 'existence', Kant will simply take further Berkeley's reference of this to perception, insisting that the *only* meaning we can attach to the notion of an object *quâ existing*, is that our concept of such an object in general—as which it must be 'thought only as conforming to the universal conditions of possible empirical knowledge' (i.e. cognitive judgments) in general—should (when thinking it as existing') 'thereby [have] obtained [only] an additional possible perception.'[3]

Only here, therefore, is the world at last centred on 'experience', in the sense of 'possible experience'; Berkeley has only begun this process of re-location of the basic concepts of philosophy. Nor would it be possible for him to give an adequate account of the cognitive basis of 'dynamical principles' until this notion had been logically 'anaesthetised' by Kant, as mere 'analogies', e.g. in the concept of causality (physical dependence), viz. as 'succession [necessarily] subject to a rule'[4]—thus taking (so to speak) the 'push-pull' element out of efficient causality. And so in the end, just as the notion of 'law imposed from without' is in Kant replaced by that of 'law self-imposed', so the concept of 'system', instead of mirroring a harmony designed by the Creator for the optimum good of man,[5] is here viewed as the creative organisation of man's knowledge of nature in accordance with certain 'regulative principles' (including those of a teleological kind) likewise self-imposed; something which employs the 'image' of God only as our method of *viewing* the teleological organisation of nature.

We have here looked ahead, because the accomplished stage of a later period makes it easier to appreciate the true significance of the earlier developments. At any rate we have now shown in more detail

---

[1] Cf. below, ch. VIII, sects. 4(c): (vi–vii).
[2] K.494; cf. below, ch. VIII, sect. 4(a).   [3] K.506.   [4] K.212, 185.
[5] At P.105 Berkeley speaks of the objectives of the scientist being the discovery of analogies and harmonies; at P.107 'the whole creation' is said to be 'the workmanship of a wise and good agent', so that scientists should 'employ their thoughts . . . about final causes of things'.

the development of the model (nomothetic model) of law-like connection, as well as the transition from the idea of explanation through substance and force to that of the systematic connection of laws (the systemic model of connection):[1] the true *raison d'être* being Berkeley's attempt to achieve a clearer separation of physics and metaphysics, at least as these concepts are understood by him.

9. THE PROBLEM OF 'ABSOLUTE SPACE AND MOTION'

The curious interplay between factual considerations (bearing on the choice of 'substitution-terminology') and metaphysical objectives, to which we have drawn attention,[2] plays its part also in Berkeley's important theory of space, time and motion; it explains how purely metaphysical issues can cast their shadow on scientific formulations, particularly where conceptual matters are involved. After what has preceded, the broad outlines of Berkeley's theory here are as expected.[3]

Let us first consider motion. Here, in the first instance, Berkeley applies the results of his argument against abstract ideas. The point is made in terms of what it is possible for us to 'conceive':[4] it is not possible 'to conceive motion without . . . [conceiving] two bodies, whereof the distance or position in regard to each other is varied'.[5] The concept of motion thus 'does necessarily include relation'. The contrary supposition depends upon false 'abstraction'.[6] So Berkeley arrives at the exciting conclusion that 'all motion is relative'.[7] However, in order to secure this conclusion, he also conjoins an empirical argument, involving the notion of 'perception' rather than 'conception'. According to this, 'motion by itself is never perceived'.[8] That this is a novel argument becomes clearer at M.63, where we meet with an early anticipation of an 'operational approach' towards meaning, when Berkeley writes that 'no motion can be recognised or measured, unless through sensible things'.

I think that Berkeley is here proposing the view that non-relative motion is a meaningless concept ('cannot be understood')[9] in virtue

---

[1] Cf. above, ch. II, pp. 51ff., 55ff.   [2] Cf. above, p. 305.
[3] We shall not discuss his treatment of 'time', which receives only cursory mention. (Cf. P.97, 98, 111.)
[4] We have noted in our section on abstract ideas the systematic ambiguity surrounding this approach. Cf. above, pp. 280–5.
[5] P.112. Indeed, it may be the case that we need at least *three* such bodies.
[6] P.111. Cf. M.43.   [7] M.58, P.112.   [8] M.43.   [9] M.58.

of the fact that its *magnitude* cannot be *measured*. Thus he contends that it is not possible to 'comprehend' that an 'actual motion can exist in' a single body, in a universe from which all other bodies have been removed. Perhaps this is an open question. Someone may object that it is surely difficult to believe that a body, at first *known* to be in motion by reference to some other physical system, should *cease* to be in motion merely in virtue of the annihilation of that (and any other possible) system.[1] But Berkeley's answer would be that such a supposition involves a vicious 'abstraction'—at least in virtue of the *esse-percipi* principle, if nothing else; the point being that without such a reference-system the notion of motion is simply undefined. It is not that we have to decide whether the body is still in motion (after annihilation of the reference system), but rather that the assertion of *either* motion *or* rest is now meaningless.[2] If we do not appreciate this it is simply because in such cases we unconsciously introduce a hidden reference-system whose *possible* existence *might* define the motion, and even 'produce' it.[3] Nevertheless, it seems that the conclusion is at least partly dependent on the invocation of the *esse-percipi* principle, and its attendant metaphysics—a salutary reminder in the light of parallel modern doctrines which imply that the concept of 'sharp simultaneous position and momentum' of sub-atomic particles, not being operationally defined, is meaningless.[4]

There are two difficulties for Berkeley in this definition of motion. First (as he himself admits)[5] it seems against 'common sense' to speak of any body being in motion simply because some other body is moving relative to it—as appears to be implied by his theory. His answer is that this is *not* a necessary consequence, and that anyhow 'common sense knows' when to speak of motion. In a more serious vein, he suddenly introduces the proposal that we should speak of motion only in such cases where, in addition to relative change of

[1] And it would certainly not be open to Berkeley to plead an essential interconnectedness of physical objects with one another, in the light of his professed 'atomistic' theory of knowledge!
[2] This fact is concealed by the Cartesian contention that motion is a 'state of the body'. On the other hand, the reader will note that Berkeley's move from nonperceptibility to meaninglessness involves the *esse-percipi* principle.
[3] P.115. Here we meet again with the introduction of the subjunctive.
[4] There is, however, a difference, for the 'non-existence' of the sharpness is also the implication of an axiom of that theory and of its non-commutative algebra. Berkeley has no corresponding scientific axiom, unless the first law of motion is viewed in this way. But apart from a casual remark that the laws of motion hold, whether we introduce absolute space and motion or not (M.65), this is not worked out.
[5] P.113.

position, some 'force or action' either has, or is imagined to have produced that motion.[1]

The employment of force may astonish the reader accustomed to Berkeley's inveterate hostility to this concept. However, as *De Motu* makes quite clear, he would probably say that such forces are merely 'mathematical hypotheses', in other words, theoretical concepts entering the axioms of mechanics; purely 'mathematical entities [that] have no stable essence in the nature of things'.[2] However, it is by no means clear whether, if such a 'reduction' is carried out, Berkeley's reference to 'force' can still be used in the way proposed. We cannot pursue the matter; certainly any reductionist account would require a far more complicated apparatus of mechanical theorising (such as later appears in the works of Hertz[3] and Einstein) than was at Berkeley's disposal. In the absence of such an apparatus, his proposals (though historically influential) are at this stage no more than a piece of hopeful optimism.

There was, however, a second and more interesting difficulty for Berkeley, which has powerfully affected the history of the subject. Briefly, this is Newton's contention that whilst—consistent with the First Law of motion—we cannot distinguish operationally between absolute and relative *uniform* motion (*sc.* velocity), there are ways of making absolute circular motion manifest, through certain forces which are set up in such cases.[4] Berkeley's reply to this is two-fold. On the one side he argues that 'centrifugal' forces set up in the rotating body are as such fictitious, so that absolute circular motion can hardly be recognised in this way.[5] On the other side, more

---

[1] This is a criterion which creates the distinction between 'apparent' and 'real' motion, and does not affect that between 'relative' and 'absolute'. According to P.112, both apparent and real motions are relative.

[2] M.67. Cf. above, pp. 286, 288f, 312, for a discussion of this 'instrumentalist' approach. Besides, as has just been suggested, once metaphysical misunderstandings were out of the way, Berkeley would presumably have no qualms at using 'force' in a physical sense. In fact, his invoking the concept in the present context provides some support for my interpretation of Berkeley's intentions.

[3] Cf. H. Hertz, *The Principles of Mechanics Presented in a New Form*, (1894), Introduction (Dover ed. 1956), especially pp. 26–8. (Published nine years after Mach's *Mechanics*, it owes much to the latter's ideas.)

[4] This refers to Newton's famous rotating bucket experiment. There is no need to go into the matter in any detail. For a discussion, cf. E. Mach, *Science of Mechanics*, ch. ii, sect. 6; for a modern discussion, cf. D. W. Sciama, *The Unity of the Universe* (London, 1959), ch. 7, pp. 94–7. For Newton, cf. *Principia*, op. cit., pp. 10–11.

[5] M.60. Actually, in the *Principles* Berkeley also reiterates his point that the real motion of the water in the bucket can be distinguished from its initial motion (when only the bucket was revolving) by virtue of its not being acted on by any forces.

fundamentally, every argument which could prove the existence of some effect associated with circular motion, could equally be (and in fact is) described by reference to the frame of fixed stars.[1]

This conclusion is for Berkeley also decisive against 'absolute space', implicitly thus treating the notions of 'absolute' and 'relative' space as corresponding to the distinction between absolute and relative motion.[2] But before turning to this, let us reflect further on the relation between the general metaphysical and conceptual contentions of Berkeley on the one hand, and their relevance to scientific theorising on the other; remarks which at the same time may be taken to have also some bearing on the topic of the relations between philosophy, philosophy of science, and theoretical science in general.

We may discuss this most usefully by asking the question: Why did Berkeley have so little immediate influence on the physicists of his and even later days?[3] It is trivial to reply that he was perhaps 'ahead of his time'. Part of the answer lies of course in the very success of Newton's *Principia*, and the resulting intellectual atmosphere, which did not require discussion of the foundations of that system. But there is a deeper reason. Berkeley, with all his criticism, leaves untouched and unattacked the actual content of the physical system. It had after all been his contention that this system could stand as well without the notions of absolute space, time, and motion, as with it. With the knowledge of hindsight—with the results of the arguments from Einsteinian relativity in mind—we can see that the significance of the denial of absolute motion was not really appreciated; or rather: the various possibilities which exist when the denial is further specified were not investigated; conceptual reformulation (under the guidance of metaphysical interests) ran ahead, preventing fruitful contact with the bases of the Newtonian system.

Consider one of the interpretations given to this principle in the Special Theory of Relativity. First, it generalises a corollary of Newton's dynamics, to the effect that the laws of nature are not altered in form for systems moving in straight lines, whatever the value of their *uniform* velocity may be ('inertial frames'). Secondly, it employs the observed phenomenon that the velocity of light is constant with respect to every inertial system by interpreting this as

[1] This constitutes part of what later came to be called 'Mach's Principle'; cf. Sciama, op. cit., p. 97.   [2] Cf. M.52.
[3] Mach's frequent mention of Berkeley makes it likely that by the end of the nineteenth century his ideas *were* beginning to be noticed.

implying an impossibility of observing the motion of any physical system with respect to an 'ether' ('absolute space').

It will be clear that, to say the least, such formulations are very *specific* interpretations of Berkeley's *general* 'relativity' principle. The situation here resembles that of Descartes' formulation of a 'conservation of motion' principle, which too, allows of several interpretations.[1] All this shows that there is a certain 'looseness of fit' between the general conclusions derived from philosophical arguments (whose special aims are broadly—and certainly here—of a 'metaphysical' nature), and those conceptual investigations into the foundations of a science which become imperative at times of crisis. 'Looseness of fit' means that there is no sharp dividing line between the two. A moral of this story is that we cannot predict, antecedently to the event, what relevance some particular philosophical argument may have for some given branch of science.[2]

These last remarks have special relevance for Berkeley's arguments concerning space. For although these—as far as we have seen—are drawn from considerations of a relatively technical nature (the question of absolute motion), they are supported by (and in their turn support) a different and more general set of considerations. For the argument against 'absolute space' is supplemented by another, directed against what Berkeley calls 'pure space exclusive of all body'.[3] In fact, we may regard Berkeley's arguments against Newton as being motivated by an attempt to resist the claim that his general objection to the conception of 'pure space' *is in conflict* with the implications of Newtonian science. Once this claim has been successfully rebutted, he can then turn to general philosophical matters without fearing any conflict with the scientists, though—as I have suggested—the connection between the two notions of non-relative space may not be as close as Berkeley believes.

On the whole Berkeley's views on space amount to a certain clarification and tightening of some of Locke's rather fluid, and even contradictory views; and it is especially where these views had maintained themselves through the weight of the associated terminology

[1] For other examples, cf. above, ch. II, pp. 40–1; and ch. III, pp. 149–50.
[2] Our studies of Leibniz's and Kant's philosophy of science will offer further support for this generalisation.
[3] P.116. Sometimes Berkeley seems to treat 'pure' and 'absolute' as interchangeable terms. Thus he asks at M.54 whether 'it is possible to form any idea of that pure, real and absolute space continuing to exist after the annihilation of all bodies'. But they are really quite distinct.

that Berkeley calls the bluff. In this way he dismisses without mention Locke's manoeuvre of calling space a 'simple idea'.[1] Similarly, he rejects Locke's tentative and occasional adoption of the Newtonian and Neo-Platonic device of making space (and time) dependent on, or connected with, God's presence.[2] Let us turn to his own account.

Basically, 'pure space, exclusive of all body', being 'unperceivable to sense', and indeed 'abstracted from our senses'[3] is taken to be a conception of something non-spiritual, putatively 'existing without the mind',[4] and for this reason already not able to fit into Berkeley's inventory. Upon investigation it turns out to be 'nothing positive', simply amounting to 'the absence of bodies';[5] it seems to mean no more than the 'absence' of 'resistance' to some other body (such as my arm); or, if we imagine all bodies annihilated, a *possibility* for my limbs to move without resistance.[6] It may indeed *seem* that our sense of sight furnishes us directly with the idea of pure space; but this Berkeley denies: 'distance or outness' is 'only suggested' by ideas which themselves are not thus characterisable.[7]

We cannot subject this contention to any lengthy critique. It involves a remarkably restricted interpretation of the nature of those ideas of sensation which the mind 'perceives', and the way in which it 'perceives' them. But the comparison with Kant is again instructive. This we shall discuss in detail in its proper place. Let us only say that Kant like Berkeley denies the existence of any empty self-subsistent space, if by this be meant something subsisting independently of the

---

[1] Cf. above, ch. IV, pp. 254, 258, 260.

[2] At M.54 Berkeley simply dismisses without argument this theory of space being 'a participant in the divine attributes', and at P.117 it is called a 'pernicious and absurd' notion. Why? Presumably, because any conception of a non-spiritual element in God, and any mingling of 'matter' and 'mind', is basically forbidden in his system. Berkeley is silent on the development of this theme in the controversy between the Newtonians and Leibniz.

[3] P.111. P.116 calls it an 'abstract idea'. Unlike Locke in some of his moods, to be able to distinguish between body, extension and space does not argue to any ontological separateness. (For Locke, cf. above, ch. IV, pp. 256ff.)

[4] P.116.

[5] M.55. This had already been one of Descartes' criticisms. Moreover, such a conclusion was certainly 'in the air', witness Locke's 'earlier' theory of space. (Cf. above, p. 258.)

[6] P.116. Cf. Locke's similar suggestion in the *Journal*, above, ibid.

[7] Cf. P.116, P.43, which refer to Berkeley's earlier *Theory of Vision*. Being only 'suggested', space and time do not (as in Locke) take on a quasi-independent existence: the opposition to abstract ideas makes this impossible. (Cf. against this the crucially different approach in Locke, above, p. 254, passage quoted from *Essay*, ii.12.8. Also above, p. 294n.5; and p. 286, for the remark quoted from P.116.)

perceptual situation. Again, according to Kant, that which is 'given *a posteriori*' (and as in Berkeley, 'passively received') does not in its own right, so to speak, possess characteristics which we would want to describe as '*its own* space and time', neither basic nor, however, as in Berkeley, even derived. For where Kant differs is in denying that these characteristics could be inferred from, or 'suggested' by, and hence 'reduced to' the qualitative aspects of the *a posteriori*, which in Kant's eyes would make space something 'illusory', or an 'imaginary entity'.[1] (The Kantian '*a posteriori*' must therefore not be identified with Berkeley's 'ideas'!) On the contrary, the space 'in which' what appears is perceived 'out there' is basic and non-derivative, being simply an '*a priori* presupposition' for anything to be as such perceived at all. In short, *like* Berkeley's space, it is not 'given in itself', but *unlike* Berkeley's, it is *not reducible* to the qualitative aspects of what *is* given.

Now this is not a thesis which can be supported by any ordinary empirical (e.g. psychological) evidence. This shows that both the Kantian doctrine which prevents, and Berkeley's which permits, the reduction of spatio-temporal characteristics to the other qualities, are in some way the outcome of metaphysical 'decisions', a conclusion which confirms our previous appraisal of Berkeley's discussion of space. And we further note that prominent feature of much philosophical doctrine which amounts to a differential 'weighting' of the various epistemological components involved in the logical structure of experience and of things; it is a question of where each philosopher places the centre of gravity of the different structural elements, in coming to a decision on 'what there is'.

At any rate (to return to Berkeley), space (and time) are not for him ontologically basic. Any positive account which he may have is exceedingly shadowy: we can speak of space only in the context of 'body and motion', e.g. in the context of extended bodies, which have certain qualities and mutual *relations*. Some passages, e.g. M.52, suggest that he was simply adopting Newton's definition of 'relative space' as 'some movable dimension or measure of the absolute spaces, which our senses determine by its position to bodies',[2] with Newton's reference to 'absolute space' omitted.[3]

---

[1] Cf. K.89, 244. For a more detailed discussion, cf. below, ch. VIII, sect. (5c), p. 549.
[2] *Principia*, Bk. i, Scholium 2 to Def. viii; op. cit., p. 6.
[3] Both Leibniz and Kant will have a far more detailed account of the nature of 'relative space'. It is worth noting that although Kant adopts Newton's viewpoint as quoted in this passage, he rejects the concept of 'absolute *motion*'.

The result is to place an ever-greater emphasis on those 'ideas', especially 'sensory' ones, which form the raw-material for reception into the mind. The progressive critique which we have followed has completely destroyed the importance for ontology of the initial Cartesian conceptions of mathematics, space and extension; and with Locke's shadowy world of substances and external objects and real essences gone, the tasks for Berkeley's finite and infinite mind to provide a substitute have become far too demanding, since they lack that articulation and integration within the limits of experience which would enable them to reconstruct the universe within the framework supplied by the growing empiricism.[1] Moreover only an uncritical retention of the conceptions of causation and deity for the moment prevents the collapse of the whole edifice. To a critique of these assumptions we must now turn. But the reader should not misconstrue such a seemingly censorious analysis of the historical situation. For in the end, as I have suggested, the 'reconstructions' still only amount to an alternative critical analysis of the controversial notions involved, producing alternative 'metaphysical' placings of weight and emphasis.

[1] The latter understood as being directed against dogmatic and authoritarian postulationalism.

VI

HUME: THE CRITIQUE OF CAUSATION

I. 'EMPIRICAL' AND 'METAPHYSICAL' LEVELS: CONTRASTS BETWEEN 'METHODOLOGICAL' AND 'EPISTEMOLOGICAL' ARGUMENTS

Hume is the heir to the tradition whose growth we have followed in previous chapters: the theory of ideas; particularianism; suspicion and eventual rejection of the 'external object', and of the 'essences' of things; rejection of 'substance' *quâ* substratum; criticism of the distinction between primary and secondary qualities; of 'innate ideas' as well as 'abstract general ideas'. The major influences on him are Locke and, to a lesser extent, Berkeley; though in fact Malebranche rather than Berkeley. And later we shall also have to add the name of Hutcheson. But there is also the emphasis on 'experience' and 'experiment'; on generalisation, and the importance of the 'laws of nature' obtained by induction from the phenomena; and here the influence is Newton. From Newton also comes the suspicion of the employment of hypothesis.

However, the consequences of these doctrines are worked out in sharper outline; above all, the supplementary scaffolding which had been erected to safeguard the building is now deprived of its ancient props and comes under similar attack—in particular, the conception of the uniformity of nature, supported either by God's activity, or—as in the case of the postulated acceptance of the principle of causality—as a deliverance of reason.

Hume's 'scepticism' is however more subtle; there is a clearer indication of what the whole enterprise of epistemology is about. His primary interest is in the credentials which philosophers present, together with those operative concepts just mentioned: external object, inductive inference, causation. Perhaps to make his position

325

seem more paradoxical, perhaps to aid the force of his 'positive' teaching, Hume never denies the position of these notions in human experience. 'I know that inductive conclusions are always inferred, and that they are justly inferred',[1] he tells his reader in so many words, the question only being, what entitles us to make such inferences. And, lest we think this a contradictory statement, let us anticipate, and realise that this is turned by Hume into the quite different question, why *do we think* we are entitled; a question whose sting is felt only if we realise that the answer proposed is given in terms of *how we think* in these circumstances, i.e. what the 'experimental'[2] conditions are surrounding the inferences which we make. Again, he holds, we cannot question for a moment 'Whether there be body or not?'; this is something we must all take for granted.[3] Instead, we can only be concerned with the credentials of that belief, where this once more becomes: 'What causes induce us to believe in the existence of body?'[4] Similarly, despite Hume's doubts about induction, he is quite firm that

> one wou'd appear ridiculous, who wou'd say, that 'tis only probable the sun will rise to-morrow, or that all men must dye.[5]

On the contrary, this is quite certain; it is a datum with which we must start; we must not—as Locke had done—let a certain notion of 'knowledge' (and its implied idealisations) force us to misunderstand the logic of inductive generalisation.[6]

Moreover, let us be clear from the start, that Hume (like Berkeley) is very much aware of the central part played in contemporary science by the methodological structure of scientific laws and theories. His paradigm is once again the *Principia*, and the whole approach which this embodies. His primary examples of laws of nature are the laws of

---

[1] Cf. *An Enquiry concerning Human Understanding* (1748) (ed. Selby-Bigge, Oxford, 1951), p. 34, hereafter referred to as E.

[2] This term in Hume's time covers both 'experiential' and 'experimental' in the modern sense; when only the former is intended, Hume speaks of 'observation of phenomena'.

[3] *Treatise*, op. cit., p. 187. (Subsequently referred to as T.)

[4] Ibid. A question whose significance is generated by Hume's position that strictly speaking we have no acquaintance with 'body' but only with sensations.

[5] T.124.

[6] Locke had talked as though one could distinguish only between 'knowledge' and 'opinion'. Hume proposes to distinguish between 'knowledge', defined as arising out of the comparison of ideas, 'proof', i.e. causal or inductive arguments, and 'probabilities', i.e. where the evidence is of a conflicting or of a statistical nature. (Ibid.)

motion, rules of impact, and the like.[1] Nor is his universe of 'proof' a single-level structure, concerned with nothing more than direct generalisations from experience. It is the task of 'human reason', he writes (echoing Newton and Berkeley),

> to reduce the principles, productive of natural phenomena, to a greater simplicity, and to resolve the many particular effects into a few general causes, by means of reasonings from analogy, experience, and observation.[2]

In the *Treatise* he shows a similar awareness of a 'dimension' underlying the observed phenomena when he remarks that exceptions to the normally observed course of events are properly explained by the scientists when they search in such cases for the 'secret operation of contrary causes' (rather than just postulating an upset in the general uniformity of nature), the reason being that

> almost in every part of nature there is contain'd a vast variety of springs and principles, which are hid, by reason of their minuteness or remoteness.[3]

Of course, if Hume's philosophy amounts to a ruthless development and critique of the primary tenets of the doctrines of Berkeley and Malebranche, whose main criticism had been directed against the notion of 'force' or 'activity' or 'efficacy', then phrases like 'the operation of causes' will require a special interpretation; but then this is the aim of Hume's work: not to proscribe such phraseology, but to prevent it from evoking misleading associations and views.[4]

Again the example of Newtonian gravity focuses the discussion. The imperfectly resolved attitude of Hume towards this will illustrate with particular elegance two approaches, the recognition on his part of what I shall call 'two levels'; his approach towards one being positive; the other receiving the full weight of his logical critique. Berkeley and Malebranche had criticised the notion of bodies '*acting*' on one another, locating all efficient causality in God. We shall later turn to Hume's epistemological arguments against this notion when he contends that we have as little 'comprehension' of God's power as

---

[1] Cf. for instance, E.29, 31, 73.
[2] E.30; the example being the axiomatic, hierarchical structure of the *Principia*.
[3] T.132. In the context, Hume refers to the usual example of the mechanism of a watch. The phrase 'minuteness and remoteness' occurs literally, as we have seen, in Locke's *Essay*, iv.3.24–5, where the reference is to the 'insensible corpuscles, . . . the minute constituent parts of matter'.
[4] 'Misleading': i.e. in collision with Hume's metaphysical atomism.

we have of that of matter.[1] For the moment, let us take his response for granted, and consider only a passage in which he seeks to meet possible objections to the universal denial of 'force and activity'. If we deny the applicability of these concepts what are we to make of conceptions like 'inertial force' and 'gravitational force', 'so much talked of in the new philosophy'? Now Hume's first reply to this question is not too unexpected; it has a rather Berkeleyan flavour:

> We find by experience, that a body at rest or in motion continues for ever in its present state, till put from it by some new *cause*; and that a body *impelled* takes as much motion from the *impelling* body as it acquires itself. *These are the facts*. When we *call* this a *vis inertiae*, we only mark these facts, without pretending to have any *idea* of the inert power; in the same manner as, when we *talk* of gravity, we mean certain effects without *comprehending* that active power.[2]

A passage like this might suggest that terms such as 'impelling' and 'force' are nothing more than summary *linguistic* devices, the non-linguistic aspects being exhausted in the description of the kinematical uniformities involved.[3] Against such an assumption, I want to claim that Hume's approach is far more 'positive' than appears at first sight; reinforcing a tendency I have claimed already implicit in Berkeley, when I drew the distinction between 'force' considered physically and metaphysically.[4] His remark about our lack of 'any idea' of 'inert power', of any 'comprehension' of 'active power', must not be construed (at least: not for the most part) as being directed against anything located on a 'physical' or 'empirical' level.[5] Rather, it relates to a denial of the 'objective' status of those connections (Hume's 'necessary connections')[6] which form the main butt of his atomistic epistemology.[7] On the other hand, when Hume speaks of the

---

[1] T.160; E.72. Cf. below, pp. 353, 377.

[2] E.73n.; italics mine. Hume evidently follows Newton's claim that the law of inertia is 'derived from phenomena'.

[3] And it matters little in the context whether Hume has in mind a reductionist or an instrumentalist account of the concepts of force and inertia. (Cf. above, ch. V, sects. 3 and 8.)

[4] Cf. above, pp. 310, 311 and 319n.2.

[5] A term like 'empirical' is systematically ambiguous. Its significance may range from the unanalysed commonsense of the physical scientist, *via* Hume's use of 'experience' as designating the level of 'impressions', to that which it is given in the Kantian notion of 'empirical reality', when contrasted with 'transcendental ideality'. In the present context, I intend something like the first of these meanings, with overtones of the third.

[6] Cf. T.77–8, 87, 155–7, 162–6, 405; E.33, 63–6, for Hume's use of this term.

[7] I shall consider the nature of this denial below, pp. 362ff, although in the pages

'fact' of one body 'impelling' another, this seems to me no *mere* manner of speaking. To appreciate this, we need to refer at some length to the rest of our passage from the *Enquiry* (E.73n.), which to those who imagine Hume to be opposed to dynamic action *univocally* must seem extremely puzzling.

Before turning to this, we must anticipate again a result of Hume's *general* argument, viz. the denial of the 'comprehensibility' of *any* kind of active power, be it spiritual or material; of any kind of connections in nature. That this is no ordinary denial becomes clear as soon as we find that Hume does not, as we might have expected, welcome the concept of gravity, understood as 'action-at-a-distance', which—as we saw in our discussion of Locke and his contemporaries—was deemed so incomprehensible as to require either explanation by way of the action of an intervening medium, or outright rejection;[1] unless, as with Locke himself, 'incomprehensibility' is used as a mark of the logical contingency of the fact of gravitation; a somewhat eccentric mixing of different kinds of logical and conceptual matters, as previously noted. But Locke's option is not a relevant consideration for Hume; contingency for him is defined on different grounds; formulated as it is (so we shall see later) by means of a metaphysical atomism not unlike what we have already encountered (though in an insufficiently clarified and generalised form) in Descartes, Locke and Berkeley. Yet, the metaphysical discontinuity which is the very basis of his epistemological doctrine, and which we may paraphrase through the slogan that 'everything is loose and separate',[2] is totally

---

that follow I have taken some of those later results for granted. I have nevertheless thought it advisable to emphasise first Hume's positive or realist teaching, because this seems to me the correct approach in attempting an adequate demarcation between the empirical and the metaphysical sides of the matter.

[1] Cf. above, ch. IV, pp. 269ff. In the Appendix to this chapter I have given some additional relevant historical details concerning the question of gravity, including Newton's own reactions. The literature on this topic is voluminous. See for instance, I. Bernard Cohen, *Franklin and Newton*, 1956, Part II, especially pp. 125–47, 163–72; (note particularly the classification of the varieties of the Newtonian use of the term 'hypothesis' pp. 138–40); also A. Koyré, *Newtonian Studies*, 1965, ch. II, especially pp. 29–40, 45–52; Appendices A–H, especially B, C, E; ch. VI. Newton's views on this are exceedingly complex. For our purpose it is only necessary to estimate what, in the light of contemporary discussions, Hume's own estimate of Newton's position may have been.

[2] Cf. '[All simple] ideas are loose and unconnected' (T.10); 'all distinct ideas are separable from each other' (T.79); 'there is no object, which implies the existence of any other if we consider these objects in themselves' (T.86); 'any thing may produce any thing' (T.173). The implications and significance of the doctrine implied in these passages we shall develop as we proceed (cf. below, pp. 366ff).

distinct from Hume's appraisal of the *physical* situation, where he does by no means countenance a similar discontinuity.

Here we need to note first of all that Hume is on the side of those who hold that action-at-a-distance as a physical conception is incomprehensible, if only because he finds the idea of a 'vacuum' (empty space) inconceivable.[1] Now it is true that this does not prevent him, any more than it had prevented Locke, from affirming the inductive truth of the law of gravitation; as is implied by the end of that part of the passage from E.73n, cited before, where Hume says that though we do not 'comprehend that active power' of gravity, we may denote by it 'certain effects'.

Nevertheless, he does not leave it there. For he finds it necessary to express his conviction that if we want to make the phenomenon of gravity intelligible, we need to postulate *physical continuity*; it is apparently after all not sufficient to reduce the phenomenon simply to law. We need to conceive it as involving the apparatus of genuine physical causal action; and causal action—as we shall note later—includes for Hume the criterion of spatial contiguity.[2] Only if this criterion is satisfied can we begin to make the phenomenon comprehensible. In short, the satisfaction of the criteria for causal action, in the *positive* sense of this term, must be kept quite distinct from those parts of Hume's arguments which postulate *metaphysical* discontinuity.

This brings us to the second part of our passage from E.73n. Hume has previously affirmed once again his 'metaphysical' doctrine (as I have called it) that genuine 'connection', or what he calls 'the energy of the cause' of any effect, is 'unintelligible' in *all* cases, be they familiar or altogether uncommon.[3] Some philosophers, dimly perceiving the truth of this, e.g. Malebranche and Berkeley, though 'they

---

[1] Cf. T.39ff., sect. IV. For instance: 'It is impossible to *conceive* either a vacuum and extension without matter, or a time, when there was no succession or change in any real existence' (T.40; my italics. Note that *conceptual* considerations do even for Hume sometimes have *physical* relevance!)

[2] T.75.

[3] E.70. Though I shall take up the matter in detail later, let me say that I speak of 'metaphysical' here because Hume's arguments against causation as a 'necessary connection' *in nature*, which purport to establish the *logical impossibility* of such a conception, are characterised as leading to the conclusion that 'we cannot penetrate into the reason of the conjunction' (T.93); that the 'ultimate springs and principles are totally shut up from human curiosity and enquiry' (E.30). My reference to 'metaphysical' for the moment means simply that Hume characterises a logical point by means of empirical or logical *images*. (The *empirical* use of 'hidden springs' we have already noted above, p. 327 (T.132). For this contrast, cf. below, p. 336.)

rob nature, and all created beings, of every power', then proceed by attributing the only seat of power to the Deity.[1] Hume counters by insisting that this 'power', in the sense in which he is *here* using that concept, is as unintelligible in one case as in the other.

However, he goes on, these philosophers have gone altogether too far. Provided we give a satisfactory explication of the concept of cause,[2] we are entitled, and indeed ought not to deny, causal activity to matter and created beings. Nay more, so the passage presently to be cited implies, the physical mechanisms through which we model causal activity should reproduce precisely those features whose true bearing we had misunderstood when previously we had thought of causality as some hidden metaphysical 'necessary connection'. And it is here that lies the interest of our passage. These philosophers, says Hume, were not at all entitled to seek support for their confused metaphysical doctrine—which had talked as though there *was no causal activity in the world, simpliciter*—from the Newtonian theory, interpreted as involving the postulate that one body's presence can have a determining influence on another *without* causal continguity. For, he writes with obvious approbation,

> it was never the meaning of Sir Isaac Newton to rob second causes of all force or energy; though some of his followers have endeavoured to establish that theory upon his authority. On the contrary, that great philosopher had recourse to an etherial active fluid to explain his universal attraction; though he was so cautious and modest as to allow that it was a mere hypothesis, not to be insisted on, without more experiments.

Obviously the dialectical web is here quite complex. There is *a* sense, in which all 'powers and energies' are 'incomprehensible' or 'unintelligible'; this is the 'metaphysical' sense of force. But there is also a second sense of force,[3] where, if such a concept seems unintelligible because it conflicts with the postulate of contiguous action, we must try, and moreover, may succeed (at least *in principle*; it was for Newton of course only 'a mere hypothesis') to render the phenomenon intelligible, by demonstrating it to be an instance of genuine causal

---

[1] E.71.

[2] As we shall see, this Hume purports to give in experiential terms.

[3] At E.67n. Hume speaks of this type of force (in the positive sense) which we feel ourselves 'exert', as a '*nisus*. or strong endeavour'.

activity.[1] Moreover, this shows that the presence *or* absence of causal action in the world, in the *positive* sense, cannot be used as an argument by metaphysicians like Malebranche or Berkeley in favour of the presence or absence of causal activity in the metaphysical sense. Quite generally, we find in Hume a more sharply focused division than hitherto admitted between the physical and the logical, as well as the metaphysical levels; our passage makes it clear that according to him we should not use a logical or metaphysical position to settle a physical question, and *vice versa*.[2]

Taking stock, we are beginning to see something of Hume's basic intentions. He will insist that we cannot question, in one sense, things like external objects, inductive inference, the hidden explanatory structure of nature, and the forces of nature. At the same time, he *will* dispute the possibility of showing these things being supported at a special, 'deeper', metaphysical level; or better: that among the elements of the 'basic inventory' of his reality there are any which could supply a support. Misleadingly such a view is often put in the form of a complaint that certain things do not exist—at least: not 'strictly speaking'; e.g. that strictly speaking there are no forces, no necessary connections. But secondly, and perhaps surprisingly, he also talks as though a kind of 'support' for these things had after all to be provided;[3] or at least (according to Hume) ought to be sought. For him, this support will for the most part be found in the forms and nature of human belief and imagination. We may expect that this will lead to confusion; that (for instance) causal inference considered on the one hand as 'descriptive fact' and on the other, as 'self-justifying' support, will often collide. So in what follows, it is important to be on our guard, and to remain aware of the distinction between these different layers in Hume's doctrine: first, the 'metaphysical level' (especially as found in his teaching concerning 'impressions' and

---

[1] We can now see that Locke confused these two sorts of 'unintelligibility'; cf. above, p. 270ff.

[2] As we saw, this had frequently been Locke's procedure, particularly in connection with gravity.

[3] If Hume were consistent, the notion of 'support' should be otiose, corresponding to his 'positive' approach towards the things here in question. But then, as we shall see, to speak of support is an archaic move; despite appearances, Hume will not provide a 'support', but instead replace the missing feature by a peculiarly descriptive approach, consonant with his reading of the Newtonian ultimate attitude to 'gravity'. Those who are surprised at this development, and accuse Hume of 'begging the question' or not providing what was required, are clearly hankering after older, Aristotelian, solutions.

'ideas', and the related doctrine of 'metaphysical atomism'), as well as the sharpened demarcation of the domain of 'demonstration and knowledge'; secondly, the methodology of 'reasonable inference', causation, interpreted as a curious mixture of 'judgment' and 'imagination', both of which are gathered together under the portmanteau notion of 'belief'; and finally, the level where these concepts of belief, imagination, instinct, provide a descriptive framework in terms of which we, if not justify, at least attempt to make intelligible, the functions of human reasoning.

2. THE USES OF 'HYPOTHESIS'

To express some of these doctrines, Hume will often use the jargon of his time, particularly where it is part of the methodological pronouncements of the scientists, and above all, of Newton. Consider for instance the notion of 'hypothesis'.[1] Here again, as in the question of gravity, the situation is not only complex and fluid; we must distinguish between scientific realities and received slogans. It will be clear already that Newton did not think the employment of hypothesis to be invariably otiose. It had a proper place in his theory of the ether for instance, not to mention those cases in the *Principia* where he makes certain assumptions concerning the resistance of mediums to the motion of bodies, or the immobility of the centre of the universe, or again concerning the size of the repulsive forces acting between the particles of a gas.[2] At the same time, in the General Scholium, Newton does insist that hypotheses have no place in 'experimental philosophy', which should be limited to those propositions obtained as inductive inferences (generalisations) from the phenomena. With a little charity, we might understand him to say that hypotheses concerning the underlying explanatory structure of things *would* be admissable if there were a chance of their being eventually confirmed inductively in some fashion or other.[3] However, the firmness of their exclusion in the Scholium is no doubt due to an attempt to avoid a consideration of the tedious multiplicity of theories of the 'causation'

---

[1] For this, see again the references given for the topic of 'gravity' above, p. 329n.1 and Appendix.

[2] Cf. *Principia*, op. cit., pp. 235, 244, 300–1, 419.

[3] Cf. in particular the case mentioned in Newton's *Opticks*, where he suggests an 'hypothesis' about the structure of the medium which could explain certain phenomena of refraction, concluding that he will for the time being not 'consider . . . whether this hypothesis be true or false' (op. cit., p. 280).

of the phenomenon ('existence') of gravity.[1] Indeed, the language used is interesting:

> Hypotheses, whether metaphysical or physical, whether of occult qualities or mechanical, have no place in experimental philosophy.[2]

There is clearly here a slide from a rejection of hypotheses as being untested, or (a different case) as lacking direct, or at least, independent observational or inductive support, *via* opposition to what are only plausible 'models' (as we should call them), to the quite different case where the hypothesis—for one reason or another—is believed to be untest*able*. This last case would be particularly relevant where we are dealing with something which, whilst parading as a scientific hypothesis, really belongs to quite a different level of considerations, to what Newton calls the 'metaphysical or occult'—such as Descartes' God, or even his theory of 'matter' in so far as that is anchored in a purely conceptual framework. Finally, there plays into all this the question whether the notion of force, which (as Newton claims) within the framework of the *Principia* need only be regarded as a purely 'mathematical' entity, or at least, no more than the expression of a law-like kinematical relation between physical bodies, stands in need, and is capable of, what we would now call 'a physical (or 'realist') *interpretation*'—Newton's reference to 'physical hypothesis'.

Evidently, the concept of 'hypothesis' conceals many intellectual conflicts and cross-currents, as well as multiple ambiguities and philosophical intentions. This being appreciated, we have a clue for an understanding of Hume's conflicting attitudes towards this notion. Like Newton he freely employs 'hypotheses' and 'suppositions' which he believes to be experimentally testable, either in fact or in principle. His theory of passions mentions them frequently, with the suggestion that they are all 'experimentally' testable.[3] Similarly, both *Treatise* and *Enquiry* recognise that form of 'hypothetical argument' in which we 'infer effects from causes'.[4] Significantly, Hume applies to this form a stringent condition of validity: we must here be able independently

---

[1] A brief account of these is given in the Appendix to this chapter.
[2] Op. cit., p. 547. At the end of Query 28 of the *Opticks*, Newton asks whether it does not clearly 'appear from the Phaenomena' that there is a God. Evidently, the *Principia* passage therefore involves an additional sense of 'hypothesis', in part signifying 'philosophical doctrine', intended to explicate the 'nature' of gravity. (Thus Newton thought he could provide an explication by reference to providential design—the application of a teleological point of view. Cf. *Principia*, op. cit., p. 546; *Opticks*, Qu. 28, op. cit., p. 369.)
[3] Cf. T.285, 289, 290, 345.   [4] T.82-3.

to 'establish the existence of these causes', if we wish to use them for the purpose of genuine explanation. A passage at E.144 more explicitly insists on the need for *empirical* acquaintance with the cause which will enable us to predict phenomena other than those on which the hypothesis was originally founded.[1] This obviously echoes one of Newton's grounds for objections to hypothetical explanations of gravity, and will be an important lead for Hume's epistemological approach.

All this incidentally highlights the injustice of those comments which castigate Hume for an insufficient comprehension of the nature of *theoretical* reasoning in science, and for having been excessively, if not solely, pre-occupied with 'direct generalisation'. The emphasis placed on the latter has in fact little to do with Hume's assessment of science, but stems from his enquiry into the logical basis of induction, considered as a form of inference.[2]

On the other hand, we naturally expect Hume to make use of the pejorative use of 'hypothesis' to make plain his epistemological position, all the while retaining the traditional link with the joint concepts of 'experience' and 'experiment'. The use in question occurs prominently in the Introduction to the *Treatise*:

> Thou' we must endeavour to render all our principles as universal as possible, by tracing up our experiments to the utmost, and explaining all effects from the simplest and fewest causes, 'tis still certain we cannot go beyond experience; and any hypothesis, that pretends to discover the ultimate original qualities of human nature, ought at first to be rejected as presumptious and chimerical.[3]

Here, the formal similarity with the logical situation as presented in the *Principia* (and its problems!) is so obvious as to require little comment. What is important is a material difference: the proscription of any hypothesis concerning the 'ultimate qualities of human nature' is now intended in a special sense, not to be confused with those warnings against hypotheses which we found to have a fairly obvious

---

[1] Interestingly, Hume adds that where this requirement is ignored, we have 'mere conjecture and hypothesis' (E.145), obviously now using the narrower meaning of these terms.

[2] What we *can* say is that Hume's insistence on direct independent acquaintance with the 'cause' shows (as with Mill, who in this follows Hume, like many other Newtonians) insufficient awareness of the place of 'indirect' theoretical reasoning in science, and an understanding less subtle than we encountered in writers like Descartes and Berkeley.

[3] T.xxi.

meaning. We can see this best where we meet with a clash of meaning or use, when identical terminology is employed. We have noted already[1] the passage from T.132, in which we read of Hume's sympathetic reception of explanations in terms of 'concealed causes', of the 'hidden springs and principles' of nature. But now compare this with an identical wording in the sequel to a passage in the *Enquiry* (E.30), which I also mentioned in that context. Hume, we noted, had said that it is the task of human reason to explain the variety of phenomena in terms of a few general causes. He then goes on:

> But as to the causes of these general causes, we should in vain attempt their discovery. . . . These *ultimate springs and principles* are totally shut up from human curiosity and enquiry. Elasticity, gravity, cohesion of parts, communication of motion; these are *probably* the ultimate causes and principles which we shall ever discover. . . .[2]

Clearly, Hume is here more radical. Though the tone is that of a Lockean sceptic and borrows Newton's wording, the meaning is that of an *impossibility*[3] of as much as conceiving of any such 'springs and principles'. The force of the injunction against hypothesis is that which proscribes the invocation of an 'occult quality'.[4] The scientist's highest level explanatory laws and principles (following Newton, Hume refers to them as 'causes') no less than those at the lower rungs of the deductive ladder of theories, are one and all (as to their logical status) inductive generalisations. It is behind them, Hume will insist, that we cannot peer.

Here we must issue a caution: just because Hume will ultimately have difficulty in making clear the nature of his insistence on the *impossibility* of 'discovering' any 'link' between the terms of these generalisations, a difficulty due to his inability to explain *clearly* against what metaphysical entity he is here inveighing, he will fall back on 'extraneous' characterisation. Thus his purely *logical* argument, roughly, the contention that empirical propositions are not connected by any *logically necessary* ties, though it cleared much lumber from the philosophical stage, will really be too thin to support the importance he claims for his doctrine. For this reason, he will erect additional props. The philosophical grammar of 'ideas' (his 'impressions and ideas') on the one side, and on the other, the semantic associations of the Newtonian notion of hypothesis, with its

---

[1] p. 327 above; also p. 330n.3   [2] My italics.
[3] Of a logical kind to be discussed presently.   [4] Cf. T.224.

consequent muddle of conflicting meanings and positions: these will be used to give plausibility to Hume's metaphysical anti-hypothetical bias, which will eventually settle as the doctrine that although empirical generalisations *assert* the inseparability of the conjunction of their terms, they do so at the level of experience, something connected with the *manner of asserting* them. What we 'cannot' do is to 'penetrate into the reason of the conjunction'.[1]

3. INDUCTIVE PROCESS AND INDUCTIVE INFERENCE

Before examining this doctrine in detail, we need to scrutinise further the background of Hume's methodological approach, in order to enlarge on two of the contentions put forward so far, (1) that Hume accepts the reality of inductive inference, albeit with a 'basis' different from his predecessors; (2) that this type of inference assumes the usual structure of theoretical and/or causal reasoning. I will discuss these topics in turn.

(1) The first topic connects once more with Newton. Perhaps the most influential pronouncements made by Newton (outside the sphere of science proper) were his four 'Rules of Reasoning in Philosophy', prefixed to Book III of the *Principia*. The first of these rules enshrines a principle of simplicity, the second postulates that to similar causes there correspond similar effects. The third rule, on 'the analogy of nature', sanctions the inference from observed qualities known to attach without diminution to all bodies in our experience, to 'all bodies whatsoever', the inference being intended to cover both the step 'in depth' from the realm of the observable to that of the unobservable (insensible particles: a kind of 'analogical inference'), and the step 'in breadth' from local regions of space to all regions. The last rule really supplements the third, introducing a 'provisionality clause': 'propositions inferred by general induction from phenomena' are to be regarded as 'true, . . . till such time as other phenomena occur by which they may either be made more accurate or liable to exceptions'.[2]

Now the phrase 'inferred by general induction from phenomena' actually covers two things. First, it signifies the *process* through which

---

[1] T.93; cf. T.169.

[2] Op. cit., pp. 398–400. The omitted section insists that 'contrary hypotheses' that may have been imagined are not to count against the induction. This was meant to be directed against those who had opposed Newton's theories of colour and gravitation by bringing in certain hypotheses concerning the structure of light and ether; or whose philosophical views appeared to proscribe gravitational forces altogether.

we determine the particular form of the causal or 'functional' relation, or even of the explanatory hypotheses that appear to account for the particular observations. This process is never terminated in principle, as implied by the 'provisionality clause' of Rule IV. On the other hand, and secondly, throughout this process there is also a logical aspect under which we *hold*[1] the law or hypothesis, arrived at by causal or theoretical investigations, to be universally applicable; and this is the *inference* proper.[2] It is of vital importance to note the distinction between these two features; the process defines an aspect of actual *method*, involving scientific investigation; the inference is spontaneous, and concerns a logical function.[3]

Now it is clear that these two rules, especially the third, play a central rôle in the fashioning of Hume's thinking. The Third Rule is indeed once referred to explicitly as 'Newton's chief rule of philosophising', when he argues (at E.204) that since 'public interest and utility' are the foundation of the virtue of justice, 'therefore' they are the source also of many other similar social virtues. This mode of arguing, he writes,

> is entirely agreeable to the rules of philosophy, and even of common reason; where any principle has been found to have a great force and energy in one instance, to ascribe to it a like energy in all similar instances (*Principia*, Bk. iii).

Clearly, this is an instance of an inductive projection (to use Nelson Goodman's term),[4] and an implicit reference to a form of 'uniformity of nature' principle. Such a principle is referred to by Hume in many places, both in the *Treatise* and the *Essay*, and there is a close

[1] Cf. Newton's comment on Rule III: 'For since the qualities of bodies are only known to us by experiments, we are to *hold* for universal all such as universally agree with experiments . . .' (op. cit., ii, p. 398; my italics).
[2] For a discussion of this distinction between inductive process and inference, see my 'Semantic Sources of the Concept of Law', in *Boston Studies for the Philosophy of Science* (op. cit.), vol. III, 1968, pp. 272ff., especially pp. 281–4; and 'Inductive Process and Inductive Inference', cited above, p. 67.
[3] Whewell's comment on the two rules reveals an interesting and very common misunderstanding. On the Third Rule he remarks that no 'arbitrary maxim' can authorise a generalisation, but only 'the reason of the case'. But with the Fourth Rule, he is surprised that Newton, after having sanctioned the generalisation, should allow the possibility of it being overthrown: how can the latter possibility be reconciled with the former assertion? Clearly Whewell did not understand the import of Newton's intentions; which is the more ironical since he eventually puts forward a similar view, though couched in terms of his theory of 'Ideas'. (Cf. *Philosophy of the Inductive Sciences*, 1840, vol. 2, xii, 1, 13, pp. 450–4.)
[4] *Fact, Fiction and Forecast* (London, 1954), chs. 2–4.

affinity between it (and the problems which it raises) and his discussion of causation, since the 'necessary connections', supposed by him to hold between the terms of causal propositions, are also just the ones which support inductive inference.[1]

Now let us note our main point—that Hume, like Newton, treats the projective 'act' of inductive generalisation as a spontaneous element, an intimate part of the whole nature of scientific thinking, which either follows or even accompanies the various determinations of the 'facts', involving a more or less extended survey of the determining circumstances. Moreover, in the light of the 'provisionality clause' of Rule IV, one could regard the aspect of generalisation as a purely grammatical, even ceremonial, feature. For that clause makes it as clear as could be desired that induction has nothing to do with 'knowledge of the future', nor with having to stipulate conditions which could guarantee such knowledge. Induction is about *inferentially predicting* the future on the basis of the present total evidence[2] which could reasonably be expected to be in the hands of the scientist. It has nothing to do with knowing now what the future will be like.[3] One could, therefore, have regarded the inferential step as a purely 'formal matter', equivalent to the scientist's locution about 'adoption of hypotheses in universal form', employed subject to further tests. But this is not how it is usually regarded, the question commonly asked being: what *justifies* the principle of uniformity?

It is remarkable that Newton is still uncommonly reticent about considerations of *this* question. The 'basis' of his Rules is never discussed. The most we get is a reference, in explanation of Rule III, to 'the analogy of Nature, which is wont to be simple and always consonant to itself'.

And Hume? Well, one can understand his doctrine to be not too unlike that implied by Newton: treating the inductive projection

---

[1] Cf. T.89. We shall return to the matter presently. References to the principle of uniformity in a variety of formulations may be found at T.104–5; E.33, 35, 38; e.g. E.35: 'our experimental conclusions proceed upon the supposition that the future will be conformable to the past'.

[2] This is not to say that there are no problems surrounding a definition of 'total evidence', and the logical criteria which such evidence has to satisfy in order to give us 'satisfactory' results. But this aspect does not concern us at the period with which we are dealing.

[3] Cf. Goodman, op. cit., p. 65: 'If the problem is to explain how we know that certain predictions will turn out to be correct, the sufficient answer is that we don't know any such thing . . . [it would be like] asking for prevision rather than philosophical explanation.'

process (the 'inference') as something that is altogether 'natural';
where all that is needed is a survey of its genetic conditions, the
human frame of mind when it thus operates on the results of experiment and observation. Note that the term 'inference' has been put in
quotation marks, since in the end Hume will deny that there *is* any
inference; or at least, a *'rational* move', proceeding by virtue of a
principle of sanction. In other words, his is a presuppositionless form
of induction, involving an approach whose significance was not really
grasped until our own century.[1] Instead of inference in the normal
sense of that term, we have a 'customary transition', from inductive
antecedent to consequent (e.g., from cause to effect). Nor is this doing
more (according to Hume) than 'to point out a principle of human
nature, which is universally acknowledged, and which is well known
by its effects'.[2] Sometimes Hume describes this as 'the imagination'
carrying 'us on to the conception of the usual effect' that has been
observed to occur.[3]

'But', so the normal objection runs, 'to give an account of inference
in terms of a description of what we, when acting as scientists, are
normally, and 'instinctively'[4] led to do is surely quite irrelevant, is to
ignore the real problem, which is, what *entitles* us to make the
inference?'

(2) This objection will take us to the heart of the second of our two
contentions mentioned on p. 337. To discuss this, it is important to
take the sting out of this criticism by reminding ourselves once more
of the distinction between the inductive process and the inductive
inference. For the objection sounds too much as though imputing
that Hume had replaced honest scientific investigation by merely
habitual association of ideas; the more so since he often describes the

---

[1] For this result, cf. below, p. 382. We shall there find that the principle of the
uniformity of nature is not for Hume a presuppositional condition, but only a
device descriptive of the logical nature of the inductive projection, or the causal
inference. (In this, he already foreshadows Mill, whose views on this matter have
unfortunately been misapprehended only too frequently.)

[2] E.43.

[3] T.148; cf. T.93. There is a complication, to be considered later (cf. below, pp. 342,
346f.): Hume normally distinguishes here between 'imagination' and 'judgment';
but the point is not relevant for the present.

[4] Cf. T.179: 'Reason is nothing but a wonderful and unintelligible instinct in our
souls, which carries us along a certain train of ideas, and endows them with
particular qualities, according to their particular situations and relations. This
instinct . . . arises from past observation and experience; but can any one give the
ultimate reason, why past experience and observation produces such an effect, any
more than why nature alone should produce it?'

nature of the 'attraction' between the ideas of cause and effect as an 'association of ideas'.[1]

However, with our distinction in mind, we shall be able to differentiate Hume's account of the nature of the causal relation as such and the corresponding notion of inductive inference (or projection) from that process which involves an investigation of 'what goes with what', in so far as this is capable of scientific or pre-scientific study. For it is quite wrong to object to Hume when, as we shall see presently, he speaks of 'deriving' the causality of certain events from their previously observed 'constant conjunction', that such a procedure constitutes a very crude way of determining the causal conditions of events. Such a complaint parallels the objection to Hume's account of induction that it is preoccupied almost solely with direct generalisation; that it pays insufficient regard to the actual methods of science, its testing of hypotheses under a variety of conditions, not to mention its integration within a system of theoretical principles.

I believe that most of these objections miss the point. Hume was perfectly aware of the need for learning 'to distinguish the accidental circumstances from the efficacious causes',[2] however much the former were observed 'constantly to be joined' to the member-events of a certain sequence. Indeed, no one is more emphatic than Hume in insisting that

> there is no phaenomenon in nature, but what is compounded and modify'd by so many different circumstances, that in order to arrive at the decisive point, we must carefully separate what is superfluous, and enquire by new experiments, if every particular circumstance of the first experiment was essential to it.[3]

This must mean that the relevance of 'constant conjunction' is only for the 'philosophical question', the definition and nature of the causal relation as such; but this comes *after* the scientist has 'deduced' his 'principles from the phenomena': for only *then* are they 'made general by induction'.[4]

---

[1] Cf. T.13; E.23–4.   [2] T.149.

[3] T.175. But, as we shall see later, this admission *will* be an embarrassment at a different stage of the argument. (Cf. below, pp. 683–4.)

[4] Newton's phrase, in a letter to Cotes. (Cf. Thayer, op. cit., p. 6.) Nothing could express more clearly the two components of 'induction' in the general sense. 'Deducing' is here, of course, used in a general popular sense, in which scientists speak of inferring the form of a quantitative law from the values obtained in experiments. Sometimes we *have* logical derivation, as in the case of Newton's law of gravitation, which in the *Principia* is actually obtained *deductively* from Kepler's

This, then, is the second of the two points I wanted to establish.[1] Hume is very alive to the 'complexity' of nature[2] and the need for care when generalising from instances, where—as he points out—sometimes even 'one single experiment' may be sufficient.[3] If a criticism is appropriate it is that he showed little awareness of the additional problems surrounding even those principles (or hypotheses) supported according to their form by an experimental or theoretical investigation. Thus, he paid little regard to the possibility of alternative hypotheses fitting the same data at any given time. Hume's response would no doubt be that, first, we can always continue in our search for additional 'crucial experiments' which would decide between these alternatives; and that, secondly, such a process has a good chance of reaching stable equilibrium eventually, thus testifying to the seventeenth and eighteenth century optimism about the power of science.

Such an optimistic assessment was engendered partly by the rigid distinction, itself built into Hume's epistemology, between theoretical and data-language. Once that is abandoned, such finitist optimism would most likely come to grief also. Another reason for Hume's optimism is more paradoxical. It springs from the very critique to which he submits the notion of inductive *inference*, with the concomitant seemingly sceptical view that 'reason' cannot 'guarantee' the future.[4] It seems that Hume felt that, having established the open-endedness of inductive *inference* in this sense, he could leave the result of the inductive *process* safely looking after itself. In this way, the metaphysical level did after all cast its shadow on the positive aspect, with the result that the problems which have bothered later philosophers of science, particularly in our own day, concerning the definition and status of simplicity, law-likeness and uniformity in science receive little attention.

Moreover, it is true that the problem of the complexity or 'contrariety' of causes is sometimes viewed by Hume as a quasi-psychological matter of the balancing of conflicting beliefs or rules; though elsewhere this is described as a conflict between 'imagination' and 'judgment'.[5] Counter-instances, though we may explain them by

---

laws using the laws of motion and the geometry of Euclid. Quite a different case is that of a 'causal' investigation, using say, Mill's Canons, where from the absence of a factor in the presence of an 'effect' we 'conclude' that it is not causally relevant; this is precisely the case Hume mentions in the passage just referred to (T.149).
[1] Cf. above, p. 337.   [2] Cf. his discussion of 'contrariety', T.131–7; 148–9.
[3] T.131.   [4] Cf. below, pp. 363ff.   [5] Cf. T.148.

reference to 'hidden springs and principles', are then viewed as their quasi-psychological effect which is that of 'weakening' our belief in the constancy of the conjunction,[1] a position which involves a somewhat fanciful psychology. But for the moment our main point must be that Hume's philosophy of the nature (and genesis) of the causal concept and causal maxim should be detached from any considerations of the empirical structure of nature.

We have here the clearest instance of the two levels which I have noted.[2] At the 'metaphysical' level (the joint resultant of Hume's adoption of the theory of ideas and of his interpretation of the 'contingency' of propositions concerning matters of fact) we meet Hume's atomistic claim concerning the separateness of all things, the absence of any 'real' causal connections. On the other hand, at the 'positive' level, the level at which critical inference is a perfectly appropriate function, we find him ascribing more than mere 'linguistic' causality to nature. Now these two levels occur side by side dramatically also in the very section of the *Treatise* which enunciates a number of 'rules by which to judge of causes and effects',[3] to which we must now turn.

4. HUME'S CANONS OF CAUSAL REASONING

Hume in this section has just completed his argument to show that 'without consulting experience . . . there are no objects [which] we can determine to be causes of any other. . . . Any thing may produce anything'.[4] Moreover, he has explained that ascription of 'causation' is entirely a matter of 'constant conjunction', which (in a manner to be considered presently) leads to the establishment of a certain belief, custom, habit, a 'determination of the mind'. For the moment, we may understand this to be an explanation of the genesis of the '*idea* of causation' (or rather, necessary connection), as well as of the *principle* of causality or uniformity—Hume eventually does not distinguish very sharply between these.

Now, as we have noted, it is sometimes suggested that this is all Hume offers us to establish the formulation of a particular causal law or connection. This interpretation, though wrongheaded, is unfortunately on occasion supported by Hume himself. When—as just remarked—he is faced by an obvious 'contrariety of events', he

---

[1] Cf., for instance, the remarks at T.132, 150. Cf. below, p. 347.
[2] Above, pp. 330–2.   [3] I.3.15, T.173–5.   [4] T.173.

embarks on a sort of psychological calculus of the strength and weakness of our beliefs resulting therefrom. But side by side with this approach, he also follows another, which is the assumption that we ourselves are in the possession of the idea and principle of causality, and that we only need to seek for instantiations or applications of these. Naturally, putative instantiations would in principle have to satisfy the 'constant conjunction' postulate, though subject to the supplementary considerations of the complexity and fortuitousness of events, and strengthened on the other side by the detection of certain 'analogies'.[1]

Be that as it may, we have here a theory which at the metaphysical level denies connection, and which suggests that the meaning of such connections is exhausted by reference to certain 'subjective' aspects of our own experience, our own minds. However, immediately thereafter, Hume continues as follows:

> Since therefore 'tis possible for all objects to become causes or effects of each other, it may be proper to *fix* some general rules, by which we may know when they are *really* so.[2]

Again we confront Hume's positive approach, his 'empirical realism': he is not just saying that we *call* those things causes which satisfy certain rules; on the contrary: an application of these rules reveals when anything (satisfying these rules) '*really*' is a cause. Moreover, we see that Hume's metaphysical reconstruction of causality in terms of belief, custom, instinct, imagination, is evidently an analysis or definition of what it is for some event (shown to satisfy these rules) to be 'really' a cause.

To consider the relation between these two levels, let us take the simplest case, that of Hume's third rule: 'there must be a constant union betwixt the cause and effect'.[3] Few would question that, *ceteris paribus*, if two classes of events ('species' is Hume's term) satisfy this rule, then this does or may indicate a causal relation. What is still open to discussion is the nature of the connection involved, of which the constant conjunction is (according to this rule) the sign.[4]

Similarly the seventh rule, part of which defines the 'concomitant variation' indicator of 'cause' and 'effect': this again would under certain circumstances 'signify' a causal connection. According to the sixth rule, if an effect is not forthcoming in the presence of some

---

[1] For the latter, cf. T.142.   [2] T.173; my italics.   [3] Ibid.
[4] Let us mark well: *here* constant conjunction is only the *sign* for a causal connection, and not its genetic *source*, as it is in the philosophical argument.

putative cause, this must be due to the absence of a factor which is part of the object hitherto believed to be the cause. Clearly once more, this canon is an indication of a causal relation. Methodologically, we here see, there is no question of concentrating merely on constant conjunctions.[1] The latter is pertinent only because Hume (as we shall see later) uses this also as a genetic factor involved in the definition of what it is to *have* a causal connection.[2] But these are two quite different uses, corresponding to our distinction between inductive process and inductive inference.

To sum up: the *epistemological* account of causation which supplies us with the analysis and status of the concept is a basis entitling us to label the result of a *methodological* investigation in accordance with our rules as a discovery of 'real causes'.[3]

Hume enunciates, in all, eight rules but it would be tedious to go through them in detail. They are somewhat indebted to hints provided in Bacon's *Novum Organum*, and they will be found further developed later by Mill and other writers. But the fourth rule, roughly: 'same cause, same effect', deserves separate attention. 'This principle we derive from experience', Hume tells us, 'and is the source of most of our philosophical [*sc.* scientific] reasonings'. 'Derivation' is a reference to the previous complex argument (which we have yet to study)

[1] Goodman's judgment on Hume is therefore very peculiar, when he writes: 'The real inadequacy of Hume's account lay not in his descriptive approach but in the imprecision of his description. Regularities in experience, according to him, give rise to habits of expectation; and thus it is predictions conforming to past regularities that are normal or valid. But Hume overlooks the fact that some regularities do and some do not establish such habits; that predictions based on some regularities are valid while predictions based on other regularities are not' (op. cit., p. 81). But it is quite clear that Hume has overlooked no such thing, at least in his methodological account. What can be challenged is only his epistemological account: whether the 'habit established', when considered as an *analysis* of the causal relation (as against the question of the method of discovering causes) is satisfactory. But that is another question.

[2] Note that whilst the causal *methods* can be used as signs *indicative* of causal connections, constant conjunction when regarded *definitively*, as *generating* the idea of necessary connection, is more than this. For this necessary connection is not 'discover'd by a conclusion of the understanding' arising out of a contemplation of constant conjunction, but is (*quâ* being generated in the mind) itself 'a perception of the mind' (T.405–6). Cf. also the remark added to the constant conjunction rule (3.): 'Tis chiefly this quality that constitutes the relation [of causation]'; 'chiefly', because we have to add the 'mental determination' *arising* from the constant conjunction.

[3] It is therefore a profound error when we read, as we frequently do, that Hume replaces causation by constant conjunction. Such a remark does scant justice to the subtlety of the interplay between the various levels of his discussion to which I have been wanting to draw attention.

which has attempted to explain the genesis of the 'idea' and 'principle' of causation; and is thus a reference to the 'epistemological' aspect of the matter. On the other hand, the rule, however it may actually have lodged in our 'experience', is now spontaneously applied in the manner suggested by Newton's Third and Fourth Rules. As Hume explains, a single 'clear experiment' may have aided us to 'discover the causes or effects of any phenomenon'; our rule is then applied to 'immediately extend our observation to every phenomenon of the same kind, without waiting for that constant repetition, from which the first idea of this relation is derived'.[1]

Hume, we saw before, speaks of 'fixing' these rules. But such phraseology does not imply that he thinks of them as 'conventions' of a more or less arbitrary kind.[2] What then is their status? Partly they (e.g. the third and fourth rule) are no doubt supposed to result from the very same process which also generates the idea of 'necessary connection'; and in this way, the epistemological argument would once again provide a basis for the methodological structure. However, as we shall see, it is not always so, for the same can hardly be said of rules like that of concomitant variation, nor those rules which attempt to deal with the complexity of causes (e.g. the fifth, sixth and eighth rules). That they are not conventions is also suggested by Hume's comment that they are 'the natural principles of our understanding'.[3] Is this perhaps a reference to those 'natural habits' which arise from the effect of 'constant conjunctions', the essential begetter—as we shall see further presently—of the '*idea*' of causal connection? Hume here begins to betray an uneasiness; the tightness of his fundamental scheme is beginning to show certain cracks produced by internal tensions which can be plastered over only by a vagueness in the terminology employed. It is true that 'custom' invariably 'carries the imagination' from some objects to others 'constantly conjoined' with them, thereby resulting in the establishment of a 'belief'.[4] However, we only too frequently meet with a 'complication of circumstances';[5] and we then realise that what looked like a causal factor is only an 'accidental circumstance'. But what is the nature of this realisation? Hume wavers; sometimes he views it

---

[1] T.173-4.
[2] They are 'conventions' only for those who are satisfied with nothing less than 'deductive sanction'. And it may be that there is just the hint of a shadow thrown by this ideal even on Hume's choice of this term.
[3] T.175.   [4] Cf. E.48-9; T.148.
[5] Cf. our quotation above, p. 341, of the passage from T.175.

only through its supposed effects, as 'weakening our belief'. But at other times, he suddenly introduces the term 'judgment' in a novel way, *contrasting* judgment and imagination, when previously he has maintained that there is no such distinction. Judgment is then singled out as that faculty of the understanding which operates peculiarly under the influence of our 'general rules'. And

> these rules are form'd on the nature of our understanding, and on our experience of its operations in the judgments we form concerning objects.[1]

Are the rules a function of the general operations of our understanding leading to belief, or are they dependent on a special faculty, i.e. judgment in a special sense? Hume wavers again. Wanting to limit his basis, and keep it to fundamentals, one moment he says that even the crude and erroneous generalisations which the 'imagination' leads us into are in some sense due to the 'influence of general rules'— presumably the 'constant conjunction' and the 'same cause same effect' rules (the third and fourth). However, immediately thereafter, he describes the result of 'contrariety' of causes in these words:

> But when we take a review of this act of the mind, and compare it with the *more general and authentic* operations of the understanding, we find it [*sc.* imagination] to be of an irregular nature, and destructive of all the most establish'd principles of reasonings; which is the cause of our rejecting it. This is a second influence of general rules, and implies the condemnation of the former.[2]

But terms like 'more general and authentic' should really have no place in this context, where we expected our rules of reasoning to be built upon a merely natural and instinctive basis. Evidently, the notion of the 'understanding' smuggles in a more 'reflective' type of reasoning or mental operation, borrowing its meaning from the consideration that—as a matter of empirical fact—our initial hasty generalisations stand usually in need of modification in the light of further causal investigation. But this presupposes a kind of 'causal thinking' which is set apart from any 'instinctive process', has emancipated itself from this basis; and this is implied also by the use that Hume makes of the causal principle when he views it as being 'applied' to the case of 'one single experiment', particularly when it is a 'clear experiment', and one 'duly prepared and examin'd'.[3]

[1] T.149; cf. T.175.   [2] T.150; my italics.   [3] T.131; 175.

Evidently here is a 'soft centre', or 'opaque core', in Hume's scheme. The notion of 'the understanding' covers a multitude of intellectual operations, and the resulting ambiguity is slurred over by suspending it at one moment on such 'technical' terms as 'the understanding', 'authenticity', 'clearness', whilst at the next supporting it on the conception of 'imagination' and 'belief'.[1]

The general upshot is this. There can be no question but that Hume was perfectly aware of the complexity of nature, and that he realises that the laws of nature[2] are not a matter of a 'blind' process of generalisation. On the other hand, the internal split which this consideration generates, and which comes to the fore in the acknowledged distinction between 'imagination' and 'judgment', presupposes a causal principle that stands in need of greater independence from the instinctive process of generalising on given individual occasions. For, in fact, his own scheme can give a firm foundation to this principle only when described as that natural form of belief which is an item in the totality of those 'operations of the understanding' which indifferently include both the reflective processes of the scientist and the more involuntary generalisings of primitive commonsense. Nevertheless, it remains an important fact that Hume's scheme was an interesting response to the background of scientific thought of his time, as crystallised in the methodological disquisitions of Newton. Hume, in a peculiar and left-handed way, introduces the conception of the 'autonomy' of scientific reasoning; it no longer is made to depend upon external, metaphysical sanctions. In this way he takes further the task begun by Locke.

---

[1] Clarification here can be gained by a contrast with the way Kant will later develop these themes. The principle of causality will there be given an independent basis as a presuppositional ingredient of the understanding. With this 'independence', Kant can then with greater consistency speak of 'applying' the principle in the process of building up a scientific scheme of causal laws which provide instances of its application.

[2] For this term, cf. E.29, 144. It is a myth that Hume views science in terms of causal sequences only. His avowed concentration on 'causes' is due, partly, to the use of 'cause' where this means 'explanatory scheme', in which form it appears, for instance, in Newton's First and Second Rules of Reasoning (cf. also the nineteenth century expression 'induction of causes'); partly to the fact that Hume was primarily concerned with the concept of 'necessary connection' which for him made up the essential element of causation; and obviously this is an element which is the essential aspect of laws in *any* form, whether causal or not. Hume's disquisitions on causation hence can easily be extended to cover all forms of functional and law-like relationships.

5. THE STATUS OF 'IMPRESSIONS'

This thesis of the emancipation of causal reasoning needed however independent defence, springing peculiarly and unexpectedly from the existing philosophical atomism of Locke and Berkeley and Malebranche, which had originally supplied the sceptical attack on this form of reasoning, further sustained by Locke's distinction between 'knowledge' and 'opinion'. All this was, however, supplemented and transformed by the introduction of an altogether new element, emphasising the importance of 'feeling' and 'belief' (for instance, in accounting for the *relations* between aesthetic forms or practical actions and our responses to these): an element which Hume borrowed from certain British philosophers of his time, especially from Hutcheson.[1]

As we have anticipated, Hume develops the idealist thesis more ruthlessly; correspondingly, the incipient atomism in that thesis is drawn in sharper outline, whilst Locke's sceptical tone, with its implied suggestion of our 'ignorance' being a merely contingent matter, though not totally abandoned, is counterbalanced by the greater importance attached to the conceptions of belief, imagination, instinct and nature,[2] and thus considerably modified. This is coincident with the move to deny still further to the notion of 'knowledge' any operative force of the kind it had possessed in the writings of Descartes and even earlier rationalists. We will discuss these themes in turn.

We have seen that Hume insists that we cannot ask 'whether there be body or not';[3] we can only enquire how we come 'to believe in the

---

[1] This influence has been traced in detail in N. Kemp Smith's *The Philosophy of David Hume*, London, 1941. We may note in particular Hutcheson's denial of a 'necessary connection' between certain 'forms and actions' and the aesthetic pleasure which we experience in their presence; likewise, our recognition of moral facts, as experienced through a 'moral sense' which is not due to any 'innate ideas' or any peculiar sort of 'knowledge', but merely to 'a determination of our minds to receive amiable or disagreeable ideas of actions, when they occur to our observation'. (Quoted from Hutcheson's *Inquiry* in op. cit., pp. 32, 34. Hume, of course, extends this from aesthetics and morals to theoretical reasoning.)

[2] Because of this, the tension between scepticism and belief reaches breaking-point in the writings of Hume; if reason is derided, instinct and nature are given an all the more prominent place, with the final judgment on the outcome of his reflections on the causes of our belief in 'external objects': 'Nature is obstinate, and will not quit the field, however strongly attack'd by reason' (T.215), and the corresponding judgment in the field of 'practical action', that 'reason is, and ought only to be the slave of the passions' (T.415).

[3] Cf. above. p, 326.

existence' of bodies, external objects. The advantage of the Humean formulation lies in making it clearer that we are not concerned here with an empirical question. But at the same time, to pose the question in Hume's way makes it sound too much as if belonging to the field of psychology, whereas in fact it is more like a request for an 'analysis'; for spelling out what it is one *means* when using the locution of an external world of bodies.

However, even this interpretation is not literally correct. For Hume's question is, after all, just as Locke's and Berkeley's had been, due to certain views about the ultimate inventory of the world; it really asks, how is it *possible to talk* of a cognition of 'body', when the latter is, *strictly speaking*, never its object?[1]

Evidently, Hume's problem is simply posed in virtue of his adoption of the idealist position. The clearest statement occurs perhaps at T.67:

> We may observe, that 'tis univerally allow'd by philosophers, and is besides pretty obvious of itself, that nothing is ever really present with the mind but its perceptions or impressions and ideas, and that external objects become known to us only by those perceptions they occasion.[2]

This sounds still very Lockean, as though Hume assumed a realm of external objects *behind* our perceptions. There are, however, a number of variants in his formulation, betraying a lack of clarity about the decision to describe the reference of all cognitive situations as an 'ideal' object. The echo of the Lockean confusions between the empirical, theoretical and metaphysical motivations for this theory are made evident precisely by the lack of a single clear formulation. How, then, does Hume characterise these 'impressions' and 'ideas', the offspring of Locke's 'ideas' (a label which Hume criticises for being too general).[3]

Let us ask first of all, 'What is the difference between "impressions" and "ideas"?' Now, impressions of sensation, as well as our emotions and feelings, are simply the basic material of our mind (or 'soul'); by contrast, 'ideas' are 'derived' from the impressions; they are 'copies' of them. But how distinguish between them? This question is impor-

---

[1] Cf. below, p. 352n.3.

[2] A similar statement occurs at E.152. Note that this position is declared to be 'obvious', a clear indication that there has been some kind of linguistic (equivalent to 'metaphysical') *decision*—as explained at length in my account of Locke.

[3] The term 'idea' has been 'perverted by Mr Locke', he tells us, by being 'made to stand for all our perceptions' (T.2n.).

tant, since, although both 'impressions' and 'ideas' are said to be '*in our* minds', it is the former which stands for that element which is, despite Hume's language, presented *to* our minds; through it we are meant to have access to what is 'other' than ourselves. Sometimes Hume is as basically cryptic as Locke when the latter had written[1] that we are 'invincibly conscious' of the difference between perceiving the sun and thinking about it at some subsequent time. Similarly Hume: we all know, he tells us, 'the difference betwixt feeling and thinking'.[2] It is not unlike the difference between the 'ideas' themselves, which we note in 'that act of the mind, which renders realities, or what is taken for such, more present to us than fictions', which—for Hume—is the difference (in this context) between what is merely 'imagined' and what is 'believed' to be the case; a difference which itself can again not be further explained, being no more than a '*manner* of their conception, a 'feeling to the mind'.[3] In a similar way, Hume holds that impressions are more forceful and lively, more vivacious than the ideas, which are 'fainter'.[4] But one feels instinctively that this (at least as regards our 'impressions of sensation'), is too weak a criterion. Impressions, like ideas, may be 'internal', but throughout Hume's *Treatise* they stand for tokens of a world set over against ourselves, our minds, pains and emotions. (As we shall see presently, the term 'impression' is continually being equated with the term 'object'.)

To ease the tension, therefore, at other times we get an 'external' criterion. The suggestion is then that impressions are 'caused', though it is not clear where Hume means to place the cause. Thus at T.7 we are told that impressions of sensations 'arise in the soul originally, from unknown causes'. At T.84 their cause is said to be 'perfectly inexplicable'; but there is also a hint about the sort of causes Hume has in mind, for he adds that it is 'impossible to decide with certainty, whether they arise immediately from the object, or are produc'd by the creative power of the mind, or are deriv'd from the author of our being'.

Clearly, the history of the concept in the writings of Descartes, Locke, Berkeley and others has provided here the model, and thus

---

[1] iv.2.14; cf. above, ch. IV, pp. 237n.1, 241f.   [2] T.2.
[3] E.49; italics in text. I.e. the difference lies *only* in the manner in which we conceive the 'ideas', and *not* their 'nature or order' (ibid.). Hume is not always consistent in his account of the difference; cf. below, p. 355 and n.2.
[4] T.2.

conveyed a meaning, where—if we took Hume's epistemology seriously—none ought to be possible; a fact which Kant was later to make
the core of his own doctrine. Hume, however, is not at all interested
in the question;[1] he will never fall victim to the temptation of offering
arguments through which to 'break out' of the circle of our impressions and ideas, as is shown by the rhetorical passage in which he tells
his reader . . . that whether we

> chace our imagination to the heavens, or to the utmost limits of the
> universe; we never really advance a step beyond ourselves, nor can
> conceive any kind of existence, but those perceptions, which have
> appear'd in that narrow compass.[2]

However, such a passage suggests also quite a different attitude
towards the 'causes' of our impressions, and one moreover that makes
the 'external criterion' of the difference between impressions and
ideas impossible to use. For our text suggests that the 'causes' in
question are not so much 'unknown', as 'inconceivable'. Thus in the
same context Hume writes:

> Since nothing is ever present to the mind but perceptions, and since all
> ideas are deriv'd from something antecedently present to the mind; it
> follows, that 'tis *impossible for us* so much as *to conceive* or form an idea
> of any thing specifically different from ideas and impressions.[3]

This reminds us of Berkeley's theory, which it strongly resembles.
But it is more radical. The whole weight of 'reality' is thrown upon

[1] Cf. also a very Kantian passage like T.638: 'As long as we confine our speculations
to the appearances of objects to our senses, without entering into disquisitions
concerning their real nature and operations, we are safe from all difficulties. . . .'

[2] T.67. But there are occasional inconsistencies. At T.275 we are told unblushingly
that impressions of sensation 'arise in the soul, from the constitution of the body,
from the animal spirits, or from the application of objects to the external organs'—
very much a Lockean account. Indeed, Hume adds that these impressions 'depend
upon natural and physical causes,' an investigation of which would require our
going 'into the sciences of anatomy and natural philosophy'.

[3] Ibid.; my italics. This is Hume speaking with a metaphysical voice, for, as we have
already mentioned, he will subsequently give a very full account of the reason why
we *believe* in the existence of independent objects. This account does not belong
to our text. It is fashioned *via* an explanation of the coherence and constancy of
our impressions (T.195), involving an act of the imagination (T.198, 205), even
though philosophical 'reflection' tells us otherwise (T.215). The trouble is (as Hume
in the end came to see) that, according to this system, the 'mind' likewise is 'nothing
but a heap or collection of different perceptions' (T.207), so that it is difficult to see
how its imaginative processes could ever result in the 'unity' which represents the
notion of an 'object'. (T.635–6.) It is worth remembering that Kant later *starts* with
the presupposition of the unity of the mind's reflective processes; a supposition
which is 'proved' from the existence of our experience of objects; an interesting
variation on Hume's theory (of which Kant was, however, ignorant).

the 'impressions' and 'ideas': 'To reflect on any thing simply, and to reflect on it as existent, are nothing different from each other.[1] And no other 'reality' is introduced; even the 'self' (mind) is ultimately resolved into impressions (cf. p. 352n.3); and the whole theory of divine action so elaborately constructed in Berkeley's and Malebranche's philosophies is shrugged off as 'fairyland'.[2] For Hume will insist that since there is no 'power', at least in the sense required by this argument, we must also refuse to endow the Deity with this 'energy' exactly as in the case of the conception of 'matter'.[3]

Such a tremendously narrow basis involves Hume in difficulties, which we cannot elaborate in detail; it seems *prima facie* unlikely that a proper definition of external existence could ever be articulated consistently in this way. Partly the difficulties are concealed, because the notion of 'impression' carries rather more than its ostensive criteria (strength, vivacity, etc.) would seem to allow, no doubt owing to the fact that the rejected *sceptical* point lingers on, i.e. that impressions are due to 'unknown causes' (the latter being evidently something 'other than ourselves').[4]

Moreover, Hume is quite unscrupulous in using the terms 'impression' and 'object' interchangeably; witness for instance T.5: 'I present the objects, or in other words, convey . . . these impressions'; T.19: 'no objects can appear to the senses; or in other words, . . . no impression can become present to the mind . . .'.[5] But are not 'impressions' constituents of minds, expressly said to be 'internal and perishing existences',[6] quite unlike what is suggested by the grammar of a 'presented object'? Clearly, this fluidity of usage enables Hume to bypass any lengthy consideration of impressions and ideas regarded as 'representing' something 'other'; it is a lack of distinction which allows him, for instance in his discussion of the nature of space and geometry, to identify ideas of minute extension with minute

---

[1] T.66–7. Note, however, that according to T.73, conceiving something clearly and distinctly only 'implies the *possibility* of existence'.

[2] E.72.   [3] T.248–9; E.72–3. Cf. above, pp. 327ff.

[4] Apparent inconsistency thus functions as a lever which bestows supplementary meaning. But it is, of course, a question whether these *are* inconsistencies; we can never be sure whether Hume would not have rebuffed the charge by saying that most of the time he was only offering characterisations of our modes of thinking or believing; and that 'ultimately' the notion of the object apart from our impressions is otiose.

[5] Cf. also T.20: 'To form the idea of an object, and to form an idea simply is the same thing'.

[6] T.194.

ideas.[1] Nor should it be replied that Hume is here talking loosely; that although impressions do not represent 'any thing beyond', he has given ample explanation of how they assume this representative character.[2] This is not denied, but my point is that Hume's 'loose language' conceals the highly *technical* use of the term 'impression', endowed with a logical grammar for which no independent defence is provided.

In a similar vein, examples of impressions that are 'conveyed by the senses' are invariably the *qualitative* aspects of objects, their 'figure, bulk, motion and solidity for instance; their colours, tastes, smells, sounds, heat and cold';[3] but beyond repeating Locke's notion that in our idea of 'substance' our ideas of such qualities are 'united by the imagination',[4] we are not given any further details; the term 'object', so frequently used in these contexts would again conceal the awkwardness of the position.

'Impressions', then, like Locke's 'ideas of sensation', are the raw material of experience, and indeed, of what exists. But, unlike Locke, instead of a 'metaphysical argument' (as we called it)[5] concerning the 'reality' of our simple ideas of sensation, only the ghost of a reference to anything external to mind, or rather, to our 'perceptions' (impressions and ideas) remains. Hume argues that 'every impression' must be 'considered as existent',[6] in explanation of his point that the idea of existence adds nothing to the idea of what we conceive to be existent;[7] which again shows up the ambiguous status of 'impression': evidently 'impressions' are at one and the same time both our

---

[1] Cf. T.29. Elsewhere we are told that 'our senses offer not their impressions as the images of something distinct, or independent, and external . . . , and never give us the least intimation of any thing beyond' (T.189). Again, as in Berkeley, our impressions of sight do not 'inform us' of 'distance or outness (so to speak) immediately and without a certain reasoning and experience' (T.191). Compare this again with Kant, who (starting with the same logical status of sensations, i.e. lacking 'outness') explicitly adds the 'form of space' as a presuppositional aspect: 'In order that certain sensations be referred to something outside me . . . the representation of space must be presupposed' (K.68). Here, once again, a vagueness in a predecessor is made explicit in the progress of philosophy. (Cf. above, ch. V, pp. 294n.5, 322, for Berkeley and below, ch. VIII, sect. (5c), pp. 549–50 for Kant.)

[2] In T. I.4.2, entitled 'Of scepticism with regard to the senses'. Cf. footnote 1 above, p. 352.

[3] T.192. Cf. T.5 for further examples.

[4] T.16. Elsewhere, 'substance' (*quâ* substratum) is, for Locke's and Berkeley's reasons no doubt, called an 'unintelligible chimera' (T.222). For 'united by the imagination', cf. above, ch. V, p. 294n.4.

[5] Cf. above, ch. IV, pp. 251–2.   [6] T.66.   [7] Cf. above, pp. 352f.

'sensings' and *what* it is that is sensed, or what we remember as having sensed. Of course, it is Hume's intention that we cannot ask justificatory (but only causally explanatory) questions reaching beyond 'impressions'; the question whether anything corresponds to a sensory perception is not to arise.

But is this true also of those 'perceptions' which Hume calls 'ideas'? It ought to be, if their status differs from that of impressions only in being 'fainter copies'. Yet, Hume likens the having of 'ideas' also to 'thinking' of the corresponding object, originally given *quâ* impressions; it is then more like a concept of that object, in the normal sense of the term. Does 'existence' then add nothing to a concept? Certainly Hume is right, in the sense that it adds nothing to the *meaning*; existence does not predicate anything, e.g. a sensory quality. But it surely makes a difference whether a concept denotes an existing entity or not? This is, however, a difficulty for Hume; he must needs refer us back to impressions, in explanation of existence; he cannot allow some realm of 'external objects', apart from these impressions. But are these not qualitatively similar to the ideas? Are they not just 'internal and perishing existences'? Hume bypasses the difficulty and suggests that what we mean when we here refer to something existing, the referent of the idea, is that we merely conceive our 'perceptions' as having 'different relations, connections and durations'.[1] Elsewhere he distinguishes 'realities' from 'mere fictions of the imagination', by the criterion of our impressions being formed into 'a kind of system'; and such a system may be extended further, *via* the various inferences which we may draw from them, either deductively or inductively.[2]

This sketchy account of the matter will at least suggest this much: Impressions are the raw material of reality, as they are the raw material of our experience, of our knowledge (in a general sense) of such a reality. Here there can be no error (at least *in principle*): for

---

[1] T.68. This is then expanded later into his account of the genesis of our ideas of external objects (I.4.2).

[2] T.108. But he is a little uneasy about such a 'coherence' account of 'reality'. Elsewhere (E.49), as already alluded to above (p. 351), he tells us that 'in philosophy, we can go no farther than assert, that *belief* is something felt by the mind, which distinguishes the ideas of the judgment from the fictions of the imagination'. And this 'belief', as against 'mere imagination', consists 'not in the peculiar nature or order of ideas, but in the manner of their conception, and in their feeling to the mind'. Here, as in many places, we therefore get a mixture of criteria; those that appeal to the *system* of ideas, the others that bear on the phenomenological differences of the two situations in question.

here there is 'direct access'. Hume puts the doctrine graphically at T.190:

> For since all actions and sensations of the mind are known to us by consciousness, they must necessarily appear in every particular what they are, and be what they appear.[1]

At the same time, such a doctrine can present itself as an account of reality only by indulging in the luxury of considerable obscurities, inconsistencies and, above all, allusions from associated contexts, injected into its formulation. It is enough here to indicate broadly the radical extension of the older account; its excision of an external world from that realm in which 'impressions' alone are allowed to exist. It is the latter which will form the pivot of Hume's logic, and the limits of his world. For in discussing and elaborating the concepts of that logic, he will repeatedly apply the surgical question, viz. 'Given such and such an idea, is there a corresponding impression from which it is derived?'[2] Impressions, so to speak, are there in order to bestow ontological credentials. *Per contra*, a decision concerning the constituents of their realm, and of their logical grammar, will determine the outcome of our investigation into the status (and 'reality') of the critical concepts in question. The central point is that henceforth any comments made on the logical relations between 'objects' will necessarily be couched in terms of those holding between 'impressions' (or their corresponding 'ideas'). At T.29 Hume writes:

> Wherever ideas are adequate representations of objects, the relations, contradictions and agreements of the ideas are all applicable to the objects; and this we may in general observe to be foundation of all human knowledge.

---

[1] Here is again the influence of the Lockean doctrine of pure passivity in sensation; cf. T.73: When objects are present to the senses, their perception is 'a mere passive admission of the impressions thro' the organs of sensation'. Error can only come in at the level of ideas, and of causal reasoning, or more likely, the effect of the play of the imagination. (Cf. T.148, 150, 225.) But as in the case of the distinction between Hume's general doctrine of causation, by contrast with the scientific determination of causes, this again is only a matter of 'principle', as just mentioned. Naturally Hume is aware of the fact that sensory perceptions are open to correction; cf. T.632: 'The understanding corrects the appearances of the senses . . .'.

[2] This will be the vital lever, for instance, when turning to the idea of 'necessary connection'; cf. T.77: 'In order to discover the nature of this necessary connection, . . . [we must] find the impression, or impressions, from which its idea may be deriv'd.' (Be it noted here that it is not vital for our account whether in fact every idea *is* derived from a corresponding impression. Once again, it is only Hume's metaphysical *intentions* that concern us.)

The immense bearing of this assumption will be obvious, in virtue of the fact that—as we have noted at length—Hume in the end treats 'ideas' (or 'impressions') and objects as equivalent. The identification (for philosophical purposes) of object and impression will then ensure that there is no second ('constitutive') world beneath the 'epistemic' level; in short, that Locke's epistemic atomism will no longer be mollified by any ontological postulations.

6. THE CLASH BETWEEN THE TRADITIONAL AND HUME'S APPROACH TO INDUCTIVE REASONING

With these preliminaries we can now turn to the *epistemological* account of causation, previously sketched scantily when contrasting it with the methodological situation provided by the 'rules of causal reasoning'. How then does Hume build up his world from these foundations? To answer this, consider his distinction between the two grand types or 'objects of human reason': Relations of ideas, and matters of fact—foreshadowed already, we found, in Locke's *Essay*.[1] In the *Treatise* Hume describes the distinction in terms of the types of *relations* involved; those that depend entirely on the ideas related (such as resemblance, or degrees in quality, or proportions in quantity or number), and those which can change without any change in the ideas concerned (such as spatio-temporal relations and, above all, the relation of cause and effect).[2] The former yields 'knowledge'; but this is now so circumscribed as to be a mere shadow of its former self in Locke, let alone Descartes. It gives us what is 'either intuitively or demonstratively certain',[3] the most important instances being algebra and arithmetic, whose 'chains of reasoning' preserve 'perfect exactness and certainty'.[4] (Geometry, which likewise is given as an example in the *Enquiry*, is excluded in the *Treatise*, on the ground that here we have no exact standards determining equality or straightness, in a field where ideas must ultimately be 'drawn from the general appearance of the objects'.)[5]

---

[1] E.25, 35. Cf. above, ch. IV, p. 186. Interestingly, at T.463, Hume divides the 'operations of human understanding' into 'the comparing of ideas, and the *inferring* of matter of fact' (my italics).
[2] Cf. T.13–15; 69–70.   [3] E.25.   [4] T.71.
[5] T.71; cf. T.42–52. There are interesting hints here that the axioms of geometry are synthetic (cf. T.51), a view partly generated by treating mathematical points as *minima visibilia* (cf. T.40, 42; E.156n.), and the consequent attempt to build up a specification of equality and straightness in 'operationalist' terms, which entirely

By contrast, the second set of relations yields only 'probability'. As already noted, this is 'philosophical' jargon; propositions based on causal or inductive arguments in *a* sense *do* yield 'proof' and 'certainty';[1] by contrast the notion of probability is now intended to mark such propositions off from those which supply intuitive insight, rational transparency. Hume, like Locke, resists the temptation to allow mathematical certainty to infiltrate the knowledge yielded by physics, and experience in general, as had happened in the philosophy of Descartes. The task of mathematics is strictly limited to drawing out deductively the consequences of those laws (e.g. the laws of motion) which have themselves been 'discovered by experience'.[2]

The second type of propositions bearing on questions that appertain to 'matters of fact' is then those involving relations concerning which nothing can be determined by a consideration of the terms alone (the 'ideas') which they contain; they involve 'no exercise of thought', but 'a mere passive admission of the impressions thro' the organs of sensation'.[3] And the relations involved 'may be changed without any change of ideas.'[4] It seems, therefore, that the assignment of any relation of this type, and of its detailed specification, must be entirely a matter arising out of a consultation of 'experience'.

Let us note once more that 'probability' (in its 'philosophical sense'), or even more strongly, the purely 'arbitrary' nature of the conjunction—another metaphorical way in which Hume is found to characterise the logic of the situation[5]—and 'experience' are coupled here primarily in virtue of the absence of 'intuitive transparency', the

---

rejects the rationalist fiction—as we found in Descartes—of a Deity, that putatively supports 'the supposition' of 'a perfect geometrical figure' (T.51; cf. T.48–9). (Cf. above ch. III, sect. 1(e) especially pp. 111ff., 117.)

[1] Cf. above, p. 326, and the reference to T.124.

[2] E.31. Cf. above, ch. IV, especially p. 185, for Locke's formulation of this position, and the reference there to the Newton passage.

[3] T.73. We have seen how important this element of 'passivity' is in the scheme of Berkeley's philosophy, and the picture it offers in terms of which to describe the 'atomicity' of ideas. Also, as in the cases of Locke and Berkeley, all impressions and ideas 'are particular in their nature' (T.24).

[4] This is an extreme case of the rigid distinction between a basic observation language and an over-layer of theoretical language. We shall take this up below, pp. 396–70. But one might insist on the one hand that *some* distinction between levels has to be made in any theory of knowledge; and on the other, that Hume's ultimate naturalistic account of inference and its methodological ingredients is quite compatible with any degree of fluidity of transition between these layers.

[5] Cf. E.29.

ideal standard of 'knowledge'. Probability is not assigned because of any such feature as the 'unreliability of experience', of the 'vagaries of nature' or the 'complexity of the facts'. For all this, as we have laboured to show in the earlier parts of this chapter, belongs to the field of the methods of causal investigation, of searching into what goes with what.[1]

It would be bizarre to use the results of such investigations to argue inferentially to the *unreliability* of inductive generalisations *in principle*, for that would be to reason in a circle, as much as when we seek to base the methods of inductive inference on their 'proven reliability' in the past. And indeed, this is not at all the intention of Hume's argument. That will be devoted primarily to showing why the notion of 'intuitive transparency', or of the 'methods of demonstration and reasoning' (as he usually puts it) are in principle here inapplicable. And they will be inapplicable because of the intrinsic 'atomicity' of our ideas; something which has little to do with our 'inability to know the future'.[2] The problem which Hume really sets himself to solve, and which he positively believes he has solved, is the reconciliation of the implications of a language of causation and of causal inference, which had traditionally been held to imply the existence of 'necessary connections in nature',[3] with a philosophy according to which no such connections can exist. Now, as we noted before, when the existence of these connections is disclaimed it is not very clear what it is that is being denied, since the locution 'existing in nature' is systematically vague; it covers, for instance, both those levels—the 'positive' and the 'metaphysical'—which we have sought to distinguish in Hume. The importance of Hume lies in having sharpened this notion, making it clear that his denial extends solely to the 'world' of impressions and ideas (or indeed, as we shall see, of impressions only)—Hume's terms for the 'ultimate furniture' of objective reality.[4] Moreover, his reading of the logical grammar of 'impressions' and 'ideas' is such that the denial of necessary connections practically follows with necessity.

[1] Note for instance, that the passage where Hume refers to the 'contrariety of events', and the resulting 'uncertainty of nature' occurs in the section where the *methodology* of causal inferences is being discussed (T.131), under the general title of 'probability of causes'.
[2] Cf. above, pp. 339–41.
[3] This was R. B. Braithwaite's phrasing, quoted before, ch. II, p. 19n.1
[4] It is therefore not in the least evident that a logic of induction is incompatible with Hume's philosophy, which is what is frequently asserted. (E.g. K. R. Popper, *The Logic of Scientific Discovery*, op. cit., p. 29, and *passim*.)

The crucial feature of Hume's philosophy is the revolutionary explicitness with which he insists on the dichotomous nature of the classification of types of propositions. The closing words of the *Enquiry* have indeed become a classic banner under which later philosophers, such as the school of logical positivism, were to sail—without, alas, doing justice to the subtlety of the remainder of Hume's approach:

> If we take in our hand any volume; of divinity or school metaphysics, for instance; let us ask, *Does it contain any abstract reasoning concerning quantity or number?* No. *Does it contain any experimental reasoning concerning matter of fact and existence?* No. Commit it then to the flames: for it can contain nothing but sophistry and illusion.[1]

This must not be shrugged off in the light of our discovery that every serious philosophical work (including Hume's) contains at its core a variety of basic claims of a primarily metaphysical nature.[2] What Hume is trying to bring to light is the essential lack of clarity in the employment and application of models of necessity such as we found so abundantly illustrated in the philosophy of Locke, with its flirtation with 'intuition' as a satisfactory approach towards the problems of knowledge; just as he resists the application of the theological model to the solution of the problem of *a priori* propositions which are believed to have a bearing on the foundations of empirical science. His dichotomy is meant to emphasise the insufficiency of these attempts to provide 'justifications' or 'foundations'. If we were to put this in Kantian terms, we should say that he raises the problem surrounding those '*a priori* judgments' which at the same time claim to be synthetic; but even this is an insufficient characterisation. Obviously there are plenty of quite unpretentious candidates for such propositions, witness Locke's instance of colour incompatibility. But Hume's protest is directed against the use of any propositions of this kind for the purpose of securing the 'foundations of knowledge', e.g. the principle of causality; the statements of geometry and arithmetic or algebra; the conception of substance as substratum. And his claim is that any justificatory feature which is not rooted in 'experience' is a feature which is simply idling—where for 'experience' must be read Hume's 'basic inventory' of impressions and ideas, and the framework

---

[1] E.165.
[2] Witness the whole doctrine of 'impressions', and, for instance, the nature of Hume's denial of the Aristotelian position, below, pp. 361–4, below.

of consciousness in which these come to be experienced—something rather more complex and controversial than the simple reference to 'experience' would at first sight seem to suggest.

Let us follow the way, then, in which Hume's dichotomy bears on the problems raised by the processes of scientific reasoning. And here there will at once arise the puzzle. The most important member of the set of relations belonging to the second type is that of 'causation'. But 'causation' (under which label there are included—as we have seen—really all the inductive varieties of reasoning) Hume describes as that relation *in virtue of which* we are able 'to go beyond what is immediately present to the senses'.[1] Causation—by whatever methods (using the 'general rules') it may have become manifest—claims to manage this feat, because it 'produces a connection'[2] between objects, 'an union among our ideas, [so] that we are able to reason upon it, or draw any inference from it'.[3]

This language, though couched in the grammatical form of the categorical indicative is really that of a claim; our idea of causation can only *claim* to give such a support. If Hume puts it in the indicative form, this is done not only to present the resulting difficulty more paradoxically, but also because the naturalistic solution which he eventually gives will precisely treat the relation in this way; it will be equivalent to ourselves (or some function of our minds) producing the desired union. That is to say, it will shift the focus from preoccupation with causation as that relation *in virtue of which* we make inferences, to that relation as considered a function of the mind *which is involved in* every such inference.

This is not, however, what the reader is at first led to expect. Hume's formulation of the problem is put in an altogether traditional form that has a very ancient history; which is not to say that we may not still encounter it in our own day. It occurs already in the *Nicomachean Ethics*, where Aristotle characterises the situation of empirical reasoning as follows:

> We all suppose that what we know is not even capable of being otherwise; [for] of things capable of being otherwise we do not know, when they have passed outside our observation, whether they exist or not.

---

[1] T.73, 89; E.26.   [2] T.73.

[3] T.94. Hume calls relations which involve such a union, 'natural relations', by contrast with those which concern no more than a mere 'comparison' of ideas; the latter are denominated 'philosophical relations'. (Cf. T.15, 94, 170.) The causal relation may be viewed as either (T.94).

Therefore the object of scientific knowledge is [*sc.* must be] of necessity.[1]

The main feature of Aristotle's argument of interest to us in this passage is that it aims to establish the need for a justificatory feature (the so-called 'commensurate universal'), a necessitarian tie, which is supposed to sanction the transcendence beyond the present moment. But this is precisely the way in which Hume presents the case. His definition of the causal relation is well known. Apart from the 'essential'[2] criteria of spatio-temporal contiguity and temporal priority of cause and effect, 'there is a NECESSARY CONNECTION to be taken into consideration',[3] as forming an 'essential part' of the idea of causal relation. This is not surprising. We saw in our second chapter how philosophers usually feel driven to acknowledge that the analysis of causal laws, and laws in general, leads to such a conclusion. But we saw there, also, that the existence of such a feature, the existence of a 'universal of law', can be taken in two ways, (1) as an expression of what we *mean* when we claim that a certain relationship is law-like; (2) as expressing something which bestows upon our relation 'validity in an objective sense'.[4]

Now it is in this latter respect that Hume at first presents the relevance of the necessitarian feature as providing the intrinsic ingredient of causal inference, his problem being how to interpret it, and how to integrate it with his general principles. As he understands it, his rationalist predecessors had made *two* assumptions: (1) that the postulation of the existence of such a necessitarian tie at the ontological level of either particulars or universals[5] had some definite meaning; (2) that in some form or other it was possible to assign meaning to a mental or epistemological process or function— the most prominent of which had been 'intuition'—in virtue of which one could gain some knowledge of such a necessary connection, some 'rational insight' into the relation. Aristotle certainly had postulated

---

[1] vi, 3: 1139b 20–22. As we saw in ch. IV, it is because Locke could not 'see' the necessary connection in question, that he denies 'knowledge' and 'certainty' to anything transcending 'the present testimony of our senses' (*Essay*, iv. 11.9); cf. above, p. 201. For a discussion of Aristotle's position as given in this passage, and others which occur in the *Posterior Analytics*, see my *Induction and Necessity in the Philosophy of Aristotle*, op cit., sect. II, pp. 18–27. For a recent Aristotelian approach, cf. above, ch. II, p. 42.

[2] T.77.   [3] T.77 (Hume's capitals); also T.87.

[4] In Johnson's formulation. Cf. above, p. 29.

[5] Needless to say, Hume, who follows Berkeley in his approach towards 'abstract ideas', has a maxim 'that every thing in nature is individual' (T.19).

the connection in the form of the 'commensurate universal'; and he often talks as though 'intuition' could yield the required insight.[1] More tantalisingly, Locke had introduced the notion into his philosophy of science, but with a sceptical caution attached. Here also, as was shown at length in chapter IV, the necessitarian requirement still haunts the scene as the ideal of 'knowledge', though it is seldom realised. Moreover, through the notion of 'real essence' and 'substance$_m$', it in some sense occurs also as a postulated bond in a realm intrinsically inaccessible at the epistemic level of 'ideas'. In other words, though Locke appears to subscribe to the first of the above two assumptions, he wavers sceptically about the second.[2]

When Hume talks of 'reason' and 'demonstration' as being those functions which his predecessors had believed to be the *sine-quâ-non* of the foundation of causal and inductive inference, it is those assumptions and notions mentioned in the previous paragraph that should be his terms of reference; notions which are obviously far less well defined (as our discussion of Locke showed) than his argument, involving largely considerations of 'logical necessity', would seem to imply.

Consider the language used where he discusses the problem of the inference from the past 'constant conjunction' of impressions to their future conjunction.[3] 'If *reason* determin'd us', he writes,

> it wou'd proceed upon that principle, that instances, of which we have had no experience, *must* resemble those, of which we have had experience, and that the course of nature continues uniformly the same.[4]

But, he goes on (for reasons to be discussed presently), 'there can be no *demonstrative* arguments to prove' such a principle. (And he rightly includes in his rejection the alternative of 'probable' conclusions whose inference, as he makes quite clear, similarly presupposes a connection between those instances of which we have experience, and those which transcend it.)[5]

---

[1] Though we cannot be certain of such an interpretation; cf. my *Induction and Necessity etc.*, sect. iii, pp. 28–34, especially pp. 32ff.

[2] It is important to see that there *are* two assumptions, and not just one, i.e. the second, since philosophers critical of Hume have often contended that his denials touch only the latter, and not the former; that his denial of 'necessity' touches only the question of 'lack of insight', a contention which clinches the issue only if the matter is understood as a denial of 'logical necessity'. I hope that I have presented the issues so as to show that Hume's net is cast rather wider.

[3] T.88–9.   [4] T.89; my italics. Cf. E.35.   [5] T.89–90; E.35–6.

Such an argument again presents us with an instance illustrating the very essence of 'change' and 'progress' in philosophy. It is quite clear that a considerable haze had surrounded the older notions of 'reason', 'demonstration' and 'intuition', whilst at the same time it was just they that formed the metaphysical core of the doctrines in which they appeared. And they achieve this quite often by way of sheer 'assertion': the making of a metaphysical claim which I have called 'semantically opaque'.[1]

Thus, that we can and do 'intuit the universal' is tantamount in Aristotle to a bare insistence, an expression of a conviction, that induction is a possible achievement notion.[2] But where there is sheer assertion, the reaction will also be no more than simple denial, even though such denials may be accompanied by a variety of arguments and empirical considerations. And such denials become real through a reinterpretation, a narrowing and sharpening of the older terminology, and the introduction of novel considerations on the bearing of certain logical principles, and the employment of new metaphysical key-terms such as Hume's impressions and ideas.

So Aristotle and his followers *could* be taken as merely *insisting* on the necessity of a postulate guaranteeing the inductive inference, a postulate concerning the 'existence' of a necessitarian tie, whose significance is not clearly defined, but whose 'metaphysical' power is to give expression to what must be granted if such an inference is to be possible. And this would be a correct *historical* assessment of the operative expressions of 'reason' and 'demonstration'.[3] Against this, when we turn to Hume we find the corresponding denial. Why then can there be no 'demonstration', no 'reason', in Hume's sense of these notions?

---

[1] Cf. above, ch. I, p. 3.

[2] Cf. my *Induction and Necessity*, op. cit., for a defence of such a reading.

[3] We should remember that Aristotle never claimed that scientific conclusions can be based on 'demonstration' alone. On the contrary, they presupposed 'induction', which for him was equivalent to 'intuition of the universal', and this involves a very different sense of reason. I should add that I am not *defending* here this sense of 'reason'; it is merely that I want to insist that this is a wider, vaguer, and more general sense, than that to which Hume sharpens it. But *per contra*, it should not be thought that all Hume's contentions amount to, is that causal and inductive reasoning is *not deductive*, or cannot be 'reduced' to that form of inference. For that would not have been a very important result, and no more than a strawman, seeing that no reputable philosopher has ever held such a view.

7. CONSEQUENCES OF THE 'LOGIC OF IMPRESSIONS': ARGUMENTS FOR AND AGAINST NECESSITY

This brings us back to the relevance of the place in Hume's philosophy of the doctrine of 'impressions and ideas', and the parallel elimination from his universe of any supplementary 'objects', by means of the virtual identification of object and impression. The logical features, and the implications for theory of knowledge, and, indeed, ontology, of these basic elements will cover therefore all possible assertions concerning the logical features of objects in general.

To bring out the significance of this step it is useful to contrast it once again with Kneale's treatment of Hume's 'impressions' (which he interprets as 'sensa' or 'sense-data'), and 'ideas' (which he treats as equivalent to images), whilst introducing in addition a third realm, that of 'perceptual objects'.[1] Such a step is of course quite foreign to the Humean enterprise, even though an understandable development, considering the contortions surrounding Hume's use of 'impressions'.[2]

At any rate, the interesting point here is that Kneale tries to save the notion of 'natural' or 'nomic' necessity by insisting that we are free to endow the 'perceptual object' terminology with 'rules of use' quite different from those which operate for the level of sensa.[3] For this assertion Kneale never, any more than Locke before him, offers any real argument, apart from pointing to the *fact*, which is of course never questioned by Hume, that the laws of science have a necessitarian logic. Once again this presents us with an interesting example of the metaphysician's activity of 'asserting' certain justificatory features, supplemented by the suggestion of empirical contexts: in the present case, of certain linguistic aspects, not so far removed from the Aristotelian position to which we drew attention before.[4]

---

[1] To avoid confusion, let us remember briefly our account of the various levels we have distinguished. Before, in this chapter, we had the two-fold division between the 'positive' and the 'metaphysical' levels, a division due to the 'methodological' and 'epistemological' accounts, respectively. But note (as we found earlier) that the latter involves itself the distinction between the 'epistemic' and 'constitutive' realm (the 'constitutive' component deriving from the 'ontological viewpoint', in Johnson's terminology). At first, 'epistemic' and 'positive' are, however, confused, as are the methodological and the ontological account; a confusion which will not be disentangled until we come to Kant.

[2] For Kneale, cf. *Probability and Induction*, § 18; and above, ch. IV. pp. 189–91, 198n.1; ch. II, pp. 29–34.

[3] Op. cit., pp. 80, 88.   [4] Above, p. 364.

Certainly, Hume would have contended that the only rules justified are those which operate at the epistemic level; which for him is the level of 'impressions' and 'ideas'. And his identification of 'impressions' with 'objects' might be taken as an expression of his insistence that this is the only meaningful level; i.e., that the rules which govern the usage of terms for objects must involve a necessary reference to the conditions of experience. In other words, if it can be shown that at the level of 'impressions and ideas', 'necessary connections' are not 'demonstrable', then this is a remark which must extend to all 'objects' of which in any sense whatever we can be said to have knowledge (in the general, and non-technical, sense).

Still, Hume likewise does not supply any *arguments* which would have the effect of *proscribing* the assumption of a supplementary realm of what Kneale calls 'perceptual objects'; an expression with 'constitutive' use. For that we have to turn to Kant's proscription of 'things-in-themselves', made possible only by a more complex epistemological structure, which distinguishes between 'perceptions', phenomenal objects and noumenal things-in-themselves.

With these remarks, we may consider briefly the consequences of Hume's logical appraisal of 'impressions' and 'ideas'. As foreshadowed, the most prominent feature is Hume's relentless insistence on the atomistic nature of impressions and ideas; moreover, here in particular the language of 'impression' is totally interchangeable with that of 'object'. At one extreme, we get references to 'acts of the mind': each of these, we are told, is 'separate and independent'.[1] Farther along, we have all 'ideas entirely loose and unconnected'.[2] At T.86 we get the classical formulation in terms of 'objects': 'There is no object, which implies the existence of any other if we consider these objects in themselves'; and a little after, the related point: 'reason can never shew us the connection of one object with another'.[3]

---

[1] T.140. Some of the relevant passages were cited above, p. 329n.2.   [2] T.10.

[3] T.92; similarly at T.97. This has its reflection in some modern writings, though transposed to the concept of the elementary propositions. Cf. Wittgenstein's *Tractatus*: 'One elementary proposition cannot be deduced from another. There is no possible way of making an inference from the existence of one situation to the existence of another, entirely different situation.' Wittgenstein concludes that 'there is no causal nexus to justify such an inference. We cannot infer the events of the future from those of the present'. Even Hume's positive doctrine we find here formulated, though only approximately, and doubtfully labelled 'justification': 'The procedure of induction . . . has no logical justification but only a psychological one. There is no compulsion making one thing happen because another has happened.' (Op. cit., 5.134–6; 5.1361; 6.363–3631; 6.37.)

In the same vein, 'every effect is a distinct event from its cause',[1] so that in this sense, and at this level, 'anything may produce anything'.[2] Clearly, the threat contained in such locutions is meant to 'soften up' the reader and to make him the more eager to accept Hume's subsequent positive doctrine. In fact, Hume is simply denying the suitability of the notion of 'necessary connection' inherited from his predecessors. Having sharpened the older notion of 'knowledge' and anaesthetised it by reduction to perception of *logical* necessity or incompatibility (e.g. the 'contrariety' between 'existence and non-existence'),[3] he dismisses forthwith all those different 'opaque' claims traditionally associated therewith that we have noted, and of which Aristotle's doctrine was a paradigm-case; in particular, the two claims formulated on pp. 362–3 above.

Now as such it is no more than a tautology to say that one object cannot logically entail another; if one object being 'independent' of another *means* 'logically independent', clearly the assertion of the existence of one can never entail that of the other. However, the importance of this is that it provides Hume with a kind of standard in terms of which to test whether the 'necessity' claimed for the causal relations can possibly meet the requirements of 'reason' and 'knowledge'; and we shall not be surprised to find that it cannot.

These requirements mostly amount to the contention that in all such reasonings we claim (the rational or intuitive emergence of) a necessary connection between one idea and another, where the association between such ideas is in the first place due to the suggestions of 'experience'. Now we have seen that all ideas are 'loose and unconnected'; all are 'distinct' from one another. Being distinct entails their being 'separable'.[4] Here Hume introduces a simple test to prove their separability: Take any two such ideas whose connection is suggested by experience. Then ask: 'Is it possible to *conceive* the thing to be otherwise (to conceive its contradictory)?' For instance, take the principle that 'whatever begins to exist, must have a cause of existence'. Since it is possible 'to conceive any object to be non-existent this moment, and existent the next', the ideas of cause and of a beginning of existence are separable; therefore the actual separation is possible, which implies the absence of any necessary connection

---

[1] E.30.
[2] T.173. Cf. above, pp. 330–1, 343ff., 365n.1, for the distinction between the two 'levels'.
[3] T.70.   [4] T.79.

and our total inability to 'demonstrate' a 'law of causation'.[1] And the same goes for the principle of uniformity of nature, construed as an entailment: we can 'conceive a change in the course of nature; which sufficiently proves, that such a change is not absolutely impossible'.[2] The impossibility referred to here is of the logical kind; for Hume adds that the denial of the connection, and the assertion of separability, 'implies no contradiction'.[3] Or as the *Enquiry* puts it: 'The contrary of every matter of fact is still possible.'[4]

It will be evident from all this that Hume is now offering a logical criterion defining the atomism of impressions and ideas. He has sharpened the previous rather vague notion of necessity; he presumably takes it that this is the only meaning of necessity which has ever been traditionally in question.[5]

Of course, it is a little too easy to dismiss the complexities of this revolutionary development by a brief reference to 'logical necessity', a notion itself beset with difficulties of a clear definition, and in any case at this stage of the history of the subject in process of emerging from a much broader and vaguer background. Nor is it possible here to go into this matter in any detail; it is amply dealt with in the great number of commentaries that exist on the philosophy of Hume, not to mention the vast literature that exists on the topic of logical necessity, analyticity, and the *a priori*. We have space only for a very few remarks, and these assuredly controversial.

According to Hume, the test of the *possibility* of a happening is its 'conceivability', but this can hardly be a psychological matter. Yet, if a reference to a *logical* feature is intended, is Hume doing more than *repeating* that the happening is possible; that being conceivable *means* being possible? By what tests, then, shall we decide that a certain relation, e.g. a sequence of ideas, is *impossible*? One might here refer to the concept of an 'analytic proposition', or again, to what is and what is not permissible in accordance with the rules of some language game or calculus, e.g. of arithmetic. But the concept of 'analyticity' is extraordinarily plastic, and its demarcation from its opposite, 'synthetic', difficult in the extreme, quite apart from the fact that analytic propositions hardly exhaust the class of 'logically necessary', let alone necessary propositions *simpliciter*. (The nature of the putative necessity of the propositions of simple elementary arithmetic, for instance,

---

[1] T.79–80. A similar argument is repeated at T.161.   [2] T.89.   [3] T.80.   [4] E.25.
[5] Cf. the point made above, pp. 336f. and 364ff., concerning the significance of this 'logical 'interpretation of the connection between ideas or impressions.

is still under dispute.) Such considerations make Hume's test of necessity seem rather less imposing, although for many purposes it no doubt has a great therapeutic and clarificatory value.

Let us consider an objection to Hume which, whilst it cannot ultimately claim the force of a final verdict on his doctrine, does throw light on the rather rarified and theoretical nature of his argument. Let us ask: 'Can we conceive, is it possible, that flowers should grow on the moon?' It might be thought easy to conceive this possibility, and thus to allow it. However, it might be objected that the flowers which are here imagined are not the flowers of the botanist, indeed, not 'real flowers' in any sense. It is easy to construct a cartoon film showing flowers growing on the moon, but this surely is not what is required in order to prove this possibility. For the flowers which are relevant are organic structures which (for instance) require carbondioxide (CO_2) for 'respiration' and growth. Now the moon lacks CO_2. Hence 'real' flowers *could* not grow, not just because it is a general physical matter of fact which Hume would of course admit, but rather because by flowers we *mean* those physical structures requiring CO_2 for growth, and by growth the process involving the utilisation of CO_2, and because the moon's surface is taken to lack CO_2.

However, Hume's answer to this objection is obvious: our objector has now made the requirement of CO_2 a *conceptual matter*. Therefore, 'growth' by way of the utilisation of CO_2 is now a logically necessary, or at any rate, a conceptual requirement.

There is, however, a reply to this answer. For we must remember that though what is involved here is our *concept* of flower, this concept is a 'plastic' one; by this I mean that although the concept has developed by reference to empirical enquiry, it now is part and parcel of a complex theoretical system of science, at the intersection of a complicated network of observation, theory and laws. And all these in turn are modifiable, so that the boundaries of our 'concepts' also may vary from time to time. So we cannot determine by mere inspection what is possible and what is not; the answer will always be delayed, *and may be indefinitely delayed*. If Hume misses this point, it is simply a reflection of his assumption of the sharp distinction between observational and theoretical predicates.

However, on this a comment may still be that although in fact a decision can be delayed, it cannot be *in principle* delayed indefinitely. For in that case we should end up with the result that—just as we would say that a flower necessarily breathed CO_2—so whatever happened happens necessarily, at least provided it passed certain tests (cf. Descartes' old test of 'clarity and distinctness'); a position not unlike that at which Spinoza eventually arrived, identifying physical law-likeness with necessity—when it is just this concept of necessity that is in question.

At any rate, not all concepts have the plasticity which we have ascribed to those of 'flower', 'growth', 'moon', etc. For instance, the concept of 'thinking' (to take a very basic one), and many other quasi-transcendental concepts, are not as easily handled in this way, and the same is true of many of the more common concepts designating fundamental qualities, e.g. colour, smell, etc. Here, Hume's test seems appropriate.[1] On the other side, there are certain concepts, e.g. that of a vacuum, where the test is not so easy to apply. At first sight one might hold that, in so far as we are concerned, say, with the question of the *existence* of a vacuum, this was simply one of empirical evidence and fact. And there were many seventeenth- and eighteenth-century scientists who treated it as such. On the other hand, in Hume's hands, as we have already noted,[2] the criterion that becomes decisive in his argument depends (as he admits himself) 'upon the *definition* of the word, touch'.[3]

All this shows that the test of the 'conceivability of the opposite' cannot be used mechanically, and that much depends upon what concepts are involved. The fact that the concepts which enter into the basic laws of nature, for instance, do not easily or obviously fall under this rule explains perhaps why 'necessity' has so frequently been ascribed to them. On the other hand, obviously this kind of defence does not so far supply us with a *principle* by which to extend 'necessity' indefinitely to all generalisations whatever, let alone to statements concerning particular matters of fact;[4] so far, this seems possible only as a matter of metaphysical postulation, as we find exemplified by the case of Leibniz. All we can say at present is that Hume's rule applies to some concepts and not to others. That it applies to all, as he claims, is simply the result of a sharpening of positions, couched in the terminology of 'impressions' and 'ideas',

---

[1] The metaphysical equivalent of this fact is, of course, Hume's notion of 'impression'!

[2] Cf. above, p. 330n.1.   [3] T.638; my italics.

[4] I add this case because it will figure prominently in the philosophies of Leibniz and Kant.

whose philosophical logic is so construed as to generate the atomicity of all concepts.[1]

8. FINAL CRITIQUE OF NECESSITY

Let us revert to Hume's basic tenet that the possibility of 'separation' of one idea from another (or from some impression, either of sensation or memory)—however frequently the corresponding impressions may have been conjoined in our experience—is fatal to the claim of necessity in any 'objective' sense. And let us ignore possible escapes from this position by way of postulation of a 'constitutive realm', or by regressing to a position such as Locke's, which involves the ideal of a perception of necessity in some intuitive fashion. We are then left only with the case where the hypothesis of law with necessitarian import is based on the dual foundation of inductive evidence coupled with those 'consolidative' features which we studied in our second chapter, and as a result of which we end up with the structural network of concepts mentioned at the end of the last subsection. Moreover, when we assume that the ascription of nomic necessity is 'based' on this dual foundation, we shall assume that this is not a reference to any sort of '*logical*' implication, either deductive or inductive. We are not to suppose that the necessity claimed for the law *follows from* this foundation. The most we can say is that the ascription of necessity is a rational response to such a foundation. But it does not much matter for our purposes whether the 'rationality' in question is interpreted in 'psychological' or 'conventionalist' fashion.[2]

---

[1] In Kant, the distinction between an empirical and a conceptual question will be articulated in a far more complex manner, partly by discussing the latter in terms of the notion of 'real possibility'. Thus, if action-at-a-distance means action across a 'vacuum', we need first exhibit the real possibility of such a notion, both in respect of 'empty space' and of 'action without contact'. If we take the common Newtonian conceptions of matter and space, Kant holds, then it is not permissible even to *hypothesize* concerning 'a force of attraction without any contact' (K.613). On the other hand, if matter is *defined* through the criteria of forces of repulsion and attraction, as well as 'moving force', all in accordance with categorial principles of possible experience, then action-at-a-distance is a 'possible' concept. (Cf. *Met. Found.*, ii–iii, and below, ch. VIII, sects. 6, 7, 9.) All this suggests a far more intimate connection between conceptual foundations and contingent principles or scientific hypotheses than allowed for by Hume.

[2] Indeed, when we come to Kant, we shall find that nomic necessity *is* attributed to laws *in virtue of* the presence of this consolidative background (the systemic and regulative structure of science); without this there would be no laws with a necessitarian logic. But again, the necessity doesn't *follow* from the background; rather it is just *another way of saying* that such a background is presupposed.

We are then in a position where it is simply *claimed* that in the statement of a certain causal law, some element A necessarily connects with some other element B. All this quite fairly represents the way in which Hume envisages the question of necessity as emerging. And we shall try to discuss it here by and large within a Humean framework, partly since our task at present is to illuminate only the implications of *Hume*'s doctrines, partly because a general discussion would take us too far. The following, then, are questions which Hume either does ask, or might well have asked, if not in precisely these terms; and so, of course, are the answers.

The questions which arise are as follows: (1) What is here to be understood by 'element'? (2) Is it possible, even in principle, to claim the existence of necessity? (3) What is the meaning attaching to the notion of necessity here involved?

(1) The elements, by exclusion, do not belong to any constitutive realm (on the lines of a Lockean 'real essence', or Kneale's 'perceptual objects'). According to Hume, the *only* interpretation we can give is that they have the status of 'ideas'; impressions are only relevant in so far as they are 'present to consciousness' at the moment at which they occur; and here they are no more than 'passively received'. Any logical remarks that can therefore be made about permanent *relations* supposed to hold between 'elements' must be interpreted as remarks about 'ideas'.

(2) But can one even *claim* a necessary connection between ideas, without possessing rational insight into this necessity? We have already noted one attempt to render such a suggestion significant. We sometimes may frame, and seek proof for, an hypothesis that a given mathematical theorem holds. Here, we do *not yet know* whether the theorem holds, and by implication, whether its terms relate with necessity, though we may assume that they either do or do not so relate.[1]

Most people would, however, want to reject this example. It assumes that nomic necessity may be equated with the kind operative in mathematics. Furthermore, our ascription of necessity in this case is on an epistemic basis. Hence, once I assert ('claim') that the theorem is true, this assertion would be inconsistent with the *possibility* of its being false. Whereas in the case of natural laws, to assert their truth—based as it is on grounds in a sense extraneous to the question of

---

[1] Cf. Kneale's example of Goldbach's theorem, in connection with Locke's point that there are mathematical cases where we have not yet made the necessary connection 'visible'; above, ch. IV, p. 198n.1.

necessity—is supposed not to be inconsistent with their possibility of falsehood.[1]

But is there any other sense (or meaning) in which one could speak of a *claim to necessity*, albeit epistemically opaque, between ideas? Evidently, if there is such a sense, it must be a fairly 'weak' one, since it is a claim to necessity which is supposed to envisage the possibility of its own non-existence.

(3) This brings us to our third question, concerning meaning. Now we have already seen, once or twice, that when questions are posed concerning the 'meaning' of a certain concept, this is an oblique request for the presentation of the 'object' which (it is believed) will correspond to such a concept.[2] But this is precisely the way in which Hume at first proceeds. Part of his apparent scepticism in fact arises from the fact that whilst he is an implicit adherent of the theory that causal laws involve the ascription of necessary connection, when he enquires into the meaning of this concept he formulates this as the demand to exhibit 'the impression or impressions from which its idea [of necessity] may be derived'.[3]

Formulating the question in these terms has the immediate consequence of suggesting that the necessity here demanded ought to be an 'objective feature'—something like a quality. Moreover, it is taken for granted that the logical strength of there being such a property is equivalent to that of a logically necessary connection, so that it would not be logically possible to entertain its negation without self-contradiction.

Now the 'quality' in question ought to have the status of an 'impression'. However, the logical grammar of 'impressions' is simply that of 'ideas', even though it is the latter that are said to be 'copies' of the former. Therefore, whatever we can say about our 'impressions' must be under the guidance of the corresponding 'ideas'. But according to Hume it is a logical property of any 'idea' that it is always logically separable from any other idea, in accordance with the principle of atomism.

---

[1] This objection could be avoided only if it were maintained that the assertion of law-likeness connects with the truth-claim in an *essential way*, which is certainly not the situation Hume ever envisaged, though it is possible to maintain that this would come close to the Kantian position.

[2] This was, for instance, noted by Berkeley when he criticised the assumption that 'every noun substantive stands for a distinct idea'—meaning by 'idea' an ontologically basic feature such as was represented later by Hume's 'impression'. Cf. above, ch. V, pp. 285f. And this was also the core of Wittgenstein's comment on the traditional theory of 'necessity', in the passage quoted in ch. II, p. 20.

[3] Cf. T.77; cf. T.163–5.

It follows from this peculiar combination of the ontological function of impressions,[1] coupled with the fact of their logical properties being those of 'ideas', that so long as our question is framed in Hume's original form, the answer to the problem of the *claim* to necessary connections between our ideas, even where opaque, must be in the negative.

It is clear that an escape from these difficulties is only possible if the nature of necessity is not construed in this ontological fashion, on the model of 'impressions'. This is suggested, moreover, by the fact that the ascription of necessity is in any case not 'grounded' on the inductive (and for Hume, not even the consolidative) basis of the natural laws in question. Remarkably, though Hume continues to frame his problem in terms of the search for an 'impression', going so far as to couch his eventual answer in this terminology,[2] the importance of his answer lies in the fact that it introduces the theory of a 'weak' form of necessity, not construed in 'ontological' but 'psychological' fashion. And the answer will supply, not so much an 'impression', but rather a sketch of those psychological or 'natural' conditions which obtain when we claim necessity of connections. 'Meaning', from being a quest for ontological objects, becomes a description of the 'manner' in which certain law-like locutions arise.[3]

9. NEW FOUNDATIONS FOR THE 'CAUSAL TIE'

Hume, as just mentioned, does not himself give up the necessitarian position. But in the face of his own destructive criticism, he attempts to assign a place to necessity which will escape his own condemnation; and this is possible only by renouncing the notion of a justificatory employment of the causal tie, placing it at a non-constitutive level, an aspect or 'manner' of believing certain propositions.[4] He thus replaces the question, 'What is our justification for stipulating a necessitarian tie?' by another, 'How is the idea of such a tie produced?', or by the still different question, 'Whence do we get the impression of which

---

[1] This should not be confused with the 'constitutive function' of Locke's and Kneale's world of 'objects' *behind* the impressions and ideas.

[2] Cf. T.165, where the impression is said to be 'internal', an 'impression of reflection'.

[3] This is the part of the answer which Kant was later to take over, reacting against Leibniz's 'ontologising' procedure. But instead of construing necessity as a 'psychological' feature, he makes it transcendental.

[4] Cf. E.49.

this idea is the copy?', the last question—being, as we have just noted, two-edged—acting as a kind of link between the other two.

Hume's procedure, then, will be to describe the conditions under which these 'beliefs' arise. Here, Hume employs again a Newtonian paradigm. The beliefs will be 'explained' by reference to certain pervasive 'forces'[1] which 'unite' ideas when certain conditions are fulfilled. It is these 'forces' that provide the missing 'medium', relating what—as a matter of logic—had previously been unrelated.[2]

Discussing the 'inference' that we 'in fact' make from past conjunctions of qualities to future ones, Hume writes: 'There is required a medium, which may enable the mind to draw such an [inductive] inference'.[3] Again: 'Where is the medium, the interposing ideas, which join propositions . . . ?'[4] At T.10 it is described as a 'bond of union' among our ideas. Similarly at E.29 we are told of the 'tie or connection between the cause and effect, which binds them together'.[5] This was what Locke had searched after, using the deductive ('demonstration') model, under just this name of 'medium'.[6] But the previous analysis has shown that the search for such a medium, even to speak of such a search, is a task beset with confusing perplexities. To name it a 'quality' suggests that it should have the status of an 'idea'. On the other hand, if an 'idea', this means that there should be a previous 'impression' of which it is a copy: a picturesque way of saying that it should be a detail in Hume's basic furniture. Yet, as we saw, *quâ* idea it could hardly be expected to do what it was introduced for: to 'bind' our other ideas together, since *all* ideas are loose and separate.

Sometimes Hume uses an 'inspectionist' device: 'Let us cast our eye on any two objects, which we call cause and effect, and turn them on all sides, in order to find the impression, which produces an idea of "connection".'[7] And we are not surprised at his inability to discover one, since this is no mere contingent matter. We know already that candidates such as *experienced* forces, or the 'animal nisus', are held by Hume to be incapable of supplying any 'bond' in the sense

---

[1] Though 'gentle'; cf. T.10.

[2] Here we again confront this important concept of 'tie', 'bond' or 'medium'. This question will be all the more pressing for Hume, since the easy theological solution which Berkeley had employed for his characterisation of 'laws of nature' is no longer available.

[3] E.34.   [4] E.37.   [5] At E.63 this medium or tie is called a 'quality'.
[6] Essay iv.3.4; cf. iv.2.2–3.   [7] Cf T.75.

required,[1] even though such a 'nisus' is an experience, an 'impression'.[2]

But though Hume will *claim* in the end that he has discovered an 'impression' or 'uniting principle'[3] for our 'idea' of necessary connection, closer inspection will reveal this to be a rather misleading characterisation.[4] That he manages the feat of describing it in these terms is partly due to his using the chief model provided by the Newtonian heritage when he likens it to the place and status of another 'uniting principle', viz. gravitational attraction. For the connection or 'bond of union' that exists between our ideas (the occasions or conditions of the causal relation are the subject of protracted investigations over many pages of the *Treatise*, involving features like 'resemblance', 'spatio-temporal contiguity', and 'constant conjunction'),[5] is (writes Hume)

> a kind of ATTRACTION, which in the mental world will be found to have as extraordinary effects as in the natural, and to show itself in as many and as various forms.[6]

But at this point we are reminded of the complex problem of the status of gravity in the Newtonian corpus.[7] It leads a somewhat nebulous existence; sometimes it is said to be 'real', at others it is known only by its effects, a matter of the regular and law-like uniformities which inductive investigation uncovers. Sometimes it is thought of as an ultimate 'explanans', at others there is an insistence that some causal mechanism must certainly be responsible for it. However, in the end, the reductionist approach prevails, and Hume can use for his own 'gentle force', which unites the mind's ideas, the language and corresponding appraisal used during this period by many when they try to characterise the place and status of a 'force of attraction':

---

[1] Cf. above, p. 331 and n.3, with its reference to E.67n. A similar point is made at T.632. We have discussed this in ch. IV, when considering Hume's reply to Locke, pp. 264–5.

[2] We could not know *a priori* what '*follows*' upon the exertion of force: thus the passage at E.67n. This shows that it is insufficient to produce an 'impression' for the 'idea of force or energy' (ibid.). Thus Hume's argument demands that we produce an 'impression', yet the very logic of 'impression' necessarily prevents the solution of his problem. This is the usual sign of a philosophical doctrine, whether negative or positive: its terms are so adjusted as to produce a kind of circularity; it is a form of metaphysical description, which leaves the empirical surface unchanged.

[3] T.10. [4] Cf. below, pp. 379f., 381f. [5] Cf. T.11, 87, 92–3.

[6] T.12–13; Hume's capitals.

[7] Cf. above, pp. 329ff. and Appendix to this chapter.

Its effects are every where conspicuous; but as to its causes, they are mostly unknown, and must be resolv'd into *original* qualities of human nature, which I pretend not to explain. Nothing is more requisite for a true philosopher [e.g. scientist], than to restrain the intemperate desire of searching into causes, and having establish'd any doctrine upon a sufficient number of experiments, rest contented with that, when he sees a farther examination would lead him into obscure and uncertain speculations. In that case his enquiry wou'd be much better employ'd in examining the effects than the causes of his principle.[1]

This ceremonial Newtonian phraseology is here evidently used in order to argue the case that explanation should not be construed in terms of underlying metaphysical powers but by reference to general law-like 'principles' which relate the phenomena. It supplies Hume with the tools for sharpening the Newtonian position which had only uneasily banished the search for 'underlying causes'. For the causes that Hume wants to proscribe are the 'secret powers and forces', the 'power and efficacy', which all his predecessors had claimed to be located in one place or another, be it in the 'primary qualities' or the 'real essence' of a Locke, or the efficient causality of a Berkeleyan Deity.[2] 'We never have any impression, that contains any power or efficacy', he writes in a characteristically Berkeleyan vein; and 'we never therefore have any idea of power', generalising this contention —as we have already noted—to the concept of God's power.[3]

The logical, or ontological, status of what is here *denied* is evident if we contrast this with the favourable reception accorded to Newton's 'hypothesis' of an ether, understood by Hume as the employment of the notions of 'force and energy' to explain attraction—but at what we called the 'positive level'.[4] And just as gravitational attraction is removed from its logically subterranean position and elevated to the level of the 'phenomena', through corresponding laws of uniformity, so the connection between ideas is placed at the level at which they themselves are operative, and will thus become a descriptive feature of the regularities which they exhibit. And this is the significance of the replacement of 'reason' by 'custom or a principle of association'. It is

---

[1] T.13.
[2] Cf. E.36–7; T.160–1. For Hume's reduction and relegation of the 'secret powers' of 'matter', cf. E.33, 37.
[3] T.161. Hume, of course, treats 'power' in this sense as equivalent to 'necessary connection'; cf. T.156–7; 162.
[4] E.73n.; cf. above, pp. 329–32.

important to note carefully the precise language in which Hume expresses his position:

> Reason can never satisfy us that the existence of any one object does ever imply that of another; so that when we pass from the impression of one to the idea or belief of another, we are not determin'd by reason, but by custom or a principle of association.[1]

We have already seen what we are to understand by this reference to 'reason'.[2] Here we only note that what replaces it is a 'natural feeling', which 'determines us' or 'the mind' to pass from impression to idea.

'But how can this be said to *justify* the passage?': thus goes the usual objection. Hume replies by placing the 'justification' precisely where the Newtonian scientist (in his 'positivist', or rather: positive guise) had placed his 'explanations': at the point where the passage regularly occurs. 'Why do all bodies always and invariably attract; or better: mutually accelerate towards each other?' The first answer is: in virtue of the force of gravity, or gravitational attraction. But for our Newtonian, this is simply to state the law-like uniformity, though regarded as a 'principle' which articulates *all* the phenomena of dynamics through a deductive system. In a similar manner, Hume states the conditions in terms of which to describe the presence of the epistemological connection, though interpreted as a quasi-psychological feature.

The details of this account may be summarised very briefly, since they are well known, and have been commented on and criticised *ad nauseam* by most commentators on the philosophy of Hume, involving as they do necessarily complex discussions of Hume's theory of 'association of ideas' and 'belief'.[3] A few critical remarks is all that is perhaps needed. The final account occurs at vii.2.[4] One caution, or rather reminder, is, however, in place before proceeding. We are not about to give an explanation of how, according to Hume, we discover causes, or causal relations. For this, as we have seen at length, belongs to the methodological enquiries of the scientist, following the variety of 'general rules', the 'natural principles of the understanding'.[5] Rather, we are seeking to delineate what we *mean* by saying that one

---

[1] T.97.
[2] Above, p. 364. To the denial of 'reason' here corresponds, in Hume's eyes, the replacement of metaphysics by physics.
[3] Cf. *Treatise*, 1.3.7–10.   [4] T.163–6; E.74–76.   [5] Cf. above, pp. 343–6.

event causes another, that one objective situation is causally connected with another. However, asking for the 'meaning' is, as I have remarked, only too often taken to be equivalent to asking for the 'object' or 'idea' which corresponds thereto; and through this locution we invariably seek for a feature providing 'justification'. To determine the 'meaning' of cause is then expected to reveal the 'feature' which *sanctions* causal reasoning. Hume's naturalistic and descriptive account reminds us of alternative possibilities of discussing meanings, and he ought not to be accused of not providing in his account any 'justification' for causal and inductive inference.

In a somewhat puzzling fashion, Hume's account appears simply to trace the *genesis* of the idea of cause or necessary connection, arising spontaneously on such occasions where previously we have encountered repeated 'perceptions' of constant conjunctions of a set of impressions. There is here perhaps a weakness, for this account squares badly with all the tasks of 'causal reasoning' which his *methodological* approach tries to accomplish. His epistemological account ties causation to a basis which leaves little chance for this idea (and the corresponding principles of causality and uniformity of nature) to emancipate itself from its narrow mechanical foundation, so as to make it possible to 'apply' it as a directive or formal principle of inference. Let me explain.

Hume's basic contention is of course that the idea of necessary connection, and in general all notions of causal relationships, must be 'founded on experience'.[1] The notion of 'experience' here involved is, however, rather complex. The operative condition in all causal reasonings from one class of objects or events[2] to another is the fact that the corresponding impressions are *constantly conjoined in our experience*. Now it is a fact, says Hume, that under these conditions, we (our 'imagination') spontaneously claim to have the idea of a necessary connection between the conjoined terms. However, since these terms (impressions) as such are subject to the atomistic postulate, neither a single conjunction, nor any number of them repeated indefinitely, can give rise to this idea, since no new impression is discovered.[3] To what

---

[1] Cf., for instance, T.88.

[2] As Hume often points out, we can only reason concerning 'species' or classes of objects that 'resemble' each other in some given respect. This 'resemblance' is itself quite basic; it is simply perceived, and we are not to suppose that it depends upon the existence of 'underlying universals'; for that would make Hume's argument clearly circular, if not beg the whole question. (Cf. above, ch. V, p. 284n.3.)

[3] T.88, 163–4; i.e. discovered as an object of perception.

then do we owe the idea? The answer is this: though nothing is added to the '*object*', 'yet the *observation* of' the 'resembling instances . . . produces a new impression *in the mind*, which is its real model'.

> For after we have observ'd the resemblance in a sufficient number of instances, we immediately feel a determination of the mind to pass from one object to its usual attendant, and to conceive it in a stronger light upon account of that relation. This determination is . . . the same with power or efficacy. . . . The necessity or power, which unites causes and effects, lies in the determination of the mind to pass from the one to the other.[1]

Hume's claim to have 'founded' causation 'on experience' thus splits up into two parts. First, there is our experience of constant conjunctions. Secondly, there is the fact that we find ourselves 'determined' to pass, when presented with one instantial member of a conjunction, to the other, in spontaneous fashion. It is not clear whether this 'determination to pass' is also supposed to be 'experienced', or whether it is something whose existence 'reflection' exhibits to the self-conscious observer. Still, though these two aspects of 'experience' are very different, Hume undoubtedly meant to convey his conviction that this 'determination to pass' from antecedent to consequent was a 'natural' fact, albeit of an 'internal' or 'mental' kind.

Has Hume explained 'necessity'? It might be objected that to speak of the mind as 'determined' is either a synonym for 'being necessitated', and yields only a tautologous explanation; or that, if it is not, it is not strong enough to give us what we want. But remember that Hume is not trying to 'justify' the causal connection. He is only telling us that on those occasions on which we postulate causation, we 'feel the determination to pass' from 'cause' to 'effect'. The feeling is not meant to 'guarantee' that the future will be like the past. He is only characterising our belief, described as a peculiar 'manner' of the way in which the ideas in question 'feel to the mind'.[2] If it is described as a 'customary transition of the imagination from one object to its usual attendant',[3] we must remember that Hume does not consider it to be a deficiency that, as he says, the 'gentle force' in question is one that only 'commonly prevails'.[4] Although it is a 'principle of union', it is not an 'infallible' one, for thought often moves along very irregularly,

---

[1] T.165-6; Hume's italics. Instead of 'determination to pass', Hume also speaks of a 'propensity, which custom produces, to pass from an object to the idea of its usual attendant' (T.165).
[2] Cf. E.49.   [3] E.75.   [4] T.10.

'may leap from the heavens to the earth' and 'without any certain method or order'.[1]

Of course, it is true that Hume feels uneasy about whether such a description completely explicates causation. To compensate for this he adds remarks which affirm that the activities involved in causal reasoning are very basic, albeit still human activities. The 'justificatory power', so we might say, reappears here under the guise of the basic needs of rational man. Though 'reason' has tumbled, it is now replaced by nothing less than 'a wonderful and unintelligible instinct in our souls'.[2] The activity is likened to breathing and feeling:

> Nature, by an absolute and uncontrollable necessity has determin'd us to judge as well as to breathe and to feel.[3]

Indeed, we find eventually that the 'customary transition from cause to effect' is a

> principle which [is among those that] are permanent, irresistible, and universal ... , the foundation of all our thoughts and actions, so that upon their removal human nature must immediately perish and go to ruin.[4]

Here again we see the wrongheadedness of the impression often given that Hume did not 'believe in causation' but only 'constant conjunction'. Such a judgment is the denigration, indeed, of all Hume stood for, which was that a quality or relation is not removed just because no metaphysical foundation for it can be produced.

It was perhaps peculiar to claim that the idea of necessary connection was 'based on experience'. True, it takes its rise from the observation of constant conjunction. Stripped of misleading talk about the 'production of a new impression', what Hume really posits, given the condition of constant conjunction, is that the mind finds itself 'determined to pass' from impression to conjoint idea. But this characterisation of the mind's state is, as we have already noted, that of a second observer (of the philosopher reflecting on the regular processes of the mind); that it is described as an 'instinct' suggests that the operative feature is simply that of the *existence* of a *process* of 'reasoning' (in the popular sense of this term, including inductive reasoning).[5] Indeed, the close relationship between the 'necessary

---

[1] T.92.　　[2] T.179.　　[3] T.183.　　[4] T.225.

[5] One may, however, question whether the content of the idea represented by the claim of the *existence* of a necessary connection, as was discussed above, pp. 370–3, may be so easily identified with the *process* of causal and inductive inference, as

connection' (equivalent to the mind's determined passage from 'cause' to 'effect') and the relation that supports an inductive inference is expressly asserted. The original and traditional position, we have seen, was to base induction on necessary connection, or some relation of entailment between putative 'cause' and 'effect'.[1] At T.88 Hume rhetorically wonders whether this should not be inverted: 'Perhaps . . . the necessary connection depends on the inference, instead of the inferences depending on the necessary connection.' The answer comes eighty pages or so later: they are to be identified.

> The necessary connection betwixt causes and effects is the foundation of our inference from one to the other. The foundation of our inference is the transition arising from the accustomed union. These are, therefore, the same.[2]

This is the answer to the question: where is the 'medium' or 'tie' or 'bond'? For Hume, when speaking in this mood, there *is* no additional 'detail' corresponding to the idea of the bond. Intuitive 'reason' had promised to produce such a bond, and failed. It is replaced by the notion of a 'belief', acquired under a certain set of circumstances, e.g. a series of constant conjunctions. But belief is only a 'particular manner' of turning to an associated idea; it is only a 'particular way' of 'conceiving our object'. So we need not have 'recourse to a third to serve as a medium betwixt them [*sc.* two ideas]'.[3]

This then is Hume's notion of a presuppositionless induction, as foreshadowed already by the suggestions contained in Newton's Third and Fourth Rules of Reasoning. Correspondingly, the whole apparatus of scepticism has once again provided—as in Locke—a weapon for a back-door advance towards 'naturalism'.[4] At the same time it becomes clear that the causal connection, like its cousin, inductive

---

characterised by Hume. That Hume is uneasy about this identification is clear from the fact that necessary connection is at one moment thought of as a certain mode of mental *activity*, and at the next as a mental 'impression'.

[1] Cf. above, pp. 361ff. For an example, cf. the passage from Ewing, quoted above, ch. II, p. 42, exemplifying the 'Aristotelian' position.

[2] T.165.   [3] T.97 and 97n. Cf. Kant's reference to a 'third thing', below, p.489.

[4] Cf. above, pp. 338–40. It is, however, important to remember, as we have already seen, that Hume is unhappily forced to admit that although *here* 'necessity' is the *effect* of experience of constant conjunction plus habit operating on the imagination (T.165, 265), and though 'judgment and reasoning' are not in any way different from the mere 'conception' of the 'unity' thus achieved (T.96–7n.), yet sometimes 'judgment' may have to be set against the 'imagination' (T.148, 150). This is clearly an unresolved conflict. (Cf. above, p. 347.)

inference, is not so much something 'based on experience' as an essential ingredient of all scientific reasoning. It is not so much the result of some specific pair of (mental) events constantly conjoined 'producing a new impression', itself observable in turn, as a spontaneous formal activity in which we find ourselves involved; an approach consonant with that expressed in Newton's Rules of Reasoning.[1]

Unfortunately, Hume never resolves these conflicting positions. And yet this is curious. We have already seen that he allows generalisation on the basis of a 'single experiment', provided it is 'clear', 'duly prepared and examin'd'.[2] And more, does he not even admit that this whole story of constant conjunctions is a myth, when he expressly tells us that we practically never meet any series of constant conjunctions that could serve as premises for causal inferences since they are almost always encrusted in a vast variety of surrounding, causally irrelevant, circumstances?[3]

Here we meet again this feature, which is really foreign to his epistemological account, and which Hume characterises as the 'exercise of judgment', pitted *against* the 'imagination', which we have noted already at length.[4] And we might dismiss the apparent contradiction simply by distinguishing between our two levels, that of positive causal reasoning, and the other, metaphysical one, when Hume attempts a characterisation of causation as such: corresponding to the methodological and epistemological strands of his argument. But there clearly exists a tension between these layers. When under the sway of the constant conjunction approach, regarded as 'producing' the custom or habit of association, all is 'psychology', mechanical response. Hume then describes the occurrence of contrary

---

[1] Cf. above, pp. 337ff.
[2] T.173, 131; above, p. 347. This exposes the hoary criticism that not all constant conjunctions are causal relations; witness Hume's direction to distinguish between accidental and essential circumstances (T.149, and above p. 341). But neither does it seem to be a necessary condition. To meet this, we must either employ Hume's balance sheet of conflicting forces, or his distinction between judgment and imagination; or we must assume that 'constant conjunctions' produce the 'idea', in the sense that they lead to the formation of the concept of necessary connection; which concept may then find application on proper occasions. At any rate, Hume's theory on the whole would be that constant conjunction first gives rise to the hypothesis (as we might put it) of a causal connection, but that this needs criticism in the light of further experience; except that the mind does not *feel* it to be an hypothesis but rather experiences it as a 'habit'.
[3] Cf. T.175, and above, p. 341, for the relevant passage.
[4] T.148, 150; above, pp. 346–8 and 382n.4.

instances as only 'weakening' the custom, creating an 'imperfect habit'.[1] The mind is then only under the sway of conflicting forces. But it is really rather difficult to see how the mind could on that basis end up with a principle of uniformity, viz. that 'what we have found once to follow from any object, we conclude will for ever follow from it'.[2] We are merely *told* that this 'supposition, that the future resembles the past' is 'entirely derived from habit'. But it is also described as a 'determination to transfer the past to the future', as an 'impulse of the imagination'—an expression which includes, as we can see, both the habit approach as well as that which makes allowance for 'judgment'.[3]

However, whilst we may doubt the psychology, we can still appreciate the general intentions. For neutral to the two accounts is the contention that inference is a form of spontaneous projection; that it involves no more than an 'extension of our observation' from the field of the observed phenomena to 'all phenomena of the same kind', as the fourth rule had expressed it; a rule which we have seen Hume describe quite simply as a 'natural principle of our understanding'.[4] Here is a theory which makes inference 'self-wrought', and not dependent on any underlying or over-arching principles and features. True, it is given psychological trappings; and there were perhaps alternative ways of coming to a similar conclusion, the most illustrious of which is of course Kant's. Kant will concentrate on the causal *principle*; since he gives up the ostensively genetic account, there is no motive for discussing individual geneses of the ascription of causal connections. But this causal principle will, in a manner reminding us of the Humean approach, be 'proved', or better, shown to be an essential aspect or part or component of cognitive experience. The difference is that Kant treats it as an ingredient of any particular judgment of change or happening as such, i.e. arising at a level prior to explicitly 'causal reasoning' as this has been construed so far. This will have the advantage of giving Kant a lever to employ the principle with greater freedom than made possible in Hume's account, using it as a 'postulate' for all scientific enquiry, though of course with no greater justificatory power than in Hume's theory.

---

[1] T.132.   [2] T.131.

[3] Although in the passages where this occurs, T.134–5, the account is again given in terms of the 'weakening' of the habit, of 'dividing and afterwards joining in different parts' the 'perfect habit', in virtue of the contrary instances. But this is fairly fanciful psychology, constituting no more than an *ad hoc* description of the facts of the case in terms of a psychology of forces.

[4] T.175; above, p. 347.

To this we shall turn in the last chapter. Before doing so, we must follow, in the development of the philosophy of Leibniz, a more searching attempt to construe the 'medium' or 'link' between the terms of propositions than we have met so far. The problem which (if we except a few hints concerning Locke's problem of the concept of substance$_p$) we have so far encountered is this: the atomism of ideas has led our philosophers to search after links which would return to us the unity of our uniformities of nature. What has not occurred to them is to bring up the question whether a similar search for links, and like problems, may not arise when we consider the nature even of the individual or particular proposition or judgment. This problem emerges for the first time in Leibniz. But for a clear separation of the two questions, the links which generate uniformities, and the links which generate individual singular judgments, we have to await Kant, with the distinction between 'understanding' and 'reason', between principles that have 'constitutive', and those that have 'regulative' force.

APPENDIX TO CHAPTER VI
(pp. 333ff. above)

A NOTE ON THE CONCEPT OF GRAVITY IN NEWTON AND HIS IMMEDIATE DISCIPLES

Owing to the attacks of the Cartesians, of Malebranche, Leibniz and Berkeley, together with the various responses to these attacks from the camp of Newton and his followers, the subject of the interpretation of gravity bristles with difficulties. It exerted, however, such an immense influence on contemporary thinkers (we have witnessed already something of this in Berkeley) that some very brief notes on the historical situation of Newton ought to be added, to be able to appreciate Hume's reaction to it.

In a Scholium to Proposition 69, Book I of the *Principia*, Newton speaks of attraction as an 'endeavour . . . made by bodies to approach each other', whether this be due to the bodies themselves directly 'agitating each other by spirits emitted', or to the action of a circumambient 'ether' or some other 'medium'.[1] According to the General

[1] Op. cit., Dover ed., i, p. 192.

Scholium, as well as Cotes' Preface, both added to the second edition (1713), and also a letter by Cotes, gravity is a 'power' which 'does really exist', though it is left open by Cotes, and denied by Newton, that it is an 'essential' property of matter. At any rate, writes Cotes, if it is an 'ultimate' mechanical cause, no further mechanical explanation ought to be expected of this property.[1] This was partly a thrust against those who insisted that Newton's explanation through gravity amounted to postulating an 'occult quality': Bodies have a tendency to be mutually accelerated in virtue of an inherent power to do so. Newton's reference to 'endeavour' comes very close to this.

On the other hand, Newton for the most part did think 'gravity' in need of 'explanation'. Thus in the famous letter to Bentley (1692–3), he writes:

> It is inconceivable, that inanimate brute Matter should, without the Mediation of something else, which is not material, operate upon, and affect other matter without mutual contact, as it must be, if gravitation ... be essential and inherent in it. ... That gravity should be innate, inherent, and essential to matter, so that one body may act upon another at a distance through a *vacuum*, without the mediation of anything else, by and through which their action and force may be conveyed from one to another, is to me so great an absurdity, that I believe no man who has in philosophical matters a competent faculty of thinking can ever fall into it. Gravity must be caused by an agent acting constantly according to certain laws, but whether this agent be material or immaterial I have left to the consideration of my readers.[2]

Bentley (1693) and Clarke both argued that the cause of gravity must be something 'immaterial', partly on the ground that this force penetrates all matter indifferently in depth. Bentley explicitly called it 'a divine energy and impression'.[3]

When more explicit on the subject, such as in a letter to Boyle (1678–9), and in Queries 21 and 28, appended to the *Opticks* edition of 1717, Newton opts for the 'hypothesis' of a 'subtle', 'elastic', 'etherial substance', having varying degrees of density, which 'impels bodies from the denser parts of the medium towards the rarer, with all that power which we call Gravity'.[4] However in the *Opticks* he

[1] Op. cit., i, p. xxvii, ii, pp. 547, 400; Appendix to ii, pp. 634–5.

[2] Cf. *Isaac Newton's Paper & Letters on Natural Philosophy* (ed. I. Bernard Cohen, Harvard, 1958), p. 302.

[3] For Bentley, cf. Cohen, op. cit., pp. 332, 338, 340–1, 343. For Clarke, cf. the latter's edition of *Rohault's System*, 1723, i, p. 54n.

[4] *Opticks*, Dover ed., p. 350–1; for Boyle's letter (which appeared in the Birch edition of Boyle's works of 1744), cf. Cohen, op. cit., pp. 250–1.

adds that the ether cannot be an ordinary 'dense fluid' since this would put up too much resistance to the motion of bodies and their constituent particles—leaving it open to the reader to guess at the constitution of his 'rare Aethereal Medium'. Indeed, on the next page we are back again to the suggestion that perhaps neither this hypothesis, nor that of the atomists or any other such hypotheses are the 'business of natural philosophy', in which we should 'argue from phaenomena without feigning Hypotheses, and to deduce causes from effects, till we come to the very first cause'.[1]

Clearly, for his more 'serious' writings on the 'mathematical principles of natural philosophy', the intellectual difficulties raised by all these possibilities: action-at-a-distance, essential property of matter, occult property, hypothetical medium, divine action, would make these somewhat unattractive, and it is therefore not surprising that in his '*ex cathedra*' position Newton shelters behind the alternative—already mentioned in chapter V—of treating gravity as a kind of 'mathematical hypothesis'. Thus, in the Scholium to Def. VIII of Book I of the *Principia* he remarks that in that book he will consider forces,

> not physically, but mathematically; [the term force is not to be taken] to define the kind, or the manner of action, the causes or the physical reason thereof; ... [nor are we to] attribute forces, in a true and physical sense, to certain centres ... [when speaking of] attractive powers'.[2]

Evidently Hume selected Newton's ether option both because it avoids the employment of empty space or a '*vacuum*' (rejected by Hume on *conceptual* grounds); and because (whilst denying, it is true, any 'active power' to material bodies) it operates with the notion of 'activity' on the part of a medium where 'force' or 'power' is employed in its 'uncritical', physical sense: that denoted by Newton's term 'endeavour'. On the other hand, in order to characterise his view that special 'media', employed for justificatory and metaphysical purposes, need to be excluded, Hume adapts Newton's 'mathematical approach' in the earlier part of the passage in the *Enquiry* (E. 73n), above, p. 328.

---

[1] Op. cit., pp. 368–9. It appears indeed from Newton's unpublished writings that he did not believe in bodies acting by impact directly on one another any more than by action as a distance; here, too, 'forces' were needed (repulsive forces), in turn requiring their own ether. It seems that in the end Newton's ether turns out to be not much more than a 'substantialisation' of 'force'; the 'subject' for the verbs 'to undulate', 'to attract', 'to repel'; a substantialisation, whose rationale is supplied in terms of the action of divine providence. (Cf. above, p. 334n.2.)

[2] Op. cit., i, pp. 5–6.

VII

LEIBNIZ: SCIENCE AND METAPHYSICS

I. THE MAIN THESES OF LEIBNIZ'S PHILOSOPHY

We now turn back in time, to Leibniz's creative period during the seventies and eighties of the seventeenth century, when Cartesianism was at its height, and ending only when Newtonianism was becoming the ruling orthodoxy, during the second decade of the eighteenth century. Leibniz was intensely responsive to the great intellectual movements of his time, and particularly those stimulated by the recent advances in science; and he in turn moulded these influences in his own image, by a synthesis of prevailing currents (especially Cartesianism) with an older basically Aristotelian tradition still operating with 'substantial forms' and 'final causes', though always interpreted by Leibniz in a special and idiosyncratic way; 'substantial form', for instance, becoming a metaphysical equivalent of physical 'force', labelled 'entelechy'.[1] His writings are encyclopaedic, though mostly contained in smaller essays, tracts and letters to contemporaries. Statements of his complete 'system' are to be found for the most part only in a number of rather summary accounts, intended for occasional purposes of a semi-popular nature, such as his *Principles of Nature and Grace* and the better-known *Monadology*, both of which were not published until 1714, near the end of his life.

One impediment to the understanding of the full sweep of his thought in his time was the fact that many of his most important writings were not published (or not published in full) during his life; thus the first statement of his mature views, *Discourse on Metaphysics* (1686), was not printed until 1846, and his commentary on Locke's

---

[1] Cf. his general remarks in *Specimen Dynamicum*, printed in L. E. Loemker (ed.). *Leibniz, Philosophical Papers and Letters*, 2 vols. (Chicago, 1956), pp. 712–13. (Referred to hereafter as L.)

Essay, the *New Essays concerning Human Understanding*, completed around 1709, appeared only in 1765, when it had such a profound effect on Kant, second only to the latter's somewhat earlier reading of Hume's *Enquiry*. All these delays not only frustrated a proper grasp of the ramified strands of Leibniz's thinking (concealed owing to the already somewhat petrified style of the late summaries), but by the time its significance was appreciated, what philosophers saw in it (as is usual in such cases) was often simply accommodated to their own current interests; witness Russell's emphasis on Leibniz's logic and the subject–predicate structure of his language, in his *Critical Exposition of the Philosophy of Leibniz* (1900).

Some of Leibniz's writings on 'scientific method' have only been discovered in recent years;[1] written in the early 1680's, at the most crucial period of Leibniz's development, and evidently influenced by Boyle, they include an interesting account of the place of hypothesis in science and its relation to the deductive method, forcing us again to the realisation—as in the case of Descartes—that a 'rationalist' approach to knowledge and metaphysics does not entail a lack of awareness of the existence of hypothesis, let alone of experiment, in science; it is rather in his estimation of the relevance of this feature and its relation to the interests of logic and metaphysics that the 'rationalist' philosopher differs from the empiricist. It is not true that for him physics is replaced by metaphysics. Even Descartes' boast that the basic axioms of physics are deducible from the principles of metaphysics we have seen to be in the end simply equivalent to a vague recognition that these axioms are not ordinary physical hypotheses but that they function by defining the conceptual framework of the science in question, the 'metaphysics' acting as a kind of consolidative device. In the case of Leibniz even more, science, far from being suppressed, plays a central rôle in suggesting a 'way of seeing' the complex totality of empirical phenomena, by providing certain privileged 'empirical analogues', acting as a mirror which reflects the underlying metaphysical situation, or a form of philosophical spectacles through which to view this situation; although by a queer ontological reversal the latter comes to be regarded as the 'true foundation' of the former, once more in effect acting as a consolidative device. This again resulted in the usual tensions; and if

[1] Cf. the 'Introduction' to his *Elementa physicae*, translated in Loemker under the title *On the Elements of Natural Science: An Introduction on the Value and Method of Natural Science*, L. 431–47.

this had manifested itself not only in Descartes, but also in Locke, Berkeley and Hume, preoccupied as they were, ostensibly at least, with the problem of perception, it will be all the more intensified in Leibniz, whose thought moves more boldly, and almost with reckless abandon, at the metaphysical level.

The temptation, as well as the ability, to avail himself of a number of such analogues was the greater for Leibniz, since his interests and creative activities ranged over a wider field, stretching from logic and mathematics (where he helped to lay the foundations of the modern symbolic approach to logic, as well as of the Infinitesimal Calculus), *via* penetrating researches in dynamics and optics and the fostering of certain biological theories of the time, to writings on jurisprudence, ethics and theology, though with his philosophical writings always at the core—the focus of all his efforts. He was the last great polymath, and the number of writers whom he cites, and on whom his critical comments have a bearing, is enormous: Bacon, Gassendi, Galileo, Hobbes, Paracelsus, van Helmont—the list stretches endlessly; as does that of his correspondents: Arnauld, Foucher, Oldenburg, Conring, Tschirnhaus, Huygens, Varignon, Bernoulli, De Volder, with the classical exchange of letters between Leibniz and Clarke at the very end; and in his letters Leibniz often developed the most significant parts of his statements. He wrote, chapter for chapter, commentaries on the chief writings of the most important thinkers of his age, on Descartes' *Principles*, Spinoza's *Ethics*, and (as mentioned) Locke's *Essay*. This variety is matched also by a peculiar habit of adapting his language, like an intellectual chameleon, as nearly as possible to the thought and preconceptions of the writers with whom he was engaging in controversy, often criticising, but as often trying to reconcile the main principles of his own philosophy with theirs, be it Cartesian Dualism or some Jesuit priest's worries about the meaning of the Eucharist.

All this makes the task of critical evaluation more difficult than usual, and here we can do no more than select those aspects of Leibniz's work which contrast with what has gone before and place him as a key-figure in our guiding theme, as the vital bridge to Kant. Unfortunately, such a partial treatment is prejudicial to a writer whose tone is so different from that of the philosophers we have looked at so far. One way to rectify this is to preface the critical remarks of this chapter by a brief list of the chief tenets of Leibniz's system. I avoided this approach in previous chapters because it is artificial and does scant justice to the most important aspect of philosophy, which is that of response to given tensions; and I have preferred to let whatever was selected emerge 'from within'. If an excuse is needed for presenting Leibniz's thought in a summary fashion, it is that (as mentioned) Leibniz himself several times attempted such summaries, particularly in the *Monadology*; and it might even have been the best tactic to refer the reader to the latter at this juncture. Unfortunately, the form of that work does not quite meet our purposes, since it pays too little respect to some of Leibniz's own major logical principles and points of view. The list of theses which follows tries at least to preserve these; but it is no more than a list; and in the remainder of this chapter I shall develop only some of its items by critical analysis. But the list will at least bring home to the reader forcefully the change of tone and interest in Leibniz's thought when compared with the philosophers we have studied previously.

Th. 1. The *principle of continuity*: No transition is made through leaps, whether it be from place to place, form to form, or state to state. Nothing is born, nothing dies, everything is always changing, and differs from everything else only 'by degrees'.

Th. 2. All connections exist primarily between the attributes of each individual substance ('monad') in which they are founded.

Th. 3. Corresponding to the last principle there is a logical *dictum*, according to which the predicates of all true propositions inhere in the subject (*'praedicatum inest subjecto'*). In the case of 'logically necessary' propositions ('truth of reason'), this can be shown by means of a finite analysis; in the case of 'logically contingent' propositions ('truths of fact'), only an 'infinite analysis' could show it; the predicates are in the subject only 'virtually'.

Th. 4. Strictly speaking, there are no attributes 'external' to any substance (no 'extrinsic denominations'); and in that sense, there is no 'external world' of objects; all is 'internal' to each monad. The distinction between the real and the imaginary lies only in the types and orderliness of the connections between the phenomena, and the latter (though 'well founded') are only appearances that have a dependence on the mind. The distinction between the real and the imaginary lies only in that of the types and orderliness of the connection between the phenomena.

Th. 5. Substances are centres of activity and passivity, at three levels that differ in respect of their greater or less degree of 'confusedness' or 'distinctness', or of 'expression': (a) force; (b) 'perception' and 'appetition'; (c) apperception. (Sensory qualities are 'confused perceptions'.)

Th. 6. The notion of force defines the reality of substance; that of economy or perfection defines existence or actuality (as contrasted with essence or possibility); that of infinite analysis of its predicates defines the nature and boundaries of the evolution of each substance. There is always a 'pressure', 'exigency' or 'pre-tension', under God, of possibility towards actuality or existence which, in so far as it is 'realised', is subject to the principle of perfection.

Th. 7. Matter (whether in the Cartesian or Newtonian sense) is mere abstraction, and an incomplete notion, although it, like space and time, being an 'appearance', is a *'phenomenon bene fundatum'*. But space and time, moreover, are reducible to an order of and relation between bodies. Hence, there can be no such thing as empty space, no atoms, nor action-at-a-distance; this is the 'principle of plenitude'.

Th. 8. The logical connections between attributes are governed by the *principle of contradiction* ('truths of reason'). The factual connections between attributes, though not 'demonstrable', exist nevertheless with certainty, in virtue of the *principle of sufficient reason* ('truths of fact'). From this, given Th. 1, 3, 6 and 10, it also follows that there can be no mere numerical difference between things: *'the principle of the identity of indiscernibles'*.

Th. 9. Since each substance is wholly separate from every other, what is normally called 'interaction' between substances is 'properly' or 'metaphysically' speaking only a 'correspondence' between them, each acting in conformity with the *'principle of pre-established harmony'*, 'expressing' the world to different degrees of perfection, and from different points of view. This applies to the mind–body relation also.

Th. 10. The source of the 'reasons', of 'perfection', and of 'realisation' (creation and conservation) is God, whose nature that has always foreseen and always aimed at perfection, and always does, also guarantees the (pre-established) harmony between the distinct substances.

Such a list is far from complete, but it will be sufficient to indicate that Leibniz's philosophy constitutes an explicitly formed system, differing in this from our previous writers. It may also be helpful in presenting the different strands of Leibniz's argument in a more unified perspective. But it is only an 'explicit system'; we should not be misled by such a form of presentation to imagine that all the parts of Leibniz's thought do in fact hang together like a deductive chain; on the contrary, his work is a very untidy colossus, and great effort was employed in trying to reconcile the different parts, which do not cohere very naturally.

Also, in another respect such a list is too definitive and magisterial, an effect which is heightened by the unusual choice of terminology, relying as it does heavily on the force of analogy and metaphor, which is the most intrinsic and pervasive feature of Leibniz's philosophical method. Actually, Leibniz's philosophical style creates a feeling of estrangement from and a discontinuity with the problems and methods of our previous philosophers which is far more apparent than real. There is, moreover, ironically, one respect in which our list misleads less as regards the character of Leibniz's procedure than might appear at first sight; for it is characteristic of the latter that it often asserts rather than argues.[1] Nor is this surprising in a writer with such intensely metaphysical interests, who is thereby prone to adopt suggestive analogies instead of arguments, be they mathematical, mechanical or biological. Possibly, this is an important part of all creative thinking, and especially, 'metaphysical thinking'; but it is such an intimate feature of Leibniz's philosophy that we can argue such a suggestion most effectively by presenting the development of Leibniz's main theses as extensions of such empirical analogies.

Thus, the theory of mechanical energy will give us Th. 5, the doctrine of substance as activity. Optical theory will provide the vital clue for the pervasiveness of perfection, Th. 6. The 'Infinitesimal Calculus' as well as the critique of Descartes' rules of motion will lead us to the principle of continuity, Th. 1. 'Analysis', and the concept of 'irrational numbers' will support Leibniz's doctrine of contingency, and his interests in formal logic support his doctrine of the relation between subject and predicate, Th. 3 and 8. Finally, biological discoveries will suggest his concept of the immortality of monads, and his general 'evolutionism'. Such a view of the matter gives also a far more balanced idea of the actual development of Leibniz's thought than, for instance, Russell's suggestion that the major part of it was

---

[1] Leibniz sometimes recognises this. To Bernoulli's complaint that he has 'given definitions rather than explanations', he replies, 'would that definitions might always be given, for they virtually contain the explanations' (L. 830). And the lack of truly systematic exposition he recognises when he writes to De Volder in 1706: 'I cannot yet order everything in such a way as to present the demonstration conveniently to the eyes of others' (L. 879).

deduced from Leibniz's notion of the subject-predicate form of propositions—even though Leibniz does seem occasionally to succumb to such a deductive approach.[1]

In the next section we shall therefore concentrate on the significance of analogy for the development of Leibnizian metaphysics, thus incidentally illustrating the rôle played by analogy in metaphysical thinking in general: analogy here forming the last great link which connects that thinking with the empirical field. Where that link is snapped (as in the Kantian system), there can be no sympathy left for such an analogical approach, and the two realms will henceforth part company for good.

2. ANALOGY AND LEIBNIZ'S METAPHYSICS

Our first example is the most interesting, although unfortunately it presupposes concepts which can become properly intelligible only in the light of results of the later sections of this chapter; it illustrates Leibniz's method of using analogy for the purpose of definition of a 'difficult' philosophical concept, viz. that of existence. That there is a difficulty will be clear if we remember Hume's point that the idea of existence adds nothing to the idea of any thing as such,[2] and Kant's similar contention that existence adds nothing to the concept of a thing.[3] In Leibniz, the problem comes up in connection with his conception of substance. Being strongly opposed to the Cartesian form of the dualist dichotomy of mind and matter viewed as pure extension, he hotly denies that pure extension can be the candidate for material substance, extension being both infinitely divisible, consisting of mere geometrical points, and being moreover 'purely passive'.[4] Leibniz has a strong sense of what has sometimes been called 'the mystery of being' or 'existence': 'Why is there something rather than nothing?', as he puts it in his *Principles of Nature and Grace*.[5] His answer to this question is two-fold.

First, whilst there are an indefinite number of logically *possible*

---

[1] Cf. for instance, *First Truths*, L. 412–17, where the '*praedicatum inest subjecto*' principle is claimed to entail the Principles of Sufficient Reason and of the Identity of Indiscernibles; that there are no 'extrinsic denominations'; the nature and separateness of substances, the body–soul parallelism, the purely 'phenomenal' nature of matter, as well as the impossibility of atoms and of a vacuum.

[2] Cf. above, ch. VI, pp. 353–4.   [3] K. 504.

[4] Cf. L. 741. The same contention can be found throughout Leibniz's writings.

[5] L. 1038.

worlds (consistent with the truths of logic and mathematics), the one that *actually exists* does so in virtue of God's 'free decrees', regarded as the 'principal sources of existences or facts'.[1] In defence of this contention we are offered, as we might expect, a version of causal principle, the 'principle of sufficient reason'; we need not enter into the details of this argument.[2] Existence seems thus defined in terms of God's creative activity; but how are we to understand this? Ignoring the notion of 'decree', we at once note here the deployment of analogy. In the version of the *Discourse*:

> Created substances depend on God who conserves them and also produces them continually by *a kind of* emanation, *as we* produce our thoughts.[3]

The second part of Leibniz's answer is connected with the first. For God's creative activity is mirrored in the fact that each individual substance is 'a being capable of action',[4] or 'a natural force ... implanted by the Author of nature, ... which is provided with a striving or effort'.[5] But again, this is surely difficult to understand; our two preceding chapters have illustrated in full the difficulties surrounding the notion of 'endeavour' in material bodies, when used for the purpose of metaphysics. In Leibniz, instead of 'scepticism', we get analogy.

At the back of this there lies the most important of Leibniz's discoveries in mechanics, the principle of *vis viva* (conservation of mechanical energy; note that *vis* means 'force'), through which he believed he had discovered an element which, unlike Descartes' 'motion' (volume times speed)—which he interprets as entirely 'passive'—did provide at least indirectly an essential feature of 'activity',

---

[1] L. 509. Decrees which (we shall see later) are executed in accordance with the principle of perfection. (Th. 6, 10.)

[2] Cf. L. 1038–9. Cf. L. 349: 'Nothing happens without a cause, or there is nothing without a reason'. L. 605: 'It is indeed certain that every phenomenon has some cause.' Note the equivalence in these passages of 'cause' and 'reason', although Leibniz sometimes is aware of the difference, as when he realises that in the case of geometrical truths, for example, we can only determine 'reasons' and not 'causes' (L. 790). Cf. also *New Essays*, iv.17.1: 'The *cause* in things corresponds to the *reason* in truths. This is why cause indeed is often called reason, and particularly final cause.' (ed. Langley, op. cit., p. 556. Future page references will be to this edition.)

[3] *Discourse on Metaphysics*, sect. 14 (ed. Lucas & Grint, Manchester, 1953), p. 18; my italics. Other quotations from the *Discourse* will be taken from this edition which has a very faithful translation; thus, the present passage would be referred to as DM. 14, p. 18. [4] L. 1033. [5] L. 712.

and this in a sense other than entirely 'physical', namely 'metaphysical'. In section 3 a fuller account of the various scientific analogues will be given in their proper setting, and their detailed relation to Leibniz's 'metaphysics' will be explored; here let us only note the result, through a comment Leibniz makes in the *Discourse* on the principle, and notion, of *vis viva*:

> This consideration of force distinguished from quantity of motion is very important, not only in natural philosophy and mechanics for finding true laws of nature and rules of motion . . . , but also in metaphysics for understanding the principles better.[1]

However, as we shall find, force in the sense required, viz. as genuine activity, can play its part in Leibniz's metaphysics only analogically, as 'active primitive force', and not as 'active derivative force' (= *vis viva*, under which form alone it occurs in dynamics).[2] But again, the extension of the notion to cover all being, animate as well as inanimate, is achieved once more by means of analogical reasoning, supported by considerations of 'continuity' (Th. 1). For the element of 'striving' which Leibniz ascribes here to matter is also asserted to be 'analogous to soul',[3] or again as 'something analogous to sense and appetite . . . so that we must think of them [*sc.* substantial forms] in terms similar to the concept which we have of souls'.[4] The neutral term for this kind of substance in Leibniz's terminology is 'monad'; and thus 'a monad . . . contains perception and appetite, *as it were*'.[5]

In the present context, all these analogical references are due to

---

[1] DM. 18, p. 31.

[2] Cf. below, sect. 3(a): ii, pp. 415ff. For concerning the latter, Leibniz will be as puritanical as Hume; if anyone wanted to regard *it* as a 'cause', he would be met by the very Humean remark that 'what we call causes are in metaphysical rigour only concomitant requisites' (L. 415), there being no more 'true interaction' between substances in the philosophy of Leibniz, than there is between the atomistically conceived 'impressions' of Hume.   [3] L. 416.

[4] L. 741. As shown later, Leibniz, again with the help of the principle of continuity, likens force and action, as 'actions of the soul' (L. 429), to what he calls 'appetition', which he regards as a sort of dynamic 'tendency' in virtue of which the individual substance or monad passes from one perception to another (L. 1034, 1046), a bit like a writer who *labours* to pass from thought to thought. Perception, here, is likewise used in a somewhat unusual sense, as explained to Arnauld, where it is said to be a species of 'expression', a term here meaning 'representation', *on analogy with* the representation of one geometrical figure by another *via* topological transformation. (For the geometrical example, cf. L. 521; also *N.E.*, ii.8.13, p. 133.)

[5] L. 820; my italics; the 'as it were' once again reminding us of the analogical nature of the argument.

Leibniz's attempt to supply a significant explanation of 'existence'. We have seen the difficulties encountered over this concept by Berkeley and Hume, where eventually it comes to be closely associated with 'perceptions' and 'impressions'. Given this basic position, it was an easy step for Hume to declare (what we just noted) that the idea of existence adds nothing to the idea of any thing as such. And Berkeley had defined it as actual or possible perception on the part of some mind. But such approaches are quite foreign to Leibniz's viewpoint. He is no mere Humean phenomenalist, despite all his teaching concerning 'appearances' and *'phenomena bene fundata'*, for 'behind' these there is a world of 'real' substantial entities.

On the other hand, the contrast between the phenomena and such a world might lead the reader to expect something on the lines of Locke's distinction between 'ideas' and 'external objects', or again, a 'reality' to which the ideas 'conform'. But such an interpretation, too, is far from Leibniz's intentions. As we shall see later, the contrast between a 'self' and an 'other', in the shape of a world of external objects in Locke's sense, does not even begin to arise.[1] On Locke's basic contention that 'the idea' is 'the immediate object of our thinking', Leibniz comments that this is correct, 'provided . . . that this object is an *expression* of the nature or the qualities of things'.[2] But since (metaphysically speaking) monads in the end are all that is real (under God), notions like 'expression' and 'correspondence'[3] are again to be understood in a special way. The different substances or monads simply *'are* . . . the perspectives [and hence, expressions] of a single universe according to the different points of view of each monad'.[4] Indeed, he at least once goes so far as to assert that 'phenomena' *are* simply 'appearances which exist in my mind';[5] but such a phrase must be read in conjunction with the rest of Leibniz's philosophy.[6]

Similarly, when Leibniz comes to the crucial passages in Locke's *Essay*, iv.4. 1–5, concerned with the question of the correspondence of

[1] Cf. for instance, DM.14, p. 23.

[2] *N.E.*, ii.1.1, p. 109; my italics.

[3] Ibid. For 'expression', cf. p. 396n.4

[4] As Leibniz puts it in the *Monadology*, sect. 57; L. 1053; my italics. (Th. 9.)

[5] L. 603; cf. Th. 4.

[6] Berkeley, with whom Leibniz's later years overlapped, is mentioned only once, with the remark that he does not seem 'to explain his position sufficiently', and seems to 'want to be known for [his] paradoxes', in attacking 'the reality of bodies' (L. 993)!

things to our 'ideas' of them,[1] he barely grasps the point, seemingly under the impression that Locke is still discussing the question of 'certainty'. And here his comment is significant. Simple ideas of sense, being merely 'confused perceptions', are a very insecure basis for knowledge, the true foundation for which is the 'universal and eternal truths' which are 'originally in our minds'; only if the 'phenomena of the senses' are 'rightly united as the intelligible truths demand' can we be satisfied: an example, or rather analogy, being the way in which 'the phenomena of optics are explained by geometry'.[2]

This is looking ahead; but we see already that Leibniz's approach to the problem of the external world is often simply through the usual criteria of coherence, of 'the connection of phenomena'.[3] As L. 605–6 shows, such a notion applies, however, merely to 'phenomena' or 'appearances'. Metaphysical considerations, which relate to the causal *foundation* of phenomena, will—so we shall find—lead Leibniz to assert that all existing things, considered as substances, are in one sense each 'a world apart',[4] whilst in another sense they are necessarily 'in intercourse with each other'.[5]

So far Leibniz does not appreciate the point of the question so central for Berkeley: 'How do epistemological limitations affect the concept of meaning?' On the other hand, he is too scrupulous a thinker to allow the notion of 'reality' to stand undefined, a mere limiting contrast with that of 'phenomenal appearances' in the manner of Locke. And in consequence, any difficulties which he experiences with meaning (such as that of 'existence') are resolved by means of the employment of analogy and metaphor.

For reasons such as these Leibniz will seek a definition of exis-

[1] Cf. above, ch. IV, pp. 251–2.

[2] *N.E.*, iv.4.4, p. 445, and iv.2.14, p. 422. We have met the notion of 'confused modes of thinking' already in Descartes (above, ch. III, p. 179), and we shall see how central it is in Leibniz's system. Leibniz's optical example cannot be taken seriously unless it is treated as an analogy; in fact he is far less prone than Descartes to certify factual truths by reference to mathematics; what he wants to say is that we are in need of a metaphysical foundation, but this does not strictly certify. The question of the relation of empirical knowledge to mathematics and metaphysics will be taken up in a later section. (Cf. below, pp. 458ff.)

[3] *N.E.*, iv.2.14, p. 421. L. 603–4 gives the criteria of 'vividness', 'complexity', 'coherence', and 'power to predict', to distinguish the 'real from imaginary phenomena'; cf. Th. 4, also L. 357.

[4] L. 518; cf. below, p. 401.

[5] L. 607. A seeming contradiction which Leibniz tries to resolve by his doctrine that whilst monads do not stand in relations of mutual transeunt causality, they do 'represent' the world, and thus each component monad, from their own particular point of view.

tence in something connected with the character of selves or minds or monads and their 'expressions', understood as a sort of 'activity'. Moreover, by the principle of continuity, *anything*, in so far as it can approach the level of 'reality', must be a centre of 'activity'. Here would be another reason why 'ideas' of the Locke–Berkeley–Hume variety can have no place in Leibniz's basic universe, since they are (as we have abundantly seen) entirely 'passive' and 'inert', on the lines of Descartes' 'matter', which Leibniz usually singles out for attack in respect of this feature.[1] The corresponding constraints on the notion of existence through its relation with the possibility of experience (the having of perceptions in the Locke–Berkeley sense) therefore do not arise.

The whole difficulty, as we anticipated—one which gives Leibniz's pretended solution its interest—is due to the fact that existence, in the sense here required, has no conceptual content; as Kant was later to point out: 'being is obviously not a real predicate';[2] it adds nothing to the concept of a thing. Thinking of a thing as actually existing is to think of it 'as belonging to the context of experience as a whole'.[3] Perhaps Leibniz would, however, reply that he is concerned with the concept of existence as such, i.e. of a *possible* existence, since he specifically points out that one cannot simply add existence to the predicates of a substance (e.g. God), without first showing that the addition as such is possible.[4]

Here Leibniz is none too clear. In the passage just mentioned he suggests that we have only to show that there is no logical contradiction, and this is how Kant was later to understand him. Yet there is no doubt that, in his mature thinking, the concept of God's perfection plays an important part, given through the mutual adjustments of the different monads, which in some sense 'express' the world, a process whose cause ultimately is founded in God. The notion of 'possibility' then relevant is that of 'compossibility': a reference to what either has existed, is existing, or will come to exist, subject to the principle 'determining existences . . . that the more perfect shall exist';[5] an idea through which Leibniz gives us some third term, lying between mere logical, and completed actual existence. In that case, we are, however,

[1] Cf. L. 416, 709; cf. below, pp. 408ff.
[2] K. 504. See also above, ch. V, pp. 353–4, for Hume's doctrine of existence. For Kant, cf. below, ch. VIII, sect. 4(a), pp. 487–8.
[3] K. 506; not unlike Leibniz's connection of phenomena.
[4] Cf. his comment on Descartes' *Principles*, i.14; L. 634–5; also L. 354–5.
[5] L. 262; cf. L. 1075 for 'compossibility'.

no further advanced in our understanding of existence, being thrown on the analogy of perfection, which ultimately indeed does seem to be what defines existence for Leibniz, at least in its essential part.[1] Once more reference to Kant will give us our perspective. Kant who declared that Leibniz had failed to 'comprehend *a priori* possibility', gives an account of the notion of a concept of an object in general, *not* (like Berkeley) in terms of possible or actual experience, but of the '*universal conditions* of possible empirical knowledge in general' (i.e. space, time and the categories), to which we 'think' the concept of the object as conforming.[2] Thus, where Kant offers us a (transcendental) logic, Leibniz employs analogy (here: that of perfection), a different 'way of seeing'.

Other important analogies which we shall meet later in this chapter are '*maxima* and *minima*' principles in optics which illustrate 'a kind of divine mathematics or metaphysical mechanism',[3] a reference to his view that these principles illustrate the existence of 'finality' (teleology) in science, and hence in nature. 'Analogies in anatomy' bolster up the view that plants are a kind of 'imperfect animals', which is in line with the principle of continuity of forms.[4]

Then there is the important analogy between finite and infinite series on the one hand, and necessary and contingent truths on the other, the analysis of the subject in the latter case, though a significant possibility according to Leibniz, not in fact being performable, since the series is infinite, just as the ratio of the area of a circle and that of the circumscribed square, though a definite number, is irrational, having the value $\frac{\pi}{4}$, which Leibniz had himself determined to be equal

---

[1] Above, p. 395n.1, reference to L. 509. I say 'in part', since in his later writings (*Radical Origination of Things*, L. 791, 793; *Monadology*, sects. 43–4, L. 1051) Leibniz comes more to emphasise the 'pressure' of possibilities towards existence, a tendency whose 'realisation' he postulates to be 'founded' in God, as an essence that necessarily includes existence. In other words, through God there is pressure towards existence, in accordance with the principle of perfection (Th. 6). However, the difficulty is not removed; such a locution once more presupposes a previous understanding of the notion of existence. Cf. also below, p. 403–4.

[2] K. 507, 506; and below, ch. VIII, pp. 487–8. Moreover, in the Kantian system this means that the notion of 'happening and coming to be' cannot arise in connection with 'substances'; the 'principle of sufficient reason' (K. 226) applies only to *states* of substances (K. 248). Kant later allows that we may *think* of an 'intelligible ground of appearances', but such a thought transcends the limits of possible experience (K. 481–2). He thus sharpens the distinction between the metaphysical and the physical realms, where the limits are never clearly drawn in Leibniz.

[3] L. 791–2; cf. section 3(b), and pp. 430, 432 below.   [4] L. 531.

to $1 - \frac{1}{3} + \frac{1}{5} - \frac{1}{7} + \frac{1}{9} - \dots$, *ad inf.*[1] Leibniz's description of the function of this analogy is interesting; it is introduced, so he tells us, 'in order to fix our attention . . . so that our mind will not wrestle with vague difficulties'.[2] The difficulty in question, so we shall see, is to accept the significance of the concept of infinite analysis of the subject of contingent propositions (Th. 3), which Leibniz requires for his most central contention that every predicate of a true contingent proposition inheres in its subject, as it does in the case of necessary propositions (e.g. those that are analytic)—the relation between the two types in turn being evidently again mediated through analogy.[3] We see therefore that the analogue which Leibniz offers carries more weight than his reference to its mere need 'to fix our attention' would suggest.

The presence of analogy should not be so surprising to our reader. Already when discussing Berkeley's concept of causation and his 'doctrine of signs', we noted the importance that we must attach to a 'substitution terminology' as a more correct explanation of some given situation than that which in uncritical moments we might imagine to hover behind the scenes.[4] Berkeley wanted to say that the causal nexus is properly described in terms of the analogy of 'signs' and 'laws', rather than that of 'invisible chains'. In quite a similar way, Leibniz who, for reasons yet to be considered, also holds that 'a particular substance never acts on another particular substance',[5] and that thus 'each individual substance or complete being is as a world apart, independent of every other thing but God',[6] proposes that where normally we speak of a substance exerting power and thus acting, we must more properly 'say that the substance . . . passes to a greater degree of perfection or to a more perfect expression'.[7]

Now Leibniz is quite aware of the status of such a description; it is, so he expressly remarks, 'remote from ordinary usage',[8] and in the following section he refers to the need 'to reconcile metaphysical language with practice', through the consideration that the 'expression' of one substance may 'hinder or limit' that of another, and 'in this sense . . . so to speak' act on the other. 'Hindering and limiting' is here however an ambiguous terminology; it may mean, (1) the

---

[1] *N.E.*, iv.3.6, p. 424, and note; L. 408–9, 421.   [2] L. 408.
[3] Cf. below, pp. 451ff.   [4] Above, ch. V, pp. 304–5.
[5] DM.14, p. 24. Cf. also the 'Humean remark' cited above, p. 396n.2   [6] L. 518.
[7] DM.15, p. 26; cf. L. 794, 817; also Leibniz, *Basic Writings* (ed. Montgomery, Open Court, 1962), p. 183; abbreviated as 'OC'.   [8] DM. 14, p. 24.

universe is best arranged when the existence of one substance is compatible only with less perfection on the part of the other;[1] (2) one substance directly prevents another from realising its full potentialities, in the normal sense of preventing. Here we see how analogies produce 'illumination' through the employment of ambiguous locutions. But then, there is built into language a vast variety of such ambiguities; and whilst logic and science may demand precision, philosophy, which sails closer to the winds of an ambiguous reality, must navigate more closely to language in its living context.

Indeed, it is at the point of choice of the operative analogies that the philosophical schools divide. Where the Berkeley–Hume tradition emphasised atomicity, under the influence of the analogue of the individual 'sensation', Leibniz slides into the opposite direction, the assertion of continuity, and connectivity—though the latter within each substance only. For this reason, the analogue of 'force', dismissed by Berkeley and Hume as 'irrelevant' in supporting the contention of the existence of a 'nexus', will on the contrary play a vital part in Leibniz's thinking. As early as 1671 (in the *New Physical Hypothesis*) he speaks of bodies when at rest (after having been stopped), as 'striving to move',[2] of impressing a '*conatus*' on other bodies, and so on: a notion later used to articulate the concept of force. Locke, too, had been searching for connections, but enshrined in his sceptical voice; in Leibniz they appear as 'metaphysical' constructions, shored up by the various empirical analogues. Of course, it may be said that Berkeley, too, was a full-blooded 'connectivist' at the metaphysical level, expressed *via* God's efficient causality. But the difference is that connectivity is here an extrinsic element; 'force', as we know it, does not provide anything of significance for our understanding of things as they are. Divine causality is simply inserted because the original basis was too narrow. In Leibniz's philosophy there is no sharp cut of this kind; and the connective link —or in logical phraseology: 'the nexus of terms or the inclusion of the predicate in the subject'[3]—becomes the original centre of gravity of the whole system. To this we shall return in a later section.

---

[1] In accordance with the principle of 'pre-established harmony' (Th. 10), and cf. our reference to the notion of 'compossibility' above, p. 399.　　[2] L. 218.

[3] L. 407. It is a mystery why Leibniz stopped at 'many subjects' or a plurality of substances, where—as already noted—atomicity momentarily re-appears, only to be made a unity again through the principle of pre-established harmony; and why he did not end—like Spinoza—with just one substance. When he poses the question, he simply points to the *fact* that there are evidently other minds.

Of course, one cannot expect more of an analogy than that it should point to identity and differences. Leibniz wants the metaphysical insights to stand on their own feet; but for the most part, he needs the guideline of the analogues, to provide a significant direction for this process. A remark which he makes in connection with yet another analogy, this time from scientific methodology, and which likens the perfection of the world to that of a theory or hypothesis, brings this out very clearly. Already in a letter of 1678 he had explained the usefulness and power of hypotheses, promising at least 'physical certainty' even though the hypothetical mechanism may not be known directly, provided they account for the greatest number of phenomena by means of the fewest additional assumptions, and that they be relatively 'simple to understand'.[1]

Now in the *Discourse* this is applied analogously to God who becomes just such a scientist, always choosing that world from among those logically possible,

> which is the most perfect, that is to say the one which is at the same time the simplest in hypotheses and the richest in phenomena, as a geometrical line might be, of which the construction was easy and the properties and effects were admirable and of great extent.

Here Leibniz is quite conscious of the nature and purpose of such analogies, for he adds:

> I make use of these comparisons, however remote they may be, in order to sketch some imperfect resemblance of the divine wisdom, and in order to say something which may at least elevate our spirit to conceive in some sort that which cannot be adequately expressed. But I do not claim to explain thereby this great mystery on which the whole universe depends.[2]

We are again led to a consideration of the ubiquitous notion of perfection. Here we must note that with Leibniz the language of a creator-God can be misleading if it suggests an original author, creating the world once and for all. As emphasised in the Clarke–Leibniz correspondence, 'God preserves everything continually and nothing can subsist without him', his 'conservation is an actual preservation and continuation of beings'[3] (Th. 10). And this includes the growth of things, as well as their coming into being. Leibniz (in connection with his attempts to explicate 'existence') often looks at

[1] L. 288.   [2] DM. 6, p. 10.   [3] L. 1103, 1111.

this in another way, as though God had placed into the universe an 'exigency towards existence in possible things . . . a pre-tension to exist . . . or that all possible things . . . tend towards existence . . . in proportion to the degree of perfection which they involve'[1] (Th. 6). Here again, the whole centre of gravity rests on the notion of 'perfection'. And yet, as we have noted, to build perfection thus into the notion of existence is surely extremely precarious? One might not object at the equation as such; we have agreed that Leibniz, in the light of the difficulty of defining 'being', must needs resort to analogy. But it is surely queer to maintain the universal evidence of perfection? Perhaps there is yet another analogy lurking?

In answer, Leibniz occasionally employs crude evasions, as when he suggests that the apparent 'confused chaos' in 'the government of mankind', though suggesting nothing so little as perfection, 'when we look more deeply, the opposite can be established', and that, as regards 'afflictions, especially of good men . . . we must take it as certain that these lead to their greater good', being 'short cuts to greater perfection'.[2]

But behind this there must lie deeper convictions. And these are, as we might by now expect, more pervasive analogies, flowing both from the methodological end, as in the case of Leibniz's doctrine of hypothesis, and from certain physical principles, exhibiting either the cause–effect relation (principle of *vis viva*), or the principle of 'least effort' in optics,[3] or, as he once claims, 'the best of all examples', which comes from 'ordinary mechanics [statics of equilibrium] itself', according to which all heavy objects tend to descend, so that any motion would produce the greatest possible descent of these weights: this, once more an analogy, to show that the 'world [which actually] is produced [is one] in which a maximum production of possible things takes place'.[4]

---

[1] L. 791. Cf. above, p. 399, and p. 400n.1 for the relation to 'perfection'.

[2] L. 795, 797.

[3] Cf. Leibniz's general comments on these in *Specimen Dynamicum*, L. 723; *Tentamen Anagogicum*, L. 778–82, 787–8; *Principles of Nature and Grace*, L. 1039–40, and several other places.

[4] *Radical Origination*, L. 792. Cf. also L. 791: 'There is always a principle of determination in nature which must be sought by *maxima* and *minima*; namely, that a maximum effect should be achieved with a minimum outlay, so to speak.' This is in part illustrated by the example from optics (to be discussed presently), in part by the problem of the catenary, solved by Leibniz, Huygens and others in the year previous to which this was written (1696). See E. Mach, *Science of Mechanics*, op cit., pp. 85–9. Note that Mach states that the principle of maximum descent more or less summarises the various problems of classical equilibrium statics; and

The problem of the relation between the empirical analogues and the metaphysical positions, the pull between 'identity and difference', apparently so neatly resolved through the concept of the analogical relationship, with its overtones of 'suggesting without proving', never quite disappears from the Leibnizian scene. It manifests itself in a certain vacillation, Leibniz sometimes stressing the immunity of his metaphysics from physical principles of enquiry and even its results, sometimes the great relevance of the latter for the former: a difference partly due to a development in Leibniz's thinking under the stress of certain discoveries not only in the foundation of his Infinitesimal Calculus, but also in statics and dynamics and optics, as mentioned in the last footnote. To these we must now turn because only a more detailed exhibition of the factual basis of these analogues can bring out graphically the actual 'feel' or 'tone' of Leibniz's expansion of this basis into the realm of metaphysics.

3. SCIENTIFIC DISCOVERIES AND THEIR REVERBERATION IN LEIBNIZ'S PHILOSOPHY

Our discussion of the place and function of analogy, as far as it has gone, indicates the importance of the 'empirical analogues' which recur periodically in Leibniz's writings. The pull between these analogues and the metaphysical positions, to which we have drawn attention, may be discerned in the way in which Leibniz deals with a number of related questions: (1) Are physical phenomena fully explainable by reference to purely 'mechanical' (Cartesian) concepts, such as figure and motion alone,[1] or do we require additional concepts, and if so, what is their status? In particular, what is the relevance here of the earlier scholastic notions of 'substantial forms', and of 'purpose', i.e. of teleological models of explanation? (2) This leads to the connected question, whether and in what sense physical concepts and principles require supplementation by reference to metaphysical ones. (3) More general, though still connected with the

---

he moreover declares it to be 'instinctively familiar' (p. 86; cf. p. 94). Its effect on Leibniz is therefore easily understood.

[1] A question still answered in the affirmative before 1680, in the letter to Conring, where Leibniz insists that Aristotelian forms of theorising about light do not contribute in any way towards an explanation of the phenomena of reflection and refraction, and that our only hope lies in their explanation 'by mechanical laws' (L. 289–91). Our section on optics will show how far Leibniz was to change his views subsequently.

first two, is the question of the relation between propositions of fact and propositions of reason,[1] on which bears once more the problem of the nature of Leibniz's metaphysical position.[2] This last question not only presupposes a settlement of the first two but also an appreciation of the significance of Leibniz's dictum *'predicatum inest subjecto'*, his view that one can significantly assert, so to speak, the existence of 'individual essences' (Th. 3).

(a) *Matter, Force and the Conservation of Mechanical Energy*

Let us take, then, our first question, which connects with Leibniz's views of 'substance' and 'matter', and the relation between them. And here Leibniz differs in the strongest possible terms from those that immediately precede him (Cartesians as well as Malebranche and his disciples, the so-called 'occasionalists'), and also the main representatives of the 'empiricist' school, such as Boyle, Locke and the Newtonians, and (the later) Berkeley and Hume. Generally their central assumption is that matter (material substance) is 'passive'—the very term 'inanimate matter' denotes this general presupposition.[3] There is no need to explain this here; the whole sequence of Descartes' metaphysics as well as the 'atomism' of Locke, Berkeley and Hume have this assumption built in as an integral element of their philosophical doctrines. Moreover, the epistemological outcome of this view had been—as we have seen in detail—that none of these philosophers could find a place for 'force' and 'activity', in any but a

---

[1] For this terminology, cf. *N.E.*, iv.11.13, p. 514. Alternative expressions are 'necessary and contingent truths' or propositions; also 'truths of reason' and 'truths of fact'.

[2] Thus, for example in *N.E.*, loc. cit., Leibniz dissents from Locke's statement that truths of reason concern nothing but our perception of the relations between our ideas; in the spirit of Descartes, he asserts on the contrary that these truths would hold even in the absence of any finite minds, since their 'ultimate ground' is God. Since elsewhere he says that the 'reason' for truths of fact is likewise ultimately grounded in God, there is obviously *some* connection between the two types of truths for Leibniz which did not exist for Locke.

[3] This will also be Kant's view, in *Metaphysical Foundations of Science*, where he comments that 'matter as such is inanimate', a fact which he takes to be equivalent to the law of inertia. Matter is there likewise contrasted with those aspects of substance which concern desire and thought (Leibniz's 'appetition' and 'perception') which belong to *animate* substance. But where Kant insists on a rigid distinction between these two kinds of substances, Leibniz—who cannot fit matter into his ultimate scheme of things—must assign matter to a merely 'ideal' realm of 'phenomena', or else saddle it, in so far as it is 'real', with the possession of a core of activity. It is this view which Kant explicitly opposes, labelling it 'the death of all philosophy of nature' (op. cit., iii, Prop. 3, Observation).

sense either 'positive' or 'instrumentalist'. As with Descartes, extension, figure and motion are their ultimates, brought to a focus all the more sharply by the radical cut made, on the one hand between 'passive matter' and 'active mind', on the other, between the merely orderly passage of 'ideas', in accordance with laws of nature viewed as 'rules', and the causally efficacious activity of God (Berkeley) or 'mind' (as in the case of Hume).

In all this, let us, however, remember the important proviso, which we tried to highlight in connection particularly with Hume: it is not that the denial of force or activity concerns anything straightforwardly 'factual'; on the contrary, 'force' (and in Berkeley's case, 'matter') is eliminated only as something 'metaphysical'. If now Leibniz reintroduces the notion of force as a vital component of matter[1] (at least in so far as it constitutes something substantial), the importance of this must therefore lie in the fact that we find the concept operating here at the purely 'metaphysical level'. In other words, there is no empirical clash.[2]

What confuses the issue is that although Leibniz makes a very definite distinction between physical and metaphysical levels, it is an important facet of his philosophy that in it physics is decidedly not isolated from metaphysics, and that the relevance of their respective 'findings' for each other is assumed—a fact which our exposition of his analogical mode of argument will of course have led us already to expect. The interesting point is indeed that the influence does not just operate from the side of physics on metaphysics, but also in reverse, for Leibniz's 'metaphysical needs' (if the expression may be

---

[1] Here, an important caution must be observed. Leibniz, when speaking 'metaphysically', holds that one substance cannot act on another. Action, as he understands it, must hence denote a process of 'striving' (*conatus*, cf. L. 580 n. 286) whereby a substance is 'urged' to develop its capacities, and to 'realise' them. The operative analogy here is evidently 'life'. Thus, in the letter to Wagner (1710) he says that 'this active principle, this first entelechy, is, in fact, a vital principle' (in P. P. Wiener (ed.), *Leibniz Selections*, New York, 1951, abbreviated as W; p. 504); and similarly throughout many other writings. For the reconciliation of the metaphysical and physical modes of speech, cf. letter to Arnauld, OC. 183; also L. 749; and below, pp. 456–7.

[2] Moreover, to avoid confusion, let us anticipate by noting that 'force', in so far as it enters as the *physical* equivalent of transeunt causality between substances, is likewise denied except in the sense of a purely 'phenomenal existence', expressed in terms of the *rules* of 'derivative force' (*vis viva*) and 'directive force' (momentum)— very much on the lines of Berkeley's 'mere mathematical hypotheses'—governing the impact between material bodies. It is only the *metaphysical* equivalent of this derivative force (the so-called 'primitive force') which has for Leibniz any basic 'reality'. (Cf. sect. (ii), below, pp. 419ff.)

permitted) in their turn quite evidently did much to dictate the direction of his enquiries into the foundations of physical science, in particular, of dynamics. Nor should such a situation surprise us, for the 'mechanistic' viewpoint of the 'anti-activist' school of his own century had similarly guided scientific enquiry and the formulation of its primary hypotheses, both in the physical, and more importantly, in the biological sciences—as it has done ever since.

Leibniz is quite explicit in his insistence both on the central importance of the concepts of 'force', 'action', *conatus*', and 'entelechy' for his understanding of substance, and its connection with his discoveries in dynamics. In the *Acta Eruditorum* of 1694 he writes:

> The concept of substance which I offer . . . is so fruitful that there follow from it primary truths, even about God and minds and the nature of bodies. . . . The concept of forces or powers . . . for whose explanation I have set up a distinct science of dynamics, brings the strongest light to bear upon our understanding of the true concept of substance.[1]

In a similar vein, in the *Essays on Dynamics* (so he tells us) he has meant to single out the central feature that the 'concept of corporeal substance' involves 'force of action and resistance rather than extension, the latter being merely a repetition or diffusion of something prior to it, namely, this force'.[2]

(i) *Matter and Phenomena*

Before coming to the details of Leibniz's dynamics, in order to test whether, and how far, this supports his conception of force, we must first, however briefly, look at some additional background considerations which directed him to an 'activist' position in the first place, parallel with the related doctrine of the '*phenomena bene fundata*'. One of these considerations is his reminder that 'extension is an abstraction from the extended', since extension must be 'the extension of something'.[3] Now we have noted before the difficult problem of interpretation of Descartes' views on 'extension'; his slide (if slide it was) from 'extension' as a 'property of substance' to something that is 'substantial' in itself.[4] Such a slide would be aided and abetted by the problems (brought to light very clearly by Locke) that surround the concept of 'substance' *quâ* substratum: such as our supposed inability

[1] L. 709.   [2] L. 752.   [3] L.874.
[4] Cf. above, ch. III, pp. 89–91.

of 'knowing' anything about it.[1] In that queer sceptical ambivalence of his, Locke, when characterising our thinking about substance$_m$, had referred to it as an 'unknown cause of the union' of the sensory qualities: a thought immediately submerged under the interests of the epistemic approach.[2] Leibniz, against this, presents the other extreme. He yields freely to the temptation which such a picture (of causal *activity*) holds out, asserting it as fact that

> taking action in metaphysical strictness as that which takes place in substance spontaneously and from its own depths, that alone is, properly speaking, a substance which is active.[3]

In the tract *On Nature Itself* Leibniz refers to Boyle's *Notion of Nature*, which we have previously mentioned in chapter II, noting that Boyle there defines the concept of nature as 'a cosmical mechanism' comprised of 'figure, size and motion' and subject to 'the laws of motion'.[4] But, Leibniz objects, the notion of 'mechanism' (and the laws that describe its mode of behaviour) is incomplete unless we add also certain 'principles', a reference once more—to judge by the analogies he mentions, of the weight and the spring which drive a clock mechanism—to his 'force or action'. Moreover, such principles, he declares, do not have a material or mathematical origin, but possess 'a higher and, so to speak, a metaphysical source'.[5]

From the beginning, Leibniz was impressed by this extra feature which—so he believed—we must add to pure extension. Partly (reverting again to the purely physical level) it appears as the 'impenetrability' of one solid body to another.[6] Again, in the early seventies

---

[1] Cf. above, ch. IV, sect. 4 (b), especially pp. 215–20. Leibniz, in his commentary, interprets Locke's epistemic denial—like most commentators—to imply total rejection, and therefore all the more forcefully argues that to insinuate that we should come to 'know' this substance, apart from knowing 'it' through its attributes, 'is to demand the impossible. . . . This opinion of our ignorance arises from that which demands a kind of knowledge of which the object does not admit' (*N.E.*, ii.23.2–5, pp. 226–7). On the other hand, Leibniz's positive account is no clearer than Descartes'; he confines himself to the assertion that 'reflection suffices to discover the idea of substance within ourselves' (where it is 'innate') (i.3.18, p. 105). No doubt this sanguine approach is due to his fundamental principle, pressed already in the Preface, 'that activity belongs to the essence of substance in general' (p. 60).
[2] Cf. above, ch. IV, pp. 219f., 221–2.
[3] *N.E.*, ii.21.72, p. 218. Though Leibniz sometimes notes that 'action and power are different things' (L. 815), he normally uses them interchangeably.
[4] Above, ch. II, p. 54; also p. 44.   [5] L. 810–11.
[6] L. 159; cf. also L. 645 (critique of Descartes' *Principles*, ii.4), where he objects that Descartes omitted to show that this property *is* eliminable or reducible to

he had spoken of a moving body when colliding with another, as 'striving' (in the sense of incipient motion) to propel the latter,[1] the term being retained and gaining greater significance after he had realised that some dynamical function of this '*conatus*' (what he calls 'derivative active force' (*vis viva*)) is conserved in collision-phenomena. In the slightly later *Elements of Natural Science* he notes that, in addition to extension and motion, we must consider 'resistance or impact'—'resistance' being distinguished from 'impenetrability' as 'resistance to motion';[2] a 'force' which, so Leibniz claims once again, cannot 'be derived in any way from extension alone'.[3]

But Leibniz has additional reasons, stemming from a different quarter. 'Matter in itself', he often tells his readers, 'is only a phenomenon or well-founded appearance (*bene fundatum*), as are space and time also.'[4] To say that it is a '*phenomenon bene fundatum*' is to say that it is subject to the laws of nature, and thus part of the order of nature; it is not an illusion.[5] On the other hand, it also means that it does not possess the status of 'substances' ('monads'). Unfortunately, the precise relation which phenomena have to substances Leibniz never made clear. In the *New System*, the correspondence with

---

extension. Whether Descartes would have pleaded guilty to the charge is an open question. No doubt he would have insisted that unless such a property *was* reducible, it would not be 'clear and distinct', and hence would contribute nothing towards scientific explanation. But this is precisely what Leibniz asserts also (cf. L. 444–5), only adding that we must not therefore ignore the 'metaphysical' sources lying behind. As we shall see, Leibniz, no more than most of the other thinkers of our story so far, resolves the pull between the 'purely phenomenal' (e.g. secondary qualities), the 'mechanical' and the 'metaphysical realms'. Yet he is not prepared to cut them loose from one another in the fashion of a Kant.

[1] *The Theory of Abstract Motion*, sect. 11 (L. 218). In this tract Leibniz still held that 'a moving body impresses upon another, without diminution of its own motion, whatever the other can receive without losing its own earlier motion' (sect. 20, L. 221). It was this erroneous theory, employing nothing but the concepts of 'magnitude, figure, position, and their change', which his later work tried to amend, made necessary by the fact that it conflicted with the law of action and reaction and was moreover in conflict with 'experiment' (L. 719–20; 839). Cf. below, pp. 415ff.

[2] L. 427. Sometimes (L. 818) this is called 'natural inertia', a term and concept borrowed from Kepler. It has a very confusing place in Leibniz's scheme. The Newtonians will object that a body displays only resistance to *change* of motion (if by motion we understand uniform velocity); a fact to which they refer by the term *inertia*, though with them (as Hume had also echoed) this is an oblique reference to the 'first law of motion'; cf. above, ch. VI, p. 328, and the passage from E. 73. However, Leibniz is quite aware of the difference (cf. L. 818). We shall discuss the matter further below, pp. 421 ff.

[3] L. 839.   [4] L. 528/OC. 222.

[5] But for the limited force of 'law', cf. below, sect. 5(b), especially pp. 462–3.

Arnauld, and the examination of Malebranche's system, he stresses the following aspects.

Matter, in so far as it is regarded as extension, involves abstraction as well as certain logical difficulties. On the former point, he insists that 'nature' is infinitely complex, so that none of the *precise* relations posited by geometry are to be found in it.[1] Figure, size and motion are not 'real and determined qualities outside thought'; they are 'phenomena like colours and sounds' even though they 'involve a more distinct knowledge'.[2]

In this connection, Leibniz makes a second point, namely, that the conception of matter regarded as geometrical extension inextricably entangles us in 'the labyrinth of the composition of the continuum'.[3] On one side, a line cannot consist of mere points, since there will always be gaps, and in any case, points cannot create a continuous line.[4] On the other hand, a line is infinitely divisible, so that we must come eventually to extensionless 'mathematical' points, which are merely 'ideal' or 'modal'. And if matter had this character, it would follow that it was made up of nothing.[5]

Such a conclusion contributed powerfully to Leibniz's view that 'the real' (substances) cannot be 'material' in this sense. Nevertheless, the age-old dream of 'analysis into simples' makes him side with the

---

[1] L. 529/OC. 222. This does, however, not prevent him from asserting elsewhere, in a letter to Varignon, that 'though continuity is something ideal and there is never anything in nature with perfectly uniform parts, the real, in turn, never ceases to be governed perfectly by the ideal and the abstract. ... [It is] *as if there were* material atoms, although there are none, because matter is actually divisible without limit.' On the other hand, though the division never proceeds to infinitely small particles, 'the rules of the infinite apply to the finite, *as if there were* infinitely small metaphysical beings, although we have no need of them. This is because *everything is governed by reason*; otherwise there could be no science and no rule. . . .' (L. 884; my italics.)

We shall find Leibniz insisting throughout, both that our *knowledge* of physical reality is not in need of metaphysical concepts and principles, and (at the same time) that the former 'ultimately' rests on the latter, which may even cast their shadow on its actual formulation.

To delineate the regions with fresh strictness was the Herculean task which Kant set himself. I shall return to the whole question in section 5(a).

[2] Ibid.   [3] L. 529/OC. 223; cf. OC. 163.

[4] A resolution of this difficulty had to await the labours of nineteenth and twentieth century analysis of the continuum. It involves the formulation of a concept which allows that between any two points there will always be others, *and* that there is never '*a* point' *next to* some given point. (Cf. B. Russell, *Our Knowledge of the External World* (London, 1914), ch. V, p. 140).

[5] This argument resembles the Thesis of the Second Antinomy of Kant's *Critique* (K. 402).

'discontinuity' solution; as he tells one of his correspondents, he had indeed in his youth, after 'freeing' himself 'from the bondage to Aristotle, . . . accepted the void and the atoms, for it is these that best satisfy the imagination'.[1] The trouble was that they were still composite; the kind of 'atoms' he is after (Leibniz sometimes calls them 'metaphysical points' or 'atoms of substance')[2] must be truly indivisible, and thus simple, whilst yet not being just 'nothing', like the mathematical points.

The argument for 'metaphysical atoms' is, however, not based merely on the considerations just mentioned. Cutting across, there is another, concerning the logical nature of the 'composite', which leads Leibniz once more to the conclusion that such composites or 'aggregates' are 'mere phenomena'. Put briefly, the argument asserts that the elements of a composite whole, i.e. of a plurality of things, are related only by what one might call 'external relations'.[3] Indeed, in his account of these relations between the elements of a composite Leibniz is quite as uncompromising as Locke had been about those qualities of a 'substance' (substance$_p$) which are defined by its 'nominal essence',[4] the 'necessary connections' between which had been claimed as ineluctably unknown.[5]

As Leibniz poses the question, it is: What makes any plurality a unity? And what is the logical status of this unity? His answer is that the relations between the parts of an aggregate are purely ideal, or 'imaginary', in the sense of being 'mind-dependent'.[6] Two separate things do not become one however much they may be cemented together by physical ties. It is true that there are composites which constitute in some sense a unity to varying degrees, depending upon whether they have more or less mutual connections (Leibniz refers us to the examples of a pile of stones, of a nation, an army, a society or a college), but the relations involved in so far as they give us anything 'real', *depend* 'simply on the mode of being of the' constituents.[7] To talk of 'connection' in any other sense (and this, we shall see, covers also the case of those connections between bodies in physical impact, subject to certain laws) 'is only a means to abbreviate our thinking

---

[1] L. 741.　　[2] L. 745.

[3] By contrast, as we shall see, the qualities of 'monads' are related by an 'internal principle' (cf. *Monadology*, 11; L. 1045).

[4] Cf. above, ch. IV, sect. 4 (a), e.g. pp. 208, 213, 216.

[5] Leibniz is not, of course, concerned with the epistemic deficiency, but with the constitutive character of such relations.

[6] Cf. OC. 161.　　[7] OC. 191.

and to represent the phenomenon'. This ideal status of the connections is therefore a further strand in the account which considers complex physical things, and any aggregates in general, 'mere beings of reason' or of 'the imagination or perception, that is to say, phenomena'.[1]

Does it follow that there are any 'true unities'? Certainly Leibniz asserts this without question: 'A multitude can derive its reality only from the true unities';[2] 'every being by aggregation pre-supposes beings endowed with true unity, because it obtains its reality only from the reality of the elements of which it is composed';[3] and there are many such passages. One might unkindly object that this is defining 'reality' by 'reality', itself a blank term. It gets substance only from the denial of the 'reality' of the 'ideal' relations holding between the constituents of what is thereby regarded as a 'mere aggregate'. More than likely, Leibniz simply joins the results of the previous considerations to the present argument, all of which are intended to contrast the unreality of 'matter' with some more genuine candidate for reality.

Leibniz here often deceives his reader by suggesting the existence of a systematic argument or inference, when only a number of different considerations are joined extraneously. Thus in the *New System*, after having rejected material atoms, as containing diversity, he concludes:

> It is only atoms of substance, i.e., real unities, that are absolutely destitute of parts, *which are* the sources of action and the absolute first principles out of which things are compounded.[4]

But it does not at all follow from their simplicity, that substantial atoms can be characterised as 'sources of action'; this must stem from some different consideration. At the head of this paragraph, Leibniz indeed offers us a vital analogy; for 'by means of soul or form there is a true unity corresponding to what is called "I" in us'. More openly, in a letter to Arnauld, he says that a true unity can be found, for

---

[1] OC. 190; i.e. to *see* things in spatial or physical contact no more makes the connection a genuine one, than it had been in the case of Locke, in reference to the qualities defining the 'nominal essence'. It should be noted that, late in life, Leibniz tried to introduce the idea of composite things, such as physical bodies, being more than mere phenomena, their parts unified by what he calls a 'substantial chain' (*monadum substantiale vinculum*), but it is an open question how definitive are his tentative remarks to Des Bosses in which this theory is outlined.   [2] L. 741.

[3] OC. 189; cf. OC. 220: 'I hold that every multitude presupposes unity.'

[4] L. 745; my italics. These 'atoms of substance' are, of course, the monads.

instance, 'in a soul, or substantial form, *such as is* the one called the self'.[1]

After the discussions of section 2 of this chapter, we shall not be surprised at being offered an analogy at a vital stage of the argument. The question is merely whether the conception that results is sufficiently enlightening and powerful to support the complete account that Leibniz desires. For we see already that the 'true unity' which results will be fashioned on the lines of the mental analogue—the mind being the most powerful example of a 'multiplicity in unity', the paradigm of the 'internality' of relations. So we shall not be surprised to be told (in the uncompromising words of the *Monadology*, sect. 14) that 'the passing state which enfolds and represents a multitude in unity or in the simple substance is merely what is called perception',[2] where 'perception' means a 'state of representation of the external in the internal', though not—as we shall see—'caused' by the external.[3] The crucial problem is whether such a narrowly conceived model of reality, viz. mind, can make contact with the world of physical relationships. This is what Leibniz believed, despite the fact that there is considerable uneasiness in his writings whenever he has to explain the status of the phenomenal world, its own relationships, and its relation to the substances.[4] His self-assurance in this matter can only be due to the power exerted by the analogy of 'force', derived from physics. If this analogical bond is fractured, we shall need a reshuffling of the cards, a re-arrangement of the philosophical pieces, as happens in Kant's reaction to Leibniz.

We have thus come full circle. In order to show what supports Leibniz's conception of substance as activity, we found it necessary to

[1] OC. 161; my italics. At OC. 225 Leibniz admits explicitly that this is an analogical argument: 'Every corporeal substance must have a soul or at least an entelechy which has an analogy with the soul, because otherwise the body would be only a phenomenon.' For this reason, 'it may be said in general that they [corporeal substances] are alive' (OC. 221).

[2] Cf. Kant's criticism of this analogy, in the important 'Note to the Amphiboly of Concepts of Reflection' (K. 281ff.), especially pp. 284–5. Kant complains that such an approach makes it necessary to 'intellectualise all appearances', including those belonging to physical nature.

[3] For this definition of 'perception', cf. W. 505; L. 521. Cf. also above, p. 396n.4

[4] Thus, being spiritual, it follows, as he admits, that 'accurately speaking . . . matter is not composed of these constitutive unities but results from them, since matter or extended mass is nothing but a phenomenon grounded in things, like the rainbow or the mock-sun'. (L. 873; cf. also L. 983: 'Monads are not really ingredients but merely requisites of matter.') But neither the abstract terminology, nor the quite insufficient analogy here employed, can explain the relationship.

enter into his strictures on the concept of substance *quâ* matter or extension; yet, in the end, to explain that rejection requires in turn the re-introduction of the concept of (purely internal) 'sources of action'. Two complexes thus require to be investigated further, that of 'action', and that of 'internality'.

(ii) *The Doctrine of Forces*

We have mentioned before[1] that in his *Theory of Abstract Motion*, Leibniz had held that a body in motion interacting with another can impart to that body (however large) its own velocity without losing any of its own. He seems to have derived this from the idea of a body's 'indifference to motion and rest', a fact expressed quite correctly in Descartes' law of inertia,[2] which says that a thing once at rest (or in uniform motion) will always remain at rest (or in motion in the same direction) if left to itself.[3] Now obviously there was here a *lacuna*; the size of the body impelled did not enter as a factor into the total situation, and it seems to have occurred to Leibniz quite soon that such a situation conflicted with the principle of the 'equality of cause and effect'.[4] There must therefore enter into our calculations entities which make the physical relationships involved uniquely determined, in order to avoid saying that 'anything could be accomplished by anything'—'pure chaos';[5] but 'our minds look for conservation', as he puts it in his *Essay on Dynamics*.[6]

The existence of such entities in physics was not unknown, the most prominent being the Cartesian concept of 'quantity of motion',

---

[1] pp. 409f. and 410n.1   [2] Cf. above, ch. III, p. 150.

[3] *Theory of Abstract Motion*, sect. 809; L. 218.

[4] Cf. L. 429, 721 for the introduction of the principle in this context; also L. 726: 'The whole effect is always equal to the full cause'. The correlative principle applied to the concept of 'force' in general is to be found in DM. 17, p. 29: 'It is very reasonable [obvious] that the same force should always be conserved in the universe.' It will be noted that this is a peculiar, substantialised version of the causal principle. This reading was to be dominant also in J. R. Mayer's extension of the principle of energy; cf. above, ch. II, pp. 40f., for Mayer's cognate 'rationalist' approach to the relation of this principle to 'experience'. An interesting discussion of this occurs in E. Meyerson, *Identity and Reality* op. cit. ch. 1, pp. 28–29; ch. 3, pp. 191, 194, 206.

[5] L. 720, 839. At W. 101, the consequences of the *Theory of Abstract Motion* are said to be 'entirely irreconcilable with experiment', or, as he writes to De Volder, with the 'two tests upon which I always depend—success in experiment and the principle of order' (L. 839). For the idea and interpretation of the 'two tests' (taken from the realms of experience *and* of reason) forming a *cumulative* support for a given scientific fact or law, cf. above, p. 411n.1 and below, pp. 440ff.; and section 5(a), pp. 458ff.   [6] In *N. E.*, p. 657.

whose conservation Descartes had connected with ('derived from') God's perfection.[1] Unfortunately, as we have seen, Descartes' concept was specified as a purely scalar quantity, as volume—later mass—times *speed* (ms), rather than speed in a given direction, or velocity (a vector quantity); and only the latter is in fact conserved. Under the stimulus of studies on the mechanics of collision, particularly by Huygens, Leibniz began an extended critique of Descartes' writings on dynamics, showing that in certain collisions it is not the scalar but the vector quantity of motion (what he calls 'directive force')[2] that is conserved. But further, there is another kind of 'force' which is likewise conserved, and which he calls 'living force' (*vis viva*). This is a quantity (scalar) which is a function, not of v, but of v^2, and of the mass m of a body.

The direction of Leibniz's thought is most interesting. Huygens before him had already found that, in the case of colliding bodies, the sum of the products of m and v^2 remains constant.[3] Again, in his work on the compound pendulum, he had employed the 'most certain axiom' (implicit already in the work of Galileo) that the centre of gravity of a number of bodies falling through a height h cannot rise subsequently more than h; and the mathematical development of this work had involved relations of h and the square of the velocity reached by the falling bodies which imply the energy equations of Leibniz. But the conception of energy or work is never made quite explicit, and apart from its mathematical representation it occurs only fleetingly in this special context as 'elevating force'. I say 'only fleetingly' because Huygens himself was initially hostile to Leibniz's generalisation of the corresponding conservation principle on its first appearance.[4] It looks as though the concept first needed 'peeling out of its shell', something that neither the study of the phenomena of mechanics nor of their mathematical form could achieve.[5]

What facilitated the breakthrough here was clearly Leibniz's

[1] Cf. above, ch. III, pp. 147–50.

[2] L. 717. This is equivalent to what we now call 'momentum'.

[3] Cf. L. 583 n. 331; also R. Dugas, *Mechanics in the Seventeenth Century* (Neuchâtel, 1958), ch. 10, sect. 5, p. 300.

[4] Cf. Dugas, op. cit., pp. 300–5; and ch. 14, sect. 6, pp. 468–9.

[5] Mach's comments on Huygens' achievements seem rather more applicable to Leibniz, when he writes: 'What Huygens asserted, therefore, no one had ever really doubted; on the contrary, everyone had *instinctively* perceived it. Huygens, however, gave this instinctive perception an *abstract*, *conceptual* form.' (*Mechanics*, op. cit., II.ii.17, p. 212.) Clearly Mach does not appreciate the need for a 'substantialisation' of the concept involved for its initial recognition.

achievement of 'substantialising' the product of m and v², no doubt under the guidance of his search for a suitable candidate for conservation, since from the very start he freely speaks of 'forces' that he does not regard as purely mathematical entities. The problem only turns on the correct quantitative 'definition' of such a force.[1] This shows that if his discovery of the principle of *vis viva* eventually came to be a kingpin in the analogical body of support for his metaphysical position, the latter in turn influenced the search for suitable analogues, and would, after its discovery, contribute vitally to what I have called the 'consolidation' of such a principle.[2]

Leibniz's *Brief Demonstration of Descartes, Error concerning a Natural Law* appeared under (roughly) this title in 1688,[3] and the gist was repeated in several places, including the *Discourse* (sect. 17), the commentary on Descartes (sect. 36) and *Specimen Dynamicum*. The basic argument is simple. Let us consider two bodies, A and B, weighing 1 and 4 pounds, respectively, falling through heights of 4 metres and 1 metre respectively. Leibniz now makes two assumptions which he claims would be admitted by Cartesians and all other contemporary scientists: (1) a body falling from a certain altitude acquires the same force which is necessary to lift it back to its original altitude—Huygens' axiom; (2) the force necessary to raise A through 4 metres is the same force as that required to raise B through 1 metre —an assumption already admitted by Descartes.[4]

---

[1] Cf. Leibniz's comment on Descartes' choice of volume times speed, in his *Critical Thoughts*, L. 648–9, quoted above, Ch. III, p. 149. Note, incidentally, that Leibniz at first claimed that with his discovery he had 'at last achieved pure mathematics ... which contains only numbers, figures, and motions' (L. 422). The transition came *via* the consideration that his principle showed the impossibility of 'perpetual mechanical motion' (ibid.; cf. OC. 164–6; DM. 17, p. 29).

[2] For 'consolidation', cf. ch. II, pp. 36ff. Planck's attitude to 'consolidation' in this context is an interesting later example: 'However overwhelming the number and significance of inductive proofs may seem, nobody can be such an inveterate empiricist that he would not feel the need for another [kind of] proof which, resting on a deductive foundation will permit the principle [of conservation of energy] to emerge from certain and more general truths in its total comprehensive meaning as a single complete whole.' (*The Principle of the Conservation of Energy* (Leipzig, 1908), pp. 149–50.)   [3] L. 455–63.

[4] The 'force acquired' by the body at the end of its fall through a height h is what we now call 'kinetic energy'; the 'force' imparted to it after it has been 'lifted back' through the same height h would be the 'potential energy' (considered as a function of h). Assumption (1) postulates their equality. On the other hand, the equality of the 'forces' claimed in assumption (2) is really the equality of what is now called 'the work done', defined as the product of Newtonian 'force' (in the modern sense' Leibniz calls it 'dead force') and the distance h. Clearly the use of the term 'force; was an expression of insufficient conceptual clarity.

It follows from our assumptions that A and B should acquire the same force at the end of their fall, ignoring air resistance and other interfering factors. Now our two bodies being then in motion, the question is whether the Cartesian product mv measures this force. But it follows from Galileo's work on the velocity of falling bodies (to which Leibniz here refers explicitly) that the velocity is proportional to the square-root of the height; a fact which implies that in the present example the velocity of A will be twice that of B, so that mv will not be constant in the two cases: a contradiction! The argument also shows that if we measure the force by means of 'the quantity of the effect which it can produce, e.g., from the height' directly,[1] then it must be estimated by the product of the mass and the square of the velocity, mv², which is the quantity that measures the *vis viva*;[2] and it is this which is *constant* in the cases where the force to raise the bodies A and B is the same (i.e. for constant 'work').

It may help the reader if we represent the argument that follows in its equivalent and more general Newtonian terms. Newton's second law of motion says that $F = m\dfrac{dv}{dt}$. If we now multiply both sides by dx, where $v = \dfrac{dx}{dt}$, we have $Fdx = m\dfrac{dv}{dt}dx$ $= mdv\dfrac{dx}{dt} = mvdv$. But the total work done between x_1 and x_2 is given by the integral of $Fdx = \displaystyle\int_{x_1}^{x_2} Fdx = m\int_{v_1}^{v_2} vdv = \tfrac{1}{2}mv_2{}^2 - \tfrac{1}{2}mv_1{}^2$. If finally we take the difference in potential energy U (a function of x) between x_2 and x_1 as the negative of $\displaystyle\int_{x_1}^{x_2} Fdx$, we get the i.e. the mathematical formulation of the conservation of energy principle, sum of potential and kinetic energy at x_1 and x_2 is constant.

Let us also note that the *time*-integral of F, i.e. $\displaystyle\int_{t_1}^{t_2} Fdt = \int_{t_1}^{t_2} m\dfrac{dv}{dt}dt = \int_{v_1}^{v_2} mdv = $ $mv_2 - mv_1$. Now mv is the momentum (F and v being vectorial quantities). Using Newton's Third Law, it can easily be shown that the total momenta of two colliding bodies, before and after collision, are constant.

The main point to note is that energy relates to the space integral, momentum to the time integral of Newtonian force. Both obey conservation laws. For accidental reasons Leibniz takes *vis viva* to be given by mv², i.e. twice our kinetic energy $\tfrac{1}{2}mv^2$.

[1] L. 457; this clearly implies the concept of 'potential energy'!

[2] Since $v = \sqrt{2gh}$ (where g is the constant acceleration of gravity), $\tfrac{1}{2}mv^2 = mgh$ = work done, where the l.h.s. is the 'kinetic energy'. Leibniz uses more round-about calculations in *Specimen Dynamicum* where he first clearly defines the new product, as what we should call the time-integral of mv, getting mv². *Vis viva* is thus double the k.e. There is no space to go into the details of Leibniz's dynamics here.

It will be seen that from the mathematical point of view we have here nothing but certain numerical law-like equivalences; it would have been quite possible to take a purely 'reductionist' or 'instrumentalist' attitude *vis-à-vis* such a result, although as noted in our contrast with Huygens' attitude, it is unlikely that its full significance would have been grasped without the 'substantialist gloss', and its significance for Leibniz's general philosophical position: a fact which he rightly claims in *Specimen Dynamicum*.[1] The relation between that position (and its metaphysical core) and the law of *vis viva* is viewed by Leibniz in two different ways, one 'direct', the other 'indirect'. The first occurs occasionally when Leibniz argues that since the traditional Cartesian mechanical approach to nature, involving purely '*geometrical*' concepts like quantity of motion, is insufficient and leads to contradictions, 'we perceive that we must add to them some higher or *metaphysical* notion, namely, that of substance, action, force'.[2] Here Leibniz appears simply to *identify* 'metaphysical' and 'non-geometrical'. But surely such a conclusion does not follow, since although Leibniz has shown that it is mv^2 rather than mv which is conserved, this could all be understood in relation to purely spatiotemporal ('geometrical') quantities. Evidently, Leibniz must be treating mv^2 as an independent physical entity, by contrast with 'motion', which he takes to be nothing but a purely quantitative (and 'abstract') function of other entities such as mass and velocity—entities which we have already seen, moreover, he treats as 'mere phenomena'. On the other hand, even if it be independent, why should mv^2 not simply take on a purely 'phenomenal' (if physical) sense? This leads us to Leibniz's second or 'indirect' way with this question.

When employing this second approach, Leibniz admits that the 'forces' which he has so far considered do indeed act only at the 'phenomenal' level, manifesting themselves on the occasion of what we normally *call interaction* between different bodies. However, since his ultimate scheme does not allow for interaction, all substances being each 'a world apart' (Th. 3, 9), entities defined in the context of interaction can at best be no more than 'short-term' or 'abbreviating' representations of something, which will then have to be of a purely 'internal' nature.[3] In short, guided by the semantic suggestiveness of

---

[1] L. 719. [2] W. 101.

[3] Cf. above, p. 398. Thus, to De Volder Leibniz writes: 'The forces which arise from mass and velocity are derivative and belong to aggregates or phenomena' (L. 863); i.e. they belong only to a purely ideal order of things, a view which corresponds to the sceptical approach to force on the part of Berkeley and Hume. For this reason,

the aspect of 'internal endeavour' in the concept of force, Leibniz will postulate an entity that is only *like* a force, 'internal' to each substance, from which the forces manifesting themselves in impact are *in a sense 'derived'*. This sense (like the relation between the two kinds of forces) is however by no means made clear, the vagueness and ambiguity surrounding it being indicative of the slide from physics to metaphysics.

In *Specimen Dynamicum* and elsewhere Leibniz distinguishes two pairs of forces (echoing the old Aristotelian distinction between 'matter' and 'form'), the first being called 'active' (A), the other 'passive' (P). Moreover, each member of this pair is subdivided into 'primitive' (p) and 'derivative' (d) force, thus resulting in a four-fold scheme, Ap, Ad, Pp, Pd.

Here, Ad is *vis viva*. Now behind Ad lies the primitive active force Ap, a force 'inherent' in substances as such which though 'it can be understood distinctly,[1] . . . it cannot be explained by sense perception'.[2] Nor is it of any use in physical investigations; it is not the sort of thing (being a 'general reason') by means of which 'to explain phenomena'.[3] Leibniz refers to this force as 'the basic entelechy', which 'corresponds to the soul or substantial form' of older philosophical schools.[4] Now primitive force is clearly altogether 'inward', non-manifest; constituting 'the inmost nature of the body';[5] only its equivalent, the derivative force, enters into calculation and experiment, though the name suggests that the latter in some way derives from the former, which is its 'ground'.[6]

Leibniz would also have been quite unruffled by D'Alembert's comment, about fifty years later, that the word 'force' does not express 'a hypothetical entity which resides in the body, [but that] we use the word only as an abbreviated way of expressing a fact'. (*Traité de Dynamique*, 1743, Prel. Disc.; in Magie, *Source Book of Physics* (New York, 1935), p. 56).

[1] For later reference we note that according to L. 607, substances have 'active [power] in so far as they express it distinctly', whilst 'passive power' they express 'confusedly'.

[2] L. 814. Though this is seemingly contradicted by L. 712, where 'this nisus sometimes appears to the senses', showing how Leibniz will switch from the metaphysical to the physical use.

[3] L. 714, 445; cf. OC. 163. This is not to say, of course, that their 'analogical shadow' may not aid physical explanation; a fact which may explain some of Leibniz's inconsistencies on this point.

[4] L. 714; cf. 429, 530, 712, 815, 818, 863, 1015; *N.E.*, ii.21.1, p. 174.　　[5] L. 712.

[6] Ap should not be confused, as is sometimes done, with 'potential energy', which for Leibniz occurs only accidentally in his argument concerning *vis viva*, and which can be *estimated* from the 'quantity of the effect' which a body falling from some given height produces (L. 457). Cf. above, p. 418.

But in what way? Leibniz's vacillation betrays his uneasiness. At L. 714, 818 he speaks of Ad arising from Ap through the 'limitation' or 'modification' of the latter when bodies interact; at L. 863 he calls Ad 'modifications and echoes of' Ap. But is this more than simply asserting that though Ad is indeed subject to all the sceptical strictures of the Newtonians and their philosophical disciples, we are nevertheless entitled to postulate a corresponding Ap which has the required features of 'inward striving or effort'?[1] Yet, if the relevance of the argument from the dynamics of Ad is exhausted by nothing better than the mere existence of a vague 'echo', why place on it such heavy emphasis?

Let us see whether we can throw further light by considering the second pair of forces (passive), Pp, Pd. Pd is more difficult to interpret because it does not seem to correspond to anything in Newtonian dynamics, being rather a fusion of a number of different elements found therein. Leibniz names it the 'passive force of resistance', by which he understands sometimes impenetrability, but more usually also what he calls the 'repugnance or resistance of matter to motion'; occasionally it means only the latter.[2] To understand the latter, it is best to call to our aid Newtonian concepts, making it possible to view Leibniz's scheme with the knowledge of hindsight.

When one body collides with another at rest, the total Newtonian momentum is conserved, so that the first will lose what the second gains. This follows from Newton's Third Law of motion: 'action is equal and opposite to reaction'. A similar conservation principle applies of course to energy. Also, it follows from the first and second laws of motion, that Newtonian force (to distinguish it from his other 'forces', Leibniz calls it 'dead force' (df)), has to be employed to *alter* the state of motion (including zero-motion or rest) of any given body. (Note that the time-integral of df is Newtonian 'momentum'.)[3]

Now Leibniz quite correctly insisted that 'every action involves a reaction',[4] which inferentially he describes as resulting in the 'loss of force' by an impelling body to one impelled, this manifesting the

---

[1] Cf. L. 712.

[2] Cf. L. 818, 839. He then calls this the 'natural inertia' of a body, after Kepler (L. 839). It differs considerably, as we shall see, from Newtonian inertia. Cf. also above, p. 410n.2.

[3] Cf. above, pp. 417–8n.1.

[4] Cf. L. 721; this was his important improvement on his earlier *Theory of Abstract Motion*; cf. above pp. 410n.1, 415.

supposed 'resistance' of the impelled body to a change of motion.[1] In other words, he slides from a body's *'indifference' to its state of motion* (a force being needed only to *change* that state), to its *not being indifferent to such a change*; and interpreting this as having 'a force and an inclination, as it were, to retain its state and so to resist motion.'[2]

To turn now to Pp: this is simply again the equivalent of Pd, at the non-phenomenal level. But surprisingly, Leibniz here at last finds a function also for Ap. As he explains to De Volder:

> Since matter in itself therefore resists motion by a general passive force of resistance but is set in motion by a special force of action, or entelechy, it follows that inertia also constantly resists the entelechy or motive force during its motion.[3]

Interestingly, the two forces—as befits metaphysical entities—cancel at the phenomenal level. Indeed, this Leibniz himself admits under further pressure from De Volder:

> Properly and exactly speaking, perhaps we should not say that the primitive entelechy *impels* the mass of its own body but that it is merely *combined* with a passive primitive power which it *completes*, or with which it constitutes the monad.[4]

So we see that through a certain amount of conceptual confusion, Leibniz manages to maintain a precarious bond between the two levels: the employment of 'analogy' with a vengeance! In this way, he

---

[1] Instead of 'change of motion' Leibniz more usually just puts 'motion', thus abetting the resulting confusion. For the Newtonians would say that 'inertia' was not really a 'force' at all resisting *motion*, being 'fictitious'; an expression of the law-like fact that 'force' ('dead force') has to be exerted in order to effect a *change of motion*. On the other hand, the *resistance* of the body arises from the phenomenon connected with the *third* law in the context of interaction. The fluidity of the concept of force has here resulted in Leibniz confusing the (fictitious) 'force of inertia', defined by the first law alone, with the 'momentum' (the time-integral of Newtonian force), a confusion facilitated by Leibniz's calling momentum 'directive *force*' (cf. the popular surplus meaning of 'momentum', and the frequent misunderstanding of the nature of 'centrifugal force'). Kant draws attention to this confusion in 'Observation 1' to 'Proposition 4' of his *Met. Found.*, iii, after having himself been subject to it during his earliest period.

[2] L. 839. This is the confusion, mentioned in the previous footnote, between the 'resistance' (in the sense of 'reaction') accompanying an 'action' during impact, in accordance with the third law, with the fact that a force has to be exerted to change the uniform motion of a body, in accordance with the first and second laws of motion, where the notion of 'resistance' is not relevant.

[3] L.839. The slide is concealed by using the term 'inertia' here to cover both Pp and Pd!   [4] L. 862; my italics.

will be able to use the indications of physical conservation-phenomena for his ultimate metaphysical insistence that 'the substance of things itself consists in the force of acting and being acted upon',[1] although one can never be sure whether and how far the separation between the two realms has been achieved.[2] His constant insistence that although 'all phenomena are indeed to be explained by mechanical efficient causes but that these mechanical laws are themselves to be derived in general from higher reasons'[3] is obviously both too strong and too weak as a characterisation of his procedure. Clearly there is no *deductive* development from the causal principle to the law of *vis viva*, and still less is this true of the step from *vis viva* to the principle of the constancy of primitive force (i.e. natural inertia and entelechy). On the other hand, by using the term 'metaphysical' also as a corrective for 'purely geometrical' notions of matter, current amongst Cartesians and Newtonians, the result constituted a real change in the formulation of the current dynamics, leading to a new emphasis on and understanding of the concept of mechanical energy. The two approaches were mediated by Leibniz's search for 'inward forces' rather than 'outward phenomena'. The trouble was that, apart from sheer assertion that these are opposite sides of the same coin, his actual argument involves—at least at the physical level—a considerable conceptual confusion, but a confusion which is understandable since 'confusions' of this type regularly accompany the birth-pangs of any scientific advance.

Another matter that plays into this is that his *'phenomena bene fundata'* are, as we have seen, attached to their substances only in the sketchiest of ways. Having only mock-anchored them in the order of nature, the precariousness of their reality becomes an added element of pressure, assuring Leibniz that the derivative forces in some way *do point* to their 'inner' equivalent. Thus, right at the start of his *Specimen Dynamicum* we meet with the confident assertion that 'like time, motion, taken in an exact sense never exists', the reason being

---

[1] L. 815; cf. *Monadology*, 18.49; also *Principles of Nature and Grace*, 2, 11, for the metaphysical image of these conceptions (L. 1034, 1039, 1047, 1052).
[2] Cf. to De Volder: 'But in phenomena or aggregates every new change arises from an impact according to laws prescribed partly by metaphysics, partly by geometry; for abstractions are necessary for the scientific explanation of things' (L. 864). Of course, the 'metaphysical' aspect might be said to exhaust itself in the consideration of the causal equivalence, and its application to *vis viva*, but this is not the bearing of Leibniz's remark; he inverts the image, and the metaphysical forces become the 'grounds' and 'justifications', of which the physical ones are the 'echoes' (L. 863).
[3] L. 722; cf. DM. 18, p. 32, and elsewhere.

that both involve a succession of extraneous parts, so that they are in need of some 'momentaneous state'; a state which '*must* consist of a force striving toward change'.[1] Therefore, 'whatever there is in corporeal nature besides the object of geometry, or extension, *must* be reduced to this force.'[2] *Per contra*, if, as in Kant, 'phenomena' are given a status *independent* of considerations touching the 'internal nature of substances', the 'inner [side] of things',[3] this slide towards 'the inner in things-in-themselves'[4] will be arrested or constrained.

Of course, Leibniz is not concerned with 'knowledge' and its conditions in the Kantian sense, not with the epistemic but with the constitutive aspect; and no shadow of any epistemic considerations falls on his postulate of a constitutive (ontological) order of things;[5] still less is he embarrassed by the notion of a semantic critique of the very concept of a constitutive nature (of 'things-in-themseves') as carried out by Hume and Kant. In *Specimen Dynamicum* Leibniz writes (in an attempt to develop his generalised concept of force or entelechy):

> God has always put into things themselves some properties by which all their predicates can be explained.[6]

Similarly, in *On Nature Itself* he insists that without certain 'intermediaries' through which what is remote in time and space can operate here and now 'anything can be said to follow from anything', so that 'some connection, either immediate or mediated by something, is necessary between cause and effect'. Once again, the need is met by a 'vestige' which God leaves in things, viz. 'a certain efficacy ... a form or force ... from which the series of phenomena follows. ...'[7] Here, the reference to 'explanation of the predicates' from the

---

[1] L. 712; my italics; a conception very Aristotelian in tone. Only in the case of substance, he tells De Volder, does the 'present state involve its future states and vice versa', and only the primitive force which characterises it constitutes 'a basis for the transition ...' (L. 879). Note that by comparison Kant will locate the basis for succession and simultaneity in a 'time' (considered as 'duration'), the conceptual 'analogy' for which is 'substance'; but the latter is now considered as a *presupposition of the possibility* of 'knowledge as such' of succession and simultaneity. (Cf. K. 213.)

[2] Ibid.; my italics.　　[3] K. 286–7.　　[4] K. 285.

[5] Locke's operation with the concept of such an order, whilst at the same time taking due note of epistemological considerations had, it will be remembered, produced his peculiar sceptical tone.

[6] L. 722. The immediate point is directed against the followers of the 'occasionalists', who make God's actions *directly* responsible for every happening.　　[7] L. 813.

'properties of substances' no longer bears any sort of methodological sense but has an ontological significance. This is a notion of 'force' which is meant to provide the analogues for the 'exigency' (Th. 6) or pressure in virtue of which the properties of a substance (all virtually contained therein from the start) will be 'evolved' and thus 'realised'; to use Leibniz's biological analogue: 'All generation is nothing but its unfolding [evolution].'[1] But if the source of that explanation were totally cut off from our world it would lose its interest; explanations in terms of hidden entities have a charm only if they have some relevance to the observable given order of things. Because Kant asserts that Leibniz's hidden order *is* quite cut off it does lose its interest;[2] Leibniz retains his precarious hold because his reading of that hidden order proceeds parallel and by analogy with the phenomenal order of matter and mind. A consideration of how all this affects Leibniz's views of the status of natural laws in general will have to be postponed until a fuller account of his metaphysical position and its relation to experience has been given.[3]

(b) *The Teleological Analogue from Optics*

If dynamics provided an image for activity, optics will supply Leibniz with an analogue of the specific directedness of that activity; and the teleological mode in which he reads the situation will supply a further meaning for his contention that the evolution of predicates from the subject (the logical equivalent of which is the doctrine of the *praedicatum inest subjecto*) has an underlying unbroken continuity of the kind for which teleological causation provides the model. We have noted already that for Leibniz 'existence' is defined in part *via* that realisation of possibilities which aims at economy or perfection (Th. 6).[4]

Leibniz argues for the 'utility of final causes' at three different levels: (a) theoretically, in the employment of scientific hypotheses exhibiting 'optimal form'[5] (i.e. '*extremum* principles' as we should call them); (b) methodologically, in the employment of teleological

---

[1] L. 832. Teleology will supply another prop for this conception; cf. below, p. 433.
[2] But for a restriction on this appraisal of Kant, cf. below p. 524n.2.
[3] Cf. below, section 5(b), pp. 461ff.
[4] Cf. above, pp. 399f. The close connection between these teleological aspects, 'exigency', and the physical (as contrasted with either mathematical or metaphysical) necessitation of things is particularly stressed in *Radical Origination of Things*, L. 789ff. [5] L. 780; DM. 21, p. 37.

forms of explanation in science, e.g. optics and biology;[1] (c) ontologically, the assertion that the world obeys, or exists in accordance with, principles of perfection, as fashioned by God.[2] 'Everything can be explained architectonically, so to speak, or by final causes', Leibniz tells us; or again: 'nature is governed architectonically', i.e. by an 'architectonic principle', which is 'that it acts in the most determined ways'.[3]

As usual, Leibniz is intent on reconciling opposites of different philosophical traditions. His views in dynamics had enabled him to add the *'metaphysical'* concepts and *principles* of force and action to the merely *'mathematical principles'* of the Cartesians, thereby reintroducing the notion of 'entelechy' or 'substantial form', though admittedly with a change of meaning. Now he seeks to add *moral principles*, or 'principles of the good',[4] in the shape of the concept of the 'final cause' of the Aristotelians. He frequently refers to his mode of reconciliation as 'the middle science', 'the middle way' or 'middle term',[5] by which he means the possibility of a 'double demonstration'[6] of the phenomena of nature, through 'mechanical' or 'efficient' as well as 'final' causes.

But there is a slight shift to be discerned in the course of Leibniz's intellectual development, which testifies to the influence of the specific scientific analogues. At first (as we already saw) he begins by an uncompromising insistence that everything has to be explained in science by the application of mechanical (here, 'geometrical')

---

[1] L. 778–80; DM. 19, pp. 32–34; DM. 22, pp. 38–39. Leibniz is aware of the step from (b) to (a), cf. L. 779–80: 'The most beautiful thing about this [teleological] view seems to me to be that the principle of perfection is not limited to the general but descends also to the particulars of things and of phenomena and that in this respect it closely resembles the method of optimal forms, i.e., of forms which provide a maximum or minimum.' Note again the power of analogy—here: the mathematical analogue. The history of science shows that the most revolutionary breakthroughs often depend on logical penetration coupled with the application of the results of one science to another.

[2] Cf. the passages from *Discourse* 6 and *Radical Origination*, L. 791–2; above, pp. 403–4.

[3] L. 780, 787; cf. 782. This, including the use of the term 'architectonic', will reappear in the teleological sections of Kant's first *Critique* (K. 560ff., 566f., 653ff.) as well as the third *Critique, passim*. But there the 'end' is not defined by reference to the direction of God's will, but by what is contained in the 'scientific concept of reason' (K. 653). The search for a purpose is now characterised as an 'idea' which has a merely 'regulative use' (K. 561), something which emerges in virtue of the 'speculative interest of reason' (K. 560); cf. below, ch. VIII, sect. 4(c): vii.

[4] For the three classes of principles, cf. L. 778–9.

[5] Cf. L. 409, 446, 779.   [6] DM. 21, p. 37.

principles, just leaving room for the abstract postulate of 'substantial forms' ('real essences' as they were in one of Locke's uses of this term).[1] At the second stage of his development, before 1682, he adds the consideration that in addition to 'mathematical science', 'metaphysics provides existence, duration, action and passion, force of acting, and end of action, or the perception of the agent'.[2] Here, finality is still merely 'extrinsic', a mere ontological, if not psychological postulate.

After 1682, the year of Leibniz's publication of his paper on refraction,[3] not only does he hold that God's wisdom appears in the laws of motion in general,[4] as witnessed by his remark that there is more in bodies than mere 'geometric necessity', viz. conservation of living and directive force—though 'force' in the metaphysical sense, as we have seen, enters into the actual theoretico-scientific complex only very indirectly; but he now claims confidently that the actual effects of nature can be 'demonstrated' not only by a 'consideration of efficient' but also of 'the final cause'.[5] And by 1696, in the *Tentamen* he insists that

the principles of mechanics themselves cannot be explained geometrically, since they depend on more sublime principles which show the wisdom of the Author in the order and perfection of his work.[6]

But what 'proves' all this is once again only an 'example',[7] viz. the optics paper (UP) just referred to, and its extension to the more complex cases of the *Tentamen*. Let us look very briefly at the essential points of UP.[8]

In an earlier chapter we considered at some length the principles of

---

[1] L. 289.
[2] L. 447. Loemker ascribes the date of *The Elements of Natural Science*, where this passage occurs, to the years 1682–4; but there is little in the text to justify such a late date, and at least one other editor dates it 1677–80 (W.v. Engelhardt, *Schöpferische Vernunft* (Marburg, 1951), p. 488, n. 1).
[3] 'On a unitary principle of Optics, Catoptrics and Dioptrics', *Acta Eruditorum*, 1682; in L. Dutens (ed.), *Leibnitii ... Opera omnia*, 1768, III, 145. A German translation is in Engelhardt, op. cit., pp. 287–98; references are to this (UP).
[4] The reference to God's 'wisdom' is always to the principle of perfection, just as 'God's power' is a symbol for creative activity, and his 'knowledge' for his possession of ideas.
[5] DM. 21, p. 37. The *Discourse* was written in 1686.
[6] L. 779.　　[7] DM. 21, p. 37.
[8] I shall only discuss Leibniz's account of 'dioptrics', i.e. refraction. The principle is meant to explain the straight-line propagation and reflection of light as well.

Descartes' theory of refraction.[1] Some of Descartes' contemporaries, especially Fermat, expressed dissatisfaction with this, their reasons being partly that it involved reference to 'invisible particles', partly that it used some very artificial and implausible arguments, including the assumption of the direct proportionality of the velocity of light to the density of the medium.[2] Fermat had evolved a very different procedure, deriving Snell's Law by the use of the 'variational approach', from the principle that the *time* of the light's passage was a minimum, thus avoiding reference to the particular theoretical structure of the medium.[3] Leibniz's proof leans on this, though with some significant differences, the chief ones being the first practical application of his new differential calculus, the replacement of 'time' (or the ratio of path-distance to velocity) by the 'path-difficulty', defined as 'distance times resistance of the medium', and—as in the previous case of Leibniz's adoption of the ideas of a scientific contemporary, with his expansion or generalisation of Huygens' ideas on conservation—a more penetrating emphasis on the conceptual aspects of the subject. The argument which Leibniz employs is roughly as follows, references being to the accompanying figure (which is taken from UP. 288).

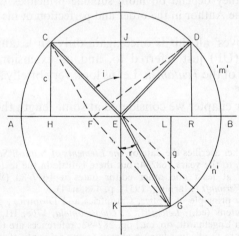

---

[1] Above, ch. III, sect. 2(d), pp. 136ff. See especially the details of Descartes, 'proof' of Snell's Law, pp. 142–3.   [2] Cf. above, p. 137.

[3] This is a *technical* case of the distinction between 'efficient' and 'final' causality; an instance of a difference which turns up again in later times as the distinction between 'abstract' and 'hypothetical' (or 'phenomenological' and 'atomistic or statistical') theories. Cf. Nagel, op. cit., ch. 6, sect. II. For the details of Fermat's

Let C be a source of light, and G an illuminated point. We ask: what is the path by which the light-ray reaches G; or rather, where lies the point E on the interface A-B between the media, m, n? Leibniz now assumes that this point must 'be situated in such a way that, of all the possible ones, this path is the easiest'. The path will be the easiest if the 'path-difficulty' is the smallest, i.e. if the total path times resistance is a minimum.

Let HL = h

EH = y

EL = h − y

m, n = the coefficients of resistance in the two media, say air and water respectively.

Then,

Total path-difficulty = s = m $\sqrt{c^2 + y^2}$ + n $\sqrt{g^2 + (h - y)^2}$ = minimum. Using the 'method of minima and maxima' (details were not published until 1686), which amounts here essentially (as shown more explicitly in *Tentamen*)[1] to differentiating and putting the result equal to zero, we get,

$$\frac{ds}{dy} = m \frac{2y}{2\sqrt{c^2 + y^2}} - n \frac{2(h - y)}{2\sqrt{g^2 + (h - y)^2}} = 0;$$

or: m.sin i − n.sin r = 0; from which we obtain:

$$\frac{\sin i}{\sin r} = \frac{n}{m} = \text{constant},$$

which is Snell's Law; the demonstration affording us, in the words of the *Discourse*, an 'example [of] the decree of God always to produce his effect in the easiest and most determined ways'.[2] Moreover, as the *Discourse* notes, in UP Leibniz goes on to give a second 'proof' of the law by using Descartes' particle model,[3] i.e. 'by consideration of efficient causes'.

---

proof, cf. R. Dugas, *A History of Mechanics* (London, 1957), Part III, ch. 5, pp. 254–9; A. I. Sabra, *Theories of Light from Descartes to Newton* (London, 1967), chs. 4–5.

[1] L. 784–5.   [2] DM. 21, p. 37; cf. also DM. 22, p. 39.

[3] His special motive for this is technical, since it supposedly enables him to show, using considerations from the theory of *vis viva*, that the velocity of light is proportional to the 'resistance' of the medium; a fact which would make the velocity larger in the denser medium, in line with Descartes' and Newton's theories, though opposite to those of Fermat and Huygens.

Once more Leibniz thus wishes to appear as the great conciliator of the two main competing approaches towards causal explanation in science,[1] insisting, against Descartes, that in addition to 'efficient causality', 'final causes' are perfectly acceptable even in natural science,[2] where they offer us a chance

> to discover the properties of those things, whose nature is not as yet known sufficiently clearly, so that we could use the more proximate efficient causes for the explanation of the internal mechanism, which the creator has used in order to produce the effects and realise his purposes.[3]

I have given a relatively detailed account of Leibniz's paper, because only by thus reproducing the scope of the technical foundation can we appreciate the nature of the expansion by which Leibniz reaches his general scheme—his view of the 'divine mathematics or metaphysical mechanism;[4] a move which is intended eventually to explain his concept of 'physical necessity' and 'contingent existence', in accordance with the principle of perfection, and which Leibniz in *Radical Origination* links directly with his 'principle of determination which must be sought by *maxima* and *minima*'.[5] And the technical core is so important because it is that which in the end supplies the ounce of meaning for the ton of mere semantic plausibility.

Our example makes clear one thing: the 'way of final causes' has little to do with any conscious purposes on the part of matter, or any goal-directed behaviour in general, although this aspect may be useful in biology. 'The light-ray does not reflect how it may most easily get to the point E or D or G', Leibniz expressly notes in UP, 'rather it is that the creator of things has fashioned light in such a way that this beautiful effect arises from its nature'.[6] Nor should we think that Leibniz is here linking the concept of 'minimum difficulty' with purely economic considerations, since he is quite aware that

---

[1] Kant will inherit this role, though with a systematic attempt to make the possibility of a reconciliation more plausible, by linking the two ways to the different 'regulative' approaches of 'reason'. (Cf., for instance, *Critique of Judgment*, par. 70 and 78.) Where in Leibniz there is still a suggestion that different explanation patterns imply different physical or metaphysical organisation, Kant replaces what looks like an ontological incompatibility with a purely methodological dichotomy.

[2] For Descartes, cf. *Principles*, i, 28: 'We shall reject entirely from our philosophy the search for final causes.'

[3] UP. 291. This is use (a), distinguished above, p. 425. *Discourse*, 22, p. 39, makes the similar point that teleological approaches are often easier than 'the way of efficient causes', adding however that the latter 'is indeed deeper and in some fashion more immediate'!   [4] L. 792.

[5] L. 791. The full passage is quoted above, p. 404n.4.   [6] UP. 290.

extremum principles 'provide a maximum *or* a minimum'. What matters is that the path of light, and phenomena in general, 'hold to the *most determined* [way], which can be the *simplest* even when it is a maximum'.[1]

The thought behind this seems to be as follows. The comparison with which Leibniz is here concerned is primarily that between 'geometrical' and 'physical necessity'.[2] Geometrical determinations allow an indefinitely large number of physical possibilities; the ones realised obey those of an 'architectonic' nature, which introduce 'uniqueness' and 'simplicity', or 'determinateness', being technically subject to the calculus of *maxima* and *minima*.[3] Now one might object that not only do such considerations depend on very ill-defined notions of simplicity and determinateness; they seem positively to be no more than Leibniz's amazed reaction that his calculus should so powerfully apply to the selected range of physical phenomena which he has specifically examined by way of a theoretical systematisation of the relevant laws. Besides, one might urge, Leibniz simply insists dogmatically that to every teleological explanation there must always correspond an equivalent one of a mechanistic kind, without any proof of this contention.[4]

But one could invert this argument, and point to explanations—as in the instance of optics—involving reference to underlying structure

---

[1] Depending on whether, e.g., a reflecting mirror is plane, convex or concave; cf. *Tentamen*, L. 780, 781; also 782.

[2] The significance of this concept will become clearer in the next section when we turn to the '*praedicatum inest subjecto*' (*P.i.S.*) doctrine.

[3] Note that Descartes, when presented with the same problem, had responded by the methodological comment that we require intermediate hypotheses between which to choose by way of crucial experiments, in order to determine the corresponding secondary causes responsible for observed effects. (Cf. above, ch. III, p. 121.) Leibniz's proposal so far corresponds to the introduction of a further constraint upon the *type* of hypotheses likely to be the most basic. But, as we shall see, his metaphysical expansion of this point will be that not only are these hypotheses governed by special principles of simplicity and determination, but all the contingent facts springing from them are also. The former does not, of course, entail the latter; this required the addition of the *P.i.S.* principle as well.

[4] This was subsequently demonstrated for mechanics by the researches of Lagrange and Hamilton where we get the deduction of minimum principles from the differential equations of motion, themselves derived from the Newtonian dynamics of force-systems. As Mach later put it: 'Since the properties of the motion of bodies or of their paths may always be defined by differential expressions equated to zero, and furthermore the condition that the variation of an integral expression shall be equal to zero is likewise given by differential expression equated to zero', it follows that the integral expressions need not possess any particular physical, let alone metaphysical, significance. (Cf. *Mechanics*, op. cit., p. 463.)

which can be replaced by others of a completely 'abstract' nature,
that are 'governed', so to speak, in so far as they yield unique law-like
solutions, by nothing but this powerful calculus of *maxima* and
minima; a consideration which Leibniz clearly generalises in his
reference to the 'divine mathematics or metaphysical mechanism',
which governs nature in general.[1] Such a reading however denotes a
highly specialised interpretation of teleology, quite at variance with
the supporting model of biological enquiry, where Leibniz likewise
advocates that we should proceed both by way of mechanical as well
as final causes, interpreting the latter as an investigation of the
purpose of the animal organism.[2] Still less clear is the identification of
this expression of 'God's wisdom' with 'perfection'; how can Leibniz
manage to jump from the 'proof' that some regions of physics are
interpretable in terms of *extremum* (also called 'stationary') principles, to his contention (even in one and the same essay) that physical
reality ('physical necessity') 'is determined in such a way that its
contrary would imply *imperfection or moral absurdity*'?[3]

Perhaps some light on the slide from calculus to God's wisdom is
provided by the *Discourse*. He there argues that the mechanistic
approach admittedly shows that nature is subject to law (some
principle of causation). But this, as the analogy of fitting some
mathematical function to a geometrical line, however irregular, illustrates, is not sufficient; *some* order and regularity the world is sure to
exhibit, provided no restrictions are placed on the degree of complexity. What the *minima-maxima* calculus shows is that there are
unique and privileged cases, and moreover, that this calculus and the
considerations associated with it govern a surprisingly large number
of phenomena. The world is like one that has been constructed by the
use of the 'simplest hypotheses' covering the largest number of
phenomena,

> as a geometrical line might be, of which the construction was easy and
> the properties and effects were very admirable and of great extent.[4]

And in the section prior to this Leibniz identifies 'God's wisdom'
with this very feature, adding that

> *reason desires* that multiplicity of hypotheses or principles should be

[1] L. 792. Leibniz in this place also gives further examples where, from among many
possibilities, a *maximum* occurs as the unique state.
[2] Cf. DM. 19–20, pp. 33–7.   [3] L. 793; my italics.
[4] DM. 6, p. 10; cf. above, p. 403, for this passage.

avoided, in almost the same way as the simplest system is always preferred in astronomy.[1]

We can say that in this way the central analogue of perfection is at least made specific in outline. And this is important since it will have to supply the life-blood, so to speak, for the other two levels of teleology (b) and (c) mentioned on p. 425f. above, the methodological and the ontological. Leibniz labours hard to make the transition from (a) and (b) to (c) plausible, especially in *Radical Origination*,[2] without, it must be said, achieving much success. And in any case, whatever force his considerations from ethics, politics and jurisprudence might here possess is borrowed quite obviously from level (a). Indeed, one can say that analogue (a) is stretched by Leibniz almost to breaking-point. In the next main section we shall study his argument which leads to the identification of the contingency of existence (particular as well as law-like) with the physical or moral necessity of the determination of whatever comes to exist: the 'pre-tension' or 'exigency toward existence in possible things',[3] itself founded on the 'free decrees' of God's choice,[4] which in turn expresses 'God's wisdom,' *via* certain 'principles of the good'.[5] Now levels (b) and (c) obviously cannot offer anything strong enough to prevent the 'goodness of the choice' from degenerating simply into 'whatever happens', particularly since Leibniz argues his case for the relevance of (c) in such a way that it becomes totally incorrigible.[6] So only (a) offers us something sufficiently specific as well as 'corrigible' to provide semantic force for Leibniz's conclusion.

In defence of Leibniz, one might aver that any characterisation of the basic inventory of the whole of Nature, as well as its supplementary constructions, *can* only offer 'analogies', if one is to avoid simply a *recital* of 'all that is'. An appreciation of this fact will prevent too facile an acceptance of criticisms like those of Voltaire, for whom Leibniz's 'universal optimism' provided an easy butt (in *Candide*). For the core of Leibniz's contention, that ours is 'the best of all possible worlds' is really contained in the analogues from physics and

---

[1] Op. cit., p. 9; my italics. This is clearly one of the origins of Kant's 'reason'; cf. K. 537: 'Reason presupposes the systematic unity of the various powers, on the ground that special natural laws fall under more general laws, and that parsimony in principles is not only an economical requirement of reason, but is one of nature's own laws.' Cf. below, ch. VIII, sect. 4(c).   [2] L. 794–8.
[3] L. 791 (Th. 6).   [4] DM. 13, pp. 20, 21.   [5] L. 779.
[6] *Radical Origination* argues that all visible 'imperfections' are required to result in the greatest possible 'universal harmony'! And cf. above, p. 404.

optics which make the contention 'meaningful'. Nor should Leibniz be over-criticised for employing a method of philosophising which, on examination, we have found universally prevalent among *all* the philosophers of our survey, and which we may be certain prevails among most major philosophers of any period. At the same time, such a view offers us at once a critical task; for the essential thing to criticise is the sufficiency, if not cogency, of the analogues, the question of their essential strength or weakness. This shows that philosophical criticism, at heart, is not so much an investigation of minor logical slips, for the deductive garb in which these arguments parade is often not much more than ceremonial dress. Metaphysics is more of an 'empirical' and 'descriptive' enterprise than an exercise in formal logic. The seeing of the world through the spectacles of one's major technical, or professional pre-occupations is a deep-seated habit of the philosophical mind, and often dictates the very forms of philosophical thought.[1]

(c) *The Principle of Continuity*

Both our previous analogues, of 'force' and 'determination', testify to Leibniz's strong sense of the 'interconnection of all things'.[2] There is a third concept, that of 'continuity', which yields a still more explicit recognition of this, and which Leibniz likewise based on 'architectonic grounds', viz. 'the law of continuity'.[3] Moreover, it also, like the previous two, depends on a specific argument from physics, arising out of Leibniz's criticism of some of the special 'rules of motion' enunciated in Descartes' *Principles*, ii. 40ff., in the course of which Leibniz 'sets up' the 'law of continuity' as 'a general criterion or touchstone'.[4]

It is not necessary to go into the details of Leibniz's critique, which arose out of the need to defend his earlier criticism of the concept of quantity of motion against the Cartesians and which he extended further to an attack on the attempt of Malebranche to defend the latter.[5] The matter is summarised in the first paper Leibniz published on this question, in 1687. There, he singles out Descartes' first and second 'rules of motion' governing impact. According to the first rule, two bodies, B and C, of equal mass, moving before impact with

---

[1] Cf. the long array of examples from physics and engineering in Wittgenstein's writings, testifying to his primary education as a mechanical engineer.
[2] Cf. L. 794.   [3] L. 788.   [4] L. 655.
[5] Cf. above, ch. III, sect. 2(e), p. 150.

equal but opposite velocities, will also do so after impact, with their velocities reversed. According to the second rule, with the same conditions applying, except that B is larger than C, only C is deflected, and B continues, both with their earlier speeds, but now in the same direction.[1] Leibniz objects that this conflicts with the law of continuity, since supposing B to be only vanishingly larger than C, the former would change *suddenly* from behaviour in accordance with rule (1) to one with rule (2), despite the fact that the mass of B might have been increased only infinitesimally. Leibniz opposes to this a new principle: when 'data approach each other continuously ... it is necessary for their consequences ... to do so also',[2] this being *one* of the technical statements of the principle of continuity.

In his *Thoughts on ... Descartes* he proceeds to give a qualitative account which avoids this 'leap': B being made slightly larger, we must assume that it is repelled with a slightly smaller speed though still in reverse direction; with increasing mass of B there will come a point at which B is actually stationary after impact, and only then, with still further increase, will B not change direction after impact at all. (In the limit, with C = 0, this is obvious.) In this way, the discontinuity of the 'consequence' is avoided, the search for continuity leading to the statement of the correct rule.[3]

Again I have given the details of Leibniz's argument in order to emphasise the gap between the technical case which originally led to the idea of continuity becoming *established*, and its generalisation by extrapolation to a much wider field.[4]

Having its genesis in Leibniz's studies of the infinitesimal calculus, from which it is subsequently transferred to physics, where it served him, he tells Varignon, 'as a principle of discovery ... and as a convenient test to see if certain proposed rules are good',[5] such a 'principle of order', by an immediate extension, suggested to Leibniz 'how the true physics should in fact be derived from the source of divine perfections', thus demonstrating that 'it is God who is the ultimate reason of things'.[6]

[1] L. 540; cf. L. 657.   [2] L. 539.   [3] L. 658–9.

[4] I say 'established', for, once again, historically and methodologically, the idea of continuity presumably served as an aid drawing attention to and confirming the initial pinpointing of Descartes' error. (There were, of course, many others working on the same lines, e.g., Wallis, Wren, Mariotte, Huygens, etc.)   [5] L. 886.

[6] L. 541. Once more we see that 'reason' in Leibniz has primarily to do with this planned teleological determination of God's actions, whilst 'cause' relates more directly to the 'source of activity' placed within each substance. (On 'cause or reason', cf. above, p. 395n.2.)

The true significance of the coupling of such a 'general criterion or touchstone' of scientific laws with a metaphysical ground can be appreciated better within the context of a discussion on the status of principles like that of continuity in the philosophy of Kant, who was deeply influenced by Leibnizian approaches in this field. Kant's contention is that such criteria are not 'arbitrary', nor even purely empirically testable hypotheses, but that they supply a quasi-creative foundation for the construction of scientific theories. In its demand for system, 'reason does not here beg but command'.[1] Laws like that of continuity, he insists, are not empirical, nor simply 'tentative suggestions', not even 'mere methodological devices'; rather, they are expressions of a demand of 'reason', in the form of 'maxims', which though 'they contain mere ideas for the guidance of the empirical employment of reason', yet *claim* 'objective but indeterminable validity'—a claim which reason necessarily imposes upon itself, in its 'regulative employment'.[2]

However, here Leibniz and Kant seemingly divide. For if such principles are treated, not as maxims, but 'as constitutive', as though to the object which they *postulate* there corresponded something, *per impossible*, with the status of an object of science, given through 'observation and [theoretical] insight'[3] then 'reason would run counter to its own vocation, proposing as its aim an idea quite inconsistent with the constitution of nature';[4] it would be subject to the famous 'transcendental illusion'.[5] In other words, 'reason' can only *seek* system in nature, *as if* such systematic unity were given; it must not treat any success in that direction, let alone its postulation, as constitutive of some 'existing' object; as though, instead of postulating that there 'must' be economy and continuity built into nature, one asserted that there *is*. (This is an aspect of the contrast between Leibniz's 'transcendental *realism*' as against Kant's 'transcendental *idealism*'.)

Now Leibniz, by contrast, frequently *talks* as though just this situation *did* provide 'proofs' for the 'existence of God'. I think, actually, that this difference is more often apparent than real. At any rate, Kant's greater sharpness in the making of distinctions brings

[1] K. 539; cf. below, ch. VIII, sect. 4(c): ii–iii, pp. 506–12.
[2] K. 544–5; cf. also K. 537. In the *Critique of Judgment*, op. cit., pp. 20–1, the maxims mentioned include 'nature takes the shortest way (law of economy)', 'nature makes no leaps (law of continuity)', and that of minimum assumptions ('Occam's Razor'). [3] K. 533, 548. [4] K. 538.
[5] Cf. K. 298; cf. below, ch. VIII, sect. 4(c): vii, pp. 523–30.

into the open the nature and limits of the arguments which Leibniz is in fact already employing. What both share is the placing of a heavy emphasis on the importance for science of systematisation and of the part which higher-order methodological principles, like those of continuity, economy and causality, play in this work of *construction*: a feature which, in all the previous philosophers whom we have studied, was either suppressed or given relatively incidental attention.

It is, of course, likely that a 'realist' approach to concepts like continuity and order can contribute greater boldness in extrapolation, and the discovery of new directions in scientific research and elsewhere. Thus, Locke's fleeting suggestions in the *Essay*, iii.6.12 and iv.16.12,[1] seemingly inconsistent with his 'metaphysical' views, concerning the interconnectedness of all things, their 'gradual connection by insensible degrees', finds an immediate home in Leibniz's ontology as reflected by the enthusiastic reception it gets in the *New Essays*.[2] For the same reason Leibniz scoffs at Locke's conjecturalism concerning the 'internal [insensible] parts of our bodies', commenting that we can 'hope that we shall go beyond conjecture on many occasions'.[3] And not unexpectedly he warmly embraces Locke's reference to the need for 'analogy' when reasoning in science, particularly in the regions of 'plants, insects, and the comparative anatomy of animals'.[4]

All these are applications and illustrations of the principle of continuity, of the idea that 'everything goes by degrees in nature, and nothing by leaps';[5] and it will be noted how admirably it fits in with Leibniz's universal method of analogical reasoning and the employment of empirical analogues in his metaphysical constructions, where it will enable him to extend the notion of 'life'—as we have already noted—from animate to inanimate nature.

This is stated most explicitly in the letter to Varignon:

> Men are therefore related to animals, these to plants, and the latter directly to fossils which will be linked in their turn to bodies which the senses and the imagination represent to us as perfectly dead and formless ... [The law of continuity demands] that all the orders of natural

---

[1] Cf. above, ch. IV, pp. 214, 260f.   [2] *N.E.*, pp. 333–5, 550–3.

[3] Op. cit., p. 552. We shall presently see how the metaphysics of 'internal continuity' of the relation of subject and predicate affects in general Leibniz's approach towards the epistemological troubles of Locke.

[4] Op. cit., p. 553. Leibniz's attitudes here were to find an echo in many later generations of biologists.   [5] Op. cit., p. 552.

beings *form but a single chain* . . . so that it is impossible for the senses and imagination to fix the exact point where one begins or ends.[1]

We may say that, for Leibniz, ultimately differences in kind always reduce to differences of degree, the most important epistemological instance being the distinction between primary and secondary qualities, a difference construed by Leibniz as one only of the degree of 'confusion' or 'distinctness'.[2]

In the same way, since for reasons that will appear, the only thing 'that can be found in simple substance' is 'perceptions and their changes',[3] perception has to be claimed even where there is no 'consciousness', a supposition which leads Leibniz to the celebrated 'small perceptions', a notion which he extended to all 'organisms'. But once again, this 'constructive approach' of the transcendental realist reacts on his more ordinary methodological approaches in science and life.

So, as in the previous case of his *extremum* principle, we find the argument here proceeding at three levels of specific scientific formulation, applied first, (a), to a special case in dynamics; then, (b), methodologically extended, providing new points of view in the realm of biological studies; and finally (c) applied to ontological constructivism, in which shape, moreover, it offers us a further illustration of 'divine perfection', and the importance and validity of the notion of 'final causes'.[4] And as in the previous cases, the essential semantic power, the technical meaning, which prevents the conception from degenerating into mere re-enumeration of all that exists, or from remaining altogether 'opaque', is centred on the first of these levels.

---

[1] W. 187; my italics. Leibniz believes this continuity of species, however, to be only a possibility. In actuality, men find that 'natural classifications' can be constructed in accordance with their own chosen criteria, and that further, there is a relative and observable constancy of species, with many missing links between them. Yet this is in the end only a matter of convenience. Whilst in opposition to Locke, he insists that visible properties are based on an 'internal, essential and immutable nature', he agrees with him that there are no immutable *borderlines* carved into that nature; for this would be incompatible with the principle of continuity. (Cf. the classic discussion of Locke's position on 'real essence' and 'species' in *N.E.*, iii.3.14, p. 314; iii.6, pp. 332–61; cf. below, p. 444.)

[2] To this we shall return. Cf. *N.E.*, iv.6.7, pp. 458–9; and L.944; below, pp. 440–3.

[3] *Monad.*, 17, L. 1046.

[4] L. 541–2. The concept of 'final cause' is therefore interpreted technically both through the principles of *extremum* as well as of continuity.

4. THE NATURE OF SUBSTANCE: THE DOCTRINE OF 'PRAEDICATUM INEST SUBJECTO'

(a) *From 'External Characteristics' to 'Internal Nature'* (*the debate with Locke*)

We have so far briefly hinted at the division in Leibniz's thought between the 'external characteristics' and the 'internal nature' of things.[1] The distinction between 'phenomena' and 'substances' mentioned in an earlier section[2] was another aspect of this. Leibniz's analogical approach enables him to avoid a complete break between these two realms. Instead he tries to avail himself of the best of both worlds, working with the basic assumption of the relevance of physics to metaphysics, as witnessed by his views on the foundations of dynamics and optics, whilst professing a sanguine assurance that epistemological difficulties, which we have seen to worry the philosophers of our previous chapters, are not productive of that fatal scepticism encountered in Locke. At the same time, there are a number of intermediate levels in Leibniz's epistemological views which seemingly are aimed at narrowing or even closing the gap just noted, or rather: concealing it. Partly the different levels correspond to Leibniz's gradual emancipation from Cartesian points of views, partly they are due to the pull between empirical and rational (inductive and deductive) approaches met already in Descartes. At any rate, both the earlier Cartesian tendency and the later metaphysical position prevents Leibniz from ever clearly facing the problems of Locke and the other 'empiricists'.

As regards induction, he frequently, though, as we shall see, not invariably, labours on its 'uncertainty'. As we may expect, this attitude is shaped by way of contrast with a certain ideal of rationality. In the early *Preface to Nizolius* we are told that induction can only yield 'moral certainty', and even that only on condition of assuming a principle of causation in the form 'same cause—same effect'. But 'perfect certainty' as found in geometrical axioms we can never reach in this way.[3]

The same 'uncertainty complaint' recurs thirty-two years later in the letter to Queen Charlotte (1702), where we are given a (possibly misleading) geometrical example (asymptotic approach) to explain how extrapolation beyond experience can mislead; and Leibniz now

---

[1] *N.E.*, iii.6.14 makes the distinction in these terms. Cf. above, p. 438, n. 1, for our reference to the Leibniz–Locke debates on 'real essence'.
[2] Above, pp. 410ff. [3] L. 201–2.

makes the contrast with those 'universal and necessary truths' which we draw from 'what is in us.[1]

The relation between mathematics and experience is, however, not clearly indicated in these examples. This is faced, on rather Cartesian lines, in the *Elements of Natural Science*, where it is discussed in terms of the need for an explanation of sensory attributes like warmth and colour. And Leibniz states it as an ideal that these— being 'confused'—should be, and in principle at least 'can be reduced' to or 'resolved' into metric attributes involving matter in motion. Only by means of such 'distinct' attributes can we hold out the hope of 'making apparent . . . their connection with each other'; otherwise, he insists, 'we do not understand, but merely experience'.[2] In this way,

> the application of mathematics to physical science consists in such consideration of the distinct attributes which accompany confused ones.[3]

For 'we perceive nothing distinctly in matter save magnitude, figure, and motion'; however, fortunately, 'everything confused is by its nature resolvable into the distinct, even though it may not always be in our power to do this'.[4] The engineer may be content to operate with the unanalysed notion of gravity, but the scientist's 'under-standing' is only satisfied by

> an explanation, sufficiently understood, which, when we have compre-hended it, will enable us to demonstrate with geometric certainty that gravity must necessarily arise from it.[5]

All this looks very like the Cartesian idea of '*mathesis universalis*'; and in the letter to Varignon, from which we have already quoted, the relevant passages[6] give evidence of Leibniz's occasional extreme rationalist approach. However, from what has preceded in earlier sections, we know that such pronouncements do not state the position

---

[1] L. 895–6.

[2] L. 443; Leibniz (very much like Locke) cites the case of green arising out of a mixture of yellow and blue, which we observe but do not understand, contrasting it with a mathematical example.

[3] L. 444.   [4] L. 445, 444.

[5] L. 445. Leibniz never accepted gravitational attraction as an ultimate property of matter; cf. *N.E.*, Preface, pp. 54–5, 60–2; Clarke-Leibniz, iii.17, iv.45, v.118–23 (L. 1116, 1123, 1167); to Arnauld, OC. 246–8. (Cf. above, p. 370n.1

[6] Above, p. 411n.1; cf. also the *Reply to the Thoughts of . . . Bayle*: 'The actual phenomena of nature . . . can never violate the law of continuity and all the other exact rules of mathematics' (L. 950).

adequately. For one thing, Leibniz is here dealing with 'phenomena' and the attributes of 'matter'. And so immediately following the passages in the *Elements* which we have just noted Leibniz continues that not only can things be explained by reference to 'final causes' just as well as by 'efficient causes', but the very laws to which allusion has just been made have their 'metaphysical reasons, [which] arise from the divine will or wisdom'. Whilst 'mathematical science' provides magnitude, figure, and motion, 'metaphysics provides existence, duration, action and passion, force of acting, and end of action . . .'.[1]

This refers us back to our earlier account of force and teleology, and we can see clearly that for Leibniz the connections supplied by 'mathematical science' are not sufficiently deep or ultimate; at best the ideal of explanation which was implicit in the examples of the *Elements* is now regarded as a *model* and not the real thing.

The pull from, and the peculiar assurance against, the attacks of Lockean scepticism provided by the fields of mathematics and metaphysics are nowhere more evident than where Leibniz faces Locke explicitly, for instance in the comments on the nature of 'species', and on Locke's views concerning the 'extent of human knowledge' and the range of certainty of universal propositions regarded as inductive generalisations (in the *Essay*, iii.6 and iv.3, 6 and 11). Here we should remember the conclusions from our study of Locke, where I showed that his scepticism is mainly rooted in his inability to reconcile his 'theory of ideas' and its logical consequences with a simultaneous adherence to a realm of 'real essences', $(O_M$, arising out of $O_T)$ and 'external objects' (O_C).[2]

The argument between Locke and Leibniz here concerns, (1) our lack of knowledge of relations between secondary qualities, and between the latter and the primary qualities (or alternatively, the internal structure of bodies);[3] (2) our lack of knowledge of this internal structure itself; (3) our lack of knowledge or certainty of empirical generalisations, and of the 'constant properties' of substances.

Let us take Leibniz's comments in turn. (1) The grand answer to the first is that our 'ideas of sensible qualities are confused'. Now it

---

[1] L. 446, 447.
[2] Let us repeat that this form of 'deep scepticism' should not be confused with the kind which consists in the simple *denial* of these realms, as in Berkeley and Hume.
[3] Naturally the immense fluidity and vagueness of Locke's position lends itself here to all sorts of misunderstandings and evasions. Cf. ch. IV, sect. 5(a).

will be obvious to the reader by now that 'confused' (or 'opaque') is really a technical term, introduced precisely to push home a metaphysical theory concerning the reducibility of secondary to primary qualities and, more important, of both to intellectual conceptions and purely mathematical relations, e.g. figure, size and motion as in Descartes.[1] Continually Leibniz couples 'confused knowledge' of this kind with what can be known 'only by experience'.[2] These two responses to Locke's position concerning the 'unintelligible gap' between the different realms give Leibniz openings for meeting the difficulty, by falling back on the power of analogical cases. The first is to contrast purely 'empirical' knowledge with an understanding through theoretical explanation, e.g. of the colours of the rainbow, where theory predicts and explains why just these colours occur and in what order.[3] This amounts to a subtle move of making explicit Locke's slide from O_T to O_M, from theoretical structure to 'real essence': the former having its natural home in the hypothetico-deductive framework of scientific theorising, the latter being its 'metaphysical extension', as I have maintained in chapter IV.[4] Leibniz here chooses to ignore Locke's metaphysical intentions even though at the heart of his own philosophical method he employs exactly this slide.

Indeed, some pages later, when coming upon the same Lockean complaint, Leibniz subtly slides towards the metaphysical level, coupling it moreover with his second response, which amounts to the example of a literal model for a bridge between 'confused' and 'distinct' perception—the case of 'a confused resultant of the actions of bodies upon us'.[5] The example is the production of a transparent image when a toothed wheel rotates at high speed. When stationary, we are capable of perceiving clearly the structure of the wheel; its rapid rotation tends to 'confuse' (blur) the individual images, leading to the purely 'artificial' effect. In actual practice Leibniz comments, it is neither 'reasonable nor possible' to obtain such analyses, the 'confusion' being due to the 'too great multitude and minuteness of mechanical actions which strike the senses'. This is using the

---

[1] Let us once more remember Kant's comment on this position, amounting to a straight denial that the distinction is one of degree only, instead of one in *kind*. (Cf. particularly K. 84, 286, and the passage in the *Prolegomena* previously referred to.) *Per contra*, in accordance with the principle of continuity, all differences, as we have remarked are, for Leibniz, differences of degree.

[2] *N.E.*, iv.3.8, p. 432.   [3] Ibid.   [4] Above, ch. IV, sect. 3(a).

[5] *N.E.*, iv.6.7, pp. 458–9.

scientific origins of Locke's epistemological troubles with a vengeance. And as suggested, Leibniz, just like Locke, slides from the scientific to the metaphysical, though on the other side of the fence, adding that

> if we had reached the internal constitution . . . we should see also how they *must* have these qualities, which would themselves be reduced to their *intelligible* reasons.[1]

Leibniz's comment on the second complaint, (2), of our lack of knowledge of internal structure, is simply once again that the new discoveries in 'infinitesimal analysis' are a powerful means of 'uniting geometry with physics'; and he furthermore points to recent advances in dynamics, all of which may enable theory to secure a great advance in an approach towards the corpuscularian interior of bodies.[2] Here, again, Locke's metaphysical difficulties are ignored.

(3) The reply to Locke's complaint that we do not know with certainty that 'gold is fixed', because of our ignorance of 'real essence', is entrancing. First of all, says Leibniz (who by now betrays considerable impatience, if not irritation, with Locke's continued sceptical refrain) we 'can be assured of a thousand such truths regarding gold'. Nor does the fact that some day we might discover a new species of gold argue against the existence of an 'internal essence', for when faced by such a discovery, 'we shall have no doubt that the internal essences of these two species are different'. But what, we may ask, entitles Leibniz to speak so confidently regarding 'internal essence'? His answer is simply that

> it will always be allowable and reasonable to assume, by means of a reciprocal proposition, that there is a real essence belonging [to its respective species].[3]

Of course, quite apart from the fact that Leibniz again chooses here to consider merely the inverse-deductive move to the hypothetical structure, and overlooks the metaphysical aspect of 'real essence', he totally ignores Locke's contention that it is only from a study of 'nominal essences' that we can arrive at any kind of classifications of species; and that any notion of a sharp division between, if not the very existence of, real essences is shot through and through

---

[1] p. 458; my italics. For Locke's almost identical position, cf. above, ch. IV, sect. 4(a), especially pp. 208–9.
[2] *N.E.*, iv.3.24–25, p. 438.   [3] *N.E.*, iv.6.4, pp. 454–7.

with the very doubts which spring from the 'opacity' of experiential knowledge of the sensory qualities of the nominal essence. From where then does Leibniz get his assurances?

The clearest answer occurs at the end of this chapter, where once more Leibniz comes across Locke's insistent complaint that since we lack any knowledge of the 'connection' between the observable qualities of bodies, we cannot know 'with certainty the truth of this proposition, that gold is fixed'.

Leibniz's reply is two-fold, though the two points are connected. (a) He now admits that this is only

> an experimental certainty, and of fact, although we *do not know* the bond which unites the fixity with the other qualities. ... [But the] specific essence, although unknown in its interior, makes these qualities emanate from its depths, and makes itself known confusedly at least by means of them.[1]

We meet a similar position in Leibniz's comments on Locke's views of the reality of species.[2] For, despite the fact that he appreciates and agrees with Locke's point that classification is not by way of insight into 'real essences', but by reference only to observable and convenient criteria; and despite the fact that he even more explicitly than Locke realises that the distinction between an 'artificial' and a 'natural classification' is not the same as that between artificial and 'essentialist',[3] he holds with an assurance transcending any doubt whatever, to the 'reality' of internal essences, though these cannot even begin the work of delineation between species, since it is always possible that 'the passage from species to species may be insensible'.[4] For, provided that we stick to our classifications based on observable characteristics, we may be sure that they 'will be founded in reality';[5] a remark which *might* of course be a harmless reference to the basis of a 'natural classification' were it not for the passage where the latter is *distinguished* from one that is 'based upon the essential or immutable'.[6] Or rather, we ought to say that there *is* simply the slide here again from the 'natural' (and the 'theoretical') to the supra-natural or metaphysical: that which is meant ultimately to provide the 'justification' and 'basis' for the former.

---

[1] *N.E.*, iv.6.8, p. 460; my italics.
[2] Cf. above, p. 438n.1, where some of the details are anticipated.
[3] Cf. *N.E.*, iii.6.39, pp. 360–1; also 6.14, pp. 336–7. For Locke's position in regard to classification, cf. above, ch. IV, pp. 214n.4, 215n.2
[4] *N.E.*, iii.6.27, p. 353.   [5] Op. cit., p. 336.   [6] Op. cit., p. 360.

It is here that the ways of Locke and Leibniz divide; that of the epistemologist and of the ontologist. For both Locke and Leibniz the metaphysical is ultimately set apart from the physical. But that fact is expressed by Locke through his scepticism—with its consciousness of the opacity of the epistemic level—whereas for Leibniz it serves as a suggestive element in his studies of the realm of nature. He is more prepared (albeit unconsciously) to accept 'slides' with good grace, indeed as a creative device, the analogies which are strewn all the way including the very powerful method of theoretical analysis or hypothesis which supplies a meaning for the metaphysical basis; just as previously the physics of force and of the principle of the *extremum* were images of substance *quâ* action, and substance *quâ* end or purpose of action. Locke, just like Leibniz, thinks of 'real essence' as a basis, but is worried by the epistemological barrier. On the other side, Leibniz is so concerned with the structure of the foundation that his assurance of its 'reality' prevents his epistemological qualms from even beginning to affect the issues. It is, as we shall now try to show, because Leibniz is so preoccupied with the nature of the bond uniting the qualities, so concerned to eliminate any gaps in the foundation, that he cannot have any truck with those whose analysis of the limits of meaningful statements (cf. Berkeley's doctrines) throws doubt upon the structure. For Leibniz is here simply not concerned with 'knowledge', but with 'foundations'; with what supports 'facts' or 'truths', not with what guarantees our *knowledge* of them, irrespective of their being based even on the best of observations.

It is not always so clear that Leibniz is concerned with 'metaphysical justification' rather than knowledge. For at times, this is—confusingly —expressed as a difference in the definition of 'knowledge' itself. This becomes evident when we turn to Leibniz's reaction to Locke's sophisticated definition of 'knowledge' as our perception of the agreement or disagreement between our ideas. His comment here is telling: whilst it is quite correct to say that

truth is always grounded in the agreement or disagreement of ideas, . . . it is not true in general that our knowledge of truth is a *perception* of this agreement or disagreement. For when we know truth only empirically, from having experienced it, *without knowing the connection of things and the reason there is in what we have experienced*, we have no perception of this agreement or disagreement, unless we mean that we feel it in a confused way without being conscious of it. But you always demand a knowledge in which one is conscious of connection or opposition, and this is what cannot be granted you.[1]

It might be said that Leibniz's remarks here bear simply on the idiosyncracy of Locke's narrowing of the conception of 'knowledge'; and this makes the comment indeed seem plausible, if not somewhat trivial. But behind it there is concealed something more significant. For Leibniz means to imply that there is *a* sense in which one can speak of 'knowledge' even where the connections are not 'perceived', as is done, for instance, in the case of 'confused perception', i.e. experiential knowledge.[2]

But such an admission, harmless for a sophisticated empiricist, carries immensely important overtones for a rationalist like Leibniz. For as the italicised portions of the passage just quoted make clear, Leibniz is sanguine about the extension of the term, because he accepts without any sceptical provisos the universality of the existence of the underlying connections between things; where we must add again the reminder that for Leibniz the realm of these 'connections' stretches beyond that of the 'analogies' of science and theory to that of the underlying metaphysical bond. It is because he feels certain that there *are* connections, without qualifications of the kind which Locke's epistemic worries had brought to the fore (and which subsequently became central in the thought of Berkeley and Hume), that he is prepared to speak of 'knowledge' even in the 'opaque' or 'experiential' cases. The hidden connections are for him, we might say, the *condition* which guarantees truth, however much their existence may be concealed as a matter of epistemic principle; and even though the ways in which we in fact arrive at the 'truth' are determined solely by observational and theoretical factors.[3]

Possibly, Locke's definition of knowledge, deliberately excluding consideration of such *hidden* connections, betrays a somewhat sceptical approach also towards the potentialities of scientifico-theoretical explanation, by contrast with Leibniz's willing extension of the meaning of knowledge, as just explained. Needless to say, for purely *logical* purposes (deductive as well as inductive) neither philosopher

---

[1] *N.E.*, iv.1.2, p. 400; my italics. (Cf. above, ch. IV, sect. 7, pp. 268–9.)

[2] Thus he wholeheartedly agrees with Locke's grudging extension of the appellation of knowledge to that of 'the existence of things without us' (*N.E.*, iv.2.14, p. 420).

[3] The situation is here in some respects like that implicit in Kneale's contention that unless there were necessary connections the scientist would have no reason to search for laws, even though he can in fact obtain his laws only by means of straightforwardly inductive methods. Cf. above, ch. IV, pp. 189–90 for this point.

can hope for this to affect the pursuit and achievements of science.[1]
It is only in so far as any *belief* in them spills over into the analogical
power of Leibniz's models that they can exercise practical and pronounced influence; an influence which may be both methodologically
and psychologically powerful.

(b) Such a reading is further supported if we turn to the second part
of Leibniz's reply to Locke's complaint of our lack of certainty in
'experiential' ('confused', i.e. opaque) knowledge.[2] For he eventually
admits that our certainty is not of the kind Locke has in mind; it does
not end in 'rational assurance'.[3] As regards propositions which are
'learned by experience', he tells Locke, only 'moral or physical
certainty, but not the necessity (or metaphysical certainty)' can be
established.[4]

This is, however, not the whole story. It is true that in the *New
Essays* Leibniz says that such 'propositions of fact', when general, are
made so 'by induction or observation'; and 'this is not a perfect
generality, because we do not see its necessity',[5] as in the case of the
propositions of mathematics and logic. But here he is speaking only
from the epistemic point of view. Far more important is the constitutive aspect, which insists, as noted a moment ago, that, even as regards
propositions of fact, there is always an underlying bond, due to the
existence of a 'reason' which is 'sufficient' to relate any attribute to its
respective substance; and as the *Monadology* puts it, 'without there
being a sufficient reason ... *there can be found no fact that is true or
existent*, or any true proposition ... *although we cannot know these
reasons in most cases*.[6]

Briefly, the 'reason' here referred to is that which governs the
'activity' of any substance so that it has or evolves the attributes
which it possesses, and has the direction ('end') of the evolution
which it exhibits, and is what it is. And although the models for

[1] In the *Critique* Kant as usual sharpens the distinction, by drawing explicit attention
to the confusion which is involved in treating the contrast between 'sensible world'
and 'intelligible world' (which is really one between 'phenomenal' and 'noumenal',
or 'positive' and 'metaphysical') as equivalent to that between 'observational' and
'theoretical'. (Cf. K. 273.) To put it in his language: analogies have no 'constitutive
implications'—neither for the positive nor the metaphysical level. (Cf. below, pp.
547–8.)

[2] Cf. *N.E.*, iv.6.8; above, p. 444.

[3] Locke, we saw, never makes the nature of the 'certainty', after which he is hankering, quite clear. Here is therefore an opening for Leibniz to interpret Locke in a
novel and idiosyncratic way that will suit his own philosophical interests. It is of
such stuff, I have maintained throughout, that progress in philosophy is made.

[4] Op. cit., pp. 461–2.   [5] iv.11.13, p. 514.   [6] *Mon.* 32; L. 1049; my italics.

'activity' and 'end' are those taken primarily from science in the way we have noted, and although the conception of a mediating 'reason' borrows, as we have also just seen, from the methodological framework of scientific theorising, the present conception presents a vast extension of these models; above all, they are broadened to cover, not only contingent propositions that are *general* in form, but also *'particular'* propositions.[1]

The slide into metaphysics becomes manifest in the following way. That there *is* such a 'reason' for every fact which links predicates that appear on the surface merely to 'coexist', including the case of particular propositions, Leibniz expresses by saying that this 'sufficient reason' must be 'outside the series of contingent factors', in a 'necessary substance . . . that we call God'.[2]

The significance of this point can be illustrated by referring back to the form of the systemic model discussed in chapter II.[3] The 'reason' for the fact expressed by the conclusion of a syllogism, in the form 'g is m', appears as the middle term M which enters into the two premises, 'g is M', 'M is m'. What Leibniz insists on is that *every* premise, such as M is m, requires a middle term, linking subject and predicate. But to stop this from going on *ad infinitum*, and because otherwise 'one makes no progress',[4] we need recourse to something lying 'outside the sequence' of the infinite series of 'contingent factors'.[5] Moreover, for Leibniz, since a reference to an 'outside' ground is necessary in the case of *this* particular premise, it will be necessary always, and for every proposition. Thus unlike the empiricist, who asserts that a basic proposition reporting an individual particular fact, such as 'this piece of sugar tastes sweet', *has* no mediator, Leibniz will have to maintain that such a proposition, too, has a ground.[6] (Even though, since the 'ground' lies 'outside', there is no genuine 'middle term'; this is only an analogue.)[7]

So the optimistic assessment by Leibniz of Locke's sceptical worries concerning 'propositions of fact' (together with the other group: 'propositions of reason') is misleading. Locke's uneasiness *vis-à-vis*

[1] Cf. L. 412; below, p. 452.   [2] *Mon.*, 37–38; L. 1050.
[3] Above, pp. 55–7.   [4] *Mon.* 37; L. 1050.
[5] Ibid. Cf. Kant's criticism of this argument, K. 511 (in his discussion of the 'cosmological argument'), and by implication also in his discussion of the Fourth Antinomy, K. 417–21.
[6] A feature which will reappear in the Kantian version of the transcendental ground of *every* proposition, including contingent ones. The change consists simply in the move from 'transcendent' to 'transcendental'. (Cf. K. 159.)
[7] Cf. below, p. 450n.1.

the class of factual propositions was precisely that he had qualms about the *existence* of what for him constituted the only basis for *knowledge*: 'necessary connections'. In so far as his uneasiness is realised to be incurable in principle, he (and if not he, then certainly the Humeans after him) will have to say that such propositions state just facts, *simpliciter*. But for Leibniz, there are no 'just or mere facts', mere coexistences. He will be prepared to speak of connections even where these are in principle unknowable (although he hesitates and becomes ambiguous about the case of 'confused perceptions'); but this does not make the whole notion for him a meaningless one.

So his conception of truth as such is not anchored in what can be *known*, nor in what just happens to coexist (the actual qualities), but rather involves the doctrine that what coexists does so in virtue of an underlying ground, which *makes* the truth something in itself, something timeless, or time-indifferent. It is for this reason that Leibniz's '*mere* physical certainty' is not even the beginning of a sceptical admission. For behind physical certainty (or its possibility) there lies the assurance of what Leibniz calls 'physical necessity', i.e. the predetermined direction of evolution of the life-history of every individual substance, in accordance with the principle of perfection.[1]

Such a conception bears some analogy with those presuppositions of inductive logicians, who require the formulation of some abstract propositions (e.g. Principle of Limited Variety, etc.) which would (if true of the world) guarantee induction to be at least possible in principle, even though not certifying the actual truth of any given inductive generalisation.[2] Let us then turn to a study of the form in which the Leibnizian version of the 'bond' or 'nexus' linking subject and predicate, or substance and attribute, presents itself in his philosophy. For it is here, as suggested in chapter II, that the most consistent, as well as relentless, manifestation of this search for a bond makes its appearance.

(b) *The bond of predication*

Leibniz does not follow the 'occasionalist' model of connection according to which everything that happens is directly caused by God, with its consequent denial of transeunt activity anywhere short of God's. For, as we have seen, 'activity' is very much part of Leibniz's

---

[1] Cf. DM, 13, pp. 18–21; *Radical Origination*, L. 793.
[2] The difference is that inductive logicians limit their case to that of law-like generalisations; Leibniz extends it to that of non-general propositions as well.

conception of a substance. Yet, this activity is limited to the field of all that can happen to any individual substance; the whole thing, as we have noted, being governed by the push *a tergo*, and the determination *a termino*. 'Connection' will therefore have to be located here. It is as though Leibniz held that if we say 'this is such and such' (the standard form of the subject–predicate proposition), the ground or reason for this proposition, of the fact which it states, must be sought for within the domain of the subject S, and nowhere else; or rather, within the field of that which links S with P.

This mode of looking at the matter is common; we have found all our other philosophers constantly recurring to the theme of the questionable 'medium' or 'middle term' which links predicates, or qualities or ideas; and in quite a similar way we find Leibniz searching for the 'middle term to connect something with itself', or for 'the nexus of terms or the inclusion of the predicate in the subject'.[1]

Russell's contention was, as mentioned, that Leibniz's whole philosophy has its source here, and can be derived from it. I have suggested that this is an overstatement, and that a vast number of different considerations (studies and the models they yield) contributed to Leibniz's position concerning the relation of subject and predicate.[2] What is true is that his doctrine respecting the latter once more provided a vastly influential model in terms of which to view the relations between things or their attributes. Moreover, the study of this problem was for Leibniz of particular importance, for this reason: we have seen that his Cartesianism exerts a considerable influence on his view of the reduction of opaque to transparent predicates and principles. The Locke-Hume distinction between propositions stating the relations between ideas, and those stating matters of fact (in Leibniz's terminology: between necessary and contingent truths) is not a possible one for him on their lines, since in the end he holds *all*

---

[1] L. 348, 407. The nexus, however, will be so intimate, the connection so tight, that no genuine 'middle term', placed between S and P, will occur; what determines P will be placed within the interior of the subject itself!

[2] The search for a tenable view about 'connections' was always with him from the start, even before he formulated it in terms of a new principle, about 1679. Four years earlier he writes to Foucher: 'All our experiences assure us . . . , first, that there is a connection among our appearances which provides the means to predict future appearances successfully; and, second, that this connection must have a constant cause' (L. 239). The 'connection' sought here would evidently provide a kind of 'inductive ground'. And the two dominant scientific analogues, dynamics and optics, are obviously just as important factors in shaping Leibniz's general metaphysical position as is its logical crystallisation through the *P.i.S.* doctrine.

predicates to be interrelated.[1] Here, his views on the S–P relationship opened to him (as he saw it) a new way of classifying these two types of propositions. We can sense his excitement at the discovery, when he writes of it:

Here lies hidden a wonderful secret which contains the nature of contingency or the essential distinction between necessary and contingent truths.[2]

The discussion of his new doctrine occurs in many places; it figures prominently in *On the General Characteristic*,[3] *On Freedom*, *First Truths*, *Discourse on Metaphysics*, and in the correspondence with Arnauld which resulted from the latter, and *On Nature Itself*. The original model (and let me stress once more that it is *only* a model) is the 'logically necessary proposition', the primary and basic example of which is 'identical propositions', in the form, A is A. Others must be reducible to this form by means of 'an analysis of their terms'. Thus, given a proposition 'A is B', then if A can be shown by means of a 'definition or analysis' to be equivalent to 'A&B', we shall have reduced it to 'A&B is B'; and such a proposition will be logically necessary (a 'truth of reasoning'), since 'the opposite involves a contradiction'.[4]

Now Leibniz at once proceeds to his astonishingly bold generalisation of such a model. For it is this which supplies him with the general maxim:

No matter how often a predicate is truly affirmed of a subject, there must be some real connection between subject and predicate, such that in every proposition whatever, such as A is B (or B is truly predicated of A), it is true that B is contained in A, or its concept is in some way contained in the concept of A itself.[5]

Moreover, 'every propositions whatsoever' means what it says; for as Leibniz points out explicitly in *First Truths*, that the 'predicate or consequent always inheres in the subject or antecedent'

is true, moreover, in every affirmative truth, universal or singular, necessary or contingent, whether its terms are intrinsic or extrinsic denominations.[6]

---

[1] Cf. above, p. 406n.2.　　[2] L. 412.
[3] I am using Loemker's titles.　　[4] L. 348–9.
[5] Ibid.; cf. L. 412; DM. 8, pp. 12–13; also to Arnauld: 'There must always be some foundation for the connection between the terms of a proposition, and this must be found in their concepts' (L. 517/OC. 132). This is the formulation of the *'praedicatum inest subjecto'* (*P.i.S.*) principle.
[6] L. 412. For the equivalent passage at L. 405, cf. above, ch. II, p. 74.

This theory is full of difficulties, discussed at length in many standard commentaries; lack of space forbids extended critical discussion; here, as elsewhere, my main concern is with the general sweep and connection of the various aspects of the teaching of a particular philosopher. Thus it may be objected that the 'is' of predication should not (as here seems the case) be confused with the 'is' of identity. However, the important point lies in Leibniz's emphasis that the subject *contains* the predicate or that 'the predicate or consequent . . . always inheres in the subject or antecedent';[1] the logical paradigm is only a partial model; as often as not, Leibniz has in mind a biological image, that of the 'evolution' of the attributes, of their emerging from the substance, like the oaktree from the acorn.[2]

This is obvious if we remember that our present model is meant to cover the case of 'contingent propositions'. For, as Leibniz is quick to point out, 'though the predicate inheres in the subject', in the latter case we can 'never demonstrate this': the proposition cannot be 'reduced to an equation or an identity'. Nevertheless, Leibniz adheres to his general theory by the doubtful introduction of a supplementary model, taken from the field of the differential calculus, and its notion of the limit. For this shows that our inability to complete an analysis, e.g. in the case of a series, is not fatal, since it may nevertheless express a definite number.[3] So in this sense, we may say that we have 'demonstrations even about infinite series', and we may consider the possibility of an 'infinite analysis' in the case of the subject of contingent propositions in this light. Leibniz thus insists that contingent propositions reduce in *a* sense to the case of the logically necessary type; yet, at the same time, there is a difference which is expressed by saying that only God is capable of completing the analysis by a kind of 'infallible vision'.[4] On the other hand, for finite minds, the subject (containing the relevant collection of predicates) is 'opaque'.

We noted earlier Leibniz's insistence that unless there is 'some connection between cause and effect', . . . 'anything can follow from

---

[1] L. 412. Notice here the extension from 'categorical' to 'hypothetical' propositions, and to the corresponding 'causal' situations. Cf. above, p. 424.

[2] Cf. also to De Volder: 'It is essential to substance that its present state involves its future states and *vice versa*. And there is nowhere else that force is to be found or a basis for the transition to new perceptions' (L. 879).

[3] Cf. the example of the expansion of $\frac{\pi}{4}$ already mentioned, above, pp. 400–1.

[4] L. 407–9.

anything with equal right'.[1] We now see that the constitutive ground of the connection is being placed in the subject itself, where everything is always already given—in Leibniz's words—either 'actually' or 'virtually'.

This is how the matter is expressed in the *Discourse* where, after giving the gist of the doctrine in terms of 'subject and predicate', an equivalent version occurs in terms of substance:

> We can say that the nature of an individual substance or of a complete being is to have a notion so complete that it is sufficient to comprise and to allow the deduction from it of all the predicates of the subject to which this notion is attributed.[2]

Here again Leibniz admits that since in the case of contingency the predicate inheres only 'virtually' in the subject, the deduction of all the predicates cannot be effected by finite minds. Nevertheless, the existence of the predicate has its 'reason', a kind of 'proof *a priori*' in such cases also.[3] It is in this sense that 'nothing happens without a reason' or, in the words of *First Truths*: 'There is nothing without a reason, or no effect without a cause'.[4] That there *is* such a reason is what Leibniz eventually expresses by the concept of God.

The contention that the *P.i.S.* principle applies to individual propositions just as much as to general ones, as Leibniz's contemporaries were not slow to notice, made the connection very tight indeed, for it evidently entailed that everything that ever would and could happen to a substance (in the sense of being predicated of a subject) had to

---

[1] L. 813; above, pp. 424–5. This is just what Hume will say in the *Treatise*; to mend the defect he offered us his 'rules'!

[2] DM. 8, p. 13. The *General Characteristic* also speaks of the 'deduction' of contingent truths from the analysis of the concept (though this may not be within human power). (Cf. L. 349.)

[3] Cf. L. 349. It will be seen that, in *this* terminology, contingent propositions would be 'analytic *a priori*', though relative to us *a posteriori*. Only the notion of the 'synthetic' is here missing, though its equivalent occurs, given by the theory of the 'infinity' of the analysis of the concept here required, visualisable only to God, in whom the unitary combination or synthesis of the predicates is achieved. In the Kantian reshuffling, the latter turns up as a function of the logical framework of human experience, providing the ground for predication, giving in this way a new meaning to the notion of '*a priori* synthesis'. In other respects, such propositions are *a posteriori*, as for Leibniz. But they are never analytic; the predicate in synthetic propositions is not even *virtually* contained in the subject; any 'unity' that it has with the subject is, as just mentioned, purely formal, in virtue of the transcendental framework (the 'transcendental unity of apperception'—though the 'principle' of this unity *is* analytic: cf. K. 155).

[4] L. 413. This formulation links it at once with the context of the theory of forces.

happen *with necessity*! Now there is a peculiar logical difficulty in Leibniz's whole theory here. For it follows from this that we should be able to speak of 'the truth' of propositions concerning what lies in the future. But it is extremely doubtful whether the notion of truth, or indeed, 'truth-or-falsity', can be applied to future-tense statements. Our belief in such a possibility is mostly due to a confusion with the very different case of *inferring* the occurrence of future happenings from those in the present and past. Now Leibniz's notion of '*virtual* inclusion' of a predicate in a subject is a clever move facilitating the transition from the harmless to the difficult case. This connects with the basis of his doctrine which asserts that contingent propositions are *in some sense* both analytic *a priori* and synthetic *a posteriori*.[1]

The issues involved are too complex to be discussed here in any detail. Leibniz's own answer is ingenious, if not unexpected, to a reader who has followed the previous sections of this chapter. He points out that contingent propositions certainly are not necessary in any *logical* ('metaphysical' or 'absolute') sense, since their contradictory is logically possible. On the other hand, the occurrence or realisation of the predicates does proceed in accordance with those considerations of the 'exigency' of the realm of the 'possible' (or 'compossible') towards the 'actual' that we have noted before: which is that it proceeds in accordance with the principles of activity and perfection.[2] So whilst the opposite of what happens is not *logically* impossible, it is *physically* so, in the sense that the result would lead to a world less perfect. Whereas the actual world contains the least imperfection, or the greatest perfection possible.[3] This is of course an immense extrapolation from Leibniz's earlier disquisition on the place of the teleological element in science: there is no need to comment on this further.

What is of interest in Leibniz's position is that it is another and rather intriguing version in the long history of attempts, which our previous chapters have traced, at a definition of 'physical necessity'.

---

[1] Cf. above, p. 453n.3; or to put this criticism in a more superficial form, in which Kant sometimes misleadingly expresses it: Leibniz wants to operate uncritically with the concept of contingent propositions that are *synthetic a priori*.

[2] Cf. above, sect. 2, pp. 399f., 404.

[3] Cf. L. 409; DM. 13, p. 21; note that the concept of the 'best possible world' does not enable us to predict or determine the actual amount of imperfection it will or can contain; this is quite 'untestable'. But then, it is not put forward as an empirical principle. Whilst Leibniz is meticulous in his attention to the empirical bases of science, such considerations necessarily cannot attach to statements concerning 'bases' and 'presuppositions'. They are at best guided by empirical analogies.

For Leibniz, this is done in terms of his doctrine that the 'contrary' of a physically necessary state of affairs would imply, '[not] a contradiction or logical absurdity . . . [but] imperfection or moral absurdity'.[1] Now Leibniz's leading analogue had been the principles of determination which we meet in science as principles of *extremum*. So ultimately, physical necessity is defined by referring us to special cases of basic laws: which is not so different from Wittgenstein's remark in the *Tractatus* that if one *could* formulate a general principle of causation in nature (which one cannot) one would have to express this by saying that there *are* such things as laws of nature.[2] The difference is that the notion of physical necessity in Leibniz stretches from that of general uniformities to embrace also individual occurrences; Leibniz wants to say that, if anything happens, it must be connected with something else on the analogy of law-like or causal connections; the model being 'necessity' of a physical or moral kind, although its ground is really located in that 'necessary basis' which causes all substances to develop or 'extrude' the predicates which they do.

Here once more the comparison with Kant is instructive. Hume's search for connections had stopped short of relations between the terms of causal laws or law-like connections. Leibniz's extension of this search to individual occurrences constitutes an intensification of this effort, but at the same time also its logical 'degeneracy'. ('Whatever happens is determined' has clearly no empirical 'cash-value'.) Kant will attempt to reconcile these two approaches by way of an anaesthetisation (formalisation) of the nature of the 'ground' in virtue of which two events can be said to be 'connected'. This ground, as mentioned in footnote 3, to p. 453, will be his notion of the 'transcendental unity of apperception', and its special forms—here: the *concept* of causation. However, this ground will, as in Leibniz, link the members of *every* type of proposition, contingent or necessary, universal or particular. Actual laws will come separately, imposed, as it were, under the *guidance* (but not *guarantee*) of the causal principle.

5. SOME CONSEQUENCES OF LEIBNIZ'S PRINCIPLES

Turning to our list of 'theses', we find that we have covered the more substantial part of the doctrines which they summarise. It is not

[1] L. 793.
[2] Op. cit., 6.36. Wittgenstein adds: 'But of course this cannot be said: it [only] shows itself'.

possible here to develop all the consequences which (for Leibniz at least) follow from these basic starting points; nor to do justice to all the various facets of his complex philosophical *persona*. I shall conclude by a consideration of only some of the consequences of Leibniz's principles: the 'atomic isolation' of his monads, and the stresses which result from this, manifest in the complex interplay between 'empirical' and 'metaphysical' language; connected therewith, a final appraisal of Leibniz's approach to the relation between 'experience' and 'reason'; and the influence of this on his philosophy of space and time.

(a) *The status of 'experience', and the meanings of 'reason'*

The most important of the consequences is that since the ground of the development of the predicates from a subject lies in the subject itself, it follows that, in this sense (in this 'metaphysical language', as Leibniz puts it in the *Discourse*), 'each substance is like a world apart, independent of any other thing save God'.[1] So, as already explained, there is no interaction as such between any substance and any other. In order to 'save' the language of interaction between phenomena, Leibniz has therefore to introduce a compensating principle, whose ontological level corresponds to the metaphysical position of the 'ground' or activity of any substance. This is the principle of pre-established harmony (Th. 9). The principle provides the metaphysical basis for the restoration of the possibility of normal speech. As Leibniz explains to Arnauld, although

> everything happens to each substance in consequence of the first state which God gave to it in creating it . . . [and there is only] conservation of the substance itself conformably to its preceding state and to the changes which it carries in itself . . . nevertheless we have the right to say that one body pushes another; that is to say, that one body never begins to have a certain tendency excepting when another which touches it loses proportionately, according to the constant laws which we observe in phenomena.[2]

It is fascinating to find Leibniz even here still suggesting empirical models of the internality of action accompanying apparent interaction; when one body repels another, this is really because elastic

[1] DM. 14, p. 23.
[2] OC. 183. This approach applies also to the relation between mind and body which are said merely to 'correspond' to each other; 'metaphysically speaking' there is no action of body on mind, but the 'states' of one 'represent the states of the other' (OC. 181–2).

forces are set up within each body which cause the subsequent motions. (The fact that these internal forces are triggered by the collision is silently omitted!) At any rate, metaphysically the intention is that all internal adjustments occur as though there were interaction; interaction as such is a 'phenomenal matter', one to which the metaphysics of the *P.i.S.* principle is indifferent.

At this point we see the 'constitutive' level being driven 'underground', so to speak, loosing its fragile contact with empirical reality, which the method of analogy had so precariously tried to preserve. One should not be surprised at this result. If things are viewed through the spectacles of a principle which says (in effect) that everything is 'really' internal to substance, then it is impossible to extract anything genuinely external from the result. 'Phenomena', particularly those possessing sensory qualities, will be 'external' to each other and 'unrelated' only in so far as they are 'merely experienced.' But this will be regarded as merely a half-way house; for beneath the 'just experienced' all is still related and internal. The realm of the sensory and *a posteriori* differs from that of intelligibility only through being more 'confused', and being one whose relations (or 'reason') are not known. But unlike Hume, Leibniz does not regard this 'ignorance' as fatal or as impugning the semantic significance of his metaphysical constructions. Hume's approach, so we have seen, consisted in paying special attention to the significance of the 'merely empirical' or 'passive' realm of sensory qualities; and his logical manoeuvres had the result of removing 'connections' from the constitutive realm altogether, converting them into a function of experience, at least as regards causal law-like uniformities. But the point of my remarks is this: such approaches do not constitute 'refutations' of a vision like that of Leibniz. It is merely a question of how to interpret the relevance of the *a posteriori* factor. And this, as we shall see in the next chapter, becomes at once clear in the basic reaction of Kant to Leibniz, which embodies a new respect (under the influence of Hume) for the *a posteriori*.

Above all, we should desist from any indictment of Leibniz for not being sufficiently 'empirical'. Categorical divisions like those of 'rationalist' and 'empiricist' are quite misleading when we draw distinctions between philosophers on such lines. Leibniz, Hume and Kant are all *equally* aware of the existence of the empirical factor. It is merely a question of how they interpret its place and relevance, when measured against non-empirical factors, and its function in the construction of the philosophical edifice. Kant's starting-point here will consist simply in *asserting* that the *a posteriori* is an independent, separate and irreducible realm: its 'origin' having to do with 'the mode in which we are affected by objects'.[1] But clearly, such a position can by no manner of means be construed as anything but an aim to provide autonomously new ontological placings; it is certainly not deducible as such from anything empirical (psychological or physical); nor can it come from any considerations of formal logic. It constitutes a creative construction of a new philosophical framework by means of which the different elements of everything given are fixed and apportioned.

All this will enable us now to appreciate more clearly Leibniz's position concerning the relation of 'experience' to 'reason'.[2] Time and again we have found Leibniz seeking support for his views by a locution, of which a typical instance occurs in *Principles of Nature and of Grace* (sect. 6), in the following way. We have seen that the monads are simple, and not subject to external influences. Leibniz concludes from this that they are indestructible, as well as non-created (not unlike Locke's 'simple ideas');[3] and as usual, he refers us to an illustrative, if not probative, empirical situation, pointing out that recent biological research has shown that the unborn animal is already 'preformed' in the 'seed'.[4] And 'just as [with] animals', he writes, so it is with 'souls' in general: 'generation' is not really birth, and to perish does not really imply 'what we call death'. So we see: the empirical analogue of the biological fact supports the metaphysical doctrine of the monads. But more: Leibniz actually introduces this whole analogy by inverting the reasoning:

> The investigations of the moderns have taught us, and reason confirms them, that the living beings . . . come from the transformation of living beings existing prior to them.[5]

To appreciate his intentions here, compare this with a case like the one cited before, from *New Essays*, where Leibniz tells Locke that an observation is a very insecure basis for knowledge, since it is always necessary to employ in our scientific reasonings the 'universal and

[1] K. 65, 83.
[2] Cf. Leibniz's remarks on this to Varignon and Bayle, already quoted above, pp. 411n.1, 415n.5, 440ff.   [3] Cf. above, ch. IV, p. 251.
[4] A reference to the celebrated doctrine of 'preformationism', which was to lead a running fight with the competing idea of 'epigenesis' throughout the 18th century.
[5] L. 1037.

eternal truths . . . originally in our minds'. Now we already noted the peculiarity of the example he offers in this connection, being that of an explanation of 'the phenomena of optics' as 'explained by geometry'.[1] And here, at first sight, it might be said that this interpretation of 'reason', in the sense of a scientifico-theoretical basis, is one with which even Hume would not disagree; on the contrary (as I have argued at length), Hume is aware of the theoretical basis of science; his whole concern is only to disengage this sense of 'reason' from another, which links it with metaphysics (though we have seen that this does not, of course, exclude metaphysics at a different point of his argument).

On the other hand, it is very difficult to avoid sliding onto the metaphysical level, which is precisely what Leibniz is here doing. Indeed, in Leibniz's hands, the optical example is a particularly slippery one, since—as he so often insists (though not at this point when conversing with Locke!)—there is an explanation of optical phenomena (by *extremum* principle) which (as shown) was for him the supreme illustration of the principle of finality in nature. So that optical phenomena in this way gain support from his general metaphysical standpoint concerning the interpretation of 'existence' in terms of 'the best possible order'.

Now just the same situation holds also for the first example, of the indestructibility of substance. Biological theory might well confirm the microscopical observations of Swammerdam, Malpighi and Leeuwenhoek,[2] concerning the part played by micro-organisms in generation. But, at the same time, these theoretical principles are for Leibniz *images* of a more fundamental order. In sum, the reference to 'reason'[3] covers, (a) mathematics and logic; (b) the latter, regarded as 'eternal truths', and hence possessing special powers of validation;[4] (c) scientific theory; (d) the latter, as involving privileged basic principles, like those of conservation and determination; finally, (e) the metaphysical story of the monadology, which throws all this into a systematic formulation.

[1] Cf. above, p. 398. Cf. also L. 1048, where 'mere empiricists' who infer inductively by enumeration of instances that the sun will rise tomorrow are contrasted with the 'astronomer' who 'concludes it by reason'.

[2] Cf. L. 742 for a reference to their work on 'transformationism'.

[3] And remember here, how 'reason' is built by him into the very definition of 'contingent facts'!

[4] Cf. Leibniz's comment on Locke's conception of 'mere relations of ideas': 'necessary truths being anterior to the existence of contingent beings, must be grounded in the existence of a necessary substance' (*N.E.*, iv.11.13, p. 512).

As I said at the start of this chapter, this is a 'way of seeing things'; but it represents perhaps more faithfully the intellectual form of much scientific thinking, although such thinking does not always explicitly affirm its metaphysical presuppositions.[1] Nor should this be misunderstood: metaphysical presuppositions do not (despite appearances to the contrary) state special sorts of facts, which '*exist*' in some super-sensible realm of reality. They are extreme, or (if you like) 'degenerate' assertive generalisations of ways of 'seeing' (in mode (e)) the special facts just classified under headings (a) to (d). Their life-blood is the empirical analogies. But the older metaphysician inverts the order, and sees metaphysical 'truths' as generating the empirical and theoretical order. We, as critical commentators, can redress the balance, and, having become more self-conscious of the mode of philosophical argument, see 'metaphysics' in its proper place.

It is in this light that we should understand Kant's famous 'constraint of reason'[2] (giving us further perspective on Leibniz's use). Kant's 'reason' is deprived of its 'constitutive' powers, and left with a purely 'regulative' force.[3] Such a change became necessary when the last connecting link between reason and experience, viz. analogy, was snapped. (Descartes' optimism regarding the power of clear and distinct perception to generate *a priori* the major principles of dynamics had already cut little ice with Leibniz.) In consequence, the notion of a 'constitutive foundation' had to be channelled into and apportioned over a number of different compartments, dividing into the purely 'metaphysical' (or 'transcendent'), the 'regulative' (as in Kant's treatment of the principle of continuity), the epistemological presupposition (the 'transcendental' framework; for instance the conservation of matter, the law of inertia, and the law of action and reaction, whose essential components are the three analogies of experience applied to specific analyses of the concept of matter), and finally, the purely scientifico-theoretical basis. In the thinkers prior to Kant, all these still diffuse into each other and their particular confusion lies at the heart of both Locke's and Leibniz's doctrines, and continues even to pervade the philosophy of Hume. Their complete distinction is first found in Kant, though that philosopher's forbidding

[1] Maupertuis almost certainly discovered the 'Principle of Least Action' through the suggestiveness and guidance of the teleological type of approach inherited from Leibniz. He makes the relation quite explicit.   [2] K. 481.
[3] The detailed explanation of this term is left for the next chapter. Also, 'constitutive' has for Kant a somewhat different sense. But the difference does not affect the present context.

architectonic has, by and large, prevented his work from being seen in this light. Of course, the making of such distinctions occurs, as I have already suggested, at a price. Kantian philosophy produces, if clear, yet arid minds, partly because its distinctions result in taking the 'drive' out of what is left. Leibniz already could write (concerning conservation of *vis viva*), 'our mind *looks for* the conservation of something absolute' (he means: for a basic candidate for conservation);[1] i.e. its basis is an intellectual *demand*. Now this clearly foreshadows the form of Kant's 'regulative approach' at its source. But unless one believes it to be a reality (the 'constitutive' leap which Kant proscribes) one is likely to draw back and lose interest, although this is admittedly no more than a psychological consideration.

(b) *The status of 'laws of nature'*

The peculiar interpenetration of the physical and metaphysical levels in Leibniz's thinking affects also his views on the scientists' laws of nature, the most important of which we encountered as the law of conservation of *vis viva*. Now, however basic this law may be, we shall not be surprised to find, in the light of Leibniz's dual approach to 'force' (as 'primitive', and as 'derivative', where *vis viva* belongs to the merely 'derivative' order), that in the end he does not accord the law of *vis viva* any place in his ultimate, metaphysical, order of things. And in this refusal he again draws (though only like a mirror-image) close to the teaching of some of the more empiricist philosophers.

The whole account is displayed most clearly in the *Discourse*, where there is an explicit recognition that the laws of nature, including the one of *vis viva*, have relevance only for the phenomenal level, being at best only an *analogy* for 'the general order of things';[2] an order which manifests itself *properly* speaking through the predicates involved in the notion of the individual substance; predicates that are realised in accordance with God's 'general will', which permits only the best of all possible worlds. Strictly speaking, it should be noted, this world is a unique individual sequence; the principle of the identity of indiscernibles is enough to forbid the recurrence of absolutely identical states of affairs, on the lines of the instances of

[1] *N.E.*, Appendix IV, p. 657; my italics.
[2] Cf. DM. 7 and 16, pp. 11–12, 27–28, dealing with the question of 'miracles'. Miracles clearly constitute a *prima facie* difficulty for Leibniz (who does not wish to dismiss the concept as such) since everything that happens and that physically speaking can happen, takes place as the result of God's will acting to perfection.

our natural laws; and although Leibniz refers to the 'general order' sometimes as governed by the 'most general laws of God'[1] (and not allowing of any exceptions), the notion of lawfulness is here itself only a metaphysical or analogical way of expressing the fact that the members of the sequence are linked by natural necessity.[2]

By contrast, the laws of nature are labelled by Leibniz as merely 'subordinate maxims'. It is true that God has 'established' these also. But, says Leibniz, exceptions to them are possible. So they are 'established' only in a manner of speaking; not in the 'iron' way in which a Newtonian might understand this. How is this possibility of exceptions to be conceived? Leibniz's answers are given as usual at different levels.

(1) Epistemically speaking, he wants to say that these laws are generalisations resulting only from our own limited point of view; they do not include the full and rich individuality of the 'general order'. They only 'abbreviate our thinking'.[3] He does not expand on this, but if pressed he would presumably employ his notion of 'abstraction', as for instance in his disquisitions on 'matter'.[4] He gives, however, a constitutive argument which supplies a more explicit answer.

(2) This connects with the distinction between 'distinct' and 'confused' predicates. First of all, we should remember that although substances, phenomenally considered, 'interact', when translated into Leibniz's 'metaphysical language', this amounts to substance being 'limited' in principle by expressing its various predicates to different degrees of 'distinctness' or 'confusedness': more perfectly, when it 'acts on another substance' (phenomenal language), less perfectly, when it is acted upon.[5] Moreover, this general 'limitation' results in our perceiving 'distinctly' what goes on near or in our own personal field, e.g. our bodies; i.e. the individual substance, although bearing within itself all the 'traces' of everything that goes on in the rest of the universe, for the most part is not conscious of this; almost all its knowledge is 'confused' (opaque). This has the result that we cannot predict all that is going to happen, and our very laws are framed only within this personal and narrow field. So everything that is *called*

[1] Cf. the notion of 'general reason' on p. 420 above.
[2] A useful example of this is certain modern cosmological laws, which purportedly govern only the whole universe, which itself is a unique individual.
[3] For this locution, cf. above, p. 412, in a letter to Arnauld.
[4] For matter as a mere 'abstraction', cf. above, p. 411.
[5] For 'interaction', cf. above, pp. 401–2, 419, 456–7.

'*natural*' is defined by these subordinate maxims (laws of nature) only; nature can be understood by us only through these laws.

This means that our 'maxims' or 'laws' can never get a perfect grip on reality; they live only on the relative surface of things.[1] This fact Leibniz also expresses by saying that we may liken the 'nature of things' (defined by the maxims) to a 'habit of God'. To call it a habit might imply that God can dispense himself from it, and thus suspend the maxims, previously established by him in this sense also—a locution which Leibniz actually employs. This is, however, to be understood as mere metaphor. It is we whose limited view of nature pictures a God acting on habit merely.

We can understand this best if we remember the contrast between 'laws' regarded as mere rules describing observed uniformities, which may—for all we know, and certainly for all we can 'see'—be mere accidental uniformities on a very large scale; and laws regarded as expressing necessary connections within the species covered by a given rule. For this difference is also rather like that between a man acting on mere habit, and a man reflecting on his actions, and acting for a reason, on every given occasion. In the former case, he just does it, no matter how many times; in the latter case, there is a 'reason', which, in so far as it is operative, *determines* the action.[2] But this picture indicates at once the view Leibniz must be holding of laws, *quâ* subordinate maxims, viz. that they state no more than mere uniformities, which grip at best only the selected surface of phenomena and not the underlying 'order'.

On the other hand, it is significant that he nevertheless continues to operate with the notion of 'a general order'; nor is this for him an empty matter since it gives a sense to the notion of miracle compatible with a view according to which all that can exist—physically speaking—does exist in accordance with God's unique decrees. Miracles on this view are simply occurrences which, whilst they involve a suspension of 'subordinate maxims', in the sense of transcending the '*natural* order', and thus are 'supernatural', are nevertheless always in conformity with the universal law and the *general* order.[3] We may add that this move is not as surprising and arbitrary

---

[1] Or as we put it above, p. 410: It is only the *phenomena bene fundata* that are subject to the laws of nature.

[2] We will overlook the troublesome objection that the force of the 'reason' could be suspended just as well, for we know already that, for Leibniz, the ultimate reason is exempt from this possibility!   [3] DM. 16, p. 27.

as it seems, since the very notion of existence we have found to be defined in terms of God's perfect will.

It is important to see that the 'limitations' involved in the framing of our subordinate maxims are not just accidental, due to the insufficient power, empirically speaking, of our intellectual abilities. They are due to the ineluctable fact that most of our perceptions are beset by the feature of 'opaqueness'. It follows that if this notion is made absolute, i.e. something basic, as in Kant, and not just—as in Leibniz —a matter of degree, the 'limitation' itself becomes something extremely general, and indeed extremely queer. For we are then totally cut off from the 'general order', and of course the whole conception (clearly Kant's realm of the 'things-in-themselves') becomes altogether inoperative. And at the phenomenal level, the only contrast remaining is that of the laws as uniformities of experience, and the individual happenings, which occur as instances of these laws.

Nevertheless, the shadow of the general order remains. For it is difficult for those who do not wish, in the fashion of Boyle, to operate with the Newtonian notion of laws imposed upon nature absolutely, to express law-likeness adequately without losing the important sense of the notion of natural law. Thus those who, like Kant, retain the idea that laws belong only to the phenomenal order (their law-likeness being in some sense only a deep convention, a necessary analogy, not of a 'general order', but of human experience), still retain the contrast with the realm of things-in-themselves, however much the latter may be totally cut off (at least epistemically) from the realm of phenomena. On the other hand, without such a notion, the concept of 'law as convention' becomes somewhat shadowy, and it is not difficult to understand why most philosophers should have protested spontaneously against the insufficiency of this metaphor (e.g. as advanced by Poincaré), without however putting much else in its place.

(c) *Space, Time, and the 'Monadology'*

Not unexpectedly, metaphysical preconceptions react upon a philosopher's viewpoints and analyses, particularly of complicated notions like those of space and time, just as in the last section we noted this influence on Leibniz's conception of force. If space and time, being inconsistent with the mode of being of monads, have to be 'driven out of' the realm of the basic inventory of things, we shall expect corresponding attempts also to give an account of these concepts compatible with such a viewpoint. Just such a result we get in Leibniz's relational account of the spatio-temporal framework.

As Kant later pointed out, nothing external, no 'dynamical relations', indeed no 'outer relations' (space and time) of any kind, can have a place in a scheme like that of Leibniz which assigns to substances only an 'internal nature', modelled on the analogy of thinking ('perception').[1] We shall therefore expect Leibniz to relegate space and time to a merely 'phenomenal' order of things—and this is indeed the case.[2]

The problem of space and time is argued most fully in the Clarke–Leibniz correspondence, although Leibniz does not there state his most basic objections in detail, being intent (as is usual with him) on selecting those aspects of the problem, and using the language, which fit in best with that of his correspondent. Basically he cannot find a means of accommodating the concept of 'outward', since all that metaphysically speaking happens is internal to a substance. That is why he has to translate everything spatial (like interaction) into the language of comparative degrees of perfection and imperfection of the states of substances,[3] states which ultimately can be defined only as either 'thoughts' or 'perceptions'.[4] Monads cannot thus be said to dwell in space, or have any spatial relations.[5] Any such relations can only be a kind of 'phenomenal emergence' out of those qualities of substances relating to 'force'; but this aspect Leibniz does not consider except to point out that without the concept of (primitive) force, since all bodies are only in relative motion, we would not be able to say what is 'really' moved[6]—clearly a phenomenal matter. To Des Bosses he writes that 'space is the order of coexisting phenomena, as time is the order of successive phenomena', where 'coexistence' and 'succession' are presumably qualitatively or independently definable —how, Leibniz does not explain. In *New Essays* he opts for Locke's

---

[1] Cf. K. 284–6; cf. ch. VIII, sect. 7(a).
[2] L. 417; 528/OC. 222.
[3] Cf. Mon. 49; L. 1052; DM. 14–15, pp. 24–6; cf. above, p. 401.
[4] Cf. DM. 14, p. 14. But, as mentioned at the start, 'perception' is for Leibniz a very broad term, including 'representation' of topological relationships. The difficulty on such a view is to see *what* is being represented; it is as though there was only structure, but Leibniz was silent on the material components. To say anything he would need 'confused perceptions'; but that is to allow into the philosophical edifice something which ought to be 'reducible'!
[5] Cf. L. 983.
[6] Cf. DM., 18, p. 32. However, '*derivative* forces' and their laws cannot 'determine absolute motion mathematically', since everything is here relative (L. 729, 750).

alternative of space being 'a relation, an order, not only between existences but also between possibilities'.[1]

The *Correspondence* defines place, and through it space, essentially in terms of the notion of 'relation of coexistence' which one thing may have with a number of others, the latter regarded as 'fixed', where a 'fixed existent' is defined as one 'where there has been no cause of any change of the order of their coexistence with others'.[2] If the relations of two things, A and B, 'agree' in respect of such a group of 'fixed existents', then they are said to be in the 'same place'; and Leibniz adds slyly that this is not so much to explain the notion of 'place', as of 'same place'.[3] And 'space is that which results from places taken together'.[4] Leibniz then points out that all this is only an abstraction, since no two individuals, such as A and B, can have identical 'affections'.

> But the mind ... looks for an identity ... , and conceives it as being extrinsic to the subject; and this is what we here call place and space. But this can only be an ideal thing, containing a certain order, wherein the mind conceives the application of relations.[5]

Moreover, Leibniz notes that the definition of space and time as relations involves the postulation of something which is 'out of the subjects' related, falling somewhere between the things related; which again makes it 'a mere ideal thing'.[6]

Understandably, then, Leibniz does not manage to say anything detailed about the *kind* of relations spatial relations constitute, what *kind* of quantity[7] they measure, although at the start of sect. 47 he speaks of the order of coexistence of things as being an order of their 'distance'. Now it will be remembered what few qualms Locke experienced when telling his reader how we 'come by this idea of distance'. Leibniz's comment on this in *New Essays* is of interest. For he there makes an operationalist move, insisting that we should, when referring to distance between two fixed things, speak 'more distinctly', for instance in terms of 'the length of the shortest possible line that can be drawn from one to the other'. And he adds the significant reminder that the length of the shortest line depends upon the kind of space or 'figure' comprised between these two things, e.g. whether it is drawn on a spherical surface, or a plane; concluding that

---

[1] *N.E.*, ii.13.17, p. 153; cf. above, ch. IV, p. 259.
[2] Leibniz's 5th letter, sect. 47, L. 1146.
[3] L. 1148.　　[4] L. 1146.　　[5] L. 1147.　　[6] L. 1147–8.　　[7] Cf. sect. 54, L. 1151.

the 'capacity' (space) of the interval is 'constituted by all the shortest lines which may be drawn between the points of each'.[1]

Now whilst this does not constitute a reductive definition of space (how does one draw a line?) we can nevertheless guess at the kind of model Leibniz had in mind, and which would serve as an analogue for the metaphysical reduction of spatial relation to something non-spatial, purely qualitative. On the other hand, such a move involves an uncomfortably close association between physics and geometry. The 'empirical approach' which results constitutes an astonishing anticipation of later developments. Kant was to object to it on just this ground,[2] and his view of space and time as presuppositional forms is in part motivated by his attempt to salvage their non-empirical ('*a priori*') nature. Leibniz, of course, was not very conscious of these empiricist consequences, and not unexpectedly argued for the necessitarian or pure or *a priori* aspect of spatio-temporal relations by his usual device of asserting that their 'truth and reality' (i.e. their governing axioms), 'like all eternal truths, is grounded in God'.[3] His final judgment on this uncomfortable definition of space and time is that

> it is sufficient to consider these relations and the rules of their changes, without needing to fancy any absolute reality out of the things whose situation we consider.[4]

As with so many philosophers, the discomfort arises from the fact that space and time are neither substances nor qualities, and that if they are regarded as relations, they seem to lack some of the 'qualitative presence' which they should possess. It is of interest to reflect that Kant managed to tread his way between the Newtonian extreme of space as an empty vacuum, and Leibniz's theory of relational space, only by the creation of a specific technical device, that of a transcendental (i.e. presuppositional) 'form of intuition'. And obviously this is not an *answer* to these other theories, but a different way of reshuffling the philosophical cards.

What then, finally, can we say of the 'monads', the substantial centres of activity and passivity, the latter a kind of reaction to action, which are 'realised' to 'different degrees of perfection'? As we have seen, they have to be 'simple'; and they are the centre of a variety of 'qualities' or 'affections', the principle of change from one to the other being 'internal'. How can we describe such a 'plurality in

[1] *N.E.*, ii.13.3, p. 149.   [2] Cf. K. 81.   [3] *N.E.*, ii.13.17, p. 153.   [4] L. 1146.

unity'? One analogy Leibniz finds in geometry, where a multitude of angles are formed by lines which intersect at the centre.[1] The other, and more dominant one, is the self and its thoughts: 'perception' being called that state which 'represents a plurality in unity'.[2] Perception would obviously be a useful analogue, because it implies representation 'within', of what is 'external' to the monad, yet without physical contact.[3] Leibniz also defines 'perception' as a

> correspondence of internal and external, or representation of the external in the internal of the composite in the simple, of multiplicity in unity.[4]

Two important provisos must here be remembered. First, as already explained, perception has a much more generalised connotation for Leibniz than the definitions just given would suggest, since it includes the case of 'unconscious representation', something like the existence of unconscious traces.[5] Secondly, Leibniz has no 'problem of perception', for metaphysically there *is* no 'external world'. The 'correspondences' of the perception are to one another, each of them locked in an internal world. Leibniz puts this cryptically, and misleadingly, by saying that each monad represents 'the world' from its own point of view or perspective (similar to the way in which for Russell's philosophy an 'object' just *is* a series of perspectival sensory visions from the observer's point of view). And the correspondences are, of course, mutually adjusted in accordance with the principle of the pre-established harmony.

So after all, not unlike Hume's world, each self is locked within its own perceptions; only, they are not 'impressions of sensation'; not *ultimately* 'sensations', for reasons we have stated already; and not 'impressed', since they are 'internal', almost 'innate'.[6] But, by contrast with Hume, there is a greater emphasis on structure, and above all, on *activity*. And finally, the greatest difference lies in this: for Hume, 'reality' is situated in the phenomena, the 'self' being dissolved into them; for Leibniz, phenomena are only '*bene fundata*', only a shadowy 'extension' of the selves or monads, which are the ultimate reality. It

---

[1] L. 1034.   [2] *Mon.* 14, L. 1045.

[3] Cf. the definition of 'perception' in the Oxford English Dictionary, 'being affected by an object without contact'.

[4] W. 505. For an alternative explication, cf. above, p. 396.n.4.

[5] Conscious perception he designates by the label 'apperception', under which form it will become a central concept in Kant's *Critique*.

[6] Although in Hume the psychological association of an external 'stamp' which, as a cause, impresses the sensation, had been rejected also. Nevertheless, Hume had retained the notion of such 'unknown causes'. Cf. above, ch. VI, p. 351.

was for Kant to try to reconcile these two extreme doctrines. There we shall meet again the 'unity of the self', but it will be purely formal, transcendental, presuppositional. We shall again meet the 'phenomena', but they will be suspended on the transcendental framework, although with the '*a posteriori*' element restored to its rightful place of independence. The 'monads', too, will re-appear, but their utter separation from the 'phenomena', which alone can be *known*, will be emphasised by their characterisation as unknowable 'things-in-themselves', capable only of being 'thought' as an 'intelligible' order. In consequence, the place and status of the 'nexus' in the doctrine of the *P.i.S.* will have to be redefined, appearing as that very 'unity' of the self which effects the relation of the phenomenal manifold, but in a purely formal or *a priori* way.

VIII

KANT: THE SHIFT FROM METAPHYSICS TO TRANSCENDENTAL LOGIC AND TO METHODOLOGY

I. INFLUENCES FROM LEIBNIZ AND HUME.
FOUNDATIONS FOR THEORETICAL SCIENCE

The Kantian *opus* is so vast, and the complexities and ramifications of its interests so enormous, that in this last chapter I shall have to be even more selective than hitherto. I shall therefore concentrate on those aspects of Kant's system which, through the shift of attitude that they display towards the problems of our earlier philosophers, will also contribute toward a deeper understanding of their thought. This will simply be an expansion of the method followed already throughout, where I have frequently emphasised possibilities of meaning that emerge in the light of such shifts in Kant's reinterpretation of the relevant epistemological concepts. As usual, I shall consider matters from the viewpoint of the student interested in the philosophy of science—in Kant's case a very natural starting-point, by virtue of the impact of his scientific preoccupations on his general epistemological or metaphysical beliefs. Naturally enough, such an emphasis results in a certain lopsidedness of presentation, for not only was Kant's thinking subject to many non-scientific influences in the fields of aesthetics, history, morals, religion; the ultimate *intentions* embedded in the eventual 'critical' standpoint were deeply influenced by his revolutionary ideas in the field of the foundations of ethics and the nature of religious thought. In what follows, by including references also to Kant's earlier writings, we shall however be able to give at least some expression of Kant's breadth of view, by giving indications of the close connection in his thinking between science and religion.

470

Despite these cautions, it remains profoundly true that science provided the basic initial stimulus for the direction of Kant's thought, as may be noted from the first ten or so publications from his hand that appeared between 1747 and 1762, which were mostly concerned with scientific matters; and Kant's remaining years (if we take into account the *opus posthumum*) were still greatly preoccupied with this field. This reference to 'science', however, should not be misunderstood, for it would be misleading to think of Kant as a creative scientist. It is more true to say that he is the philosopher of science *par excellence*, whether he be reflecting on the question of the correct definition of 'force' (in mechanics) or on the evolution of the universe in accordance with Newtonian principles, the causes of earthquakes, meteorological questions, the theory of mechanical impact or the problem of absolute space in the systems of Newton and Leibniz. Moreover, we must remember that 'philosophy of science' does not operate in a limbo. Many of the general questions and assumptions of Kant's epistemological work pervade those earlier scientific writings, if not indeed already setting the tone for later definitions and meanings. If Kant's thought developed, and changed radically, one is apt to exaggerate the extent of these changes. More often than not traditional attitudes are merely integrated within new vistas; and the latter may even be subtly modified to accord with the earlier and older framework.

With all this, however, Kant is also heir to those basic concepts and problems whose development we have traced. A question like that of the agreement between mathematics and physics naturally gained new impetus and significance in Kant's eyes, with its seeming realisation in Newton's *Principia*. Or take the Leibnizian influence: the problems of the relation between the Leibnizian physics of forces and the metaphysics of the monadology; Leibniz's two-pronged approach to scientific knowledge *via* the paths of efficient and final causation; his discovery of the distinction between necessary and contingent propositions, as well as the general metaphysical standpoint supporting that distinction; indeed the basic puzzle of the nature of the subject–predicate tie; all this makes the deepest impact on Kant, during the progress of his early intellectual development.[1]

---

[1] I say 'during the progress', for Kant did not initially come under the *direct* influence of the basic sources of these problems in the great classical writings that are the basis for this book, but received his early instruction at one or two removes. Thus the thought of Leibniz, apart from Euler, was mediated *via* the teaching of

Subsequently, the epistemological problems bestowed by Locke come to the fore, culminating in Kant's reading of Hume's *Enquiry*. Let us briefly trace a possible line of development. Given the influence of Leibniz, one will appreciate the central place held in Kant's thought almost from the start by the question which he asks towards the end of the tract on *Negative Quantities*: 'How can I comprehend [the general notion], that, because something is there should be something else?'[1] We know the general trend of Leibniz's answer to this question, referring us to the conception of the sufficient reason, in virtue of which predicates evolve from their subjects, by some kind of 'analytic' process. Kant, in his *Nova Dilucidatio* (1755), a work on the principles of contradiction and sufficient reason, had followed such an approach in general form;[2] moreover in the year in which he published NQ, he was inclined to formulate a similar 'dogmatic' solution in his *Only Possible Ground for a Demonstration of the Existence of God* (1763), God being postulated as the ultimate ground of 'existence' as such.[3]

Now it is because of the rich preparation from the side of 'rationalist' philosophy that Hume's problem not only fell on such fertile soil in Kant but also that subsequently Kant's answer to Hume received such a peculiar twist; a fact not easily appreciated by those who ignore Kant's 'rationalist' upbringing. In his eyes, Hume essentially had asked a very 'Leibnizian' question, though in a slightly narrower context, concerning the relation of causation. And the form in which Kant couches this question years later still echoes the language of NQ, when he summarises Hume as asking

> with what right [reason] thinks: that anything can be of such a nature, that if it is posited, something else must thereby also be posited necessarily;

something which so far is no more than an unjustified idea involved simply in the *concept* of cause.[4] Now it will be clear from my account of Leibniz and Hume that the thinking of both was already converging on Kant's position in the *Prolegomena*, namely that it is quite

---

Christian Wolff, which contained vital modifications of the former. Only gradually did Kant come into contact with the main works themselves.

[1] NQ. 202; cf. below, pp. 563, 570f. A list of Kant's writings used in this chapter, with the abbreviated titles, will be found in the bibliography at the end of this book.

[2] Cf. N.D. 392; Insel ed., pp. 423–5.

[3] Cf. OG. 77–83; and ND. 395 (Insel ed., pp. 433–5) for a similar argument.

[4] P. 6 (Preface).

impossible to *see* (by a kind of rational intuition) how, because something is, something else must also necessarily be.[1] Hume's work, which 'interrupted' Kant's 'dogmatic slumber',[2] could therefore only trigger off a deepening of this question, leading him to ask 'how therefore the concept of such an *a priori* connection [as such] could be introduced'[3] in the first place?

In Leibniz's 'dogmatic' approach (as has been shown at length) there had been nothing better than mere metaphysical postulation of a ground in general. Hume's answer (in his special case, *and as understood by Kant*) was that the transition from antecedent to consequent was effected as a matter of habit, produced by frequent repetition. Kant rejects this answer, though the general criticism sticks; Hume's naturalistic account seems to him irrelevant; the problem surely is one of principle: how are we to conceive in general of a justification for the introduction of a concept *considered as a ground* for a connection? Yet the echo from Hume's experiential approach intrudes into this very question of principle: the justification—if it is to avoid 'dogmatic' postulation—must be tightened by the insertion of a 'critical constraint'; and this constraint will have to be formulated in terms of something having necessary relation to experience; here, however, not so much to the *actual* experience (considered, say, on the lines of Berkeley's *percipere*), but instead to the *possibility of experience in general*; which means: to that general framework which, according to Kant, is presupposed in, or an essential part of, the formulation of any cognitive judgment whatever; a framework that is part of the presuppositional structure of both objective cognitional form and of the 'objective situation' experienced or cognised.[4]

In this way, the philosophical situation where the two great schools had left it becomes concentrated at one point. And Kant will try to combine essential aspects of both, though in the process he subtly transforms either. This shifting of positions, corresponding to a curious sharpening of problems and of associated concepts, takes

---

[1] It will be remembered that Descartes had claimed this ability in relation to some of the laws of dynamics.

[2] P. 9.   [3] P. 6.

[4] In other words: the presuppositional (what Kant calls 'transcendental') framework is *part* of the experiential or cognitive situation; a result of Kant's analysis of this situation.

'Cognitional form' is a less misleading translation of Kant's 'mögliche Erkenntnis' than the more frequently used 'knowledge'. Also, instead of 'objective situation', Kant more usually has 'object'. Finally, 'objective cognitional form' may be thought of as equivalent to 'formal structure of a public language'.

place throughout Kant's work—though frequently concealed by his habit of retaining traditional modes of expression, which are used to telescope a rich variety of overlapping concepts, unfortunately into a much too narrow received terminology. His generalisation of Hume's problem, by joining causation to substance and interaction, and treating all three concepts together in a common approach, will come as no surprise to those who appreciate Kant's Leibnizian background, with *its* generalisation of the problem of connection between the terms of propositions *to all* propositions, particular as well as general, contingent as well as necessary.[1]

For this reason, Kant's way of meeting Hume's problem is not quite what might have been expected; it is *not* simply that of proving (by means of the 'transcendental method') the existence of necessary connections for the case of putatively causal propositions of science as such. It is true that the general presentation of the subject in the *Prolegomena* (parallelled by the 'Introduction' to the *Critique*) frequently suggests such an account, particularly since Kant here emphasises his claim to have given a 'deduction' of the principle of causation; so that we might perhaps expect him to have supplied a justificatory ground for the framing of particular causal inductions; equivalent to a general justification of the law-likeness of nature. Actually, the situation is far more complex. There are two sides to it.

(1) The problem set for Kant by Leibniz might be described by asking how something like the equivalent of the *praedicatum inest subjecto* principle can be elucidated by the 'critical method'. Now to explicate a relation 'critically', for Kant, means to display it as a required ingredient of the analytic framework of *experience*; a framework which in Kant's idiosyncratic *definition* possesses (as will be shown in more detail later) three components: a sensory or perceptual, a spatio-temporal, and a conceptual component. At the same time this structure is taken to be definitive also of the structure of *what is* (or alone *can be*) 'experienced', viz. the empirical object *quâ* 'appearance' or 'phenomenon'. So Kant's problem becomes the elucidation of the propositional connective (involved in the *P.i.S.* principle) in terms of the transcendental framework of experience; and in particular, of the principal categorial concepts (e.g. substance, causation, interaction) which thus come to be explicated with equal generality, as conditions of the possibility of something being an

[1] The generalisation, it will be remembered, that had been expressed in the *Praedicatum inest subjecto* doctrine. (Cf. above, ch. VII, sect. 4.)

object, or of being objective, part of a public language,[1] i.e. of empirical, or contingent propositions, be they general or particular, contingent or law-like.[2]

Now it is in this way, primarily, that we must interpret Kant's reference in the *Prolegomena* to the 'introduction'[3] of the concept of causality: as shoring up the possibility, as being part of the transcendental framework, of the cognitive objective judgment in general. It follows at once that the '*justificatory power*' of this concept is exhausted for this purpose alone. Its relation to the formulation of causal propositions of science still remains to be made good separately; we shall find that this is largely the task of constructive scientific theorising, of what Kant calls 'reason in its theoretical employment'.[4]

(2) Kant does, however, claim also that this transcendental method supplies the framework for the possibility of a 'pure natural science'.[5] The explanation of this is now less difficult; let us for the moment state it in summary fashion, anticipating the results of this chapter. By supplying foundations for the possibility of 'pure natural science' Kant means (a), as just noted, supplying 'proofs' of certain general principles, e.g. causality, substance, interaction, as presuppositions of a public language, of verifiable observation reports; (b) by suitable 'constructions' of the object of dynamics, viz. 'matter', and its primary predicates, viz. extension and motion, ensuring the application of mathematics (geometry) to material nature; (c) by further *construing* matter as the seat of repulsive and attractive forces, and as exerting motive force, and by the applying to this concept the aforementioned principles of substance, causality and interaction, demonstrating the *possibility* of the primary laws of Newtonian dynamics.[6]

---

[1] The reference to 'public' or 'objective' provides the contrast with, say, Hume's 'impressions', whose frequently smuggled-in identification of 'impression' with 'object' is now disallowed. Thus 'public' should be contrasted with the 'privacy of sensations'; in the Kantian context the implication is not that there is a need for public *discourse*. Kant's 'objectivity' is defined without reference to other selves. It only seeks to elucidate the conditions that must be satisfied before any public discourse can even begin. The term 'cognitive judgment' is used with this meaning.

[2] That this is the logical situation, Kant occasionally registers explicitly. Thus in a crucial passage of the transcendental deduction (para. 19) he says that the 'necessary unity' of the elements of the judgment is postulated as holding 'even if the judgment is itself empirical, and therefore contingent' (K. 159; cf. below, p. 637). This fact is concealed from the reader of the 'Introduction' to the *Critique*, since Kant there talks as though the empirical judgment (which is synthetic *a posteriori*) posed no problems, contained no 'necessary connection'; as though only the synthetic *a priori* judgments required a special defence.

[3] P. 6; cf. above, p. 473.   [4] Cf. below, pp. 500ff.   [5] Cf. K. 56; P. 35, 52ff.

[6] Cf. below, sections 8 and 9, pp. 615ff. Evidently, (b) and (c) apply to a slightly

However, Kant does not limit himself to 'pure' or 'nearly pure' natural science alone; his claims extend also to 'empirical science', i.e. to our concern with those empirical generalisations of science that are based on particular experimental data. And the claims are as follows: (a) The possibility of the formulation of particular causal laws depends on the *a priori* 'injection' of the causal concept into propositions which prior to such injection state no more than constant conjunctions; (b) To make good the claim that science operates with empirical laws of a *necessitarian* kind requires the construction of theoretical systems, involving the addition of theoretical concepts in no way reducible to the purely observational basis of the empirical domain of science. The existence of such systems, moreover, is not 'given' but merely 'demanded' by the searching (or 'regulative') activity of the scientist, of what Kant calls 'theoretical reason'. In sum, just as without conceptual principles (whose domain is the field of what Kant calls 'the understanding') there would be no public language, no 'nature', regarded as the concatenation of things and happenings, nor any 'pure science'; so without the 'regulative' activity of 'theoretical reason' there would be no science in general, in the form of an interconnected system of things and happenings.[1]

This resumé will indicate something of the close relation in Kant's system between his epistemological theories and his construction of a foundation for Newtonian science and, indeed, science in general. Nevertheless, the latter should be kept distinct from the former; there is—as we shall see—considerable looseness of fit between the two parts of the total structure. Let us now turn to a preliminary development of some of the details of the system.

2. SOME CENTRAL CONCEPTS AND LINES OF APPROACH IN THE DEVELOPMENT OF KANT'S THOUGHT

The last few pages have anticipated much later discussion, but this was a necessary part of any outline of the origins, the influences and the development of Kant's thought. And our last point ended where we began: the central place of natural science, and in particular, of Newtonian science—not to mention some of the scientific topics

---

less 'pure' case, since it involves reference to 'matter' and 'motion' (cf. P. 53). But this is still quite distinct from the case of 'empirical science', which involves reference to the result of particular observational matters of fact. (Cf. also below, pp. 672f.)
[1] For this, cf. below, section 4, pp. 484ff.

which we found so closely interwoven with Leibniz's philosophy. Let us look at some of these scientific contexts in Kant's pre-critical writings, to see what light they may throw on Kant's basic assumptions and contentions in the first *Critique* and its ancillary text, the *Metaphysical Foundations of Natural Science*.

The following are some of the main ideas which emerge from such a survey:

(1) The central position of the concepts of law and systematic theory. (Section 4.)

(2) Attempts at a reconciliation of the concept of a law-like and mechanically evolving universe with the competing model of a universe subject to design. (Section 4.)

(3) The distinction between the object as phenomenal substance (matter) which has only 'outer relations', and substance possessing an 'inner nature' (which later becomes: the thing as it is in itself, has only the status of a noumenon). (Sections 5 and 6.)

(4) A corresponding contrast between the 'mathematical' and the 'metaphysical' approach. (Section 6(a).)

(5) The more straightforward distinction between a 'mechanical' and a 'dynamical' approach in physics which emerges confusedly from both (3) and (4) (though fully emancipated therefrom in the final version of M). (Section 6(a).)

(6) The distinction between the province of purely logical relations and that of nonlogical relations, through the model of mathematical (geometrical and arithmetical) magnitude. (Sections 5 and 6(b).)

(7) Corresponding to (6), the concept of 'construction', central both for the characterisation of mathematical reasoning on the one side, and of physical reasoning on the other; a concept which, together with the associated one of 'combination' or 'synthesis', becomes linked more and more with the element of 'process', or of 'methodological procedure'. (Section 6(b).)

(8) Under the influence of (6) and (7), the distinction between what is 'necessary' and what is 'contingent', or again that between what is 'sensed' and what is 'thought' (and the later finer distinctions between the 'synthetic'—'analytic' and *a posteriori*—*a priori*) all of which are affected by the notion of 'construction', linked in a complex way to the concepts of space and time. (Section 6(c).)

(9) The proper appraisal of the concept of the 'universe', regarded as a 'whole', consisting of 'parts' that are themselves independent substances, bodies, material particles; or (by abstracting from body)

the same universe regarded as a purely spatio-temporal whole. (Section 7(c).)

(10) The concepts of space and time: viewed, first, as functionally derivative from those laws which govern the structure of the physical universe; then, as the medium for 'construction' of whole from part; then, as something which is 'given' in addition to material body, non-reducible to the latter; and finally regarded as an independent 'form' (or 'manner' or 'mode') through which the conscious and constructive processes involved in Kant's definition of 'experience' and of 'phenomena' are to be understood (or 'made possible'). (Section 7.)

Although this inventory of key concepts has been compiled with a knowledge of hindsight of the part they play in the final *Critique* (Sections 8 and 9), most of them are developed, and certainly acquire much of their semantic significance in Kant's pre-critical, and especially the 'scientific', writings (in the sense of 'philosophy of science').[1] Missing there was only the emphasis which Kant, under the influence of Hume's criticism, was to put on the 'experiential' aspect of 'construction'. For this provided the clue, convincing Kant that the concept of law needed building, not only into the notion of empirical science regarded as a body of laws and theories, but, more deeply, into the notion of empirical knowledge in general, regarded as a system of cognitive judgments concerning contingent states of affairs.

Consider, for instance, in preliminary fashion, the case—as it is discussed in the *Critique*—of 'the possibility of our perception' of a set of states or objects coexisting in space.[2] The notion of such a set— according to Kant—presupposes 'synthetic construction'; and since this is a process which takes place against a background of a purely 'subjective consciousness' of perceptual states, it will (in order to display the formal character of objectivity) necessarily require a unifying concept—here: of mutual interaction, the empirical model for which is given in Newtonian physics by the third law of motion

---

[1] Although Kant's transcendental concepts and principles are intended to provide a 'foundation' for scientific concepts and principles, the meanings of the former are usually deeply entrenched in the latter, and often link up with Kant's outlying scientific interests. Moreover, the concrete contexts of many of Kant's later unspoken assumptions are found often only in the pre-critical writings; indeed, these writings are actually seldom as 'uncritical' as they may appear to the reader at first sight.

[2] Cf. K. 237.

(the law of action and reaction).[1] And a similar contention will hold for the case of cognition of a sequence of states, which will be shown to require the concept of 'ground and consequence', in its spatio-temporal version of 'necessary determination in accordance with a universal rule'.[2]

This approach, although developed under the stimulus from Hume, implies a much more sophisticated version of the old 'empiricist' condition that the notion of knowledge of objects must be defined relatively to 'experience'. Nothing is easier than to read this condition with the associations in mind of the philosophies of Berkeley and Hume. We may avoid such interpretations by seeing Kant's thought as developing against the Newton-Leibniz background to which we have alluded. One of the ways to read Kant's contention that the possibility of any capacity for empirical cognition extends only to things as they 'appear', and not as they are 'in themselves' (i.e. as regards their 'inner nature'), is to understand this as a reference to the need for a conception of 'things' that presupposes the process of 'construction' in space and time: something that must not be confused with the logically more primitive Berkeleyan doctrine concerning the requirement that the object should be actually or potentially perceivable. Even at the most Humean stage of the argument Kant is primarily concerned with the conditions that make perceptual talk itself *possible*, and not the *actuality* of perception. Again, where subsequently we come to the 'subjective' aspect of the process of construction, Kant is almost wholly concerned with the logical character of the experiential structure; how it has to be conceived to yield the notion of a public world. Only indirectly, we shall find, are the philosophical notions of the British writers, especially the notion of 'idea', thereby subjected to a subtle transformation; as are their theories concerning the field of the 'other-than-idea', whether it be Locke's external object, Berkeley's God, or Hume's 'unknown causes' from which we derive our impressions.

With this general caution, let us turn to some of the concepts summarised in our list—lack of space forbids an exhaustive treatment of more than a few. In each case we shall follow the development of Kant's thought in its transformation of the pre-critical positions

---

[1] The significance of the notion of 'subjectivity' will be explained later. For a detailed account of the Third Analogy, and of the reasoning that leads up to it, cf. below, section 7(b), pp. 580ff., and section 8(e): iv, pp. 665ff.

[2] For the 'schematised' definition of the concept of cause, cf. K. 124–5. For the 'schematisation' of causality, cf. K. 184–5.

(often inherited from the earlier tradition) under the weight of the 'critical approach'. This will enable us also, as we said, to appreciate these traditional positions and to deepen their own significance in the light of the Kantian criticism.

3. PRELIMINARIES CONCERNING 'FOUNDATIONS'

First some remarks to characterise the direction of Kant's enterprise in a general way. Doubtless, in one of his most important guises, Kant is the philosopher of Newtonian science, but it is important to realise that he embarks on the object of 'laying the foundations' for that science in a very roundabout way,[1] the most exciting aspects of which have very little to do with the foundations of *science* as such, but rather, with empirical knowledge in general. Nevertheless, one may picture the advance from the specific to the more general question in a quasi-chronological order, though even at the start Kant undoubtedly approached the whole subject in the full consciousness of the general philosophical implications of the tradition bequeathed by his predecessors.

The sequence of questions (set in motion by Kant's preoccupation with problems raised in his *Universal Natural History* of 1755) is as follows. The basic framework of Newtonian science presents us with a number of material bodies whose states are subject to certain law-like changes and which constitute a system of substantial entities coexisting in space, the 'systematic' nature of this edifice manifesting itself as a network of interlocking laws. Here arises the first question: What justifies us in representing the concatenation of material bodies as a system? What is the basis of such a system of laws? What justifies the very possibility of *a* system? This leads to the second question: What is the basis of our conception of a set of 'axioms of motion' that govern the states of the bodies of such a system? (Involved in this is also the problem of the applicability of mathematics to the world.) This question leads to a third, which involves the problem of the possibility of natural science in general, since it makes a number of basic assumptions concerning material bodies, subject to certain sequential changes of state and coexisting in space.

Now whilst the first question raises problems about the notion of

---

[1] Besides, the attempt to delineate the mechanistic approach of Newtonian science leads Kant eventually also to a characterisation of the competing teleological approach of the biological sciences of his time.

the systematic element in nature, of what Kant calls 'the order of nature', the last concerns 'nature' itself, regarded as a mere concatenated plurality of bodies in space. Hence, the question, how is natural science possible, leads to the more general enquiry into the possibility of framing cognitive judgments concerning bodies and their states as such. But clearly such a question completely transcends the interests of Newtonian science, and indeed of natural science in general, and concerns the whole body of human knowledge. Moreover, as has already been hinted, and will be shown in more detail later, Kant's treatment of our third question, which is dealt with in the sections on the transcendental aesthetic and analytic of the *Critique*,[1] as such fits only loosely into his concerns as presented by the two initial questions, the solution of the first of which he tries to settle in certain sections of the Transcendental Dialectic (and also in J), the second being for the most part the concern of his *Metaphysical Foundations*. To portray Kant as laying the foundations of Newtonian science as such in the Aesthetic and Analytic sections of the *Critique* (as is frequently done) is therefore misleading. What is true is that Kant's thinking proceeded in the order in which we have enumerated our questions, beginning with the enquiry into the basis and extent of Newtonian science.

To speak of Kant as a philosopher who lays foundations of science is misleading also in that it presents Kant's intentions in too limited a way. For it is one of these intentions to give an account of 'foundations' that will enable him to square the claims of Newtonian science with certain other concerns of human beings, especially in the fields of morality and religion. Thus, his eventual solution of the problem of the existence of an 'order of nature' (a universe of law) will be that the latter is the result of the 'projective activity' of the scientist in his continuing attempts to create systems of order, a solution which is so adjusted as to relativise the whole notion by its very 'subjectivity'. This is to enable Kant to say that a change of concern permits us to view everything from quite a different angle; to switch, for instance, from a concern with the realm of fact to that of duty and obligation. The notion of a 'critique of reason' which is here relevant is meant to constrain reason from affirming that anything corresponding to the

---

[1] References to '*Critique*' in the text are to the *Critique of Pure Reason*, also frequently called the 'First *Critique*'. The *Critique of Practical Reason* and *Critique of Judgment* will be distinguished either by their full title or by the appellations 'second' and 'third', respectively.

'source' of such a systematic order exists, or can be known, or can be 'given', independently of the context of scientific activities.[1]

In a peculiar way, a generalisation of such an approach is subsequently imported also into the 'solution' of the third question, connected with the existence of bodies in space. Here, too, we shall find that the *possibility* of 'phenomena' (e.g. material bodies in space) is caught up in certain subjective constraints. Moreover, the very framework which defines these constraints, i.e. the forms of space and time, and the categorial concepts, are thereby in turn 'tied down' to the context of 'experience' through which alone their 'reality' (or 'applicability') is defined. Beyond these confines—and it is of the essence of the Kantian approach to see the 'phenomenal' realm of empirical fact as confined in such a framework—they cannot be employed to yield anything relating to a possible cognitive judgment. If they are to be used at all, as in the case of the concept of causality, this can only be done in the context of a different range of concerns, here: of a moral appraisal of the practical activities of man. And the possibility of this new context is justified by the just-mentioned 'relativisation' of the empirical realm within a special framework. For this implies, Kant asserts, that it is at least *meaningful* to postulate alternative frameworks, as a logical basis for these different concerns.

The peculiar and characteristically Kantian circularity or mutuality of these approaches will be evident. In respect of the first question, for instance, the 'reality' of physical science is taken for granted; or more properly: the reality of the activity of the scientist, the activity of his research. What is problematic is the logical character of that research, for instance, the existence of an 'order of nature', and it is the 'possibility' of this character that has to be 'proved'. And it is 'proved' by linking it to that very activity of the scientist, to his ceaseless attempt to create a system of laws. It follows that the idea of such a system is defined only in the context of that activity.

There result thus the following relations. There is no 'order of nature' without scientific research; that research is characterised by its idea of *aiming at systematic* interconnection. Both the 'order of nature' and the 'idea of system'[2] point towards the activity of research, and are held suspended in it alone. On the other hand, that notion of research itself would be nothing unless it were characterised

[1] Cf. below, section 4(c): ii.

[2] We shall later list some of the subsidiary maxims which define this idea. Cf. below, pp. 511–2.

by these two aspects, which in turn hold each other in mutual balance. Above all, only as much 'system' can be found in nature as is imported into it by the scientist. The 'possibility' of a systematic order hinges on the logical character of the scientific act. But that act, in turn, must needs contain (as intellectual character) this idea of system.

This is the method that Kant will follow throughout, even when he answers the deeper and more general questions of the possibility of the foundations of Newton's specific laws, of Euclidean geometry, of the concept of the object in general. In this last case, for instance, as will be shown in more detail below, there is no object (*quâ* 'phenomenon') without the framework of 'experience' being presupposed, a framework which is an essential part of each individual moment of experience, and on which ultimately the whole edifice hinges. On the other side, the framework itself obtains its significance only in the context of a situation in which phenomena are actually given.

This approach is then quite general. The Kantian question, 'How can we be certain of the possibility of a cognitive account of some given subject-matter as such?',[1] is always answered by an analysis of the logical character of the cognitive situation as such. But that character is always shown to be part of what has to be fed into the epistemological situation, so that the result may become a possible object of knowledge, or perception, etc. For instance, let the question be: how can there be knowledge in principle of empirical objects or of objective happenings at all? This question is answered by claiming that we ourselves necessarily inject the logical character corresponding to this objectivity (the categories) into the perceptual situation. At the same time, the *empirical* significance of this logical character (here: the categories) is entirely linked to, and exhausted by, the injective function.

Or again, take the case of Euclidean geometry. How can we prove the possibility of the kind of knowledge (according to Kant non-analytic and yet necessary) that this geometry claims? Only, because

---

[1] It is important to note this formulation. We are no longer concerned with the question of the certainty of our knowledge of such and such a set of statements, but only of the formal conditions that have to be satisfied before questions of certainty can even be asked. This is the main shift away from the more confused situation of the earlier philosophers. Even the case of Euclidean geometry (as we shall see) should not be construed as a claim to 'knowledge' but rather to the possibility of knowledge of the kind which Euclidean geometry (in Kant's view) claimed to represent.

its essential character (spatial construction in accordance with a given set of axioms) has to be injected into the apparatus of empirical cognition, here: in the aspect of 'intuition',[1] as a condition of the very possibility of any such cognition at all. At the same time, the very meaning of the formal aspect of this cognition, which is here 'space', is tied down entirely to the context of such a cognition being assumed to take place at all.

This somewhat general and abstract sketch of Kant's approach must at this stage of the discussion necessarily appear both obscure and difficult, and we shall develop it at least in outline in what follows. Nevertheless, it is imperative to insert these preliminaries because one must always bear in mind Kant's general approach and ultimate objectives—so complex and ramified that it is easy to lose sight of its many different components. Yet without this the various specific arguments are often misconstrued. Let us trace, then, some of the original sources of the questions that have been enumerated.

4. MECHANISM, TELEOLOGY AND ONTOLOGY

(a) *Possibility and Ontology*

If we except the concerns of Kant's earliest work (LF), which deals with issues arising out of Cartesian and Leibnizian dynamics, it was undoubtedly the *Principia* of Newton that was the first great scientific initiator of Kant's thought, as well as the dominating influence as a generator of some of his most basic assumptions. For this was the work which exhibited a system of laws, formulated geometrically (rather than algebraically), managing to include apparently all the phenomena of dynamics and mechanics as well as observational astronomy in a vast edifice, and in the old Euclidean axiomatic form. The effect produced by the work shows in Kant's second major writing, whose full title eminently testifies to the direction it gave to Kant's philosophy of science. In his own words, Kant here sets out 'to discover the systematic element which links the great members of creation within the whole extent of infinity'.[2] His objective is an extension of Newtonian systematics to the evolutionary history of the whole cosmos. By 'system', as he explains, Kant does not simply mean the spatial relations of the stellar orbits, as represented by the

---

[1] This is not of course rational or 'intellectual' intuition, but sensory or 'sensible' intuition, either 'pure' or 'applied'. Cf. below, pp. 533n.3, 545n.2, 562n.3, 563.
[2] UH. 221.

Copernican theory, but rather the more 'specific relations [i.e. laws] which produce their mutual bond in regular and uniform manner', the laws of the system being linked together in a coherent manner.[1] Kant's general idea is to postulate an initial material chaos, subject from the start to the 'general laws of motion' as well as 'the accepted laws of attraction'.[2] This by itself, he claims (although the claim is partly at fault, but this cannot be discussed here), is sufficient to produce the orderly development of the universe as we now know it. The broad picture which is here introduced will rule Kant's thinking throughout his life and remains his major paradigm when in the *Critique* he characterises the 'Regulative Employment of the Ideas' as 'reason's search for unity'.[3] Moreover, the transcendental image of 'mutual bond' will reappear even more basically in the Third Analogy as a concept supporting the very possibility of a plurality of bodies coexisting in space.[4]

However, there is a second facet of this work, which shows that from the start Kant's objectives were never purely scientific, not even as regards method, but were concerned further with more general philosophical matters. I refer to the disturbing element of teleology, the question of the world's design by a creative God. We have noted how these two aspects, causality of law and causality of purpose, struggle for ascendency in the work of Leibniz. But there was a more immediate stimulus in the *Principia* itself. For not only had Newton maintained that such a system as he had described 'could only proceed from the counsel and dominion of an intelligent and powerful Being';[5] in Question 31 of the *Opticks* he had expressly claimed that it was 'unphilosophical to seek for any other origin of the world' or 'to pretend that it might arise out of a chaos by the mere laws of nature'.[6]

Newton had not tried to reconcile these conflicting positions; the resulting logical pressures produce one of the most customary traits of Kant's procedure, the attempt to reconcile seemingly contradictory elements that pervade the doctrines of predecessors and contemporaries. He accepts Newton's general teleological approach but denies that it is inimical to such a view to hold nature to have evolved from chaos in accordance with the general laws of motion and attraction. Unfortunately, in his early work, this defence of Newton is somewhat slight, as may be recognised from Kant's repetitious and rather worried tone. He admits that the natural interpretation of such an

[1] UH. 246.   [2] UH. 222, 228.   [3] K. 544–5.   [4] Cf. K. 233.
[5] General Scholium, op. cit., p. 544.   [6] Op. cit., p. 402.

evolutionary account is to draw atheistic conclusions, but rebuts the charge by simply *asserting* that the existence of matter together with its laws leads to such 'necessary beautiful connections' that this by itself proves the existence of a designing deity.[1]

But the problem will haunt Kant throughout his life: how to reconcile the assumption of a universe subject to law, both in its existence and in its development, with the postulate of a universe whose teleological aspect demands a God. He returns to it eight years later in his *Only Possible Ground*. This work shows a greater awareness of the problem surrounding any attempt at reconciling mechanism with teleology. Kant tries to meet it, on the one hand by the provision of an independent ('ontological') proof for the existence of God, on the other by a more complex articulation of the teleological argument; an argument, be it noted, which Kant never ceased throughout his life to regard with great respect. Indeed, after his subsequent abandonment of the 'ontological' proof, it was to become of central importance, a key concept for his characterisation of 'speculative reason'.[2]

Let us begin by considering the ontological proof. This proceeds (in a not altogether un-Leibnizian vein) from the contention that we must grant the conception of the possibility of things; but since all possibility presupposes something real through which it can at least be thought, this reality is necessary as such; for with its suspension, all possibility would be suspended also, which is impossible.[3]

This 'ontological argument', from the 'inner possibility of things', Kant claims, has the precision of a demonstration possessing mathematical certainty.[4] To it he again joins now the teleological approach of the 'physico-theological' proof; an argument which moves from 'what experience teaches' to both existence and attributes of God.[5]

This type of argument, however, Kant has by now realised more clearly, does not possess the 'geometrical strength' of the ontological

---

[1] UH. 228.

[2] Cf. below, section 4(c): vi, pp. 520ff. For his 'respect', cf. K. 520.

[3] OG. 78–83; cf. ND. 395 (Insel, pp. 433–5), which already sketches this argument in the same year in which appeared UH. K. 488–93 will reproduce this argument concerning the 'transcendental substrate' of 'the complete determination of things, that contains . . . the whole store of material from which all possible predicates of things must be taken', also labelled 'the sum-total of all possibilities' (K. 490, 488). However, Kant will now say that this is a 'mere fiction'; and that all we can do is to ask how and why 'reason' argues (and perhaps *must* argue) in this way (cf. K. 493–4; and below, pp. 489f.).

[4] OG. 155, 161. There are many different species of this argument. For another example, cf. ch. III, sect. 3(c) above.

[5] OG. 156.

argument; it has no more than 'a very high probability', and in any case cannot go farther than the properties of the world (e.g. its adaptive organisation), which it assumes as premises from which to draw its conclusion.[1]

Ignoring for the moment the way Kant here fits mechanism into his scheme, if we now compare all this with the *Critique*, we find that the two main arguments, regarded as proofs of the existence of God, are here criticised and rejected.[2] Or rather, if the ontological argument is rejected, the physico-theological argument can no longer be employed as a supplementary buttress; and, as regards the former, Kant is now pressing his earlier doubts concerning the logical power of that proof with greater insistence. Let us briefly consider the character of the 'critical' rebuttal and reinterpretation of the earlier argument. Broadly, this amounts to shifting 'possibility' from a purely ontological to a transcendental level—no unimportant matter since the concept of possibility has already been found to be the very core of the Kantian edifice.

As regards the ontological argument, only two points are of interest here. To the contention that the suspension of all reality in thought, and so of all possibility, is impossible, i.e. self-contradictory, the *Critique* objects, first, that if we remove all things in thought (together with their predicates), they cannot 'leave behind a contradiction'; secondly, that there 'is [already] a contradiction in introducing the concept of existence . . . into the concept of a thing which we profess to be thinking solely in reference to its possibility'.[3]

The important point here concerns the critique of the concept of possibility. We must distinguish between 'logical possibility' and 'real possibility', the former alone being the concept of what is not self-contradictory.[4] Now the need for *this* distinction Kant had recognised from the earliest years of his reaction to the philosophy of Wolff, which had tried to eliminate it (itself reacting against Leibniz's own highlighting of this distinction). And, paradoxically at first sight, nowhere more than in the *Only Possible Ground* does Kant labour the distinction, just as he there is already insisting that 'existence' must not be treated as a real predicate on the lines of the qualitative

[1] OG. 159, 160.
[2] *Transcendental Dialectic*, ch. III, sects. 4 and 6: 'The impossibility of an ontological' and 'of the physico-theological proof' (K. 500ff., 518ff.). Cf. also above. p. 486n.3.
[3] N. 503.   [4] K. 503n.a.

determinations of a thing.[1] But how then characterise 'real possibility' if we reject recourse both to logic and to experience (as Kant explicitly states that we must)?[2] How rescue a notion which falls somewhere between logical possibility and completed actual existence?

In Kant's responses to this question we meet again the varieties of metaphysical procedures to which I have frequently drawn attention. It will be remembered that Leibniz, although attacked by Kant for inattention to the distinctions involved, had actually tried to give a meaning to 'real possibility' through his conception of 'compossibility' and 'pre-tension', both involving a reference to God's perfection, accompanied by an attempt to smoothe a logical passage from possibility to actuality through an application of the principle of continuity.[3] Kant was quite blind to such a method, and therefore in his earlier work he could do no more than 'dogmatically' link 'real possibility' with the semantically opaque notion of God as its ontological ground. Now compare this with the definition given in the *Critique*: the 'objective reality', we are told, which alone prevents a concept from being 'empty' (however much it may be non-self-contradictory), must first be 'proved'; 'and such a proof . . . rests on principles of *possible experience*'.[4] Kant, we see, has immensely sharpened the division between pure possibility and full actuality; no such gradual flowing from one modal state into another as suggested by Leibniz's 'tendency for the possible to acquire actuality' is here as much as contemplated.[5] The transition from possibility to actuality requires a reference to 'experience', or—since we are here concerned with the general framework that belongs to experience—to the principles which define *possible* experience: everything, in order to become an object of experience, must first be *given* as an extensive and intensive magnitude constructible in space and time; and secondly, it

---

[1] Cf. OG. 72–3. Cf. the celebrated passage in the *Critique* which repeats this (K. 504).

[2] OG. 80–1.

[3] Cf. above, ch. VII, pp. 399f., 403f. Altogether the reader should compare Kant's 'method' here with Leibniz's use of analogy, as discussed in sect. 2 of the above chapter. It is the rejection of analogy for *metaphysical* purposes, I want to argue, that drives Kant, first towards sheer 'dogmatic' postulation, as in the *Only Ground*, and then to the introduction of the 'critical' approach which links 'real inner possibility' with the formal conditions of 'experience' or 'appearances', allowing a place for analogy only at the level of 'reason', in its 'speculative employment'.

[4] K. 503n.a; my italics.

[5] Cf. Kant's brusque rejection of such an approach, culminating in the blunt remark: 'This [alleged] process of adding to the possible [to constitute the actual] I refuse to allow' (K. 250).

must be *thought* as suspended in a framework of concepts which yield connection, viz. substance, cause and interaction.[1]

In his reply to Hume, Kant describes this notion of '*possible experience*' as a 'third thing',[2] in virtue of which we apprehend in an *a priori* manner one empirical concept, i.e. a perception, to be related to another (although for the specific content—e.g. specific degree of intensity of illumination—we need to resort of course to observation).[3] We at once realise that this 'third thing' is the last and most sophisticated of the list of models through which philosophy characterises the underlying bond uniting the terms of propositions. In its broadest aspect, we shall see later, this notion of possible experience refers us to the 'synthetic necessary unity of apperception' (whose specialised forms are the categorial principles just enumerated); and it is Kant's basic contention that what is 'intended by the copula "is" ' is the bringing of 'given modes of cognition' to 'the [transcendental] unity of apperception'.[4] It is 'the unity of apperception' which will yield that 'real connection between subject and predicate' which Leibniz had postulated in his *P.i.S.* principle, but whose *raison d'être* he had placed in God, as formulated in the principle of sufficient reason.[5] The difference is that Kant translates possibility of *being* (as representing a quite unarticulated concept) into possibility of *experience*; the 'unity of apperception' is viewed as an *a priori* ground that first generates the concept of an object, although it can *achieve* this function only in the instant of an actual or potential experience: an experience that must involve reference to causation, space, time and the categories. But 'ground' means here no more than 'principle'; a principle which is 'presupposed' as a structural component of Kant's analysis of the abstract notion of (possible) objective experience.

This explains the sense of his account of the 'Ideal of Pure Reason', where he argues that it was only through a kind of 'natural illusion' that we had regarded the 'empirical principle of our concepts of the possibility of things viewed as appearances [which has meaning only within 'the context of a possible experience'] as being a transcendental

---

[1] Cf. *Critique*, Transcendental Analytic, ii, ch. 2, sect. 3.   [2] K. 610.

[3] To avoid confusion at this early stage, let us note that although Kant in this context speaks as though what we could *apprehend a priori* was the *empirical law-likeness of the sequence* of events, what he means (as clearly implied by his arguments elsewhere) is that we have to *think* (*transcendentally*) *law-likeness* into the perceptual situation in order to speak of the sequence of our perceptions as yielding a perception of a *particular contingent sequence* of events. (Cf. above, p. 474; and below, sect. 4(c): i and v; 8(e): iii.)

[4] K. 159, cf. below, sect. 8(d), 631–37.   [5] Cf. ch. VII, pp. 452–3.

[really: 'transcendent'] principle of the possibility of things in general';
a notion of 'possibility' which Kant identifies with that of the 'ideal of
pure reason', i.e. God.[1] It follows at once that any 'dogmatic' postulation of possibility is no longer available for a demonstration of God's
existence, since non-logical possibility is linked to the components of
the structure of experience in general.

Now these structural components, which define the concept of a
possible experience, Kant calls 'transcendental'.[2] In summary fashion,
transcendentalism has here taken the place of metaphysical ontology;
space, time and the categories are Kant's substitute for the 'God' of
the ontological argument. How this is to be achieved in detail we
shall see later. But it seemed necessary to sketch first the broad trends
of Kant's many-sided 'critical approach' towards this idea of possibility in the context of the whole tradition of philosophy which he
inherited, before examining the details.

(b) *Teleology and Mechanism*

If 'real possibility' and the resulting ontology are given a new
twist in the later *Critique*, a similar reinterpretation was necessary for
the place of teleology on the one hand, and the concept of matter and
its 'necessary properties', expressed through its laws, on the other. By
comparison with the earlier cosmological treatise (UH, whose scientific parts are briefly incorporated also in the later work),[3] the distinctions between these three facets are considerably sharpened. Kant
now distinguishes explicitly between the 'necessary laws'[4] (the reference being to the laws of motion), and the general order of nature,
which he labels (surprisingly at first) an 'accidental order'. In Kant's
usage, an order here is 'merely accidental', or 'contingent' ('*zufällig*'),
if the arrangement of the facts is not sufficiently explained by reference
to the general laws or properties of matter, but in addition requires

---

[1] K. 494; cf. K. 488–9. Cf. also above, pp. 486n.3, 487f.; below, pp. 496, 507 and n.3,
536f.
[2] This is very different from 'transcendent', which denotes something that lies
beyond all possible experience, such as a God postulated as the ground of all
existence. (Cf. below, pp. 523f., where this re-emerges as an 'idea of reason'.) On the
other hand the *category* of ground and consequence, or its spatio-temporalised
version, causation, i.e. necessary sequence in accordance with a rule, when viewed
as a concept which *must be added* so that we shall be able to frame the notion of an
objective happening as such, is 'transcendental', and for this reason likewise *a priori*;
i.e. it is not 'received' by the mind as a 'passive' element, in the fashion of the *a
posteriori* component of a phenomenon.
[3] Cf. OG. 137–51.   [4] OG. 110.

reference to a purpose or purposes, i.e. to teleological explanation.[1] Thus, although individual ranges of physical phenomena certainly obey the general laws of mechanics, and are subject to efficient causality, in their joint action they often express a high degree of adaptive behaviour, sometimes appearing even especially suited for the well-being of man, and thereby suggesting a designing intelligence. Moreover, such adaptiveness or purposiveness manifests a certain economy of effort which nature displays in its dealings.

Kant realised, of course, that most of the considerations and arguments current in his time in favour of teleology had a somewhat precarious force. He himself harps on the narrow selectiveness of most of these considerations, not to mention their rather naïve anthropomorphism. But, just as in Leibniz's case, a powerful example from contemporary developments in natural science comes to serve as a paradigm possessing the necessary generalising force to affirm Kant in his stand. The example concerns the discovery by Maupertuis, a few years prior to the publication of the *Only Possible Ground*, of the so-called 'Principle of Least Action', which the latter had been the first to enunciate and apply to an explanation (in the sense of theoretical derivation) of a variety of scientific laws in the branches of optics, mechanics and statics.[2] Maupertuis' principle amounts to a generalisation of the principle of least effort that we have met in Leibniz,[3] whose distinctive feature had been its teleological twist (cf. Principle of *maxima* and *minima*). The striking aspect of Maupertuis' work was the power of the Principle of Least Action to unite subjects

---

[1] OG. 96ff. It should be noted that this is a notion of the 'accidental' which operates at the level of the 'accidental manifold of empirical laws', themselves accidental in the sense of 'contingent'. As such it must be carefully distinguished from the 'accidental manifold of perceptions' which we shall meet in the transcendental deduction of the categories (below, section 8 (d)). There is, however, the usual Kantian parallel between 'understanding' and 'reason': just as the latter 'unites' the aggregate of laws into a theoretical system (cf. below, section 4c), so the former 'unites' the individual perceptions to result in an 'objective order' (cf. 8(b)–(c)). Because most Kant commentators pay insufficient attention to this office of 'reason', and omit to note the parallelism just mentioned, they endow (or think Kant endowed) the understanding with powers Kant never intended for it. It is not surprising that one then discovers 'inconsistencies' in Kant's teaching. (For this 'parallelism', see also below, pp. 502, 505f. for the accusation of 'inconsistency', cf. pp. 661ff.)

[2] Cf. *Histoire de l'académie royale des sciences et belles lettres*, 1746, pp. 268–94: 'Les loix du movement et du repos déduites d'un principe métaphysique'; also *Essai de cosmologie*, 1751, p. 21. Cf. below, pp. 498, 521.

[3] Cf. above, ch. VII, sect. 3(b), pp. 425ff. It is not clear whether Maupertuis was directly indebted to this or whether his was an independent discovery. He was certainly under the general influence of Leibniz's finalism.

which previously had appeared altogether unrelated (laws of mechanical collision, static equilibrium, optics).[1] At the same time, in the eyes of Maupertuis, the 'explanatory power' of the principle lay in its 'metaphysical' nature, expressing for him the activity of a Maker who had fashioned nature in a manner displaying purposive activity.

Now these were precisely the aspects which made such a deep impression on the Kant of the *Only Possible Ground*. *Every* scientific system (such as that of Newton's *Principia*) imposed unity of principles upon diversity of laws; but these were laws which by themselves display strong apparent similarities (gravitational force). On the other hand, the connection between the different laws which were brought together by Maupertuis' principle was quite unsuspected; without it the laws seemed to display no more than an accidental concatenation of uniformities. By applying a principle involving the concept of purpose, here: of a designing Deity, unity of these laws had been effected. Maupertuis' formula, so Kant expresses his reaction to this discovery, manifests 'a reference to appropriateness, beauty and well-adaptedness'.[2] It supplies evidence of a unity which subordinates the otherwise seemingly 'blindly necessary' (behaviour of matter), and accidental aggregate (of different empirical uniformities) to a higher order, bestowing a unity of systematic connections that points to a perfect archetypal Being. If we keep in mind how systematisation and purposiveness here come together, we shall understand better Kant's apparently artificial bracketing of these aspects in his later writings.[3]

Nevertheless, the old problem of how to conceive of the coexistence of a set of 'necessary' mechanical laws with such a teleological order was not thereby solved. So far, we have nothing but a more systematic specification of the notion of a teleological order. Before turning to the development of the problem in Kant's later work, we must however first also consider briefly the place held by the laws of motion in the scheme of the *Only Possible Ground*, and the reinterpretation of their status in Kant's later writings. Broadly, as the general result

---

[1] Admittedly, his various derivations required a certain liberality in their interpretation of 'action'. Still, the principle, and other *extremum* principles, have subsequently been shown to display very great systematic fertility. Cf. above, p. 431, and the reference to Mach's comment on the situation; also W. Yourgrau and S. Mandelstam, *Variational Principles in Dynamics and Quantum Theory* (London, 1955).

[2] OG. 99.

[3] Cf. below, pp. 520ff. Thus, the first *Critique* often speaks interchangeably of 'the principle of systematic unity' and the 'principle of purposive unity' (K. 563); or even, in one phrase, of our search for the 'greatest possible systematic and purposive unity' (K. 567); and the third *Critique* develops the same theme at greater length.

of Kant's re-evaluation of both the teleological and the mechanical aspects, whilst the former is shifted into the realm of *methodology*, where also we find located mechanism, in the general sense of efficient causality; in its aspect of 'necessary properties of matter' (laws of motion) mechanism is placed within the context of Kant's doctrine of physical possibility and the connected notion of 'construction', leaving for physical actuality only the set of individual contingent happenings.

First, then, the question of the properties of matter. (Teleology will be taken up in the next section.) In the *Only Possible Ground* these are said to derive with 'logical necessity' from the concept of matter, the laws expressing the 'essence' of this concept. The passage in which Kant states this anticipates the position still taken in almost unchanged form in the *Metaphysical Foundations* twenty-three years later:

> The possibility of matter as such and of its essence is evident since we cannot conceive that which fills a space, or which is capable of imparting motion to matter or which exerts pressure, under any other conditions than those from which these laws derive with necessity.[1]

Although the relationship between the notion of matter and the laws which govern it is conceived in a somewhat similar fashion in the *Metaphysical Foundations*, there are nevertheless important differences. For in the *Only Possible Ground* Kant goes on to say that 'the inner possibility of matter itself', of what he calls: 'the *data* and the real', is 'not independent or given in itself' but rather is 'posited through a principle' which relates the diverse laws in a unitary fashion, in a manner which proves their contingency'.[2] Since a few pages later we are told that this kind of 'unity' which 'is founded in the possibility of things', proves 'a wise being, without which all these objects in nature are not possible, and in which, as an eminent ground, the essence of many of these objects are united in uniform relations',[3] it is clear that this 'inner possibility', i.e. the ground of existence of matter as such, is located in God. So God is here *both* the ground of the 'inner possibility' of matter and its laws, *and* of the higher-order unity of all empirical laws; although it must be said that the operations of the Deity occur (though unrecognised) at such disparate logical (and ontological) levels that this attempted reconciliation of mechanism and teleology is highly artificial.

[1] OG. 100. The coupling of 'possibility' and 'essence' should be noted.
[2] Ibid.   [3] OG. 125.

As we might expect, this diversity of levels is more clearly recognised in the *Critique*. We shall later show that the 'uniting' of the 'essence of objects' in 'uniform relations' in a ground, viewed as the source of the *higher-order* unity of empirical laws, is subsequently construed as the expression, on the one side, of a *methodological* procedure, and on the other, of an *'analogue'* which is cast in *theological* form.[1] By contrast the 'critical' approach will (as already mentioned) formulate the ground of the possibility of an individual material thing as such in terms of transcendental structure, i.e. of construction in space and time, under the guidance of the categories: an essential requirement for the definition of the 'reality of the data' being that the formal framework should contain an *a posteriori* element, i.e., be anchored in an instance of actual experience of the completed object (*quâ* 'phenomenon'). In other words, the 'inner possibility' of an object (i.e., the meaning of the concept of a real object) will exhaust itself (as in the case of the criticism of the ontological proof) in the description of the formal structure of this whole experiential process.[2] The 'ground' of the 'inner real possibility' of a thing becomes the actualisation, occurring in a given instance, of 'possible experience', the latter itself being a reference to the structural framework of this notion to which we have briefly alluded already, and which we shall unfold in more detail later. But though the 'critical account' still has to be discussed, it seems to me important to relate it (prior to such an exposition) to these early contexts in which the *functions* that it is intended to fulfil are first conceived. Let us only note the important result that, at the 'critical' stage, Kant's older ontological level is replaced by, and split up into, the tripartite set of methodological, analogical (or 'theological') and transcendental levels, with the ghost of the 'noumenon' hovering behind the scenes.

Concurrent with a redefinition of 'possibility', the *Metaphysical Foundations* will also display a change in the conception of the relation of the laws of mechanics to the concept of matter. In the *Only Possible Ground*, these laws derive from the 'essence of matter', whose ground lies in God. In the later treatise, the categorial principles (deriving from the categories of quantity, quality and relation), e.g.

---

[1] Cf. below, pp. 521ff.

[2] True, Kant will often add that it must be possible to *think* an 'intelligible ground' (the 'noumenon') for that which in the experiential moment 'merely appears'; a reference to the doctrine of this earlier period of his life. But these aspects are of no consequence in the present context.

the 'Axioms of Intuition', the 'Analogies of Experience', etc. will supply the necessary constraints, in terms of which the 'essence of matter' will be developed into the specific laws defining its possibility. That the 'necessary properties' of matter derive from considerations of its very 'possibility', providing (incidentally) a new interpretation of the place of 'necessity' for this special group of laws of science,[1] is a principle to which Kant adhered throughout; but where this possibility is anchored in God in the early period, it is later linked to 'experience', as *its* possibility. Still, the old contexts never quite die; on the contrary, it is certain that they give much-needed substance and meaning to Kant's rather more abstract later schema.

(c) *Teleology and Theoretical Science* (*Methodology*)

Only the basic laws of motion have so far been removed to the level of 'possibility'; the rest of the particular laws of physical and biological science, still seen through the eyes of a thoroughgoing mechanistic approach (the approach of 'efficient causality'), continue to live uneasily with teleology. Moreover, Kant's rejection of the ontological argument, whilst freeing him to move the mechanism of the laws of motion to another level, has also withdrawn a prop from the teleological (or 'physico-theological') argument. The resulting need to re-evaluate this argument, so crucial for Kant,[2] suggested that it, too, be shifted to another level: something which would once and for all remove any apparent clash ('Antinomy' is the expression Kant uses in the third *Critique*) between the two approaches to causality. The change of approach is worked out in the relevant sections of the first and third *Critiques*.[3]

As already noted, in the *Only Possible Ground* Kant had judged the physico-theological argument (from the manifestation of a purposive order to the existence of God) to be no better than plausible and persuasive; it lacked 'demonstrative' strength and was bolstered up only by the ontological proof. In the *Critiques* he declares it to be quite inconclusive. In the third *Critique*, even Maupertuis' work, to

---

[1] I.e. they define the near-pure part of Newtonian science, mentioned above, p. 475. The 'necessitarian' character of ordinary empirical laws will—as we shall show in the next section—be explicated in a different fashion, as a function of the deductive architecture of scientific theories.

[2] Cf. K. 520.

[3] Cf. for instance, K. 560–4; J, paras. 68, 85, as well as the two introductions to J. (For the 'First Introduction', see IJ.)

which (without mentioning names) he alludes, is declared to be irrelevant for this purpose:

> Universal mechanical laws—however strange and admirable may seem to us the union of different rules, quite independent of one another according to all appearance, in a single principle—possess on that account no claim to be teleological grounds of explanation in physics.[1]

However, the rejection of teleological causality for the purposes of a 'natural' theology did not at all lead to a corresponding rejection of the teleological viewpoint itself. On the contrary, this continues to haunt Kant's general thinking in a very important way, affecting particularly his whole approach to the nature of theoretical reasoning in science ('theoretical reason'), and the relevance of the latter for the whole meaning and organisation of the laws of science at the empirical level. The result will be that whilst teleological causality is retained, its significance for science in general, as well as biology in particular,[2] including its theological significance, is re-interpreted. For this reason, we shall not be surprised that Kant's views on the methodology of scientific theory are located in the section of the first *Critique* entitled 'The Ideal of Reason', which is a 'critique of theology', and particularly in the 'Appendix to the Transcendental Dialectic',[3] and in the introduction to the third *Critique*, together with the original introduction to that work.[4] Kant's statement of his views is prolix, and I shall condense them in a certain order, pertinent to our interests. Such an account will give us some idea of the subtle complexity of his views concerning the methodology of science, whilst indicating at the same time the intimate connection between

---

[1] J. 229 (para. 68). Kant means: the teleological form of such theories should not allow us to imagine that their explanatory power consists in their foundation on the *existence* of a God; let alone can such a form be used to *infer* such an existence. This point will be explained presently. (Cf. below, pp. 521ff., 527–8.)

[2] The biological aspect will not be treated in this chapter.

[3] K. 485ff., 532ff. Cf. above, pp. 486n.3, 489f. Kant has a double account of the 'ideal of reason', essentially like that advanced in OG. The first occurs at K.488: It is the sum-total of all possible predicates, from which everything is imagined as deriving its own possibility, and in virtue of which everything is related to a common correlate. Moreover, being represented also as an individual, it is called an archetype, or archetypal ground. The second account, of which traces again occur in OG., occurs at K. 556, 566: Kant there speaks of the ground which reason thinks, as the substratum of the systematic unity and order which the world exhibits. In the version of the *Critique*, of course, Kant's restriction is that such a concept, though 'thought' by reason, has no corresponding object in sense experience, and is hence, by definition, a 'mere idea' (cf. K. 318).

[4] Especially J, sects. iv–vii; IJ, sects. i–vii.

these views and his general epistemological and theological interests. We shall thus discover how central a part methodology plays in the whole edifice of Kantian philosophy.

(i) *The need for systematisation*

Kant begins with the position of the *Only Possible Ground*. Observation and experiment present us with individual happenings, and (by inference) with uniformities of association and sequence of such happenings, labelled 'particular empirical laws'. Now the aggregate of such laws is at first sight no more than an accidental or contingent collection.[1] But organised experience, or science, cannot rest there; it 'prescribes and seeks to achieve its systematisation' of these laws.[2] These scientific tendencies Kant also describes as a 'law of reason which requires us to seek for this unity';[3] in conformity with such a law 'we interrogate nature'; 'reason does not here beg but command'.[4] In these pages of the *Critique* the centre of gravity shifts from God as the ground of the adaptive order of the world to the constructive activity of science, or 'reason', which is, as already mentioned, self-constrained to bring about a view of the world in accordance with principles of adaptive order. Reason determines itself to seek for a

---

[1] Cf. K. 538; J. 20, 21, 23; IJ. 10. For 'accidental', cf. above, pp. 490f.   [2] K. 534.
[3] K. 538. K. 320–1 shows that Kant construes reason on the formal model of an agency which generates a logical process that either seeks to draw conclusions from a given set of premises, or which seeks to fit a set of 'hypothetical' premises to conclusions. Moreover, dropping of the experiential conditions leads reason into contradictions. As shown in the Antinomies, it is then easy to 'prove' that the world is and is not limited in space and time, that it is thoroughly causal and that it cannot be; that it has a necessary cause and that it cannot have one. Only if reason is 'constrained', its expansionist drive channelled into procedural ('regulative') directions, can its formal needs be allowed a legitimate place. For instance, instead of saying that the world *is* thoroughly causal, we must say that causal conditions *must and can be fitted* to every event, as a matter of scientific methodology.
[4] K. 534, 539. Cf. the celebrated passage in the Preface to the Second Edition, which interprets the work of Galileo, Torricelli and Stahl in terms of this methodology, speaking of the combined effect of reason and its principles, acting as a judge interrogating nature, and experiments devised in conformity with these principles, in order to answer its questions. We are here a long way from the over-emphatic insistence on patient observation suggested by some aspects of the British empiricist school, although even there we have found considerable concessions to the place of theory.
   It should be noted that in Kant's very fluid presentation, such a constructive approach actually covers *four* domains: The constructive activity of mathematics; the constructive element among the conditions determining cognitive judgment in general; the constructive activity of the scientist, in the present section our primary concern; and finally, the constructive approach of the 'metaphysics of nature' (cf. below, pp. 672ff.).

unity of principles, which will result in the creation of a higher-order scientific theory.[1]

There are a number of motives for this search for 'systematic unity'. Evidently reverting to the case of Maupertuis, Kant remarks that

> the discovery that two or more empirical heterogeneous laws of nature may be combined under one principle comprehending them both, is the ground of a very marked pleasure, often even of admiration, which does not cease, though we may be already quite familiar with the objects of it.[2]

The aesthetic ingredient as a characteristic feature, in what we may call 'the pressure of reason' towards theorising, should not be under-rated. A second factor is of a more heuristic order, and belongs to methodology. Kant points out that theorification provides us with a powerful tool for the mutual testing of the individual laws of a system; it is a vehicle which functions as 'the touchstone [Probierstein] of the truth of its rules'.[3] We should not, of course, understand this as the suggestion of a superior way of *'proving'* the truth of some given hypothesis. Rather, a theory provides a means of exposing a given hypothesis to a variety of test-situations, even though the question of 'success' may remain entirely 'problematic'.[4] Kant is far removed from any simple notion of 'truth'. The strength of an hypothesis is 'problematic', in the sense that it is a 'problem' which is only 'set as a task';[5] it is an ideal whose realisation we may seek to achieve but can never *more* than seek. We are not here concerned with a proof *of* universality, but an argument *to* universality.[6]

[1] In the first *Critique* Kant speaks of this as 'the hypothetical employment of reason' (K. 535). In the third *Critique*, 'reason' is limited to the task that *requires* particular laws to be 'subsumed under the universal laws' of a theory; and it is 'reflecting judgment' which *supplies* (as explanations) the higher-order universal laws. In this it is guided (in a way to be explained) by the teleological viewpoint. (Cf. J. 252.) Both functions are fulfilled by the 'reason' of the first *Critique*.

It should be noted that this 'power of judgment' is quite distinct from the notion of 'judgment' (simply), prominent in the first *Critique*, whose function is that of relating a predicate to a subject, or forming a disjunction between judgments of the latter sort, etc. (Cf. K. 105–6; 158.)

[2] J. 24 (Introd. vi). Newton's *Principia*, and Linnaeus' taxonomical systems supply further examples.

[3] K. 535. A chemical touchstone is any reagent which tests for a specific substance.

[4] Ibid. Cf. the analogue of such a view in K. Popper's concept of 'corroboration', as employed in *The Logic of Scientific Discovery*.

[5] Cf. K. 449 for this locution.   [6] Cf. below, pp. 509ff.

A third motive for theoretical enquiry is that it automatically 'contributes to the extension of empirical knowledge',[1] supplying it with a predictive power which is immeasurably increased by its employment of theoretical notions that designate entities which would otherwise remain entirely hidden from view, thus opening up new regions of fact and further possibilities for testing both itself and other neighbouring theories.[2]

The fourth and most powerful consideration pertains to Kant's interest in the conception of 'the order and harmony of nature'.[3] This is an important notion which must be distinguished carefully from Kant's concept of 'nature' simply.[4]

By the term 'nature' Kant denotes alternatively the following: (a) the totality [Inbegriff] of all objects of experience; (b) the existence of these objects, considered—*quâ* existence—determined according to universal ('transcendental') laws, or better: concepts; (c) more specifically: simply this conformity to law of all objects of experience.[5] By contrast, the concept of the 'order of nature' refers us to what we express through the formulation of a scientific system of particular *empirical* laws, or of a biological taxonomical system. The distinction between these two notions is perhaps the most vital as well as most contentious in Kant's whole philosophy of science and his general epistemological viewpoint. We can therefore only hope to unravel it gradually as we proceed.

To a certain extent we must here assume the 'proof' of the categories and of the 'Principles of Pure Understanding' (e.g. Analogies of Experience) as discussed below, in section 8.[6] However, since any intelligible approach to that proof presupposes in turn the argument of the present section, we can again do no more than employ our method of leap-frogging. What is important for the moment is to warn the reader not to read misleading interpretations into Kant's language. First of all, let us note that the 'universal laws' alluded to in the definition of 'nature' are the general categorial principles (or

---

[1] K. 550. [2] Cf. K. 556. [3] Cf. J. 21, 23.

[4] The distinction is indeed crucial for any grasp of the relation between the Kantian notions of 'understanding' and 'reason'. Without this, we cannot get a clear conception later of the limits of the Kantian proof of the categories, and it is partly to facilitate this that I have in this chapter inverted the usual presentation of Kant's doctrines by first discussing the doctrine of the 'transcendental dialectic' (the domain of reason) before venturing on an account of the 'transcendental analytic' (the domain of the understanding).

[5] P. 52, 54 (paras. 14, 16). Cf. K. 392n.b. [6] Pp. 615ff.

concepts) of the *understanding*. Furthermore, the expression, 'conformity to law of objects of experience', will later be shown to be a reference by Kant to his contention that the concept of an object as such, corresponding to the general concept of a cognitive experiential judgment, *contains*, as a necessary presuppositional ingredient of its *possibility*, the various categorial concepts, e.g. quantity, quality, substance, cause, interconnection, etc.

Let us exemplify this by the case whose 'transcendental' aspect we shall discuss in more detail later: the notion that the possibility of cognition, as such, of a *sequence of states* of an object necessarily contains or presupposes (according to Kant) the concept of cause, where the latter is defined as '*determined sequence* under a universal rule'.[1] It should be stressed with particular emphasis that the sequence whose notion (as an object of cognition) presupposes the concept of cause, and is thus made 'possible' (a possible object of experience) may be, and indeed as such *must* be, regarded as an altogether *contingent* event; not only contingent in the sense in which empirical laws (and their instances) are contingent (a fact which Kant assumes throughout),[2] but contingent also in being simply an individual particular happening, mentioned in some unique observation report. The question whether some observed event or change of state is an instance of an empirical law can be determined only by those inductive procedures distinctive of all scientific enquiry.[3]

It follows that we must interpret the phrase 'objects of experience conform to universal laws'[4] very carefully and sparingly, and not understand it in the *normal* sense of such a locution, viz. that all natural processes are *in principle* describable as subject to empirical laws (whatever these laws may be which subsequent inductive investigation may discover). For that there *are* any such laws is (even in principle) not determined by the concept of nature and its possibility.

Consequently, from the fact that the possibility of an empirical sequence (as part of 'nature') presupposes the abstract concept of cause, we *cannot infer* that such a sequence is an instance (even in

[1] Cf. below, pp. 648ff. It will be noted that this is a Humean interpretation of the *concept* of cause, as necessary connection.

[2] Cf. J. 19–21; cf. above, p. 490n.1, for the variety of Kant's notion of the 'accidental' or 'contingent'; also below, sects. 6(c), pp. 563ff. 8(d), 628ff.

[3] Cf. K. 173. Moreover, as we shall see presently, the concept of *empirical law-likeness* itself will likewise require a 'transcendental support', though from 'reason', and not the 'understanding'. (Cf. below, pp. 505f.)

[4] Since these laws 'define' nature, Kant often calls them 'the universal laws of nature', cf. at P. 53 (para. 15), J. 23 (Introd. vi).

principle) of some causal law or other. The most we can say (as Kant indeed once remarks explicitly) is that since the notion of 'nature' contains this concept of cause, 'the *understanding* has all it can *demand*' if in the context of scientific enquiry—employing the principle of causality that we 'here require'—we (i.e. 'reason') 'search after and formulate' the 'natural conditions of natural events'.[1] So, to explicate the existence of empirical law-likeness is a task for 'reason'; the very definition of the concept of empirical law (as an entity with a necessitarian logic) will have to be given in terms of the activity of reason.[2] And it is only a 'regulative principle of reason' (not of the understanding)

> that for every member in the series of conditions we must expect, and as far as possible seek, an empirical condition in some possible experience.[3]

The same 'looseness of fit' between the fields of understanding and reason we note in another passage, where Kant again refers to the logical character of the empirical laws of nature. *Quâ* laws, he holds, these 'carry with them an expression of necessity'. But the latter is a function of reason. For Kant goes on to say that the necessity here relevant (i.e. empirical lawlikeness) carries no more than an '*intimation* [Vermutung] of a determination from grounds that are valid *a priori* and antecedently to all experience'.[4] In other words *empirical* laws (like all empirical statements) presuppose transcendental concepts, e.g. causality. But the *relation* between this fact, and the necessitarian character of statements of *empirical laws* (involving a law-likeness which—as we shall see subsequently—is first injected by 'reason'), is not one of entailment but merely an 'intimation'; what *is* true is only that the concept of causation (necessary connection) involved in the statements of the *a priori* conditions of experience in general is the *very same concept* which is employed subsequently by reason in its generation of empirical law-likeness: a fact which explains the significance of the sentence following the passage just quoted, where Kant adds that the laws of nature 'simply *apply* [the categorial principles] to special cases in the [field of] appearance'.[5] And this shows that the categorial principles are here to be viewed not as validating grounds (in the way of justificatory principles of induction) of the possibility

---

[1] K. 471; my italics.   [2] Section (v) below will explain the details; cf. also pp. 502–3, 505f.
[3] K. 481.
[4] K. 195; my italics. These latter grounds are the *transcendental* concepts, explicated by the understanding.
[5] Ibid., my italics. For a contrary view of Kant's argument, cf. below, pp. 660ff.

of empirical laws, but (in respect of causality) supply merely the *form* of such laws.

There are other hints in the first *Critique* which construe the relation between the work of reason and of understanding analogically and indirectly. Thus, at K.568, the regulative activity of reason (*via* its 'idea' of a 'ground' of the systematic connection of laws, and hence of law-likeness)[1] is made the 'foundation for viewing the appearances, in conformity with the *analogy* [nach der Analogie] of a causal determination, as systematically connected with one another'.[2] Here, what determines the activity of reason is the 'analogy' with the situation at the level of the understanding—again the same 'looseness of fit'.

In yet another passage it is stated quite categorically that the whole 'idea of reason', in its aspect of regulative procedure, is 'an analogon of a schema of sensibility'.[3] The analogy is this: just as the categorial concepts yield a conception of the object itself, when applied *in the field* of the understanding, so, at the level of reason, the categories supply us with 'a rule or principle for the systematic unity of all employment *of* the understanding', a phrase which refers again to the notion of a *system* of empirical laws. The relation between the two realms is made quite explicit in the next sentence, where every principle of the understanding 'also *holds, although only indirectly,* of the object of experience',[4] i.e. in the field of applied theoretical science. Evidently the 'indirectness' of the relation is its character of being employed analogically, reason as the 'analogon' for understanding.[5]

All this suggests strongly that Kant's real view is that the concept of the law-likeness of nature, in its *normal sense*, is that of the '*order* of

---

[1] For this, cf. below, pp. 521, 524ff.

[2] My italics. Kemp Smith's translation has 'principle' instead of 'analogy'—an obvious misunderstanding of the meaning of this passage, presumably due to the fact that the *principle* of causality occurs in the Second Analogy, as one of the 'Analogies of Experience'. But that is not the significance of the term 'analogy' in the passage here under review.

[3] K. 546.

[4] All italics mine. 'Object of experience' is here a grossly misleading reference to the *system* of objects, in the empirical explorations of science; it is not a reference to 'possible objects of experience', a term by which Kant refers always (as for instance in a sentence quoted before, on p. 500) to the level of the understanding.

[5] For this term, cf. K. 546. It will facilitate our grasp of Kant's method in the 'analytic of principles', when we deal with the understanding, if we remember that historically Kant's treatment of the logical function of reason more than likely yielded the analogue for the far less obvious doctrine of the understanding, as a 'unity of experience'.

nature', alternatively described as 'experience as a system according to empirical laws'.[1] It is one of Kant's basic assumptions that we stand in need of systematic experience, and not just 'experience' simply (corresponding to the conceptions of 'order of nature' as contrasted with mere 'nature') in order to make good this notion of law-likeness.[2] But just like the characterisation of the relation between understanding and reason in respect of the logic of law that we have just noted, the nature of this absolute requirement of the advance from contingent to systematic experience is supported only by a number of heuristic and informal arguments.[3]

One such argument is connected with the point already mentioned: theory facilitates extension of the domain of nature beyond the realm of the observable. The importance Kant attaches to this emerges from the fact that he counts as a reason for sometimes calling the principle of systematic unity 'objective' that it allows us to 'further and strengthen *in infinitum* the empirical employment of reason', thus 'at the same time opening out new paths which are not within the cognisance of the understanding', whose domain reaches no farther than observation of given empirical particulars.[4]

A few pages later he returns to this topic, when—referring to the

---

[1] IJ. 9. There is hence an ambiguity in the term 'experience', which denotes cognition (i) of a singular objective happening; (ii) of a systematic relationship between a series of events, processes, laws.

It is vital for any understanding of Kant to determine from the context which of these two senses is being used on any particular occasion.

[2] Another reason connects with his theological interests. For God is seen by Kant no longer (as in Leibniz) as an ontological basis for, but only as a mirror-image of, theoretical reason. Hence a *necessary* advance to reason would make fast this conception of God. At the same time, however, there is Kant's need to make the relation between nature and the order of nature (understanding and reason) relatively loose, and to this end, give it merely analogical strength. For in this way Kant hopes to preserve the autonomy of his conception of God, its relative insulation from the realm of objective experience. In section (c): vii we shall sketch briefly the function of this notion of God in Kant's philosophy of science.

[3] To be sure, the more conceptualised our experience is, the more it forms a systematic whole. But this is not an aspect which Kant treats in any detail. It is true that he is the very founder of the view that all experience contains conceptual ingredients, but these are of an entirely abstract nature, i.e. the categories. On the other hand, his position with regard to the 'order of nature' *may* be taken as an *implicit* avowal that nature is conceptualised also in respect of concrete concerns.

[4] K. 556. Cf. Kant's reference to the 'objective validity' of the 'systematic unity of nature' as K. 538. The changes in meaning and function of such operative terms as 'object', 'objective', 'reality', etc. should be noted. This is a basic aspect of his procedure, and provides one of the greatest stumbling-blocks for the study of Kant.

power of reason to 'extend' the field of 'empirical knowledge'—he comments that

> this empirical [body] of knowledge is more adequately secured within its own limits and more effectively improved than would be possible . . . through the employment merely of the principles of the understanding.[1]

In other words, the notion of a fully developed empirical *corpus* of knowledge requires an extension of the domain of the general categories, to include the creative power of theoretical science, and the presupposed methodology that lies behind it.

A further argument in favour of the inevitable need for the advance to the field of 'reason' occurs in the First Introduction to the *Critique of Judgment*, where he maintains that 'particular experience which is something thoroughly coherent under invariable principles,[2] stands in need of [bedarf; the translation has 'demands'] this systematic connection of empirical laws as well'.[3]

Kant seems hardly conscious of the slide, as is even more evident a few pages later in the *Introduction*. He writes:

> Unity of nature in time and space and unity of for us possible experience are one and the same thing, because unity of nature is a concept which comprises as such only the aggregate of mere appearances . . . and can possess its objective reality solely in experience; an experience [however] which must itself be possible as a system of empirical laws, if (as one must) one thinks experience as a system. It is therefore a subjectively necessary transcendental presupposition that [since] this alarming and limitless diversity of empirical laws, and the heterogeneity of natural species, does not befit nature, the latter should rather be characterised as experience *quâ* empirical system, through the affinity of particular laws under more general ones.[4]

The notion of 'unity of possible experience' in the earlier part of this passage belongs once again to the realm of the understanding; it is its transcendental unity, whose full meaning we shall expound

---

[1] K. 550–5.

[2] The context of what has just preceded this passage suggests that this is a reference to the categorial principles, for otherwise the argument would be tautologous. It is perhaps true to say that by leaving it open for the reader to interpret this phrase *either* as a reference to abstract, *or* to concrete particular laws, the semblance of an argument is produced which on any specific choice of these two possible interpretations is either a tautology or a *non sequitur*.

[3] IJ. 9–10. This is the complementary case (in reverse) of the one where Kant spoke of the necessitarian logic of laws 'intimating' categorial necessity; cf. above, p. 501.

[4] IJ. 14–15; the translation has been modified.

later.[1] Here I only wish to note the almost spontaneous transition—and how many commentators have been confused by it, not to say bewitched—from this type of unity, to the quite different notion of 'experience as a system', which is likewise referred to by Kant as 'systematic unity',[2] though as a function of reason. Nevertheless, there can be no doubt that the transition from 'unity' in the first sense to that in the second was facilitated by the similarity of expression, and the consequent analogy which Kant aims at in the two cases.

Arguments like these show, then, that the principle of systematic connection of particular laws is no more than *suggested*[3] by the situation that prevails with respect to 'nature' and the 'understanding' as such. At any rate, as already noted, Kant throughout declares specifically that the categories do not yield any particular laws.[4] Nor do they yield empirical law-likeness. For, interestingly, we might (with Kant) reverse this point, and begin with a conception of the 'possibility of infinitely diverse [putative] empirical laws', and of an indefinite number of such laws.[5] Now it is here that the importance of the concept of system arises, for only such a concept supplies us, Kant claims, with the necessary constraints on the number of possible 'rules' of uniformity, selecting those (as laws) from among infinitely many which can be fitted into such a system. *Per contra*, only those putative uniformities which can so be fitted will be regarded as laws.[6] It is not just the case that systematisation supplies us with the *links between* empirically given laws, but this process is really the very presupposition for any empirical uniformity being regarded as law-like in any serious sense of the term. The 'law-likeness of nature', in the sense *normally* understood by this term, is thus first defined through the systematising activity of reason, and not through the categories of the understanding, whose sole function it is to form a framework for the concept of a *possible object*. Once more we meet the 'analogy' of the structure of the 'understanding' with that of 'reason', noted above:[7] as 'nature' presupposes the categories (or more generally—as

---

[1] Cf. sect. 8(c), pp. 623ff.   [2] K. 535, 567.

[3] Cf. J. 233, where Kant says that the principle of universal mechanism is *a priori* 'suggested [an die Hand gegeben] by the mere understanding' (i.e. the domain of the general categorial principles); cf. below, p. 658. It will be seen what an important part is played throughout Kant's arguments by 'suggestions', 'intimations', 'hints' and 'suspicions', contrary to the superficial impression of a tightly-knit logical architectonic.

[4] Cf. K. 173; J. 16, 19–20; and above, pp. 500–1.

[5] Cf. also the passage from IJ. 14–15 just quoted.   [6] J. 20.

[7] p. 502, in the passage from K. 546.

will be shown later—the unifying synthesis of the understanding (a 'transcendental' matter)), so the 'order of nature' presupposes its 'transcendental principle . . . of systematic unity'.[1]

(ii) *System as a 'projected unity'*

We have just noted that the concept of an order of nature is said to 'presuppose' the principle of systematisation issuing from reason. This notion of presupposition is however ambivalent. It might mean that we *assume* the *existence* of this order, or unity, in nature; or, going even farther, as Kant had done in the *Only Possible Ground*, assume the existence of an author of that order.[2] There is, however, a second interpretation, and it is this which makes up the specific transition from the earlier to the later, 'critical', position. This interpretation treats unity as a methodological recommendation, as a presupposition of scientific research. One of the consequences of such a step is that statements formulating and governing this 'unity' (Kant calls them 'regulative maxims') cease to be either true or false. Indeed, they may even contradict one another formally, as in the case of the regulative maxims of teleology and mechanism. (This will be part of Kant's solution of what in his earlier work had seemed to constitute a real contradiction.)[3]

The idea of unity (in its 'regulative' use) is treated as something which expresses a *decision* to seek systematic connections; reason is regarded, not as *assuming* ('dogmatically') *the existence* of a unity, but as something 'which [itself] *requires us to seek for* this unity'.[4] It is pictured as acting 'spontaneously'—an expression which Kant always employs to indicate that a certain concept is not derived from any

---

[1] K. 537. The parallelism is expressly acknowledged at IJ.10n, although the translation has unfortunately concealed the analogy. The relevant sentence should read:

'Now what the category is in respect of every particular experience, purposiveness or adaptiveness of nature (also in respect of its special laws) is to our capacity for [exercising] the power of judgment, according to which it [i.e. nature] is conceived not just as mechanical but technical; a concept, to be sure, which does not determine, unlike the category, the synthetic unity objectively, but which nevertheless yields subjective principles that serve as guiding-thread for our inquiry into nature.'

The reference to purposiveness will be explained below; it is equivalent to the idea of a projected system, which is hypothetically conjectured by reason (in the terminology of K) or by the 'reflecting power of judgment' (in the terminology of J). Again, 'mechanical' here denotes, roughly, efficient causality; 'technical', teleological causality.

[2] OG. 125.   [3] Cf. J. 233–4.

[4] K. 538; my italics. Cf. the reference to 'the command of reason', above, p. 497.

other source than reason or understanding;[1] and in a similar vein he speaks of 'the command of reason, that all connection in the world be viewed in accordance with the principles of a systematic unity'.[2] Granted such a conception, it would make no sense to ask whether natural processes, and the laws which describe them, do 'in fact' form a systematic whole, if this question supposes that such a 'systematic unity' is 'given in itself',[3] or (to use again the theological model) that it does *in fact* flow[4] from the act of a Supreme Being whose existence must first be postulated. For the conception of reason as commanding, and as the source of unity (regarded as the methodological or 'regulative' requirement to seek such a unity), is intended precisely to bypass the problem that would arise if such a unity were derived from an external source. We could then, as Kant remarks, never possess any assured cognitive grasp of the existence of such a unity, for the latter would then be 'foreign and accidental to the nature of things'; nor can it be cognised by means of the 'universal laws of nature' (transcendental principles), these being quite insufficient for such a task.[5] This we ensure only through the doctrine that the 'unity' lives by virtue of what it presupposes, viz. the regulative procedure of reason.[6]

[1] i.e. not sensation; nor are we to assume the existence of any 'innate ideas' in the traditional Cartesian sense. Reason and understanding are always purely 'formal'; they lack 'content'.

[2] K. 559; cf. K. 539.

[3] K. 535. Notice that this yields us a first sense in which something is or is not 'given in itself'. Whenever this phrase occurs, it is a reference to the question of presupposed transcendental forms (here 'reason' with respect to 'the order of nature') that are said to determine the issue. Later we shall find, similarly, that a thing *quâ* 'given in itself' would be something abstracted from the transcendental apparatus of the understanding; and the same goes for space. Note that at K. 490, the 'ideal of reason', in abstraction from the framework of experience, is said to be just the concept of the *thing in itself* (italics in text); cf. below, p. 536.

[4] Presently we shall see that reason may, or even *has to, pretend* that it flows from such a source. But reason *cannot assert* this to be a contingent fact, either actual or hypothetical. (Cf. below, sections (c): vi–vii.)

[5] K. 564. I have again translated 'erkennen' as 'cognise', and not 'know'. (Cf. above, p. 473n.4.) As in all such cases where 'presuppositions' are involved in Kant's argument, we are concerned only with the concept, not with what the concept comprises; thus here, only with the 'concept of unity' and not with actual instances of such unities. The distinction will be important for our subsequent account of the transcendental deduction of 'objectivity', where again we are only concerned with the presupposed 'concept of the object', and not with whatever it is that is claimed to be objective.

[6] Cf. the passage from IJ.14–15, above, p. 504, where Kant speaks of the 'subjectively necessary transcendental presupposition' of 'experience *quâ* empirical system'. Note that the presupposition is only 'subjectively' necessary, by which Kant means that it bears only on a system which itself does not have the standing of an empirical

Here we meet with an instance of Kant's revolutionary position: There can only be as much knowledge (in the sense of 'Erkenntnis', i.e. of the formal and conceptual aspects of experience, whether 'systematic' or 'non-systematic') as is injected by the spontaneous activities of the subject itself, be it reason or understanding. Only if we regard the unity of laws as 'a projected unity', not 'given in itself',[1] can we be certain that the concept is not 'foreign and accidental to the nature of things'.[2] It is because Kant *assumes* as a basic fact that it is a 'law of reason' that 'requires us to seek for unity',[3] for system, for consistency, that he does not seriously raise the question I asked earlier,[4] *why we should seek* to extend the domain of experience from the individual self-contained object to its systematic interconnection with other objects. All we can say is that just because the concept of the object is suspended on the spontaneous framework of the understanding, without which it would possess no 'life of its own', so we, through the faculty of 'reason', are entitled (*because we must*)[5] to go farther, driven by 'the analogy of a causal determination' supplied by the concept of the understanding,[6] and seek such a systematic interconnection which, though we require it, we can be certain is not given by the transcendental processes of the understanding.[7]

Before proceeding, and for the sake of what follows, I want to stress once more the importance of the Kantian phrase 'what makes knowledge [Erkenntniss] of some given X possible?' This is always

---

'object'; a *system of objects* has not the same logical status as the objects which form part of it. A system is never 'given' in the sense in which an 'object' is given at the empirical level; only heuristically speaking may we say that 'systematic unity under merely empirical laws' is sometimes met with, 'as if it were a lucky chance favouring our design' (J. 20). (Cf. below, p. 516n.3.)

[1] K. 535. For the same point, cf. J. 16; IJ. 10–11, 14–15.

[2] Precisely the same mode of argument will be met again at the level of the understanding. There, instead of the *system* of objects, what has to be made good is the concept of the object itself, the possibility of which we can 'know' only if it becomes 'self-wrought', a spontaneous function of the cognitive judgment itself. We note at once the close connection in Kant's doctrine, on the one hand, between the question of a cognitive grasp (Erkenntnis) of an object (*quâ* concept of that object as such) and the ontological status of the object on the other. (Cf. also below, p. 527n.3.)

[3] K. 538; cf. above, p. 497. This is perhaps the only 'explicit' metaphysical element or 'fulcrum' in Kant's system, in the traditional sense of the term.

[4] Cf. above, pp. 498ff., 503ff.

[5] For this phrase, in just this context, cf. IJ. 15; above, p. 504.

[6] The power of such an analogy is mentioned at K. 568; cf. above, p. 502.

[7] I have discussed Kant's motives for this advance from understanding to reason in my 'The Relation between "Understanding" and "Reason" in the Architectonic of Kant's Philosophy', *Proc. Arist. Soc.*, **67** (1967), pp. 209ff.; see especially pp. 221–6.

concerned with the notion of cognising such an X *in principle*. The question is a little like asking *whether it makes sense to speak of* knowing this X, and refers us to the enquiry as to how to construe its *concept* so that we can show it to be a possible object of cognitive understanding. It is really always an enquiry after both meaning and adequacy of the concept involved. Thus, in the present instance, Kant is not concerned with the details of the theoretical unity concerned—that is a matter for inductive enquiry—but only with the formal aspect; and his point is that we can be assured *apodictically* that 'unity' has *meaning*, or again, has 'objective reality', only if this concept is understood as the result of a methodological procedure on the part of the reasoning self, which thus establishes its 'possibility'. Only if we have injected unity ourselves can we be certain that the empirical facts 'really' indicate the 'existence' of such a unity. Thus, in a paradoxical manner, we can assure ourselves of the 'existence' of unity in one sense only by denying it in another. The same basic approach will later be followed also in respect of nature, i.e. possible knowledge of empirical objective situations. But, being more basic and abstract, that idea is also more elusive, and it is therefore well to understand it in the present more concrete context.

(iii) *Regulative and constructive ingredients of theories*

Kant, when emphasising that man does not find, but projects systematic interconnection into the world, refers to 'reason' in its 'hypothetical employment'.[1] (In the third *Critique*, as already noted, he employs instead the term 'reflecting power of judgment'.)[2] Such a terminology reminds us that he is here treating of theories whose explanatory basis has mere hypothetico-deductive support. The explanatory hypotheses, Kant points out, are not *'proved'* by their consequences; theory can only argue to, or 'approximate' to their truth.[3] Moreover, in the present context, such an 'approximation' is envisaged solely as the 'aim' of reason 'to bring unity into the body of our detailed knowledge'.[4] This is a far more sophisticated logic than the old idea of inductive support through accumulation of instances; the emphasis is all on 'extension'[5] of the domain of the theory; in

---

[1] K. 535; cf. K. 537. [2] J. 15; IJ. 16.

[3] K. 535. Cf. above, p. 498. Nevertheless, their 'probability' may 'increase and be raised to an analogue of certainty' (L. 75). For the *locus classicus* of the formulation of the hypothetico-deductive logic, cf. Descartes, above, ch. III, p. 144. Descartes does actually there use 'proof', but he means non-deductive inference.

[4] K. 535. [5] For 'extension of theory', cf. above, pp. 499, 504.

Kant's terminology: 'the hypothetical employment is regulative'; it creates the field in which the hypothesis is to be tested; and it must be remembered that such experiments as are here undertaken in their turn involve (as he points out expressly) further theory.[1] Here, the expression of 'regulativeness' always refers to the methodological, constructive or creative aspect of reason;[2] it 'indicates a procedure' for constructing a theoretical system.[3]

The constructive aspect of the 'regulative employment of reason' involves, however, reference to more than mere logical systematisation. Kant anticipates in an astonishing manner more recent consciousness of the element of 'theoretical construction' in science.[4] Instances of such theoretical constructs are found in the 'pure air, pure earth, pure water' of earlier times, or in the efforts of the chemists of Kant's time to reduce the many different chemical substances to two or three groups, e.g. acids, alkalis, earths. Nor, Kant adds, should we think of the theoretical ingredient of science as aiming at a mere increase of the 'probability' of the resultant explanation; the whole theory could never *begin* if the constructive approach were not postulated from the start, governing the form of the theory itself.[5]

Kant's remarks at K.539 make the logical status of 'construction' at the level of reason particularly clear. One should not construe, he points out, the elaboration of such constructs as a mere labour-saving device whereby 'reason' might hope 'through the unity thus attained, to impart probability to the presumed principle of explanation'. Now the hasty reader might interpret this as the claim that

---

[1] Ibid. Cf. also IJ. 6 for the remark concerning the 'theory-ladenness' of experiment, which foreshadows present-day views, from Duhem onwards: in 'experimental physics', even the 'principles according to which we perform experiments must themselves always be derived from the knowledge of nature, and hence from theory'.

[2] Cf. the expression 'creative reason' at K. 551.

[3] Cf. K. 547. Cf. K. 566, which speaks of 'the regulative principle of the systematic unity of the world'.

[4] Cf. for instance the concept of 'law-cluster terms' mentioned in ch. II, p. 33, and Kant's reference to experimental physics quoted above, n.1.

[5] Cf. K. 534, 539; and a similar reference to the example of gravitational force at K. 545. In the latter case, Kant's approach is again very sophisticated: it is a misconception to ask whether such a basic force *exists*, for it is really no more than an expression of the working of a number of regulative ideas, such as 'manifoldness, affinity, unity' (K. 544), in terms of which the analogies holding between different phenomena of dynamics are reduced to an orderly whole, by the introduction of this concept of force, whose quantitative expression, through the law of gravitation, is alone a purely inductive matter. (For the experimental status of this law, cf. below, p. 674, where it is shown that there is yet a third question, concerning the 'possibility' of attractive force. See also next subsection (iv), and pp. 565–7.)

'reason' bestows *more* than probability on its explanations, aiming at 'rational certainty'; but nothing could be farther from Kant's mind. The point is that such unitary constructs are presuppositional forms in accordance with which nature is investigated and in terms of which it is 'seen'. And if Kant adds that 'in conformity with the idea everyone presupposes that this unity of reason accords with nature itself', this is only because he holds that 'nature' (i.e. 'the order of nature') is a function of this constructive activity itself: 'Reason does not here beg but command'.[1] One thing only may seem to divide Kant from similar 'subjectivist' methodological doctrines espoused during the present century: he evidently believes that the advance of science in its constructive activity will be unilinear, resulting in ever-growing unification of concepts. But the question whether science will continue to operate with competing 'paradigms', or whether it is advancing towards a true 'unity of science' is after all still under debate.

To return: just as specific theoretical concepts govern the construction of theories, so there are a number of more general 'regulative maxims' that determine acceptable forms of such theories. This is the real status of the earlier Leibnizian principles of the economy of nature, of continuity and of the shortest way.[2] It is an important example of Kant's method of converting metaphysical principles into something possessing purely methodological force.

To call such principles 'methodological' is not, however, for Kant a pejorative gesture. Certainly, he insists, we should not think of them as merely psychological, 'for they do not tell us *what happens* . . . but how we *ought* to judge':[3] once more a reference to his basic claim that, if we want certainty concerning the order of nature, this must be viewed as a spontaneous creation of methodological reason. Indeed in his concern to prevent any undervaluing of the creative status of regulative principles, Kant on occasion is not satisfied with the

[1] K. 539; cf. above, pp. 497, 506–7. Clearly 'reason' has here broken through its old seventeenth and eighteenth century confines. The whole doctrine somewhat cryptically anticipates such recent views as those espoused in T. Kuhn's *The Structure of Scientific Revolutions*, with its basic contention that central scientific 'paradigms' are not tested, verified or falsified in any straightforward way, but govern the very direction of such tests, etc. They can only become obsolete, collapse under the stress of competing paradigms, through an accumulation of internal logical stresses or 'anomalies'. This is a clear case where modern methodological and historiographic discoveries help us to understand more clearly Kant's vision, hidden from earlier readers if only because they were expressed in too traditional a language, a language whose terminology pointed to older preoccupations and logical situations, and which could not bear the weight of such richer insights.

[2] J. 18; K. 538, 543; cf. ch. VII, sects. 3(b)–(c).   [3] J. 19; my italics.

appellation 'methodological', since such a label may seem to put too much emphasis on the heuristics of method. The latter may yield certain anticipations and corresponding hypotheses which, if and when confirmed, lend strong probability to the related maxims. But regulative maxims, Kant protests, are more than this, for they are what first creates the machinery in terms of which we look at nature and construe it. They are not so much 'methodological devices' as 'transcendental principles'.[1] In other words, they are not just *precepts of* methodology, but *principles presupposed for* the development of such methodologies, and the resulting 'order of nature'.

(iv) *Criteria for the legitimacy of hypotheses*

It may be of interest to see how the various criteria stipulated by Kant as certifying the legitimacy of an hypothesis bear on his general epistemological outlook, with which they are closely intertwined. At any rate, it will be clear by now that Kant's methodology yields powerful tools for a fuller articulation of such criteria than had so far been offered—or have been offered since until recently. Indeed, the rough classification into the inductive and consolidative components, making up the basis of an hypothesis, which were mentioned in chapter II,[2] can now be refined. The most explicit statement in Kant's writings occurs in his *Logic*, but we need to supplement his remarks there by reference to the sections in the first and third *Critique* which either implicitly or explicitly bear on the matter.[3]

Briefly, we may distinguish between criteria for the probability, the possibility, and the rationality of hypotheses.[4] Regarding the first, we have already noted Kant's view that, since we can never know all possible consequences that may actually follow from some 'adopted principle' or 'hypothesis', we can never '*prove* its universality'.[5] This is no isolated remark; throughout the *Critique* Kant gives evidence of having a clear appreciation that, as he once puts it: 'In natural science [as distinct from mathematics] ... there is endless conjecture, and certainty is not to be counted upon'.[6] Evidently, more in line with the

---

[1] K. 543–4; cf. passage from K. 539, quoted above, p. 511.
[2] Above, pp. 36–41.
[3] L. 75–6; K. 240–1; 535, 550–1; 612–13; 625–8; J. 315–20.
[4] 'Rationality', it should be noted, is not a term that Kant uses in this context. I employ it as a reminder that the considerations relevant here occur largely in connection with the concerns of 'reason' in its hypothetical employment. (Cf. also the notion of the 'functional *a priori*', discussed in ch. II, pp. 33, 40–1.)
[5] K. 535; italics in text. Cf. above, pp. 498, 509.
[6] K. 433. The principles laying the 'metaphysical foundations of science' (e.g.

tone of Locke, we have travelled a long way from the claims to
rational certainty found in the writings of Descartes or Leibniz; or
rather: these claims here find a complete reinterpretation, and a more
adequate articulation. Nor does Kant's more complex classification of
criteria (corresponding to the levels of a transcendental metaphysics
of nature, generalising induction or hypothetico-deduction, and the
systematic activity of reason) leave intact the rationalist metaphysics
behind Locke's sceptical tone. Kant's conjecturalism does not involve
a contrast with any claims to the possibility of certainty at the level of
physical enquiry into nature, but at that of the *metaphysics* of nature,
which develops only the *possible* forms of physical principles and
concepts.

Certainly, all we can demand for inductive probability is that the
facts explained by the hypothesis should 'follow legitimately from the
assumed principle'; and that one should require no arbitrary additional hypotheses to save an hypothesis threatened by disconfirma-
tion.[1] In general, if we assent to the 'truth' of an hypothesis, we do so
'on account of the sufficiency of the consequences'.[2]

The echo of the old rationalist theories of science we meet then only
in the criteria for possibility: one of the most brilliant cases of Kant's
exciting 're-locations' or 'shifts' of earlier traditions. His *a priorist*
claims never extend farther than the *possibility* of some given subject,
group of statements or concepts, whether in connection with the
axioms of Euclidean geometry, the Newtonian laws of motion and
gravitation, with causal generalisations, or straightforward empirical
statements of particular matters of fact.[3] The *a priorist* approach
towards possibility is crystallised in Kant's criterion that the 'possibility of the supposition itself' must be 'apodictically certain'.[4]

At first sight, this criterion seems to involve too strong a claim, at
least if 'possibility' has the broad sense which it sometimes bears in
Kant, when it includes reference to preferred explanation-types, or,

---

Newton's laws of motion) are, however, excluded from this. Their status will be
discussed in section 9, where it will be found to relate to 'possibility'.

[1] L. 76.

[2] L. 75. 'Sufficiency' is clearly here for Kant a relative matter; it bestows, as he has
told us, no more than 'probability'. But this may reach a degree which is an 'analogue of certainty' (ibid.); cf. above, p. 509n.3.

[3] If this is often concealed from the Kant student, it is because, in the geometrical
and Newtonian cases at least, Kant *seems* to be concerned with the actual axioms,
rather than their possibility. This, as we shall see, is a misconception. (Cf. below,
sections 7(e): ii, 9(a)).

[4] L. 76.

again, mathematical analogies.[1] I think, however, that these are really examples of quite a different group of methodological criteria, which we might name criteria of 'intelligibility'; they should be distinguished from the more basic and, for Kant, much more important criterion of 'possibility' proper, in its 'transcendental' sense: a genus which itself covers two species, first, the 'pure' transcendental principles, secondly, the application of these principles to the empirical concept of 'matter', which leads to Kant's 'metaphysics of nature'. To this we shall return; for the moment let us note only Kant's contention that an hypothesis should involve only suppositions which do not transgress the limits of categorial principles and the framework of space and time—the 'pure' case; and that it should not conflict with those more specific physical principles which are connected with the general principles, and which are enumerated in the *Metaphysical Foundations*.[2] Again it is not yet possible to discuss this criterion of pure and applied possibility, until an account of Kant's transcendental method has been given. It is sufficient, for the moment, to appreciate that, unless we keep in mind Kant's methodological views as here sketched, the significance of the transcendental approach cannot be grasped; an approach whose primary aim is precisely to supply a systematic treatment of the notion of 'real possibility'. Once again, a leap-frogging account is forced upon us.

Turning now to 'rationality', let us remember Kant's view that the empirical, inductive foundation can impart to an hypothesis no more than probability; confirmation by way of 'repeated comparison of observations' with the deductive consequences of an hypothesis, he tells us, 'is for the most part of little importance' in science.[3] It is here

[1] Thus, according to L. 76, the explanation of earthquakes and volcanoes by means of an hypothesis which assumes the earth to be 'an animal' is an impossibility, as compared with the 'possible' hypothesis of a subterranean fire. At M. 194, the geometrical properties of the sphere are employed to suggest the 'possibility' of a gravitational law of the inverse-square. Neither of these are cases of the kind of possibility which Kant is mostly concerned with, and it seems best to distinguish such cases by using a different term, such as 'intelligibility'.

[2] K. 613, for example, states that it is illegitimate to assume an hypothesis that involves extra-sensory perception, or a claim to clairvoyance, or again, the conception of 'a force of attraction without any contact'. (Cf. also K. 241.) The latter example is of particular interest, since one of the main objectives of M is indeed to devise a 'metaphysical' account that will legitimise this concept (make it 'comprehensible'). What Kant has in mind in K is only the concept of attractive force as it occurs in the 'mathematico-mechanical' atomistic scheme of Newton, as contrasted with his own 'metaphysico-dynamical' account, given in M. (Cf. M. ii; p. 201; also below, pp. 557, 566-7.)

[3] K. 627.

that the criterion of what I have called 'rationality' becomes relevant. Whilst generalising induction, as well as comparing the deductive consequences of *single* hypotheses with observation, are relatively primitive, not to say unrealistically conceived methods in science,[1] we must always seek to incorporate a given hypothesis into the more complex framework of some one or more theories.

At this point, a confusion between the aspects of probability and rationality is easy. For the incorporation of some empirical generalisation into a theory, thus linking it with all the other laws forming part of such a theory, has always been believed to add more inductive strength to an hypothesis, or to increase its inductive probability, notwithstanding the possibility of the subsequent overthrow of the hypothesis in question. But the logic of such a 'consilience of inductions' (as it is often called) is still quite unsettled. And in any case, we have already shown that although Kant's position may have gained plausibility from this conception, it was primarily concerned, not with the question of the strength of the inductive generalisation regarded as a law, but rather with the question of what made such a generalisation into a law, treated as a problem of logic and semantics rather than the pragmatics of induction.[2]

The aspect of 'rationality' thus resides at the point where the notion of 'consilience of inductions' merges into Kant's general approach involving the 'hypothetical employment of reason', already discussed in the previous sections. With that discussion in mind, it will be understood that the 'rationality' of this method is certainly not that of the older pre-critical tradition. Rather, it belongs to the 'projective' nature of the system, within which such a (law-like) hypothesis first comes to life, the methodological drive of reason's systematic activity being the presupposition of its acceptability. The centre of gravity of the hypothesis is located in the dynamical nature of reason itself, the system never being 'given in itself', but capable only of being 'regarded . . . as a problem'; being no more than 'a projected unity . . . prescribed by reason'.[3] Any 'objectivity' which such a system, and

---

[1] Cf. Kant's remark, mentioned earlier, from IJ. 6, that experimental testing itself involves principles embedded in theory; cf. p. 510n.1.

[2] This is clear also from his remarks concerning the 'probability' of hypotheses at K. 539, discussed above, pp. 509–12. In terms of ch. II, we are more concerned here with the criteria of lawlikeness; and it will be seen that Kant's criteria include some of those listed in sect. 4(b) of that chapter, viz. the 'systemic group', pp. 45–8, to which the reader should refer.

[3] K. 535. Cf. above, pp. 505–8.

with it, the individual explanatory hypothesis, may possess, is thus defined only in the sense that it derives from 'a procedure' which reason *must needs* follow.[1] In short, the hypothesis is legitimised as rational because it is a creature of reason; a reason which, when expressed as the system of 'regulative principles of the systematic unity of the manifold of empirical knowledge in general',[2] is what Kant calls a 'transcendental presupposition' of the very possibility of a theoretical system itself.[3] The subtle change in the notion of 'rationality' will thus be evident, paralleled by a similar transformation of the notion of 'objectivity'. This goes also for the associated notion of 'necessity', both of the empirical laws and that of the system of laws.

(v) *The 'necessitarian' character of empirical laws and theories*

Kant never wearies of repeating that empirical laws of science are for us (or 'our understanding') only contingent, since they can be 'discovered only empirically'.[4] Nevertheless, simultaneously, we must also 'think' these laws as 'necessary'. In the light of the discussions in the previous chapters, it will be interesting to trace Kant's arguments in favour of such a view and to consider the sense in which laws are here claimed to carry necessity.[5]

One thing is clear already: obviously it is axiomatic for Kant that no empirical hypothesis or law can acquire necessity, become 'necessarily true', on any inductive grounds, since the latter yield only

---

[1] K. 547; for 'objectivity', cf. K. 556, 538; also above, p. 503, and n.4.

[2] K. 550.

[3] For this reason he speaks in the same place of the 'transcendental deduction' of the 'idea' which such a unity pictures. We shall return to the notion of 'idea' presently. Note that this transcendental deduction of ideas should be strictly distinguished from the deduction of the categories. As already mentioned, the latter has to do with the 'possibility of nature', the former with the 'possibility of an *order of* nature'. Moreover, 'nature' is given as an 'object', the 'order of nature' is not, but entirely results from the self-wrought activity of reason. Hence the present deduction is said to be only 'subjective', whereas the categories are 'deducible' objectively. (Cf. above, p. 507n.6.)

[4] J. 21; also J. 20, 16; IJ. 10.

[5] It should be noted that in this section we are concerned with 'empirical laws', as contrasted with the laws which are the object of investigation in M. The former, being contingent, are so far synthetic *a posteriori*, although based on contingent statements, they contain, like all such statements, an *a priori* link, viz. transcendental concepts (as shown in section 8). Empirical law-likeness is akin to W. E. Johnson's 'nomic necessity', as discussed in ch. II.

On the other hand, the laws discussed in M are synthetic *a priori*, and concern the *possibility* of certain statement forms.

probability, however much such a probability may reach the status of practical certainty (an 'analogue' of apodictic certainty only), or 'truth' *simpliciter*. Nor is an hypothesis made 'necessary' by the fact that it fulfils the conditions of possibility, in the sense explained. To be sure, as we have already noted, statements of hypothesis, being empirical propositions, have a categorial ingredient; for instance, as propositions concerning some change of state of a physical object. They presuppose (in a transcendental sense) a necessitarian nexus. But this is not the context in which the necessity of laws is now conceived. At best, as we have found Kant saying, this law-like necessity carries with it an 'intimation' of that transcendental necessity to which all empirical propositions, law-like or not, are subject.[1] Necessity therefore attaches to laws simply in virtue of their being regarded as laws.[2] But why then must these uniformities be 'thought as laws'? Kant's answer is that although we can never have 'knowledge of' or 'insight into' such a necessity, we have to 'think' it—let *us* say: inject it into the grammar of empirical law-like statements—because 'otherwise these laws would not constitute an order of nature':

> Although the understanding can determine nothing *a priori* in respect of such laws (*quâ* empirical objects), it must, in order to trace out these empirical laws (*quâ* laws), place at the basis of all reflection upon the latter an *a priori* principle, viz., that a cognisable order of nature is possible in accordance with these laws.[3]

But we are clear by now that such a reference to 'the order of nature' carries with it that whole vast complex of the hypothetical or regulative employment of reason traced in the preceding sections. The argument comes down to this. Necessity is never a feature derivable from our experience of particular matters of fact, including empirical uniformities. (This was also the result of Hume's reflections.) Secondly, although necessity as a categorial element determines the possibility of experience in general, it still does not yield the necessity of particular laws, which is a necessity *not* connected with their being *empirical* statements. Therefore, only as an analogy which parallels[4] the categorial case, should we regard necessity as 'injected' into the scientific situation, to yield its specific character of systematic order,

---

[1] K. 195. Cf. above, p. 501.

[2] The equation is stated explicitly at J. 21, where Kant asserts that inductive uniformities 'must be thought as laws (*i.e.* as necessary)'; my italics.

[3] J. 21; cf. above, ch. II, pp. 32f.   [4] For this parallel, cf. above, pp. 506.

with all those 'regulative' aspects (e.g. the regulative maxims) which this notion entails.

Necessity is thus infused into statements of empirical uniformity in a somewhat indirect way. We have already noted that Kant regards the aggregate of empirical putative laws as only an 'accidental aggregate', prior to 'unification' into a systematic whole. Only when this becomes 'a system' do we obtain 'a thoroughgoing connected whole of a law-like kind [durchgängig gesätzmässiger Zusammenhang]'.[1] The necessitarianism here implied belongs so far only to the logical connections *between* the laws established by the deductive theoretical system. However, Kant slides from this also into a law-likeness (necessitarianism) of individual laws, evidently implying that, without the idea of system, we have no reason to 'postulate' any law-likeness whatsoever. The result of this slide becomes plain in a passage at J. 16:

> These laws, as empirical, may be contingent from the point of view of *our* understanding; and yet, if they are to be called laws (as the concept of nature requires), they must be regarded as necessary in virtue of a principle of the unity of the manifold, though it be unknown to us.[2]

In other words, laws (*quâ* which they are principles of necessitation) are necessary not because they *follow validly* from higher-level hypotheses, which would give only a trivial sense,[3] but also because without the concept of a theory, itself engendered by the regulative systematic activity of reason, the whole concept could not arise in the first place. Once more we meet here those exciting transformations of traditional locutions made possible by Kant's method of re-locating older concepts within a different and richer methodological structure. It is interesting also to mark the way in which Kant's treatment of law-like necessity transcends in an elegant way earlier, cruder views, as much as it does Johnson's later concept of nomic necessity, applying to the constitutive[4] level, from which derive indirectly the later versions of Kneale and Braithwaite. The equivalent of the 'constitutive point of view' in Johnson and Kneale appears in Kant at the regulative level of reason, whilst the shadow of the 'constitutive' order

---

[1] IJ. 10.
[2] Cf. above, pp. 505–6. The 'thinness' of Kant's vocabulary here easily causes confusion. For 'nature' we should read 'order of nature'; 'unity of the manifold' is a reference to the system of empirical laws, and not—as in the sections of the transcendental deduction of the first *Critique*—to the manifold of perceptions.
[3] For a modern example of this, cf. above, ch. II, pp. 46–7.
[4] Cf. above, Ch. II, pp. 28ff.

proper in this context will be allowed a place only in the fictional procedure of reason, to 'give an object to its idea'[1]—an idea which in its true function, however, only crystallises the regulative aspect of reason. On the other hand, what in Braithwaite's version is only a purely formalistic ('honorific') element is for Kant more than mere formalism; the idea of unity has more than mere psychological, logical or methodological significance.

To appreciate this, however, we must pursue further Kant's teleological version of reason (or 'reflecting judgment').

The relevance of teleology to the grammar of law-like statements is fairly clear. Many philosophers and scientists have felt that what converted a scientific generalisation into a law was not just its deducibility from higher-level axioms (hypotheses) but rather that behind this logical form there lies what alone is of genuine value in theorising, i.e. to provide an explanation. Others have gone even farther, and maintained that any such explanation, in order to be more than formal, must have a genuinely *causal* aspect. They have held that in explanation the facts rendered by the explanatory hypotheses must in some sense be causally related to the facts to be explained. But further, teleological explanation has frequently seemed here the supreme form of causal explanation, witness Leibniz's favourable attitude—not to go back to Aristotelian views.

The preferential status accorded to teleological explanation or causation arose from the fact that the 'cause' was here conceived as acting 'internally' to the 'effects', somewhat on the model of conscious purpose—neither pulling nor pushing, but 'determining from within'.[2] It was in this way that, according to Leibniz, the monads evolve in accordance with considerations that touch on the optimum possible arrangement of all. Clearly, this idea of an internal bond, provided by the final cause, would be admirably suited to supply an image for the nomic kind of necessity of empirical laws for which Kant is searching. But quite generally, it is not surprising that Kant should have sought to re-vitalise the teleological point of view. Always conservative, always trying to preserve deeply entrenched intellectual forms, he was ever ready to seek to retain as much of these forms as was consistent with his general approach, the application of his 'Copernican method'. Let us follow Kant's application of this method to teleology.

[1] Cf. below, p. 524.   [2] Cf. Kant's definition of 'purpose', below, p. 521.

(vi) *The interpretation of 'systematic unity' as 'purposive'.*
    'Physico-teleology'

We have already noted that Kant often uses 'systematic' and 'purposive' unity as alternative expressions. Actually, at first sight it seems surprising that he should have called the systematic unity of reason 'purposive'.

Consider an artefact, such as a watch, which constitutes a 'technical' system of parts and functions. The coherent working of such a system we can *understand*, if we see it as the creation or actualisation of a coherent plan (or 'concept') in the mind of the artisan who fashioned it. Also the different parts of this 'machine' will, in Kant's language, display adaptiveness or purposiveness both mutually and with respect to the plan. The plan (when first envisaged) contained the 'ground', or explanation, or 'final cause' of the eventual artefact. Without the plan, or knowledge of the 'end' or 'purpose',[1] we might describe the working of the separate parts of the structure in accordance with their mechanical laws, but we could not predict or grasp the reason of the cooperation of the different parts in the systematic whole.

Kant now employs this analogy in its application to the case of the whole of nature, viewed as such an 'artefact' or piece of natural engineering.[2] The point of the analogy is this: without a 'plan', i.e. without the interpretation of the different uniformities as members of a theoretical system, the aggregate of the laws of such a system would only form a 'contingent' (or meaningless) concatenation. *With* such a theory, constructed under the guidance of 'hypothetical reason' (or 'reflecting judgment'), this contingent aggregate is, as we have seen, converted into a systematic whole. By analogy with the technical model, Kant therefore speaks here of the 'purposiveness of nature',[3] or again, of 'a formal teleology of nature'.[4] It is *as though* the world had been fashioned by a designer, some 'archetypal intellect',[5] in accordance with a plan, specifying the details of the system through

---

[1] It will be appreciated that Kant's most general notion of 'purpose' has a close affinity with the Greek concept of 'final cause' or 'telos', although he bolsters this up also by reference to narrower interpretations such as 'function', or again 'relative purpose', as well as the concept of 'natural purpose' in the biological context of eighteenth century science, where one thinks of an embryo as containing a 'formative power' in terms of which it strives to realise its perfection as the completed member of a given species. (Cf. J, paras. 65ff.)

[2] Thus IJ.11 speaks of 'nature as art', or again, of 'the concept of the technique [*Technik*] of nature'.

[3] J. 17.　[4] IJ. 10.　[5] K. 565.

which the different parts or laws are related. In fact, of course, this plan has been fashioned only by us, to help us grasp the coherent system of laws.

This is thus again an expression of Kant's 'Copernican method'. Final causes do not 'exist': things *are not* organised teleologically. Rather, they are *viewed by us* under the guidance of the notion of a teleological organisation: this point of view acting as 'a guide for the investigation of nature'.

Here, the example of Maupertuis, which had supplied physics with such a 'strange and admirable . . . union of different rules, *quite independent of one another* according to all appearance, in a single principle',[1] provided a clinching paradigm. The different laws are derivable from the Principle of Least Action, but in this case we also come to 'understand' *why* these laws are what they are; the facts which they express are 'the result' of the economical arrangement of the universe, each individual fact being 'adapted' to such an arrangement. It is as though each of these facts were 'pervaded by' this idea of economy, an idea which determines in this paradigm the very nature of the system, very much as in Greek philosophy where the 'end' or 'purpose' or 'final cause' (*telos*) determines the being of facts; it is 'the concept . . . [which] contains the ground of the actuality of' these facts.[2] But 'ground' here means no more than 'explanation' in terms of a 'final cause', through which we 'see' and 'understand' the arrangement of the laws.[3] Such a cause is entirely 'intellectual'

[1] J. 229; my italics. (For the full passage, cf. above, p. 496.) Note that this supplies us with a *technical* case of Kant's 'contingency' of the empirical uniformities prior to their systematisation, a case which Kant means to *generalise* to the universal contention that it is the idea of 'system' which first injects, and hence defines (makes possible with apodictic certainty) the concept of a law-like nature. Kant does not mention Maupertuis' name but the phrasing is almost identical with that employed in OG, when he discusses the Principle of Least Action. Most commentators have concentrated on the biological paradigm of teleological explanation. But it is difficult to see how Kant, on the basis of this paradigm alone, could have been so bold as to regard all 'system' in nature as 'purposive', including the 'physical' cases. When Newton had claimed the evidence of design to be derivable from the system of the *Principia*, he had clearly bequeathed upon Kant the *problem* of giving a sense to this claim.

[2] This is the phrase in which Kant *defines* his use of the term 'purpose' (J. 17). Note that this is one of the re-located positions for Leibniz's Principle of Sufficient Reason, the other being the Second Analogy. (Cf. below, p. 532n.3)

[3] As so often in Kant, the choice of term enables him to smuggle in those earlier 'ontological' contexts of the concept of ground mentioned in the discussion of Maupertuis in OG. (Cf. above, p. 493.) The importance of these associated meanings will emerge presently.

and does not refer us to any physical determinant. The individual laws in our present example from Maupertuis are undoubtedly due to the action of 'efficient causality'.[1] But their logical nature as 'law-like' entities is first caught in the teleological approach, though there does not 'correspond' anything to this 'existing' in the world. We only '*consider*' the particular contingent uniformities—in Kant's terminology: 'rules'—through the spectacles of such a teleological unity which the rules (thereby becoming 'laws') would have 'if *an understanding* (*although not our understanding*) had furnished them to our cognitive faculties, so as to make possible a system of experience according to particular laws of nature'.[2]

So not only does the notion, or 'idea', of a non-human Designer serve as an anthropomorphic image in terms of which to describe the world as designed; we may also view that intellectual activity (hypothetical reason, or reflecting judgment) which yields an understanding of the world in terms of a designed organism, and thus regards it as fashioned in accordance with a plan, *as though* there had been a non-human intellect who had handed us this design.

In Maupertuis we noted that there had been a slide, a purported 'inference', from design to a divine Designer. Kant subjectivises this process, as he makes quite explicit in the third *Critique*. Just as the 'final cause' is only introduced as a device for our understanding, he tells us, so the non-human understanding, in terms of which we may instead view this matter

> must not be assumed as actual (for it is only our reflecting power of judgment to which this *idea* serves as a principle—for reflecting, not for determining). . . .[3]

Thus 'subjectified', we have in Kant the slide or movement from *our* viewing nature under the notion of a 'final cause' to that of a non-human, i.e. 'divine' understanding, and from there to the further

---

[1] Remember, for instance, Leibniz's use of his principle of optics. Most certainly, he had held, the refraction of light can and must be 'explained' also by reference to the mechanical action of Cartesian ether particles. Cf. above, ch. VII, pp. 430ff.

[2] J. 16; italics mine. This is a more general use of the term 'understanding' than that employed in the dichotomy of 'understanding' and 'reason' It in fact shares aspects of both. For the non-teleological version of this point, cf. above, pp. 505f.

[3] J. 16–17; my italics. 'To determine', in Kant's technical jargon, here means 'to deduce from *given* universal principles'; 'to reflect' means 'to subsume particular given laws under more general ones which are *not* given' (IJ. 8–9); i.e. as-it-were 'inductively' to supply 'hypotheses' from which the laws may be deduced. (Cf. above, p. 498n.1.)

notion, of this divine understanding being itself 'the ground' of the unity of the system.[1] This understanding is, as implied in the last passage, to be regarded as 'an idea', a notion which does not enshrine an 'actual' entity, but concerns merely 'a substratum, to us unknown, of the systematic unity, order, and purposiveness of the arrangement of the world'.[2] Let us consider this use of 'idea' in more detail.

(vii) *Teleology as 'Idea'. Physico-theology*

We have returned full circle to the doctrines of the *Only Possible Ground*, where a divine archetype was posited as the source of the adaptiveness and coherence of the special laws of nature. However, the ontological status of this God is altogether changed. The significance of the change is caught by Kant in the conception of the 'idea'.[3] At the end of the last section, I spoke of Kant 'sliding' from the concept of a 'human understanding' to that of a 'divine understanding'. Instead of speaking of a 'slide', Kant's intentions are, however, better described by saying that he holds such expressions to have *equivalent* logical (and ontological) force. This view is stated explicitly in the reply to his rhetorical question: 'May I regard seemingly purposive arrangements as the intentions of a divine will?'; a question to which Kant replies:

> Yes, you may indeed do this, but only provided they are *equivalent expressions*, whether one says, *divine wisdom* has disposed all things in accordance with its supreme ends, or, that the *idea of supreme wisdom is a regulative procedure* in the pursuit of scientific investigation and a principle of the systematic and purposive unity, involving universal laws of nature, even in those cases where we are unable to detect that unity.[4]

In this passage Kant clearly postulates a parallelism between 'idea' *quâ* 'subjectivist' regulative procedure which injects unity and leads to

---

[1] Note well that the divine understanding (in the present context) is only the ground of the unity of laws, not the cause of the existence of what is subject to these laws, i.e. 'matter'. For remember that 'matter' and its 'necessary properties' were incompatible with the teleological idea, a fact which Kant repeats at K. 516. But by making 'matter' a function of the exercise of the understanding, operating on the sensory data, i.e. the possibility of experience in general, and the unity of empirical laws a regulative expression of the function of reason, the contradiction is removed.

[2] K. 566. This point enshrines the very epitome of Kant's notion of a 'critique of pure reason', of a restraint on reason preventing the movement from the idea of purpose to an ontological ground, in the manner of the philosophy of Leibniz.

[3] It will be plain from the context that this 'idea' is altogether different from the 'ideas' of Locke, Berkeley and Hume. Historically, it relates more closely to Platonic ideas. (Cf. K. 309–11.)

[4] K. 567; my italics. The translation has been modified.

the creation of theory, and an equally subjectivist approach towards a divine wisdom which disposes things fittingly, which results from 'giving the idea simultaneously an object'.[1] Let us broaden this picture by noting that Kant in fact recognises four 'analogues', all mutually equivalent: methodological, teleological, theological and empirical.[2] Methodologically, we have the regulative procedure; teleologically, the notion of the 'ground' (*quâ* 'final cause'), as explained in the previous section; theologically, such a ground regarded as 'supreme wisdom', a 'being of reason', a 'substratum', a 'supreme intelligence', a 'divine understanding', a 'wise author', a 'divine will', etc.[3] Finally, the empirical analogue is to be traced in those technical paradigms of theoretical reasoning which yield unitary principles or sets of explanatory axioms or taxonomical systems, as displayed in the achievements of Maupertuis, Newton and Linnaeus. Specifically, we note the parallelism between the regulative procedure which demands and creates unity and connection, and the 'source' or 'ground' of this connection, as well as the location of such a ground in the 'divine wisdom': the 'ground' being that which unites and connects the multiplicity of forms and laws.[4]

Now the Kantian 'idea' (in the present case it is that of the 'ideal of reason')[5] is precisely intended to cover these parallel aspects. Regarded methodologically, it has merely 'regulative' force. On the other hand, we are always 'constrained'[6] to give such an idea 'an object'. When we do this, we *imagine* that the theoretical unity, previously viewed as merely 'projected', is actually completed, and

---

[1] Cf. K. 556, for this Kantian locution. K. 566 speaks of an 'object in the idea and not in [the realm of] reality'. An 'idea' may be regarded either as the expression of a regulative process or as denoting such an object. From this arise the tensions in human thought, which erupt, as we shall see presently, in a necessary 'transcendental illusion'.

[2] Although Kant does not sanction, as we have noted previously, the employment of analogy to supply foundations for metaphysical conclusions, as for instance on the lines of Leibniz's procedure, analogy is permissible provided it bears only on notions that have the status of 'mere ideas', which lack 'constitutive' force, and whose 'object' lacks the status of even 'possible' empirical existence.

[3] Cf. K. 567, 556, 566, J. 16, K. 567, etc. The reckless abandon of this rich tableau of alternative expressions implies Kant's acknowledgment of their lack of existential import.

[4] Cf. Kant's description of the 'supreme intelligence' as the 'primordial ground of the unity of the world' (K. 566).

[5] For the distinction between 'idea' and 'ideal', cf. K. 486–7. The idea, in its regulative employment, 'supplies the rule' for systematisation; *quâ* 'ideal' it supplies the notion of an 'archetypal' and 'individual' ground.

[6] Cf. K. 566, 556.

that moreover it has its source, not in the creative search of the scientist, but in nature itself, or again, in the divine wisdom.[1]

But there are two aspects to such a procedure. On the one hand, Kant wants to say that we *must* 'give the idea an object', the reason being that 'experience never supplies an example of perfect systematic unity', and hence no corresponding 'ground' of such a unity either.[2] And only such a simulated model, an 'object in the idea', is powerful enough to express both the absolute necessity or law[3] of reason to seek systematic order, and of effecting the corresponding heuristic achievement of enlarging our empirical *corpus* of knowledge:[4] 'Reason cannot think this systematic unity otherwise than by simultaneously giving to the idea of this unity an object'.[5] For the scientist, the 'idea' thus serves as a necessary device whereby the *indefinitely extendable* process of research, and the imprecisely defined rules determining its direction, are pictured as *completed*, or at least *completable*, and precise.

On the other hand, Kant is anxious to prevent us from reading any 'ontological' significance into this postulated 'unity'. The idea which simulates the 'ground of the unity' expresses only the scientist's insistence that the world should be regarded 'as if' it displayed such a unity.[6] In a parallel fashion we are enjoined not to misconstrue the status of the 'object in the idea'. It is no more than a 'heuristic fiction',[7] a *'focus imaginarius'*,

> from which, since it lies quite outside the bounds of possible experience, the concepts of the understanding do not in reality proceed; none the less it serves to give to these concepts the greatest possible unity combined with the greatest possible extension.[8]

Even though the idea may thus 'represent' the *object* as 'transcendent', it is itself in fact 'immanent' only,[9] not possessing *transcendent* status: a fact easily confused with its having *transcendental* employment (in

---

[1] These, too, are equivalent expressions: 'It must be all the same to you, whether you say: God in his wisdom has willed it so, or: Nature has wisely arranged it thus' (K. 567).

[2] K. 556.

[3] Cf. K. 538: 'The law of reason which requires us to seek for this unity is a necessary law, since without it we should have no reason at all'.

[4] K. 556–7. Cf. K. 535, which sums up the whole matter by saying that it is the 'function of reason . . . to assist the understanding by means of ideas, in those cases in which the understanding cannot by itself establish rules'.

[5] K. 556.   [6] K. 559; also K. 551, 561.   [7] K. 614; cf. 493.

[8] K. 533. This resembles the Leibnizian God of the *Discourse*, DM. 6; cf. ch. VII, p. 403.

[9] K. 532.

its guise of the representative of a methodological presupposition). The confusion between the two results is what Kant calls 'a transcendental illusion',[1] according to which the 'idea' is treated as 'the concept of a real thing'; with the result that, although lacking the 'conditions of possible existence', this 'real thing' (the 'object in the idea') becomes 'transcendent'. But such a step conflicts, of course, with the very definition of an 'idea' as 'a necessary concept of reason to which no corresponding object can be given in sense-experience'.[2]

We can see that Kant is here operating on two fronts at once. On the one hand, he is concerned to close the advance from 'idea' to 'existence', since the idea of God as a potentially *existing* entity would certainly break through those barriers which Kant has erected to protect the conditions of possible existence, or experience; still less can such an idea be entertained as a *problematical* empirical existence; if God 'exists', in the sense left (after the present critical purge) for the moral argument, His existence must be non-problematic, and apodictic.[3] Moreover, from the start of the *Critique* we have seen Kant wishing to prevent the move from the idea to existence through some form of ontological argument. For 'existence' requires a proof of its 'possibility'; and such a proof involves a reference to the formal conditions of possible experience, or of 'phenomenal reality'—space, time, categories—which in the present case are (by definition) absent; not to mention the material condition of a sensory *a posteriori* content.

On the other hand, the resulting confinement of the idea to its regulative or teleological aspects releases its methodological driving power. If Kant can make good his claim that we stand in absolute need of the idea; and more (even if only as a 'fiction'), of the 'object in the idea' which will complement the notion of 'nature' by leading us to the 'order of nature', this will—in Kant's eyes—supply us with the strongest possible argument for a God, though *not* of course for God's *existence*, in the 'phenomenal' (constitutive) sense of that term.

---

[1] K. 298; cf. K. 532, 549.   [2] K. 318.

[3] In Kant's subjective method, this existence will be demanded as a 'postulate of pure practical reason' (KP. 126); as something 'necessitated by [our concept of] morality' (J. 301n). (Cf. below p. 529.)

The whole nature of Kant's 'transcendental method' is intended, in fact, to purge 'existence of God' of its 'constitutive' use (in the sense of belonging to the 'phenomenal' order) in a way that will permit us to retain *a meaning* for the notion of 'existence of God' which is not logically otiose. And this is why the transcendental method has as its aim to 'relativise' the concept of 'empirical phenomenal reality'; an argument whose bearings we shall elucidate further in the next section (5).

'Can we, on such grounds, assume a wise and omnipotent Author of the world?', Kant asks; and he replies emphatically: 'Undoubtedly we may; and we not only may, but *must*, do so'.[1] Yet, what here we 'must' do is only to 'presuppose' an 'unknown being *by analogy* with an intelligence (an empirical concept)' for the sake of achieving a 'systematic and purposive ordering of the world' in our work as natural scientists.[2]

It is as though Kant was saying that the 'transcendental illusion' had to be incurred by us necessarily, as rational beings, whilst at the same time inviting us to remain ever-conscious of the logical nature of this illusion, in order to prevent us from endowing the 'object in the idea' with properties that must inevitably make it problematic. And this is the very core of Kant's method: the 'critical constraint', or the insulation of the basic key-concepts of traditional philosophy and natural theology from an ontological anchorage which—in Kant's eyes—does not actually have (or at least: cannot be known to have) any such foundations either in systematic experience generated by reason (as in the present case), or in elementary experience in relation to the understanding.[3]

We have noted the *parallelism* between the methodological, teleological and theological aspects of reason, jointly caught in the Kantian notion of 'idea'. What is not so clear is the degree of strength of the arguments in favour of the putative need for the movement from the methodological analogue to its teleological, let alone from the latter to its theological form. Of course, it may be that Kant did genuinely want only to stress no more than the *analogy*, with no intention of letting the old ontological ('constitutive') associations of the concept of God infiltrate into the situation. And more than once he emphasies that whilst 'physical teleology impels us, it is true, to seek a theology . . . it cannot produce one'.[4] Moreover, it is possible that what thus 'impels' the advance to the theological analogue is only

---

[1] K. 566; italics in text.   [2] K. 566; italics in text.

[3] The reference to experience is a reminder that Kant expresses the same point epistemically by saying that the assumption of an author of nature does not 'extend our knowledge beyond the field of possible experience' (K. 566), although this is meant to fix at the same time also the 'ontological' status of the idea of God, as not belonging to the 'constitutive' but only to the 'regulative' order of things. (Cf. below p. 528.)

Throughout what follows the reader should keep in mind that Kant's discussion of the conditions of 'knowledge' of anything are simultaneously meant to fix its ontological status, e.g. as being a phenomenon, or a noumenon. (Cf. above, p. 508n.2.)

[4] J. 291.

derived from Kant's older arguments (expressed in the *Only Possible Ground*). For it is evident that we here meet the echoes from, the shadows cast by, the physico-theological argument, the proof of the existence of God from the putative prevalence of design in the world. Now, as already mentioned, Kant rejects the probative strength of this argument in his later period, pointing out that not only does it commit an ontological fallacy, but it cannot even begin to postulate its premise (the existence of design or 'purposiveness') without either *assuming* a designer (which is reasoning in a circle) or reading design into nature in accordance with the regulative approach of reason (or 'reflecting judgment').[1] However, Kant holds the argument, when critically purged, to have great importance. The idea of the author of nature, which it reveals, is not only a necessary 'guiding concept' in our search for the unity of nature; but in proportion as we are successful in our researches, the concept opens up for us 'the wonders of nature and the majesty of the universe', and thus leads to the contemplation of 'the supreme and unconditioned Author' of nature.[2] In other words, we have here a historically conditioned transition from nature to God, in a fairly typically eighteenth-century guise. As usual, just as in the transition from 'nature' to the 'order of nature', from the 'idea' of the 'systematic unity' to that of 'purpose', so here again we meet a 'looseness of fit', tying 'purpose' and the corresponding concept of 'ground' to the theological conception of God as the author of nature.

Kant has, however, a further, perhaps the most important, motive for suggesting that we are seemingly 'impelled' to seek the ascent to an author of nature. For, as briefly anticipated already, paralleling the physico-theological, there is an ethico-logical argument, where God again plays the rôle of an idea projected, or more technically: 'postulated', by the processes of moral consciousness (of 'reason in its practical employment').[3] Here, considerations of the 'conditions of experience' are no longer relevant since we move now in the realm of 'noumenal freedom'. If we assume a moral nature for man, then we *must* (Kant argues) also see this morality as postulating the reality of a God, as an expression of the necessary fulfilment of the conditions that alone can realise the full exercise of moral activity.[4] There are

---

[1] The point is expanded in detail in J, paras. 75, 85; K. 518–24; 563.

[2] K. 520.

[3] Theoretical reason is concerned with 'what *is*', whilst practical reason deals with 'what *ought to be*' (K. 526). Cf. also K. 474, and below, p. 538.

[4] Cf. J. 301; also KP. 128–30. Kant's ethical and religious doctrines will not be

only two points that concern us here. In Kant's view, the 'deficiency' of the physico-theological argument (in yielding a concept that is only 'subjective') 'impels us to supplement' it with the ethico-logical proof. But further, and *per contra*, the latter, though it yields 'objective reality' (in respect of man's moral nature, which is for Kant the ultimate substrate of reality, *quâ* man as noumenon), requires the intellectual content supplied by the physico-theological approach.[1]

Needless to say, this 'objective practical reality' of the moral postulate must not be confused with the objective phenomenal reality which feeds on the 'constitutive force' (in Kant's sense) of a straightforward empirical entity.[2] But then it was precisely the intention of Kant's 'critical purging' of the physical conception of God, exhibiting it as a necessary subjective device, to enable him to speak unequivocally of 'objective reality' in the *moral* sense; having constrained the claims of theoretical reason (let alone of the understanding) to yield such a notion, in respect of God, he holds that practical reason is free to supply it without fear: once more providing testimony of this peculiar shifting of traditional meanings to new levels—a transmuting of old locutions into novel contexts.[3]

All this constitutes an interesting attempt to reconcile the claims of

---

discussed since they fall outside the bounds of reference of this book. Let it be noted that, here again, the argument is eventually given a teleological form which will increase further the suggestiveness of a new parallelism between the moral and the physico-teleological argument.

[1] For the 'unity of principles' thus achieved, cf. J. 307; also J. 12, 33. (For J. 33, see also pp. 538f. below.)

[2] Hence the famous passage in the Preface to the Second Edition: 'I have therefore found it necessary to suspend *knowledge* [Wissen] in order to make room for *faith*' (K. 29). K.32 makes it clear that the main object of the *Critique* is in fact to 'sever the root' of 'materialism' and 'atheism', as well as 'idealism and scepticism', an aim in which all the philosophers that we have discussed seem to have joined!

[3] It should be remembered that—in line with what we anticipated in ch. II (above, p. 61)—in the present chapter I am using 'constitutive' in Kant's and not Johnson's sense, which was the sense in which I used it in all previous chapters. In Kant, a concept has 'constitutive' force if it relates to what can be 'given' for a 'possible experience', and thus includes reference to the various components of experience, both 'sensory' and 'formal'. 'Regulative' by contrast relates only to the self-wrought processes of ordering the elements of experience, but there does not 'correspond' to this process anything in what is thus ordered. (For a more detailed discussion, cf. below, pp. 617ff.)

On the other hand, Johnson's 'metaphysical' use corresponds to a vague conflation of Kant's notion of a 'noumenal' order, or again, what was called 'the object in the idea' (e.g. of teleological unity), with that of Kant's transcendental ground, not to mention 'theoretical foundation'—a notion with 'constitutive' use (in the Kantian sense).

science and religion, by making both expressions of the subjective interests of reason, in its search for an 'order of nature', insulated by the method of 'critical constraint' from the existential (or 'constitutive') import of the conception of 'nature' as such. But we can already see that, on the other side, as a complementary aspect of this approach, the realm of 'nature' (understood as the aggregate of 'objects' or 'objective situations') will have to be constrained likewise, by regarding it as 'merely phenomenal', depending on the formal conditions of experience in general; thus assuring with logical force that it should stand deprived of any claims to yield (*without* the aid of 'reason') that *order* of nature which alone constitutes a fully-fashioned universe; or to yield (*without* the aid of the 'subjective' *forms* of sensibility and understanding) an independent type of 'existence' ('existence-in-itself') that could make it a serious competitor with the God of Kantian reason.

(d) *The resolution of antinomies*

We have now seen something of Kant's method of solving the problems posed by the old contradictions between mechanism and teleology inherited from traditional sources and his own earlier days. The method consists largely in prising the relevant concepts off their metaphysical location and shifting them to new levels provided by the 'regulative approach', whether to the level of 'reason', or of the 'understanding' (which has yet to be expounded). It may be useful here to survey some of these attempts towards reconciliation, without discussing them in any detail, since such a discussion would both take us too far afield and add nothing to our deeper understanding of Kant's method. Such a survey will graphically illustrate the variety of denotations which so often lie hidden beneath Kant's terminology. We will select the various notions of 'mechanism' involved in Kant's attempt to reconcile mechanism with teleology. When such a survey is entertained, the following contexts of 'mechanism' are revealed.

A. Causality at the level of the understanding. (Transcendental causality.) The necessary attachment of the concept of causality to the notion of 'experience' in general, in respect of 'phenomenal sequence'.[1]

B. Causality at the level of reason. (Empirical causality.)

[1] Cf. below, section 8(e): ii, pp. 648ff.

(1) As 'regulative principle of reason': 'everything in the sensible world has an empirically conditioned existence' (K. 481).[1]

(2) Every empirical process is subject to empirical laws.

(3) A tightened version of (2): Every such process is subject to empirical laws in deterministic form. (This version is sometimes suggested by the treatment of the Antinomies, especially the third.)

(4) As 'regulative maxim or principle' of investigation, especially in the field of biology: 'All production of material things is possible according to merely mechanical laws' (J. 234).[2]

C. The existence, in physics, of a set of laws, which as the 'necessary (mechanical) laws of matter' govern all inanimate nature.[3] These, are transcendental, as in the first case, (A), but are called 'metaphysical' because they involve the 'empirical concept of matter'. (Metaphysical causality.)

With such a rich interpretation of the concept of mechanism, we have a corresponding variety of ways in which a reconciliation of contradictions, or solution of antinomies, becomes possible. Ignoring the antinomy between freedom and determinism which concerns only the foundations of ethics, permissible relationships between mechanism and teleology are as follows.

First of all there is the 'looseness of fit' between A and B. The relations between these are (as shown above)[4] not of a strictly logical nature; at best they are 'analogical' and suggestive. It follows that the mechanism of B will be a function of the spontaneous, regulative activity of reason, engaged in the scientific and religious concerns that have been considered. In this way, for instance, B.4 and its teleological formal contradictory[5] will be declared mutually compatible, provided they are viewed as regulative principles of research, in the causal investigations of biology. At the extreme end of the scale, we get the resolution of the antinomy between the purposive arrangement of nature as an over-all system of empirical laws governed by the principles of B.2–3, and the realm of law in the sense of C. Here, the logical distance between the regulative principle of teleology and the transcendentally conceived laws of motion is at its maximum, thus removing the paradoxical situation of the earlier views.[6]

Finally, our old search for images of the notion of a 'bond' receives here a further penetrating solution. We have already briefly

[1] Cf. above, p. 501.   [2] Cf. above, p. 505; below, p. 658.
[3] Cf. above, p. 495; below, p. 672.   [4] Cf. pp. 499–506.
[5] Cf. J. 234; cf. op. cit., paras. 68, 70, 75, 77–8.   [6] Cf. above, p. 523n.

foreshadowed Kant's interpretation of this concept as the 'transcendental unity' of the understanding, part of the logical structure of possible experience, or phenomenal objectivity, or 'nature'.[1] And in the rest of this chapter, the more detailed account of these notions will be explored. But it will be evident that a second version of this 'bond' occurs at the level of 'reason', as part of the conception of the 'order of nature', of the ground of the systematic or purposive unity of its particular empirical laws; and as the generative source of the concept of law-likeness in general. This is obviously a far more penetrating and subtle classification than Johnson's dichotomy of 'epistemic-constitutive', which we noted in chapter II.[2] Moreover, we now see that Leibniz's notion of the 'ground', postulated for the sake of the *P.i.S.* doctrine, can be seen—against the background of Kant's finer structure—as a conflation of the 'unity' of reason and of the understanding, of 'nature' and 'the order of nature'.[3] We must now retrace our steps, and approach a delineation of the realm of 'nature' by reference to other sides of Kant's interests.

5. THE SIGNIFICANCE OF THE CONTRAST BETWEEN NOUMENA AND PHENOMENA

(a) *Contextual varieties*

In the previous section we have outlined some of the contexts where Kant employs freely notions like God; the substratum or ground of all connections; or again, the sum-total of all possibilities.[4] And we have noted that when Kant invokes these notions, during his 'critical' epoch, he insists that their use is permissible only if we keep in mind certain necessary epistemological constraints, the chief of which is formulated as a constraint on the limits of the use of the concept of 'knowledge': we have no 'knowledge' i.e. cognitive grasp, of such a God, or of a substratum, because these conceptions as employed here transcend the general conditions which—for Kant— define a grasp of empirical phenomena. Similarly, the 'sum-total of all

---

[1] Cf. above, p. 489; below, section 8(c), pp. 623ff.

[2] Cf. also above, p. 529n.3.

[3] Hence we may expect to meet Leibniz's 'principle of sufficient reason' twice in Kant's duplicated procedure. It occurs explicitly under this name as a synonym for the principle of causality in the Second Analogy, as what supports the possibility of objective succession in time. In its second guise, it appears as the demand of reason for a 'primordial ground of the unity of the world' (K. 566; cf. above, pp. 521, 524n.4.

[4] Cf. pp. 489f., 496, 521–4.

possibilities' is a conception which is permissible only if it is located within the boundaries of possible experience, beyond which limitation it has only a purely 'dogmatic' use.[1] Thus, even where (for methodological purposes) we invoke—as an assumption—an author of the world, 'our grasp' (Erkenntnis) is by no means 'extended' thereby 'beyond the field of possible experience'.[2]

Such Kantian declamations sound almost trite unless they are understood as so to speak 'grammatical' consequences of Kant's analysis of empirical cognition, as *necessarily* requiring a reference to sensation, spatio-temporal forms and conceptual (categorial) framework.[3] Moreover, it follows that not only can God not be 'known', in the sense required, but the notion of any ordinary empirical 'thing' or object, in so far as it is treated in abstraction from the conditions of empirical cognition, is subject to the same limitation. Here, again, Kant employs certain technical terms, at first sight suggesting familiar familiar epistemological contexts, with which they are often misleadingly identified. Things, when viewed *subject* to the above-named conditions, are called 'appearances' or 'phenomena'; by contrast, when abstraction is made from these conditions, Kant speaks of 'things as they are in themselves'. Now the latter, from the logic of the situation, are clearly not 'knowable', or 'cognisable'. For certain purposes (to be noted presently) Kant claims, however, that they can be 'thought' —a somewhat Pickwickian sense of 'thinking'; the term 'noumenon' being employed to denote the object of such a thought.[4]

---

[1] Cf. above, pp. 489f., 496. This use leads to the 'ideal of reason' as 'thing in itself' (K. 490).

[2] K. 566.

[3] In what follows, it is necessary to keep in mind that the element of 'sensation', in Kant's use of the term, is the result of an abstraction from spatio-temporal form. It refers to a specific shade of red, for instance, but not to an *extended* patch of red surface *in time*; it is a reference to the 'matter', and not the 'form' of the sensory datum, i.e. of the 'appearance' (K. 66); it should not be confused with the modern 'sense-datum'. When Kant wants to refer to what would correspond to our *apprehending* of this sensory aspect (again a purely abstract notion), he speaks of 'perception' (Wahrnehmung; not to be confused with the normal meaning of this term). 'Matter' plus 'form' of appearance are usually referred to as 'sensible intuition', which still abstracts from the aspect of categorial synthesis, that alone generates the full-blown 'appearance'. (Cf. below, p. 555n.5.) As we shall see, this 'matter' (sensation) is the only element which is altogether *a posteriori*; this again involves a considerable sharpening of received notions of the *a posteriori* as current in the Locke-Berkeley-Hume tradition. Kant's *a posteriori* elements of sensation are not to be confused with Lockean 'ideas of sensation', for instance, since these are extended in space and time.

[4] Throughout the Kantian *opus* the use of these philosophical key-terms is strictly

An expression like 'the thing as it is in itself' easily gives rise to false expectations, even if only of a negative kind. Descartes, we saw,[1] had demanded that objects should be grasped as they are 'in themselves'; and Locke, again, had played with the notion of the 'inner constitution' of things, of their 'real essence'. But for Kant, even the *hypothetical* supposition of any such 'knowledge' is otiose; indeed, interpreted as 'thing-in-itself', the very empirical meaningfulness of the concept of an 'inner nature' is impugned.[2] The 'thing-in-itself' is not a 'thing' in any ordinary sense of the term, for that would require reference to the formal conditions of 'experience'. We should not even think of it as 'existing' or 'not existing' if we use 'to exist' in the context of empirical meaningfulness as defined by Kant. It was for this reason Kant had noted specifically that questions as to whether the uniting underlying 'ground' of the phenomena had 'substance', or was 'real' or 'necessary', are *'entirely without meaning'*, since even the categories 'have no meaning when not applied to objects of possible experience'.[3] Yet, as we have already seen, Kant holds that a concept like that of a 'ground' can be 'thought' by 'reason' (in its regulative employment).

This suggests that Kant is operating with a double-sense of 'meaning'. According to the first sense, a meaning can only emerge in the context of possible experience: a forerunner of the verification-theory of meaning. The second sense occurs in the context of those interests which demand Kant's noumenal thing-in-itself. All this has, however, some interesting consequences for the Kantian noumenon. It is, as just noted, Kant's view that within a context of empirical meaning one cannot speak of 'existence' unless the conditions of experience are allowed for. On the other hand, he seems to be of the opinion that, once he has made it *clear* that one cannot speak of existence *nor* non-existence of noumena in this sense, one is free to speak (or 'think') of noumenal existence in another sense or context, as long as the concept is not self-contradictory.[4] It *is* self-contradictory, he implies, to speak of the existence of noumena in a phenomenal context; but not so if you remove this context. On the other hand, the very notion

---

'internal' to the doctrines involved, though at the same time Kant—like every other philosopher—frequently avails himself of their external commonsense meaning.
[1] Above, ch. III, p. 105n.5.
[2] Cf. below, pp. 546–7, for Kant's strictures on the Leibnizian concept of inner nature.
[3] K. 566; italics in text.
[4] Cf., for instance, K. 481–2. A similar practice we noted before in connection with the concept of 'objective reality' (p. 529).

of such a context (the transcendental framework of experience) being *removable* presupposes the contrast between phenomena and noumena itself; but since this evidently assumes that the latter conception *is* meaningful apart from (and not just within, as Kant sometimes says) such a contrast, Kant's reasoning seems circular.

A comparison may perhaps clarify the difficulty. We sometimes entertain the hypothesis of beings whose mode of existence is quite other than that of ourselves, endowed with a perceptual-conceptual framework altogether 'different' from our own. The problem is to explain this difference, since—short of mere negative—it seems necessary to use our own perceptual-conceptual framework for such a description. This would seem a self-defeating task, whose frustrating nature is only concealed by the fact that we usually envisage a comparison which employs analogies.[1]

Here it may be objected that the task can hardly be hopeless since we ourselves employ the phrase 'perceptual-conceptual *framework*', which suggests a contemplation of such a framework so to speak 'from without'. But it is not certain that this phrase has acquired its meaning from more than mere analogies, coupled with a certain *limitation* in its interpretation, which leaves enough material 'outside' the 'framework' to generate a 'boundary'. Thus, for Kant, the framework is exhausted in mentioning space, time and a limited number of categories, but he offers no proof (apart from rather unconvincing considerations of an architectonic nature) that such an inventory is either necessary or sufficient. Again, phrases like 'a world outside space and time', or 'a non-spatial and non-temporal world', have no clear meaning, unless we employ something like the theory of Leibnizian monads,[2] itself faulty in the present respect; so that it seems that such phrases presuppose that the concepts of space and time be previously explicated—a task that Kant never undertakes, since he really *assumes* spatio-temporal locutions, requiring only that we clarify the ontological implications of these locutions, or that we show how the ascription of putatively unique properties, as of space, is *possible*. Postulating a world of noumena is not so much to answer such questions but rather to pose them. In the present section we shall be able only to note the various attempts Kant makes at a positive and independent characterisation of a noumenal order, in the

---

[1] Thus, there is no difficulty in imagining men who see by means of their hands rather than their eyes; or with no sense of time, being bereft of memory.

[2] We shall presently turn to this important interpretation of Kant's noumena.

hope that one or the other of these may shed a little light on this vexed problem.

What we require, then, are contexts for the notion of a 'noumenon' which relate it to something other than what is merely *not* 'phenomenon'. Now actually, 'noumenal contexts' seem as diverse as the different occasions of Kant's arguments require. Thus, the 'idea of reason' (*ens realissimum*), God, the Leibnizian monad, the Berkeleyan 'idea', as well as the 'intelligible ground' of 'the appearances and their connection',[1] are all characterised at one time or another as 'thing in itself'. It is not even clear whether this appellation denotes a plurality or a singular individual: not surprisingly, since unity and plurality are categorial notions which can not as such apply to the concept of the noumenon. What is perhaps common to all these instances is that they involve something whose spatio-temporal and categorial aspects are *not* regarded as a framework defining the conditions of experience (as is the case for the phenomenon) but as something putatively belonging to the thing itself in question. Thus, to take the extreme case of the *ens realissimum*: this is defined as the 'substratum' of the 'sum-total of all possibility' of things in general,[2] which is, says Kant, 'just the concept of a *thing in itself* as completely determined'[3]—exactly as had been postulated in OG.

Although this seems to supply us with a meaning obtained from the context of traditional metaphysics, and which no doubt continues to linger and exert its semantic influence, Kant now suggests however that the whole notion, if it is to be regarded as something substantive, is merely a confusion of the 'concept of the possibility of things viewed as appearances' (i.e. as suspended within the *framework of possible experience*) with the concept of a '*principle of the possibility of things* in general'.[4] This suggests again that if we want to remain within the field of phenomenal meaningfulness, we can characterise the thing-in-itself only *negatively*.[5] Yet at the same time, again we see that dropping the conditions of empirical meaningfulness (i.e. the reference to possible experience) apparently does not entail for Kant the loss of meaning in the sense of absence of logical contradiction; an 'intelligible' meaning remains and can be 'thought'; enough at

---

[1] K. 471; cf. above, p. 507n.3.   [2] K. 488, 490; cf. above, pp. 489f, 496n.3.
[3] K. 490; italics in text.   [4] K. 494; cf. above, p. 490.
[5] This is the view emphasised in the second edition of the *Critique*'s chapter on Noumena and Phenomena, where Kant says that we must not think of the noumenon as an object of a *non*-sensible, i.e. intellectual, intuition, but only as *not* an object of a sensible intuition. (K. 270, 272; cf. below, pp. 539–40.)

least to attach such concerns as do not involve—in Kant's eyes—
reference to anything given as appearance; e.g. the noumenon as
moral self, as moral cause.[1]

The question however is, of course, whether it is possible for Kant to
utilise any such positive attachments for his characterisation of the
contrast between phenomenon and noumenon; and if this is not
possible, whether the contrast between the former and the latter
(regarded as merely a negation of the former) is sufficient to explain
the nature of the phenomenal order, i.e. of the status of space, time
and the categories. Yet, as we shall see more and more clearly, it is
just this status, as universal *a priori* conditions of experience (intuition
and objective cognition), that needs to be explained. If ultimately this
can be done by Kant only through the contrast with the noumenon
(contrasting for instance the notion of space and time as *a priori*
conditions of intuition with the competing notion of them as properties of things-in-themselves[2]), this contrast must fail to clarify either
notion, unless we are offered an independent characterisation of the
noumenon.

At best, so we shall find, Kant to this end does no more than use
his own characterisations of earlier philosophical positions; though
even here he leans heavily on the views he tries to establish. Perhaps
this is not surprising if—as we must—we view Kant's own system
less as a timeless critique of previous philosophical systems, but rather
as a rival system itself, operating (like its predecessors) within its own
closed sphere of philosophical terminology. At any rate, without these
'positive' meanings, Kant's whole procedure revolves in a logical
vacuum; yet, the positive meanings which Kant offers us do not really
meet his difficulty, since it is not shown that they are logically relevant,
or give us the sense of noumenon required, however much they may
have contributed towards providing a psychological basis, or 'analogies' in terms of which to picture the thing.[3]

But let us explore these further contexts where supposedly positive
(independent) meanings arise in Kant's presentation. One such context we have dealt with already in detail, as the notion of the ground
of the multiplicity of natural laws or 'forms', and I shall say no more
about it, except to note that any meaning attaching to this notion is

[1] Cf. K. 471–2.
[2] The significance of this expression will be developed.
[3] Cf. K. 484, where Kant explicitly notes that we must frequently fall back on 'the
use of analogy, by which we employ the concepts of experience to form some sort
of concept of intelligible things.'

derived from the analogy with teleological reasoning.[1] A further context, which cannot be dealt with in this book except in passing, connects with man's 'practical' concerns; with what 'ought to be' rather than 'what is'; how man *ought* to act, as contrasted with the way in which he *is observed* to act.[2] *Quâ* object of observation, man must be regarded as 'appearance', and it is the transcendental framework of the categories that gives this the logical character of an 'object'. Knowing man observationally as an object is possible only by regarding him as 'appearance'. On the other hand, we may regard man also from a different point of view, with a purely moral interest, as 'subject'. And since the epistemological status of *any* object entirely *depends* upon some given point of view (the phrases 'appearance' and 'thing-in-itself' being otherwise quite undefined) we are *free* to regard man also as a subject whose status is quite other than that of 'appearance', Kant claiming for it the character of a thing-in-itself.[3]

But now it is by no means obvious that the thing-in-itself, which is declared to be 'unknowable' with respect to the realm of objects as 'appearances' in general, may thus be identified as the subject of moral activity, even if the general lines of Kant's ethical doctrine be accepted. It is perhaps best to regard Kant as *assigning* this meaning to the notion in its present context, and to claim a right to this simply because he has *deprived* it of significance at the phenomenal level.

At any rate, Kant evidently does think that all these contexts (of the noumenon as what lies behind the individual 'appearance', as the ground of what systematic experience exhibits to science, and as the moral or rational self) converge unequivocally. In a passage of the third *Critique*[4] they are in fact actually treated as such by implication, the noumenon being referred to as the 'supersensible substrate'. *Vis-à-vis* the understanding, Kant says, our cognition of the phenomenon leaves the substrate 'quite *undetermined*'. Judgment (in its 'reflecting' or regulative activity) makes it '*determinable by means of the intellectual faculty*' (a reference to the theoretical processes studied in the preceding section of this chapter). And if the first context emphasises the notion of a ground on which things *quâ* appearances 'rest',[5] the second belongs to the notion conceived as the source from

[1] Cf. above, p. 524, and sect. 4(c): vii.
[2] For the contrast drawn in these terms, cf. K. 474, and above, p. 528.
[3] For the identification of the moral self as 'thing-in-itself', cf. K. 467–8; also G. 85–86; KP. 6.
[4] For this locution, cf. K. 467.   [5] J. 33.

which spring the natural forms and laws of the phenomenal world. Finally, 'reason, by its practical *a priori* law, *determines* it [the substratum or ground]'.[1] Clearly it is nothing more than the Kantian architectonic that assures here an identity of referent (the substrate or thing-in-itself).[2]

The term 'substratum', however, reminds us of yet another context also, viz. Locke's concept of 'substance' *quâ* substratum; of his account of the substratum as that which 'makes the whole [combination of simple ideas, of whose union it is their cause] *subsist of itself*'.[3] Moreover, though distinct, the associated concept of 'real essence' is not here irrelevant, when Locke speaks of it as 'that particular constitution which everything *has within itself, without any relation to anything without it*'.[4] Now this is precisely the characterisation Kant frequently gives of the noumenon; except that, unlike Locke, he will sharply distinguish between the epistemological and the scientific account of the 'internality' of the 'inner constitution'.[5] Thus, in the first edition of the *Critique*, he speaks of the noumenon as that which has no relation to anything, by contrast with an 'appearance', the very term 'indicating a relation to something'. Against this, the noumenon 'must be something in itself, that is, an object independent of sensibility'.[6]

For Kant, two points are here important. First, when so characterised, we can never frame any notion as to *what* the thing may be in itself, since no cognitive framework is defined to give empirical meaning to this. For this reason, it is—somewhat like Locke's substance—said to be quite 'indeterminate', except that the 'unknowability' of the substratum is made more clearly than in Locke a matter of logical principle.[7] Secondly, with respect to its cognisability, the

---

[1] J. 33; italics in the text.

[2] Thus as J. 12, the realm of the supersensible ground is asserted to be, without much argument, the 'ground of the unity' of the realms of teleology and moral freedom. Elsewhere (at J. 271, 257, 261), Kant postulates the 'coalescence' of teleological and mechanical nature. In all these cases, the 'supersensible ground' is no more than a postulated answer to the demand of reason that there *should* be such an agreement, coalescence, or 'unity'. The answer *defines* Kant's use of ground. Indeed, J. 307 explicitly affirms such a 'unity of principles', whose 'givenness' is assured simply because reason 'bids us to seek' it!

[3] *Essay*, ii.23.6, my italics; cf. above, ch. IV, pp. 216ff.

[4] Op. cit., iii.6.6; my italics; cf. above, p. 208.

[5] For Locke's rather 'fluid' treatment of this distinction, cf. above, ch. IV, section 4(a), pp. 204ff.

[6] K. 269-70.

[7] Indeed, the expression, 'the thing as it is in itself' in Kant has lost its 'ordinary

noumenon is only characterised as a 'limiting concept', in the 'negative sense' of 'not [being] an object of our sensible intuition': something which is easily though wrongly confused, Kant adds, with a putative object of a 'non-sensible intuition' (i.e. intellectual intuition, on the lines of Descartes).[1]

Now the notion of 'limiting concept' is interesting. It says: 'do not regard a thing as though it were something given in itself, but only in relation to some cognitive framework'. This seems, however, meant to provide a real contrast between *one and the same thing*, first, regarded as possessing—or rather, *being*—an indeterminate substratum, and secondly, as something cognisable. Note that, in its former respect, and in *this* context, it is not nothing, even though it be 'indeterminate', and even though—as already noted—we cannot say of it that it does or does not exist, in any empirical sense of that term. It is for Kant more like an index, or like an incomplete symbol in mathematics, e.g. the functional symbol f( ); though, as the passage from J. 33 shows,[2] it is (in his eyes) sufficiently well defined to take on fresh interpretations, either in the context of the regulative procedures of theoretical reason, or of the implications of practical reason. In the former case, for instance, we are quite entitled to 'think' of the world regarded as a system of empirical law-like processes as having an unconditioned 'ground'.[3] The mistake lies only in treating such a ground as something 'substantial' or 'real' (existing), an illusion (the already mentioned 'transcendental illusion')[4] which arises when we treat the 'object in the idea' as having existential status (existing or not existing).[5] And in the second, or moral, case, the 'existence' of the object, here: the 'ground' of the world, has in any case only the status of a 'postulate of reason'.

Nevertheless, Kant is concerned to preserve a place for the noumenon, at least in order to avoid the objection that the procedure

---

sense'. Kant is concerned to argue that we must distinguish between the empirical, i.e. scientifico-theoretical, case of the atomic constitution of a thing, of which we have inferential evidence, and its 'inner constitution' which supposedly 'exists' in independence of *any* framework of possible experience. At K. 273–4 in a similar way, Kant notes that we must not confuse the distinction between observational and theoretical astronomy, both of which belong to the level of phenomena, with the quite different distinction between the latter, and noumena. For the general point, cf. below, p. 547, the passage from K. 287, which graphically represents this situation. But Kant cannot be absolved from having played on the 'ordinary sense' in order surreptitiously to insert meaning where none perhaps existed.

[1] K. 268, 270, 2nd edition.   [2] Above, p. 538.   [3] K. 565.
[4] K. 565n.a; and cf. above, p. 526.   [5] Cf. K. 566; above, pp. 525f.

whereby we 'think' these interpretations within the framework of theoretical and practical reason is meaningless *quâ* self-contradictory. To this end, he must needs retain the locution, referred to a moment ago, of 'one and the same thing' regarded from two different points of view.

Here, indeed, the paths divide. There are two ways of formulating the Kantian position. (1) To speak of a 'thing' *is* to speak of something the concept of which makes necessary reference to the conditions of experience; this is to treat it as 'appearance'. (2) To speak of a thing *quâ* appearance is to speak of it as something which in so far as it is regarded as given as an object of cognition (in a cognitive judgment) presupposes the conditions of experience.

The first formulation suggests that 'thing' *means*, and *only* means, 'thing regarded as phenomenon', i.e. given only in the context of experience. The second suggests *two* interpretations of 'thing', 'something' which is 'mere appearance', and 'something in itself', Kant suggesting that 'the [very] word appearance must be recognised as already indicating a relation to . . . something [as it is] in itself'.[1]

Each formulation is deficient. The first conceals the fact that in developing the significance of the phrase 'conditions of experience', the contrast with the thing in itself is—as already suggested—still required. The second formulation conceals the fact that the phrase 'in itself' lacks empirical meaning (and must hence borrow it 'analogically' from ordinary usage): a fact which Kant certifies by referring to this notion of the noumenon as being 'quite indetermined'.[2]

It all suggests that Kant is subject to divergent tensions. The first formulation is intended to eradicate the thing as it is in itself from the phenomenal realm, and thus to *restrain* reason from even hypothetically entertaining the existence of anything noumenal regarded constitutively. Yet the intention of all this is to *free* the concept of the noumenon so as to give it a legitimate place at another level: i.e. that of morality, which Kant claims not only to supply an interpretation of man as noumenon but also to *entail* the postulate of the existence of God; there being no longer any possibility of confusion between *two* notions of existence, since 'phenomenal existence' (which presupposes 'real possibility') has been proscribed. In short, we are *restrained* from giving the concept of the noumenon any positive meaning at the

[1] K. 269–70, 1st. ed.   [2] J. 33; cf. above, p. 538.

phenomenal level (of either understanding or theoretical reason), so that we may *retain* it at the level of practical reason.[1]

Now we have suggested that the concept of the noumenon requires a positive characterisation as well. And it seems possible though quite arbitrary that the positive characterisation of the noumenon at the level of practical reason may have emboldened Kant to assume that at the theoretical level he was entitled to operate with the purely negative ('indetermined') notion. The need for the positive characterisation begins to show more explicitly when Kant wants to contrast the status of the spatio-temporal aspect of 'appearances' with that of things-in-themselves; where he claims that space and time are often (though misguidedly) represented as the 'properties' of the latter (e.g. 'qualities or relations').[2] How is one to understand this, so that the contrast with phenomenal appearance may be grasped?

Before answering this, let us look at a second positive characterisation of the noumenon which becomes apparent when we consider Kant's definition of the concept of the phenomenal object in the Analytic of Principles. Taking Kant's description of the noumenon as the thought of a thing in itself which lacks any 'relation to something' else, let us first of all compare this with the logical characterisation which he imposes on the concept of the object when regarded as 'appearance'. To this Kant refers as something of which 'we are asserting . . . that the [perceptual details] are combined *in the object*, no matter what the state of the *subject* may be', contrasted with the case where there *is* such an explicit egocentric reference.[3] In a second place, we find a similar contrast, when cognitional experience ('Erkenntnis') 'of objects through perceptions' is construed as involving 'a relation' where 'the existence of the manifold has to be represented in experience, not as it comes to be [subjectively] combined in time but as it *exists objectively in* time'.[4]

Now it is quite obvious that, in both cases, the concept of the noumenon, *quâ* thing as it exists apart from reference to the conditions which surround the cognitive apparatus associated with the 'conscious subject' (e.g. the observer), a public, rather than a private

---

[1] Cf. above, p. 529, for the parallel situation of Kant's concept of the Deity.
[2] For this locution, cf. K. 89; 244; 280, 286; M. 154, 180. We shall presently return to this contention.
[3] K. 159; my italics. Cf. below, pp. 632ff., for the use of this characterisation in the development of Kant's transcendental deduction.
[4] K. 209; my italics. Cf. below, p. 643, for the whole passage.

object,[1] throws its semantic shadow and serves as model for the required notion of objectivity. It is true that (according to Kant) it does this only indirectly. For as the first edition version of 'Phenomena and Noumena' points out, the object through the concept of which the 'perceptual manifold' is united in one 'thing' is a 'mere something = X'; and this 'cannot be entitled the noumenon'.[2] But the only difference is that in the case of the noumenon we think of this X as a *second* object, 'corresponding' to the 'appearance', and as being 'something-in-itself'.[3] Apart from this aspect, it still has the same logical character which later serves to formulate the 'concept of the object' of the transcendental deduction, i.e. the non-egocentric, non-subjective, non-accidental aspect of the phenomenon given in experience.[4]

The only difference between thing *quâ* noumenon and thing *quâ* objective phenomenon is that in the latter case the notion of objectivity is articulated as a logical function of the transcendental conditions of experience, through the corresponding categories, which have to be 'added' to the perceptual manifold. For, just *because* this is regarded as the component of a mere 'appearance', it lacks spatio-temporal determinations of its own which it would have (though, as we shall see presently, only *per impossibile*) if it were a noumenon.

(b) *A context from Leibniz: The Monad*

All this suggests that we shall have to look further and search among Kant's predecessors for sources from which he may have drawn further 'positive' characterisations of the concept of the noumenon, if only by way of critical reaction. Here, the description of the noumenon given by Kant especially suggests the influence from Leibniz. The noumenon is 'given' as a self-subsistent entity, apart from relation to anything else. On the other hand, in so far as we characterise its spatio-temporal aspect, we see that this must in some sense 'belong' to the noumenon, space and time being in some way its 'properties'. The first aspect suggests that the noumenon is some-

---

[1] One is of course tempted to use the model of the 'psychological conditions', or perhaps Locke's and Berkeley's 'mind'. But, as we shall see later, all these are, in an important sense, themselves 'objective', and presuppose a condition which is 'subjective' in quite a different and technical sense. To the difficulty of characterising this sense (which yields the 'phenomenon'), there corresponds the parallel difficulty concerning the noumenon. The two stand and fall together.
[2] K. 268, 271.
[3] K. 269–70. Kant implies that, at this level, it is again something only negative.
[4] For this, cf. below, sect. 8(d), pp. 623ff.

how really non-spatio-temporal, in the manner of 'thoughts', or 'universals' or mathematical and logical truths. The second suggests that, if spatio-temporal aspects belong to it, they cannot do so 'ultimately'. But this is precisely the account which we found to attach to the Leibnizian monad; and this impression is reinforced in Kant's rather more radical and harsher version of this account. It is therefore more than likely that the monads were a primary model of Kant's notion of noumena; and it follows, moreover, that no understanding of this conception will be complete without a full realisation of this fact.

We referred earlier to the difficulty of understanding phrases like 'out of space and time', and suggested that the theory of monads may have met this deficiency in Kant's eyes.[1] Kant's first published work, *Thoughts on the True Estimation of Living Forces*, still strongly under Leibnizian influence, gives evidence of this. When discussing the question of 'other worlds', he throws in the suggestion that 'things may actually exist and yet not be present anywhere in the world'.[2] And he characterises these things as 'substances [which] contain within themselves the complete source of all their determinations', and thus have no 'outer relations to anything': from which he infers that they are 'nowhere present in the world'.[3] This is obviously an interpretation of monadic reality regarded as outside space and time, and whatever meaning the theory of Leibniz may hold would also be infused into this phrase as used by Kant. And even if subsequently Kant comes to insist that we can have no knowledge, nor even any cognitive understanding ('Erkenntnis'), of such monadic substances in the empirical (phenomenal) sense of that term, since for this we necessarily require reference to 'outer relations', the contrast which this phrase requires with the contrary case would be primarily supplied by the doctrine of monadic substances whose 'self-sufficient being' contains the source of all their determinations.

Now this is in fact precisely the theory which is subsequently attacked and criticised in the chapter of the *Critique* entitled 'The Amphiboly of Concepts of Reflection',[4] where Kant seeks to settle accounts with Leibniz.[5] A glance at this section may be helpful, since

[1] Cf. above, p. 535.   [2] LF. 8, para. 7.
[3] Ibid. They are to be distinguished from the physical substances of our world which require such a framework of outer relations.
[4] K. 276–94.
[5] Note especially the explicit equation between 'monad' and 'thing-in-itself' at K. 280, 282, 287; and between 'monad' and 'noumenon' at K. 279.

a study of Kant's *criticisms* of the monads is indirectly an important aid in our search for positive meanings Kant may have attached to the conception of the noumenon itself. Here, the primary criticism concerns the position held in Leibniz's philosophy by space and time; the position which (as we have seen) makes these (as being merely 'confused concepts') ultimately collapse into (or 'reduce to') the qualitative and intellectually conceived world of monadic activity.[1]

The basic error of Leibniz, Kant maintains, was to imagine that sensibility was

> only a confused mode of representation and not a separate source of representations. Appearance was, on his view, the representation of the *thing in itself*.[2]

I have drawn attention on several occasions in previous chapters to the central place held in the philosophies of Descartes and Leibniz by the doctrine that sensation is a kind of confused thinking; and I have anticipated that Kant makes the question of the relation between sense and intellect a focal one, insisting—against Leibniz—on their total distinctness.[3] The two corresponding faculties of the mind, Kant says repeatedly, 'understanding and sensibility', are *independent* 'sources of representations'.[4] And the 'Amphiboly' is meant to demonstrate the result of our ignoring this distinction. Since the monad or thing-in-itself[5] was basically only the seat of 'internal

---

[1] Against this, we shall see later, Kant will argue for an ontologically independent position of space and time, though not, of course, as self-subsistent beings. But if they are independent of the 'perceptual' world of the monad (of 'thought', in Kant's phraseology), we shall find them equally independent of the logical level of the element of 'sense'; space and time belong neither to sense nor to thought, but are allotted a position which makes them 'some third thing' (K. 181), the 'form' or 'mode' in which sense, as articulated by thought, is turned into the concept of the object. This will be the broad trend of the argument.

[2] K. 282; italics in text. 'Sensibility' is defined by Kant as our 'capacity (receptivity) for receiving representations through the mode in which we are affected by objects' (K. 65), i.e. the *a posteriori* element of sensation. For Kant, all sensation must 'appear' under the forms of space and time, as 'intuitions'. Here he is criticising Leibniz's doctrine that both these forms and their 'matter' (sensation) are 'confused thoughts'.

[3] Cf. K. 83–4; 286; P. 47; ID. 58 (para.7) for the *locus classicus* of Kant's contention; and not directed just against Leibniz. The latter had falsely allowed sense to merge into thought; he had—as Kant puts it—'intellectualised appearances'. Locke, on the other hand, had 'sensualised all concepts', instead of which Kant postulates that 'understanding and sensibility [are] two [generically distinct] sources of representations' (K. 283). And the same goes for the formal aspect of intuition, i.e. space and time.

[4] K. 282–3.   [5] The explicit equation is made at K. 280.

perceptions', thereby making space and time only derivative entities, the logical principles that characterised it could only be those of the identity of indiscernibles and of contradiction.[1] Not being able to assign any ultimate place to 'outer relations' (of space, and the same goes for time), the monads could not be pictured, except through the model of 'thinking', as something purely 'internal' (governed ultimately by the principle of contradiction). As a further consequence, the conception of 'matter' could not be assigned a succinct place in such a scheme either. Worse still, such a philosophy, according to Kant, could not explain the possibility of a *community* of substances or monads, since for this we require either an independent and prior conception of space and time, or alternatively of a 'physical influence'. Since neither of these were available to Leibniz, he could only fall back on the doctrine of a 'pre-established harmony', through which the 'merely inward' nature of the different monads were mutually adjusted.[2] (*Per contra*, we may already note, the notion of 'phenomenon' ought to be so defined by Kant as to require necessarily a reference to 'dynamic association'. The problem haunted him, as can be seen by a glance at ID, Scholium to para. 22; its solution is presented in the doctrine of the *Critique* as given in the Third Analogy. To this we shall return.)[3]

All this supplies us with a technical interpretation of Kant's recurrent phrase that we have no knowledge or insight into things as they are in themselves, which may now be understood as a reference to the purely intellectual isolation of the monads. Moreover, this sharpening of the complaint, 'that we have no insight whatsoever into the inner

---

[1] K. 278–9; 283–4. Here we meet with a further delineation of 'noumenal reality', as that which *can only be thought in accordance with the law of contradiction*, whereas 'phenomenal reality' involves in addition the need for 'construction' of a sensory, or at least mathematical, manifold. In Kant's thought, this is really the oldest and perhaps most basic distinction, and we shall come back to it in our attempt to trace the development of Kant's doctrine towards his final 'critical' position. (Cf. sect. 6(b).)

[2] K. 279; 284–5. It will be seen how much Kant has sharpened Leibniz's conceptions. This is due to the fact that Kant is not prepared to play Leibniz's game of maintaining the uneasy relationship between the purely metaphysical or 'intelligible' and the phenomenal aspects of the system, through the application of analogical considerations. These (as we have seen) have their place, at best, only at the level of speculative reason, in the picturing of its 'ideal', and not at the level of the understanding, with its concept of the object—two of the Kantian offsprings of Leibniz's metaphysical realism. As noted—and we shall see it again—Kant himself, in his first published work (on 'Living Forces') is still in the thrall of an uneasy marriage of the two older Leibnizian aspects.

[3] Cf. sect. 7(b), pp. 584ff., especially p. 586 for the Scholium.

nature of things',[1] is paralleled by Kant's interpretation of the notion of 'inward nature' in a way that not only reveals the real meaning of the complaint of those who (like Locke) were so insistent on our ignorance of this inner nature, but also makes explicit the confused sources of such a complaint. In our discussion of Locke, we have already noted the confusion contained in the notion of the 'inner constitution' of things: 'unknowable' one moment because merely 'hypothetical' or 'unobservable', and the next, because involving some more intrinsic metaphysical ignorance (of 'real essence').[2] Kant's answer is that our ignorance of the one is accidental and curable indefinitely; of the other is simply an expression of the metaphysics of the monadology.

> Through observation and analysis of appearances we penetrate to nature's inner recesses, and no one can say how far this knowledge may in time extend. But with all this knowledge, and even if the whole of nature were revealed to us, we should still never be able to answer those transcendental questions which go beyond nature.[3]

Here Kant embarks on that division of the notion I have called the 'constitutive' realm which was foreshadowed in chapter II, placing some of its aspects in the field of the 'hypothetical', but still 'phenomenal', others in that of the transcendental, and yet others in the transcendent (noumenal) realm of the thing-in-itself.[4] And it is one of Kant's motives for insisting on our ignorance of the latter to free the way for a thoroughgoing 'empirical realism' with respect to the hypothetical entities of physical science. If later he rejects the mechanical-atomistic conceptual scheme in favour of one of a continuous matter permeated by basic forces of attraction and repulsion, the question of physical transcendence, the 'unobservability' of 'atoms', is not at issue, nor as much as raised, being regarded by him as possessing little logical interest. In the terminology of our chapter on Locke, Kant's concern was to proscribe the Lockean 'metaphysical extension' of the theoretical object O_T, whereby the latter had come to be misconstrued as O_M, a kind of metaphysical essence. Or, to choose Kant's own example at K.273, the contrast between the observed motions of the

---

[1] K. 287.   [2] Above, ch. IV, section 4(a).

[3] K. 287; cf. K. 83–4: 'Even if that appearance could become completely transparent to us, such knowledge would remain *toto coelo* different from knowledge of the object in itself.' See also below, p. 556, for the 'scientific' background of the conception of 'inner nature', as it had occurred in LF.

[4] Cf. above, p. 529n.3.

stars and the scientist's theoretical explanations of these motions by means of 'hidden' Copernican or Keplerian orbits is not an example of the contrast between the phenomenal and the noumenal realm, since both observation and scientific theorising (he insists) belong to the phenomenal realm.[1]

Here then lies one of the great deficiencies of the thing-in-itself as monad: In Kant's sharpening of this conception, it has a 'being' which is entirely 'internal' to itself, for—as we saw—'monads have no windows'.[2] The 'sharpening' in Kant's presentation of the monadology consists in rejecting the doctrine of the 'obscure and confused' attribution of a spatio-temporal order to the monads.[3] For Kant, Leibniz's characterisation of space and time, and of phenomenal objects in general, as '*bene fundata*' ('real', and yet not '*ultimately real*') is inconsistent with such an approach. Phenomena, he seems to argue, can be '*bene fundata*' only if they are altogether 'cut off' from the realm of the 'intelligible' (monads), and given an independent foundation; above all, an independent spatio-temporal framework, which will have to be built into the very conception of an object *quâ* phenomenon.[4]

(c) *A context from Berkeley's 'ideas'*

Now such an 'independent framework' of space and time has its own difficulties—as our study of Kant's predecessors has already shown, and we shall return to this below; difficulties that will result in Kant's treatment of this framework as an expression which has application (and hence empirical meaning) only in the context of possible experience, requiring in principle reference to a sensory basis. In this way, the 'necessity' of a spatio-temporal component in the very definition of the expression 'empirical phenomenon' (and, of course, of 'experience' or 'knowledge') is sharply contrasted with the theory of Leibniz's monads, and *their* metaphysical characteristics.

Quite generally, in the subsequent development of Kant's doctrine of phenomena or appearances, the contrast with these older theories,

---

[1] Cf. K. 273–4; cf. above, p. 539n.7.   [2] Cf. Leibniz, *Mon.*, sect. 7.
[3] Cf. above, ch. VII, sect. 5(c).
[4] The contrast, in these very terms, between an 'intelligible world of substances' and the 'phenomena of the senses' is already in Leibniz (*N.E.*, p. 428). But here, phenomenal 'matter cannot subsist without immaterial substances'; it 'supposes simple substances or real unities'. Unfortunately, the relation between the two is never explained. In Kant, this becomes the 'intelligible ground of the appearances' (K. 482), but this language moves at a level strictly distinguished from that of the phenomenal order of cause and effect.

whether they be Leibniz's or Berkeley's or Locke's, must be remembered. Thus in characterising objects as 'appearances', Kant often contrasts his account with that of earlier philosophers who—as he formulates this in terms of his own philosophical position—so construed the concept of the object as though the space and time in which they were placed were 'properties . . . which must be found in things in themselves'.[1] And it is then important to understand the use of such a locution in these Kantian terms. Let us get two or three of these contextual uses into further focus.

Kant may mean that space and time have the status ascribed to them in Leibniz's monadology, i.e. as 'confusedly perceived' attributes of monads, and ultimately unreal. At other times this is a reference to the theory according to which things are falsely viewed as 'given' antecedently to, and in an impermissible abstraction from, the process of a 'synthesis—as being 'constructed' from a relational manifold, because only in this way can the antinomy between the infinite divisibility of matter and the simplicity of its constituents be avoided.[2]

A third case—and one of great interest because it places the concept of the thing-in-itself in a Berkeleyan context—concerns a reference to the theory which assumes that space and time belong to a logical, or rather ontological level, similar to, and parasitic on, that of the 'things' existing in space and time: a view which Kant imputes to Berkeley, with 'things' standing for 'sensory ideas'.[3] A complete explanation and appraisal of Kant's criticism of this view will have to await the discussion of Kant's view on space proper.[4] But briefly, his criticism runs as follows. Let us assume that 'things' must be seen 'in space' in order to be cognised *as* 'things'; space is an 'inseparable condition' of things. Now Berkeley had realised 'the absurdity' of positing for this purpose 'the existence' of 'two infinite non-substantial things' (space and time).[5] His escape from this had made space and time in some sense *derived from* our primary ideas of sensation.[6] But since the

---

[1] K. 89, in reference, of all people, to Berkeley!

[2] Cf. Second Antinomy, K. 405–7; M. 180. Also K. 83. Cf. below, p. 602n.7.

[3] Cf. K. 89, 244, where Kant ascribes the view of space and time as 'properties' to be 'found in things in themselves' to Berkeley.

[4] Cf. below, sect. 7(c), pp. 588ff.

[5] K. 89. Kant agrees with this characterisation of Newtonian space, and its condemnation. Cf. K. 81, which speaks of 'two eternal and infinite self-subsistent non-entities', which K. 400 asserts to be 'nothing real in itself'. And the *Dissertation* already opposes the view of 'an absolute and boundless receptacle of possible things' (ID. 70).

[6] Cf. above, ch. V, pp. 321–2, where Berkeley is shown to hold that 'distance or

latter are merely 'subjective' perceptions, we have no means of 'explaining' our cognitive experience of bodies being 'really out there' in space and existing in time. To this we shall return in its due place.[1] What interests us here is that this is a further context yielding an interpretation of the Kantian expression, 'thing-in-itself'. For Berkeley is described in the *Critique* as holding that 'space [has to be] interpreted as a property that must belong to things-in-themselves'.[2] So evidently, even a Berkeleyan 'idea', when coupled with Berkeley's view on space, can stand for the notion of 'thing-in-itself'. This is at first sight surprising since 'ideas' are anything but hidden; they might indeed be called 'appearances' *par excellence*. It is therefore evident that the characterisation of anything as thing-in-itself concerns primarily the question of the status of space and time. It is because in Berkeley's theory the latter—as in some sense 'belonging' to the 'ideas'—lack reality, are merely 'imaginary', that 'ideas' do *not* count as 'appearances' but are instead said to be 'mere illusion [Schein].'[3]

Kant will reply to Berkeley in his usual way, that space and time are not dependent on 'things' but must rather be regarded as *independent conditions* of such things. And since their independence cannot be pictured on the lines of Newton's 'absolute non-entities', they must be anchored in something else, which already we know broadly to be a transcendental condition defined by the need to make cognitive discourse intelligible, or possible.

Here we are already beginning to see that actually Kant makes the need, under which space and time are specified as *independent* conditions, logically compulsive because of his formulation of the notion of 'appearance'. For the *a posteriori* component of such an appearance (its 'matter', i.e. 'sensation') is—unlike Berkeley's 'ideas'—*affirmed* to be incapable of yielding (even derivatively) anything spatio-temporal.[4] (It is the contrary supposition that characterises a thing as a 'thing-in-itself'. And both characterisations, it will be seen, are, as 'affirmations', in the most important sense of the term, metaphysical, i.e. in the sense which was foreshadowed at the start of chapter I.)[5] With this

---

outness' is only 'suggested' by our ideas, but is 'nothing positive' in itself. For Hume's similar contention, cf. ch. VI, p. 354n.1.
[1] For the case of space, cf. below, pp. 596–7.   [2] K. 244.
[3] K. 88. The German for 'appearance' is 'Erscheinung', and there is thus a neat play on the contrast between 'Schein' and 'Erscheinung'. It should also not be forgotten that 'to appear [erscheinen]' does not in this context have the sense of 'to seem' but rather that implied in the phrase 'X suddenly appeared on the scene'.
[4] Cf. above, p. 533n.3.   [5] Above, pp. 3ff.

reading of 'appearance' we can therefore be at once *certain* that for anything with an *a posteriori* content to 'appear' as empirical object, the space in which it is *empirically* 'seen' to appear must be a 'real' space, since it must have been 'contributed by', be part of, and can occur only in the context of, the perceptual situation, as the 'form' of a sensory intuition.[1] This space, though defined only as a transcendental condition, and thus not 'objectively real' (i.e. not a self-subsistent entity), but only 'transcendentally ideal', is thus nevertheless 'empirically real'.[2] This is roughly—as will gradually emerge—the way in which Kant's argument will proceed. It involves a logical re-interpretation of the notion of 'idea' (of what is merely passively 'received'), which becomes a purely theoretical abstraction, or rather: a 'metaphysical re-description'.

Now as a contrast which assists our evaluation of Kant's own doctrine, the opposition between thing-in-itself and appearance, framed within the context of a critique of Berkeley's theory, is undoubtedly more revealing of Kant's intentions than some of the other contrasts that have been considered. But we see at the same time that this positive doctrine is only a response to certain purported 'difficulties' in Berkeley. Kant's 'reading' of the key-terms, especially of 'idea' *quâ* '*a posteriori*', is so contrived as to supply a more satisfying explanation of common locutions. To say that space is not 'suggested' by our spatio-temporal experience or language, but is infused '*a priori*' in the context (and as part of the fabric) of that language, is really only giving a different 'metaphysical description' of these locutions. The matter is clearly not empirical; a transcendental account explicitly not so. The metaphysical centres of gravity are shifted, in line with the general method of most of our philosophers.

Quite generally, we may say that any reference by Kant to noumena comes down to a contrast of some earlier philosophical doctrine with his own, whether the 'noumenon' in question be the 'ground of all possibility', or the source of natural forms, or the indeterminate substratum of attributes, or a monadic unity of thinkings (subject only to the law of contradiction), or whether, finally, it be used to characterise some aspect of Berkeleyan idealism. The thing-in-itself is thus a reminder of the theories against which Kant is working out his own scheme and, with the conceptual framework of the former, suitably

[1] Always assuming, as Kant does, that space and time are not 'self-subsistent'.
[2] Cf. K. 72. This is contrasted with Berkeley's 'empirical ideality' of both 'things' and their derivative space.

interpreted, often worked into the fabric of his own doctrines. This is a more fruitful way of approaching the problem of the noumenon than to ask whether or not it exists. Once more, progress in philosophy emerges as a constructive dialogue of opposing schools.

6. THE CONCEPT OF CONSTRUCTION

(a) *From metaphysical to physical forces: shifts in meaning*

Kant felt deeply the need for emancipation from earlier misunderstandings and confusions concerning the distinction between the 'inner nature' of a thing as it is in itself, and the thing regarded as given through 'external relations', for this paralleled his own intellectual development which had begun with a relatively uncritical acceptance of these distinctions in his earliest 'scientific' writings, with their contrast between 'the inner and the outer',[1] which involves also the rather confused dichotomy of the 'mathematical' and the 'metaphysical' approach in LF, or again, the distinction between purely logical conflict and non-logical opposition, discussed in the tract on *Negative Quantities*.[2] Many of these distinctions and classifications, with the general presuppositions underlying them, continue to linger on in Kant's later work, and seldom are the intellectual spectacles through which these problems are approached abandoned altogether, however much they may be transformed in the gradual emergence of the themes of the distinction between the contingent and the necessary, or its finer classification into the analytic and synthetic, *a priori* and *a posteriori*. Above all, the meanings involved in the terminology of the earlier treatises are never quite lost in these transformations—a fact that makes a reference to these writings essential. In the present chapter, rather broad indications of these transformations will have to suffice; obviously a complete treatment would require another volume of the present size. Nevertheless their mention is needed in view of the perennial problem of indicating the *significance* of Kantian terminology. Throughout I have stressed the importance of focal models that serve as centres of gravity for metaphysical construction; and nowhere is this of greater importance than in the Kantian system.

We have noted before the way Kant shifts the key concepts of his philosophy to different levels; and we have already discussed the

---

[1] As this is called in the Amphiboly, K. 279; for LF, cf. above, p. 544.
[2] Cf. the survey of the relevant ideas given above, pp. 477–8.

re-location of the teleological and the ontological notions of his earliest writings. Similar shifts also happen to the key concepts of his *Thoughts on the True Estimation of Living Forces*, with its contrasting pair of 'mathematical' and 'internal' forces. In Kant's later period, with its clearer articulation of 'mathematical' (through the concept of construction), the distinction is sharpened, and the somewhat confused idea of an internal force is removed to a different level.

In LF, Kant still extensively employs the distinction between a 'mathematical' and a 'metaphysical' approach,[1] which reminds us of a corresponding dichotomy that we have met in Leibniz. Interesting here, in the light of his later constraints on the 'metaphysical' order (through the doctrine of the *noumenon*), is that in this earlier work Kant is still interpreting the distinction in quite a different way. A physical body is subject to forces of two kinds: (1) Forces due to external action, which though they may have only momentary counterparts in the body, and do not last for any finite length of time, are subject to the principle of conservation. These forces are capable of 'mathematical' treatment and have genuine 'phenomenal' significance, for they are expressed in the Cartesian formulation of the product of mass times velocity, velocity being the quantitative expression of a 'real motion'. (2) The second kind is an internal force, which arises when, after a body has been set in motion, it is *caused* to continue to move with uniform velocity if not subject to *external* forces. The source of the *internal* force resides within the body, is not subject to the law of conservation of force, and is *for that reason* (according to Kant) not capable of mathematical treatment. Nevertheless, he does provide an 'estimate' of its magnitude, which turns out to be none other than Leibniz's measure of *vis viva*, mv^2.[2] This concept of 'living force', whose task it is to preserve a body in inertial motion, evidently still follows Leibniz's doctrine according to which an 'internal force' is required to preserve such motion.[3]

But, unlike Leibniz's living force, Kant's is not subject to a law of conservation. Not therefore being a mathematical conception, it is

---

[1] Cf. LF, paras. 17, 19.

[2] Cf. LF, paras. 115, 121. Here we note part of Kant's early meaning of a 'mathematical' treatment. It is in his view not sufficient to supply (as done for *vis viva*) a quantitative estimate of a force, based on 'metaphysical' (which here means: conceptual) considerations; instead, it is necessary that the concept should enter an important *law* of mechanics, here: of conservation; moreover, this will involve, as we shall see, construction.

[3] Cf. above, ch. VII, sect. 3(a): ii, pp. 421–2.

really more like Leibniz's '*primitive*' rather than '*derivative* active force',[1] despite the quantitative specification. For, like the former, it is not only altogether 'internal' but also 'different in kind' from those 'external' forces which are subject to conservation. In the case of Leibniz, the difference involved was the distinction between forces conceived as 'physical' or 'phenomenal' and as 'metaphysical', the latter supplying the essential aspect of his notion of 'action' in respect of the monad, when considered 'confusedly' in its physical aspect. The distinction which Kant introduces instead is that between those forces which operate with respect to what he calls 'mathematical body', and those acting in the 'natural body', a distinction which is in his early thought vaguely connected with that between inanimate and animate nature, 'living forces' being modelled on animate nature.[2]

Despite, however, these Leibnizian echoes, the introduction of the distinction between the 'natural body' and the 'mathematical body' is already an early indication of a more rigid architectonic: an unwillingness to think in the Leibnizian terms of a continuous transition from inanimate, via animate, to the 'metaphysical' realm, with its analogue of human self-consciousness, or 'apperception'.[3] And these Kantian distinctions become more pronounced when subsequently the notion of the 'natural body' is eliminated from the realm of mechanics altogether, finding a place only in the teleological part of biology. (This is paralleled by a similar shift in the notion of the 'accidental' already met in Kant's teleological approach, and which we should therefore expect to find associated with 'nature' in the earlier work.[4] And sure enough, in LF, whilst Cartesian forces are said to be presupposed as the axiomatic and necessary properties of matter, the 'living forces' have a merely 'accidental' nature, are purely 'contingent', and are introduced only in a 'hypothetical' manner.)[5]

Obviously, the conception of 'living force', as here employed, contained a number of 'idling' features. In M, where Kant has embraced the Newtonian doctrine of inertia, inertial motion no longer requires any external cause for its continuation, and the concept of action in accordance with an 'internal' principle is abandoned as inconsistent with the very concept of matter. Kant's earlier 'hylozoism' is likewise rejected, it following from the very

---

[1] Cf. above, ch. VII, pp. 420–01.   [2] Cf. LF., paras. 114–15.
[3] Cf. above, ch. VII, p. 468.   [4] Cf. above, p. 491n.1.
[5] LF, para. 129. All such contexts are important for an understanding of the richness, and even unexpectedness, of Kantian meanings.

'concept of inertia, as the merely inanimate, that there can be no positive endeavour [Leibniz's *conatus*] on the part of matter to maintain its state'.[1]

In the same work, Kant will now insist that *all* forces require mathematisation, and this in two ways: (i) they must have a place in the laws of mathematical mechanics; (ii) they must in some way relate to 'real motion'; and the latter can be expressed quantitatively only through the concept of velocity, rather than of its square.[2] The notion of 'internal forces' being now abandoned, all forces act 'externally', issuing from the material points of an indefinitely divisible 'matter', and are regarded as 'causes' of potential motions, the quantitatively expressed laws of these motions *defining* the very nature of 'matter'.[3]

Now these motions are expressed as velocities; but velocity (according to the later Kant) can be made *intelligible* (in his phrase: can be shown to be 'possible') only through certain *constructions in space* (in the form of what we would now call 'vector quantities').[4] It becomes at once apparent that the defining conception of matter and force presupposes spatio-temporal representation. It follows also that Kant's insistence on the 'unknowability' of the 'internal nature' of bodies will now be expressed in terms of an impossibility to construct. Positively this becomes the doctrine that the concept of a material or 'mathematical' body presupposes this element of spatio-temporal construction through which alone it can be represented in mechanics. Moreover, construction links up with intuition: for it is 'in intuition' that 'we construct [any] concept' of a mathematical, e.g. geometrical, nature (cf. K. 577).[5] In this way, 'construction' becomes a basic

[1] M. 222–3; cf. below, pp. 676–8, for the 'proof' of the law of inertia.

[2] In LF Kant also suspects *vis viva*, for the reason, now known to be mistaken, that it is not real because it is independent of the time during which a body has moved: a confusion of the fact that velocity, as a rate of change of displacement with respect to time, involves time, with the quite different fact that momentum, i.e. mass times velocity, is the *time* integral of Newtonian force, whereas (as Kant rightly knew) *vis viva* is the *space* integral of that force. In M, Kant still adheres to his opposition to *vis viva* explicitly on this ground. In this later work, however, he has given up altogether the concept of living force. (Cf. op. cit., iii, Prop. 1, Note; M. 216–17.)

[3] M. 170. Forces are, however, clearly distinguished from the motions, and constitute for Kant irreducibly *a posteriori* and opaque entities which cannot, unlike the 'motions', be 'constructed'. They retain indeed something of Kant's older doctrine of the 'striving' of matter, except that this is now interpreted as an independent 'dynamical', and of course purely mechanical, phenomenon. (Cf. M. 200–1; also K. 230.)

[4] Cf. M, i; especially M. 157, 161–3, 164–5.

[5] To 'intuition' [Anschauung], as we have indicated (p. 533n.3) corresponds 'matter'

ingredient in Kant's definition of possible cognitive experience, of which intuition is an integral component.

Here we can see clearly at work Kant's method of re-interpreting and re-locating the concepts of his earlier period. The truly mechanical becomes the mathematical, which in turn is further articulated in the crucial aspect of 'construction'.[1] The realm of 'internal activity' that had been located in the 'natural body' is split into a number of factors. No longer being required to 'explain' inertial motion, 'striving', in so far as it is still a mechanical phenomenon, becomes the purely mechanical ('dynamical') forces of attraction and repulsion. The aspect of 'life' is shifted on to the biological and, with this, teleological level. On the other hand, in so far as the internal activity of a substance had in Kant's earlier work denoted something 'metaphysical', partly because 'non-mathematical', that notion, when reinterpreted, splits into the purely 'noumenal' aspect on the one side, and into those conceptual aspects of mechanics which Kant expounds as being governed by the categories when applied to the spatio-temporalised concept of matter.[2] The re-shaping of this concept of internal activity is particularly interesting when we compare the shift from 'inner determinations of body' in LF, para. 18, with the more clearly defined and re-located corresponding dual pair of the post-critical period, the 'inner principle of action' (viz. 'life') of M. 222–3, expelled from mechanics and moved to biology, and the 'internal nature' of substance, at K. 284–5 regarded as a mere 'state of representation', i.e. something which is 'a thinking or analogous to thinking' (K. 379), the monad or thing-in-itself.

All these shifts are, it will be appreciated, the result of certain developments, due to difficulties which Kant encountered at the start and for which he gradually sought a solution by the application of the transcendental method. Often, however, the basic element in the

---

(sensation) and spatio-temporal 'form'. Even the German does not recognise its ordinary use as a noun (though as a verb [anschauen]), which indicates its technical and abstract nature in Kant, where it mostly occurs as a noun. Thus what Kant means, in the passage referred to, is best expressed by saying that the representation of the intuitional aspect of anything involves construction. The logical properties of both will have to be developed as we proceed.

[1] And 'construction' itself, from being initially located in the realm of mathematical operations, is generalised in the body of the 'critical' argument as the process of 'synthesis' of the 'empirical manifold', i.e. of a process which operates on the material of sensation, our perceptions. In this way, 'synthesis', and with this 'construction', become transcendental elements.

[2] This is the interpretation of the term 'metaphysical' in M. (Cf. below, sect. 9(a).)

changing situation just simply consists in this shift of meaning or location of the central terms. As a result, whilst Kant appears to be arguing against his earlier positions, in some way the old meanings retain their force, particularly where the concept in its *new* environment has such a tenuous semantic hold that it requires support from the older associations. The historical development of Kant's thought can then explain (for instance) the peculiar contrast between such pairs as the 'mathematico-mechanical', and the 'metaphysico-dynamical manner of explanation',[1] which lies at the heart of Kant's theory of physical science. Or again, we can understand why Kant could employ without serious question the notion of an 'internal nature' of the noumenon, contrasted with that which possesses 'outer relations' (of space) (K. 284). It is true that the model of a mind 'in' which there reside its ideas may supply a background for this locution; but it is difficult to believe that the notion of a noumenon possessing an 'inner nature' would have retained its attraction for Kant, had it not been for the earlier quasi-scientific contexts.

At any rate, this discussion will have drawn attention to the connections which Kant postulates between the notion of empirical matter, or the empirical object, and the need for 'construction in space', this being contrasted at the earlier period with 'inner determinations' (by *vis viva*), and at the later period with a noumenon, determined solely by the activity of thinking—a thinking which is subject only to the principles of pure logic, above all, the principle of contradiction. Let us now trace the development of this doctrine in Kant's writings, in order to delineate certain epistemological aspects of 'empirical (phenomenal) reality', which will go a long way towards defining its 'contingency'.

(b) *Logical contradiction v. physical opposition*

At several stages we have already encountered the notion of construction. We must now examine this to see how it is used by Kant to articulate the idea of 'phenomenal reality' and logical contingency, contrasted with the realm of 'noumenal reality' and logical necessity.[2] The centrality of this will not be underrated if we remember the importance that attaches to the development of the notion of contingency in the philosophers that have preceded. The modern

---

[1] Cf. M. 201.
[2] The terminology of phenomenal and noumenal reality, which Kant uses to label this contrast, may be found at K. 279.

definition simply states that the contingent is that the opposite of which is not self-contradictory. But this does not tell the whole story. In Locke we found an unconscious attempt to equate the realm of the contingent both with what was supported only by 'experience' and with what was 'unintelligible' as well as 'ultimately' related only by divine *fiat*. In Hume, the element of 'unintelligibility' was sharpened (or recognised) as a form of 'metaphysical atomism', any 'relations' between the logical atoms being located in the domain of 'the mind'. In Leibniz, both the term as well as the concept of contingency had played a very basic rôle. The contingent was again that whose relations are 'opaque to the intellect'; but relations (between substance and attributes) there must *be*, even though the 'analysis' that can alone exhibit the relation cannot be effected by man, being of an 'infinite' nature. The relation is thus, literally, 'contingent on God's power' which provides the 'reason' for its 'existence'.

Finally, in Kant's doctrine of the office of 'hypothetical reason', the contingent was the 'accidental' and the 'unexplained', though there the conception extended only to the relations between the empirical laws and forms of nature. More basic is the question as to how Kant will interpret contingency at the level of the understanding, relative to the problem of the relations between the members of the perceptual manifold. And again, all the elements current in the older philosophers recur, though as usual transformed. Moreover, if we stick to our approach of interpreting the method of these philosophers as that of providing basic analogues, we shall be in a stronger position to grasp the development of Kant's attempts to formulate the notion of contingency. What we shall find is that the empirical aspect returns as the notion of the '*a posteriori*' (though with the technical sharpening already noted); its 'opaqueness' or 'atomicity' is one half of the meaning intended by the term 'synthetic', whilst the other half is to be found in Kant's re-interpretation of the Locke-Leibniz function of the divinity, through the notions of 'construction' and '*synthesis*', both of which, being regarded as 'transcendental' or 'presuppositional', are labelled *a priori*. Finally, the necessity of logic is that of the analytic *a priori*. With this broad survey in mind, we may now turn to Kant's early deliberations on contingency.

In tracing the sequence of Kant's reflections over about twenty years of his most active period, I shall stress the basic continuity, rather than the discontinuity, of his thought, which consists mostly in the heightened emphasis laid on the notion of 'experience' during the 'critical' period. This will mean once again using the method of shuttling between pre- and post-critical writings, so that each may illuminate the other. This is at times confusing but I think that only thus can we steer close to the centre of Kant's thought; and we shall be less tempted to over-rate the (perhaps) more accessible 'Humean' aspects of Kant's revolution. In laying greater emphasis on the 'Leibnizian' rather than the 'Humean' ingredients in Kant, we shall also automatically keep in mind Leibniz's generalisation of the problem of the 'link' between the terms of propositions of whatever form; altogether crucial for an understanding of the extremely generalised nature of Kant's intention, if we want to avoid its trivialisation.

Here, the relevance of Kant's early scientific preoccupations stands out very markedly, and once again helps the student to appreciate the often idiosyncratic contexts of Kantian technical locutions, e.g. those of possibility, *a priori*, construction, intuition, etc. Without a knowledge of these contexts, one is too apt to fall back either on pre-Kantian interpretations, or worse, on modern uses which frequently have outgrown their original home. It cannot be stressed too often that a philosopher's key-concepts are internal to his arguments, being a critical response to the general ambience of his thought. And for Kant this was clearly located at the start in those scientific contexts. Rather than enter upon minute logical criticism of Kant's individual arguments, let us follow then the trend of the evolution of his key-models.

Kant first began to discuss the contrast between 'phenomenal' and 'noumenal' reality (though not of course in these terms) in some of his writings of the 1760's, where he characterises the distinction between the physically real and the logical. Let us single out the important discussion which occurs in a tract of 1763, bearing the curious title: *An Attempt to introduce Negative Quantities into Worldwisdom [Philosophy]*. In this Kant sets out to delineate the difference between logical denial or contradiction, and physical opposition. Thus, given some positive character a, the assertion not-a, when joined to the former, leads to contradiction. By contrast, the algebraical addition of a to its negative (minus-a) produces, not contradiction, but a numerical result, zero. Now it is one of Kant's basic ideas to regard this as an important model, and as a way of understanding physical and psychological relationships: for instance, action and reaction by one body on another, as an example of 'real opposition'.[1] In general, if we want to understand such relationships, we must treat them by what

[1] NQ. 175, 195. I shall ignore Kant's psychological examples.

we might now call a vectorial calculus, in which we 'add' or 'subtract' positive 'qualities' or 'quantities'.

Kant goes so far as to characterise the logic of these cases by enunciating a generalised form of the principle of conservation of linear momentum![1] The 'basic rule' behind this he formulates in the statement that 'real opposition *can take place only in so far as* of two things [regarded] as positive grounds, one counteracts the effect [Folge] of the other'.[2] The condition here stated, which is supposed to express the specifically non-logical aspect of things, is evidently fashioned through the model of the vectorial addition properties in the realm of mathematics. Moreover, Kant formulates all this by saying that we can and must express the physical situation in terms of a process of construction.

Thus, suppose we want to make intelligible, and thus explain the possibility of, two forces in opposition applied to one and the same body. We shall then draw the vectorial diagram of forces, which represents their 'additive' property, and thus obtain the resultant of the two forces, which may be zero or some other number. (If the forces are equal and opposite, as in the case mentioned previously, the resultant will be zero.)[3]

This representation of forces, moreover, involves the construction of a *spatial* diagram. Only in so far as the forces are represented by corresponding lines drawn in space, actually or potentially, and in accordance with a set of axioms, can the construction be effected, the assumption being that forces can be represented by their corresponding spatio-temporal manifestation, i.e. motions,[4] e.g. a parallelogram of accelerations, etc.

Here, then, is the origin of the Kantian contention that to represent *concepts* of *physical objects*, we require a reference to spatio-temporal construction, this being for Kant the basic paradigm for distinguishing merely logical from physical relationships. In his later writings, Kant discusses this frequently in connection with the foundations of *scientific* 'knowledge', rather than 'knowledge' in general. Thus in M we find the famous *dictum* that

> in every particular branch of natural enquiry there is to be found only as much of science *proper*, as can be found therein of *mathematics*.[5]

[1] Cf. NQ. 194–5.

[2] NQ. 175; my italics. 'Folge' may mean either 'effect' or 'consequence'. Note the presuppositional form of the part of the statement we have italicised.

[3] Cf. also below, sect. 7(e): i, p. 601.   [4] As stated at K. 230.

[5] M. 140; italics in text.

Now it must be realised that not only does the reference to mathematics here denote 'construction', but as such the situation is quite general, and not just concerned with the *law-like* aspects of science proper; for all cognition of empirical objects (happenings, processes) presupposes spatio-temporal construction.[1] To grasp this difference between that part of the discussion of M which concerns the specific foundations of science, in respect of its 'pure' component, and the part which concerns knowledge in general, is however important, since it involves some of the trickiest problems for any interpretation and understanding of Kant's doctrine. For the moment we will just say this: the constructive element (despite the empirical models for it, e.g. drawing) is something presuppositional. Now in order to define a basis for all cognition of 'possible objects' that shall have 'certainty', Kant always interprets 'presupposition' as something 'injected by the cognitive subject as a spontaneous act'.[2] It is that element of the cognitional situation which is not 'found' or 'received' (as part of the *a posteriori*) but which belongs to the formal side, the *manner in* which we 'receive' the *a posteriori* elements; and this is one reason why 'construction' is said to be '*a priori*'.[3] Construction thus stands for an *a priori* formal component of cognitive experience in general.

Now, according to Kant, *science proper* has, however, itself also a more *explicit* 'pure' and '*a priori*' component, which, as we shall see later, yields its 'metaphysical' basis. It is true that *all* construction, if it is to yield a construct, must be carried out in accordance with a concept and with the rules which govern this. In respect of cognitive experience of objective situations, these concepts (categories) are *implicit* with respect to the definition of such an experience. But in regard to natural science these rules *must be made explicit, and applied*

---

[1] Naturally, the reasons for this are complex, as will be clear from our discussion of the contrast between noumena and phenomena, and from what follows also in the next few sections, especially the section on space. Here, however, we are concerned to trace the roots of Kant's choice of models at an early period.

[2] Cf. K. 577: In 'empirical intuition we consider only the act whereby we construct the concept'. The procedure of 'injecting' what we want to regard as 'certain' we have already found to be basic in the regulative and 'projective' activity of reason. Cf. above, pp. 508–9. We shall later see that it is the primary aspect in the whole definition of the 'transcendental deduction'.

[3] The element of construction is *a priori*, first, because it is a 'subjective' presuppositional element; secondly, because it 'borrows nothing from experience' (K. 577), as in the construct of a triangle, which proceeds in accordance with definitions and axioms which are introduced *a priori*. Finally, it involves the distinction between 'empirical' and 'absolute' space, which requires the former to be transcendentally ideal, and hence *a priori*. (Cf. below, p. 601.)

to the empirical concept of matter.[1] Thus, in so far as we are concerned solely with extension (category of quantity), the rules may be the axioms of Euclid, with the help of which we formulate the quantitative (kinematical) aspects of matter. And the same goes for the remaining categorial concepts, the qualitative and the relational aspects, which as rules *are applied* to 'matter'; e.g. when the concept of causality is applied to the 'empirical concept of matter', it yields the law of inertial motion.[2]

Nevertheless, although M discusses construction primarily under the aspect of its relevance for science, what it says defines at the same time Kant's general position in respect of cognitive experience as such, with its *implicit* employment of the categories. It informs us that, if we want to formulate the conditions of physical (as against purely logical) language, we cannot operate 'merely with concepts', subject only to the law of contradiction.

> The possibility of some determinate natural things [Naturdinge] cannot be cognised from mere concepts; for from these may be grasped, it is true, the possibility of the thought (that it does not contradict itself), but not that of the object, as natural thing. ... Hence, in order to cognise the possibility of determinate natural things, and therewith [to cognise them] *a priori*, it is further requisite that the *a priori* intuition, corresponding to the concept, be given, i.e. that the concept be constructed.[3]

The spatio-temporal component of things thus turns out to be an important constraint, which defines part of the notion of empirical (phenomenal) cognition, and of phenomenal reality. The vectorial model, first met in NQ, has (as 'construction') become an important

[1] Cf. M. 142: 'But to render possible the application of mathematics to a physics of body [die Körperlehre], through which alone it can become a natural science, principles of the construction of concepts which belong to the possibility of matter, must first precede.'

[2] Cf. below, sect. 9(a–b). The significance of speaking of these concepts as 'applied' will there be made clear. For the 'explicit-implicit' distinction, cf. below, pp. 603f.

[3] M. 140; cf. K. 255, where it is shown that change, coexistence, etc., require 'outer intuitions'. In the chapter of the *Critique* on the 'schematism of concepts', the same point is made though in a different language. A category concept can only be 'applied' when the formal conditions of sensibility (space and time) are given. Thus, the concept of the triangle must be regarded as a 'rule of synthesis', as 'the representation of a method', of drawing 'in thought', in conformity with the rule implicit in the concept. (Cf. K. 182.) The reader should note the peculiar equivalences, implied in the passage from M. 140, between 'cognition of *possibility*' and cognising something as *a priori*; between 'constructing a concept' and 'giving the concept in an *a priori* intuition'. There is always an easy traffic in Kant between possibility, *a priori*, intuition, and construction, and of course, space and time, as forms of intuition.

conditional element for defining 'the possibility' of phenomenal objects in general. And it will be noted that this approach of NQ (in all but the term 'construction', which it does not yet use) vastly antedates the more 'subjective' experiential and transcendental concerns of Kant's later period. Nevertheless, its origins in NQ are a helpful pointer towards an understanding of the significance of the meanings of Kantian terms, such as intuition, where they occur subsequently. Indeed, in a way the 'subjectivism' of the later period is already incipient in the earlier thought. Thus the concept of construction itself has a double use, the term standing for the 'act of construction', as well as for the 'result of constructing'. Correspondingly, intuition may be regarded as involving the spatio-temporal form of a given sensory manifold, but also as standing for the 'manner' in which this manifold is 'received' into consciousness. The second alternatives will become central in Kant's later transcendental approach.

(c) *Articulation of the concept of contingency*

For the moment, let us revert to Kant's discussion of the mathematical model for empirical possibility, or contingency, contrasted with logical necessity, which we met in NQ. Very much in Leibnizian fashion, he there generalises the problem of physical *v.* logical opposition into that of delineating the realm of logical necessity from that of contingency. We must distinguish between the 'logical ground' that governs the consequences of certain assumptions, and the quite different 'real ground' which mediates the possibility of something temporally following upon something else. His frequent question, as formulated in the text in bold letters, is: '*How am I to understand that, because something is, there should be something else?*'[1] The question is significant. It does not ask, of course, how we come to be informed empirically that something follows, or even is caused by, something else. This is for observation to decide. Rather, it is concerned with the kind and status of the 'ground', in terms of which the answer is formulated.[2] No *logical* ground is here available, Kant contends in the manner of Leibniz, since that would be to subsume the discussion under the principle of contradiction. On the other hand, he appears already to reject Leibniz's answer, which had sought the solution in

[1] NQ. 202; my italics; cf. above, p. 472.
[2] Cf. above, pp. 557–8, and, for instance, the corresponding distinction between these two questions we made in Hume. Only, Kant generalises it considerably. We are clearly already concerned with contingent propositions of matter of fact *in any form*, and not simply those couched in the causal mode.

terms of a sufficient reason, ultimately located in God. What remains in Kant of Leibniz's procedure is only the belief that the question, in the sense in which it is here asked, is significant and must have an answer. In NQ, this answer is not yet fully developed, but its outlines begin to appear. And, as usual, the models required for the elucidation of the meaning of the question are taken from physics.

Consider Kant's chief example, the phenomenon of the impenetrability of solid bodies. How are we to understand this? Only, Kant replies, by endowing bodies with repulsive forces, i.e. by reducing the conflict between two solid bodies, one of which seeks to interpenetrate the other, to a conflict between two forces, a conflict once again to be formulated in terms of vectorial addition. Only in this way can such a phenomenon be rendered 'transparent'.[1] Similarly, in a collision phenomenon, we can 'understand' how through the motion of one body that of another may be annulled, by endowing the bodies with forces. For this assures that we are dealing not just with logical contradiction but with physical conflict.[2] Yet, how are we to 'explain' real opposition as such; how are we to 'grasp distinctly' why, given that something (e.g. motion) exists, another (motion) should cease to be? This, Kant insists, can not be made 'transparent'. The forces are simply given as opaque elements of fact, however much we symbolise them graphically in space. We may analyse complex explanations into simpler ones, but we shall finish only with 'simple and unanalysable concepts of real grounds' which can not be made 'distinct' or 'transparent'.[3]

Some further facets of the later critical system here already appear again. On the one hand, there is an *a priori* element; this connects with the definition of a real, as contrasted with a purely logical order (an important mode of formulating, we have seen, the contrast between the phenomenal and the noumenal realm); this is the element of mathematisation, which supplies the model of 'real conflict'. On the other hand, that something *specific* happens in the context of certain given empirical conditions, or better: *what* it is that happens (in the later *Critique*, Kant will speak of this as 'content'), cannot be made transparent. Nor is Kant prepared any longer (as Leibniz had been) to fill the epistemic gap by the metaphysical or 'dogmatic'

---

[1] NQ. 179–80.

[2] For the sake of what follows, let us note how important here already for Kant is the conception of 'mutual interaction of forces'.

[3] NQ. 203–4. There is an echo here from Locke's philosophy of 'simple ideas'.

postulation of some principle of sufficient reason which would formulate the self-developing history of a monad, its history correlated
with that of all the others in virtue of a principle of pre-established
harmony. For the character of the monads is now beginning to be
absorbed into 'logic'—and ultimately into the noumenal realm. Kant
seems to conclude in NQ (though he does not here explicitly say so)
that the 'what' of the happenings is a matter for experience, an *a
posteriori* element. The point is certainly made with explicit emphasis
in the *Critique*, when Kant writes:

> How anything may be altered, and how it should be possible that upon
> one state in a given moment an opposite state might follow in the next
> moment—of this we have not, *a priori*, the least conception. For that we
> require knowledge of actual forces which can only be given empiri-
> cally. . . .[1]

Corresponding to the place assigned to mathematics in NQ, we
should, however, expect Kant to hold here that there is to be found in
every empirical happening an *a priori* aspect also. This, too, is clearly
enunciated in the *Critique*, where Kant continues the passage just
cited, with the following proviso:

> But apart from all question of what the content of the alteration, that is,
> what the state which is altered, may be, the *form* of every alteration, the
> condition under which, as the coming to be of another state, it can alone
> take place . . . can still be considered *a priori* according to the law of
> causality and the conditions of time.[2]

Further elucidations of Kant's main ideas concerning 'physical
conflict', first formulated in NQ, occur mostly in M, in connection
with the 'pure' parts of natural science. We have seen that one of the
main ways by which Kant delineates the notion of the empirical
reality is through the contrast with logical necessity. Thus he had
written in NQ:[3]

> How through the motion of one body that of another body may be
> counteracted [e.g. when in collision, or in mutual attraction] is another

---

[1] K. 230. For this reason, forces cannot be 'constructed'; cf. below, p. 674.
[2] Ibid. In NQ Kant had not yet reached the conviction that the categories were also
essentially involved in the formulation of the notion of an empirical realm. Moreover, in the later work, the aspect of form often uppermost in Kant's mind is 'time'
as 'form of intuition', although space is no doubt equally basic. For the moment it
is sufficient to ferret out the origin of Kant's contention that each empirical account
contains a reference to something *a priori*, regarded as 'construction'.
[3] NQ. 203.

question [*sc.* other than purely logical], since the two motions surely do not [logically] contradict one another.

Now for one body to resist another in impact raises also the question of the solidity or impenetrability of bodies. Let us first turn to this. In M Kant mentions 'Lambert and others' who had attempted to explain impenetrability as deriving from the very 'concept' of matter, as involving 'solidity'. Kant interprets this as an explanation which proceeds 'in accordance with the principle of contradiction'. And again he rejects it, with the celebrated remark: 'the principle of contradiction does not repel any material body'.[1] To understand this, we must first note that in M one of the criteria of matter is its capacity to 'fill space'. Kant then goes on to 'prove' that this can take place only by attributing to every point of matter (regarded as continuous) something like a spherical field of repulsive force. The 'proof' involves the contention that all attempted penetration into a material body is to be regarded as a tendency to motion, and so is the resistance to the latter. In this way, the whole phenomenon is again exhibited as an opposition of motions; motions which—*via* the concept of velocity— Kant has previously shown to be *constructible*. (Thus in M a 'clear' demonstration of the theorems of the composition of velocities has previously been given.[2]) The *resistance* by matter to the motion of another body (attempting to penetrate it) is then defined as an instance of a cause of motion, i.e. a force. The main point involved in this proof is that 'matter fills space in virtue of a moving force, and not merely of its existence'.[3]

Now Kant's basic objective in this mode of presentation is to show that there is an intimate, not to say essential, relation between matter and force, and through this, motion.[4] In this scheme, force is not simply 'added' hypothetically, as something 'foreign', to the concept of matter, as had been the case with the 'mathematico-mechanical' approach of the 'Newtonians', by contrast with which Kant calls his the 'dynamic form of explanation'.[5]

All this is of some importance, since Kant will show that a similar

[1] M. 170.   [2] M. 160–4.   [3] M. 170.
[4] Cf. K. 230: '... knowledge of actual forces ... as, for instance, of moving forces, or what amounts to the same thing, of certain successive appearance, as motions, which indicate [the presence] of such forces.'
[5] Cf. M. 201. Though historically independent, this in some respects resembles Boscovich's theory of matter, but more importantly it anticipated and perhaps contributed towards the development of physical field theories in the nineteenth century, particularly in the work of Faraday.

argument (the details of which need not concern us) makes *attractive* force likewise a basic aspect of matter, thus depriving all anti-Newtonian objections to 'attraction at a distance' of any basis.[1]

Kant's line of argument is basically as follows. As regards 'repulsion', this is not sufficiently 'explained', he insists, by making it an 'analytic' component of one of the 'essential' properties of matter, viz. 'solidity'; for, thus understood, this is just an 'empty concept'.[2] But neither are we entitled to 'add' to matter, defined solely by reference to extension and solidity, subsequently—even if only 'hypothetically' —the quite 'foreign' property of action without contact. Rather, in either case, we must—before forming any hypotheses—first show that matter is necessarily and exhaustively *comprehended* as *being* simply the seat of these two forces, a demonstration which moreover involves considerations of the composition of motions and the corresponding theorems. Only if this is done, Kant declares, can we 'understand [for instance] how there is *a contradiction*, that into a space which is taken up by any one thing another thing of the same kind should nevertheless enter'.[3]

All this provides us powerfully with a further explanation of the kind of 'understanding' that Kant intends by the locution, 'ability to comprehend the empirical *possibility*' of matter and force, in the context of natural science.[4] Clearly, it involves the notion of (explicitly applied) 'construction', Kant implying that his demonstration involves thus a 'first datum of the construction of the concept of matter'.[5]

All this will help us now to deepen our understanding of Kant's critique of Leibnizian metaphysics in the Amphiboly, considered earlier, and the contrast there drawn between noumenal and phenomenal reality, when Kant writes that in the former we can think only

[1] For this, cf. M. 182–94, and below, pp. 673–4. An earlier version of this approach is to be found in Kant's pre-critical CP. 18–20.
[2] M. 201.
[3] M. 171; my italics, i.e., it is 'real opposition', not logical contradiction.
[4] Cf. above, pp. 513–4, on Kant's criterion of 'possibility' of hypotheses.
[5] Ibid. It is however *only* a 'first datum' for its construction in any *a priori* sense. Force itself is a purely *a posteriori* element, as we have noted in the passage from K. 230. For this reason, the 'complete concept' of matter (for instance its particular *laws* of force) cannot be constructed, but only the 'general concept' of matter. (Cf. M. 200–1, 212.) In the end, therefore, it turns out that the 'possibility' of matter and force is established in two senses, (1) a strong sense, since it involves 'construction' of the corresponding motions; (2) a weak sense, since—although 'forces' cannot be 'constructed'—Kant has shown that the concept of matter, as understood by him, essentially involves a reference to forces, and thus, indirectly, to motions. (Kant does not, however, explicitly distinguish these two senses of 'possibility'.)

of logical conflict, whereas in the latter we have to take note of the concept of 'real conflict', *via* the 'opposition in the direction of forces'.[1] Now in the cases just considered, taken from the discussion in M, this model of 'construction' was of course employed in what I have previously called the 'explicit sense' applied to the empirical concept of matter.[2] But just as that concept is claimed to be possible only by reference to 'construction', so the phenomenal object in general is subsequently said to be possible only on condition that something corresponding to 'construction', which involves a 'synthesis' of parts, be presupposed: an equivalent locution for which is that the object be given in an intuition.[3]

It follows that our understanding of the *general* case (involving 'construction' in the 'implicit' sense), which provides perhaps the greatest stumbling-block for any understanding of Kant, is provided with a vital clue, if we realise that much of its meaning and significance is borrowed from this model of 'construction' applied 'explicitly' to the concept of matter, where it is meant to offer legitimisation of any hypotheses concerning the action of matter. And we shall get an inkling of Kant's intentions when in the Amphiboly he maintains that phenomenal reality involves the concept of construction. For this is the concept which—by acting as the central criterion—distinguishes also between the realms of what Leibniz had called logical necessity and logical contingency. To exhibit the significance of Kant's reconstruction of the concept of contingency, let us revert to our discussion of repulsion and attraction, this time in connection with Locke.

So far, we have only mentioned Leibniz. But we can get further clarification by returning to some of Locke's spoken and unspoken assumptions. It will be remembered that Locke's *Essay* had shown a certain scepticism towards 'knowledge' of these theoretical concepts. Communication of motion by impulse had been declared to be 'inconceivable', and the 'original rules and communication of motion' unintelligible, the main reason being that there seemed to be no discoverable 'natural' or 'necessary' connection between our ideas of these diverse phenomena. On the other hand, we also noted that Locke admits that the 'fact of the matter' is observed 'by daily experience'.[4] Nor had he left the matter there, for he had added that in such cases we have to ascribe our ideas, and the connection which

---

[1] K. 284; 279; cf. above, pp. 544–8.　　[2] Cf. above, pp. 561–2.
[3] For the equivalence, cf. above, p. 562, the passage from M. 140, and footnote 3.
[4] Cf. *Essay*, ii.21.4, 23.28; iv.3.29; above, ch. IV, pp. 191, 269–70.

they exhibited in experience, 'to the arbitrary will and good pleasure of the wise Architect', i.e. God.[1]

Now I have maintained that Locke was here involved in a complex reconstruction of earlier views on 'knowledge'. His ideal is still that of some form of 'necessary connection', but he feels instinctively that this is inappropriate: hence the emphasis on the 'empiricist' approach of Newtonian science.[2] On the other hand, there is a residue of the older doctrine, namely the conviction (however confusedly or subconsciously it may have been held by Locke) that beneath the empirical 'coexistence' of 'ideas' there must be concealed some kind of 'link', albeit of a 'non-logical', and certainly of a non-transparent order. And this was the significance of his appeal to God: a philosophical model which—as I pointed out in my chapter on Leibniz—is not so very far from the Leibnizian ideal of the universal connection, said to be involved in all propositions, including those that are particular and contingent.[3]

We have already seen that Kant had been emphatic from the start in drawing a sharp dividing line between formal or logically necessary and contingent propositions. But, a true heir to Locke and Leibniz, he is not content to define contingency solely by reference to 'experience' —at least not to its 'sensationalist' aspect. Nor is it simply a question of joining to this basis a logic of inductive generalisation. His only objection to Locke and Leibniz is that they had 'postulated' in purely 'dogmatic' fashion a link purporting to mediate between the otherwise 'opaque' terms of contingent propositions. It is the dilemma of traditionalist metaphysics, as Kant insists already in NQ, that it never more than postulates such links, without certifying our entitlement to this process—without, indeed, as much as asking what it would be like to produce such certification and what meaning one was to attach to this.[4]

Let us now see how Kant develops this in his Introduction to the

---

[1] Cf. above, ch. IV, pp. 212 ff. [2] Cf. iv.3.29.

[3] Locke had, of course, additional models. His complaint that knowledge of 'coexistence' does not yield insight into necessary connection was matched by his doctrine of 'substance' (*quâ* substratum) as well as of 'real essence', both of which had yielded in their own separate ways either formal or metaphysical models for the source of the putative 'union' of the qualities of a physical thing.

[4] For this complaint, as developed in the *Critique*, cf. K. 50–8 (for 'dogmatic', K. 57). It will be remembered that Hume's search for a meaning of the causal tie becomes infinitely more pressing once the problem of the meaning of the propositional link had been extended by Leibniz to contingent non-general propositions.

Critique, where he uses a finer classification of the necessity-contingency pair. One aspect of necessity stressed in the Introduction is the element of 'analyticity', analytic judgments being defined as those judgments in which the predicate is '(covertly) contained in' the concept of the subject.[1] Also such judgments are—according to Kant—from the nature of the case 'prior to experience', i.e. *a priori*. On the other hand, in contingent judgments, the relation between subject and predicate must rest on a different basis, for the predicate goes 'beyond' or 'outside' the concept of the subject—in the sense that the connection is non-analytic, i.e. 'synthetic'. Kant chooses this term because it graphically represents his view that in such judgments the predicate is 'attached synthetically'[2] to the subject. But why should it be so attached? And what makes it possible to speak of 'attachment'?

Part of the answer must be, of course, that this is 'suggested by' or 'inferred' or 'derived' from 'experience', i.e. from what we have observed—at least in those contingent judgments which involve observation, sensation, i.e. those that are synthetic and *a posteriori*. There are, however, many judgments where, according to Kant, this element of the *a posteriori* is, to say the least, irrelevant: mathematics is one such group; the basic laws of motion another; a third group is exemplified by the principle of causation; finally there are the bolder judgments belonging to traditional metaphysics, concerning the existence of God, the immortality of the soul, etc. Yet all of these are synthetic in the sense that here the predicate never attaches analytically to the subject.

Now in the group last mentioned the problem of what 'links' subject and predicate is obviously most pressing if we take the view—as Kant did—that their validity at least, if not their genesis, borrows nothing from the *a posteriori*. But though most commentators concentrate on these cases, it must be said that this needlessly particularises Kant's problem—quite foreign to the basic direction of his general intentions. It prevents us from reaching the sense of his doctrine in the light of the pre-critical and especially Leibnizian elements in his thinking. Yet it is just these writings which exhibit Kant's questions and even his somewhat sketchy answers still in their most general form.

Let us here revert once more to the question asked in NQ which we cited before: 'How am I to understand that, because something is,

[1] K. 48.   [2] K. 50.

there should be something else?',[1] asked against the background of Kant's demarcation of the contingent from the logically necessary. Evidently, he is already here searching for a model which will describe, or at least supply the formal conditions for, a link relating the terms of contingent propositions in general, so as to catch the aspect of contingency, non-logical reality.

Now in NQ, Kant could only proceed some of the way. Negatively, he rejects the theological form of the answer, whether it be offered from the side of Locke or of Leibniz. On the other hand, there is one positive result, though it is only hinted at contextually: the relation between the terms of the contingent proposition must be modelled through mathematics, so as to define the notion of physical agreement and opposition. But we saw that this theme returns in the post-critical period, e.g. in M, as the notion of construction in space and time, as being a presupposition of the edifice of 'knowledge' in general, as well as when the principles of construction are made 'explicit'.[2]

We are here beginning to see something of the deeper relations in Kant's system. The older Locke-Leibniz theological model for the link is now being built far more intimately into the very notion of physical (non-logical) reality as such: in Kantian locution, as something which makes the definition of such a reality possible. Contingent judgments concerning this reality, though we depend for our knowledge of their 'content' on an *a posteriori* element of experience, involve a further reference to something *a priori*, the forms and concepts through which such terms are linked.

So, at a general level, *both* those contingent propositions that are synthetic *a posteriori* and those which are normally regarded as synthetic *a priori*,[3] *are treated by Kant on similar lines*, as raising the problem of the nature, 'existence' or status of the link that unites their terms. And he shares the Locke-Leibniz position that the existence of a sensory (*a posteriori*) basis is here not relevant, and cannot provide a 'ground' for the relations between the terms of contingent propositions, which bestow 'unity' on the latter.

[1] Cf. above, p. 563,. This formulation still occurs in the *Critique* (K. 255).
[2] Cf. above, pp. 561ff.
[3] This is contrasting 'contingent' with 'necessary', as in Leibniz, and not as 'depending on sensory information'. Whilst all *a posteriori* propositions are contingent, the converse does not necessarily hold. It is simply the case that the negation of a contingent proposition, unlike that of a necessary one, is not self-contradictory.

Now NQ suggests already that *the basis* for such a link must lie in the *formal* character of the empirical object or process in which the latter is enveloped, thereby becoming one of its essential ingredients. This 'formal character', through which Kant defines the specifically physical, rather than logical, had been given at first, we have seen, in the model of vectorial addition of physical quantities (e.g. forces, velocities). Subsequently, it had become 'construction'. Since at the foundation of either there lies space and time, clearly Kant will end with the doctrine of space and time as presuppositional conditions of the definition of *empirical reality*.

But by way of anticipation of what follows we may go farther. In the next section we shall find that space and time will turn out, in their first aspect, to be forms of intuition, i.e. 'entities' defined solely by the context of experience, in its formal (though non-conceptual) aspect, injected '*a priori*'. This will at once lead to a parallel development of the results achieved so far. For just as 'phenomenal reality' presupposes 'construction', so our 'experience' of such a reality will contain, as a transcendental element, construction, *quâ* form of intuition, as a 'spontaneous act'.[1] Finally, Kant will argue that only when experience is defined in this way can what it yields 'correspond' to the phenomenal object. Construction, from being a physical or mathematical result, becomes transformed into the transcendental form of a 'subjective' process: subjective only, of course, in the sense of not being *a posteriori* but as contained in the definition of the form of the cognitive judgment which results from reporting our 'experience'.

Kant frequently does actually describe construction as such a subjective process representing the kind of link for which we were searching. Consider a line. We need not *actually* draw it; yet its construction as a process (in accordance with a concept, here a mathematical formula) may be regarded as the formal condition of the concept of a line.[2] This leads to the second aspect of space (and time). For, as pointed out in the Transcendental Deduction (2nd ed.), we see at once that space, *quâ* 'object' of geometry, involves more than mere 'form of intuition', but in addition also *combination* of the

[1] Cf. passage from K. 577, quoted above, p. 561.
[2] Cf. Kant's important example: 'To cognise anything in space (for instance, a line), I must *draw* it, and thus synthetically bring into being a determinate combination of the given manifold, so that the unity of this act is at the same time the unity of consciousness (as in the concept of a line); and it is through this unity of consciousness that an object (a determinate space) is first cognisable' (K. 156); cf. below, p. 622.

manifold.¹ It is as though we said that the points of the line lie *in* space, representing a spatial aspect of reality, but that the drawing of the line, e.g. in accordance with some given formula (or, speaking more generally, a concept), which determines the points to lie in this space in a specific way, results in *a* space.²

'Combination' is thus an aspect of construction which represents a further and more sophisticated step in the old search for the 'link', on the side of both Leibniz and Locke. Its significance as 'transcendental unity of consciousness' will emerge later.

Comparing Kant's pre- and post-critical formulations thus helps us to appreciate more fully, first, the nature of the questions which the later work is trying to answer by means of the concept of 'synthesis'; secondly, the nature of the answer, by tracing the roots of the notion of synthesis back to 'construction', and from there back to the mathematico-physical model of forces and their vectorial representations. It is through this form, Kant claims, that thought makes contact with the world of physical reality: an anticipation of the doctrine of the 'schematism' of the *Critique*.³

For the moment let us only note that the essential novelty of the *critical* approach lies in validating this model of a link (for the notion of 'the synthetic') by reference to the logical character of 'experience' (and of 'consciousness', as noted in the passage cited from K. 156), expressed in a cognitive contingent judgment, linking subject and predicate or, more generally, the various sensory elements entering into consciousness. For this alone makes the latter 'belong to one another, though only contingently, as parts of a whole, namely, of an experience which is itself a synthetic combination . . .'.⁴

¹ K. 170n.–171n. Moreover, it is combination in accordance with a concept; so space (and time), when fully defined, involve reference to concepts. (This is stated in more explicit form in the *Prolegomena*, para. 38. Cf. also below, p. 587n.2.)
² Kant distinguishes between these two aspects of space as 'form of intuition' and 'formal intuition'. The former 'first gives a manifold', i.e. enables us to speak of points *in space*; the latter presupposes prior 'combination' or 'synthesis', and hence also the conceptual aspect of such a synthesis, its 'unity', specified *via* the categorial concepts; and only this yields *a space*. (Cf. below, p. 587n.2.) We shall discuss this twofold aspect of space further in what follows, distinguishing them as 'spatiality' ('indeterminate', cf. P. 84) and 'determinate space' (cf. K. 156).
³ For the characterisation of 'construction' within the explicit context of geometry, cf. below, p 607.
⁴ K. 50. We need not labour Kant's basic empiricist assumption that all formal and conceptual aspects of experience are only legitimised provided they are in principle regarded as given in the context of an actual experience in general, i.e. determine the *a posteriori* sensory component as such.

7. KANT'S THEORY OF SPACE

(a) *The criticism of Leibniz's doctrine of space*

It has been shown that one of Kant's central problems was the adequate characterisation of the difference between noumenal and phenomenal reality. And the first important model had been construction. Now construction involves the notions of space and time, so we may expect that an examination of these will provide a key to an understanding of the whole enterprise. The distinct ways in which the relation of space and time to noumena and phenomena respectively is construed will have an important bearing on that distinction itself.[1]

The growth of Kant's theory of space (and to a lesser extent, time) on which we shall concentrate in this section is an immensely complex matter, into which play external influences as well as Kant's own earlier views. We have briefly discussed Leibniz's theory of space in the previous chapter; let us now consider Kant's characterisation of this, a characterisation which, as usual, bears the stamp of his own ultimate doctrine, in order to approach the latter *via* Kant's criticism. We have noted previously two discrepant elements in Leibniz's procedure: on the one hand, space was interpreted relationally; on the other, it was declared to be mere phenomenon (though *bene fundatum*), which for Leibniz means: a reducible property of the monads, not ultimately 'real'. Kant discusses the former position in the Aesthetic, where he compares the doctrines of Newton and Leibniz.[2] His objection to the first aspect of Leibniz's theory is simply that a relational theory makes geometrical propositions *a posteriori*, instead of *a priori*; the latter a status which Kant always claims for Euclidean geometry. To this we shall return. The argument against the second aspect occurs more explicitly in the *Amphiboly*, though there is also an allusion to it in the earlier place.[3]

Kant proceeds by pointing to the consequences which follow from Leibniz's conception of monads; but his account is couched again in terms of his own distinction between noumena and phenomena and thus does not so much constitute a genuine criticism as an affirmation of his own position; at best it can further an appreciation of this position by allowing us to see how Kant contrasts it with his own

---

[1] A preliminary summary of the position will be found above, p. 548; the discussion begins on p. 544.
[2] K. 81. [3] K. 280, 285; cf. K. 87.

vision of Leibniz's standpoint. Things-in-themselves, he argues, could, if at all—though this is according to Kant for man an impossibility—be known only by a kind of 'intellectual intuition', through their 'inner properties': Leibniz's 'preceptions and appetitions', which 'mirror' those of all other monads. Phenomena are knowable only through 'outer relations'.

This point we have already met.[1] It looks at first sight as though there were a commonsense model for it. We have no 'intuitive' insight into other minds, let alone into the properties of inanimate nature. For knowledge of these properties we require, as the British empiricists had insisted, 'sensation'; we are reminded of Locke's picture of the sensory organs facilitating the despatch of information from the 'outside world' into the mind, external 'motions' providing a kind of bridge, or 'outer relation'.[2]

There can be no doubt that this model lends plausibility to Kant's argument, but it cannot *be* his argument since it would lead him into the whole quagmire of the Lockean confusions and difficulties concerning the relation of mind to the external world. In reality, Kant is involved here in an epistemological analysis of how we, as philosophical analysts, should construe the notion of phenomenal reality. But the object of such a construal *includes* the 'external world', our bodies, and, in a sense, our 'minds' or 'selves' regarded as 'objects' (e.g. of psychological study or even introspection).[3] We cannot ask any questions concerning the 'mind' of the philosopher who engages in the construal. Nor can the talk of a self supposedly 'affected by the object'[4] be taken as a reference to a physical or psychological process. For the emphasis is here all on 'receptivity', a characterisation of the 'matter' of 'appearance', sensation, as *a posteriori*. Now, as already mentioned,[5] the grammar of the material element of sensation as the only element that is *a posteriori* with respect to things regarded as 'appearances' is so construed by Kant as to *necessitate* the 'addition'

---

[1] Cf. above, pp. 544–7 for an introductory summary of this position.

[2] Cf. also Kant's definition of 'sensibility '(quoted already above, p. 545n.2) as 'the capacity (receptivity) for receiving representations through the mode in which we are affected by objects' (K. 65). At K. 83, again, we are told that a 'body', *quâ* 'appearance of something', must be regarded by us only as 'the mode in which we are affected by that something'. Incidentally, the expression 'something' should not here be construed as a reference to the 'thing in itself'. Also Kant naturally assumes that other factors (formal and conceptual) are required here.

[3] Cf. K. 168–9: 'We know our own subject only as appearance. . . . My existence is . . . only determinable as the existence of an appearance.'

[4] Cf. the reference to K. 65 and 83 above, n.2.   [5] Cf. above, p. 551.

of spatio-temporal form, in order that it may yield 'experience' of 'appearances' in so far as concerns their formal 'intuitional' component. This component has therefore to be placed at the level of the *a priori*.[1]

One important reason for this is ultimately a negative one. It is, as the rest of this chapter will make clear, that no other theory of knowledge is capable of 'explaining' or 'accounting for' the status of space and time which these forms do in fact possess as basic components of empirical language. In a theory like Leibniz's, they must ultimately 'reduce' to something non-spatio-temporal; and the same goes for Berkeley's account, with the added disadvantage that on that theory things reduce to 'ideas'; and on neither theory can geometry be an *a priori* synthetic science. The most fundamental motive for Kant's *a priorism* (in respect of space and time) will, however, become evident only in his account of the relational aspects of experience. For it will then become apparent that the 'subjectivity' of space and time (as forms of intuition) is required by Kant because only in this way can an account be given of 'objectivity' that will be apodictically certain.[2]

All this means that Kant's mention in the *Amphiboly* of the *need* for 'outer relations' (space) is really an oblique reference to his own theory of things having to be regarded as 'phenomena' or 'appearances', because otherwise no 'account' could be given of the spatio-temporal component of language; a component which can sometimes be shown to be required explicitly if we want to specify all the 'relations' into which things may enter.[3] Yet, so his objection to Leibniz runs, no such account is given in the monadology, since space and time are there regarded as merely 'intelligible form of the connection of things . . . in themselves', 'sensibility' being no more than a

---

[1] Cf. K. 66: 'Whilst the matter of all appearance is given to us *a posteriori* only, its form must lie ready for the sensations *a priori* in the mind.' Kant frequently employs the psychological model of 'the mind'. What is meant however is the total context of the transcendental conditions which make possible the locutions involved in cognitive judgment—a logical, and not a psychological matter. Mind, *quâ* 'empirical consciousness', is itself 'an appearance', presupposing the transcendental apparatus. (Cf. p. 575 n.3; and K. 157–8.)

[2] Cf. below, sect. 8. For 'relational', see pp. 617–8.

[3] Cf. below, pp. 599ff., for instance, the argument concerning 'incongruent counterparts' (right-handed and left-handed glove), which to Kant proves that a full specification of empirical reality involves reference to more than merely 'internal' relations—rather a different notion of 'internal', than the purely 'logical' context of the argument against 'noumenal reality', which focuses on the aspect of logical contradiction, though in fact Kant seems to slide here from one use to the other!

'confusing and distorted' form of thinking ('understanding').[1] In short, the monadology represented the situation as 'if space and time were determinations of things-in-themselves'.[2]

Let us be clear: Kant is not denying that Leibniz acknowledges spatio-temporal locutions. What he objects to is that these have not been 'accounted for' satisfactorily, his point being that such an account *must* exhibit space and time not as 'confused modes', inexplicably attaching to monadic substances to which they must ultimately 'reduce', but as something with independent status. The quarrel does not concern any empirical, physical or psychological matters but solely the account that is to be given of the 'foundations'. And the stance taken here is not, in Kant's eyes, purely arbitrary For it connects with considerations of the status of geometry, as well as with his whole view on the construal of the foundations of physical science. But let us for a moment follow Kant's further criticisms of Leibniz's theory of space in the *Amphiboly*.

The criticism runs as follows. How is it possible for Leibniz to speak of a 'community' of monads, if by this we mean a plurality of substances in some way coexisting in space? Surely once again the notion of such a community requires reference to 'outer relations'? Here a new element enters into Kant's criticism, which we shall take up in more detail in the next section, (b). For, Kant now contends, such 'outer relations' require not only an account in terms of space in general, but involve reference to 'a concept of mutual interaction' between the monads.[3] So evidently, in order to construe the complete concept of space, we need more than mere form, spatial locution *in general*, which is quite 'indeterminate'.[4] We need (in the words of the First Antinomy) also a reference to 'things', which first 'determine space' in respect of 'its possible predicates of magnitude and relation'.[5]

---

[1] K. 286.

[2] K. 280. But only *per impossibile*: K. 163, for instance, shows that on Kant's doctrine 'things-in-themselves' are not in space and time, in line with LF, §7. (Cf. above, p. 544.)

[3] K. 285. Kant makes the same point here also with respect to the category of cause ('ground and consequence'), by which alone 'two states of one and the same thing' are 'connected'. And again the application of this category will require schematisation through space and time, (which defines it as 'following in time in accordance with a rule' (cf. K. 185)) in the context of sensory experience, so that its validity (as transcendental presupposition) may be proved.

[4] For this characterisation of space *quâ* 'form of intuition', cf. also P. 84 (para. 38); and above, p. 572f.; below 587n.2.

[5] K. 400. Kant is here, of course, assuming his own theory of things *quâ* phenomena or appearances; i.e. he is forcing his own theory of space.

However, Kant continues in the *Amphiboly*, since Leibniz does not have available an independent concept of construction, and with this, no account of space in general, such reciprocal action cannot be made comprehensible, and thus reduces to the merely logical or metaphysical notion of community, in which spatial coexistence can only be a 'confused' concept,[1] the 'ultimate' status of which has to be defended 'dogmatically' in terms of the notion of a 'pre-established harmony'.[2]

But, as we said, the question of the empirical or non-empirical status of geometry is always at the back of Kant's mind; and here it also receives another, and deepened, interpretation, which explains further some of the background to Kant's affirmation of the need for an independent status of space and time (or construction). For Leibniz to speak of space and time as *applying* to things (*via* the structure of geometry), Kant contends, was altogether dogmatic and unexplained.[3] It cannot stand for two reasons, (1) because, being an 'opaque fact', the application becomes an *a posteriori* matter—this was like the objection he had previously urged against Leibniz's relational interpretation of space; (2) since, as 'confused concepts', space and time must ultimately collapse into the merely conceptual nature of the monads or noumena, their relation to the monads is never made 'transparent', and we can hence have no clear cognitive assurance of the way space and time *do* apply to 'things'. To appreciate the bearing of this objection we must remember that Kant's theory of the 'application' of space and time to 'phenomena' is construed as a kind of 'injection' of the spatio-temporal component of 'intuition'. Since 'phenomena' presuppose this 'injection', the specific axioms governing the metric of space will likewise determine the space of phenomena; a procedure which yields

---

[1] 'Confused' in the Descartes-Leibniz technical meaning of this term.

[2] K. 285–6. Kant's language in this paragraph is apt to cause misunderstanding. He does not mean that it is wrong to represent outer relations by means of the concept of the understanding, i.e. the category of mutual interaction, but only that it is wrong to try this *merely* with the help of the understanding, without reference to space in general, as 'form'. For when this is omitted, the category cannot be 'applied', as transcendental presupposition, but can only be postulated 'dogmatically' by means of a 'third cause' (K. 285), determining all substances through pre-established harmony. (For this, cf. also below, p. 586, the argument of the 'common cause' of the *Dissertation*.)

[3] K. 286. I have italicised 'applying'; Kant says: 'applying to things *quâ* appearances'. He wants to say that, on his own theory, spatio-temporal construction applies *necessarily*, as a transcendental condition.

the account of the 'certainty' of geometry for which Kant is arguing.[1]

So this gives us an interpretation of the Kantian contention that space and time must be *independent* forms of intuition. It also follows that if we want to say that spatio-temporal forms *necessarily* apply to *things*, we shall have to regard the latter, not as Leibnizian noumena, but as phenomena which by definition require the notion of 'outer relations' (as an *a priori* addition), and the apparatus of geometrical construction as well as the 'metaphysics' of dynamical science, which yields the material for the 'possible construction' of dynamical interaction. When added to the argument against Berkeley already considered, the thoroughgoing complexity of the concept of 'phenomenon' thus readily emerges. And it is of vital importance always to keep in mind this grand complex of considerations that is fed into Kant's key notions, which ultimately are no more than a vehicle for the great variety of responses to his criticism of previous philosophical standpoints.

To return: a community of things coexisting in space, Kant has argued in his criticism of Leibniz, requires reference to 'outer relations'; but why should this in turn involve the 'concept of mutual interaction' and the concepts and laws of dynamics? Now we have already seen that there is a peculiar cross-traffic between forces, construction and spatio-temporal form. Forces and their corresponding 'motions' presuppose construction and hence spatio-temporal form. Yet the latter, in the guise of 'formal intuition' (as contrasted with 'form of intuition') itself, as already suggested,[2] requires reference to a synthesis 'under concepts', e.g. the concept of dynamic interaction. Is, then, space primary or secondary? Clearly, the answer is that in the sense of 'form of intuition' (what we may call the aspect of 'spatiality', of 'being *in* space'), it is primary; but *quâ* 'formal intuition', what we may call 'determinate space', or 'being *a* space', it is secondary, and presupposes categorial concepts. To elucidate this, we shall need to follow again the historical development of Kant's reflections on space, time and force. We shall see how to these two aspects there correspond conflicting stages in this development. And we shall see again that Kant seldom simply abandoned earlier views or positions of his own; but that rather he transmuted them by

---

[1] Cf. below, sect. 7(e): ii, pp. 606ff.
[2] Cf. above, pp. 572f., 573n.2; below, p.587n.2.

converting a feature which had appeared earlier under an 'ontological' guise, into something possessing 'transcendental' status.

(b) *The doctrine that space presupposes interaction*

Let us first turn again to the *Estimate of Living Forces*. Here, we saw, material substance is viewed as the centre of an 'active force', and to start with, as '*vis viva*'.[1] But this active force manifests itself also as an action in virtue of which the substance repels and attracts other substances, the force being transmitted in all directions.[2] Now, interestingly, in LF Kant argues that extension and space are not independently given but exist only consequent upon the outward action of substances; without the latter, 'there would be no extension, and consequently, no space'.[3] He infers, moreover, that the properties of space, e.g. its three-dimensionality, are due to the nature of these forces, as represented for instance in the form of Newton's inverse-square law of gravitation, which requires reference to a three-dimensional space. And further, since this law is not a necessary property of matter, but is contingent, it follows that the three-dimensionality of space is likewise contingent.[4] Finally, Kant concludes, contingently we can have 'no knowledge' of other-dimensional worlds; although such worlds are 'metaphysically possible', their existence is 'improbable', as conflicting with the principle of 'the greatest perfection'.[5]

By contrast, the phraseology of the later *Critique* will reformulate this in the following way:[6] Non-Euclidean space is conceded to be *thinkable*, since 'there is no contradiction in the concept' of, say, a non-Euclidean triangle.[7] On the other hand, the earlier 'improbability' now becomes a 'transcendental' or 'real *a priori*' impossibility, since,

---

[1] Above, pp. 553–4.

[2] LF, paras. 1–3, 12, 51. As already noted, this is reflected in the teaching of M, where 'matter' is defined, a.o., through the criteria of attractive and repulsive forces. The only difference is that space (as form of intuition) no longer *results* from the forces but is treated as a *presupposition* of the possibility of constructing their equivalent motions; only 'formal intuition' presupposes concepts and their application.

[3] LF. 9 (para. 9).   [4] LF. 9–11 (paras. 9, 10).

[5] LF. 12–13 (para. 11). Note again the Leibnizian tone.

[6] Note that I call it a 'reformulation'. Let us insist once more that, throughout Kant's later work, the contexts of the earlier positions linger on, like shadows, supplying hidden meanings and thus explaining the significance of the whole enterprise. It is for this reason that throughout this chapter I have tried (however sketchily) to explain Kant's thought by constantly relating pre-critical and post-critical stages.

[7] K. 240; cf. below, pp. 611ff.

so Kant asserts, it is not possible 'to construct' such a triangle in any *given* space.

More generally: the earlier doctrine of space and its properties as deriving from the forces and their laws is now transformed into a theory in which two aspects of space are distinguished. Space *quâ* form of intuition is presupposed before all forces. Moreover, there is, as we saw, imported the suggestion that this space 'must' be three-dimensional, although the interpretation of this necessity—belonging really to space in its second aspect—still needs discussion. Let us therefore consider space in this latter aspect.

Space *quâ* 'formal intuition' (determinate space) presupposes, as we saw, a reference to things and their mutual relations,[1] and with this, the concept of 'dynamic interaction', and even—when this concept is 'applied' to the notion of 'matter' (as in M)—actual forces and their laws, though only 'in general'; the specific 'content' of these laws is still a 'contingent' matter, as already noted.[2] Nevertheless, these forces, *via* the motions which 'indicate' them, involve 'construction', which clearly itself takes place *in* space *quâ* form of intuition.[3]

For the moment, let us note that the account found in LF already anticipates the close relation which Kant sees between force and space. And the gap between this and the later account is not as great as is often made out. The chief difference lies rather in the fact that in the later account both space *quâ* 'form', as well as the concept of interaction become 'transcendentalised'. Let us therefore continue for a while, under the guidance of the pre-critical writings, to gain further insights into Kant's doctrines.

The position which we find in LF is that space, and the coexistence of physical substances in space, is conditional upon (or presupposes) their mutual universal interaction. But since according to the *Critique* (Third Analogy of Experience) 'all substances, in so far as they can be perceived to coexist in space, are in thoroughgoing mutual interaction',[4] we see that this basic position still remains, except that

---

[1] Cf. above, p. 577.   [2] Cf. above, p. 510, below, pp. 672–4.

[3] This is the burden of the proof in M. 223–5, where it is shown that interaction becomes intelligible, i.e. possible, only if we construct, in conformity with the transcendental concept of interaction, the basis for the required forces as motions which are assigned to both the bodies involved in, e.g., collision. But the construction requires the drawing of trajectory lines, in turn presupposing space *quâ* form. (Cf. below, sect. 9(c).)

[4] K. 233. Kemp Smith's translation of 'Wechselwirkung' as 'reciprocity' is peculiar, since he himself translates this term as 'mutual interaction' in the 1st edition

instead of '*assuming* dogmatically' the notion of interacting forces ('*influxus physicus*', as the *Inaugural Dissertation* calls it)[1] and of 'substances' existing 'in themselves', Kant is now re-interpreting the latter as things 'perceived' (i.e. as 'phenomena'); and with this, what was earlier a physical hypothesis (i.e. 'interaction'), is now reinterpreted as a 'transcendental presupposition' of the possibility of perception, or rather, 'experience' (or 'knowledge', i.e. Erkenntnis) of such phenomena'.[2]

For a fuller appreciation we need to consult the *Dissertation* which provides vital clues for an understanding of the development of Kant's doctrine of space, lying as it does half-way between the pre-critical period and the fully developed doctrine of the *Critique*. Here, the distinction between space as form of intuition, indeterminate space or spatiality, and as formal intuition, or determinate space, is already implicit. *Quâ* form of intuition, space is now asserted to be nothing when abstracted from the context of 'outer sensations', and in this sense nothing 'objective and real'; certainly not 'an absolute and boundless receptacle of possible things' on the lines of Newtonian absolute space.[3] Space is only 'altogether true', provided it be viewed as something 'relative to all sensible things', i.e. as a universal condition (or 'foundation') of that part of experience which concerns the non-conceptual aspect, i.e. intuition.[4]

Into this aspect of space we need not enquire any further for the moment; its significance will be more properly appreciated after we have discussed (in section (c)) what I take to be Kant's reformulation of the Newtonian position. The *Dissertation* shows, however, clearly that this is not yet a complete reconstruction of space, as will eventually be attempted on the basis of transcendental idealism. Yet something concerning the assumptions and viewpoints underlying this reconstruction, including the very need for it, emerges more clearly in the *Dissertation* than elsewhere, because here the nature of this second aspect of space is still viewed in a relatively 'dogmatic',

---

version of the *Critique*. Cf. also below, pp. 666n.2, 667n.3. All such 'action' is however regarded as reducible to *laws* of motion; cf. M.188.

[1] ID. 75, 78 (paras. 17, 22). In ID Kant is critical of this 'assumption' but has not yet anything better to put in its place.

[2] For the Third Analogy, cf. below, sect. 8(e): iv.

[3] ID. 68, 70 (para. 15, A.D); cf. also above, p. 549n.5 for similar characterisations elsewhere in Kant; and below, pp. 595ff.

[4] ID. 71–2 (para. 15, E). In ID, the conceptual aspects were assigned to the realm of the '*intellectus*', which in the *Critique* became 'the understanding'.

i.e. metaphysical—not to say picturesque—fashion. And the metaphysical version is always important, since the transcendental account amounts almost universally to no more than a re-location of the earlier metaphysics. It is this which provides the primary meaning for the transcendental entities; and if this is realised we shall avoid over-misleading empirical interpretations. Or perhaps, bearing in mind Kant's use of psychological models, we have here a conflation of metaphysics and empirical models.

So far, Kant points out, 'this concept of space' which has here been proved 'denotes only the intuitively given *possibility* of universal coordination'.[1] In the clearer wording of the *Critique*, space *quâ* form of intuition concerns merely that aspect in virtue of which it is 'made possible' that 'the sensory manifold of appearance can be [i.e. is potentially capable of being] ordered in certain relations'.[2] (Note that this does not make space itself a relation; on the contrary, as the rest of Kant's exposition is concerned to show, we are here concerned with explaining the foundation that will eventually yield space as 'an infinite *given* magnitude', a 'single' and 'all-embracing space'.)[3]

Now what is it to say that we have *so far* proved only the 'possibility of universal coordination'? Kant's reply in ID is really the central clue for the rest of the critical standpoint, expounded in the *Critique*. For, Kant insists, this 'proof' does not as yet give us a foundation for the locution—hitherto believed to be obvious—that 'all things that exist are necessarily somewhere'.[4]

[1] ID. 74; clearly a reference to 'indeterminate space'.

[2] K. 66; 2nd ed. Kant altered the text of the 1st edition, in order to emphasise explicitly that this gives so far only a (spatial) basis for a *potential ordering* and not yet the material for an *actually ordered* space. And the same remarks go also for time. Cf. below, p. 621.

[3] K. 69. I say 'yield eventually', since the exposition in the *Dissertation* and the Transcendental Aesthetic (K. 68–70) is concerned only with the formal condition (form of intuition) and not with space as determinate and completed, when the aspect of synthesis and the categories is taken into account. Once this is remembered, many of the apparent inconsistencies of this exposition, with which commentators have found so much difficulty, will disappear. To mention just one such difficulty: although Kant's exposition here refers to space as a 'given whole', this would plainly be inconsistent with the teaching of the First Antinomy, which denies this. In the present context, Kant wishes merely to contrast space as 'given' with the 'faulty' doctrine of space as 'thought'. Secondly, he is assuming that the 'wholeness' of space presupposes the synthesis under concepts, which is a guided *process*, as already mentioned. So he is only saying that his characterisation of space at K. 69 is such as to make the completion *via* the synthesis result in a space of the kind here described.

[4] ID. 74 (para. 16); cf. also ID. 84 (para. 27), with a similar reference to time ('somewhen').

Kant's rebuttal is interesting because it tells us about the way he envisages the function of the postulate of absolute space: to supply a bond or scaffolding, which would relate the different substances and thereby give them a definite position in space. Space, he tells us, when regarded

> as a real and absolutely necessary bond, as it were, of all possible substances and states [would constitute] the primary condition of their possible interaction, [since on such a view] all things that exist are necessarily somewhere; [which would guarantee why things are] presented in a fixed manner, . . . being determined directly by the fact that space is a whole which includes all things.[1]

Evidently Kant holds that, if a view concerning the kind of space as here characterised could not be substantiated, a new foundation would be required to make interaction intelligible, and with this, indirectly, the notion of a plurality of coexisting bodies. The foundation for both is not given until the *Critique* (Third Analogy); for the former, it is further developed in M, involving the construction of a basis for forces, as well as for their composition, and the action of one body on another, either by way of mutual repulsion or gravitational attraction, requiring the whole apparatus of synthesis in accordance with concepts, viewed as a condition of possible experience.[2]

In ID, all this is still missing; and for the moment, at any rate, we are not yet in possession of the conditions that would allow us to postulate a coordinate system specifying the position of a single body, let alone the relations between a plurality of bodies, a possibility that had been espoused, though in Kant's eventual view, abortively, by the notion of Newtonian absolute space, regarded as a receptacle. But only such a related plurality of objects would give us what Kant calls 'space'—i.e. space *quâ* formal intuition, determinate space. He writes:

> The question, which cannot be solved save by the intellect, thus still remains quite untouched: *on what principle rests this relation of all substances, which, when viewed intuitively, is called space*? On this point turns therefore the question . . . , to make clear: in what way it is possible *that a plurality of substances should stand in mutual [dynamic] interaction*, and in this sense belong to the same whole, called a world.[3]

---

[1] ID. 74.

[2] This was the significance of the reference to the missing concept of interaction between monads in the *Amphiboly* (K. 285), above, p. 577. Cf. also pp. 566ff., and especially p. 581n.3.

[3] Ibid.; italics in text. The reference to the 'intellect' [*intellectus*] reminds us that the eventual answer given in the *Critique* involves the categories of the understanding—in this particular case, the category of 'community' or 'dynamic association',

In other words: whatever it is that makes it *possible* to speak of space (*quâ* determinate space) will be equivalent to the feature that makes it *possible* to speak of physical interaction between substances; the general notion behind this still being that space depends on physical interaction.[1]

This, then, is the sequence of Kant's assumptions: the notion of space in its determinate sense requires the concept of interaction of bodies. On the other hand, the mere existence of a plurality of such bodies does not yield interaction; 'something further is required from which their mutual relations can be made intelligible [intelligantur]'.[2] Here lies the advance from the earlier LF. For, Kant now informs us, we cannot (as in LF) simply *assume* interaction, from which in turn space would be derived.[3] To make this assumption first 'requires a special ground determining this precisely'.[4] Once again, the basic standpoint is that 'mere observation' of certain forces, for instance, is not enough. Any inductive evidence, direct or inferential, requires that the concepts involved be first made 'intelligible'.[5] Moreover, this requirement, in the case of space (as determinate space) is evidently made necessary by Kant's critical expurgation of the earlier dogmatic views on space, which were replaced by the doctrine of space *quâ* form of intuition, thereby teaching us something about the significance of this expurgation. For this new concept of space, being as such something quite indeterminate, is impotent to yield 'objective' relations or reference.[6] In the *Critique*, the required 'ground' will be sought in the transcendental synthesis under the concept of interaction. But in a way, the significance of this, as a true ground, is made clearer in the *Dissertation*, where Kant, unsure of a solution on 'critical' lines, still defines such a ground—though in very tentative language, as if he saw it as no more than a picture of traditional metaphysics—as 'a cause common to all things', a 'common principle'

---

whose application in the Third Analogy leads to the 'principle of coexistence' (K. 233).

[1] One might regard this as a forerunner of the 'physicalisation of space', which later became prominent in the outlook of General Relativity theory.   [2] ID. 75.

[3] Cf. above, p. 580. In the words of the title of para. 9, LF: 'If substances had no force whereby they can act outside themselves, there would be no extension, and consequently no space' (LF. 9).   [4] ID. 75.

[5] This point is made explicitly at M. 158 (i: Observation to Explanation IV). Cf. our discussion of the three criteria for the legitimacy of hypotheses, the first of which was 'possibility'; above, section 4(c): iv, pp. 512ff.; 6(c), p. 567.

[6] This is also the explicit teaching of P. 82–4 (para. 38).

which 'maintains all things'. He there sketches the three theories which had commonly been proposed concerning the modes in which this 'common ground' might be held to act: (1) *via* the *influxus physicus*; (2) through an harmonious internal adaptation of all substances (i.e. the Leibnizian concept of the pre-established harmony);[1] (3) through the external *fiat* of the Malebranchian system of 'occasional causes'.[2] Though not coming out finally on one side or the other, Kant states (though again with hesitation) that 'very likely' the theory of the *influxus physicus* will win eventual acceptance, i.e. the doctrine of physical interaction, provided *its* 'ground' can first be established.[3] It is to these three theories that the transcendental doctrine of the *Critique* (especially the Third Analogy) constitutes a response; an important consideration, since—as I have maintained— it is possible to exhaust the full significance of such a response only if it is seen against the background of the earlier contextual meanings.

Now, as just noted, in the *Dissertation* physical interaction is not accepted as an ultimate ground but is founded in God, a typical piece of metaphysical postulation. That Kant recognises it as such emerges also from the Scholium, appended to para. 22, where he writes:

> All outer things lie open to the mind's view *in infinitum*, only in so far as the mind, along with all other things, is upheld by the infinite power of a single cause. For this reason it senses external things only through the presence of the one upholding common cause; and space [in the sense of 'form of intuition'], which is the universal and necessary condition, sensitively apprehended, of the co-presence of all things, can therefore be entitled *omnipraesentia phaenomenon*. (For the cause of the universe is not present to each and everything because it is in their place; on the contrary, places, i.e., possible relations of substances, exist, because, it, the cause, is intimately present to all things.)[4]

A similar account is then given of time, as '*aeternitas phaenomenon* of the general cause'.[5] And Kant adds that this 'opinion' is 'little different' from that of Malebranche, 'namely, that we see all things in God'.

Kant acknowledges that this is a half-way-house, since he remarks that he is here overstepping 'a little the limits of apodictic certainty

---

[1] Cf. above, pp. 544–6, 576–8.   [2] ID. 77–8.   [3] Cf. ID. 75.   [4] ID. 78–9.
[5] Ibid. As will be seen in the next subsection, there is a resemblance here to Newton's Cambridge-Platonist doctrine, that God, 'by existing always and everywhere, ... constitutes duration and space'. (Cf. below, pp. 595ff.)

befitting critical metaphysics', and is sailing 'out into the deep sea of mystical enquiries'. Nevertheless, for us these formulations are of vital importance, as telling us something about Kantian 'questions'.

(c) The 'transcendental ideality' of space

Kant, we have seen, opposes both the relational theory of space,[1] the theory which regards space as a property of things-in-themselves (whether in their Leibnizian or Berkeleyan guise), and finally the doctrine of absolute space, under the image of the boundless receptacle. We have also had some glimpses of Kant's own responses, whose objective is to supply 'foundations' that will safeguard, or explain the locution of, external objects in space. And we have noted that the 'transcendental approach' is directed both to providing a framework for the notion of indeterminate space (spatiality), and that of a determinate space. The details of Kant's theory as regards the latter have to await an account of the transcendental deduction of the categories, for in order to supply locutions for the notion of 'determinate space' we have to discuss 'construction' in the context of 'synthesis in conformity with concepts';[2] and this will be the task of section 8. For the moment, let us concentrate on Kant's interest in the location of things *in* space—the location for which the transcendental support will be the notion of the *a priori* 'form of intuition'.

For certain purposes, at any rate, it is not necessary to distinguish between these two aspects of space. This is especially so when we follow Kant's transformation of physicalist or metaphysical conceptions of space into something transcendental. The essential question which concerns us then is always this: if space is something real, in what is this 'reality' to be anchored? The support of the thing-in-itself is gone; it cannot be a limitless self-subsistent 'non-entity' either; but if we anchor it in some metaphysical ground—the view espoused tentatively in the *Dissertation*,[3] with its suggestion of a 'common cause' (a reference to the Deity)—this is a thoroughly 'dogmatic' and non-empirical proceeding.

---

[1] In this section we shall primarily concentrate on space, though similar considerations often hold for time also.

[2] A fact which is, as has been noted, especially remarked on by Kant at K. 170n.– 171n.: 'Formal intuition . . . presupposes a synthesis [which first supplies the unity of this *a priori* intuition; and through this synthesis] space and time are first *given* as intuitions' (italics in text); cf. above, pp. 572f.

[3] Cf. above, p. 586.

The reader knows, of course, already from all that has been said that Kant will eventually anchor space in 'possible experience', and again, in the possibility of 'appearances'; and that, furthermore, the result of this move is supposed to 'secure' space as an independent element, as an 'all-embracing', 'unique' and 'given whole', not 'dependent' on things, but logically preceding them, as their 'condition'.[1] The essential part of the argument, as was noted in the discussion of Kant's criticism of Berkeley,[2] is to prevent space subsisting at the level of the *'a posteriori'*, which alone is passively received. How does Kant support this?

As already anticipated, it can only be maintained by a doctrine which so frames the notion of a thing as 'appearance' that it is one of its logical characteristics for the *a posteriori* element in it *necessarily* to lack the spatio-temporal component. Now this form of characterisation is simply the opposite of the metaphysical doctrines which it opposes (and equally metaphysical), which was that the spatio-temporal component of things is *not* ultimately a 'real' entity, whether anchored in a Berkeleyan 'idea' or in a Leibnizian monad. But since (as we shall see again in the next subsection) the only other available possibility of space and time as boundless self-subsistent receptacles is equally unacceptable, there remains at first sight no anchorage at all. True, things 'appear' as spatio-temporally processed entities. But we are not concerned with what appears, but with the 'ground' of the components of the appearance. And, as we saw, Kant refuses the 'solution' of identifying this 'ground' as God. There is perhaps a last possibility; could not space and time be *anchored in* the *'a posteriori'* element itself (sensation)? This possibility Kant rejects likewise, and most decisively, for a variety of reasons which we shall have to follow up further in what follows. It conflicts with the putatively *a priori* status of geometry; it runs foul of the possibility of giving an account of 'objective' cognition, because it comes close to the already rejected general doctrine of space and time as properties of things in themselves; and this is what an *a posteriori* sensational element would be if itself presented with 'its own' spatio-temporal framework.

At the same time, the decision to deprive the *a posteriori* of 'its' space and time (which seems to leave both these 'forms' finally anchorless) inevitably builds up the necessary pressure to locate these somewhere else, since they are undoubtedly an essential aspect of phenomenal language. If this line of argument is accepted, it becomes

[1] K. 69–70. (Cf. above, p. 583n.3)   [2] Above, pp. 549–52.

obvious that there was left only one final possibility: to locate space and time at the point at which apprehension of a spatio-temporal appearance occurs; i.e. simply to make it a *condition* of (the possibility of) this apprehension, as the formal aspect of intuition.

This transcendental re-location would seem to satisfy all the various requirements: it leaves space and time 'independent' (since 'irreducible') entities, in the sense of not being given *with* the *a posteriori*, nor being a 'confused' property of monads. Yet, it avoids the notion of self-subsistent receptacle. Being irreducible guarantees for their 'appearance' (as 'form of intuition') the status of 'empirical reality'; whilst their not being self-subsistent entities, but only conditions that make experience (spatio-temporal apprehension, intuition) possible, makes them 'transcendentally ideal', rather than 'real'. Moreover, not being part of the level of the *a posteriori*, they must be *a priori*; injected as concepts which have (outside the experiential context) no real but only a logical or ideal status.

Such, summarily, is Kant's argument. Let us pursue some of the details. Kant's actual arguments, especially those offered in the Aesthetic,[1] have often been felt to be rather deficient as 'proofs' that space and time are *independent* and *a priori*. To say that space and time are conditions of things, that they are basic, individual, infinite, does not seem to clinch the matter. Nor does the contention that space and time universally belong to all appearance, and that this can be known antecedently to all experience, in contrast with sensory qualifications. And the (for Kant) crucial argument that they are the basis for geometry, regarded as a synthetic *a priori* body of knowledge, might likewise be said to beg the question.

Finally, it is not sufficient to describe Kant's position by reference to the experiential assumption alone. Thus Kant says that

> if we depart from the subjective condition under which alone we can have outer intuition, namely, liability to be affected by objects, the representation of space stands for nothing whatsoever.[2]

But clearly it does not follow from this that the *a posteriori* sensory element ('matter of sensation') in what is given for experience might not be spatio-temporal as such, in precisely the way in which in fact it does after all actually appear. To draw this conclusion requires the additional premise that it is necessarily otiose to speak of space (like time) as a 'property of things in themselves [and]...' as 'a

[1] K. 68–70; 74–6.   [2] K. 71.

determination that attaches to the objects themselves':[1] a locution which we know now to be equivalent to treating the *a posteriori* element of the appearance as 'deprived' of space and time at the level at which it (the *a posteriori*) is 'received'. Only then have space and time as yet no place. And only then does it become *necessary* that they be supplied from another level, so that an appearance may be 'given'.

So if Kant presents his arguments in the exposition of the Aesthetic as implying that 'space does not represent any property of things in themselves',[2] we can see that although this is a crucial premise, it is hardly a 'conclusion' (as he claims) from his exposition; for this we require reference to the total background of Kant's views and to the intellectual forces that produced them.

There is, however, an important item in our presentation so far which requires further elucidation. We have often spoken of space and time being 'supplied' or 'added' or 'injected' from another level. And we have noted that this level is sometimes pictured by Kant as 'the mind'. To call space and time 'transcendentally ideal' makes it appear as though they lay ready-made in our minds, as a kind of ideal forms, to be added to the *a posteriori* on appropriate occasions.[3] This is, however, rather a misleading simile, with the added difficulty that it presents us with the gratuitous problem of 'knowing' on which occasions to add space and on which not to do so, since many of our 'experiences' are after all not 'spatialised', not perceived as 'out there'. Besides, it makes it appear far too much as though space and time were now 'properties of mind'—quite out of tune with Kant's real intentions. In fact all we can say is that the *a posteriori* elements ('matter of sensation') *are* in space and time only in so far as objects *appear* (are presented as 'appearances') spatio-temporally (in the context of experience). The question where space and time themselves *'are'* simply *has no meaning* outside this context, whether we fasten

---

[1] Ibid. We have seen that the 'object itself' might be a Berkeleyan idea, which is spatio-temporal, even though its *esse* is only given in experience ('*percipi*'); cf. above, pp. 549–51.

[2] Ibid. Kant must there also assume that they are neither self-subsistent entities, nor grounded in God, as just noted.

[3] Cf. the passage from K. 66, cited above, p. 576n.1, which thereby attempts graphically to represent the logical place of space and time as being other than that of the *a posteriori*. Moreover, 'mind' suggests that space and time are 'within', whilst the matter of sensation comes from 'without', all equally impermissible locutions, since the *a posteriori*, in abstraction from space and time, is neither within nor without— at least, in any 'phenomenal' sense.

on the *a posteriori* element by itself which is 'received', *or* on 'mind' which is aware of the appearing object when it cognises it. We cannot ask at all where 'space and time are situated', outside this context.[1]

So Kant's main point comes down to saying that space and time are 'empirically real', i.e. that '*we can speak of space*'[2] and time at the phenomenal level precisely because these 'entities' have no logical place outside this linguistic situation; that they are empirically real simply *because* we have an empirical linguistic location for them, at the phenomenal level, where it is indeed absolutely required. They are thus 'proved' by virtue of the very fact that we do have experience of an external or internal world—when we have it! Their reality would be problematic only if they could perchance hover behind the scenes, so to speak: e.g. buried in a 'mind', or in God, or in the noumenon. Modifying the passage from K. 71, quoted above,[3] it would therefore be better to say, *not*, that space is as nothing outside the condition of there being intuition, but rather that *because* space *is nothing* outside such a condition (is not a property of things-in-themselves, nor a non-substantial receptacle), it is *necessary* to supply it as an *a priori* condition, or transcendental presupposition—which 'proves' its 'empirical reality'.

Outside the intuitional context space and time are then not 'ideal forms', lying 'ready-made' in the mind, to enfold sensory reality. The terms space and time refer us only to *concepts* of 'forms' whose *designata* have empirical status solely in so far as they 'make it possible' that the sensory manifold should be intuited as 'appearance'; and they are the manner (spatio-temporal) in which intuition occurs. Kant must therefore insist that, outside the intuitional situation, space and time can be no more than concepts *a priori* or 'ideal', for only thus can he say that correspondingly the 'matter of sensation' (the *a posteriori*)—outside this context—is not 'given' either, cannot 'appear', since for this 'matter' as such, spatio-temporality is not defined.[4] This establishes the sense of the expression 'transcendental ideality', ascribed to one side of the status of space and time.

[1] The 'meaninglessness' of this question is precisely what created our earlier problem of how it was possible even to frame the notion of the noumenon, where we make just this assumption of space and time as properties of the thing, or of a thing as being set apart from space and time, etc.

[2] For this locution, cf. K. 71.   [3] Cf. p. 589.

[4] The two expressions, 'the *a posteriori*' and 'the empirical' should therefore never be confused. 'Appearances' are known empirically; but their sensory component, taken in abstraction from the conditions of intuition, where it *must* be spatio-temporalised, is clearly a purely theoretical abstraction.

We have here a double achievement. The 'transcendental ideality' of space and time is 'balanced' by the world of 'appearances', whose 'possibility' is thereby established.[1] Secondly, that world is conditional upon the *a priori injection of the concepts* of space and time, which should not be confused with the misleading picture of a quasi-physical or psychological 'injection' or 'addition' of corresponding *ideal forms*. For the metaphor of injection only refers to space and time as concepts, but says that they are *given* as *forms* in any intuitional context which makes the object, *quâ* appearance, possible.

So only at the transcendental level ('in transcendental reflection')[2] can we say that space and time (as *a priori* concepts) are 'added' to the *a posteriori* matter of sensation. But at the empirical level, from the point of view of the 'appearance', we cannot talk of space and time as independent entities, being *added* as 'forms': for there they are the forms *of* the intuitional manifold. This Kant notes expressly:

> Empirical intuition is not . . . a composite of appearances and space (of perception and empty intuition). The one is not the correlate of the other in a synthesis; they are connected in one and the same empirical intuition as matter and form of the intuition.[3]

It is in any case clear that space and time, where their conceptual (ideal) status is defined as that of components of the 'possibility of appearances', cannot be 'added' to anything; one cannot add possibilities. This procedure of 'spontaneous' injection we must therefore always read as a locution which stands for 'supplying the *a priori* concept as a condition subject to which an appearance can be given', or 'subject to which alone anything can appear to us'. But that the concept is thus *supplied 'a priori'* is 'proved' only, granted the premises, (a) that objects (as 'appearances') *are* given to us; (b) that the philosophical grammar of 'appearance' is construed as entailing that logical status of the '*a posteriori*' element which we have noted at length: an approach, as will now be clear, which precisely mirrors the procedure of 'reason', where an 'order of nature' (as a *system* of appearances) was only defined provided we supply the concept of unity as condition. And in each case, the two elements, phenomenal order and transcendental concepts, hold each other in balance in precisely the same way,

---

[1] Always understood that we must keep in mind that all this is only half the story. For an appearance to appear as an object, we require the account of the synthesis of perceptions as an act of consciousness, whose 'unity' yields the notion of an object. (Cf. section 8.)

[2] For this expression, cf. K. 207, 286.

[3] K. 398 n.b.: Note to the 'proof' of the 'thesis' of the First Antinomy.

with the difference that the 'order of nature' is never really 'given' but only 'projected', so that this unity as transcendental condition only functions relative to something that has been *postulated*.[1]

This 'bringing together' of the 'subjective' and the 'objective', of the 'act' and what results from the act, is the very epitome of the Kantian method. And just as space and time are 'tied down' in the coalescence of 'form of intuiting' with 'form of that which is intuited', so, in the transcendental deduction of the categories, we shall presently meet again such a coalescence, between the concept of the object and the object itself. In the words of the Preface to the Second Edition, Kant makes our concepts conform to the object by making the object conform to the concept[2]—each 'balancing' the other. The concept applies, and the object has empirical significance, only in so far as such a balance is achieved.

We are now also in a position to complete the account of Kant's reply to Berkeley, begun in section 5.[3] The doctrine of the 'empirical reality' of space and time implies that what is seen to be as 'out there' really is so. The reason for this is that there *is*, according to Kant, *only one kind of space*, a space which is (empirically) real; there are not *both* a real *and* an ideal space at the phenomenal level.[4] And this contention holds even in respect of those who, like Berkeley, had denied the existence of such a real space. For in ascribing 'mere empirical ideality' to space (the emphasis being on the 'mere') they had at least assumed that such a contrast *made sense*, which is what Kant denies. (All 'ideality of space' belongs to the transcendental level.)[5] For this reason, what 'appears', and might thus be held to possess the status of a Berkeleyan 'idea', must necessarily be an empirical *thing*; and by reading 'thing' for 'appearance', we reach the result that there can be no second realm of things, at least not at the phenomenal level, simulating 'ideas', of which alone we might be 'aware'.[6] To do so, as Kant points out in the chapter on Noumena

[1] The 'unity' is 'proved' or 'deduced' only 'subjectively'. (Cf. above, pp. 503, 508, 509.)
[2] K. 22. Much the same procedure lies behind Kant's doctrine of geometry, cf. sect. (e): ii.
[3] Cf. above, pp. 549–50.
[4] We are of course not concerned with cases where things are 'imagined' to be in space, or 'seem' to be in space, for these presuppose the existence in principle of a real space, with which such phenomenal locution can be contrasted.
[5] In M there is yet another conception of space, 'absolute space', which is however not a transcendental concept, but an 'idea of reason'.
[6] This is also the burden of Kant's 'Refutation of Idealism', K. 244–6.

and Phenomena, is to confuse what is really only a 'noumenon' with something which, though purportedly lying behind the scenes, ought still to have phenomenal status.[1] Yet his argument is so contrived that nothing with phenomenal status *can* lie behind appearances.

(d) *Some Newtonian influences*

Now this theory of Kant's, according to which what is 'seen' to be 'out there'—in respect of the 'outer relation' which is here involved—*is* external in a primary sense, and not something derivative, stems from many sources. Among these, there is in Kant an interesting response to some aspects of Locke's and Newton's disquisitions on space. A consideration of these is of help, since, as usual, Kant's theory simply embodies shifts in the 'metaphysical centre of gravity' of positions which, though they may represent perfectly possible pictures, display themselves in a misleading logical or theological guise.

Let us briefly consider the situation where we left it in our review of Locke's theory of space. This, we saw, comprises a large and rather bewildering variety of conflicting views. According to one version, space was 'real', *simpliciter*; a fact which had been expressed by Locke through his claim that space was a 'simple idea'. But Locke had also spoken of it as a 'boundless ocean', whose borders mind never reaches. At other times he had written that space has no positive real existence without things; that it denotes only the bare possibility of a body existing, where sometimes there may be none. And he had also echoed Newton's view, in the General Scholium to the *Principia*, that time and space exist in virtue of God's filling eternity and immensity. Finally, he had seemed to opt (though with hesitation) for a relational theory, but only after throwing out the visionary suggestion that if space could not be modelled as a substance, in the way in which God, matter or finite minds are substances, it was perhaps a fourth type of substance.[2]

Now there are many echoes in Kant of this somewhat inconsistent bundle of suggestions. He, too, holds space to be 'real' (empirically real). It is characterised in terms of 'possibility' ('condition of the possibility of appearances'). Again, it is characterised 'as an infinite

[1] Cf. especially the account given in the first edition, K. 268–70.
[2] For details, cf. ch. IV, pp. 254, 257, 258, 259, 258, in the order of the above suggestions. It is, of course, true that Kant could have known at best only of the doctrines expressed in the *Essay*. But the Journals often crystallise and repeat views dispersed throughout Locke's circle—they mirror contemporary opinions.

given magnitude',[1] although it turns out that *quâ* individual spatial whole, we require reference to a 'successive synthesis', which ultimately is viewed as a task for reason, to be attempted but never completed.[2] Equally clearly, for Kant space is neither a substance divine (anchored in God), nor material (anchored 'confusedly' in monads), nor mental (not a discursive concept but an intuition), not even a 'property of mind'. But as transcendental ground it does touch Locke's 'fourth substance'.

More important, although the Locke-Newton affirmation that space is somehow an aspect of God's immensity is on the surface irately rejected by Kant[3], it does here find an echo, even if only after undergoing an interesting transformation. We know that Newton had been deeply disturbed by the conception of 'absolute empty space'. Although it was required by methodological considerations, his ontology had no room for what in his eyes this implied: a 'self-subsistent boundless receptacle'. Certainly Kant's criticism of this conception can in strictness only apply to the later Newtonians.[4]

Newton's response to this worry owed much to the Cambridge Platonists, especially to Henry More. Space and time are given ontological support in theological guise, by anchoring both in God. Thus in the General Scholium to the *Principia* he writes:

> [God] endures for ever and is everywhere present; and, by existing always and everywhere, he constitutes duration and space.[5]

The vital question here is, of course, 'How are we to understand the way God 'constitutes' space and time?' The answer that Newton gave was that this can be made intelligible through the conception of God's *sensorium*; something which, as Clarke explains to Leibniz, may in turn be grasped by analogy with, as a 'simile' for, the *sensorium* of man.[6]

---

[1] K. 69. The meaning of 'givenness' will be explained in the next sub-section.

[2] K. 401; and above, p. 583n.3.

[3] Referring to the suggestion that 'God is comprised in infinite space all at once', he reacts with the comment that 'words cannot express the extent to which philosophers are befooled by these shadows that flit before the intellect' (ID. 85, para. 27).

[4] K. 81 refers to the 'mathematical students of nature'; and at ID. 70 (para. 15) there is a mention of 'the English view' of space, as 'an absolute and boundless receptacle' of possible things. (Cf. above, pp. 549n.5, 582.)

[5] *Principia*, op. cit., p. 545, cf. above, p. 586.

[6] It was Leibniz's misunderstanding of Newton's intentions here that was partly responsible for initiating the correspondence with Clarke. For Clarke's explanation, see his First Reply, 3; op. cit., p. 13. Cf. also Newton's *Opticks*, Questions 28 and 31; op. cit., pp. 368 and 405.

Now the notion of the 'human *sensorium*' Newton frames in terms which borrow considerably from Locke's theory of perception. We are acquainted only with our 'ideas' (Newton calls them 'images'), which alone are *in or before our minds*. At the same time, we do see things in space. We may describe this by saying that the 'idea' represents things as 'appearing in space'. Space is thus an integral aspect of the *sensorium*, which is, of course, nothing physical, and is quite distinct from the sensory organs.[1] In virtue of the *sensorium*, what is, metaphysically speaking, 'in the mind', is described as 'appearing in space'; seeing things in our *sensorium* is to see things in space.

The human *sensorium* thus yields a kind of 'ideal space'. And it is an analogy for God's *sensorium*, with that difference that God does not see the 'ideas' of things, but 'the things themselves' directly; and he sees them

> in infinite space, as it were in his sensory . . . : Of which things the images only carried through the organs of sense into our little *sensoriums*, are there seen and beheld by that which in us perceives and thinks.[2]

Evidently Newton's theory is this: 'Space' subsists in the context of 'seeing things in space' in the way in which 'ideas' *represent* things in space, out there, although there *is* no real space; it can only be ideal, since the 'ideas' themselves are not 'out there'. So *our* space is 'constituted' by our 'seeing things out there'. However, Newton at the same time affirms that this is only a 'simile' for that one real space in which—as real things—everything subsists. As a result, he argues that just as everything subsists in one space *for us*, in virtue of and in the context of our perceptions, so there is one real space, in virtue of God's perceptions; God 'seeing the things themselves', rather than their ideas.[3]

Now some of the basic features of Kant's doctrine are here already present. Space is not a self-subsistent entity; and further, it is defined in the context of 'seeing things', i.e. 'sensory intuition'. Moreover, just as the picture of God's perception assures that space is logically prior to things, so does Kant's transcendental argument. But there is this difference: the (empirical) reality of the Kantian space is assured,

[1] Cf. *Opticks*, Qu. 31: 'The organs of sense are not for enabling the soul to perceive the species of things in its *sensorium*, but only for conveying them thither' (op. cit., p. 403).
[2] Op. cit., Qu. 28, p. 370.
[3] This explains the sense of the passage quoted above, p. 586 and p. 595.

not by God's perception of things, but by the transcendental ideality of space, i.e. by its being an *a priori* condition whose empirical significance is *only* defined in the context of something being intuited as 'appearance'. It is because things are 'appearances' (whose sensory component stands *logically* in need of space, only given through intuition) that we can be certain of space being 'contributed' as a transcendental element in the moment of perceptual experience (using these terms in a colloquial sense). Newton's story is that things would not exist *in space* unless they were perceived by God. Kant's modification of this is that they would not appear unless they were given to us in intuition, where the notion of 'appearance' is so used as to *entail* that space has to be added to its *a posteriori* content, appearing as 'form of intuition'. Both Newton's 'God' and Kant's 'transcendental condition' are to make it possible for us to say that things that 'are *seen* in space' thereby also *are* in space—in contrast with the doctrines of Berkeley and Hume, according to which 'to be seen in space' is—speaking, of course, only metaphysically, not empirically—equivalent to 'merely *seeming* to be in space', since space is here regarded as a derivative entity.[1]

There is a further linguistic echo in Kant reminding us of the Newtonian locution of '*sensorium*'. For Kant often says that when we perceive things in space which stand thus in an 'outer relation' to us, we see them in virtue of 'outer sense'.[2] Such a 'sense' is, of course, no 'sixth sense', added to the senses of sight, sound, etc. It simply repeats, in a somewhat physiologising manner, the doctrine that in intuition things are seen as 'outside', suggesting through the physiological metaphor that 'outness' has empirical reality, which in turn is meant to imply that it is non-derivative.[3]

Now it is clear that this account of Kant's reaction to Newton's approach to the problem of space is not entirely fanciful, since Newton's account bears striking resemblances to certain features which we noted in the *Dissertation*.[4] There is no doubt that these Newtonian analogies greatly aided the graphical formulation of

---

[1] It is unclear whether Kant's notion of the *universal* (and necessary) presupposition of space is a sufficient analysis of our conception of a single all-embracing physical space. But presumably on his principles this is a question which has no answer. The problem of 'other minds' is not one that ever agitated Kant.

[2] K. 87, and cf. the whole discussion at K. 346–9.

[3] K. 87 also speaks of 'the representations of the *outer senses* [which] constitute the proper material with which we occupy our mind'. Here, and elsewhere, e.g. at K. 131, Kant also posits 'inner sense', whose 'formal condition' is said to be time.

[4] Cf. especially above, p. 586.

Kant's problem. Just as the scientist needs analogies for finding his way among the abstract features of his systems, so the philosopher often also employs similes, analogies and metaphors. Nor are these inevitably misleading. Knowing them to be analogies, we are less apt to interpret Kant's own solution in too empiricist or psychological a frame. Indeed, through the rôle which theological and empirical analogues play in philosophy (as images of metaphysical positions), we are reminded of the non-empirical features of philosophical thought.

(e) *Supplementary arguments for Kant's conception of space*

A considerable part of Kant's argument depends on our accepting his contention that sense ('sensibility') is generically distinct from thought ('understanding'); so that the 'formal' aspect of the intuitional component of appearance (space and time) may be kept a distinct element, not merging into the inner nature of any 'thing-in-itself'. Space and time 'in transcendental reflection' must be conceived as 'determinations [not] of things-in-themselves but of appearances';[1] and this means they must, at the transcendental level, be logically prior: independent and *a priori*. Moreover, as determinations of appearances, they subsist only in the context of the givenness of such appearances; one cannot speak of a given spatial *whole* as 'given all at once', in abstraction from the *process* of cognising appearance.

Now there are two supplementary considerations (though only worked out for 'space') by which Kant seeks to support this doctrine. The first claims that an account of the spatial determinations of any physical body requires reference to a space *in* which such a body is located, which shows that space is more than, and independent of, the 'inner relations' of body. The second consideration relates to Kant's view that space is *a priori*, which he supports by the putative fact that Euclidean geometry is a synthetic *a priori* body of knowledge. Let us very briefly consider these claims.

(i) *Incongruent counterparts*

The first of Kant's arguments centres on the so-called problem of incongruent counterparts. It is supposed to yield 'strong confirmatory evidence' for the 'proposition that space is not reducible to properties

---

[1] K. 286.

or relations of things-in-themselves . . . but merely a subjective form of our sensible intuition of things or relations'.[1] It is not at all certain that it does support this conclusion. The basic contention involved in the argument is that one cannot say everything one wishes about physical bodies, without presupposing a space in some sense independent of, therefore prior to and determining, the former. But since 'space as a whole' is a concept that needs 'critical' interpretation by reference to the theory of 'outer sense', that theory might be said to be confirmed by this argument.

If we compare a left hand with a right hand, we see that, although the relative positions of the parts of each are identical, there is still a difference between them. This difference (with respect to each taken by itself) can however be made manifest only by reference to the other, i.e. a *second body*, here: a left hand; and by *showing* (through attempted superposition) that a right hand is not congruent with a left hand (e.g. a right-handed glove cannot be made to fit a left hand). Let us, using Kant's terminology, call lefthandedness and righthandedness 'inner spatial determinations' of physical bodies, and designate them as L, R, respectively.[2] Kant's basic contention is that L, or R, cannot be explained intelligibly, nor specified by reference to the properties of any body *taken by itself*, for instance spatial relations possessed by the body which is L- or R-determined. In the present case this is obvious, since the relative positions of the parts are the same for both L- and R-determined bodies. But, quite generally, it seems that right or left, as directions 'in space', cannot be obtained by reference to any general marks, which do not ultimately involve *pointing to particular instances*; we have to say, for instance, that the direction is 'as that of the motion of the sun', or that it is 'in a direction contrary to that of the motion of the sun', etc.[3] Since L- or R-determinedness do not depend on any spatial relations internal to each of

---

[1] M. 154. The argument originally formed the basis for an earlier pre-critical tract, on *Regions in Space* (DS. 36–43), and it occurs also in the *Prolegomena*, para. 13, P. 41–3. It is not in the *Critique*, presumably because it yields only *supporting* evidence and in any case is really no more than an *analogy* of the basic contention which Kant means to support. It is nonetheless interesting because it supplies new contextual uses for the Kantian locution of the 'thing-in-itself', its purely conceptual status, as objects of 'mere thought' and the contrast with those things which presuppose space, time and 'construction' (understood as transcendental presuppositions of 'appearances').
[2] P. 42. At DS. 42, P. 42 and M. 154, Kant describes what distinguishes incongruent counterparts, such as L and R, as 'inner differences'.
[3] Cf. M. 154.

the bodies taken 'by themselves',[1] they must depend on something external.

Now in DS, which is just outside the 'critical' period, it is said to be 'absolute space' that provides the 'ground' for these 'inner determinations', such as L, R. Moreover, this space is 'primary', logically prior to the L and R. It is as though this 'absolute space' *supplied* the directions, such as L and R, through an absolute reference framework. More generally, such a space supplies the framework which makes it possible that a body may unambiguously be said to be 'somewhere'! Kant writes:

> Thus it is evident that instead of the determinations of space following from the positions of the parts of matter relatively to one another, these latter follow from the former. It is therefore clear that in the constitution of bodies differences are to be found which are real differences, and which are grounded solely in their relation to absolute, primary space.[2]

Evidently this is meant to supply a backing for Newton's 'absolute space', for Kant adds:

> A reflective reader will therefore regard as no mere thought-object [Gedankending] that concept of space which the geometer has thought out and which ingenious philosophers have incorporated into the system of natural science.[3]

Now Kant does not leave the matter there—not even in the present work, which antedates the *Dissertation* by two years. For already here he adds the important rider that 'absolute space is not an object of outer sensation', but only the concept of something that 'first makes all such sensations possible'. And he concludes—very much on the lines of the *Dissertation*—that anything which involves 'reference to pure space' must be interpreted as meaning 'comparison with other bodies'.[4]

He also hints that we have to distinguish between space in general, as a presupposition of 'outer sensations', and 'absolute space' as an

---

[1] [den Dingen selbst]: this is the locution used at M. 154. It enables Kant a few lines later to slide into the phrase 'the things in themselves [der Dinge an sich selbst]'.

[2] DS. 43. We have already seen that during the period in which he wrote ID, Kant is concerned with the problem of space and time as conditions of things being 'somewhere and somewhen' (cf. above, pp. 583-4).

[3] Ibid. The translation has been modified.

[4] Ibid. In other words, Kant here foreshadows the doctrine of the *Dissertation* which makes space as such a conditional form of intuition, and which interprets an absolute spatial whole as presupposing 'relation to other bodies'. A relation which in turn is said to presuppose 'interaction'. (Cf. above, pp. 584-5.)

'idea of reason'.[1] This again foreshadows the distinction which Kant makes later, in M, between 'empirical' or 'relative', and 'non-empirical' or 'absolute' space, where the former is regarded as form of intuition, the latter as an 'idea of reason'.[2] The motive for the distinction is once again, that without it one cannot 'construct', and thus exhibit transcendentally, the possibility and hence cognisability,'[3] of, e.g., a body being subject to two equal and opposite velocities (or two forces) simultaneously. The reason is that construction requires vector lines drawn in space, and two such lines could not be drawn in the same space. The antimony is avoided if we distinguish between two kinds of space: a process which supplies an additional explanation of Kant's insistence that 'empirically real' space is purely phenomenal, and transcendentally ideal. For only on such an assumption can we regard space under these dual aspects, being both empirical (relative) and non-empirical (absolute).[4]

It is important to realise that, whatever Kant may subsequently supply as an analysis of the concept of space (relative or absolute), as for instance in DS or K, such an analysis has always the implicit intention to reproduce also those properties of space which connect with its picture of a spatial whole *in* which bodies are situated, and relative to which they have position. Equally relevant is the fact that Kant's interpretation of the being of a body in space involves already, in DS, and even more later, a reference to the need for a *comparison* with other bodies, standing in mutual 'external relations' to each other.[5] Surprisingly, this suggests that Kant was not adverse to a relational theory of space as such, on the lines of Leibniz's analysis. What he opposed was the view that these relations *derive* from the metric constitution of the individual bodies taken 'by themselves', instead of regarding the relations as a *presupposition for* the empirical notion of a plurality of coexisting bodies.

But how can the argument concerning incongruent counterparts be evidence for the transcendental ideality of space? For this is what Kant claims in P and M. Perhaps the wording in P will help us, for it repeats the phraseology of DS, by saying that the 'inner determination

---

[1] DS. 43.   [2] M. 152, 238–9.

[3] The reader will by now appreciate the extremely broad range of Kant's enquiry into the possibility of our cognition (Erkenntnis) of things. Moreover, a relatively straightforward and technical context like the present one yields a meaning for the more evanescent case of the possibility of a cognition of objects in general, to which we shall turn in the next section.

[4] For this, cf. also above, pp. 560ff.   [5] Cf. K. 400; and above, p. 577.

of any space [i.e. its being L- or R-determined] is only possible by
determining its outer relation to space as a whole of which it is part
(the relation to 'outer sense')'.[1] This evidently connects 'space as a
whole' with 'outer sense', where the latter is a locution referring to
space as the formal aspect of intuition, or to the way the sensory
manifold has to be related in order 'to appear out there'.[2]

Now partly this is just falling back on the main argument of the
Critique, since space (if this argument is accepted) *must* be a transcendental form. But the *Prolegomena* offers further analogies which suggest what this means to Kant. In our passage he speaks of the
phenomenon of incongruent counterparts as showing that there are
important properties of a body which cannot be derived by studying
it by itself; this is true for instance of its position in space.[3] From this
Kant now 'infers' by a slide that if we regard objects as 'representations of things as they are in themselves and as a mere understanding
would cognise them', even the positional facts about such a body are
undefined. Kant here treats the case of 'body by itself' as one of
'thing-in-itself', because this is an instance where cognition of the body
(in respect of L-determinedness) *necessitates* a reference to outer
relations whilst, the 'thing-in-itself' necessarily lacks it.[4]

This may seem artificial since the notion of thing-in-itself is far
richer, and in any case concerns something of which *nothing* can be
said in abstraction from relations to space, unlike the case of physical
bodies Kant is here considering. But he does draw attention to
another characteristic which applies in the present case and which
holds of 'appearances' but not of 'things-in-themselves', as construed
by him. To understand this let us go back to the passage where Kant
says that an L-determined property is only 'possible by determining
its outer relation to space as a whole of which it is a part'.[5] But, Kant
then goes on to say, this is true only of appearances; of things-in-
themselves it is never true that 'the part is only possible through the
whole'.[6] They, like Leibniz's monads, are self-contained, or self-
determining.[7]

---

[1] P. 42.   [2] Cf. above, p. 597.

[3] There is here an interesting parallel with the 'special' theory of relativity: one
cannot determine the (uniform) *velocity* of a body by considerations which appertain solely to that body by itself.

[4] We have noted that this slide is made explicitly at M. 154; cf. above, p. 600n.1.

[5] P. 42; see above.   [6] Ibid.

[7] A similar argument is used to resolve the Second Antinomy, concerning the
problem of the division of matter into simple parts. Here, too, if we regard an
object as a thing-in-itself, we have to say that 'a whole must contain all its parts in

Suppose we accept this characterisation of things-in-themselves, does the present argument show by itself that L-determinedness proves things to be 'appearances'? We know already on other grounds that empirical cognition implicitly presupposes space (as form of intuition); i.e. that cognisability of anything presupposes a reference to external *spatial* relations, and not, as for the case of the noumenon, reference only to internal, purely *conceptual*, relations. For otherwise the specific note which belongs to the physical, as distinct from the logical, realm would not be caught. If our concepts are to make contact with the world, we require a 'third thing', viz. the formal aspect of 'empirical intuition', which here is spatial. What is distinctive in the present argument is Kant's claim that there is a *phenomenal* character (L-determinedness) which cannot be explicated except by necessary *explicit* reference to space (and not just implicit, as in the normal case).

Perhaps, as I said, Kant does not wish this argument to have more than analogical strength. Perhaps it is also true that he was not always aware of the slide which is involved in passing from an argument which involves *explicit* reference to one which posits *implicit* reference to a presuppositional feature. We noted this distinction already when comparing the implicit use of construction (and synthesis), as well as the categories, for the purpose of objective cognition in general, with their explicit employment in the foundations of science.[1] And we shall meet it again in Kant's argument concerning the concept of cause in cognitive judgments in general, as contrasted with those that are couched explicitly in causal form.[2] And the same considerations (we shall find) will apply also to the argument concerning the foundations of Euclidean geometry. There can be no doubt, however, that the powerful, because obvious, existence of the *explicit* presuppositions in these cases lent powerful support in Kant's eyes to the 'implicit' procedure of the transcendental deduction in general. It is one of the most pervasive features of the whole Kantian method.

Sometimes the inconclusiveness of the 'explicit' argument becomes quite apparent. Thus, there is an objection to Kant's argument that the existence of the L, R properties shows space to be a form of

---

advance'. Whereas with respect to 'appearances', there are only as many parts as progressive division of the whole yields. This division may *proceed ad infinitum*, but it is not *'given as infinitum'*. (M. 180; cf. K. 402–9.) A *'given'* infinity can perhaps be 'thought' but it can never be 'constructed'.

[1] Cf. above, pp. 561–2.   [2] Cf. below, pp. 653–4.

intuition and not a property of things in themselves; and the objection supplies an opportunity to review the whole compass of Kant's approach to space, as far as we have taken it. Suppose space were after all purely 'phenomenal', in the Leibnizian sense of *phenomenon bene fundatum*, i.e. not founded transcendentally, as presupposition of appearances, but grounded in the thing itself, to which it must ultimately be reducible, being no more than a Leibnizian 'confused perception'? The question need merely be asked to show that Kant's conclusion from the existence of incongruent counterparts only follows if we *assume in the first place* that the space which is to explicate the L-property is not reducible to the status of a thing-in-itself (or rather; its property), but instead to something which is logically independent from what is given *a posteriori*, an independence guaranteed by its transcendental ideality which makes it a form of intuition, i.e. the formal aspect of the way in which the sensory manifold is apprehended in order to yield appearance. And clearly Kant's specific conclusion here is compatible with either general theory of space.

So it turns out that these two alternatives (Leibniz's and Kant's) are simply different 'ways of seeing' the total logic of the situation, mirrored in Kant's basic dictum that sense and thought, sensibility and understanding, are generically distinct domains. It follows that the phenomenon of incongruent counterparts does not *'prove'* Kant's conclusion nor support it, but *exemplifies* it analogically: a situation which seems to me to present in a nutshell the logical nature of the very core of much philosophical or metaphysical (though perfectly genuine) argument.

What Kant does throughout, as a basic postulate, is to *refuse* to accept the notion of a 'confused perception'. For Leibniz it had been a technical term; Kant objects that it leaves the position of space in a fog and that the only way in which we can make it clear is to render it as an independent form of the way we ourselves 'see' things in space. Only what we inject, or what is *a priori*, or 'subjective', of that we can be 'certain', provided this is balanced by the condition that the procedure of injection is no more than a contextual notion, living only in the context of 'application' to the element of the *a posteriori*; which in its turn again *means*—perceiving 'things in space', i.e. as 'appearances'.

Kant's way of explicating the status of space is not, of course, totally arbitrary. It is a response to the experientialist approach of the British empiricists that we can only cognise what is brought into the context of 'experience'. But that, too, is re-interpreted here, under the influence of Kant's method of philosophical re-location. For the empiricist doctrine had contained a built-in confusion; it had amounted to a conflation of contradictory demands, viz. that only that can be known which is a state of the subject, whilst yet at the same time it had also to be other than the subject. Kant splits up 'idea' into 'material' and 'formal' components. By making the formal component literally a state of the subject, i.e. form of intuition (and concept of the understanding), and by making the content (or 'matter') a merely theoretical, *as such* extra-experiential, 'x', of something other than a state of the self, something 'received', which in order to be anything 'given' for us needs to be 'intuited', he can then solve this contradiction. The 'foreign' aspect of the *a posteriori* can be reconciled with the complete subjectivity of the formal element, for they operate at different levels. And they are defined as joined, through the requirement that they occur as 'appearance'.

The discussion of incongruent counterparts is also taken up again in M, in terms very similar to that found in P, except that it is couched in terms of the contrast between 'the discursive mode of cognition' (which belongs to pure concepts, or rather: purely logical relations, whose object would be the thing-in-itself), and the 'construction of the concept'. Thus, Kant tells us, L- and R-determined properties 'cannot be made intelligible' through general marks, but require *construction*: which explicates the fact that we can ultimately explain the left/right distinction only through *pointing*.[1] Now construction we already know to be the means by which Kant always develops a concept in the direction of physical significance, through the intermediary of space.[2] It provides the distinguishing mark which contrasts that which is 'merely thought' with something that can be 'given' in intuition, i.e. in space. The requirement of construction thus operates as a constraint upon concepts. The nearest equivalent in modern philosophy would be the constraint put upon the concepts of physics which demands that we should provide for these an 'operational definition'.

---

[1] M. 154. We omit a critical discussion of this claim itself, which implies the contention that the explication of these concepts involves something which *differs in kind* from most other concepts which can be explicated through ways other than pointing. In any case, if my discussion of this argument is correct, Kant can only use the putative distinction as a model for a more general theory; so a decision need not be made concerning the tenability of his example.

[2] Cf. above, pp. 572ff.

As such, it shares the later difficulties of this conception. But a discussion of this would lead us too far from our topic.

(ii) *The status of geometry*

Construction as a constraining influence upon the degree of freedom of operation with pure intellectual concepts is a basic element also in Kant's theory of geometry, as it is in theoretical physics, regarded as a synthetic *a priori* body of knowledge in respect of 'foundations'. So we are led quite naturally to a consideration of Kant's second supplementary argument for the *a priori* status of intuitional form.

Let us remove one misconception from the start. Kant's method of supporting his theory of space here consists in offering an explanation ('the only explanation') 'that makes intelligible the *possibility* of geometry, as a body of *a priori* synthetic knowledge'.[1] He is *not* trying to *prove* that geometry *has* this logical character. At best he makes this an assumption; one whose truth he may well have believed, but whose falsehood would certainly not at all affect the fabric of the general doctrine of the *Critique*. On the other hand, it would make the present argument, regarded as a support, less interesting.

Indeed it turns out, as usual with Kant, that the argument offers no more than additional contextual information concerning the significance of his key terminology. For in Kant's 'explanation' of the possibility of synthetic *a priori* geometry, the very terms, synthetic and *a priori*, automatically receive the interpretation which they have in the mature parts of the *Critique*. Or rather: the mature meanings are added to the more traditional ones. Thus 'synthetic', though originally defined as a relation between terms which is 'ampliative',[2] i.e. where the predicate is not contained in the subject, eventually comes to designate a judgment which presupposes a 'synthesis' of parts, as in the example of the line.[3] Similarly, *a priori*, introduced originally by Kant in terms of its two criteria, universality and necessity,[4] receives the additional note of being connected with a 'transcendental condition', with 'spontaneity' (as contrasted with 'receptivity', the mark of the *a posteriori*). This in turn further relates '*a priori*' to the notion of 'possibility'; witness Kant's explanation at M. 140, that 'to cognise anything *a priori* means to cognise it from its mere possibility'.[5]

Again, *a priori* cognition sometimes contrasts with 'what has not

---

[1] K. 71.   [2] K. 48.   [3] Cf. above p. 572; below, sect. 8(a).   [4] K. 44.
[5] We have already remarked on the considerable traffic in Kant between these notions, cf. above, p. 962n.3.

been borrowed from any experience',[1] i.e. what contains no *a posteriori* elements. But even here we must place this in its proper context, which, for the present discussion, is geometry. Thus, a triangle is said to be constructed *a priori* when it is constructed 'by the imagination alone', i.e. 'without having borrowed the pattern from experience'. Now such a construction is always an operation on an *individual* part of space, even though we abstract from the 'empirical' (sensory) aspect of phenomenal reality as well as from the particular characteristics of the figure e.g. the triangle, that we may happen to draw.[2] The abstract result is a 'pure intuition' (in respect of its form). Such a 'pure intuition' is said to be '*a priori*'.

However, the construction of a triangle also involves something with an *a priori* aspect in the further sense of proceeding in accordance with a concept.[3] There are in fact always two things involved in a geometrical demonstration: first, the need for construction; secondly, the concept. Kant's explanation of this[4] is that in order to prove a geometrical theorem, we require not only the concept of a triangle, we need also to 'construct', i.e. to introduce considerations of space, as well as employing the axioms which govern the demonstrations of some given system of geometry. Construction is thus *a priori*, not only because it involves the 'form' of 'pure intuition', spatiality, but also because it uses the definitions and axioms which govern the deductive development of a geometrical system, and which are *a priori* because they are injected *ab initio*.[5] Kant summarises all this, when he writes:

[mathematics considers its concepts] not empirically but only in an

---

[1] Cf. K. 577.

[2] Note that construction is always treated as operating on 'a single object' (K. 577), for only in this way can it (as the 'pure' part of an intuition) make contact with empirical reality, which is always individual. This is one of the marks which distinguish intuition from thinking, which operates with concepts *under* which fall an infinite number of instances, whereas intuition concerns an infinite number of representations *within* itself; i.e. all those for which a given triangle 'stands'. (For this, cf. K. 69–70.)

[3] Cf. K. 577: 'In this empirical intuition [of a triangle on paper] we consider only the act whereby we construct the concept, and abstract from' all its particular determinations with which the demonstration is not concerned. (Cf. above, pp. 561–2, where we considered construction in relation to its function in dynamics.) For an anticipation of Kant's point, see Berkeley's notion of 'considering' the relevant properties of individual triangles, to yield universality of an inoffensive kind; above, ch. V, sect. 2, pp. 284–5.

[4] K. 578–9.

[5] The relevance of 'construction' for the formulation of the notion of contingency has been discussed already in section 6(c). For some of the logical characteristics of construction in that context, cf. especially pp. 561–2, 572–3.

intuition which it presents *a priori, that is,* which it has constructed, and in which whatever follows from the universal conditions of the construction must be universally valid of the object of the concept thus constructed.[1]

We see then, again, that there is a considerable traffic between space, intuition, construction, *a priori,* possibility, axiomatisation and transcendental presupposition. With these explanations, we may pass to the core of Kant's theory of geometry.

Here, let us distinguish between a weak and a strong claim. According to the weak claim, all physical bodies must be subject to spatial determination. This follows from the fact, already demonstrated, that first, all 'appearances' presuppose space *quâ* form of intuition; secondly, that they presuppose it *quâ* 'formal intuition'—the notion which is equivalent to the postulate of construction, in the sense of involving a 'synthesis' which itself proceeds in accordance with a general concept.[2] This concept is the category of quantity, which, when applied to the present case, yields the principle: 'All intuitions are extensive magnitudes'; the so-called principle of the 'axioms of intuition'.[3]

Let us remember that this is a transcendental argument: the principle is balanced by the 'givenness' of appearances.[4] It holds (or applies) only in so far as there *are* appearances—and who would deny this? What *would* be problematic is only the *assumption* that these appearances are subject to a *specific* spatial metric, an assumption which—as we shall see presently—will 'balance' the corresponding *transcendental presupposition* of a Euclidean form of intuition. But, at present, we could only question the claim that physical things are 'appearances'—quite a different *kind* of criticism. Kant is at present merely claiming that we can cognise things only *quâ* appearances and that these are given only on condition of presupposing the transcendental principle, and that neither, apart from the other, makes sense. All construction is thus *a priori* synthesis in the sense that the possibility of 'appearances' is conditional upon its postulation. (It is, of course, only the first of eight such postulations.) As to the 'principle'

[1] K. 578; second italics mine.
[2] Cf. above, pp. 572–3; also P. 82–4 (para. 38). It will be noted that in the 'weak' argument, we are concerned only with the fact that *a* concept governs construction, not the specific concepts (as well as axioms) in accordance with which a geometrical figure may be constructed.
[3] K. 197. In the first edition the principle runs: 'All appearances are, in their intuition, extensive magnitudes'.
[4] For this 'balance', cf. above, pp. 592–3, 604.

lying behind the axioms of intuition: this will be '*a priori*' in just this sense, i.e. as the principle which guarantees the possibility of appearances (which involve spatial intuition); and it will be 'synthetic' in the double sense of 'governing' the *synthesis*, and of being non-analytic.

We may now pass to the strong claim. This is the claim that Euclidean geometry is privileged before all other alternatives, as uniquely applying to all things regarded as appearances. First, let us remember that the 'principle' of the axioms of intuition is *not identical* with these axioms. The latter do indeed presuppose the former, as providing a 'proof' that extensional axioms have a synthetic *a priori* status *in general*. But this does not tell us what axioms there are, nor whether there is a single and unique set of such axioms.

This looseness of fit between the special axioms and the general principle is occasionally concealed by Kant's exposition. When explaining and arguing for the synthetic *a priori* character of geometry, he often simply *exemplifies* his exposition of the general principle (instead of defending it on special grounds), and this by reference to some of the axioms of Euclidean geometry.[1] He then easily gives the impression of believing that since spatial intuition is presupposed not only in general, but also always under the specific form of *some geometrical system or other*, as a condition of a thing 'appearing' at all, that condition automatically guarantees also the *particular* specification (Euclidean or otherwise) involved. But this would, of course, be a howler. Indeed, Kant does not fail, on some occasions at least, explicitly to note the logical gap involved. Thus he introduces the exposition of the 'Principles of the Understanding' with the following general warning:

> It should be noted that we are as little concerned in one case with the principles of mathematics as in the other with the principles of general physical dynamics. We treat only of the principles of pure understanding in their relation to inner sense. . . . It is through these principles of pure understanding that the special principles of mathematics and dynamics become possible.[2]

How could Kant back the strong claim—if he indeed ever meant to do so? We have seen that geometry proceeds spontaneously by

[1] Cf. for instance, K. 70, 199.

[2] K. 197. So far, then, the general principles do not *imply* the special ones but are only *presuppositions* of the latter. How far the 'tightened' or 'special' versions of these general principles validate the subject-matter to which they are applied, merely in virtue of being determinate versions of the indeterminate transcendental principles, is of course the basic problem for our whole interpretation of Kant.

injecting definitions and axioms into the process of construction of geometrical figures. These are 'pure' intuitions, since we here abstract from the matter of sensation. Nevertheless we can be certain that there is a complete fit between pure and empirical intuition, provided we limit ourselves to appearances, for we understand by these entities whose logical grammar guarantees that the pure and the empirical cases are in them united. Now the transcendental exposition involved in the general claim amounts to the assertion that the logical character of an appearance is so defined as to 'balance' the transcendental principle of the axioms of intuition, thus making that principle synthetic *a priori*. This also entails that we can 'antecedently to experience' be certain that the general principle will hold of all appearances. Correspondingly, the 'proof' of the strong claim ought therefore to run as follows: the special axioms are 'balanced' by the Euclidean character of appearances, a character specified by the axioms of Euclidean geometry.

Now it is true that if Euclidean axioms, as special cases of the 'general principle', are *fed* into the constructive process, they will be 'synthetic *a priori*' *in the sense explained*. For it will only be on condition of our feeding these axioms into the constructive process that any '*Euclidean*' objects (*quâ* appearances) will be possible. And *if* all objects *were* Euclidean, this is what we should expect. But the argument can, of course, not demonstrate that all objects *are*, let alone *must be*, Euclidean. All that is perhaps acceptable in Kant's argument is its procedural approach: we have to feed some geometry or other into our treatment of physical phenomena, before we can even begin to reason geometrically or physically about them.[1] And this would nowadays be generally admitted. Geometrical systems are not derived from experience. They are presupposed in order to carry out empirical investigations. On the other hand, if these investigations become too cumbersome or complex, it is perfectly possible to replace one geometry by another.

Here there may enter a doubt concerning the whole question of 'applied geometry'. For it may be objected that Kant is surely concerned only with the 'pure' aspect of intuition, i.e. with pure geometry, and never with the question whether some geometrical system holds of the world.

This would, however, be forgetting Kant's basic requirement just

[1] This is a minimum interpretation of para. 38 of the *Prolegomena*; cf. below, pp. 615n.1, 674n.4.

mentioned, namely that geometrical axioms must allow of construction. Now this refers to the need to presuppose space as the form of intuition in respect of *any possible* appearance: which may, of course, be no more than a geometrical drawing. But whilst we may abstract from the sensory aspect of the drawing, it could not *be* a drawing for us, if the pure aspect of intuition were not presupposed. In this way, construction always implies a reference to (individual) appearances, and it becomes clear that for Kant the sharp distinction between pure and applied geometry does not really hold. Besides, Kant does occasionally state explicitly, for instance at K. 210, that the principle of the axioms (as well as that of the anticipations of perception) 'justify the application of mathematics to appearances'. No clearer indication of his intentions could therefore be required; our problem is what sense we can make of such explicit claims. At any rate, a large part of Kant's argument seeks only to explain how the assertion that a certain geometrical system is synthetic *a priori* can be made meaningful, and thus shown to be possible. But at times he does seem to be making the stronger claim. He then seems to offer a proof that there *is* a privileged system.

Thus, at K. 240, he notes that it is perfectly possible logically to conceive ('to think without contradiction') the concept of a figure which is enclosed within two straight lines. For such a concept involves a *synthetic* axiom—'synthetic' in the sense of being non-analytic.[1] However, Kant goes on, it is not possible *to construct* a figure in space which would correspond to such a concept.[2] So Kant admits the logical possibility of non-Euclidean geometries, but not their constructibility. Now the notion of construction immediately reminds us that the contrasting case (involving logical thought) is by Kant conceived to operate in the realm of noumena. 'Construction' moves the discussion into the realm of phenomena. Here we must again remember the aspect of 'balance' between the transcendental principles and the 'appearance'. This shows at once that reference to construction means that only those axioms are to be permitted that could yield *possible* appearances.

[1] Cf. above, p. 580.
[2] We may also apply this to the problem of the incongruent counterparts. It is logically possible (using analogy) to make L-determined objects congruent with R-determined ones, by rotating them in a fourth dimension. But Kant would object that such a rotation cannot be 'constructed'. The problem of the grounds, if any, for the privileged place of three-dimensional space (if indeed it is privileged) is still under discussion today.

Three interpretations are possible of Kant's contention in the last passage, one weak, two strong. The weak one implies that we inject only those axioms that will eventually apply to a world of physical objects, 'in fact' characterised by *some specific* geometry.[1] And it is the phenomenal character of these objects that operates as a constraining force on our freedom to conceive alternative geometries. We can frame (logically conceive) such alternatives, but it does not follow that they will apply. And only those axioms have genuine significance that *could* apply to material objects or lead to a 'successful' construction. If there are such objects, then the axioms that govern them will still be synthetic *a priori*, in the sense explained.

From the weak claim we are led to two strong claims. The first would assert that the geometry of the objects which in fact 'balances' the *a priori* construction *is necessarily* Euclidean, quite apart from the transcendental condition. This Kant cannot mean since that would give objects a non-phenomenal character. So at best he can mean that the phenomenal objects are in fact *found* to be characterised necessarily as Euclidean. Since 'necessities' cannot be discovered, this can make sense only *via* the second strong claim, viz. that the *specific* concepts which *govern* construction *a priori* are necessarily Euclidean; i.e. that we inject necessarily 'spontaneously' not just *a* geometry but a unique geometry. But this Kant cannot mean either, for in that case his concepts would not be '*constrained*' *by the result*; he would be back to the position where he would be saying that we *can* only 'think' Euclidean geometry: something we have just seen him to deny.

There is a puzzle which arises out of this interpretation: How can we be 'constrained by a result' which is of our own making? I think that the answer to this is: it is a *paradox* of our own making. Kant simply cannot be raising any claims concerning the *necessity* of any specific metric. And this is surely the case if we remember that he wants only to demonstrate the possibility of the fact—*if fact it be—*

---

[1] Strictly speaking, where we are dealing with 'pure geometry', or rather, the pure part of intuition, one cannot say this within a Kantian approach. For 'in fact' is an expression which is conditional upon a given set of axioms being fed into the construction. And these two sides of the same process, construction and the result of construction, the appearance characterised by a given metric, cannot be distinguished except as abstractions. What we can say is that if we 'find' certain objects with some metric or other, this will be because we have 'successfully' injected that metric into the constructive process. But the notion of 'success' is purely pragmatic; it means that some given system of geometry and theoretical physics gives a satisfactory account of the phenomena. Nor is this very far from the approach which we have found Kant take in respect of theoretical science.

of there being a unique geometry. For in that event it would *simply be the case* that a unique metric is injected *a priori*, which is that set which is equally '*in fact*' balanced by 'what we find', if we find it, i.e., a privileged geometry.[1] There is at least one passage in which the situation as here described is stated by Kant almost explicitly:

> The manner in which something is apprehended in appearance can be so determined *a priori* that the rule of its synthesis can at once give, that is to say, can bring into being, this [element of] *a priori* intuition in every example that comes before us empirically.[2]

It follows that this 'constraint' must be understood, not as something 'external', exerted by 'objects' which 'in themselves' have the metric which they happen to have, but as a constraint which only 'as-it-were' guides the hand that constructs in space: which is a notion of 'constraint' that operates against a background in itself bereft of all 'form'.

It may seem a peculiar procedure, to construe 'what is in fact the case' through a concept which implies necessitation of some kind.[3] Yet this is a very general aspect of Kant's procedure. We shall meet it again presently in the function of the categories (as concepts which 'carry about themselves necessity'),[4] to define the notion of empirical objective reality.

At any rate, all this is a far cry from what most commentators believe Kant to hold, namely, that *since* axioms are synthetic *a priori*, that alone will ensure that they will *ipso facto* apply to those objects that can 'appear'—namely objects that are Euclidean.

But did Kant not also *want* to make this stronger claim? We cannot be certain. I have not been able to find any decisive evidence of Kant explicitly making the claim. Certainly we should not interpret the need for construction, or for intuition, as implying that we 'intuit' space with a unique set of spatial characteristics spontaneously, by interpreting 'spontaneity' as a kind of natural or psychological fact; let alone that we are possessed of unique intellectual intuitions. The notion of spontaneity concerns a reference to the intuitional 'act' being *not a posteriori*, i.e. being *a priori*; but its *a priori* status is

[1] Here 'in fact' must be interpreted in the light of the previous footnote.
[2] K. 210.
[3] But we found Leibniz already defining empirical actuality by reference to the determining concept of perfection, which determined the result of a process without necessitating it.
[4] K. 209.

defined only as a *logical* presupposition of the possibility of certain characterisations of appearances being given; it is only a transcendental matter. So far from intuition being 'free' (however much it may be 'spontaneous'), it is constrained (held in balance) by the character of the object which it determines.

Such an interpretation has the advantage of harmonising with the earliest account of the status of three-dimensionality which Kant had given in LF, where we have found him say that this follows from the nature of the forces of the science of dynamics and the mathematical framework in which it expresses them—which is there held to make three-dimensionality a 'contingent' matter.[1] For with our interpretation we can see that Kant did not really abandon the earlier position, nor come to accept its contradictory, even though he seems to do so formally, if the claim to the synthetic *a priori* status of geometry means that the axioms of geometry *are necessarily* Euclidean. Instead, that early position is reformulated. Certainly, it is now realised that the geometry employed as part of dynamics is not *derived* from its empirical laws, e.g. the law of gravitation. What we have to say instead is that the geometry *chosen 'a priori'* is the one that will (when applied in dynamics) lead to the laws which (with such a geometry) we as a matter of fact 'discover'. But it is the existence of these laws (as characterisations of 'appearances') that 'balances' the 'choice' of the specific geometrical system: always understood that the transcendental principles which underlie such a choice, and which are required to yield 'appearances', are 'presupposed'.

So the three-dimensionality of space is not 'contingent' upon the laws of dynamics, in the sense of being *derived*, but only in being *dependent* upon their being 'in fact'[2] the laws which issue from the '*a priori*' constructive process, 'construction' being this peculiar element in which the pressure from the 'appearances' is balanced by the presuppositional element of the *a priori* principles and concepts.

Be that as it may, it does seem that Kant at most makes the strong claim apply explicitly only to a very exclusive case, when he writes that we can be apodictically certain that 'objects of the senses . . . conform to such rules of construction in space as that of the infinite divisibility of lines or angles':[3] a statement that can be argued without

---

[1] Cf. above, p. 580.

[2] For this phrase, cf. p. 612n.1. But it should be remembered that these laws, like the contingent events on which they are based, contain an *a posteriori* element, unlike the earlier case of the 'pure intuitions' of geometry. (Cf. below, p. 617.)

[3] K. 200.

involving the question of the uniqueness of Euclidean geometry. Moreover, some sections of the *Prolegomena*, a work in which Kant usually makes rather sweeping claims, heavily emphasise the interpretation of the synthetic *a priori* nature of geometry as being due to our spontaneously feeding the axioms of a geometrical system into its general concepts, in order to generate construction,[1] whilst making no claims as to the uniqueness of the result, thus interpreting the *synthetic a priori* status of geometry in terms of the 'spontaneity' of the construction, its 'axiomatic' character *simpliciter*.

To the looseness of fit between the general principle of the axioms of intuition and the special axioms themselves corresponds a similar gap between the Analogies of Experience and the Newtonian laws of motion, and finally, an even wider gap between the principle of causation in its transcendental setting and its application to empirical science. The last case we have already explored initially; both it and the second we shall discuss further in the final sections of this chapter.

8. OBJECTIVITY

(a) *The categories of relation*

The manner and sequence of our presentation of Kant's thought in this chapter has made it necessary to break through the order imposed by the architectonic of the first *Critique*. And this has also been necessary in respect of the account of Kant's theory of space, as given in Part I of that work (Transcendental Aesthetic). Thus, the discussion of the foundations of geometry required reference to a later part which first presents the principle of the axioms of intuition, viz. all appearances are extensive magnitudes—the transcendental ghost of Descartes' metaphysical dictum that all matter is 'essentially' extension. Now the formal 'proof' (we have omitted its explicit discussion) which Kant gives of this principle in the *Critique* involves reference to the unity of a 'synthesis of the manifold whereby the representations of a determinate space or time are generated'—again something only dealt with in Part II (Transcendental Logic); with its important example of the line which we 'cannot represent' to ourselves 'without

---

[1] Thus, at P. 82 (para. 38), Kant actually speaks of the understanding as 'injecting [hineinlegen] the law of the chords cutting each other in geometrical proportion when it constructs the figure' of a circle in the course of a certain demonstration; the law being injected into the concept of the circle. Space in general (spatiality) 'is something so homogeneous and in respect of all particular properties so indeterminate that there is certainly no hoard of natural laws to be found in it' (P. 84).

drawing it in thought';[1] a process which first yields, as we have noted throughout, space *quâ* formal intuition ('determinate space'), as contrasted with 'form of intuition', i.e. indeterminate space, or spatiality. Indeed, Kant insists, in cognising any appearance as extended, we presuppose not just the synthesis, or combination, of the manifold, but also the fact that we are *conscious* of this synthesis; a consciousness which itself presupposes a 'unity', i.e. the unity corresponding to the concept generating the determinate space.[2]

Nothing of this is mentioned in the section of the *Critique* expressly dealing with space (Aesthetic), although we have found throughout the need for introducing the notion of 'synthesis' in order to do justice to the wide ramifications of Kant's conceptions; the notion of construction, for instance, would be almost incomprehensible without it.

The need for such a reference may by now be obvious, since Kant's notion of space has this double implication, of an indeterminate and a determinate whole. When speaking of the latter, we noted the need to introduce the categorial concept of interaction, between different bodies said to be 'in space'. And this, too, has to be viewed as a concept which expresses the unity of a synthesis.[3]

Finally, we have throughout noted the Kantian locution concerning the conditions on which anything can become an appearance. When expressed more completely, this means, as just noted, 'appearance for a *conscious* observer'. It is a specific note of Kant's whole approach that an account of the conditions which make anything an object are at the same time also the conditions which make it possible for an observer to be conscious of this object. Thus, in respect of space *quâ* formal intuition, the consciousness of the synthesis of the manifold under the category of quantity is a condition of the cognisability of the intuitional element in any appearance, i.e. as an extended object. This aspect of consciousness, which has been ignored so far, will therefore need to be followed up in the present section. But before turning to this, it will be useful to draw attention to a peculiar difference between the ways the different categories are employed; this will also provide an indication of the status of the categories of relation (substance, causation, dynamic association) within the Kantian enterprise.

[1] K. 198; cf. above, p. 572.
[2] Cf. ibid. The notion of the 'unity of consciousness' will be taken up in section (c).
[3] The details of this 'unity' will be explained in section 8(e): iv.

Referring back to the passage from K. 210, cited above,[1] we saw Kant holding that a spatial synthesis 'brings into being' the spatial element of every appearance 'in every example that comes before us empirically'. What is thus brought into being, however, is a manifold whose members are merely placed side by side, as for example the points which constitute a line. There is no *assertion* of any connection *between* these points. More specifically, the concept of extensive magnitude does not contain the thought that the points of this line are *mutually and internally connected*, but only that they *lie side by side*, in accordance with some rule of constructing the line, going from part to whole.[2] The construction does not *assert* any *internal* connection. And a similar account can be given of the second category, which treats of the 'qualitative' aspect of perception, i.e. intensity.

In the case of other categories, especially those of 'relation', e.g. cause and effect, dynamic association (or mutual interaction), the situation is a little different. These categories involve concepts which (to cite one case), in order to formulate the notion of an objective sequence existing in time, explicitly determine the members of such a sequence ('perceptions') as being 'necessarily connected'.[3] But though in this respect stronger than the first set of categories, they are in another respect weaker. The category of quantity could be said to 'produce' or '*constitute*' the 'formal' aspect of the perceptions, thus yielding those components of 'appearances' called 'sensible intuitions', but more particularly, 'pure intuitions'. Kant therefore calls the corresponding principles 'constitutive'. They in a manner 'bring into being' the relevant aspect of intuition; which explains again the difficulty we noted before, of anything 'balancing in fact' the transcendental presupposition.[4]

By contrast, the categories of relation, whilst they bear on the relations *between* the perceptions, do not *constitute* anything further in respect of the latter. If they yield (as Kant claims they do) the notion of the *existence* [Dasein] of an object (e.g. objective sequence), they do so only provided we assume the material (sensational) content of the perceptions to be supplied *a posteriori*. Only in the context of a situation which involves (in principle at least) the presence of an *a posteriori* content do these categories therefore have valid applications. These categories hence do not 'bring into being' anything appearing intuitionally; and there is an element of recalcitrant 'fact'

---

[1] p. 613.   [2] Cf. K. 197n f.; 198.   [3] For this, cf. below, sect. 8(e): ii.
[4] Cf. above, p. 612n.1; for 'bringing into being', cf. K. 210; p. 613.

(the *a posteriori*) which is here more relevant. We require that 'a perception is [first] given in a time-relation to some other perception',[1] before any notion of objective existence in time can be formulated by means of the categories of relation.

In Kant's words: 'The *existence* of appearances ... cannot be cognised *a priori*'; and he adds that for this reason 'existence cannot be constructed',[2]—a point whose significance will not be lost on the reader of this chapter. Since the categories of relation supply only principles which function as rules of connection between members of the intuitional manifold, they are therefore said to have only 'regulative' use *with respect to intuition*. To avoid misunderstanding, it should be noted that in another sense, as Kant points out later in the *Critique*, *all* categories, including those of relation, have, of course, also 'constitutive' force, namely, *with respect to our experience of* a given object *quâ* appearance, since this is what they generate (always provided we assume an *a posteriori* content).[3]

Here, the situation is clearly different from that at the level of reason. The 'idea' of unity of the empirical laws as a system *in no sense* has any constitutive, but *only* a regulative force, since not only does it not generate any *a posteriori* content but it cannot be said to generate its 'object', which here would be the system or 'order' of nature; something (it will be remembered) which according to Kant can only be postulated or 'projected'. There is here nothing 'given' which might 'balance' the 'idea'; at best what balances the latter (and hence 'validates' it 'subjectively') is the 'need' for science to postulate system. (Whereas it is permissible to say that the categories generate 'nature', since nature (as the aggregate of objects) may be said to be 'given'.)[4]

To speak of 'rules of connection between members of the intuitional manifold', as we have just done, is a phrase which, however, requires the most careful interpretation, seeing that the relational categories have only regulative force. For this means that we must beware of supposing that these categories, as transcendental conditions of objectivity, are 'balanced by' connections that might 'in fact' (*per impossibile*) subsist between the members of the perceptual manifold, *quâ* intuitions. On the contrary (using again our main example), to say

---

[1] K. 211.   [2] K. 210; italics in text.   [3] Cf. K. 546.
[4] For this, cf. above, sect. 4(c): ii, vi–vii; especially, pp. 505f., 525ff. We employ the idea of systematic unity regulatively if we use it to construct hypothetical systems; constitutively (falsely), if we *assume that there exists* an 'object in the idea', instead of *proceeding as if there were* such an object.

that the concept of causal connection, 'injected' so as to generate objective sequence, has only regulative force with respect to the intuitional manifold is Kant's way of saying that no 'given' necessary connection 'balances' the injection at the intuitional level: which in turn means that the addition of this concept cannot imply that the members of the intuitional manifold are 'necessarily connected' at that level. It can only mean that *quâ* objective appearance they do, in virtue of the necessary connection asserted by the concept, belong together as a matter of objective (and contingent) fact. This, as we shall show presently, is precisely what Kant expressly asserts at the focal point of the transcendental deduction,[1] and thus supports the interpretation of Kant's intentions here claimed.

One general result will become clear from the various discussions of this chapter. It is extremely difficult to ascertain from Kant what precisely is 'brought into being' and what is 'found' or 'given'. In most general terms, of course, he is concerned with providing an account of the transcendental conditions of indeterminate experience (or appearance), i.e. of what is given in intuition, as well as of determinate experience, or objective appearance, whether individual or systematic. In all this, expressions like 'injecting', 'generating', 'bringing into being', 'spontaneity', 'projecting', 'givenness', are semi-metaphors. Evidently the weighting attached to the two sides of the 'balance' of which we have spoken in the last few sections varies from case to case. The emphasis on subjectivity is strongest in the case of geometry, slightly less so in connection with the notion of the order of nature, and reduced to a minimum in respect of the individual objective happening.

We shall return to this topic below in the question of the sense to be attached to the 'givenness' of objects.[2] For the moment, let us add a general remark. The whole procedure has about it something of a conjuring-trick, though it is not for this reason to be despised. There is more in great philosophical writing than sheer reporting, deducing, or inductively proving. On the other hand, if this characterisation is apposite, Kant's system is not either a clear-cut and successful argument, nor (as other hostile critics would claim) a tissue of *non-sequiturs*. Indeed, if I am right, any such appraisals err in misconstruing the nature of philosophical writing. But let us pass to the main business of this section.

[1] Cf. below, p. 637, especially the passage from K. 159.
[2] pp. 638ff.

(b) *The need for synthesis*

The necessary connection supplied by the category, as mentioned, is taken by Kant actually to form part of a process which is the 'synthesis' already alluded to previously. It will be the task of the rest of this section to explain the significance of the notion of synthesis and the place of the categories within that synthesis; where we shall now omit from further consideration those aspects of synthesis which yield the formal part in any intuition of appearances; concentrating on the need to explicate the notion of the existence of an objective situation.

Kant's rejection of the thing-in-itself has called for a new interpretation, a new structural analysis, of the concept of thing *quâ* phenomenon. The resulting exercise can be presented as a programme through which the phenomenal object is built up, piece by piece, until there results something with the required grammar of a public or 'objective' language, in which it is possible to say again: this is really such and such; this really follows that; this really co-exists with that, as contrasted with statement-forms which in the last analysis can report only on the 'sensa' that are in *my* mind. Kant's pregnant question: How is knowledge of objects [Erkenntnis, cognitive grasp of what is objective as such] possible? must be understood, not as scientifically orientated but as asking what the components of the logical structure of 'phenomena' and of 'experience' have to be so as to yield, cognitively, an object, corresponding to our concept of such an object.

As our examples show, Kant construes objectivity in a somewhat formal way. He is concerned primarily with the existence of an objective time-sequence, or with simultaneous existence of objects in space.[1] Our question then becomes, 'How are we to construe the notion of such an objective time-sequence, or again, objective coexistence?'

But what is the sense of this question? In answer,[2] we must draw on what has gone before. We are supposed to be in a cognitive relationship to an empirical ('sensory') manifold, such as is represented by an objective sequence of sensory elements. However, so far we are

[1] This is to concentrate on the material of the second and third analogies. We shall omit consideration of the category of substance so as to keep this account reasonably concise.

[2] The answer to this question is given in general in the 'Transcendental Deduction of the Categories', and in particular, in the Second and Third Analogies. In this section, we shall concentrate on the general account.

confronted by[1] no more than a set of sensory elements apprehended in the indeterminate forms of space and time. Now if this characterisation is to concern an 'appearance', this will entail that these sensory elements (in this context named 'perceptions') are characterised only in respect of degree of intensity.[2] Furthermore, the space and time of which we are here speaking, so far concern only the 'form of intuition', which was, as has been seen, no more than a reference to 'that which has to do with [the fact] that the manifold of appearance may [potentially] be ordered in certain [spatial and temporal] relations'.[3] This means that these perceptions are not so far represented as *being* ordered; the requisite considerations which are to yield that aspect of the 'appearance' are still missing. This was the import of my previous remark,[4] that the component with respect to which we are 'purely receptive', i.e. the *a posteriori* sensory content, does not arrive *with* its *own* space and time.[5] It contains no feature that would as such supply something corresponding to a putative objective temporal or spatial order.[6]

Nor do the 'forms of intuition' supply such an order, for they provide merely the basis on which such an order may be constructed. If we could use a picture, we might say that what is missing so far is a spatial and temporal clamp, evidently corresponding to 'formal intuition', which would yield the notion of a determinate objective order. Negatively, we might also say that the sensory elements are in Kant's philosophy 'loose and separate' in a much more radical sense than in Hume's. This is due to the fact that Kant has sharpened the

[1] Kant says 'affected by [affiziert]'; cf. K. 65, 83.

[2] 'Perception' [Wahrnehmung], it will be remembered, is Kant's technical term for the aspect of *consciousness* of the 'intensity' component of a sensation, subsisting in indeterminate space and time (form of intuition). It is a pure abstraction, a theoretical component of the structure of Kantian 'experience'. Cf. above, p. 533n.3.

[3] K. 66. I have modified the translation to render Kant more literally. Cf. above, p. 583.

[4] Cf. above, p. 588.

[5] It will be appreciated that this is not a remark with an empirical basis. We are discussing the transcendental structure or components of the concept of 'appearance'.

[6] I say 'putative', in order to indicate that we are concerned with the notion, the possible 'meaning', of an objective order as such and not the *empirical* question of what order *is* objective. We are here merely sketching the logical conditions under which a *question* about objective order could even be *asked*. Kant's frequent question, especially when couched in English, translating 'Erkenntnis' as 'knowledge', viz. 'how is a knowledge of objective situations possible'? calls forth misleading expectations, unless a reader has already learned what Kant means by the term 'possible'.

notion of the *a posteriori*, under the impact of the criticism of the thing-in-itself. The Kantian *a posteriori*, as has been shown at length, cannot even begin to simulate an item of experience in the sense represented by Locke's or Berkeley's 'idea'.[1] The notions of 'experience', as well as 'object' (both in the phenomenal sense) are still missing.

The degree of abstraction required here is difficult to realise. Normally, when we conjure up an image of space and time, we spontaneously use the picture of these entities in their 'determinate' form, exemplified perhaps by the image of a line, not to speak of a Newtonian absolute space.[2] But a geometrical line, we have already seen, corresponds to a (pure) 'formal intuition'. And this may be generalised to include also 'empirical intuition', where we are dealing with a given set of sensory elements. A line is not just an 'indeterminate space'; it represents a set of points, 'held together' and 'tied down' to a given locus in accordance, say, with some algebraic formula, a model for Kant's generalised notion of a 'rule', or 'unity'. Kant, we saw, speaks of the fact that the possibility of such a line presupposes that it be drawn.[3] Now this example of a 'drawing' stands also for something quite general: before we can speak of an 'order of sensory elements' we require the notion of ordering; and this in turn requires that the elements should be collected or 'combined', whether the 'combination' denote mere 'placing together' or 'inter-relating'.[4] (In saying 'this comes before that' I in some sense 'unite' both this and that in a common act of consciousness.)[5] In short, we here presuppose a synthesis of the sensory manifold, whose primary aim is to initiate the relating (in general) of the elements. Note, moreover, that, just as for the case of space and time, all synthesis must ultimately be *a priori*, for if it were not 'added spontaneously',[6] sceptical doubts of the kind

---

[1] As noted on p. 551, the *a posteriori* must by no means be confused with anything like the Lockean 'idea' or the Humean 'impression'. Kant proscribes Hume's easy verbal identification of 'impression' and 'object', which we noted to be such a dominant feature of Hume's presentation.

[2] In what follows immediately, the case of the line is not now meant to refer us back merely to the first two principles of the understanding (axioms and anticipations), but stands for what is general in Kant's procedure. But it seemed best to begin once more with this case, since we are by now familiar with its basic features. Besides, Kant himself employs it for such a general purpose, e.g. at K. 156.

[3] K. 156; cf. above, p. 572. This idea lay also behind the notion of construction.

[4] Cf. above, p. 617.

[5] Cf. K. 131–4; 153. It will be remembered that Locke had reserved this 'act of combination' for his 'complex ideas'; Kant generalises this process and deepens it.

[6] At K. 151 Kant speaks of combination or synthesis as 'an act of spontaneity' to

affecting our knowledge of things-in-themselves, would again arise.[1] It is this which produces the need to supply a logical (transcendental) structure for the notion of the phenomenon as 'object'. Kant's account, it is true, employs a great number of psychological models, particularly in the version of the first edition of the *Critique*; and it may be that without these it is difficult to *picture* the logical situation. Since this chapter is already too long, I shall desist, however, from these refinements.

Now we have spoken of a synthesis of the sensory, or rather perceptual, manifold (since we are dealing with the consciousness of the members of a sensory manifold given in an intuition). If for the moment we continue to employ the important paradigm of the line, we realise that the very drawing of a 'line' *ipso facto* proceeds in some determinate way or other, whether we are conscious of this or not. All synthesis has a determinate 'direction'; my hand is 'guided' by the formula for the line. Kant here employs the more general term of 'unity' of the synthesis, which (as he notes) is actually included in the very notion of a synthesis:

> The concept of combination *includes*, besides the concept of the manifold and of its synthesis, also the concept of the unity of the manifold.[2]

We have now already come very close to a complete reconstruction of the notion of object (*quâ* phenomenon). For as Kant defines it:

> Cognition consists in the determinate relation of given representations to an object; and an *object* is that in the concept of which the manifold of a given intuition is *united*.[3]

(c) *The transcendental unity of apperception*

Now we could go on from here to show that the categories are in fact the specific determinations articulating the 'unity' involved in the definition of the 'object'. Before doing so, it will however be best to define the significance of Kant's procedure by setting it again in the

---

indicate that it does not belong to the level of the sensory material, in so far as this is a component of the object *quâ* phenomenon.

[1] Cf. the corresponding approach at the level of reason, above, pp. 505–9; and the same approach was of course taken in Kant's account of geometry. And cf. p. 561.

[2] K. 152; my italics. The notion of 'unity' thus involves reference both to the process of combining, the result of this process, and—as in the case of the line, literally—the concept or 'rule' determining the result. The logical box to which Kant assigns all this is 'the understanding', an echo of Locke's and Hume's 'understanding'. At K. 147, the understanding is in fact defined as 'spontaneity of cognition', as 'faculty of concepts', and finally, as 'faculty of rules'. (Cf. below, p. 634n.3.)

[3] K. 156; italics in text.

older epistemological tradition and its problems, particularly those met in Locke: the problem of the 'two objects', object as 'idea', and object as the 'thing' to which the idea is problematically supposed to conform. In a single chapter it is not possible to elucidate the full complexities of Kant's transcendental deduction; and there are, more-over, a large number of commentaries in which this has been attempted, from the time of Kant down to the present.[1] We shall therefore limit ourselves to very brief indications.

To appreciate Kant's answer to Locke (and by implication also to Descartes, Berkeley and Hume, who all in their way had retained the notion of 'idea' as contrasted with a non-ideal realm),[2] let us remember once more that Kant is concerned only with providing an account of the notion of an object as such, of cognition of objects in general; and again, with the possibility of having cognitive experience of such objects. He is not concerned with the empirical question whether what we observe to be the case (in virtue of the data of experience) is in fact the case.[3] Moreover, this is a question which, in the light of his answer to the general logical question, can hardly be asked in any but the most trivial sense.

Kant's answer to Locke (and also Berkeley) is complex, attempting, as it does, to sharpen some of the ambiguities of the latter's exposition. The first half of this answer we have already discussed, when we showed that Kant's doctrine of the transcendental ideality of space is intended to proscribe the fully-fledged 'ideas' of the Lockean kind (purporting to 'represent' object in space and time) accompanied by a second realm of 'things' to which problematically they might correspond.[4] Our very ability to 'see' things in space already vouches for the 'empirical reality' of space (as form of intuition), provided the

---

[1] Unfortunately, by seldom attending to the vast ramifications of Kant's thought, especially in its scientific directions, or by paying insufficient attention to the models for his later language that are found in his earlier work, many commentaries fail to do justice to the astonishing synthesis which Kant's work represents. More-over, since the earlier work yields particularly important indications for Kant's ideas on space, and the accompanying re-interpretation of the notion of the *a posteriori*, the main tenets even of the transcendental deduction are frequently not understood.

[2] Hume, in the *Treatise*, had indeed sought to eliminate the doctrine of 'double objects', by offering an 'explanation' of our view that there are 'external objects' in addition to our 'perceptions', thus hoping to provide a reductive analysis of the former. Nevertheless, he retained the uncritical locution of private impressions and ideas, as the only 'things' with which we are directly acquainted.

[3] Cf. above, p. 621n.6.   [4] Cf. above, pp. 593–4; and p. 543 for Kant on 'correspondence'.

reference is always to 'appearances'. But the price paid for 'transcendentally subjectivising' space (and time), is that this of necessity carries in its train the urgent need for an account of 'objectivity' which will not lean on a space (and time) that are 'objectively (or transcendentally) real'. That such an account is required, we have indeed anticipated, by noting that the second aspect of space (determinate space, or formal intuition) presupposes categorial concepts, which—as the transcendental deduction shows—first yield the completed notion of an object. And it is with this that we are now concerned.

Locke had moved uneasily between many different senses of object: object as idea, as theoretical entity, as metaphysical entity, as cause of the idea, and, finally, as that beyond which, through the notion of idea, man 'intends' to point to a realm of 'real' things to which ideas may or may not 'conform'.[1] In so far as the object *quâ* 'idea', and *quâ* cause of our idea, denotes phenomenal entities, Kant's answer is that there can be only one object.[2] Further, in so far as the object as cause and the object as 'conforming' slide over into the 'metaphysical' realm, they must be treated as noumena. Finally, we must resist the slide (which we often suspected in Locke) from theoretical entity to metaphysical foundation (as in the concept of the seventeenth century corpuscles). Theoretical constructions like this, or like Newton's gravitation, do not (Kant explicitly insists) treat of any purely 'intelligible' or 'noumenal' world. We have indeed found that such constructions, though springing from the level of the understanding, involve the regulative employment of theoretical reason. But we must beware of confusing this with any insight into a non-phenomenal realm.[3]

Let us now turn in more detail, first, to the answer that at the phenomenal level there is only one kind of object. Kant does this by showing that the transcendental conditions which have been enumerated as components of the structure of the phenomenal object are the very same components which are required to define the notion of possible experience.[4] For experience requires, and is defined as, consciousness or awareness of objects. But such consciousness would

---

[1] Cf. ch. IV, p. 251; for the various senses of object, cf. also pp. 242ff.
[2] Cf. K. 268, 1st ed.
[3] Cf. Kant's discussion of this at K. 273–4; and above, p. 548.
[4] This again we have already seen to be the situation in respect of space, blended in the dual aspect of 'form of intuition', as form of *intuiting* and of that which is *intuited*.

not be possible if there were not involved that very same synthesis of perceptions to which we have already referred, and hence also the accompanying unity of that synthesis. Only if such a unity is granted is consciousness of objects defined.

In the second edition Kant distinguishes 'empirical consciousness' which as a kind of attentive awareness *must 'accompany'* each individual element of the perceptual manifold, from that consciousness—still empirical—which unites, or in which are united, the different elements. In so uniting them, we must, however, also assume that a unitary and identical something, which Kant designates as 'I think' (a sharpening of the old Cartesian *Cogito*) 'accompanies' each element of perceptual awareness in virtue of which all the different individual parts of consciousness are parts of one and the same consciousness, the unity which in fact constitutes consciousness of an object.[1] The argument is correlative, in true transcendental fashion: if we want to construct the notion of an object, i.e. of consciousness of the object, we must presuppose the 'I think', and its function of unity of synthesis. *Per contra*, that such a unity of the 'I think' is presupposed is proved by the fact that we do have consciousness of objects.[2]

It is important to keep this condition in mind. The unity of consciousness is 'proved' only 'transcendentally' in the sense that it explains the cognition of objects, which themselves contain in principle a reference to perception, and thus an *a posteriori* element of sensation. Outside this context, this transcendental unity is nothing; with such a context it is as 'objective' as its correlative, empirical objectivity (in the phenomenal sense). Unity *quâ* form is thus objective, however much the psychological model of Kant's discussion may revolve round the subjective features of empirical consciousness. It replaces Locke's conception of the 'mind' in which are lodged its 'ideas'; once more the familiar spectacle of a transformation from physicalism or psychologism to transcendentalism.

Kant refers to the 'I think' also as 'apperception'—the old Leibnizian term for self-consciousness. Since the unity which it invokes is presupposed for the possibility of empirical consciousness of objects —in which it is indeed mirrored—it is easy to see why it bears the celebrated title of 'transcendental unity of apperception'.

Being transcendental, this unity is *a priori*, and since necessity is one of the marks of the *a priori*, Kant frequently speaks also of the 'necessary unity of apperception'. For the sake of what follows, let us

[1] K. 151–4.   [2] Cf. K. 155–8.

note that this ascription of necessity covers two different aspects. The unity is a necessary one because it is transcendentally presupposed. It is, however, also said to be necessary because it involves an ascription of necessity, in the sense of some rule-like determination, at least for the case of the second, or 'regulative' group of categorial concepts on which the unity is modelled. Thus the concept of causality for Kant denotes something that determines a sequence in accordance with an absolutely universal rule; a special version of necessity.

Here, then, is the proof that there is only one object, not two. One cannot speak of the manifold of intuition as related, without presupposing the synthetic unity of apperception, which is a function of the understanding, and not given *together with* the intuitional manifold. (This follows from the fact that we are dealing with *phenomenal* entities.) But this is the very same unity which is presupposed also for that empirical consciousness described as awareness of objects. Kant concludes:

> The synthetic unity of consciousness is, therefore, an objective condition of all cognition. It is not merely a condition that I myself require in *cognising* an object [*quâ* object], but is a condition under which every intuition must stand in order *to become an object for me*.[1]

In other words, outside the conditions of experience, it makes no sense to speak of objects at all. This leads at once to the critique of Locke's metaphysical objects. For, *a fortiori*, we cannot now logically ask whether the object within experience, i.e. the object as 'appearance', '*corresponds*' to something which is not in itself appearance, to use Kant's characterisation of this question in his chapter on Phenomena and Noumena.[2] The emphasis is here on 'corresponds'; and the point is that we cannot attach to any such 'noumenal' object any 'positive sense'.[3] Only if we make a distinction between the '*concepts*', phenomenal object and noumenal object, and understand the latter 'in a merely problematic sense', as a conceptual box, so to speak, which may be used, as it turns out, in the context of morality, are we entitled to this language; but we have thereby removed the linguistic context in which Locke's sceptical question can be asked.

[1] K. 156; first italics mine.
[2] K. 269, 1st ed. version. Nor can there be any question of an hypothetical inference to such an external object, from the 'contents of my internal consciousness', e.g. 'ideas'; cf. Kant's 'refutation of idealism', appended to the 2nd edition, K. 245. The 'refutation' *assumes* of course the doctrine which has transformed the old empiricists' 'ideas' into 'phenomenal objects'.
[3] K. 272.

(d) *The accidental and the necessary*

To appreciate what is to follow we need to take stock here of the basic outlines of Kant's novel construction, by summarising what has preceded. It is clear that the core of Kant's whole position is summarised in his insistence that we regard the object as 'appearance'. At first this seems only to repeat the doctrine of constraint which, in the philosophy of Locke, Berkeley and Hume, had been imposed on the concept of empirical object: the empirical object must be given in a possible experience (e.g., must be perceived), *via* a cognitive judgment whose logical structure it must thereby come to satisfy as well.

This first step yields the 'autonomy' that we noted as resulting from the procedures of Locke and Berkeley, of converting the object into an 'idea'. Berkeley and Hume had shown that this move had eliminated the possibility of postulating hidden ontological connections; connections which Kant proceeded to transfer to the realm of the thing-in-itself. Against this, the philosophical grammar of the object of experience is to be traced within the context of that experience itself.

Again, not unlike Berkeley in his criticism of Locke, Kant insists that the thing as it is in itself has no cognitive status in the context of possible experience. The worlds of phenomena and noumena do not subsist at the same level, and become altogether incommensurable. This removes the basis for Descartes' and Locke's assumption of two worlds, 'ideas' and 'things' to which 'ideas' might problematically conform.

However, Kant goes farther than Berkeley. For the location of the object as appearance within the domain of cognitive experience, by proscribing a world of non-noumenal transcendent objectivity altogether, cannot banish the distinction between subjectivity and objectivity altogether. These notions thus require new explication. And here Berkeley's uneasy resort, on the one side to the laws of nature as a distinguishing criterion of objectivity, on the other to God as the source of sustaining activity, had offered models that are either irrelevantly naturalistic, or impermissibly arbitrary and artificial.

The Berkeleyan impasse (and in Kant's eyes the similar impasse of Hume) thus suggests the need for an important advance: a further subdivision of what is 'given' within experience; a dichotomous division of the logical character of 'appearance'. This is the step taken by Kant. Basic linguistic associations which accompany the notions of the 'given' and of 'appearance' are now articulated in a new way.

Thus the notion of something being 'given' for an observer suggests that it comes 'from without'. On the other hand, the notion of 'appearance' which subsists as such only within the framework of our experience, implies—like Berkeley's 'idea'—an altogether subjective and private subsistence. This is a 'paradox' which Kant resolves by taking up an alternative metaphysical stance, recognising *within* the 'appearance' both an aspect of 'subjectivity' and of what is other-than-self—incorporating both aspects in the character of the 'given' (appearance).

As a result, what is 'given' is no longer *identifiable* with what is passively 'received'; this is only one of its component elements, viz. what corresponds to the 'contents' of specific sensations. (The way for this view was also paved by the criticism of Descartes and Leibniz, which makes the domains of sense and thought generically distinct.) It follows that 'receiving' has here an altogether Pickwickian use, denoting the fact that the content *cannot be anticipated* prior to experience, and is thus *a posteriori*. (However, *that* specific contents have different degrees of intensity, e.g. *some* degree of magnitude of light illumination, *can* be anticipated, and is thus *a priori*.) The Pickwickian use of this notion of *a posteriori* is evident since it does not include any reference to anything described normally as 'coming from without', and 'in time'—unlike Locke's ideas of sensation. Since they are, however, also characters belonging to what is 'given', Kant concludes, as we have seen, that they are *a priori*.

To describe anything as an appearance is thus *to say* that the *a posteriori* element in it as such lacks the spatio-temporal frame, which, on the contrary, as an aspect of the 'given', is contributed *a priori*. This is the first essential change introduced into the Locke-Berkeley construal of the 'idea' regarded as what is 'received', Kant describing this by saying that they treat the idea as a thing-in-itself.

It follows further that the character of the 'given' as 'objective', i.e. as other-than-self, will likewise have to be construed as a contribution made *a priori*. The character of the object as being 'other-than-self' will have to be located, not in what is received, but in what is spontaneously contributed by the cognitive subject in the context of judging. We might describe this by saying that the 'self' is here regarded as the seat of the *a priori*, and of what is 'other-than-self': objectivity being located *within* the framework of possible experience, instead of residing in a realm that transcends experience (as the external object had transcended Locke's 'mind'), as something to which fully-fledged 'ideas' conform. However, if this model of the self is employed, we need also to make a further dichotomous division within the notion of the self, as 'empirical consciousness' and as 'transcendental apperception'. And the subjective nature of the former turns out to be no more than a model for the logical tasks of the latter. In this way, the *a posteriori* element, in order to form a component of the given appearance in consciousness, requires to be 'grasped' (in a synthesis) in accordance with certain logical (*a priori*) concepts (the categories) whose function it is to supply a construal of external objectivity.

The autonomy of the critical philosophy thus goes much farther than the idealistic doctrines of Locke and Berkeley. It claims the right to regard the character of the *a posteriori* as *deprived* of its spatio-temporal framework, the latter being removed to the level of the *a priori*, in which guise alone it promises to yield (together with the categories) the character of intra-experiential objectivity claimed for 'appearance'. Kant's step, whereby the object is regarded not as thing-in-itself but as located within experience, and whereby what is 'sensed' must always emerge as an element of consciousness ('perceptions'), has thus a double function. It peels the object off its foundation in the ontologically conceived 'sum-total of all possibility' (as in the old doctrine of OG) and places it within experience (much as Locke had placed his objects as 'ideas' in the mind).[1] But further, it claims the right thereby to strip the sense-elements also of their spatio-temporal 'clamps', and thus goes farther than Locke and Berkeley. For us the essential point in this reconstruction is that the reference to re-location in consciousness is in fact only a model for Kant's logical step of depriving the sense-elements of 'their' objective spatio-temporal order. If Kant nevertheless continues to speak of the sensory elements as occurring in the form of an ordered plurality, it will be obvious that his appraisal of this order as 'subjective' and 'accidental' (to which we shall turn in the sequel) is a Pickwickian means of indicating that the true logical status of these elements consists in their not having any order at all; that the latter is only bestowed on them in virtue of the transcendental processes associated with the notion of experience as objective cognition. 'Subjective' and 'accidental' are thus no more than expressions borrowed from psychological and realist modes of speech—a fact to be remembered in what follows if we are to make sense of Kant's general approach at all.

[1] Cf. above, sects. 4(a–b); especially pp. 486n.3, 496n.3; 486–90.

Finally, it will be seen that 'subjectivity' is a concept which does here duty twice over. First, it characterises the as yet non-processed concatenated pseudo-order of the elements of the *a posteriori*. Secondly, it characterises the 'spontaneous' *a priori* contributions of the 'self', where the self must be regarded not as an empirical entity but as the logical seat for the transcendental functions of judgment. In neither respect does the reference to subjectivity therefore quite mean what it says, and in both respects it is meant to serve as providing a contrast with the *metaphysical* characterisations of earlier philosophers. On the one hand, it wants to say that the element of sensation as such lacks spatio-temporal form; on the other, that the element of objectivity must be defined not by reference to transcendent reality (the character of the thing-in-itself) but through the logical character of certain functions of judgment, i.e. the categories. This re-location of metaphysical centres of gravity must be strictly kept in mind if we want to understand the significance of the proofs of the principles of the Analogies discussed in this section.

Let us now develop some aspects of this exposition, by asking once more: what gives the empirical manifold of intuition, when transcendentally unified, the right to be called 'objective'? Now there are really two quite separate sets of considerations which Kant employs, though they are eventually brought together in an interesting way.

(1) The first set starts from the conclusion of the argument that there is only one object, and not two, which derives from the fact that, without the synthetic process, the notion of phenomenal object is undefined. Kant is quite explicit on this when he declares that the unity of apperception is 'entitled objective', because it 'is that unity through which all the manifold given in an intuition is united in a concept of the object';[1] and that, *per contra*, 'an object *is* that in the concept of which the manifold of a given intuition is united'.[2] This sounds like a pun on the term 'objective'; it certainly involves a relatively narrow connotation of the term.

(2) The second set relates to the form of the indicative judgment which involves the copula 'is'. Consider the example of a categorical indicative contingent non-general judgment, 'the body is heavy'. This makes a publicly testable claim concerning an external object, viz. that subject and predicate are 'combined in the object', quite irrespective of, and in abstraction from, what goes on in the subject which makes this statement.[3] But this language of 'combination in the

[1] K. 157.   [2] K. 156; my italics.   [3] For this argument, cf. K. 159.

object' reminds us of the previous account, (1), which says that an object is that in the concept of which the manifold is combined and united: the concept of the object being the mirror-image of the necessary unity of apperception. This evidently motivates the form of Kant's definition of a judgment as 'the manner in which given cognitions are brought to the objective unity of apperception'. He takes a judgment to 'assert' that the members of an empirical manifold 'belong to one another in virtue of the necessary unity of apperception in the synthesis of intuitions'.[1]

Let us explain this puzzling expression. The crucial point concerns the relation which the judgment asserts to hold between the members of some perceptual manifold—a relation which has a certain logical character. Now this character is *incompletely specified* by saying that the relation holds in virtue of the *process of combining having brought together* the perceptions. For this would be an 'accidental' combination, since the combination is only the result of a 'subjective' process, effected on the part of the cognitive subject; remember that the 'act of combination' is held to be 'spontaneous'.[2] But clearly the logical character of the relation asserted by the judgment says more. It claims that the perceptions occur together as they do (in a putatively true judgment) in virtue of their *'really existing'* together.[3]

Now here lies the crux of the argument: from where are we to take the notion of 'really existing together', if this is *not* to involve claiming a cognitive relation to the thing *as it is in itself*? The only way to do this is to add to the process of combination, so far conceived 'subjectively' (or 'spontaneously'), a further *logical* characteristic, though introduced equally 'spontaneously'.[4] In other words, the notion of an 'objective reality' requires a complement, as logical characteristic, additional to the 'subjective' and 'accidental' combination of the manifold, in which 'transcendental reflection' has so far conceived the matter. Evidently, it suggests itself that we should envisage the

---

[1] K. 159.   [2] Cf. K. 151: 'Combination . . . is an act of spontaneity'.
[3] For this locution, cf. K. 209.
[4] I.e. this is a 'transcendental' matter. Remember that this 'spontaneity' is only contrasted with 'receptivity' in respect of sensation. It is not 'arbitrary', but 'added' as defining objective cognition. It is the critical philosopher who construes this as an 'addition'. In fact it is simply part of the logical characterisation of objective cognition. To say that it is 'logical' and 'spontaneous' is to say that it is not a *material* detail, but only a manner of characterising the nature of objective cognition, and of the logical intentions of the cognitive judgment. The 'ground' of the world has become 'transcendental form'. Besides, we should keep in mind the remarks on 'spontaneity' made above, p. 619.

nature of this logical characteristic as *determining* the combination in a manner which *says* that it is what it is, not 'accidentally' or 'arbitrarily', but, so to speak, '*necessarily*'. This locution borrows the case where we say of a happening in the world that it did not take place accidentally, but that it was, for instance, necessarily determined, in the sense of causally necessary; or more generally, that it was subject to some rule or law.[1]

Emphatically, this is only a model taken from the empirical realm. Berkeley had attempted to distinguish the 'objective' by reference to the orderliness of the laws of nature; Hume by noting the characteristic of the peculiar 'manner' in which we cognise what is 'real'. Just so Kant will characterise the 'objective' basically by reference to the criterion of 'order', but the Humean reference to 'manner' as well as Berkeley's 'order' are caught by shifting both to the 'transcendental' level.

In sum, we must add to the transcendental model of the 'accidental combination' a further concept which *determines* the combination, thus preventing it from being accidental. But the concept we want here is precisely that which says that the 'unity' belonging (as we have noted) to every combination, should be a 'necessary unity'.[2] Like combination, like unity, so the necessary unity is transcendental. And to say that it is transcendental is again to say that it is a condition upon which alone anything can be conceived (cognised) as empirically real, as an object; whilst at the same time, as again noted already, it is a condition of ourselves being *consciously* cognisant of any such objects.[3]

This is precisely, as Kant emphasises in his discussion of Noumena and Phenomena, where all previous philosophy went wrong. In its search for a model of objective reality, it regarded that which corresponds to the necessary unity of apperception as a kind of 'transcendental object'. Rightly viewed, however, this is nothing more than the

---

[1] Cf. also the model of the accidental set of empirical laws, requiring the unity of reason; above, pp. 490–1. Besides, the German word for 'accidental' is 'zufällig', which stands also for the English 'contingent'. The contrast between the contingent and the necessary thus suggests itself. But it should be remembered that Kant is here characterising the transcendental logic of *contingent* propositions; it is *they* which are now said to contain an element of necessity! (Cf. below, pp. 636–7.)

[2] K. 159.

[3] The transcendental unity of apperception is, to note again, 'necessary', both in the sense that it denotes a necessary determination of the combination, and that it is a necessary condition for both empirical consciousness and for objects *quâ* phenomena. (Cf. above, p. 627.)

'correlate of the unity of apperception'. But searching, as we do, for something 'which, even apart from the constitution of our sensibility . . . must be something in itself', we frame the notion of 'an object independent of sensibility', and 'there thus results the concept of a noumenon'.[1]

Kant's negative strictures can now be seen also to have a more positive side. I have argued before that one of the tasks of the concept of the noumenon is precisely to serve as a model for the 'transcendental object', or better, for the logical characterisation of the task effected by the unity of apperception.[2] So, by a peculiar inversion, Kant's strictures betray the sources of the model which serves for his characterisation of the unity of apperception. By converting the logical character of the noumenon regarded as a metaphysical entity into a description of the unity of apperception, he moves it from the metaphysical to the transcendental level, as a condition of our experience of *phenomena*, which—as in themselves lacking the characterisation of things in themselves—stand in need of the transcendental unity.

In all this, it will be noted, we are not concerned with the empirical truth (or objectivity) of the judgment, but solely with an analysis of what is contained in the very claim or notion of such a judgment. The crucial nerve of the argument lies, of course, in the Kantian view of the 'accidental' nature of the combination, prior to being processed 'conceptually' through the notion of the necessary unity. What, then, is it to say that the combination of perceptions, in some order or other, prior to the imposition of the conceptual mould, is 'accidental'?

The choice of the term itself we have already seen to be motivated by Kant's own general procedure. Since the combination requires the addition of a concept which is modelled on that of a 'rule' (for instance, a causal rule), it follows that without such a rule an order is accidental. This, however, explains only the choice of the *term*, not the justification for the logical characterisation of the combination as 'accidental' as such, since it assumes that the latter *requires* the imposition of a rule, i.e. necessary unity.[3] The basic significance is, however, this. We are here dealing with perceptions. But we have seen —as a result of Kant's critique of space and time—that perceptions

---

[1] K. 268, 270.   [2] Cf. above, pp. 542–3.

[3] The conjunction of unity, concept and rule is Kantian. Thus we have already noted that, at K. 147, the 'understanding' is 'defined' as spontaneity of cognition (in distinction from the receptivity of sensibility)', further 'as a faculty of concepts, or again of judgments', and finally, 'as the faculty of rules'. (Cf. above, p. 623n.2.)

must be viewed as given in an intuition, the form of which is space and time. Now not only are these forms 'subjective' modes (in the sense explained at length in the last section[1]), i.e. not properties of things-in-themselves; they are also so far only the *indeterminate* aspects of space and time (the factors of spatiality and temporality). For instance, no *determinate* 'before-and-after', and no *determinate* simultaneity, attaches to them *quâ* members of an intuitional manifold. Only the *possibility* of cognising them as 'before-and-after' or as 'simultaneous' is so far *given*—the argument runs parallel with that for space. *Determinate* space and time, as a kind of 'conceptual clamps', must still be supplied.

It is therefore clear that the notion of 'accidental' is not intended to characterise the phenomenal object (e.g. an objective succession or coexistence). We are here concerned solely with what only a 'training in transcendental reflection'[2] makes evident, viz. the status of space and time and the corresponding transcendental characterisation of the perceptual manifold, in respect of its *a posteriori* content, as matter of sensation. It all hinges on Kant's characterisation of the object as 'appearance'.

Yet, this characterisation of space and time is a difficult one, as our previous sections have made plain. For this reason Kant (perhaps unawares) seeks support for his transcendental analysis of the accidental character of the perceptual synthesis by means of a psychological approach, not to say empirical model. This is not surprising. Kant presents the transcendental process as something that relates to the experiential structure of the conscious subject. The transcendental elements of experience thereby easily come to be viewed as a 'hidden realm' lying beneath the surface of empirical consciousness. Being hidden, we are compelled to use empirical (psychological) analogies to make our meaning clear.

Intuitions thus are treated as primitive states of consciousness, not to say a kind of pre-conscious states. Perceptions, Kant writes, are 'modifications of the mind', and thus 'belong to inner sense', whose 'formal condition' is time.[3] 'Inner sense', like 'outer sense' (whose formal condition we have seen to be space) bears again, however, that double interpretation, transcendental as well as psychological, which attaches to outer sense.[4] In 'inner sense', i.e. in time, perceptions 'must

---

[1] Cf. also below, pp. 639–40.   [2] K. 207.
[3] K. 131. This is the more 'subjective' and psychologising rendering of the first edition of the *Critique*.
[4] Cf. above, p. 597.

all be ordered, connected and brought into relation'.[1] Of course, the 'time' here in question does not *produce* the order as is so easily and so frequently imagined to be Kant's meaning; it is only, so far, a *condition* of their *being* ordered; it offers us a *potentiality* for order. Now the psychological account which suggests itself at once contrasts 'internal experiences' in the mind with 'external events' in the world. It is as though we compared what *seemed to us* to be the case with what *is really* the case; or our images with reality. More probably, we are face to face with another model, viz. the old Lockean distinction between our 'ideas' and the 'things' problematically conforming to the former.

Every reader of this chapter will realise that this cannot be the contrast which Kant's intends. Some, however, do read him this way; and it is then not surprising if Kant's method of converting the subjective order into an objective one, by 'adding a concept', is felt to be quite insufficient, not to say bizarre. And yet it is not enough to reply that this contrast is only a *model* for Kant's contrast between the pre-conceptualised and the post-conceptualised state of consciousness. For the question ever-present is whether the contrast which he wishes to make *can* be made meaningfully without these models.

Thus, in the passage discussing the nature of the cognitive judgment,[2] Kant contrasts the judgment 'the body is heavy', said to be 'objectively valid', with a locution purporting to denote only 'subjective validity', viz. 'If I support a body, I feel an impression of weight.' Now it is true that to call the first judgment 'objectively valid' is intended only as a description of its logical form; we are not concerned whether the judgment *is* objectively valid; that would be a question of material truth. And the same goes for the 'subjective' case. However, it is clear that Kant's 'subjective' example could not be *understood*, on his own theory, unless it were already 'categorised', and thus contained those conceptual ingredients which are supposed to bestow 'objectivity' in Kant's special sense of that term. So the most we can allow is that Kant, through this example of a judgment explicitly involving reference to a subject, wishes to describe analogically something which—given the basic features of his transcendental argument —is as such 'unspeakable'.[3]

---

[1] K. 131.
[2] K. 159. Note that obviously even a contingent particular proposition requires transcendental necessity.
[3] Elsewhere, especially in the *Prolegomena*, these subjective forms are referred to as

To return to the notion of the 'accidental'. If this, as we have noted, is not intended to characterise the phenomenally 'given', which is itself the *result* only of the 'addition' of the transcendental concept to the perceptual 'accidental' order, no more does the characterisation of 'necessity', when added as concept to that order, make the resultant objective situation a necessary one. In short, at the level of empirical consciousness, i.e. empirical contingent reality, there is no necessity. The necessary unity of apperception does not yield necessary connections at the empirical level. This Kant points out expressly, when discussing the form of any judgment as such:

> [The copula 'is'] indicates the relation [of some given perceptual representations] to original apperception, and its *necessary unity*. This holds good even if the judgment is itself empirical, and therefore contingent [or accidental: 'zufällig'], e.g. 'Bodies are heavy'. I do not here assert that these representations *necessarily* belong *to one another* in the empirical intuition, but that they belong to one another *in virtue of the necessary unity* of apperception in the synthesis of intuition, i.e. according to principles of the objective determination of all representations, in so far as this can yield cognition; principles which are all derived from the transcendental unity of apperception.[1]

Evidently, therefore, transcendental necessity does not touch the question of any putative necessary connections that may hold between members of the empirical manifold as such at the level of 'intuition', in those cases where we are dealing with empirical laws which Kant regards (as we have seen before) as having a necessitarian logic.[2] Any empirical necessity will have to be made good separately.[3]

'judgments of perception', which are contrasted with the 'objectively valid' 'judgments of experience'. Many commentators, e.g. Kemp Smith, Cassirer, have argued that this is a notion quite at variance with Kant's fully developed critical view, according to which such a contrast ought not to be possible. My account which claims that 'subjective' is no more than an analogy for the transcendental or Pickwickian notion of 'accidental' may perhaps mediate between this interpretation and those commentators who operate with the notion of judgment of perception without qualms.

[1] K. 159, italics in text; cf. above, pp. 475n.2, 619. (The translation has been modified). The principles referred to towards the end of this passage are the Axioms of Intuition, Anticipations of Perception, Analogies of Experience, etc.

[2] Cf. for instance, above, sect. 4(c): v.

[3] Of course, this 'level of intuition' is an abstraction; for in order to represent anything 'empirical', it would have to subsist with 'consciousness' and thus 'in experience', so that transcendental necessity would *have* to be added. But, so our passage claims, this does not affect the question of any necessary connections that may obtain between the parts of the intuitional manifold, regarded (in virtue of the

The question of empirical necessity has already been discussed, and its relevance in the present context will be pursued below. For the moment, let us note that Kant's argument neatly solves Leibniz's problem of the 'analytical bond' or connection between the terms of contingent, and even particular propositions, by making a finer distinction of levels. It constitutes a reformulation of the *P.i.S.* doctrine.[1] For Leibniz, the connection had been opaque to the intellect, though postulated as a metaphysical condition, yielding a kind of necessity that he had been unable to characterise, labelling it sometimes moral, sometimes physical; a connection which was explained through the model of the infinite analysis of the subject of the contingent proposition, completable only in the sight of God. In Kant, opacity is caught in the synthetic character of the *a posteriori* element of contingent propositions, whilst Leibniz's 'analytical' connection is moved from the metaphysical level (itself a sharpening of the 'physical', i.e. phenomenal) to the transcendental level. Here it is sustained, not by the teleological activity of God, but by the logical need to provide the conditions for experience (cognitive judgment) of an objective world, regarded as appearance.[2]

Let us stand back and view Kant's procedure for a brief moment in its totality. The objective validation of the transcendental unity is provided by its constituting a contribution towards the possibility of an experience of objects, of the fact that objects can be 'given' at all. Here again a puzzle similar to the one met in our discussions of geometry, and of the question of 'balance' in general, faces us.[3] For it may be objected that since 'objects' are 'generated' by the contribution from apperception, how can the 'fact' that these objects are given supply a validation? Such a validation would carry force only if the objects were given—so to speak—independently. Yet this is precisely what is not the case; the notion of the object is 'bracketed' by virtue of the transcendental condition!

The reply that Kant might give to this is as follows. The givenness of objects is a fact of experience.[4] The only question which concerns

---

necessary unit) as being given as phenomenal occurrences in empirical reality. (Cf. also sect. 8(a).)

[1] Cf. above, ch. VII, sect. 4, especially pp. 453–5.

[2] Cf. above, ch. VII, p. 453n.3.   [3] Cf. above, pp. 612n.1, 612–13, 619.

[4] This argument shows that the Kantian view of the 'conceptual ingredient' of what is empirically given is framed in a very narrow and architectonic way. At the empirical level, the 'objects' of Kant's world are the straightforward things of commonsense, their grammar is not indebted to any scientifico-theoretical

us is how this 'givenness' is to be interpreted so that its 'possibility' may be demonstrated with apodictic certainty. And the answer is, of course: through the procedure of 'bracketing' the notion of the object, by which we regard it as 'appearance'. Now this answer shows that Kant must be using the notion of 'givenness' twice: first, as what is given for uncritical consciousness; secondly, as that givenness whose possibility can be 'demonstrated' if the object thus 'given' is bracketed as phenomenon—an entirely Pickwickian sense of the 'given'. The claim, that it is permissible to move from one use to the other, is one of Kant's deep metaphysical assumptions. The positive result yielded by Kant's metaphysical appraisal is that it places older philosophical problems in a new light, by reformulating their questions. It re-locates the philosophical grammar of their key terms and shifts the philosophical centre of gravity in new directions; one of the conclusions which this book was meant to elucidate.

This is well illustrated by the shift in the grammar of the Kantian 'subjectivity'. We have already touched on the debatable contrast mirroring the difference between the pre-categorised 'accidental' order and its categorised successor. Here, we may limit ourselves to the subjectivity of the transcendental apparatus of Kant's epistemology. First of all, let us note that space, time, the unity of apperception, are if anything only transcendentally subjective. There is a temptation to think of them as subjective because they do not belong, at the transcendental level, to the *a posteriori*. But the *a posteriori* should not be confounded with what is 'objective'; so Kant's intentions must connect with the contrast between *a posteriori* and *a priori* as such. Space, time and the categories are supplied, in the context of experience, as *concepts*, whether they be interpreted as concepts that designate intuitional form or as those which specify logical relations (as do most of the relevant categories). But there is no reason to locate concepts, just because they are logical entities, in a realm of subjectivity. It is true that they need *application*; and this application takes place in the context of 'experience', but all that is here relevant is the *notion* of experience, i.e. experience *quâ* 'possible'. Moreover, all

---

components. For the further conceptualisation and incorporation into theoretical contexts, we have to move on to the level of 'reason', in its 'constructive' employment. (Cf. above, particularly the account on p. 510.) In the philosophy of Hegel, and again, in modern developments, as in the writings of a neo-Kantian like W. Sellars, these sharp divisions are smoothed over, though with corresponding loss of logical compulsiveness.

this is equally true of the *a posteriori*. It, too, acquires significance only as a component of possible experience; outside that, it is nothing for us.

In truth, the subjectivity in question here is nothing psychological or empirical. It contrasts with the notion of 'objective reality' which belongs to the thing-in-itself. Both are notions which are sharpened technical equivalents of confusedly-held ideas on the part of earlier philosophers and uncritical commonsense. But out of their normal context they become difficult to characterise. This is, after all, the crucial difficulty of the thing-in-itself. And just as we saw that the notion of the latter can at best be articulated only in a variety of complex considerations, in answer to conflicting pressures, so its logical complement, the element of transcendental subjectivity, can likewise be understood only in a similar way.

This 'sharpening' of Kantian meanings is graphically illustrated in putting Kant's main point by saying that the 'objectivity' of the phenomenal object is not 'objectively given' but wrought 'subjectively'. This at once makes it clear that the second 'objectivity' in this account no longer has a normal sense. Nor is this surprising. For, figuratively speaking, one cannot characterise the limits of language and of our world by contemplating them from without, through the use of linguistic tools taken from this very world, without radically altering their sense. At best it can be done, if at all, only through this process of sharpening, and through the negative criticism of which the *Critique* itself is a supreme example. Here, an insistence on 'ordinary usage', as found in recent philosophy, is as misplaced as it is incapable of appreciating many of the procedures of classical philosophy.

To return: in the passage from K. 159, quoted on p. 637, Kant implies that the general principles of the understanding (and the categories which they embody) can be derived from the transcendental unity of apperception. How is this to be achieved? Let us first consider the case of the categories. We saw that it is a logical function of the judgment of experience to bring the manifold into unity under concepts. Now these concepts are the different categories which, Kant claims, are given by, and run parallel with, the different headings under which judgments are classified.[1]

[1] Cf. K. 160. Thus, to categorical judgments corresponds the category of substance; to hypothetical, causality; disjunctive, dynamic association or interaction; to quantity and quality, the corresponding categories which run by these names; etc.

Many commentators think this 'deduction' rather artificial, but there is no need to enter into this matter nor into many other subsidiary discussions, such as the question of schematisation of the categories. The abbreviated form of our exposition forces us to concentrate on the essential thread of the Kantian argument. What we shall do instead is to attend in slightly more detail to Kant's separate 'proofs', in which (again omitting 'substance') the concepts of causation and interaction are shown to be necessary components of judgments concerning the *sequence* of states of a thing, and the *coexistence* (in space) of such states or of different things, respectively. This will enable us, finally, to place in their correct position the function which these concepts have, on the one hand, with respect to ordinary contingent judgments concerning particular matters of fact, and on the other, the more specifically law-like statements of physical science.

(e) *The Analogies of Experience*

(i) *The general argument*
We must now proceed to a discussion of the way the ideas of causation and interaction are fitted into Kantian 'experience', for only thus can the Kantian stratification of levels, the increase in philosophical 'fine structure', be understood. Here we shall be concerned with that culminating construction of the *Critique*, which, as the reader will have realised by now, is the Analogies of Experience, the first dealing with substance, the second (causation) involving the core of Kant's answer to Hume, the third ending where the *Estimate* of 1747 began, and where the criticism of Leibniz's monadology had continued it, with its explanation of the relation of interaction and coexistence—basic, as we already know, for Kant's notion of determinate space.

Such an account of the Analogies is needed also because it will reinforce our understanding of the sense in which the transcendental method is meant to yield the objectivity of the intuitional manifold, shifting as it does the discussion from the notion of an 'object' in general, to that of 'objective sequence' and 'objective coexistence'. Thus, the Second Analogy purports to show how experiential judgments concerning an objective sequence (as an object of a potentially

---

And each of these categories can then be shown to yield corresponding general principles. (Cf. K. 107 and 113 for the two Tables. I omit discussion of the categories of modality, as not central to any of our topics.)

publicly reportable experience) is possible; the third, how perception of objective coexistence of things in space is possible.[1]

The argument of the Analogies, and particularly of the second, hinges on the conception of time. So far we have only alluded very briefly to time. Parallel with the Kantian locutions for space, time, as the 'form of inner sense', is that in virtue of which things are placeable in a temporal order of endurance, *or* succession *or* simultaneity.[2] (K. 209, which also speaks of time as the form of inner sense, equates it with 'the time-order of the complex empirical consciousness'.) We need not argue the problem of its status, since what we have said of space, holds of time also (indeterminate as well as determinate), except that time is in a sense more basic than space *quâ* presupposition; it concerns not only 'outer things' but happenings internal to us, i.e. in our minds, as well.[3] Time, like space, is transcendentally ideal, not real. And, as already pointed out, as *form of* intuition, it only yields the notion of the *possibility* of things being placed in an order of succession or simultaneity—in short, yields a temporal language. *Quâ formal* intuition, we furthermore require (as for space) synthesis and its unity (under concepts); and in neither aspect should time be pictured on the lines of some kind of absolute clock that, being in some way attached to the phenomena as a 'property' (which would be so only for things-in-themselves), can tell us which of them precedes and which succeeds the other.

---

[1] Note that sequence can take place in the *same* place or space, but things can exist simultaneously only in different places of an extended space, for which reason space does not enter explicitly into the discussion of the Second but only the Third Analogy, whilst time is an important determinant for both.

[2] K. 87, 209. In his psychologising account, Kant frequently characterises the form of time also by the remark that all our sense-contents are 'apprehended' always 'as successive' (cf. K. 213, 219), something that reminds us of Locke's attempt to define time in terms of the successive order of our ideas. But this would be a misunderstanding. Both succession and simultaneity can only be expressed as happenings *in* time, characterised *through* time, which is—whatever it may be—still presupposed. (The First Analogy argues that it is the concept of substance which expresses this time in general; but this contention has never satisfied Kant's critics.)

[3] K. 131. This 'absolutely fundamental' assumption (as it is here called) displays again graphically that curious intermixture of the phenomenal and the transcendental sense of the 'inner-outer' dichotomy. The 'inner', in the phenomenal sense, in which my sensations, for instance, are 'internal', can here serve as no more than a model for the transcendental sense of 'inner' required. Transcendentally speaking, *everything* is 'outside' as being given for or to consciousness, and *everything* is 'internal' in that it has to occur in time. However, phenomenally speaking, only *some* sensations, as experienced qualities of bodies (including my own) are 'outside', whereas others, such as certain emotions or feelings, also pain and pleasure, are not in that sense spatialised.

The various proofs of the three Analogies are preceded by a general proof which (in the version added to the second edition)[1] contains probably the most succinct summary of Kant's position, reproducing more clearly the main features of the transcendental deduction, though assuming the point that consciousness and the concept of object require one another mutually. The purpose of the general argument is to show more specifically than the transcendental deduction had done that the 'unity' of the synthesis which lies behind objective relations of time is specifiable through the categorial concepts of relation;[2] i.e. to show that, in respect of the *relation between* the members of the intuitional manifold (the perceptions), the concepts are those of substance, causality and interaction.

The proof begins by reminding us that experience involves a synthesis of perceptions, as well as the unity of this synthesis, which 'constitutes the essential in any cognitive grasp [Erkenntnis] of the *object* [as such] of the senses . . . as distinguished from mere intuition'. This again summarises the equation of unity and objectivity, found already in the transcendental deduction. There follows, similarly, the confrontation of 'accidental order' and 'necessary determination':

> In experience . . . [if we abstract from the element of unity which it contains] perceptions *come together only in accidental order*, so that no *necessity determining* their connection is or can be revealed in the perceptions themselves. For apprehension is *only a placing together* of the manifold of empirical intuition; and we can find in it no representation of any necessity which *determines* the appearances thus combined to have connected existence in space and time.[3]

The vital clue here is, of course, again the concept of the 'accidental order' of perceptions. Remember that we are concerned with the task of constructing the formal notion of the object and not with the material content (sensation). To appreciate this, let us consider an objection which arises spontaneously at this point. We want to say that the order of perceptions, e.g. a green patch, followed by a blue patch, followed by a yellow patch, surely occurs in the order in which it is 'passively received' by us. True, it may be 'accidental', for instance in the sense that no one intended it; or in another sense, that there is no empirical law determining the order. Or again, it may be said, whilst the order may be other than it *seems* to us, perhaps in virtue of a psychological defect on our part, or an optical illusion,

[1] K. 208–9.   [2] For the status of these 'relations', see also above, pp. 618–9.
[3] Ibid.; my italics.

such deceptions can surely not be mended by any formal manipulations, such as is represented by the notion of the *unity* of the synthesis of perceptions.

The reply must be that the objector has totally misconstrued the nature of Kant's problem; this is just the misunderstanding which our lengthy discussions of the concept of the noumenon and the theory of space and time were intended to meet *ab initio*. For let us repeat: Kant is solely concerned with investigating what is involved in claiming that the experienced order of perceptions is an objective one, and what is involved in expressing this through the judgment-*claim*: 'These colour patches *do as a matter of fact* follow in such and such an order', always understood that the objective situation concerning which we make this claim is 'mere appearance'. But in respect of anything regarded as appearance, nothing *can* be or has so far been said about any order at all! Only the notion of 'combination in accordance with concepts' will do this, and this has not yet been introduced.

To say that we are here dealing with an 'appearance' is to say that the 'perceptions' which 'enter into consciousness' do not carry with them any objective time-order. True, things-in-themselves might do this; but Kant's argument has been that we can have no objective cognitive conception of these, as objects of possible experience. So the concept of the object, representing the claim to an objective order, has still to be constructed. And it is, as we saw, for Kant a concept which '*determines necessarily*' the order of the perceptions. This is the logical paradigm that will 'explain' or 'define' the notion of objectivity.

Now the order is emphatically not determined merely by the fact that the perceptions have been combined as the result of a synthesis. So far, says Kant, this is

> only a placing together of the manifold of empirical intuition; and we can find in it no representation of any necessity which determines the appearances [perceptions] thus combined to have connected existence in space and time.[1]

So the abstract account, as far as we have it, does not yet include a reference to any objective time order. It is difficult to render such an abstraction in ordinary language. Kant therefore has to fall back once again on the subjective model, speaking as though it were ourselves who do the placing together, not *constrained* by what is 'really the

[1] K. 209.

case', but only by an *inner* activity. Or it may be that there are here echoes of the more sophisticated model employed in Locke's conception of complex ideas and nominal essences, the boundary-lines for which we saw to be drawn only in an 'arbitrary' manner, because we lack knowledge of the 'necessary connection' between the ideas which form it, and of any 'real essence' which may be thought to be their ground. At best, the 'activity' of combining the simple ideas was here constrained by the teaching of nature.[1]

As to the first, or psychological, model: if this is misconstrued as more than a model, and instead as some form of empirical account, it must be palpably absurd. Surely Kant cannot be referring to the arbitrary grouping of *images*. On the other hand, if he is concerned with perceptions (supposing the ordinary sense of that term!), Locke's account of 'simple ideas of sensation' would surely hold: do we not combine or 'blend' these in the order in which they are passively received by the mind? Nor can Kant mean that since it is we who do the grouping, taking hold so to speak of the 'ideas', it is always possible to make mistakes of perception: for that would once again be an empirical question. So the psychological case can only be an analogue, through which Kant seeks to contrast the 'internal' with the 'external' situation. Yet even this is misleading, since according to him there *is no such external* order, given in itself. That order, as it is 'in itself', can no more than throw its logical shadow, to guide us in our construction of the concept of the object.

Locke's more sophisticated model mentioned a moment ago brings this out more forcibly. For there the 'arbitrariness' of the combination was already clearly no empirical notion; and the contrast between it and the 'real essence' showed the latter up as a vanishing limit-notion. Moreover, even so far as we, in constructing the complex ideas of particular substances, consult experience, this had in Locke's view no bearing whatsoever on the fact that the combination of the simple ideas into complex ones was still in some sense 'arbitrary'. So this model represents more usefully the point that empirical considerations are not relevant if we want to construe Kant's contention that the

---

[1] Cf. above, ch. IV, pp. 214–5, 248–50. Kant, of course, has further sharpened what was already a philosophical appraisal in Locke. In Kant, for 'ideas' in the present context we must read (Kantian) 'perceptions'; for 'arbitrary combination' read 'transcendental synthesis'; and for 'real essence' we now have the double-conception of thing-in-itself whose logical 'reflection', as we have noted already, reappears in the definition of the tasks of the unity of apperception, itself not a metaphysical but a transcendental matter. (Cf. above, pp. 540–3, 634–5.)

notion of 'mere combination' (in abstraction from its 'unity') yields only an accidental order.

Kant's 'placing perceptions together in an accidental order' is hence no empirical matter, but a remark concerning the absence of time-determination, itself nothing empirical, but connecting with his rejection of time as a property of things-in-themselves. How Kant envisages this contrast is made clear in the sentence following the passage just quoted from K. 209:

> But since experience is a cognitive grasp of objects through perceptions, the relation [involved] in the existence of the manifold has to be represented in experience, *not as it comes to be put together* in time *but as it exists objectively* in time.[1]

Notice that Kant is concerned only with conceptual analysis, as asking how we are to 'represent' something in 'experience'. Here, the fact that the synthesis is pictured as a 'doing' (putting together) on the part of the subject is useful in order to contrast this with a situation which is *not* under the control of such a subject (as represented by the notion of objectivity—that notion standing for 'exercise of external constraint'). The 'doing' is part of the process of 'construction' in the sense considered earlier, e.g. the drawing of a line, except that in the present argument it is considerations of temporal, and not spatial, order that are uppermost in Kant's mind. Also, viewed from the side of the concepts (which as we already know will have to be introduced presently), the act of construction is required so that the concept may be applied or get a grip, since (to use our example of the line) only if you actually draw (even if only 'in imagination') can you do so in accordance with a formula or rule or concept.[2]

All this leads thus to the final stage of the argument. This is simply that, as the drawing of the line requires the guidance of a rule, so synthesis in general requires a concept, whose logical property it is (in Kant's view) to impose a 'necessary determination' upon the order of construction or synthesis:

> Since *time*, however, *cannot itself be perceived*, the determination of the existence of objects in time can take place only through their relation in

---

[1] K. 209; my italics. Kemp Smith's 'knowledge' has been replaced as usual by 'cognitive grasp'.

[2] In the chapter on Schematism, Kant refers therefore to this feature of construction as a 'third thing' (K. 181), placed between the sensation (perception) and the concept. It presupposes space and time as 'forms of intuition', and is halfway in reaching 'formal intuition'—which however requires in addition the concept.

time in general, and therefore only through *concepts* that connect them *a priori*. Since these *always carry necessity* with them, it follows that experience is only possible through a representation of necessary connection of perceptions.[1]

That time (and space) cannot be perceived is of course simply the result of Kant's general theory of space and time. To say that time cannot be perceived follows from the fact that it is not 'given' (as might for instance be the Newtonian receptacle) as a self-subsistent absolute framework into which are fitted the perceptions. For it is only in the context of an empirical language, i.e. when regarded as forms in which perceptions (actual or potential) occur, that the expressions 'space', 'time', have a meaning; and this holds even in the case of intuitions that are 'pure', i.e. where we abstract from any sensory determination or presence.[2] However, as mere forms, space and time do not yield a reference frame which would enable us to *define* the objectivity of a spatio-temporal order; and thus give sense to the expression: 'Whatever is, is somewhere and somewhen'.[3]

So it is only the concepts (categories) that supply the missing item ready to serve as the model for objectivity, in the sense of external constraints placed on the construction: a peculiar sense of 'external', intended to explicate the non-subjectivity of the form of the judgment of experience. Moreover, the 'necessity' which is here claimed for concepts is, of course, a feature which Kant believes to attach to rules. We speak of 'following rules', for instance the rule of counting integrally from one up to any number. 'Having the concept of counting' means being *determined* to count in the order demanded by the rule. But whether this model of the concept exhausts what we *mean* by objectivity is another question, which I shall not here discuss.

This, then, completes the general argument of the Analogies. Just as the argumentation against the thing-in-itself had shown that only by regarding things as appearances, is objective cognition of them through concepts possible, so the present 'proof' completes the demonstration that this possibility *is* real because, *quâ* appearance, the object *logically requires* the addition of such concepts.

---

[1] Ibid.; my italics. At K. 226 Kant says that 'absolute time is not an object of perception'.
[2] They are not 'transcendentally' but only 'empirically real'. Here is a useful mnemonic device: Pure intuitions are not empty intuitions. It is only the latter that are proscribed. (Cf. the passage from K. 398, cited above, p. 592.)
[3] This the *Dissertation* had already singled out as one of the chief 'surreptitious axioms', requiring a critical purge; cf. above, pp. 583–4, 600.

(ii) *The Second Analogy*

The principles of the general argument may now be applied to the special cases of the relation in time as a successive series, and as simultaneous existence.[1] The Second Analogy provides the 'proof' that the concept which articulates objective succession in time is the concept of causality, in the form in which it can be applied to a temporal situation (i.e. schematised).[2] In this form, Kant defines it as something 'real upon which, whenever posited, something else always follows',[3] with the occasional specification that what thus follows does so in accordance with a rule or, as he sometimes says, 'an absolutely universal rule'.[4] And there is, as I have maintained, only as much 'necessity' in this as is implied in the notion of a rule as such. It is the latter that replaces Hume's abstract concept of necessary connection.

Consider now a particular contingent state of affairs, such as the two successive positions of a ship, A, B, sailing down a stream (one of Kant's examples).[5] Let us single out, from among many, two perceptions which we will mark A', B', in order to indicate that, as perceptions, they are not yet members of an objective time-order. If indeed we could speak of any order of such perceptions as such, we should say, as Kant does, that the order is purely accidental.

In this context, Kant offers us another 'empirical analogy' meant to contrast 'subjective' with 'objective' order. He cites a case where the order of 'our perceptions' is palpably and known to be other than the real order of what is perceived. We may inspect the parts of a house, letting our eyes pass from roof to door and back to windows. Now the parts of the actual house are 'really' simultaneous, yet the 'perceptions' are successive. This may be contrasted with the succession of perceptions of the ship's positions which, putatively, happens to reproduce the actual order of succession.

---

[1] Cf. K. 213, 236. In the present section, we shall concentrate on the Second Analogy; section (iv) will take up the discussion of the Third.

[2] Unschematised, it is really the concept of ground ('Grund') and consequent, i.e. 'something from which we can conclude to the existence of something else, (K. 262, 113). The expression 'ground and consequence' in the context of an unschematised, here: noumenal, context, occurs at K. 285.

[3] K. 185.

[4] K. 124–5. Kant does not specify the rule, and it is not clear whether, for instance, he would have accepted statistical rules. Clearly the 'universality' here involved is *not necessarily* the universality of 'all-or-none' rules. That the *laws* of science should be deterministic or mechanistic is in any case for Kant governed only by a regulative maxim of reason to which contrary maxims might easily be added.

[5] K. 221.

After what has been said about the wrong-headed attempt to render the contrast between accidental and non-accidental by empirical analogies, there will be no need to appraise in detail the inappropriateness of the comparison, which would only needlessly extend the length of our account. This is not to dispute that the example has a certain graphic plausibility, provided we do not take it too seriously. What it does is to offer us instances of 'objects' (house, the history of the sailings of the ship) whose conceptual ingredient is taken to 'tie down' the perceptions in 'a determinate order' and on the basis of which we build up the relevant objective states.

The problem of the Second Analogy is solely to ascertain which concept will convert the accidental *sequence of perceptions* into one that is 'externally' determined; into the *perception of a sequence*, as Kant puts it artfully; for only this will satisfy our notion of an 'object'—which is what we seek. We do not have available for this the image of an 'objective time-clock'; only a logical concept will satisfy the requirement. And Kant's claim is simply that the concept in question is that of causality. For this concept says, briefly, that each of the states of the ship, for instance B′, will be 'objectively placed' in a sequence, if we add a concept which says: B′ follows upon some other position, say x′ (which may *but need not be* A′), in accordance with a universal rule. The concept, so to speak 'ties down' the order of perceptions in a determinate way, as required by the general argument of the Analogies. The 'subjective' or 'accidental' order of the ship's positions, A′, B′, is hence converted into an 'objective' one, by 'thinking' or conceiving their order in such a way

> that it is thereby determined as necessary which of them must be placed before, and which of them after, and that they cannot be placed in the reverse relation.[1]

Note at once a most important point: the 'thought' or 'concept' of the irreversible time-order is 'quite indeterminate'.[2] It only demands that for any state, such as B′, we posit *some* preceding state or other,

[1] K. 219. This 'thinking' is the predominant task of the understanding. When unconstrained by the need for the presence of an *a posteriori* element of sensation, the same 'thinking' becomes the province of 'reason' in its 'speculative' employment —a misapplication of its regulative function. The difference formally is that the *understanding* fits *concepts* to the empirical manifold, whilst *reason* involves *inference* from or to a set of *propositional* premises. Such, roughly, is the architecture of the first *Critique*.

[2] K. 225. This must be distinguished from the indeterminateness of time as 'form of intuition'.

which *as such* is never specified (is quite indeterminate), some x′ (which we *may* be exemplified by A′), whose whole function is simply to enable us to express the thought that B′ lies in a determinate ('irreversible') time order. Only the general concept of a determinate relation is added. And *in its transcendental employment*, the concept *cannot* be instantiated. (I cannot say: A′ *is*—or even *'perhaps is'*—a cause of B′.)

Let us be clear at once that Kant is not saying that the order A-B *is itself* irreversible. He is only positing that in so far as I think this order as an objective cognition, I think each of its members, A′, B′, tied down in accordance with the *model* of the concept of causality. This is obvious from the choice of Kant's example, which is that of a contingent happening; clearly we are not to imagine that the ship's sailing downstream is necessarily, if ever, an instance of a law-like happening, or that it is *as such* determined by preceding or underlying causes. This is what is expressed in Kant's explicit reminder at K. 159,[1] quoted above, which stated the position in general terms: in a contingent judgment (such as would be represented by A-B) we do not assert that A and B 'necessarily belong to one another in the empirical intuition, but [only] that they [*quâ* A′, B′] *belong to one another in virtue of* the necessary unity of apperception in the synthesis of intuitions'; i.e. the members of the sequence A, B are cognised as following one another *objectively*, by thinking them (*quâ* A′, B′) as subject to a conceptual clamp (here, causality).[2] This 'thought' is however no more than part of the analysis of the notion of 'cognition of objective sequence'; it is what Kant means by the reiterated reminder that the category is 'proved' only as a condition of possible experience. Its function is meant solely to indicate that in representing to ourselves the notion of an 'objective' cognitive judgment, this is insufficiently characterised as a 'synthesis', since the latter (without the categorial concept) can be regarded only as 'subjective', i.e. as the subject's 'spontaneous'—and hence as such 'unconstrained'—act; so that, as judgment of experience, we need to include a reference to a logical feature expressing 'objective validity'. The *picture* here is that since the 'subject' is 'free' to reverse the order of perceptions, we need to add—in order to speak of objective cognition—a 'constraint', i.e.

---

[1] p. 637; and cf. also p. 619. Remember also the explanation added on p. 637n.3!
[2] Regarded in abstraction from the synthetic process, the perceptions are represented by primed symbols; viewed as parts of the sequential experience, they are shown unprimed. Both, of course, concern the same sensory content. But only *quâ* A′, B′ are they 'causally related' (at the transcendental level), and not *quâ* A, B.

the 'thought' that we 'cannot' reverse this order, which is provided by adding the concept of causality.

Now this, and only this, is what is meant by saying that all succession (or 'alteration') in time takes place in conformity with the concept[1] of causality. It does not mean that the sequence A-B can thereby be claimed, even in principle, to be a member of a causal situation or nexus, still less, that it is as such an instance of a putative causal law. We might put this by saying that the logical strength of the concept lies *solely* in (and is exhausted by) its providing the model for the 'thought' that the erstwhile accidental ('subjective') sequence is really *not* under my control, but is guided by a rule of succession which prevents this succession being different from what the judgment in fact expresses. The judgment itself does not *assert* that the sequence which it states is a *necessary* one; rather, in order to assert that things are *in fact* as it states them to be, we must conceive the accidental order as prevented from being accidental by its subjection to a determining rule.[2] This is all that is meant by saying that the concept (or 'law') of causality is a transcendental component of the notion of experience, as expressed in the form of the cognitive judgment. It follows that only in this sense is it true that Kant has shown that 'nature' (as the sum-total of all objects of experience)[3] is subject to the 'law' of causality, viz. as a presupposition of possible experience.

(iii) *The 'looseness of fit' between transcendental and empirical levels*

Certain ambiguities in the general tenor of Kant's presentation of the doctrine of causation can however not be denied. For frequently, after stating his position by insisting that the transcendental argument shows the concept or law of causality to be only a presupposition of the notion of objective succession, he goes on to say that ('only')

---

[1] Kant alternatively uses here the term 'law', to denote the notion of the 'rule' contained in his use of this concept, but he thereby causes much confusion. For this, cf. above, sect. 4(c): ii, especially pp. 500–1.

[2] This is in fact precisely why Kant employs the term '*Analogy* of Experience': the concept of a causal connection (determination in accordance with a rule) between perceptions A', B' *is an analogy* for the objectivity of the time relation (sequence) between A and B. (Cf. K. 211.) Moreover, the concept of a causal relation between A, B is as such an analogy for the causal relation between, and hence time-determination of, A', B'.

[3] Cf. above, p. 509, for this definition. In this context it is particularly necessary to remember that the German original for 'object' is 'Gegenstand', which in addition to 'thing' may also denote 'any subject matter', 'state of affairs', 'happening'.

observation and inductive investigation can inform us what actual empirical laws of causality there are. But this seems to suggest that he believes himself to have shown nature to be in principle law-like, in a sense different from that so far 'proved'; as though he was implying that he had shown every empirical sequence to be subject to some empirical causal law or other, in the sense of being an instance of such a law, or at least of entering into some theoretical nexus. However, the 'law' of causality has *not* been 'proved' as a condition of the *law-likeness* of nature, but only of nature in general. We should not therefore assume without further ado that Kant is actually claiming that the law of causality, so far proved only transcendentally, may be employed *directly*, by simply resorting to experience, to indicate which empirical sequences *are* causal. A careful reading of those passages where Kant may seem to suggest this will in fact not really support such an interpretation. Thus Kant writes:

> Particular laws, even though they are all and one subject to them, cannot be completely derived from the *a priori* principles of the understanding, since they concern appearances that are empirically determined. In order to obtain any information concerning the particular empirical laws, we must resort to experience; but in regard to experience as such, and as to what can be cognised as an object of experience, the *a priori* principles alone can instruct us.[1]

Now the emphasis in this passage is on the place and function of the *a priori* principles, e.g. the transcendentally proven law of causality. Furthermore, Kant reminds us that empirical laws, since they treat of 'nature', and are thus based on judgments concerning the contingent evidence that supports them, must be subject to the categories. But the passage does not bear out the contention that these 'laws' (i.e. principles) can be applied forthwith, and even without introducing further 'assumptions'; as though by themselves they provided a justificatory basis for the empirical law-likeness of nature. True, Kant does actually say that we need to resort to experience. But I have shown abundantly (in section 4(c): i) that his use of this term is ambiguous: it may denote experience as rendered by the particular contingent judgment—and here the categories are presupposed as principles which define *possible* experience—or it may mean 'experience as a system'; thus resulting in a slide which we noted in our reference to the passage in IJ.14–15.[2] The passage from the categories to empirical

---

[1] K. 173. The translation has been modified.   [2] Cf. above, p. 504.

laws, if indeed it is a passage at all, would then be circuitous and not direct.

What further muddles the issue is that all this is easily confused with the fact, stated by Kant in the *Prolegomena*, that a causally formulated language cannot be derived from propositions stating mere regularities of sequence, without presupposing the causal concept.[1]

A study of Kant's arguments in the *Prolegomena*, and the examples which he gives, shows that he has there in mind the following situation. Take any empirical regular sequence such as a rise in temperature of a stone which always follows its exposure to the rays of the sun. He now contrasts this 'judgment' with the 'causal' proposition 'the sun *warms* the stone'. And he contends that such a proposition could not be formulated without the injection of the concept of cause. Moreover, the concept *has* to be injected 'from without', so to speak, i.e. '*a priori*', for 'no matter how often I and others may have perceived' the original sequence, it will never 'contain' the 'necessity' of the causal judgment.[2] And that causal judgments *do* contain necessity is of course assumed by Kant as it was assumed by Hume and every other philosopher of our period.

In the argument of the footnote to P. 64 Kant refers to this causal judgment as a 'judgment of experience', and contends that without the injection of causality we cannot formulate such judgments; the 'necessity' involved cannot be 'learned through experience'; 'on the contrary, experience is only generated when the concept of the understanding (of cause) has been added to the perception'.[3] I think this passage affords an admirable example of the ambivalence and fluidity of Kant's position. Clearly, we must distinguish between two cases of 'judgments of experience', (i) a judgment concerning a contingent sequence; (ii) a judgment concerning a causal sequence. In accordance with Kant's way of thinking, the concept of causality needs injecting into *both*, in order to generate their 'possibility'. Nevertheless, the situation is clearly different in the two cases. Causality is a *transcendental* presupposition of the notion of a *contingent* sequence which thus contains causality only *implicitly*; but it is a logical or conceptual or formal presupposition for the notion of a *causal* sequence, which

---

[1] Cf. P. 60n., 64n. (paras. 20, 22).
[2] Ibid. Evidently Kant ignores here Hume's quasi-psychological reconstruction of such a case.
[3] Ibid.

involves the causal form *explicitly*. Moreover, we can see that Kant is using considerations belonging to these quite different cases, in order to let each of them support, and give meaning to, the other. The injection of the concept of cause, converting mere sequence of *events* into a causal connection, is used as a model for the injection of causality into the sequence of *perceptions*, to generate the 'perception' (meaning now experience) of a sequence, in the objective sense. On the other hand, his argument that without the injection of the category we cannot talk of a judgment of experience as such, in the sense of *contingent* experience, is used to add a quasi-demonstrative strength to the contention that without an *a priori* addition of the causal concept no judgments of (causal) experience as such are possible.[1]

So at most we can only conclude that the example from the *Prolegomena* serves as a model—derived from the locutions of scientific (causal) language—for the situation which prevails in the argument of the Second Analogy. That is to say, the concept of causality is presupposed transcendentally and *implicitly* for the possibility of contingent judgments, in a manner in which it is presupposed formally and *explicitly* for the possibility of causal judgments. And the most that Kant could (and would?) claim is that the 'law' of causality, 'proved transcendentally', finds its *application* at the level of science, where it is 'presupposed' in the merely 'formal' sense.[2]

In interpreting the limits of Kant's transcendental argument it is always vital to keep in mind the details of his 'proof' and to use language carefully for its characterisation. In order to cognise a sequence of perceptions, A', B' as objective, I have to *think* the perceptions connected in a way for which the *concept* of a causal nexus is the required model. But the concept is used only 'indeterminately'. This use is emphatically not to be construed as implying that I think the perceptions connected causally, in the sense of the resulting sequence being a possible or even putative instance of an empirical, and hence contingent, causal sequence. Moreover, since this is not demanded by the transcendental argument, the latter—in so far as it succeeds—cannot be strong enough to provide a 'proof' of the existence in principle of contingent causal connections, posing

---

[1] I add again 'as such', since Kant does not of course claim, even in this latter case, that the additon of the concept can guarantee the *empirical truth* of the causal (necessary) judgment.

[2] We have already seen that this switch from 'implicit' to 'explicit' use is a pervasive feature of Kant's thinking. (Cf. above, pp. 561f., 603f.)

thereby as a proof of something like a principle of inductive justification.

This is therefore all that can be asserted when Kant says that nature (in respect of objective sequences) is subject to the 'law' of causality, i.e. 'contains', or 'involves' the concept of a causal connection. So whilst it is quite correct for him to claim (as his 'answer to Hume') that this principle is not derived, nor derivable, *from* experience, but is on the contrary a transcendental presupposition of its possibility, what is thus established is not empirical causality, even in principle. It is only that the notion of 'nature' (i.e. particular contingent happenings) essentially involves the concept of causality. If subsequently I *apply* the concept, and the corresponding law, *to* my enquiries into the (putatively law-like) *processes* of nature, I can do this only postulationally, though I can, of course, continue to claim that what I thus apply has a transcendental *footing* in my *conception* of 'nature'.

Kant's general argument, as an answer to Hume, must therefore be construed as follows. The principle of causality cannot be derived from 'experience'. If by the latter we understand the system of the law-like or causal propositions of science, these do presuppose the formal concept of causality; but its function here is simply to provide the framework for a causal *language*, and this does not touch the question of any validation of the claim that nature is law-like. This fact can certainly not be derived from experience of constant conjunctions, any more than the causal form. But if by 'experience' we understand what is expressed in statements concerning contingent matters of fact, then one can 'prove' that the concept of causality is a transcendental presupposition of any such statements, expressed as 'judgments', being *possible*. What we cannot do is to present Kant as having simply and unambiguously claimed that in order to cognise an empirical (contingent) state of affairs I have to 'assume', or 'think' it to be causal; or more generally: theory-laden, *at* the phenomenal level of reality.

In fact, much of the detailed discussion of Kant's theory of science in section 4,[1] which was seen to involve so intimately a reference, not just to the domain of the understanding, called 'nature', but to the domain of theoretical reason, called 'the order of nature', was precisely intended to show that Kant never makes this 'simple and unambiguous' claim; and to prepare us to grasp the scope and limits of the

[1] The reader should refer back to the discussion of this above, pp. 474f., 500–6, 516–9.

transcendental argument of the Analytic more adequately. We there saw that Kant is at great pains to establish that the notion of empirical law-likeness presupposes the office of reason (or 'reflecting judgment') in its regulative employment.[1] There is thus a considerable 'looseness of fit'[2] between the understanding and reason, between the law-likeness that governs 'nature' and that which expresses the 'order of nature'. And it is a 'looseness' which, as we showed in the pages referred to, is mirrored in Kant's idea that there are certain analogies and parallels characterising the structure of these two realms, whereby the transcendental unity indicates and even 'demands', 'though only indirectly', the structure of an 'order of nature', which in turn yields the notion of empirical law-likeness.[3] It is these analogies which make Kant say that the principles of the understanding hold, 'though only indirectly', also at the level or reason;[4] just as they cause him to say that the necessitarian character of empirical laws 'intimates' a transcendentally necessary relation at the level of the understanding.[5]

Moreover, and again, just as the causal language of science presupposes the causal form, so the practice of science requires that we search for causal laws.[6] But though the concept of empirical law-likeness forms part of the theorising activity of the scientist, *that* there are causal laws is still not thereby guaranteed. And it is evident that Kant did not think that the Second Analogy had proved it, when we find him saying (as already reported) that the understanding can only 'demand' that reason employ the principle which it 'requires' for the construction of causal laws.[7]

Let us for a few more moments pursue the theme of the 'looseness of fit' between transcendental and empirical causality, and extend some of the considerations in its favour offered in sections 1 and 4, considerations which may now be placed more succinctly within the body of the transcendental argument.

---

[1] See in particular the discussion on pp. 505f. and 517f., and the important passages from J. 20 and 21 there cited.

[2] Cf. p. 501.

[3] Cf. pp. 502, 505, 508. For a discussion of these analogies, see also my 'The Relation between "Understanding" and "Reason" in the Architectonic of Kant's Philosophy', op. cit., pp. 209–26.

[4] Cf. p. 502.   [5] Cf. p. 501.

[6] It will be appreciated that for this argument such laws need not be in explicitly causal form. It is sufficient to consider law-likeness in general, as formulated for instance in the differential equations of physics. All that matters is that they should be regarded as carrying 'nomic necessity'.

[7] Above, p. 501.

It is, I think, fairly obvious that Kant will have to construe the relation between reason and understanding pretty loosely, because he wishes to retain the freedom, either to disregard the 'point of view' of objective, or scientific, experience, as in the case of morality; or to avoid certain other contradictions discussed in the Antinomies of the first and third *Critiques*.[1] Consider for instance the Fourth Antinomy, which discusses the contradictions which arise if we operate 'uncritically' with the notion of a cause, supplied as an explanation or the result of some change of circumstance. The details of this conflict do not concern us here. Alone relevant is Kant's contention that we speak 'uncritically' if we *assert* that there always *exists* an actual cause or condition for every such change. Rather, we should say

> that for every member in the series of conditions *we must expect*, and *as far as possible seek*, an empirical condition in some possible experience.[2]

Two things are here of importance. First, the existence of causes is made a function of the self-imposed activity of *reason* to expect and to search for causes; this is what is labelled its 'regulative principle'.[3] Secondly, there is the clear implication that the 'proof' of the causal principle, as a principle of the *understanding*, is not here relevant, for we are now concerned with the fitting of '*empirical* conditions'. No single such condition can be *asserted* as 'given', much less must some first or last member, some 'maximum of the series of conditions', be regarded as 'given'; all this 'can only be set as a task that calls for regress in the series of conditions'.[4] The situation of 'reason' and 'understanding' are thus quite different. Kant indeed remarks quite explicitly:

> That everything which happens has a cause is not a principle known and prescribed by reason. That principle makes the unity of experience possible, and borrows nothing from reason, which, apart from this relation to possible experience, could never, from mere concepts, have imposed any such synthetic unity.[5]

---

[1] Cf. for example the Antinomies, where Kant contrasts man when his character 'can be studied, if . . . we are simply *observing*' with the situation 'when we consider these actions in their relation to [practical] reason'. Here we do not 'explain' a man's actions (which is the task of 'speculative [i.e. hypothetical] reason'), by showing why something '*has* happened', but consider whether it perhaps '*ought not to have happened*' (K. 474; italics in text). Kant's main contention, as usual, is, that the antinomy between what *does happen* and what *ought to happen* (which may be other than what does happen) can be resolved only by 'bracketing' man, locating him at different levels, regarding him both as phenomenon and as noumenon.
[2] K. 481.   [3] Ibid.   [4] K. 449.   [5] K. 306.

The 'looseness' of the principle of causation in respect of science (which presupposes the activity of reason) is here quite evident, since we are expressly told that, relative to reason, universal causation cannot be known, but is at best only prescribed.

In a similar way, this 'looseness' is employed by Kant to resolve the antinomy between mechanistic determinism and teleology.[1] Moreover, the 'analogical inspiration' of the principle of causation as a principle of the understanding comes here to the fore. Consider a relevant passage from the third *Critique*. There Kant once again repeats the point that although the 'universal laws of material nature in general' are 'given . . . through the understanding', when we turn to the 'particular laws', they 'can only be made known to us through experience'. Moreover, the multitude of laws is such that we require their unification into systems, on the lines explained earlier, where we already noted the requirement of a number of regulative maxims (continuity, parsimony, etc.) to define a possible scientific approach to theorising.[2] But now Kant adds that in order to achieve such a 'contingent unity of particular laws',[3] we stand in need of *further* 'guiding threads', if we are 'to hope for connected empirical knowledge and understanding according to a thoroughgoing conformity of nature to law'. Such a thread is provided by the two maxims of mechanism and teleology. For our purpose here, the decisive point is Kant's characterisation of the relation of these to causality as a principle of the understanding, and to teleological phenomena putatively appearing in nature, respectively. Kant phrases it thus:

> [The mechanistic principle] is *a priori* suggested [an die Hand gegeben] by the mere understanding, but [the teleological principle] is prompted [veranlasst] by particular experiences which bring reason into play, in order to form a judgment upon corporeal nature and its laws in accordance with a particular principle.[4]

So we see: the principle of mechanistic, deterministic, explanation is not entailed by the causality of the understanding; its application is only 'suggested' by it; and lest anyone quibble with this translation

[1] Cf. above, pp. 530–1.   [2] Cf. above, pp. 505f., and sect. 4(c): v.

[3] 'Contingent': it will be remembered that this label in such a context occurred first in OG, where the reference was to Maupertuis' Principle of Least Action, which collects the otherwise merely 'contingent unity' of the laws of mechanics, statics and optics under one general heading, and converts them into a system. (Cf. above, pp. 490ff.)

[4] J. 233 (para. 70); cf. above, p. 505.

of 'an die Hand gegeben' (it is Bernard's translation), the relation of 'prompting' which leads from 'evidences' of design in nature to the methodological principle of teleology makes it clear that the relation in the former case can be no tighter.[1]

We should not be surprised at the looseness of the relation between the findings of the understanding and the activities of reason. The whole purpose of the *Critique* had been to give a 'proof' of the general principles (substance, causation, interaction) which would assure them merely transcendental status, precisely so that 'reason' could make no pretensions to operate with the categorial concepts 'constitutively'. On the other hand, this looseness of fit at the same time is taken by Kant to 'liberate reason', in its 'necessary' attempts to construct a unified idea of nature. The subjectivisation of this unity gave Kant the means both to resolve the older 'antinomies' between conflicting approaches (e.g. mechanism and teleology), to claim a 'right' to speak of a systematic unity of nature, and to extend this to the realm of theological locution, just *because* the understanding could not transcend the unity which bounded the mere isolated object, i.e. objective situation.

Altogether, Kant's transcendental argument can obviously retain its logical tightness only on condition that this separation between science (as a body of laws) and the world of commonsense objects be made complete. For we have seen it to be a central feature of the transcendental argument that the 'validity' of its 'proofs' involves a 'balance' between the transcendental principles and the logical character of the phenomenal world, as 'appearance'. It is only because this character is (for Kant) unquestionable—since no one would, according to him, deny the givenness of a material world of objects, or a psychological world of human empirical consciousness—that it can provide an anchor for Kant's argument that it is only on condition of such a world being 'given' that the transcendental principles are 'validated'.[2] And it would therefore be quite wrong to interpret Kant as saying that in order to guarantee the possibility of an experience of

[1] The different readings of some commentators are, however, understandable. One *could* translate 'an die Hand geben' as 'proving' or 'demonstrating'; and similarly, 'veranlassen' does occasionally mean 'to cause'. But then, we should expect the logical fluidity of the relation which ties reason to the understanding to be expressed by a correspondingly 'fluid' language. One can only marvel at Kant's ability to choose terms which contain such ambiguities to express his lack of logical resolution.

[2] Always granted Kant's identification of the two senses of the 'given'. Cf. above, p. 639.

objectivity, we must regard this 'experience' as law-like. For it is the quite unproblematic character of contingent, and more explicitly, *non-law-like* objectivity, that is supposed to 'balance' the transcendental concept of law-likeness at the level of the understanding.

It is however easy to see why the mistaken interpretation of Kant should so spontaneously insinuate itself with many of his readers. For, in the case of Euclidean geometry, there is—as we have seen—a possible interpretation according to which the *specific* character (i.e. Euclidean) of the transcendental concept (extensive magnitude) is balanced by the character of that geometry. And again, in the case of the foundations of 'pure natural science' (to which we turn in the next section) it may be said that to the transcendental ('metaphysical') principles there *corresponds* the synthetic *a priori* character of the laws of Newtonian dynamics (laws of motion). But even if these two cases were accepted at their face-value (and I argue that they need not be), they would constitute no more than *applications* of the transcendental principles. The latter function in these contexts as no more than 'transcendental explanations'[1] of the putative logical character of a given body of science or mathematics. This is the method which Kant employed in the *Prolegomena*, referring to it as 'the analytic method',[2] which must be interpreted as the procedure that *deduces* the logical character of science and mathematics from 'super-induced', i.e. 'hypothetical' premises, here: the transcendental principles. Kant contrasts this with the 'synthetic method' of the first *Critique*, which derives these principles from an account of the structure of experience, and of phenomena (as appearances) itself. True, in a sense this is 'analytic' also, since it may be said to deduce the logical structure of experience from the super-induced transcendental framework. But the difference between the two cases is that the structure of appearance, or experience, *is not explicitly law-like*. It only 'contains' the law-like character, which has to be read into it, so that the possibility of objects as appearances (which, however, possess *explicitly* a character that is *non-law-like*) can thereby be 'proved'.

It may be useful to conclude this section by contrasting the position I have taken *vis-à-vis* Kant's argument concerning causality with some alternative interpretations. And almost universally, commentators have overlooked the gap which divided Kant's notion of 'experience' from that of 'systematic experience', and the corresponding distinction between transcendental and empirical causality, the

[1] Cf. K. 70. [2] P. 13.

former founded on the objectively transcendental argument in the context of 'experience in general', the latter on the subjectively transcendental argument which anchors causality only as a presupposition of the possibility of systematic experience. Similarly, many critics have failed to recognise the different senses connoted by the Kantian locution of 'nature' being 'subject to law'.[1]

As a consequence, it is frequently assumed that Kant is arguing that experience in general would not be possible unless we assumed it to be causal or systematic: and, *per contra*, that the transcendental requirement of causality putatively proves the world to be organised in a causal manner: as though the Second Analogy argued *systematic* experience to be impossible without the presupposition of causality.

Such an interpretation evidently proceeds in ignorance of Kant's doctrine of science, with its central contention that the systematic, and by implication, law-like element in nature (the 'order of nature') is not 'given' but merely 'projected'.[2] For that doctrine implies at once that for Kant the causality which *reason* injects into the plurality of empirical laws must be 'balanced' by an aspect of 'appearance' that is merely 'projected', not 'given', and which thus yields only an explication of empirical causality. By contrast, the objective proof of the Second Analogy is intended to balance causality against the realm of contingent particulars which *are* 'given', and whose givenness yields in this Kantian sense an 'objective' proof of causality. It can hardly be Kant's intention to mix the arguments from the domains of reason and understanding.

An interesting modification of the traditional Kant interpretation is to be found in a recent commentary on the first *Critique* by P. Strawson.[3] Unlike many others, Strawson does recognise that Kant's various arguments involve a shift in the ' "application" of the word "necessary" ',[4] namely from what we have labelled 'transcendental' to 'empirical' causality. He holds (as we have done) that the argument of the Second Analogy really only defines the notion of an objective sequence. However, he claims that in fact Kant's intentions went far beyond this, and that Kant moved from the premise that each *member-event of a sequence* is 'tied down' in time by a 'tie' construed through the model of the causal relation, to the conclusion that the

---

[1] Cf. above, pp. 499–501. A recent instance may be found in E. W. Schipper's 'Kant's answer to Hume's problem', *Kant Studien*, **53** (1961–2), pp. 68–74, which conflates the two senses of 'experience' here distinguished.

[2] Cf. above, sects. 4(c): i and v.

[3] *The Bounds of Sense* (London, 1966), pp. 136ff.   [4] Op. cit., p. 138.

event called 'the sequence' is itself necessitated, in the sense of being necessarily conditioned by some indeterminate antecedent event: an inference which, if it were Kant's, would indeed involve the latter in what Strawson calls, 'a *non sequitur* of numbing grossness'.[1]

In detail, Kant is made to argue that the fact of our having to conceive the perceptions A', B' as necessarily determined (in order to yield the notion of objective sequence) 'is equivalent to conceiving the transition of change from A to B as *itself* necessary'.[2] True, as even Strawson has to admit, we should not take Kant to mean that A causally necessitates B, but (Strawson claims) Kant does conceive 'the change from A to B as causally necessitated by *some* unspecified antecedent conditions'.[3] He concludes, summarily, that Kant must argue that

> any succession of perceptions is a perception of objective change only if the order of those perceptions is necessary; but the order of the perceptions can be necessary only if the change is necessary, i.e. causally determined.[4]

Now it must be admitted that the *language* which Kant uses in certain sections of the first *Critique*, and even more in the *Prolegomena*, is consistent with this interpretation.[5] Nevertheless, in the light of the account that we have given of Kant's view of the function of theoretical reason (reflecting judgment) this view is implausible, and it is much more likely that the idiosyncrasies of Kant's far too tenuous terminology have misled his critics.

Strawson's account is implausible on several additional grounds. To start with, it must be noted that, if we adopt his interpretation, Kant is involved not only in a *non sequitur*, but in a contradiction. For he is made to say that it is a condition of our regarding a *contingent* sequence as objective that we should regard it as necessary. Now it is true that Kant sometimes claims certain statements to be both contingent and necessary; this is for instance his view concerning the logical status of scientific laws.[6] But even here, when the complexities are unravelled, we find that contingency and necessity operate at different levels, the latter being a function of the transcendental

---

[1] Op. cit., p. 137. For a possible confusion of these two distinct senses of event, cf. below, p. 664.
[2] Op. cit., p. 138; italics in text. [3] Ibid.; italics in text. [4] Ibid.
[5] Cf. especially our discussion of P. 64n., above, pp. 653–5, and our resolution of the confusion.
[6] Cf. above, pp. 516ff.

(regulative) processes of reason, the former belonging to the constitutive realm. Similarly, the 'necessity' appertaining to contingent judgments of matters of fact belongs to the level of the transcendental unity of apperception—a point we have found Kant at pains to stress specifically.[1]

That transcendental necessity cannot be intended to 'balance' any 'necessity' of contingent judgments at the phenomenal level, and still less that the latter could be a condition of the former, is evident from the fact that it supplies only the missing factor of 'absolute time'. It no more than reconnects the 'broken' order of perceptions which—because of our inability to supply an absolute time-clamp—has to be conceived as 'accidental'. Its function is merely to reconstitute the notion of 'possible experience', and of a 'determinate', i.e. 'objective', appearance.[2] Hence the objective sequence can itself be rendered 'necessary' in no stronger sense than would result from its being anchored (*per impossibile*) in a formal framework of absolute time. But such a sense is obviously not strong enough to support the idea of putative empirical causality in general, and Kant's references to absolute time at K. 226 must be taken to absolve him of any intention in the Second Analogy to claim the stronger doctrine insinuated by Strawson.

It is easy to forget the exceedingly technical connotation of Kant's constant emphasis that the proof of the categories bears only on 'possible experience in general'. For this implies not only that we ignore here the element of actual observation but that, in order to reach the level of empirical causality, we require reference to the conditions of the possibility of *systematic* experience, which is only supplied by the regulative principles and ideas of reason. The only relevance of the understanding in this context—as Kant notes at J. 20—is the fact that it is for the latter 'a necessary aim or need' to advance to systematic experience: presumably because connections between the processes of nature (which yield its 'order') mirror the connections postulated by the understanding as present within the individual 'objects' that make up nature. But, as we noted in an earlier section, a 'need' must not be confused with a 'presupposition'; it cannot yield any 'proofs' of empirical law-likeness.

There is perhaps an explanation of Strawson's assumption that Kant moves from the necessary determination of the members of a

---

[1] Cf. above, p. 637, especially the passage from K. 159; also p. 619.
[2] Cf. K. 226.

sequence to the necessitation of that sequence. For both the members of the sequence and the whole sequence itself may somewhat ambiguously be labelled 'events'.[1]

Now there is a similar ambiguity in Strawson's use of the term 'event', when he speaks of the 'event of change . . . as preceded by some [causal] condition . . .', where previously he had defined the sequence of the states A, B as 'a single event'.[2] This gives the impression that Kant has argued that the 'event' (*sc. succession*) A-B must be conceived as necessitated, since he does speak (as in the passage from K. 225 cited below) of an 'event' (though here meaning 'member of succession') being placed in a determining relation with a preceding state.

Now actually, when Kant in the Second Analogy speaks of 'event' (for which he uses the German synonyms Wirklichkeit, Begebenheit, Ereignis[3]), he is invariably referring to states such as A, or B (*quâ* perceptions, i.e. A′, B′) separately. Thus, in the passage which Strawson seems to have in mind (since he speaks of '*some* unspecified antecedent conditions' which according to him Kant demands as necessitating 'the change from A to B'[4]), Kant says that in entertaining the notion of an objective sequence,

> there is [created] an order in our representations in which the present, so far as it has come to be, refers us to some preceding state or other as a correlate of the event [Ereignis] which is given; and though this correlate is, indeed, indeterminate, it none the less stands in a determining relation to the event as its consequence [Folge], connecting the event in necessary relation with itself [*sc.* the preceding state] in the time series.[5]

Here it is evidently the 'event' B, *quâ* B′, which is determined, whether by A′, or some 'indeterminate' x′; there can be no question or arguing that the 'event' A-B itself is claimed to be causally determined by some antecedent (or internal) condition x′! For if that had been Kant's intention, he would have spoken, as he indeed does in another context of this Analogy, not of event, in the sense of 'Ereignis', but of 'Begebenheit', used to denote the sequence A-B, and which is

---

[1] In the same way, both are described as 'appearance', although Kant usually refers to the members of the sequence by using the plural form 'appearances'. Thus, at K. 219–20 Kant moves from 'succession of appearances' and 'manifold of appearances' to 'the manifold of appearance', where the singular use of 'appearance' denotes the appearance of succession.
[2] Op. cit., p. 136.   [3] K. 221, 225.
[4] Op. cit., p. 138, italics in text.   [5] K. 225.

very properly rendered by Kemp Smith through the term 'happening'[1]. Strawson's expressions 'transition of change' and 'event of change'[2] thus conceal an ambiguity, when in the argument as stated they should properly be used only to refer to a *member* of any succession (regarded as 'subjective' perception) and not to the succession itself. Kant is only arguing that, *given* some such succession, the general concept of such a happening requires that we *think* each of its members when regarded as 'perceptions' (whose order requires a transcendentally founded relation) as determined by some conditioning factor —a factor which (like the causal relation itself) must always be conceived in an entirely indeterminate fashion, and which, *within* this argument, is never capable of concrete specification.

So once more: we must not forget that it is an essential requirement for the preservation of the strength of Kant's transcendental argument that no question shall arise concerning the putative existence of any causal determinations of the individual sequence A-B, understood to occur at the phenomenal level, and regarded as a contingent happening, since it is precisely this—in its contingent character assumed to be 'given'—which alone is strong enough to 'balance' the transcendental concept of causal necessity. Kant's transcendental arguments are never intended to furnish more than an explication of what, on other grounds, he assumes as 'given', whether it be the synthetic *a priori* character of Euclidean geometry, the 'pure' laws of Newtonian dynamics, the contingent facts ('objects') of 'nature', or the empirical laws which supply that 'order of nature' which the scientist is driven to 'project' into nature. It would therefore have been peculiar if on just one occasion Kant had wanted to go beyond the general character of his philosophical procedure, and to mix, so to speak, the levels of his argument; and Strawson's suggestion of any explicit manoeuvres on Kant's part in the one instance of causality (as distinct from his somewhat ambiguous language) must therefore be rejected. It seems at any rate to indicate some lack of awareness of the existence of the different 'levels' that has been shown to be such a prominent feature not only in Kant, but in most of the philosophers whose writings we have scrutinised.

(iv) *The Third Analogy*

The 'looseness of fit' between the transcendentally proved principle of causation and its application at the empirical level is not confined

[1] K. 221.   [2] Cf. above, pp. 662, 664.

to causality alone. We have noted previously that there is a logical gap between the principle of the axioms of intuition and the actual Euclidean axioms. Moreover, the whole question requires a survey of the use Kant makes of the transcendental principles in his work on the metaphysical foundations of Newtonian dynamics. To this end, there remains however an item still to be discussed, the Third Analogy. This is in any case necessary because, when touching on this Analogy in our discussion of Kant's views on space in the *Dissertation*, we saw how central it is for Kant's views on the aspect of space as 'formal intuition', and for his belief that the 'physicalisation' of space requires the concept of interaction.[1]

In the *Critique*, the Third Analogy is formulated, under the heading of 'Principle of Simultaneous Existence, in accordance with the Law of Mutual Interaction, or Dynamic Association', as: 'All substances, in so far as they can be perceived to exist simultaneously in space, are in thoroughgoing mutual interaction'.[2]

This Analogy is meant to do duty in respect of two quite separate topics. The first arises in the context of Kant's attempt to give a transcendental account of an objective *time* relation, in line with the general trend of the argument behind the architectonic of the Analogies. In this way, the Third Analogy is meant to articulate the notion of objective simultaneous existence as such, whether our concern be with bodies separated 'in space', or simply with the question of coexistence of different 'states', which may belong to a single substance.[3] But there is also the old problem of a plurality of substances coexisting in space, a notion—it will be remembered—which, in accordance with Kant's critical approach, is not 'clear' so long as the concept of interaction is not built into this locution, owing to the fact

[1] Cf. above, pp. 582–6. As a reminder, cf. Kant's explicit statement in the First Antinomy: 'Things, as appearances, *determine space*, that is, of all its possible predicates of magnitude and relation they determine this or that particular one to belong to the real. Space, on the other hand, viewed as a self-subsistent entity, is nothing real in itself' (K. 400); my italics.

[2] K. 233, 2nd ed. version. Kemp Smith's translation is here gravely deficient, due to the influence of his own interpretation of this Analogy. Our translation is more literal, whilst yet adhering to Kant's intentions, which admittedly require that we should bear in mind his general approach as it first emerged in the *Dissertation*. In particular, I have translated 'Gemeinschaft' as 'dynamic association' (and not 'community'), in accordance with Kant's own explanation at K. 235. (Cf. also below, p. 667n.3.)

[3] At K. 237 Kant explicitly mentions this second case, when he writes that 'in the manifold which is coexistent *the states* coexist in relation to one another in conformity with a rule and so stand in dynamic association' (my italics).

that the 'dogmatic' assumption of a 'given space' (e.g. a Newtonian receptacle) is no longer available, and that Leibniz's theory of monads has no room for it at all.[1] There are therefore certain confusions and weaknesses in Kant's proof which we shall not dwell on, but which are mostly due to the fact that 'interaction' seems not at first sight very relevant to the question of simultaneity, whilst its undoubted relevance to the question of space is not pointed out explicitly in the argument of the Third Analogy at all. Only the resemblance of the wording with the discussion in the *Dissertation* shows the relevance of the problem of space to the Analogy.[2]

Another reason for the initial strangeness of the Third Analogy is due to the fact that most readers of the *Critique* would expect Kant to offer them—after the manner of the two first Analogies—a concept belonging to the field of logic, as represented by the concepts of substance and cause, whereas suddenly they are presented with a concept ostensibly belonging to the armoury of physics.[3]

I think that part of this reaction betrays a lack of grasp of Kant's general method. *All* these categorial concepts serve only as models which Kant believes to represent some feature capable of simulating 'objectivity'. And further, whatever the model, it 'lives' within the Analogies at the transcendental level only, as a transcendental component. Thus, causality when 'added' to the synthesis of perceptions yields phenomenal objective sequence; but *at the level* of phenomenal reality, as has been argued at length, causality is, figuratively speaking, 'invisible', and only implicit. Similarly for the case of the Third Analogy: dynamic interaction yields coexistence in space; but it remains 'invisible', buried at the transcendental level, beneath the

---

[1] Cf. above, pp. 546, 585ff.

[2] This in turn pushes the discussion back to the even earlier set of questions, and to the terms in which Kant saw these, found in his first work (LF); whilst at the same time it looks forward to the problems of M. No one not fully cognisant with this side of Kant's interests can therefore hope to penetrate the full significance of the *Critique*. Indeed, it is not too much to say that a great part of the transcendental deduction, without an awareness of these aspects, must remain incomprehensible.

[3] This explains perhaps why many translators render Kant's 'Gemeinschaft' as 'community', which incidentally does indeed make some attempt to link the concept to its model in the Table of Judgments, viz. 'disjunctive judgment'. On the other hand, at K. 235 Kant explicitly remarks that in the German language the word 'community' is ambiguous, meaning either *'communio spatii'*, i.e. 'local community', or *commercium*, which 'signifies a dynamical community', which is the use Kant himself intends. In the Table of Categories, after Gemeinschaft, he in fact adds in parentheses, 'mutual action [or interaction] between agent and patient' (which Kemp Smith renders once again as 'reciprocity'—far too 'non-dynamic' a translation).

phenomenal surface. And this was after all the reason why the application of the principles of the Analogies raises the problem of the 'looseness of fit'. Without such an interpretation, one would indeed get the 'uncomfortable' result from the Third Analogy that everything was *in fact* connected by forces!

It is true that Kant sometimes construes interaction simply as mutual or reciprocal causation, which makes this case parallel to the previous one, though commentators have not been slow to harp on the difficulties of such an interpretation. But the whole significance of the argument of this Analogy requires that we understand the concept to be *modelled* on more straightforwardly physical (dynamical and mechanical) contexts. As such models, attractive and repulsive forces are indeed uppermost in Kant's mind, but there is also another which simulates a 'relation' established between bodies *via* their mutual exposition to light radiation.[1] I call these 'models', because Kant does not mean *his transcendental argument* to signify that there are hidden transcendental forces beneath the phenomenal level. It is only the *concept* of such forces that is required. What *is* true is only that the concept of interaction, thus proved transcendentally, is subsequently *applied at the phenomenal level*, to yield the law of action and reaction.[2]

Let us now turn to the temporal aspect of the Third Analogy. The second had contrasted 'subjective' with 'objective' sequence and had claimed that the transition from the former to the latter is only possible, granted the injection of the concept of a causal nexus. The Third Analogy seeks to mediate a similar transition from 'subjective succession' to 'objective simultaneity'. Here already, however, there is a difference. We are not now contrasting subjective with objective simultaneity. There is for Kant no 'subjective' simultaneity, because throughout the *Critique* he has pictured 'time', regarded as form of intuition or 'inner sense', by means of the story of our 'perceptions' being run through the mind in successive order.[3] And, as mentioned, this psychological picture is forced on him as a necessary concession in virtue of our inability to characterise the 'accidental order';

---

[1] Cf. K. 235. Again these are literally 'analogies' for the 'objectivity' of the time relation of 'coexistence'.

[2] This interpretation of his argument parallels that of causation at the transcendental and the empirical levels respectively, except that in M, the 'application' of these concepts does not involve reference to empirical information, but merely to an 'empirical concept', matter. (Cf. below, section 9.)

[3] Cf. above, p. 642.

something which is, of course, not 'really' an order at all, since it is part of the 'synthesis' of perceptions as yet not subject to concepts— a mere abstraction not expressible in language. But obviously, Kant could not characterise the 'subjective' as 'subjective *simultaneity*' of perceptions in the same place, or in different places.

The subjective account has therefore itself to be stated in terms of yet a different model, viz. reversibility of perceptions. In detail: suppose two bodies, e.g. moon and earth, to coexist in space. My 'perceptions' of moon and earth, though still successive, could then occur in any order, the perception 'moon' preceding or following the perception 'earth'. Unfortunately, such a situation can still not yield objective *simultaneity* of moon and earth, since there is nothing to show that the moon which preceded is the 'same' moon that followed, albeit in the reverse order.

The difference between this case and that of the Second Analogy is striking. In the latter, 'subjective sequence' could *simulate* objective sequence (as in the case of the moving ship). In the present case, reciprocal perception cannot even *simulate* objective coexistence. The argument which insinuates the absolute need to add a concept is hence not as compelling; too much, so to speak, is required from the concept: one reason why we may suspect that the considerations from the context of the arguments concerning space play a hidden part in this Analogy.

To explain the significance of the need for the addition of the categorial concept, let us take again Kant's empirical example or model, though with modern embellishments.[1] No matter in what order our perceptions of earth and moon occur we could not from this advance to the '*hypothesis*' of their simultaneous existence, unless we added the postulate that light is transmitted instantaneously,[2] and mutually, between these two bodies. Only the addition of this postulate makes the assumption of absolute simultaneity physically meaningful. Now we may take the Third Analogy to 'generalise' this reasoning, by locating the required 'postulate' of 'mutual interaction'

[1] K. 235.

[2] The instantaneous transmission of light is a fiction which, though held by many scientists and philosophers before the late seventeenth century (Euclid, Hero, Aristotle, Kepler, Descartes), would not be held by Kant. However, as is well known, the framework of Newtonian dynamics can be presented (though quite unhistorically) as incorporating such an assumption: a fact which the Special Theory of Relativity makes explicit, by entailing the consequence that the assumption of absolute simultaneity can make *logical* sense only if light moves with infinite speed.

at the 'transcendental level'.[1] Just as the postulate of possible *physical* instantaneous interaction bestows *physical* significance upon the putative fact of simultaneously coexisting bodies in space, so the transcendental concept of interaction yields the notion of a 'possible perception' of bodies coexisting in space.

It only remains to add that this is of course the required transcendental concept which establishes the possibility of space (as formal intuition), by transcendentally 'deducing' the concept of interaction. The *metaphysically* conceived 'common cause' of the *Dissertation*, which had yielded the 'possible relations of substances',[2] has now become the 'thoroughgoing dynamic association (i.e. mutual interaction)' which, as *transcendental* condition, enables us to formulate the categorical indicative judgment that certain substances 'exist simultaneously in space'.[3]

Here again, it may be useful to contrast our interpretation with the account offered by Strawson. In his discussion of the Third Analogy there seems to be involved a treble misconception. (1) He sometimes seems to confuse the 'subjective' level of 'reciprocal perception' (he speaks of the order-indifference of perceptions) with the categorial concept of 'mutual interaction', possibly under the stimulus of Kemp Smith's misleading translation of 'Wechselwirkung' as 'reciprocity'.[4] Thus he says that just as the Second Analogy argues to the concept of a 'necessary order' (*sc.* causal connection) presupposed by the notion of objective change, so 'order-indifference' is required for objective coexistence; implying that 'order-indifference' (like causality) is a categorial concept.[5] Whereas Kant explicitly states that order-indifference is revealed in 'the synthesis of imagination in apprehension',[6] a term he reserves for a pre-categorised process;[7] so that in order to explicate coexistence, an *additional* concept is clearly required, viz. 'dynamic association or mutual interaction'.[8]

---

[1] And it should be noted that Newtonian gravitational force also 'travels' with infinite speed!
[2] Above, p. 586.
[3] Cf. 1st ed. wording of Third Analogy, K.233. In LF, the forces of interaction had still been 'physical'.
[4] Cf. above, p. 666, n. 2.   [5] Op. cit., p. 137.   [6] K. 233.
[7] Cf. K. 144, where this is stated more or less explicitly.
[8] K. 234. Kemp Smith translates this as 'community or reciprocity', thus involuntarily almost equating the concept with that of 'reciprocal sequence' (*wechselseitige Folge*), instead of making clear the difference, which is that between temporal sequence and dynamic interaction. Possibly Strawson's identification is influenced by this accident of translation.

(2) Strawson then claims[1] that Kant 'equates' reciprocal perception (i.e. 'order-indifference') 'with the mutual causal influence of co-existent objects'. Now if order-indifference were just another word for the category of 'reciprocity' (mutual interaction), this would indeed be tautologically true; on the other hand, if order-indifference is a 'subjective' matter (analogous to the case of subjective sequence of perceptions in the Second Analogy), then it is false; and in fact (as just shown) Kant argues that order-indifference is *insufficient* to yield coexistence, and that for the latter we require a *separate* concept, i.e. mutual interaction.

(3) Finally, Strawson's formulation (on p. 139, op. cit.) of Kant's argument as a claim to interaction *between coexistent objects* is (to say the least) a very misleading rendering of Kant's contention that coexistence *presupposes* interaction at the *transcendental* level. Here it is even more obvious than in the case of the Second Analogy that Kant cannot posit the existence of connections at the *phenomenal* level. The concept of interaction exhausts its logical strength in providing a *transcendental* foundation for the notion of coexistence in a determinate space, and has thus a purely logical (more literally: analogical) use, as a 'model' in terms of which to fashion the notion of space. Clearly, this cannot by itself yield *physical* interaction in such a space. The transcendental concept of interaction is thus one which for the purposes of science, even 'pure science', must be 'borrowed'[2] from the transcendental argument, and *freshly* applied at the phenomenal level, where we 'construct' the notion of 'physical interaction' (action and reaction).

It will be seen again that an interpretation such as Strawson's suffers from an insufficient realisation of the existence of different 'levels' in Kant's argument. True, the notion of 'levels' is itself metaphorical, since it is only shorthand for Kant's theory of the 'accidental' order of perceptions, requiring 'addition' or 'injection' of the categorial concepts: a notion which in its turn presupposes acceptance of the Kantian teaching concerning the thing-in-itself. The Kantian system, like most philosophical systems (whether they be so explicitly or implicitly) is a self-contained linguistic enterprise, and we must always beware of introducing concepts from positions with a different range of philosophical ideas before first making the required re-translations.

---

[1] Op. cit., p. 139.
[2] This is in fact the term explicitly used by Kant in order to indicate the logical relation involved; cf. below, p. 679 and n.3.

9. THE METAPHYSICAL FOUNDATIONS OF NEWTONIAN
DYNAMICS

(a) *Transcendental principles and the 'metaphysics of nature'*

We have seen how Kant solves the problem inherited from his earlier writings, leading to an interpretation which connects with his teleology. As was made clear in an earlier section, there is considerable fluidity in Kant's notion of 'mechanism'. Let us briefly summarise the results of our earlier survey.[1] First, it may denote a reference to the regulative employment of the 'maxim' of a mechanical explanation of natural (especially biological) phenomena, where this itself denotes either the determination to study structure rather than function, or the more extreme claim that deterministic explanations are to be explored for preference. Moreover, this carries with it an interpretation of 'causality' as 'efficient causality', and it is this form of causality which is prominent also in the formulation of the concept in the Second Analogy. So the notion of 'mechanism' may sometimes have the bare minimum interpretation that nature is subject to causality in either the transcendental or the empirical sense.

There was, however, still another version, left over from Kant's earlier writings, leading to an interpretation which connects with his conception of the 'necessary laws of matter', defining the very 'essence' of 'matter'.[2] The reference was here not so much to the 'empirical particular laws' but to Newton's laws of motion, and perhaps the law of gravitation. This is the domain which Kant in his later writings refers to as 'pure natural science'; and it is largely this whose 'possibility' the *Critique* begins to establish in the Transcendental Analytic; rather than the theoretical apparatus of science in general, which, as we have shown, is dealt with in the Transcendental Dialectic and the *Critique of Judgment*.[3]

Even here, as already anticipated,[4] Kant makes certain distinctions. He distinguishes between that part of natural science which is wholly

---

[1] Cf. above, sect. 4(d).   [2] Cf. above, pp. 490, 493, 494f., 531

[3] K. 56–7 and P. 35 state the relevant questions: (1) How is pure mathematics possible?; (2) How is pure natural science possible?; (3) How is metaphysics, both in general, and as science, possible? Here, the complexity of Kant's treatment of gravitational attraction is particularly interesting. Its 'possibility' (as part of 'pure' science) is established in M, whereas its 'rationality' is formulated as a task of reason in K, whilst its factual truth as an inductive generalisation is again asserted in M. (Cf., M. 182–7, K. 544–5, M. 212; and above, pp. 510–6, especially pp. 510, 515f. For attraction, cf. pp. 566–7, 581f., 674.

[4] Above, p. 475n.6.

pure, and where we are concerned with the 'principles of a general physics' which are quite universal, e.g. the principles of causation and dynamic association.[1] Apart from these 'universal laws of nature' there are, however, laws and concepts which involve reference to something 'not wholly pure and independent of sources in experience', e.g. the concepts of motion, of impenetrability, or inertia, etc.[2] Unlike the empirical laws, there does not attach to these basic laws of Newtonian science that 'uncertainty and conjecture' which Kant has claimed to pervade the rest of science.[3] On the contrary, they are 'principles of necessity', not only in the formal sense in which all empirical laws are asserted to carry necessity, but in the sense that they are principles of the 'metaphysics of nature'.

Now this metaphysics of nature, by Kant also called 'special metaphysics', is to be distinguished from 'universal' or 'transcendental' (or perhaps less misleadingly: transcendent) metaphysics. For the latter, by hypothesis, omits reference to the formal conditions of experience, and can hence have no place in any programme concerned with 'knowledge' or science.[4] On the other hand 'special metaphysics of nature', while still not requiring reference to particular observations (the realm of particular experience) does involve an application of the universal transcendental principles to an 'empirical concept'. For physics, this concept is 'matter', which, by the way in which it is defined (as 'the movable in space', as 'what fills space', and 'what possesses moving force'), allows mathematical construction to be applied; a classification guided once again by the table of categories.[5] Moreover, the reference to construction as well as to the categories shows at once that Kant thereby believes himself able to establish that the resulting principles will yield the *possibility* of the concept in question. What is not so clear is how far the establishment of *possibility* makes contact with the *actual* content of pure science, as formulated in its basic laws: something belonging to the empirical level. As usual, the transition from the possible to the actual is smoothest at the level of the purely spatio-temporal concepts, e.g. velocity,

---

[1] P. 53 (para. 15). To call these 'principles of general physics' is apt to mislead, since they clearly govern human experience in general. Besides, if my interpretation of them is correct, Kant here conceals the 'looseness' of the relation between these transcendental principles and their application to physics proper.

[2] Ibid.   [3] Cf. above, section 4(c): iv, pp. 512–6, especially, p. 512.

[4] Cf. M. 139–40. Kant therefore has transferred it, as we have seen, to other levels, either of 'ideas' in their regulative employment, or of morality.

[5] Cf. M. 144–7; also M. 150, 169, 214 for this formulation of the criteria of matter. For an introductory characterisation, cf. above, pp. 561–2.

where Kant purports to establish the theorem of composition of velocities from the general considerations of construction in space of velocity as such.[1] On the other hand, in the case of the laws of repulsion and attraction (the latter equivalent to the universal law of gravitation), Kant—contrary to what most commentators suppose —makes no such claim. As already noted, the law of gravitation itself is expressly declared *not* to be an '*a priori* conjecture', but to need 'inference from the data of experience'.[2] The reason for this is that forces as such cannot be 'constructed', even though we may attempt '*possible* constructions'[3] of their laws through the use of geometrical considerations; but these constructions, though conjectural possibilities, are only 'thinkable', i.e. purely 'logical' explorations and have no empirical (inductive) significance, since they do not involve a *real* (as distinct from an assumed) possibility of construction, as do spatio-temporal entities.[4]

(b) *The Law of Inertia*

Let us now turn to two of the laws demonstrated in the third of the four sections of M, 'Mechanics'.[5] We shall limit ourselves to a

---

[1] Cf. above, e.g. pp. 555, 560, 566.   [2] M. 212; cf. above, pp. 510, 581.

[3] M. 194. This is 'possible' in the sense of 'conjectural', and 'logically possible'.

[4] Ibid., p. 201. Para. 38 (P. 82–3) of the *Prolegomena* should be read in the light of these provisos if its claims are not to be misunderstood. For although Kant there claims that certain geometrical analogies appear strongly to suggest an inverse-square law of gravitation, and that this law finds such powerful theoretical use as to suggest that 'no other law of attraction . . . can be thought up as suitable for a cosmic system', Kant's main intention here is only to illustrate the point that just as in empirical science we approach the subject with hypotheses in mind, which are used to organise the physical enquiry, so in geometry we can only extract as much knowledge from any system as we feed into it by way of axioms.

But, needless to say, the physical case can be treated here as no more than a partial analogy for Kant's characterisation of the procedures of geometry. Moreover, I have already suggested that even the geometrical case need not receive so 'strong' an interpretation as is sometimes supposed. (Cf. above, pp. 606ff.) And to complete the confusion, the model from science (spontaneous injection of hypotheses) here applied to geometry is itself taken from Kant's characterisation of the procedure of reason, which establishes, not the 'possibility', but the 'rationality' of hypotheses. (Cf. above, pp. 672.n3, 510n.5.)

[5] Obviously, an exhaustive treatment of the *Foundations* would require another volume. Also, we have already indicated something of Kant's procedure and intentions in the second of the four main sections of M, entitled 'Metaphysical Foundations of Dynamics', and whose object it is to establish the possibility of 'matter' acting in space through forces of attraction and repulsion. This section considers individual parts of matter by themselves, whereas the third section, on Mechanics, is concerned with the interaction between different parts of matter, and the laws of force which here apply. (Cf. above, pp. 566–7.)

consideration of the laws involving inertia, and action and reaction (and which correspond to Newton's First and Third Laws of motion). And again we may ask whether we can interpret Kant as establishing only the *possibility* of certain formulations, creating thereby the 'foundation' for this part of Newtonian mechanics, rather than 'proving' their inductive truth.

The section on 'Mechanics' corresponds to the Analogies of Experience in the *Critique*; and the general method applies once more, viz. an 'application' of the relevant categorial concepts to the concept of 'matter'. Moreover, if the principles of the Analogies are synthetic *a priori*, so will be the theorems of the Mechanics, with the proviso that the concept of matter is an *empirical* concept. The sole question is whether the general principle thus applied is *a priori quâ* transcendental presupposition of experience in general, or whether its logical force is *a priori* only as a postulate antecedently (*'a priori'*) injected into the mechanical situation: the result being synthetic *a priori* only in the sense that the general principles (be their foundation what they may) are *applied* in 'constructive' fashion. And I have argued that only the second alternative is available to Kant, whether he realised this or not.[1]

However, this second alternative, far from being an archaic throwback, is astonishingly 'modern', differing as it does from the purely inductivist views of many of Kant's contemporaries, especially in England; views which were to dominate the philosophy of science throughout the eighteenth and nineteenth centuries, and down to recent times.

Modern philosophy of science (as explained in chapter II)[2] has made us familiar with the view that Newton's laws of motion, and other such basic principles of science, are not straightforward 'generalisations from experience', but rather express certain definitive characteristics of the science in question. Though they are not entirely devoid of an empirical ingredient, they define the 'natural assumptions' and the actual technical vocabulary of the science in question; for instance,

---

[1] As I stated earlier, this weaker interpretation is also one which helps to establish more reasonable meanings for the passage on Galileo and Torricelli in the Preface to the Second Edition, which can now be seen as a joint allusion to the approach of the Transcendental Analytic in respect of experience in general, to the regulative procedure of reason, and finally to the Kantian metaphysics of nature, thus letting a single example stand for quite separate levels of approach. (For the reference to the Galileo passage, cf. above, p. 497n.4.)

[2] Cf. above, pp. 33ff.

what we are to understand by 'force', 'inertial motion', and the like.[1] Such laws will be 'necessary' in the sense of being part of the 'foundations' of a science, i.e. of its vocabulary, which they 'define implicitly'. Being 'presuppositional', such laws, in a very loose sense, might therefore be called 'transcendental'. What Kant adds is the suggestion that these laws are also 'transcendental' in the more precise sense of being specially 'tightened' versions of principles, which in their general *form* have been established as presuppositions of 'nature in general', i.e. of 'possible experience' of nature, although, as I have suggested, this procedure involves an 'ambiguity' in the use of the 'middle term' ('nature is subject to the categories') which Kant does not explicitly acknowledge, if he in fact ever noticed it.

Consider now the First Law. We are here concerned with the notion of those 'changes in matter' which result from the application of forces, a fact which (by involving an empirical unobservable, i.e. 'force') must, as such, be epistemically opaque. Let us then ask: how far can this phenomenon be made 'intelligible', i.e. how can we understand its 'possibility'? Now the proposition which Kant actually purports to prove is not the First Law of motion but rather the following principle: 'All change of matter has an external cause'. Only *in brackets* does he state the actual law, viz. 'Every body continues in its state of rest or motion in the same direction and with the same speed, unless it is compelled to relinquish that state by an external cause'.[2] Does he mean to imply that the bracketed addition is not 'proved' with the same strength, but is rather only an illustration of the general principle?

Let us consider this 'proof'. Since it is short, we may cite it in full.

(From general metaphysics we have as a basis the proposition that every change has a cause; here we need only prove in respect of matter that its change has always an *external cause*.) Matter, as mere object of outer senses, has no other determinations but those of external relations in space, and hence is also subject to no changes, except through motion. In respect of this, [*understood*] *as* change from one motion to another, or to rest, and conversely, it must be possible to trace its cause (according to principles of metaphysics). This cause, however, cannot be internal, for matter as such has no inner determinations and grounds of such deter-

---

[1] Cf. B. Ellis, 'The Origin and Nature of Newton's Laws of Motion', op. cit., pp. 29–67; also my 'Science and Logic: On Newton's Second Law of Motion', *Br. J. Phil. Sc.*, 2 (1951), 217–35.
[2] M. 222.

minations. Hence all change of matter is based upon external causes (i.e., a body continues, etc.).[1]

First, let us note that the argument depends upon a rejection of the notion of 'internal forces', still prominent in LF. Kant in fact explicitly *identifies* the law of inertia with the proposition that all 'matter as such is inanimate'.[2] As a result, the 'causes' mentioned in the 'metaphysical' principle *must* be external.

Kant requires, however, further assumptions. He has to identify 'change of motion' with 'change of velocity'; and by 'velocity' he has to understand *uniform velocity*. For, without these assumptions, it is not obvious why velocity itself should not as such be regarded as a 'change', nor why, when there is no change of velocity, the latter should continue to be uniform. Peculiarly, 'Explanation 2' of section (i) ('Phoronomy') defines the 'motion of a thing' as '*change* of its external relations with respect to a given space'.[3] And this is by implication repeated in the second sentence of the above 'proof'. It is only in the next sentence that Kant slips in that he wishes his argument to be understood as being in reference to *change of motion*.

Now it is true that already in the first edition of the *Critique* we find a note in the argument of the Second Analogy, to the effect that when he speaks of change, Kant wishes to be understood as referring, *not* to 'change of certain relations in general, but of alteration of state'; and he adds explicitly that when a body moves uniformly, its state is not altered.[4] This was, however, by no means an assumption involved in the argument of the Second Analogy: the sequence of positions of the ship A′, B′ does not involve explicit reference to velocity as such; nor does the argument demand that the ship should be sailing with a variable velocity! In short, very much in the manner of Descartes' proof of the law of inertia, Kant's involves many additional assumptions, although the latter makes a greater number of these explicit.[5]

However, one can never be sure how far the purely deductive train of the argument reaches in Kantian 'demonstrations'. We have noted this fact through the interpretation we gave of the relation between transcendental and empirical causality. Are the words 'understood as',

---

[1] M. 222; second italics mine. The translation has been revised.
[2] Ibid., p. 223. He presumably means that this is the heart of the law. Cf. above, pp. 554f.
[3] M. 152; my italics.   [4] K. 230 n. a.
[5] Above all, Kant has made explicit his assumption that space is Euclidean. For Descartes' treatment of inertia, cf. above, ch. III, pp. 150–4.

italicised in my quotation of the 'proof' of the First Law, claimed to be logically entailed by anything that has preceded? Or is Kant, with the knowledge of hindsight, supplying the required interpretation of 'change of motion' as change of velocity? The answer partly depends again on whether we assume that Kant is offering proofs of the actual truth of the law, or whether we take seriously his declaration that metaphysical foundations are no more than demonstrations of *possibility*—to be understood perhaps in the loose sense of displaying how much in these laws mirrors the general concepts and principles of a science, previously shown to be embedded in the general structure of consciousness of experience of 'nature'. Such considerations could then not be intended as inductive support for the laws, but solely as architectonic devices—not so very different from the procedures we noticed operative at the level of 'reason'.

No doubt we are here moving more than ever in the flickering twilight of the Kantian ambiguity, and one's interpretation will depend on what later philosophical developments have shown to be possible in respect of the foundations of physics. Kant certainly thought it a vital achievement to have exhibited putative links between the formulations of Newtonian mechanics and the transcendental principles of experience in general; moreover, the former historically no doubt conditioned the general construction of the architectonic of the *Critique*. At the same time, it is clear that such 'anticipations of nature' have no direct binding inductive force. It should not be thought that for that reason they are worthless. For purely inductive considerations, as we noted in our second chapter, are never sufficient, and the notion of inductive truth is itself so wide as undoubtedly to require additional constraints. But these constraints have very circuitous connections with the inductively formulated hypotheses. It is Kant's merit to have drawn attention to the complex structure of hypothesis formation, formulated afresh in our section on the 'criteria of hypotheses'.[1]

(c) *The Law of Action and Reaction*

If the application of the Second Analogy to the concept of matter had to employ additional assumptions involving a knowledge of hindsight, in order to yield the First Law of motion, Kant's 'surreptitious sideglances in the direction of certain observations and experimental results', cast in order to 'arrive miraculously' at the desired

[1] Cf. above, section 4(c): iv, pp. 512ff.

result,[1] seem even more prominent in connection with the Third Law of motion, the law of action and reaction. For here the concept required, mutual interaction, is precisely the very same that Kant has used to articulate the notion of coexistence in space of a plurality of bodies. And, once again, the usual logical gap between the concept regarded in its transcendental employment and the same concept applied to 'matter' is obvious.

The law that Kant seeks to prove is formulated as follows: 'In all communication of motion, action and reaction are always equal to one another'.[2] The 'proof' of this theorem begins by 'borrowing from general metaphyiscs the proposition that all external action in the world is mutual interaction'.[3] It is then shown that mutual interaction is tantamount to 'reaction'. The details of the proof are too complex to be included here. Its core consists in the demonstration that the action of one body on another, with which it stands in 'dynamic association', can be 'constructed' in space in such a way as to show also that the action is always equal and opposite to the reaction. Using the case of collision, this is proved by apportioning the momenta between colliding bodies equally between them: a feat which is achieved by a 'relativisation' of 'empirical space', suspended in a framework of 'absolute space'. All this is then also extended to 'attraction'.[4]

The emphasis is therefore once again on showing that (at least part of) the concept in question can be 'constructed'. But the context is not that of construction (or synthesis) involved in the transcendental argument of the Third Analogy, but the more explicit context of physical mechanics. Once more the dissociation between the two contexts is obvious. The function of the transcendental argument of the Analogy had been only to show that the concept of interaction is an intrinsic component of experience in general. But the move from that to the context of the metaphysics of science is not direct. On the contrary, the logical outlines are far more complex, the divisions finer and the logical complexities more subtle.

---

[1] Kant's own critical characterisation of '*a priori*' approaches towards empirical questions; for the whole passage, cf. above, ch. III, p. 153.

[2] M. 223.

[3] This 'application' of the Third Analogy shows at once how closely it relates to Kant's concerns with space, and how much less to time. Note that this proposition is 'borrowed'. Is this an intimation that it is thereby placed in a *new* context, thus supporting my general reading of the logic of the situation?

[4] M. 227. As stated before, 'absolute space' is for Kant a mere 'idea of reason'.

Certain it is that Kant's *a priori* approaches to science mirror the general advance in sophistication which it was the object of this book to clarify. We must therefore be careful not to misconstrue the introduction to the 'Observation' which Kant appends to his 'demonstration' of the Third Law:

> This, then, is the construction of the communication of motion, which at the same time carries with it as its necessary condition the law of the equality of action and reaction, which Newton did not trust himself to prove *a priori*, but for which he appealed to *experience*. . . .[1]

For, if we read on, we soon find that as usual Kant couples the notion of 'construction' with that of 'possibility'. His main concern is that of demonstrating the 'possibility' of the phenomenon of 'communication of motion', in the course of which the law of action and reaction is likewise derived (albeit by importing from 'experience' the *existence* of certain forces). This Kant makes perfectly plain, when he writes that his 'problem', in the metaphysics of mechanics, is only 'to make this possibility [of communication of motion] . . . comprehensible'.[2] So evidently, we are not moving at the level of inductive, but 'metaphysical', foundations; at the level which tries to establish the foundation for the 'possibility', and not the 'inductive truth', of laws. What is true is that to the extent to which the formulation of possibilities coincides with that 'derived from experience', it is more than likely that the latter has borrowed as much from the former as has the former from the latter: a conclusion which modern ideas would again no doubt support.

In the end, as the concluding words of the Preface to the Second Edition point out,[3] it must be realised that no understanding of any *part* of the *Critique*, such as the section on the Transcendental Deduction of the Categories, is possible without reference to all the other parts of Kant's system, and we may add, to his pre-critical writings. When this system is thus viewed in all its astonishing subtlety, we may better understand not only the displacement of philosophical concepts of which Kant's thought is the most supreme instance, but also in general see that this displacement is the over-arching feature of the whole development of philosophical thought as I have tried to

---

[1] M. 227; italics in text.
[2] M. 229. It will be remembered that Locke had found this phenomenon 'incomprehensible'. (Cf. above, pp. 968–9, and the reference to Locke, pp. 568n.4. 569n.1.)
[3] K. 37.

sketch it. And not only of *its* development; it is the very key to an understanding of philosophical thought in general. The reader will perhaps now find it easier to search the philosophical positions of Kant's predecessors, as well as his successors, especially in our own time, to discover something of this 'displacement'. Here it has been sufficient to show how some of the key conceptions of Descartes, Locke, Berkeley, Hume, or Leibniz have become transfused in the critical thinking of Kant.

BIBLIOGRAPHY

Only books and articles used or referred to in the text are included.

Primary references

Berkeley, G.:
An Essay towards a New Theory of Vision, 1709
P: *A Treatise Concerning the Principles of Human Knowledge*, 1710
I: Introduction to the *Principles*
Three Dialogues between Hylas and Philonous, 1713; all in *A New Theory of Vision and other Select Philosophical Writings*, Everyman ed., A. D. Lindsay, London, 1929
M: *De Motu* (*Of Motion*), 1721, trsl. A. A. Luce, in *Works*, iv, ed. A. A. Luce and T. E. Jessop, Edinburgh, 1951
Siris: A Chain of Philosophical Reflections and Inquiries Concerning the Virtues of Tarwater, 1744, in *Works*, v

Descartes, R.:
A. & T.: *Œuvres de Descartes*, ed. C. Adam and P. Tannery, 12 vols., Paris, 1897–1910
Anscombe: *Philosophical Writings*, trsl. and ed. G. E. M. Anscombe and P. T. Geach, Edinburgh, 1954
H. & R.: *The Philosophical Works of Descartes*, trsl. E. S. Haldane and G. R. T. Ross, 2 vols., Cambridge, 1967
K.S.: *Descartes' Philosophical Writings*, trsl. and ed. N. Kemp Smith, London, 1952
L.A.: *Discourse on Method, Optics, Geometry, and Meteorology*, 1637, trsl. P. J. Olscamp, Libr. of Liberal Arts, New York, 1965
Descartes' Discourse on Method and other writings, trsl. A. Wollaston, Harmondsworth, 1960
The Geometry of René Descartes, trsl. D. E. Smith and M. L. Latham, Dover Publ., 1954

Hume D.:
T: *A Treatise of Human Nature*, 1739, ed. L. A. Selby-Bigge, Oxford, 1946
E: *Enquiries Concerning the Human Understanding and Concerning the Principles of Morals*, 1748–57, ed. L. H. Selby-Bigge, Oxford, 1951

684   *Bibliography*

Leibniz, G. W.:

L: *Leibniz. Philosophical Papers and Letters*, 2 vols., trsl. and ed. L. E. Loemker, Chicago, 1956

OC: *Leibniz. Basic Writings*, trsl. G. R. Montgomery, Open Court Publ., La Salle, 1962

W: *Leibniz. Selections*, ed. P. P. Wiener, New York, 1951

UP: 'Ein einziges Prinzip der Optik, Katoptrik und Dioptrik' ['On a unitary principle of Optics, Catoptrics and Dioptrics'] 1682, in *Leibniz: Schöpferische Vernunft*, trsl. and ed. W. v. Engelhardt, Marburg, 1951

DM: *Discourse on Metaphysics*, 1686, ed. P. G. Lucas and L. Grint, Manchester, 1953

N.E.: *New Essays Concerning Human Understanding*, 1705, first publ. 1765, trsl. A. G. Langley, La Salle, 1949

The Leibniz-Clarke Correspondence, 1717, ed. H. G. Alexander, Manchester, 1956

Locke, J.:

An Essay Concerning Human Understanding, 1690, ed. A. C. Fraser, 2 vols., Dover Publ., New York, 1959

An Essay Concerning Human Understanding, ed. A. S. Pringle-Pattison, Oxford, 1934

An Early Draft of Locke's Essay, 1671, ed. R. I. Aaron and Jocelyn Gibb, Oxford, 1936

Kant, I.:

References to Kant's works are in chronological order, and are to the translations where these are stated.

LF: *Thoughts on the True Estimation of Living Forces, and criticism of the proofs propounded by Herr von Leibniz and other Mechanists in their Treatment of this controversial subject, together with some Introductory Remarks bearing upon Force in Bodies in General*, 1747; *Kant's Works*. Akad. ed., 1910 *et seq.*, i; paras. 1–11, 114–15 are translated in J. Handyside, *Kant's Inaugural Dissertation and Early Writings on Space*, Chicago, 1929

UH: *Universal Natural History and Theory of the Heavens, or Essay on the Constitution and Mechanical Origin of the entire Structure of the Universe, treated according to Newtonian Principles*, 1755; *Works*, i

ND: *A New Elucidation of the First Principles of Metaphysical Knowledge* (*Principiorum primorum cognitionis metaphysicae nova dilucidatio*), 1755; *Works*, i; German trsl. in Insel ed. of Kant's works, Wiesbaden, 1960, i

OG: *The Only Possible Ground for a Demonstration of the Existence of God*, 1763; *Works*, ii

NQ: *Attempt to Introduce the Concept of Negative Quantities into Philosophy*, 1763; *Works*, ii

CP: *Enquiry concerning the Clarity of the Principles of Natural Theology and Ethics*, 1763; *Works*, ii; trsl. G. B. Kerferd and D. E. Walford, *Kant. Selected Pe-Critical Writings*, Manchester, 1968

DS: *On the Prime Ground of the Distinction of Regions in Space*, 1768; *Works*, ii; trsl. G. B. Kerferd, *op. cit.*

ID: *Inaugural Dissertation on the Form and Principles of the Sensible and Intelligible World* (*De Mundi Sensibilis atque Intelligibilis Forma et Principiis Dissertatio Pro Loco*), 1770; *Works*, ii; trsl. G. B. Kerferd, *op. cit.*

K: *Critique of Pure Reason*, 1781; 2nd ed., 1787; *Works*, iii; trsl. N. Kemp Smith, London, 1953

P: *Prolegomena to any Future Metaphysics that will be able to present itself as a Science*, 1783; *Works*, iv; trsl. P. G. Lucas, Manchester, 1953

G: *Fundamental Principles of the Metaphysics of Morals*, 1785 (sometimes called *Foundations* or *Groundwork*); *Works*, iv; trsl. T. K. Abbott, London, London, 1940

M: *Metaphysical Foundations of Natural Science*, 1786; *Works*, iv; trsl. E. B. Bax, *Kant's Prolegomena and Metaphysical Foundations of Science*, London, 1883

KP: *Critique of Practical Reason*, 1788; *Works*, v; trsl. L. W. Beck, New York, 1956

J: *Critique of Judgment*, 1790: *Works*, v; trsl. J. H. Bernard, New York, 1951

IJ: *First Introduction to the Critique of Judgment*, 1790; *Works*, xx; trsl. J. Haden, Libr. of Liberal Arts, New York, 1965

L: *Logic*, 1800; *Works*, ix; trsl. T. K. Abbott, *Kant's Introduction to Logic*, New York, 1963

Secondary references

Alexander, P. & Hesse, M. B., 'Subjunctive Conditionals', *Proc. Arist. Soc.* Suppl. xxxvi, 1962, 185–214

Austin, J. L., *Sense and Sensibilia*, Oxford, 1962

Bachelard, G., *Les Intuitions atomistiques*, Paris, 1933

Baumrin, B., ed., *Philosophy of Science. The Delaware Seminar*, i, New York, 1963

Beck, L. J., *The Method of Descartes*, Oxford, 1952

Bennett, J.:
(1) 'The Status of Determinism', *Br.J.Phil.Sc.*, **14** (1963) 115ff.
(2) 'Substance, Reality, and Primary Qualities', *Am.Phil.Quart.*, **2** (1965)

1–17. [Reprinted in ed. C. B. Martin and D. M. Armstrong, *Locke and Berkeley*, London, 1968.]

Boyle, R., *Works*, ed. Th. Birch, London, 1744

Bradley, F. H.:
  (1) *Essays on Truth and Reality*, Oxford, 1944
  (2) *Principles of Logic*, 2 vols., Oxford, 1928

Buchdahl, G.:
  (1) *The Image of Newton and Locke in the Age of Reason*, London, 1961
  (2) *Induction and Necessity in the Philosophy of Aristotle*, Aquinas Pamphlet No. 40, London, 1963
  (3) 'Science and Metaphysics', in *The Nature of Metaphysics*, ed. D. F. Pears, London, 1957
  (4) 'Science and Logic: On Newton's Second Law of motion', *Br.J.Ph.Sc.* **2** (1951) 217–35
  (5) 'Descartes' Anticipation of a Logic of Scientific Discovery', in ed. A. C. Crombie, *Scientific Change*, London, 1963
  (6) 'The Relevance of Descartes' Philosophy for Modern Philosophy of Science', *Br.J.Hist.Sc.*, **1** (1963) 227–49
  (7) 'Theory Construction. The Work of N. R. Campbell', *Isis*, **55** (1964) 151–62
  (8) 'Semantic Sources of the Concep of Law', in *Boston Studies in the Philosophy of Science*, iii, ed. R. S. Cohen and M. W. Wartofsky, Dortrecht, 1968

Braithwaite, R., *Scientific Explanation*, Cambridge, 1955

Bunge, M., *The Myth of Simplicity*, New York, 1963

Burnet, J., *Early Greek Philosophy*, London, 1930

Burtt, E. A., *The Metaphysical Foundations of Modern Physical Science*, London, 1932

Campbell, N. R.:
  (1) *Physics: The Elements*, Cambridge, 1920. Reissued as *Foundations of Science*, Dover Publ., New York, 1957
  (2) *The Principles of Electricity*, London, 1912

Cassirer, E.:
  (1) *Das Erkenntnisproblem*, i–ii, Berlin, 1922
  (2) *Determinism and Indeterminism in Modern Physics*, Yale, 1956
  (3) *Substance and Function*, Chicago, 1923

Chisholm, R.:
  (1) 'The Contrary-to-Fact Conditional', *Mind*, **55** (1946) 482–97. [Reprinted in ed. H. Feigl and W. Sellars, *Readings in Philosophical Analysis*, New York, 1949.]
  (2) 'Law Statements and Counterfactual Inference', *Analysis*, **15** (1955) 97–105. [Reprinted in ed. E. H. Madden, *The Structure of Scientific Thought*, London, 1960.]

Cohen, I. B.:
  (1) *Franklin and Newton*, Philadelphia, 1956
  (2) ed., *Isaac Newton's Paper and Letters on Natural Philosophy*, Harvard, 1958
Crombie, A. C., *Robert Grosseteste and the Origins of Experimental Science*, Oxford, 1953
Dijksterhuis, E. J., 'La Méthode et les Essais de Descartes', in *Descartes et le Cartesianisme Hollandais*, Paris, 1950
Dugas, R.:
  (1) *A History of Mechanics*, London, 1957
  (2) *Mechanics in the Seventeenth Century*, Neuchatel, 1958
Duhem, P., *The Aim and Structure of Physical Theory*, trsl. P. P. Wiener, Princeton, 1954
Ellis, B., 'The Origin and Nature of Newton's Laws of Motion', in ed. R. G. Colodny, *Beyond the Edge of Certainty*, Englewood Cliffs, 1965
Euler, L., *Lettres à une Princesse d'Allemagne*, ed. M. E. Saisset, Paris, 1843
Ewing, A. C., *The Fundamental Questions of Philosophy*, London, 1951
Feyerabend, P. K., 'Explanation, Reduction, and Empiricism', in ed. H. Feigl and G. Maxwell, *Minnesota Studies in Philosophy of Science*, iii, Minneapolis, 1962
Galileo, *Dialogue Concerning the Two Chief World Systems*, trsl. Stillman Drake, Berkeley, 1953
Galileo, *Dialogues Concerning Two New Sciences*, trsl. H. Crew and A. de Salvio, Dover ed., New York, n.d.
Galileo, *Discoveries and Opinions of Galileo*, trsl. and ed. Stillman Drake, New York, 1957
Gilbert, W., *De Magnete*, trsl. P. F. Mottelay, Dover ed., New York, 1958
Goodman, N., *Fact, Fiction and Forecast*, London, 1954
Grossman, R., *The Structure of Mind*, Madison and Milwaukee, 1965
Hall, A. R. and M. B., *Unpublished Scientific Papers of Isaac Newton*, Cambridge, 1962
Hall, M. Boas, *Robert Boyle on Natural Philosophy. An Essay with Selections from his Writings*, Indiana, 1965
Hartmann, N., *Philosophie der Natur*, Berlin, 1950
Heath, T. L.:
  (1) *A Manual of Greek Mathematics*, Oxford, 1931
  (2) *The Thirteen Books of Euclid's Elements*, Cambridge, 1926
Hertz, H., *The Principles of Mechanics*, Dover ed., New York, 1956
Hesse, M. B.:
  (1) *Forces and Fields*, London, 1961
  (2) *Models and Analogies in Science*, London, 1963; Notre Dame, 1966

688 *Bibliography*

Hobbes, T., *Human Nature, English Works*, iv, ed. W. Molesworth, 1839–45

Johnson, W. E., *Logic*, iii, Cambridge, 1924

Kemp Smith, N.:
(1) *New Studies in the Philosophy of Descartes*, London, 1952. [Abbr. as K.S.N.S.]
(2) *The Philosophy of David Hume*, London, 1941

Kepler, J., *Ad Vitellionem Paralipomena, Works*, ii, ed. W. v. Dyck and M. Caspar, München, 1939

Kneale, W.:
(1) *Probability and Induction*, Oxford, 1949
(2) 'Natural Laws and Contrary to Fact Conditionals', *Analysis* **10** (1950). [Reprinted in ed. M. Macdonald, *Philosophy and Analysis*, Oxford, 1954.]

Koyré, A., *Newtonian Studies*, London, 1965

Kuhn, T. S., *The Structure of Scientific Revolutions*, Chicago, 1962

Lakatos, I., 'Proofs and Refutations', iv, *Br.J.Phil.Sc.*, **14** (1964) 71ff.

Landé, A., *From Dualism to Unity in Quantum Physics*, Cambridge, 1960

Lewin, K., 'The Conflict between Aristotelian and Galileian Modes of Thought in Contemporary Psychology', *Journ. Gen. Psychol.* **5**, 141–77. [Reprinted in *Dynamic Theory of Personality*, New York, 1935.]

Lewis, C. I., *Mind and World-Order*, New York, 1929

Lovejoy, A., *The Great Chain of Being*, New York, 1960

Mach, E., *The Science of Mechanics*, trsl. T. S. McCormack, La Salle, 1960

Magie, W. F., *Source Book of Physics*, New York, 1935

Malebranche, N., *Recherche de la Vérité*, 3 vols., Paris, 1962

Mandelbaum, M., *Philosophy, Science, and Sense Perception*, Baltimore, 1964

Maupertuis, P.-L.-M. de, *Essai de cosmologie*, Berlin, 1750

Mayer, J. R., *Die Mechanik der Wärme*, ed. J. J. Weyrauch, Stuttgart, 1893

Meyerson, E., *Identity and Reality*, London, 1930

Mill, J. S.:
(1) *Auguste Comte and Positivism*, London, 1865
(2) *Examination of Sir William Hamilton*, London, 1889
(3) *A System of Logic*, 2 vols., London, 1879

Milhaud, G., *Descartes Savant*, Paris, 1921

Moore, G. E., *Commonplace Book 1919–1953*, ed. C. Lewy, London, 1962

Nagel, E., *The Structure of Science*, New York, 1961

Newton, I., *Opticks*, Dover Publ., New York, 1952

Newton, I., *Principia* ed. F. Cajori, 2 vols., Berkeley, 1962

Pap, A.:
(1) *The A Priori in Physical Theory*, New York, 1968
(2) *An Introduction to the Philosophy of Science*, Glencoe, 1962
Planck, M., *Das Prinzip der Erhaltung der Energie*, Leipzig und Berlin, 1908
Poincaré, H., *Science and Hypothesis*, London, 1905
Popper, K., *The Logic of Scientific Discovery*, London, 1959
Putnam, H., 'The Analytic and Synthetic', in *Minnesota Studies in Philosophy of Science*, iii, ed. H. Feigl and G. Maxwell, Minneapolis, 1962
Quine, W. V., *From a Logical Point of View*, Cambridge, 1961
Russell, B.:
(1) *The Philosophy of Leibniz*, London, 1937
(2) *Our Knowledge of the External World*, London, 1914
Ryle, G.:
(1) *The Concept of Mind*, London, 1949
(2) *Dilemmas*, Cambridge, 1954
Sabra, A. I., *Theories of Light from Descartes to Newton*, London, 1967
Schipper, E. W., 'Kant's answer to Hume's problem', *Kant Studien*, **53** (1961–2) 68–74
Schlesinger, G., *Method in the Physical Sciences*, London, 1963
Sciama, D. W., *The Unity of the Universe*, London, 1959
Scott, T. F., *The Scientific Work of René Descartes*, London, 1952
Sellars, W., *Science and Metaphysics*, London, 1968
Strawson, P., *The Bounds of Sense*, London, 1966
Thayer, H. S., ed., *Newton's Philosophy of Nature*, New York, 1953
Waismann, 'Verifiability', *Proc. Arist. Soc.*, Supp. vol. xix, 1945, 132. [Reprinted in ed. A. G. N. Flew, *Essays in Logic and Language*, First Series, Oxford, 1951.]
Whewell, W., *Philosophy of the Inductive Sciences*, 2 vols., London, 1840
Wittgenstein, L.:
(1) *Philosophical Investigations*, trsl. G. E. M. Anscombe, Oxford, 1953
(2) *Tractatus Logico-Philosophicus*, trsl. D. F. Pears and B. F. McGuinness, London, 1961
Yourgrau, W. and Mandelstam, S., *Variational Principles in Dynamics and Quantum Theory*, London, 1955

Post, N.
(1) The Theory of Equipoised Theory, New York, 1955
(2) An Introduction to the Philosophy of Science, Glasgow, 1982
Planck, M., Das Prinzip der Erhaltung der Energie, Leipzig und Berlin, 1887.

Poincaré, H., Science and Hypothesis, London, 1905.
Popper, K., The Logic of Scientific Discovery, London, 1959.
Putnam, H., The Analytic and Synthetic, in Minnesota Studies in Philosophy of Science, iii, ed. H. Feigl and G. Maxwell, Minneapolis 1962.
Quine, W. V., From a Logical Point of View, Cambridge, 1961.
Russell, B.
(1) The Philosophy of Leibniz, London, 1937
(2) Our Knowledge of the External World, London, 1914
Ryle, G.
(1) The Concept of Mind, London, 1949
(2) Dilemmas, Cambridge, 1954
Sabra, A. I., Theories of Light from Descartes to Newton, London, 1967
Schappe, L. Y., Kant's answer to Hume's problem, Mind, vol lxv, 53

Schlick, M., Gesammelte Aufsätze, Wien, 1938.
Schlipp, P. A.

Scriabinger, G., Maxwell and the Physics of Newton, London, 1961
Simon, D. W., The Origins of Metaphysics, London, 1949
Sion, T. E., The Scientific Work of René Descartes, London, 1952
Sellars, W., Science and Metaphysics, London, 1968
Strawson, P. F., The Bounds of Sense, London, 1966
Tarski, H. S., ed., Works, Philosophy of Vienna, New York, 1953.
Watkins, Verifiability, Proc. Arist. Soc., Supp. Vol. 1945, 1952.
[Reprinted in ed. A. G. N. Flew, Logic and Language, First series, Oxford, 1951.]
Whewell, W., Philosophy of the Inductive Sciences, 2 vols., London, 1840
Wittgenstein, L.
(1) Philosophical Investigations, trans. G. E. M. Anscombe, Oxford, 1953
(2) Tractatus Logico-Philosophicus, trans. D. F. Pears and B. F. McGuiness, London, 1961.
Woolgan, W., and Mandelbaum S., Sensational Principles in Dynamics and Quantum Theory, London, 1953.

INDEX

691